47th Reloading Handbook

Edward A. Matunas, *editor*

IBM 9817000
UPC 0115169700

First Printing: September, 1992
Second Printing: March, 1993
Third Printing: April, 1994
Fourth Printing: February, 1995
Fifth Printing: February, 1997
Sixth Printing: October, 1998

Published by:
Lyman Products Corporation
475 Smith St.
Middletown, CT 06457
J. Mace Thompson, President
Printed in the United States of America

Warning

The publisher and editors have no control over how reloading is conducted by the individual or with what components and dies. Every change in equipment, procedure, and component lot will effect ballistics and/or the safety and usefulness of a load. Therefore, no warranties are implied or expressed by the data and copy contained in this book. We specifically disclaim any warranties of fitness for any and all particular purpose and specifically disclaim any and all liability for consequential damages of any kind.

The individual assumes all risks for the safety of reloaded ammunition. Improperly loaded ammunition, or the failure to follow all necessary precautions, may result in serious personal injury and/or death to the shooter or bystanders.

There are many precautions to which the reloader need adhere. This volume deals with many of these but cannot possibly foresee or include all possible cautions and caveats.

National Associations of Interest

A person's enjoyment of a hobby can often be enhanced by membership in an organization composed of like-minded individuals. The early tangible benefit to these memberships is receipt of the newsletter or magazine, usually filled with articles and interesting tidbits. Later, with increased involvement, friendships are formed that can last a lifetime. One or more of the following organizations will interest most reloaders. Write for membership information:

American Single Shot Rifle Association
L.B. Thompson
987 Jefferson Avenue
Salem, OH 44460

Cast Bullet Association
4103 Foxcraft Drive
Traverse City, MI 49684

International Benchrest Shooters
RD #1
Box 244A
Tunkhannock, PA 18657

International Handgun Metallic Silhouette Association
127 Winthrop Terrace
Meriden, CT 06450

International Practical Shooting Federation
P.O. Box 811
Sedro Wooley, WA 98284

National Bench Rest Shooters Association
2027 Buffalo
Levelland, TX 79336

National Reloading Manufacturers Association
Suite 300, One Centerpointe Drive
Lake Oswego, Oregon 97035

National Muzzleloading Rifle Association
Box 67
Friendship, IN 47021

National Rifle Association
1600 Rhode Island Avenue, N.W.
Washington, DC 20036

National Shooting Sports Federation, Inc.
555 Danbury Rd.
Wilton, CT 06897

Acknowledgements

Lyman would like to thank the following companies for their assistance in the production of this handbook. Without their strong support this effort would not have been as complete. Our sincere thanks and appreciation to all.

A-Square Co.
1 Industrial Park
Bedford, KY 40006

Accurate Arms Co.
P.O. Box 167
McEwen, TN 37101

Alliant Techsystems
100 Howard Blvd.
Kenvil, NJ 07847

Barnes Bullets
P.O. Box 215
American Forks, UT 84003

Blount, Inc.
Box 856
Lewiston, ID 83501

Federal Cartridge Co.
900 Ehlen Drive
Anoka, MN 55303

Hodgdon Powder Company
P.O. Box 2932
Shawnee Mission, KS 66201

Hornady Manufacturing
P.O. Box 1848
Grand Island, NE 68802

IMR Powder Company
Plattsburgh, NY 12901

Marlin Firearms Co.
100 Kenna Drive
North Haven, CT 06473

Nosler Bullets, Inc.
P.O. Box 671
Bend, OR 97709

Olin/Winchester
427 North Shamrock
East Alton, IL 62024

Remington Arms Co., Inc.
1007 Market Street
Wilmington, DE 19898

Sierra Bullets
1400 Henry Street
Sedalia, MO 65301

Stoeger Industries
55 Ruta Court
S. Hackensack, NJ 07606

Sturm Ruger
One Lacey Place
Southport, CT 06490

Thompson/Center Arms Co.
P.O. Box 5002
Rochester, NH 03867

U.S. Repeating Arms Co.
275 Winchester Ave.
New Haven, CT 06511

Weatherby Inc.
2781 Firestone Blvd.
South Gate, CA 90280

Edward A. Matunas, editor

LYMAN 47TH EDITION RELOADING HANDBOOK DEVELOPMENT STAFF

Publisher: J. Mace Thompson

Editor: Edward A. Matunas

Technical Editor: Thomas J. Griffin

Managing Editor: Richard J. Ranzinger

Staff Associates:
Edward J. Schmitt
Karen E. Hall
Stuart R. Bennett

Outside Contributors:
Raymond I. Cowles
J. Michael Nicastre

Ed Matunas has been a shooter and reloader all of his life. His entire career has been spent within the shooting sports industry. Part of this time was at Lyman Gun Sight, now Lyman Products Corp., where Ed served as Director of the Ballistics Laboratory and worked on the production of Lyman's technical publications.

In 1982 Ed turned to writing full time and has since written 12 books and over 600 magazine articles. He resides in Moosup, CT.

Thomas J. Griffin, technical editor

Photography credits:
John Giammatteo
Edward Matunas
Nyle Leatham

Meet the Lyman Management
Let us help you with questions or problems.

Karen Griffin **Mgr. Customer Service**	**Tom Griffin** **Mgr. Ballistics Lab**	**Joe Discepolo** **Engineering Manager**	**Ed Schmitt** **Mgr. Product Planning**
Rae Johnson **V.P. Operations**	**Rick Ranzinger** **V.P. Sales**	**Tom Andersen** **V.P. Fin. & Adm.**	**James Mesick** **Manuf. Engineer**

Table of Contents

EDITOR'S FOREWARD

People have been reloading longer than most of us realize. Our forefathers were reloading when they poured black powder and rammed patched lead balls down the bores of their muzzleloading rifles. The pioneer spirit of our ancestors created life styles that demanded they become reloaders. Their tools were relatively crude and hand operated, giving rise to the term handloading. Even so, they were able to get the job done without major difficulty. When loaded metallic cartridges started to become available, reloading continued to grow in popularity among shooters of the late 1800's and early 1900's.

Many people lived a great distance from a source for commercially loaded ammunition, however, it was common for black powder and lead, for casting bullets, to be available. The early Ideal handloading tools found an untapped market with these shooters. In fact, some of the early Ideal handloading tools are equivalent to today's popular Lyman 310 Tool, the main exception being a bullet mould that was built into the handles of some of the earlier tools.

While most of the basic principles remain the same as they were in the earliest years, reloading has truly gone through a renaissance. Today, shooters have state of the art reloading equipment, an endless supply of top quality components to select from and a broad base of information to assist in the reloading effort, including computer programs. Simply stated, "there has never been a better time to reload, or get started in reloading."

Reloading is an exciting and satisfying experience. While the cost saving aspects will always be there, most people that reload today do so out of pure enjoyment. Reloading is also perfectly safe as long as it is undertaken by responsible people who are willing to learn it the right way from the beginning. This requires the necessary time to read a quality reloading manual and all the information provided with the equipment and components that will be used in the reloading process. While a friend can often answer basic questions, we would caution that there are reloaders who may be misinformed or simply have not learned the correct procedure for the various aspects of reloading.

Be safe, not sorry. Know what you are doing before you do it! Follow our guidelines exactly. When a question arises and you are unsure of what to do next - STOP - and go no further. Call the component manufacturer if it is a question on components, or call the equipment manufacturer. Most companies within the reloading community have technical departments that are happy to assist, most with toll free 800 numbers. The Lyman Service Department will be pleased to assist you with any questions you might have or refer you to another source in cases where more information is required.

Serving as editor for the Lyman 47th Edition Reloading Manual was one of the most satisfying, yet challenging, endeavors I have ever undertaken. During the period in which I worked for Lyman, in the late sixties and early seventies I had considerable involvement in other Lyman Handbooks and, in fact, served on the Technical Staff for both the 44th and 45th Editions of the Reloading Handbook. The initial question I posed to myself when asked by Lyman to work on this new manual was "How do you improve upon what many consider to be the finest handbook for reloading ever written?" This was the task that the Lyman Handbook Staff and I were presented with.

Since the 46th Edition was published, many changes had taken place in the market in terms of new calibers and components. The extensive line of popular powders from Accurate Arms, for example, are totally new to the Lyman Handbook series. Lyman Handbooks have always been the most universal. Their loading data includes powders and jacketed bullets from all the major manufacturers. In addition, the Lyman manuals always include loading information for their extensive line of cast bullets. Needless to say, updating the data section for the 47th Edition was a major undertaking unto itself.

The objective in the editorial section was to provide contemporary information that will be of value to every type of reloader and shooter, whether beginner or expert. This is extremely challenging considering the many shooting disciplines which exist today and the different requirements of each. The needs of the Bianchi Cup shooter, for example, are undoubtedly different from those of the benchrest shooter. The new shooter who simply wants to load a few rounds to hunt with in the Fall would not be concerned with the latest information on bullet lubricants which might be important to silhouette shooters using cast bullets.

All in all, it is impossible for any single volume to be all encompassing, to answer every question that a reloader might have. This handbook does, however, meet the basic criteria as well as any I have ever read.

As in all Lyman handbooks of the past, we place special emphasis on the basic reloading and casting operations. These are presented in great detail in The 47th Edition and safety is stressed throughout. Beyond the basics, we offer the experienced reloader accuracy tips and suggestions on upgraded equipment which can add speed and efficiency. As always, we have commissioned outside experts to write about special topics which can add to our knowledge of reloading or, perhaps, to introduce us to an element of shooting that we have not yet experienced. Arnt Magne Myhre, for example, in his article on "Reloading for Competition" provides a fine introduction into action pistol shooting.

For those who have struggled to achieve accuracy with cast bullets, Ed Schmitt's article on the subject is especially relevant. Ed relives his own experiences in getting his cast bullets to shoot better groups. Ed's story is both personal and at times humorous, but emphasizes how far a person can go in reloading when driven by accuracy and a healthy enthusiasm.

Ken Oehler, Ph.D, is one of the foremost authorities on the subject of chronographs and bullet velocity in the United States. He was, literally, a pioneer in this field and the company that bears his name is the world leader in chronograph technology and development. Ken, in his article, shows not only how to use a chronograph, but how to process the information this instrument gives us.

Since many reloaders are also hunters, the Handbook staff felt that an article on hunting with reloads would show the satisfaction that can be experienced in bagging a trophy with one's own handloaded ammunition. In this case we took that a step further with an article on African game hunting with reloads. This article was especially enjoyable to me since, being chosen as the author, I was able to relive some very fond memories. I believe we left the reader with an expanded level of confidence in hunting with his or her own ammunition.

The last series of chapters in The 47th Edition deals exclusively with the subject of ballistics. As with other parts of the manual, we start at the beginning with a historical overview of the field of ballistics. This gives the reader some perspective on the origins of ballistics as they relate to the shooting sports. This chapter is followed by a totally new article on "Trajectory Variations". Here, we discuss how weather conditions and elevation can have an effect on the path of a bullet. This is particularly important to the target shooter or hunter that sights in a rifle at one location and travels to a different location for a hunt or a match. The authors selected specific cartridges as examples and show, through the use of ballistic tables, what changes are important to the shooter, and why.

Finally, the authors move on to an in depth discussion of air drag on bullet flight in the chapter entitled "Air Drag Effects and the Ballistic Coefficient". Here, through the use of specialized shadow graph photography, they show the correlation of bullet shape as it relates to air drag. The authors then discuss how ballistic coefficients are derived from this information. Associated ballistic coefficient tables are provided for most Sierra and Lyman bullets.

All of the articles in the ballistics section were written by Bill McDonald and Ted Almgren. This is simply because there are no two people anywhere who are as qualified to write on the subject. Ballistics is a passion with both of these men and they have devoted much of their lives to ballistics research, often writing for Lyman and Sierra.

Bill McDonald holds a doctor's degree in instrumentation sciences from MIT. In his main professional capacity, Bill was Director of Advanced Systems in the Advanced Development Center of Rockwell International. Ted Almgren has a degree in mathematics from California State Polytechnic College and has also been employed by Rockwell in the Autonetics Division as a Systems Engineer. Both are lifelong shooters and together have authored many articles and papers on exterior ballistics.

This 47th Edition of the Reloading Handbook is clearly the best and most comprehensive ever produced by Lyman. It is literally packed, from cover to cover, with the most up-to-date articles and information on handloading and shooting that you will find anywhere. We hope you enjoy it and tell your friends about it.

As always, the Lyman Staff encourages your comments and suggestions on how to improve upon this and other upcoming Lyman publications.

Edward A. Matunas

THE LYMAN RELOADING HANDBOOK

The first publication associated with Lyman is the original Ideal Handbook, published in the 1800's by John H. Barlow, founder of the Ideal Tool Company. This edition was the first reloading reference book of its type and is, without question, the longest-lived shooting reference title ever published. The Reloading Handbook series was continued by Lyman and over the years has set the standard for manuals of this type. Along the way the book has been edited by such prestigious shooters as Townsend Whelen, Charles Askins, Earl Naramore, Lysle Kilbourn and, most recently, Ed Matunas.

The Lyman Reloading Handbook has always been unique among the many manuals which are available to shooters. This is due to the Lyman/Ideal tradition of representing all brands and weights of jacketed bullets, all of the major brands of smokeless propellants, plus the most extensive amount of loading data for cast bullets offered anywhere. This tradition continues today.

This 47th Edition of The Reloading Handbook series is our first new manual in more than 10 years and is, by far, the most comprehensive ever offered by Lyman.

It is the result of rigorous testing and measurement in our advanced ballistics laboratory in Middlefield, Ct. over the past three years. The extensive data section includes all the newest calibers and components. The editorial section offers complete reloading instruction for the beginner, accuracy tips for the expert and contemporary articles on shooting and ballistics by well known writers.

More than ever, the 47th exemplifies our goal of offering you, the shooter, the most accurate and reliable data possible. We hope you enjoy it. Our technical staff stands ready to assist your reloading efforts.

1

AN INTRODUCTION TO RELOADING

CHAPTER 1
WHY RELOAD?

Today, reloading is a great deal more sophisticated than it was a century ago. Modern cartridges develop much higher pressures and are loaded with an array of propellants greater than could have been imagined by our forefathers. The level of precision built into modern reloading tools and procedures allows the reloader to assemble ammunition that is a great deal more accurate than any of the factory loaded types. Indeed, the most demanding accuracy sport, benchrest shooting, is the exclusive domain of very carefully assembled reloads. Many of those who enjoy long range target shooting, varmint hunting or big game hunting, will turn to reloads as the best, if not the only source for the level of accuracy they require.

The first reason most shooters become reloaders is to save money. Indeed, it is possible to save from 40 to 80% of the cost of factory ammunition by reloading. The exact amount saved will be reflective of the caliber(s) involved, the specific components selected, as well as where and in what quantities you purchase components. The primary reason why most of these shooters later become deeply involved in reloading is accuracy. It is not uncommon for some lots of standard factory ammunition to supply about 2-inch, or larger, five-shot groups as fired from a bench rest at 100-yards using an accurate bolt action rifle. The premium grade ammo assembled by many factories can of course improve upon accuracy levels. Reloads can, in many instances, reduce group sizes to half or less of factory ammo groups.

If a shooter cannot obtain groups of 1-1/2" or less with carefully assembled reloads it will simply be a sign that the rifle being used needs tuning or is not capable of fine accuracy. Working with loads to obtain groups of under an inch is realistic for rifles that are of the quality and tuning to make this degree of precision possible. Accuracy may not be the first reason shooters begin to reload, but it quickly becomes very important as a new reloader sees what can be accomplished.

Over four million shooters have become reloaders and annually produce as much ammo as all the major ammunition factories combined. It is debatable if the majority of these shooters really have an absolute need for the increased levels of accuracy supplied by reloads. Nevertheless, all take pleasure from it. It has been shown that many reloaders reach a point where the saving of money is no longer the primary reason to reload. Money is not a major concern to a reloader seeking a higher level of accuracy. Thus, we see that another reason for reloading is the ability to enjoy an increased level of shooting activity. This is how marksmen are made - they practice and practice and practice, often for no more cost then expended by a casual user of factory ammo. Reloading makes it possible to enjoy all the pleasures of shooting a lot of ammo and doing so frequently.

Cost Comparison of Factory vs. Handloaded Ammunition

		Handloaded Ammunition Jacketed Bullet - Rifle			Handloaded Ammunition Jacketed Bullet - Handgun	
	Factory Ammunition	W/New Cases	W/Recycled Cases	Factory Ammunition	W/New Cases	W/Recycled Cases
Cartridge	.30/.06	.30/.06	.30/.06	.357 Mag.	.357 Mag.	.357 Mag.
Bullet wt., grs.	180	180	180	158	158	158
Cartridges per box	20	20	20	50	50	50
Cost per box	$17.57	$13.06	$4.81	$18.55	$10.17	$5.35
Cost per cartridge	$.879	$.653	$.241	$.371	$.203	$.107
Savings per box over Factory	—	$4.51 (25.7%)	$12.76 (72.6%)	—	$8.38 (45%)	$13.20 (71%)
Savings per cartridge over Factory		$0.23	$0.64	—	$0.17	$0.26

		Handloaded Ammunition Cast Bullet - Rifle			Handloaded Ammunition Cast Bullet - Handgun	
	Factory Ammunition	W/New Cases	W/Recycled Cases	Factory Ammunition	W/New Cases	W/Recycled Cases
Cartridge	.30/.06	.30/.06	.30/.06	.357 Mag.	.357 Mag.	.357 Mag.
Bullet wt., grs.	180	170	170	158	158	158
Cartridges per box	20	20	20	50	50	50
Cost per box	$17.57	$7.53	$2.53	$13.25	$6.64	$1.81
Cost per cartridge	$.879	$.377	$.127	$.265	$.133	$.0362
Savings per box over Factory	—	$10.04 (59%)	$15.04 (86%)	—	$6.61 (50%)	$11.44 (86%)
Savings per cartridge over Factory	—	$0.50	$0.75	—	$0.13	$0.23

Note: All handloads loaded to equivalency of the given factory loading.
All prices based on average 1992 over the counter prices.

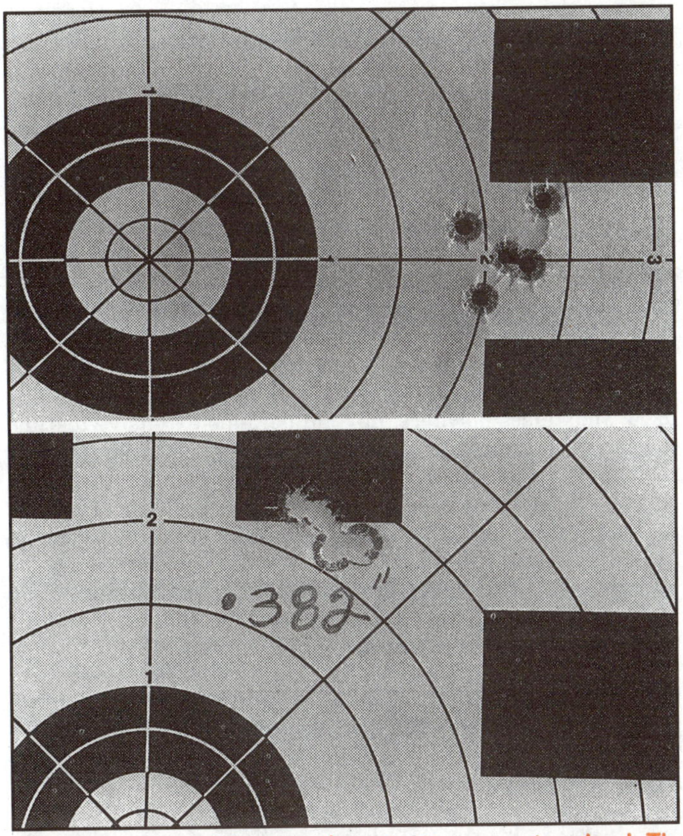

Accuracy is one of the most important reasons to reload. The top target is the result of shooting factory ammo at 100 yards. The bottom target was fired with the same rifle and bullet weight using reloads.

Reloading can also provide a deep sense of personal accomplishment. When the match is won or the trophy is taken with ammunition that was self assembled, the accomplishment has a greater and deeper meaning to the shooter.

Reloading as a hobby also adds to our enjoyment. Serious collectors of stamps or coins know that there is a great deal more to their hobby than the accumulation and cataloging of a collection. There is a constant opportunity to learn and to create a means of self stimulation of the mind. Reloading is not different in these aspects. Indeed, reloading is a hobby that can lead to test programs that afford countless hours of enjoyment while one seeks ways to enhance ammunition performance.

It is fine to vary loads and methods within the framework of published data and methods. Such testing with various components and methods can become a hobby unto itself. It can also be the stepping stone to become a deadly serious shot. Always avoid any "experimentation". Without a complete ballistic lab there is no way to "experiment" without incurring serious risks.

Reloading is what makes the sport of benchrest shooting possible, wherein a group measuring 1/4 inch from the centers of the two most widely separated shots won't be good enough to get the shooter an honorable mention at a serious match. Reloading is what makes it possible to cleanly bag a trophy that might be standing a distant 1/4 mile away without having to trust to chance whether the shot will strike the vitals or merely maime the critter. Shots taken at 400-yards with some lots of factory ammo can be less than sporting due to accuracy limitations.

There are, of course, many other reasons to reload. The reasons are as varied and diverse as are the shooters who participate in the undertaking. The fact that you are reading these pages suggests that you do not need us to supply reasons for reloading. However, you may need to be cautioned that reloading requires a very mature level of reasoning and safety awareness. Novice and veteran alike will find a great many cautions in the following pages. They are as essential to the undertaking of reloading as were the very first thoughts that brought you to this point.

Many years ago, a learned gentleman wrote what was then a very concise text book on reloading. The author was Philip B. Sharpe and the book was Complete Guide To Handloading. Some 18 years later, Mr. Sharpe felt it necessary to revise his effort with a supplement in the early 1950's. Included in that supplement was a profound effort to caution reloaders. The caution was not in the first book. It is doubtful that Mr. Sharpe ever dreamed of the number of individuals who would ignore the instructional material and data of the original volume. Indeed, he felt the caution was so required at the time of the revision that he added a chapter called "And The Angels Sing". The content of that chapter was aimed at cautioning the reader to follow all guidelines given.

Long range shooting means a real need for the use of reloads.

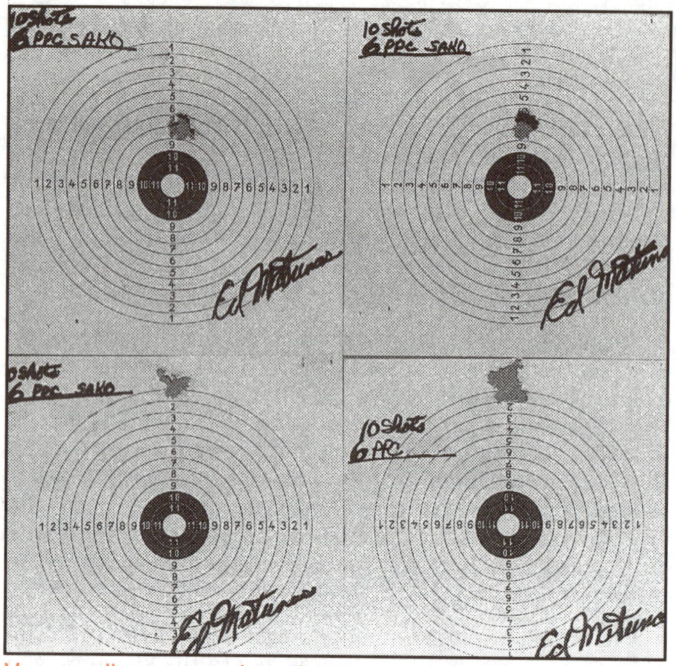

Very small groups such as these are consistently possible only with carefully assembled reloads. (10 shot groups by the author at 100 yds.)

Mr. Sharpe's words would probably not be published in this day and age as they are harsh and negative. Reloaders should keep in mind that the data in loading manuals *is not* purposely kept on the conservative side, with respect to pressure.

Reloading can provide you with safe and highly accurate ammunition. The proof is that there are more than 425,000,000 centerfire metallic cartridge reloads assembled each year. For the most part they all add a great deal to the shooters pleasure.

Almost always, new would-be reloaders want to know the physical dangers of keeping powder and primers in the home. Only a few days ago I received a call from a reloader who lived in a rented home. The landlord had not maintained the wiring in the home and there was a minor problem. When the repairmen reported to the landlord that he had seen explosives stored in the home, the renter was ordered to remove the explosives - 5 pounds of powder and about 1000 primers. The truth of the matter is that reloading components stored in reasonable quantities and in a prudent manner are no more hazardous than many common items stored in and around many homes. Indeed a half can of gasoline kept for a lawn mower is far more hazardous than several pounds of powder and a quantity of primers.

Most hunters do bag trophies. When it is done with a reload the sense of pride, for some folks, can be almost overwhelming. (Caribou taken at 440 yards with 270 cal., 130 gr. reloads)

Normal household cleaning solvents, pressurized spray can products, and paint are all debatably more hazardous than smokeless propellants and/or primers. Still each of the reloading components must be handled and stored with the necessary caution and respect. Primers, after all, are designed to cause ignition of the propellant charge by exploding. They are, however, as packaged by the manufacturer, more than reasonably safe to handle and store. The propellants used, smokeless powders, have undergone extensive testing to receive a flammable solid classification. They simply are not explosive if treated as intended by the manufacturer. All the television nonsense to the contrary, bullets cannot be pulled from loaded cartridges. The powder contained cannot be poured out in a heap and be ignited to form the nucleus of an explosion. The necessary cautions for the handling and storage of components is discussed in the chapters dealing with primers and propellants.

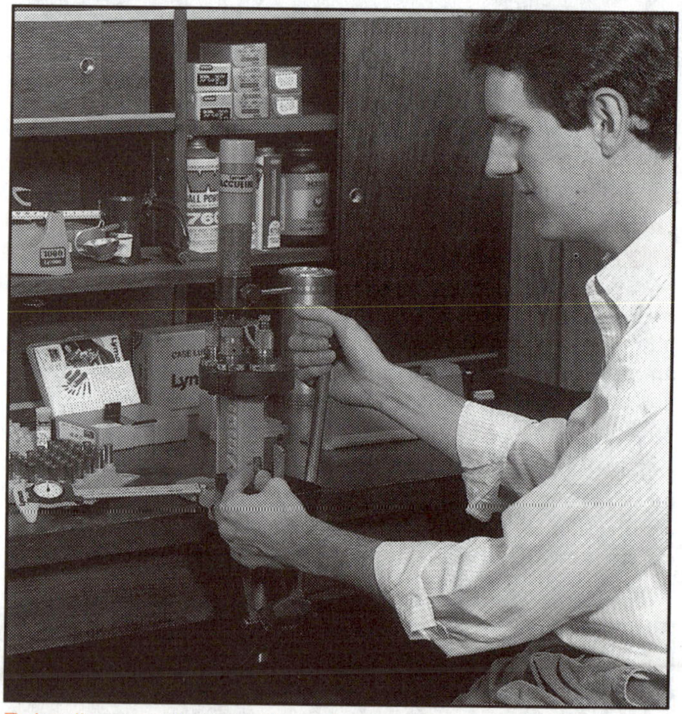

Reloading can be a safe and rewarding undertaking if the individual carefully follows all of the essential cautions.

CHAPTER 2
GETTING STARTED IN RELOADING

What does it take to get started in reloading? This question is, of course, often asked to give the interrogator some idea of the type of equipment and amount of money it will require to begin to assemble one's own ammo. But we would be remiss if first we did not mention some of those items that one does not purchase at the local gun shop.

It takes an honest measure of common sense to become a reloader. You would not expect a golfer to necessarily be qualified to make golf balls. Neither is every shooter a candidate for reloading ammunition. One would expect that everyone who shoots is a thoughtful and conscientious person and able to handle firearms in a safe manner. This kind of mature thinking is a necessity for the responsible shooter. Such maturity is also a big step in the direction of becoming a reloader. However, more is required.

Folks who are conscientious shooters are generally good candidates for reloading their own ammunition.

Reloading means following the recipes. Some cooks follow the recipe for a favorite venison sausage to a tee. Oh, they might make the recipe a bit sweeter or cook it a bit longer, but the critical ingredients will be adhered to exactly. Other cooks will vary everything possible in the recipe and justify it all as experimentation. Often such experiments fail completely. A safe reloader will follow the recipes exactingly. Anything less and the ammunition may fail with the results being someone getting hurt, or far worse.

One brand of bullet may be substituted for another or a change in primer brands might be made. But such changes are always approached with caution and the load worked up carefully. There is never a justification for experimenting when loading ammunition. Maximum loads are always adhered to and minimum loads are equally respected. If you feel you can respect the wisdom, judgement and testing of others, reloading may well be right up your alley. But if you think all loading manuals are conservative, that maximum loads can be exceeded, or that specific cautions can be ignored by a savvy person, then I promise you that reloading is not for you.

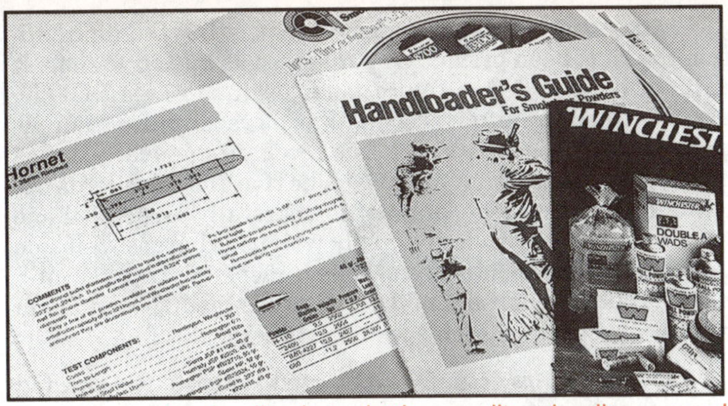

The first step in reloading is to obtain a quality reloading manual and read all the basic information from the start to finish.

Reloading should be undertaken only by persons who can appreciate that any change in components, procedures, or loaded round dimensions will create a measurable change in the ballistics of the cartridge. The appropriate candidate for reloading will also be a person who generally thirsts for additional knowledge as applicable to any specific undertaking.

A well lighted area, free from all distractions and potential hazards is essential. This aspect of getting started cannot be overstressed. You must be able to give full, undivided attention to the reloading task. Keep distractions such as TV, children and pets away. The work bench needs to be situated away from any potential household traffic flow. There must not be any unprotected lights above the work area which could be broken and thus drop sparks on the flammables with which you will be working. There shouldn't be any other potential ignition source located so as to present a hazard.

Do not overlook simple potential problems such as heavy objects so stored that they could fall onto the work area and perhaps cause ignition of primers, etc. Be extra cautious in selecting and setting up a work area. A remote storage area for components is also part of the common sense approach to reloading.

There are more than four million people who have gotten started in reloading. Most of these folks became safe reloaders and now assemble ammunition at a substantial savings with respect to cost per round. It is unlikely however that any of them actually save money. Most simply reinvest their poten-

Getting started in reloading your own ammo usually means duplicating ammunition previously purchased as factory loads. An almost immediate benefit of reloading is enhanced accuracy.

tial savings into more ammunition for more shooting. If practice makes perfect only reloaders will be able to practice enough to become highly proficient shots.

Getting started usually means duplicating the ammunition that has been previously purchased as loaded rounds. It will not be long before the reloader begins to note that the assembled ammunition is somewhat more accurate than his or her factory loaded ammo. This recognition usually comes quickly. If it does not, it might well be that the reloader is not doing a first class job. Ammunition carefully assembled one round at a time, with individually weighed charges which do not vary, and loaded to appropriate dimensions is inherently more accurate than ammo that is assembled thousands of rounds at a time on high speed machinery.

A few reloaders actually get started because they wanted a load for a specific purpose that was not available in factory ammunition. An example of this might be an off-season varmint load for a large caliber big game rifle. Another example might be the need for a highly accurate varmint load, to score hits on tiny targets at ranges that would exceed factory ammo capability by a hundred yards or more. This is also true for the long range big game hunter. Most factory ammo is capable of about two inch groups at 100 yards, though the premium types often do better. This converts to a theoretical (but seldom achieved) 6" group at three hundred yards. This is not good enough for the serious varmint hunter. Someone wanting to take a 350-yard shot at an elk, goat, or antelope will find that ammunition which does no better than 10" at that range is justifiable reason to question the worth of factory ammo for the intended purpose.

Long range shooting requirements almost always demand the use of reloaded ammunition.

Factory ammo takes a lot of game each season, but when it comes to long range shooting it takes accurate reloads to insure clean kills. Only ammo that has proven highly accurate should be used for shots past 200-yards. The shooter who needs long range accuracy will usually be satisfied only with top quality reloads. Thus, in addition to saving on the cost of each round of ammo, the need for accuracy is a real part of wanting to get started in reloading.

If the savings and accuracy requirements are to be met, the reloader should select durable equipment, tools and accessories that will allow for accuracy at a level better than factory ammo. Equipment that breaks down, falls apart, or fails to perform as hoped, will be costly as it will soon have to be replaced. It is a certainty that the cheapest equipment is not the best road to follow with respect to a reloading venture. Start with reasonable equipment and it will provide enough loaded ammo to wear out a half dozen barrels and ensure a lifetime of extensive shooting.

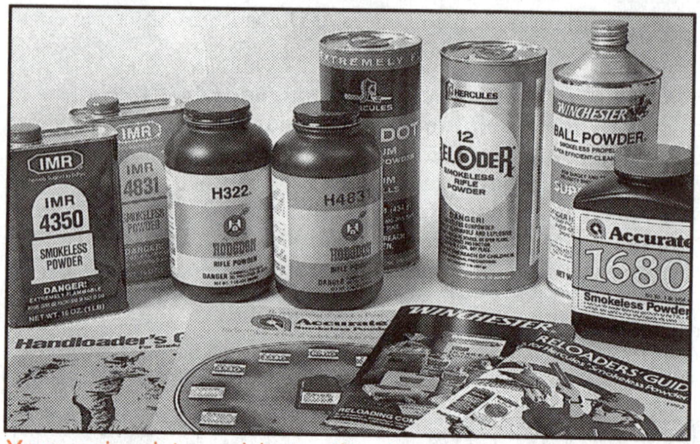

Your major data and instructional source should always be supplemented by one or more other data sources. The free information published by the propellant manufacturers is an ideal secondary source of data.

There are certain basic items that every reloader must have. You already have at least one of these - a basic reloading manual. The complete list of necessary equipment is as follows:

1. Primary data and instructional source
2. Secondary data reference (often from propellant manufacturer)
3. Reloading press
4. Shellholder
5. Reloading die set
6. Case neck brush
7. Case lubricant (not required when using a carbide die)
8. Case lubricant pad
9. Priming unit
10. Primer tray
11. Powder scale (graduated in grains avoirdupois)
12. Powder trickler
13. Powder funnel
14. Two loading blocks
15. Dial indicating caliper (graduated in 0.001" increments)
16. Case trimmer with pilot
17. Deburring tool
18. Cartridge headspace and length gauge
19. Chamber headspace gauge

These 19 items are absolute essentials. To omit any of them will insure a reloading operation that is not 100% safe.

Even the seemingly insignificant powder trickler is essential. The last few granulations of propellant can be added to the scale by using a spoon and sprinkling on the required powder. This approach is less than exact and can lead to errors made in haste. It has happened.

Using only one loading block, instead of two, will be a small cost savings but it will eliminate a safeguard against multiple powder charges in a single case or no powder charge at all in another. When adding powder, an empty case is removed from one block and charged with powder. The charged case is then placed into the second block. A seemingly simple thing such as this can be a major asset in safety. A reloader who uses a two block system is taking an important step to avoid double charged, or no-charge problems. Two loading blocks can also help prevent skipped steps in the reloading process.

A brief discussion of each listed essential tool is appropriate at this point.

1 & 2. Data And Instructional Sources

A secondary data source is always recommended. Looking up a load in two sources helps insure that you did not read the load wrong in the primary source. It also will help establish that no inadvertently published error will be used as a guideline for reloading. While every possible precaution is taken, and published errors are rare, it can happen. It is possible that a specific set of components, tooling and procedures could result in test ammo producing differing pressures to those listed in the data source. Determining that two separate sources list similar data helps assure that no errors will be made. The second source for data is ideally the manufacturer of the propellant to be used.

Remember that a data source is only as good as you are a reader. If you do not take the time to read all the pertinent material you may well find reason to regret such poor judgement. Data tables are only a part of the knowledge needed to assemble safe ammunition.

Since you have this book you already have one of the finest informational sources available. However, if you do not read it completely, but only as necessary, it will be of limited value. If you use this publication solely to look up "maximum" loads and then proceed to load them, you are indeed an accident waiting to happen.

3. The Press

The press selected is the foundation for the loading bench. Most shooters correctly settle on a well made single station press having a "C" or "O" shaped frame for their first loading tool. Such tools offer the most precise alignment of dies and shell holder as the die station hole and ram support are machined in a continuous single operation. Since there is little to wear out with respect to moving parts or alignment, such tools generally offer a lifetime of use without need for replacement. The Lyman Orange Crusher Press is an example of this type loading tool. It has the added strength of an "O" frame (compared to a "C" frame tool) and has a work opening that is 4-1/2 inches high. This is large enough to handle any normal reloading effort, unless you are looking to load 50 caliber Browning Machine Gun cartridges.

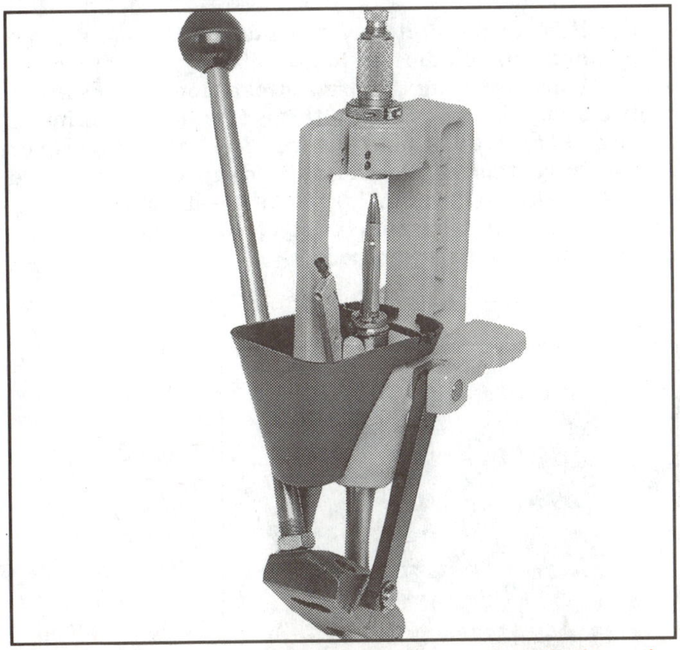

The Lyman Orange Crusher is a large, heavy-duty, single-station press that is built to last a lifetime. Its "O" frame construction gives it added strength as compared to a "C" frame press.

The Lyman Orange Crusher press package includes a priming unit and a primer catcher. Lesser priced tools are available for the shooter on a tight budget. The Orange Crusher is a universal tool as its operating handle can be set up for left or right handed manipulation.

All good reloading presses allow for the necessary mechanical advantage to squeeze (resize) fired cartridge cases back to factory-like dimensions. The ease of reloading on the Orange Crusher is a reflection of the quality of the tool.

Shell holders must be made to exact tolerances. An incorrectly dimensioned shell holder can cause a dangerous altering of bottleneck cartridge headspace.

4. Shellholder

This is a simple device designed to allow the cartridge case to be aligned with the various dies used to reload it. If it is not properly dimensioned it may fail to grip the case securely enough to pull it from a full length resizing die. The shellholder must be exact to properly align the case and primer arm during the sensitive repriming operation.

5. Reloading die set

Die sets may consist of two, three, or four separate dies. The first is a resizing die designed to return the fired and expanded case back to factory-like dimensions and it usually removes the fired primer. Without appropriate resizing, cases will not necessarily rechamber without difficulty. Some sizing dies will resize only the neck portion of the case. These are useful for special purposes which will be discussed later in these pages.

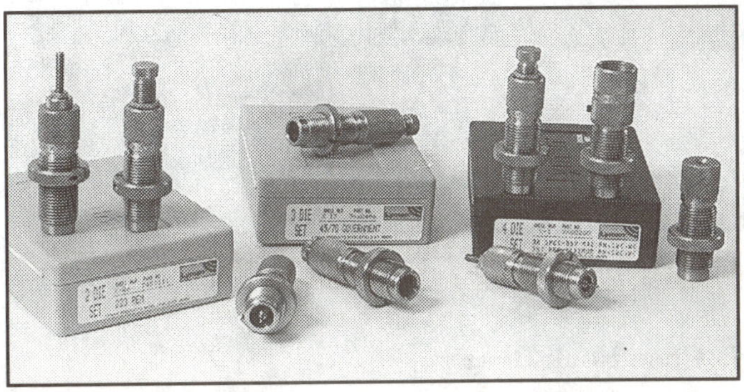

Reloading die sets come in many configurations including sets having 2, 3, or 4 dies.

Sizing dies for bottleneck cases may also contain the expanding button used to insure that the inside of the case neck is of proper dimension to accept and hold the bullet. Straight walled cases must be neck expanded in a separate die. This means that all straight cases will require a three die set.

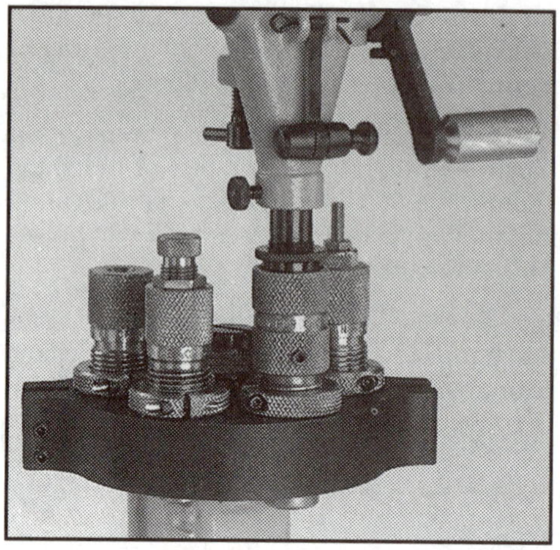

Lyman's 4-die carbide pistol set includes a taper crimp die and allows powder charging directly through the neck expanding die, a feature that saves one complete step in the reloading process.

The bullet seating die does the job for which it is named. It can also crimp cases to the bullet when required such as with ammo loaded for use in most revolvers or in guns with tubular magazines. Ammo for use in pump or semiautomatic rifles is also normally crimped. Some shooters find that crimping is more precise and less nettlesome if done separately from bullet seating. Therefore a fourth die is sometimes used. Taper crimping (as used for some reloading of handgun cases which headspace from the mouth) is always done in a fourth die.

A worthwhile variation of the standard sizing die is the type that uses a carbide ring to resize cases. Such dies are generally available only for straight cased handgun cartridges. The use of a carbide die eliminates the need to lubricate cases before resizing and also the need to wipe all the lubricant from the case after resizing.

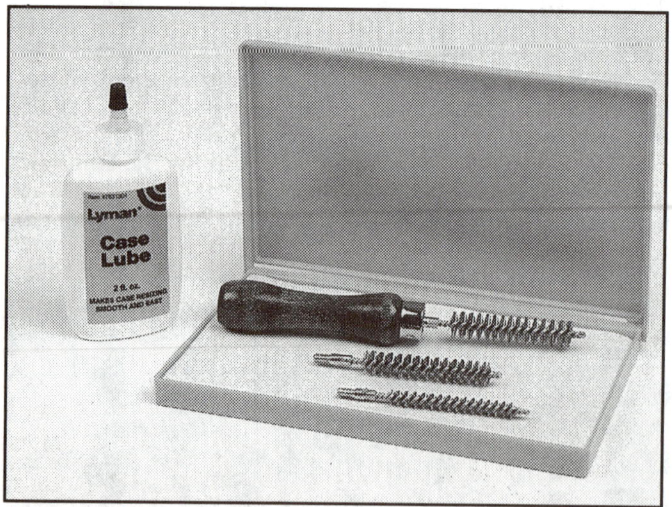

A case neck brush, case lubricant, and a lubricant pad are each important to properly assemble ammo.

6. Case Neck Brush

These handy items are used to rid the inside of the case neck of accumulated carbon and other powder burning residue. The 3 or 4 brush strokes used to wipe the accumulated crud free will greatly reduce the friction of the expanding button when being withdrawn from the case. This can help reduce case neck stretching and eliminates any need to lubricate the inside of the case neck, an accuracy destroying practice. Bronze brushes work best.

7. Case Lubricant

Fail to lubricate a case before it is resized and you are almost always going to have the case jam solidly in the die. Without lubricant, when force is applied to the press handle to free the case from the sizing die, its rim will be torn off and the case will remain frozen solidly in the die. Some reloaders will lube only every second or third case, depending on fired case dimension compatibility with the sizing die. These are the same folks who buy a lot of stuck case removers and replacement sizing dies. To avoid grief lubricate every case, or where appropriate, purchase a carbide ring resizing die to eliminate the need to lubricate.

8. Case Lubricant Pad

Some folks simply put a tiny bit of lubricant on their finger and lube each case by sliding their fingers over them. However such a practice invariably results in getting too much lubricant on the case, inside the case, or on loading tools and accessories that should not be contaminated (such as primer seating equipment). Keep lube where it belongs and only in the required amounts - use a lubricating pad to transfer sizing lubricant to your cases. Lyman also makes a case lubricant in a handy spray can applicator.

9. Priming Unit

Priming units, often called priming punches, priming rams, or priming arms are an important part of the loading equipment. Various models are equipped with priming tools that accomplish the task of seating new primers quite well. The new primer is often seated at the completion of the down stoke as the sized case is withdrawn from the resizing die (or in the case of a straight walled case as it is withdrawn from the neck expanding die). The shooter interested in the most uniform primer seating possible may wish to replace standard priming units with one of the Ram Prime Die systems.

The Ram Prime units are mounted at the normal die station and can be adjusted to give a very uniform primer seating depth by using the press stroke stop as a means to control primer insertion depth (this is the very best priming method). For those who like to use the "feel" method of primer seating (various primer seating methods are discussed in the chapter dealing with primers) the "at-the-die-station" position of this unit transmits a lot of additional "feel" not available in some other methods.

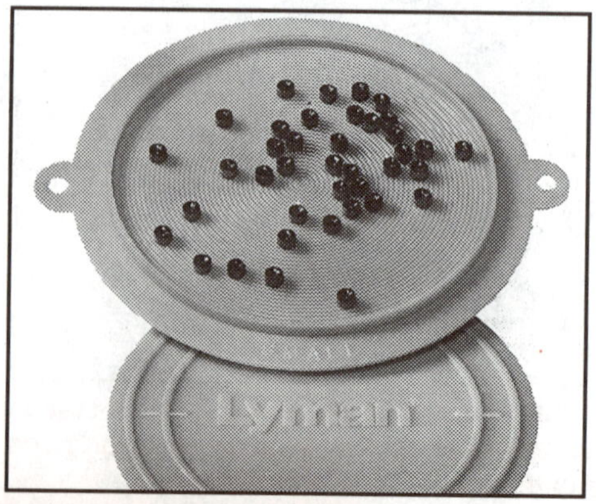

There are several style priming units available. A primer tray is also required as basic equipment.

10. Primer Tray

A primer tray is used to orient all primers one side up to insure that as each one is picked up and placed into the primer post it is correctly oriented. A primer seated upside down puts a damper on the successful firing of a cartridge. This unit will help eliminate such a possibility. It will also help keep primers free from any contaminants.

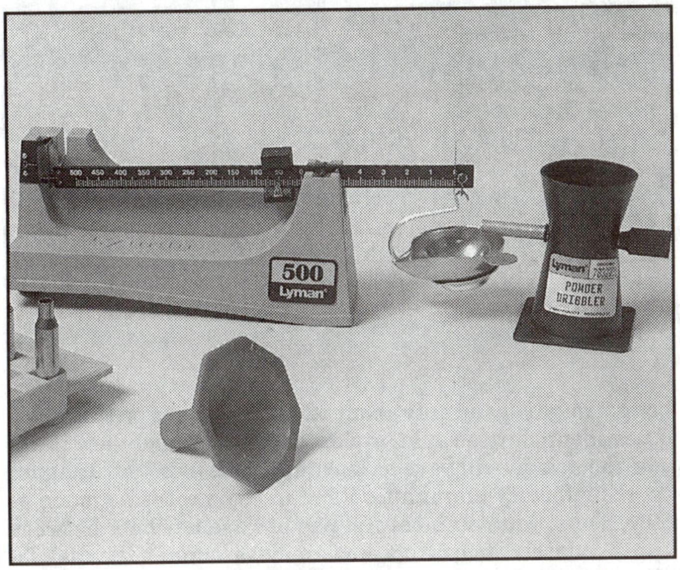

A powder scale, trickler, and funnel are each essential to any reloading effort.

11. Powder Scale

Obviously, the only way to insure that a precise amount of powder is placed into each case is to weigh every charge on an accurate scale such as the Lyman Model 1000, Model 500 or AccuScale. Powder scales are also required for the correct adjustment of any powder measure. The Lyman Model 1000 has a 1005 grain capacity and is the best choice for someone who will need to weigh shot charges for shotshell loading, arrows, or heavier than normal bullets. Both the Model 500 and the AccuScale have a capacity of 505 grains, enough for the reloading of most metallic cartridges.

For the reloader wanting the last word in powder weighing equipment, there are electronic powder scales. Lyman's LE-1000 is one such unit and has a capacity of 1000 grains, digital display, and automatic tare weight compensation.

12. Powder Trickler

As stated earlier these make an otherwise nettlesome job go quickly and accurately. The trickler can add powder at a one-granulation-at-a-time rate to the powder scale pan to bring a propellant charge exactly to weight.

Powder tricklers are available in hand turned models for about ten dollars or in the superb electric types for about forty dollars. The electric types are not suitable for use in conjunction with delicate electronic powder scales.

13. Powder Funnel

Without a powder funnel there is no practical way to get the propellant from the scale pan into the case. A simple but effective funnel will allow use with cases as small as 22 caliber and as large as 45 caliber. Lyman's funnel will do this and is designed not to roll away when it is placed on the bench.

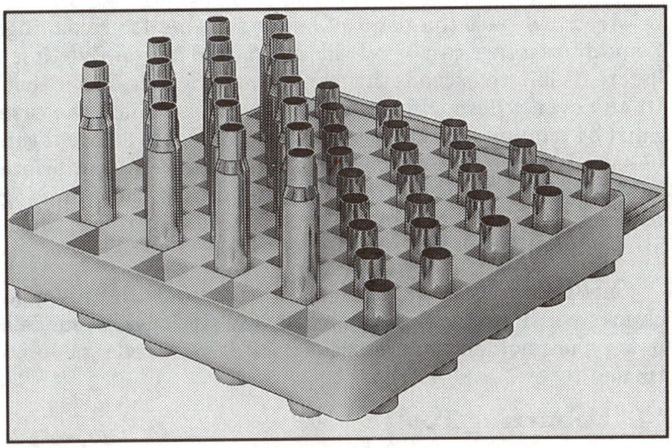

Loading blocks are indispensable in maintaining an orderly progression of the reloading process. Two loading blocks can be a safety enhancement when assembling ammunition.

14. Loading Blocks

Loading blocks will hold the cases on the bench in an organized manner - in either an upright or inverted position. The safest reloaders use two loading blocks. As an example: one block is placed to one side of the tool holding the primed cases in an inverted position. As cases are picked up and charged with powder they are placed in the other loading block in a right side up position.

Lyman's loading blocks will accept most cartridges from 22 to 45 caliber.

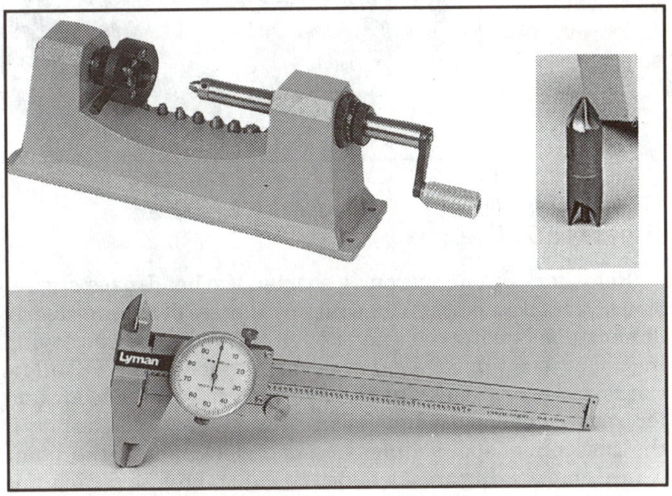

A case trimmer, deburring tool, and a caliper are "must-have" items for every loading bench.

15. Dial Indicating Caliper

This is a very essential tool. It will allow the reloader to measure the depth of seated primers, to check case lengths before and after trimming, to maintain an appropriate cartridge overall length and to discover the source of many difficulties. Calipers are available as inexpensive plastic types or stainless steel models that sometimes outlast the reloader.

16. Case Trimmer

Cases stretch when they are fired. Case bodies are often shortened when resized but their necks tend to stretch during this process. After as little as one firing and sizing, a case may

A selection of quality powders, bullets and primers, such as these shown above, can be sourced from the data section of this manual.

be stretched past the point of being suitable for reloading. Sometimes a case can be reloaded and fired four or five times before its length exceeds the maximum allowable dimension. In any event there will always be a point at which the case must be trimmed back to a workable dimension before it can be reloaded. Trimmers require the use of a pilot of appropriate caliber to keep cases positively aligned during the trimming process. Some case trimmers require the use of an appropriate shell holder.

Case trimmers are available from Lyman with universal chucks which accommodate almost any size case, in models using shell holders, in an electrically driven model or hand turned types.

17. Deburring Tool

After trimming, a case will have a lot of burrs on the inside and outside of the case mouth. A deburring tool quickly gets the case smooth and ready for reloading. Cases should be deburred for the first reloading, even if they are not trimmed.

Headspace gauges for both cartridge and chamber are two of the items required to keep ammo assembly at a 100% level of safety.

18 & 19. Cartridge Case And Chamber Headspace Gauges

Reloaders can dangerously shorten the headspace dimension of a bottleneck cartridge case by the use of an improperly dimensioned sizing die and/or shell holder, or by employing improper techniques. A cartridge headspace gauge will allow the reloader to quickly inspect a resized case or loaded round for correct headspace dimension. This gauge will also show if the case's headspace dimension has elongated to the point where chambering the loaded cartridge might be difficult or impossible. Some cartridge headspace gauges can also be used to visually reference minimum and maximum case lengths.

The use of cartridge headspace gauges on a regular basis might also enable the reloader to detect a load that was producing abnormally high pressures.

Firearm chamber headspace is not something that normally increases. However, if a load produces excessive pressure, the headspace of any firearm can lengthen slowly or rapidly past the limits of safety. Chamber headspace gauges might allow such a condition to be detected before a dangerous condition can occur.

The cost of getting set up with the necessary essential tools will vary depending upon where the equipment is purchased and exactly what model tools are selected. A person on a tight budget might spend only about $200. The return on the initial investment can be rapid or slow depending upon how much you shoot. You will save at least 40% of the cost of standard sporting factory ammunition and in some cases as much as 80%. Thus, you will be at full pay back somewhere between approximately 30 to 50 boxes of ammo (assuming they sell for about $15 per box). The more expensive the factory ammo you use the quicker a full pay back will be realized from reloading.

Naturally, you will also require a supply of primers, powder and bullets to begin reloading. The cost of these will vary depending upon the caliber you wish to reload as well as the type of ammunition you want to assemble. In almost all instances the components used will cost you no more than half the cost of similarly loaded factory ammo. The brass case that you use repeatedly is an expensive component that gets used but once with factory cartridges. Your ability to repeatedly use this expensive component is a great part of the cost savings associated with reloading.

When you are reloading, there will come a time when you will benefit from adding to your basic tools. Accessory items are available that can help improve accuracy and/or make the loading process faster.

Some of the most useful and effective accessories include the following discussed items:

1. Powder measure and stand
2. Primer pocket cleaner
3. Case cleaning equipment
4. Primer catcher
5. Scale weight check set
6. Bullet puller
7. Remote priming
8. Turret press
9. Progressive press
10. Stuck case remover
11. Case neck turning equipment
12. Flash hole uniformer
13. Primer pocket reamer/uniformer

A discussion of each of these accessories follows.

1. Powder Measure And Optional Stand

A powder measure adds speed to the reloading process.

Powder measures allow the reloader to dispense a near exact powder charge with but a single motion. Powder measures are best used in conjunction with a scale. In this application the powder measure is adjusted to dispense a powder charge that will come as close to the desired charge as possible. Because charges thrown from a measure will vary, the measure is set slightly below the desired nominal charge so that no variation will bring any specific charge over the desired weight. Then the charge is brought into exact balance by adding the last few granulations of powder with a powder trickler. Without a measure, the reloader must purchase or fashion a scoop to place the bulk of the powder in the scale pan. Scoop use tends to be less than uniform and hence a slow process. A measure speeds up the powder weighing operation by a worthwhile amount.

Some powder measures can be clamped to a bench top or shelf. Some can be installed on the press. Some, such as the Lyman 55, can be mounted either way. Still, the most convenient use of a measure, when loading on a single station press, occurs when it is mounted in a stand specifically designed for the application.

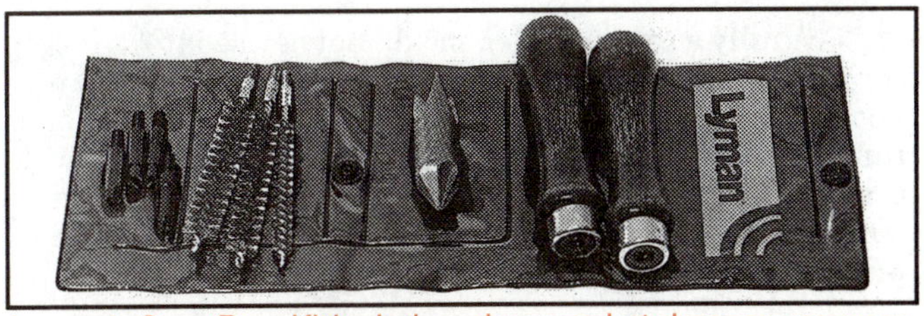

Lyman's Case Prep Kit includes primer pocket cleaners, reamers, a deburr tool and case neck brushes.

2. Primer Pocket Cleaner

As a cartridge is repeatedly reused, an accumulation of residue will occur in the primer pocket. When inspecting the primer pocket after decapping you will note a hard, crusty black deposit. If allowed to build to excessive amounts, this deposit can interfere with uniform primer seating and thus cause inaccuracy. This deposit is easily removed with a few twists of a primer pocket cleaner.

Case polishing machines can get old brass looking like new.

3. Case Cleaning Equipment

A dirty looking case unto itself will be of no consequence to the accuracy or ballistic worth of a loaded cartridge. But a dirty case might keep a defect from being noticed by the reloader. For example, the first sign of pending incipient case separation might go unnoticed on a dirty case, and the case might again be reloaded when it should be thrown away.

Dirty cases can scratch dies and cause premature die wear. A dirty reload is not something that instills personal pride or the admiration of others.

A careful reloader can maintain a quality appearance for assembled reloads by wiping each case individually. However, the finished product won't have the gleam and good looks of factory ammo. A case tumbler can take some very dirty brass and return it to looking like new while the reloader tends to other matters. At one time case cleaners were for the most part quite expensive and remained a luxury item. But today case cleaners are available to fit almost every need with respect to the quantity of cases to be processed and the wallets of almost every reloader. For example, the Lyman Tubby

Tumbler is quite inexpensive and it will process up to 125 handgun cartridge cases at a time.

The use of a case polishing machine also requires that the cases be decapped before putting them in the tumbler. A Universal Decapping Die will allow the reloader to decap almost any cartridge from 22 to 45 caliber without the need to first lubricate the cases.

Naturally a case tumbler should not be used to "shine-up" corroded brass. Any case with corrosion should be discarded. Loaded ammunition, and primed cases, should never be tumbled. There is a risk of the primer pellet or powder granulations breaking down and causing unintended problems. It is also possible that ignition of the loaded ammo or primed cases could occur.

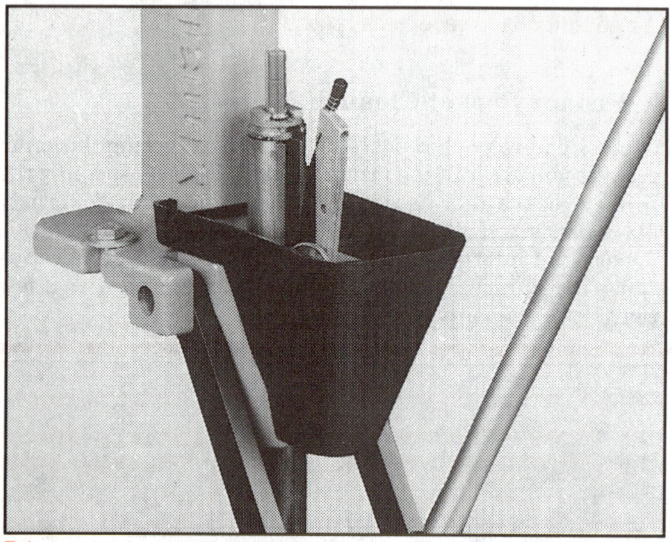

Primer catchers will keep the floor and loading bench free of decapped primers and the gritty residue that accompanies them.

4. Primer Catcher

Primer catchers are little trays that fit onto the loading press to catch the decapped fired primers instead of letting them fall to the floor. Many reloading presses come with primer catchers included in the price of the tool. A catcher will keep you from tracking primers all though the house after they become imbedded in your shoes.

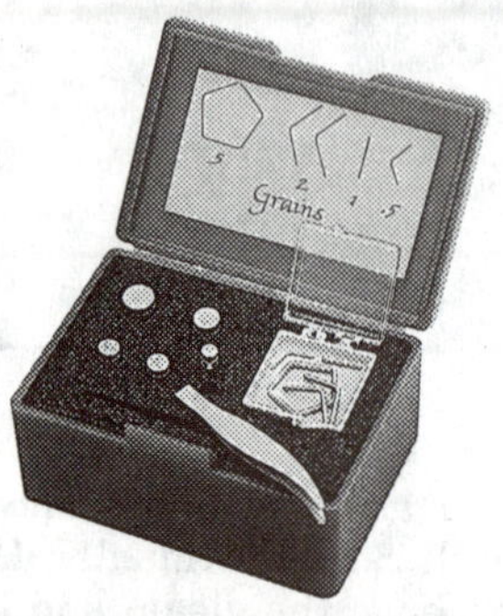

A scale weight check set will enable the reloader to verify the accuracy of a powder scale from time to time.

5. Scale Weight Check Sets

Scales are delicate instruments. When a scale receives an unplanned jolt, or worse, the logical precaution is to determine that its accuracy has not been compromised. Without a scale weight check set there is no realistic way to accomplish the needed evaluation without returning the scale to the manufacturer. Unless you are the rare person who never subjects a scale to any form of abuse, sooner or later a weight check set becomes the only alternative to factory inspection of the scale.

6. Bullet Puller

Owning a bullet puller could be misconstrued by some as an admission of making potentially dangerous mistakes while reloading. Why else would one need to pull bullets from loaded rounds?

The truth is a safe reloader will not need to disassemble ammunition and reclaim the powder and primed case. (Seldom are pulled bullets left undamaged to a point where a serious reloader would want to reuse them.) However, in the search for the most accurate ammunition possible, it is likely that certain assembled loads will not perform as hoped. There is certainly justification to then pull the bullets from this ammo to reclaim the propellant and primed cases. Should an error be realized before leaving the loading bench, the time spent pulling bullets will highlight the need for a better loading system (see the chapter on reloading safety). It is also possible to pull bullets that are seated too deeply when setting up the bullet seating die.

The best and safest bullet pullers are those that use a collet mounted in a bench press. Pulling bullets in this manner supplies ample leverage via the press linkage to make the job easy and avoids the need to subject loaded ammunition to the high force impacts of inertia type pullers.

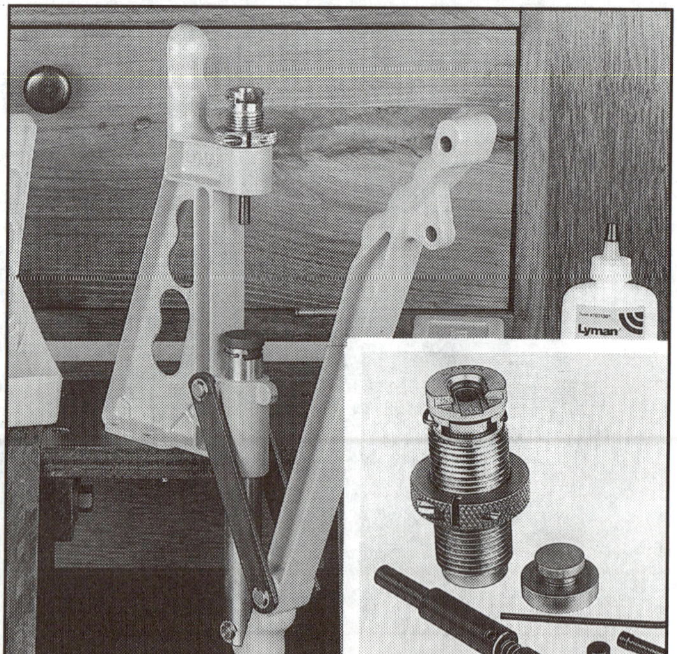

Remote priming is a popular option with many reloaders. Shown is the Lyman Acculine press with a Ram Prime Die system.

7. Remote Priming

Many reloaders find that remote priming is both convenient, free of potential primer contamination, and precise. Instead of priming on the principal loading tool, the use of an inexpensive and remote press set up with a Ram Prime system is becoming quite popular.

There are also remote priming methods which use small hand held tools. These were originally designed for benchrest shooters. Additionally, there are small bench mounted tools which are designed solely for the purpose of remote priming.

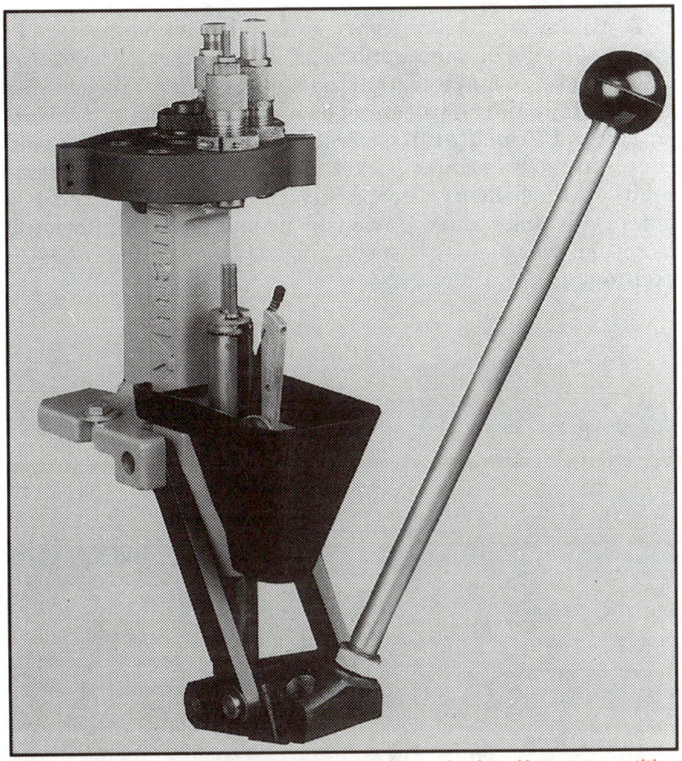

Turret presses are popular with reloaders who load large quantities of ammo. Shown here is the Lyman T-Mag with removeable turret.

8. Turret Presses

Turret presses have more than one die station, usually three to six. Such tools have a die station head which can be rotated to bring each die station into alignment with the shell holder. The big advantage to these presses is that a reloader can place a case into the shell holder and fully load it, without taking the case off the shell holder, by simply rotating the necessary dies and powder measure into position over the case. This method of reloading is often referred to as semi-progressive. This method has a speed-up advantage and is most often employed by handgun cartridge reloaders as they often shoot great quantities of ammo in a single range session. On some turret presses it is possible to load as many as two hundred handgun cartridges per hour.

Turret presses also allow the reloader who loads cases by the usually preferred batch method to set up all the dies and leave them in position on the press. Depending on the type of dies used, and the specific press selected, as many as three sets of dies can be installed in a turret. The Lyman T-Mag press, for example, has six die stations allowing for the use of up to three 2-die sets. Extra turrets are available for this press. This allows the reloader to keep as many die sets installed in turrets as desired.

9. Progressive Presses

Progressive presses can differ in basic format. What they all share in common is multiple die stations and multiple case stations. Each will produce a loaded cartridge for one complete stroke of the press handle, up to 500 or more shells per hour.

Progressive reloading has disadvantages which outweigh its speed for all but those who use very large quantities of ammunition. Even then, extraordinary care must be exercised when loading. As a result progressive presses are less than ideal for the majority of reloaders.

10. Stuck Case Remover.

Eventually, most reloaders will forget to lubricate a case, or will simply use too little lubricant causing the case to become solidly jammed in the sizing die. A reliable method is required to remove a stuck case from the die. Effective stuck case removers come in several varieties. One type requires the drilling and tapping of the case base. This method works but is not as easy or as quick as some might like.

Another method uses simple parts which allow the withdrawal of the decapping rod assembly from the die, after the jammed case is driven out of the die with a steel rod and hammer. This is fast, easy and effective - if you are using dies that allow the withdrawal of the decapping rod/expander button assembly through the top of the die.

11. Case Neck Turning Equipment

Serious accuracy buffs, such as benchrest shooters and varmint hunters, have found that neck turning, which makes the inside and outside of the case neck concentric, improves accuracy. The turning of case necks can lead to neck sizing problems that cannot be solved with standard dies. Neck turning for enhanced accuracy is best undertaken when using custom chambered rifles and neck sizing dies with custom inserts.

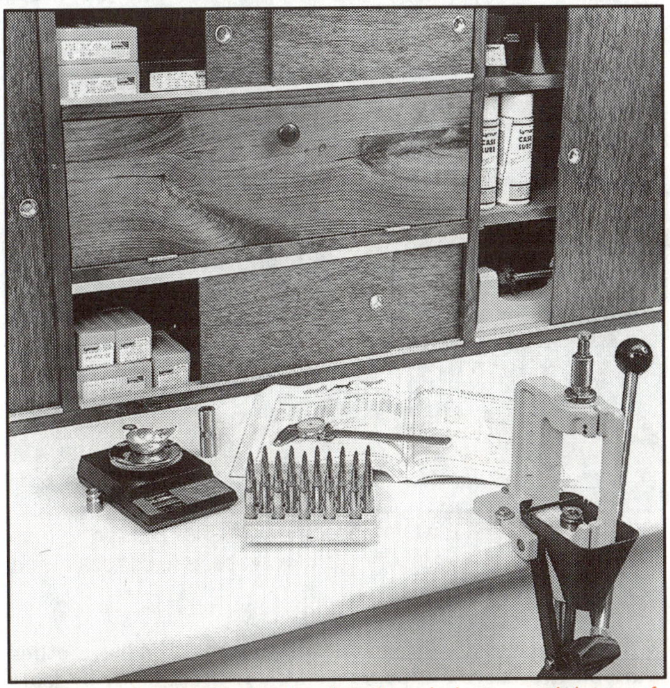

A neat and organized reloading bench is essential to safe reloading. Powders and primers should be stored separately, away from the loading bench.

Neck turning is often essential when resizing cases to a different caliber. In this instance, necks are turned only to keep cartridges at or slightly below the maximum allowable diameter and no special sizing dies are required.

Neck turning is generally performed only by experienced reloaders. More on turning necks is discussed in the chapter dealing with accuracy.

12. Flashhole Uniformer

This little handheld reamer will bring all flash holes to a uniform diameter and remove the burr at the terminus of the flash hole. This is an accessory that was originally used by serious bench rest shooters. Today many handloaders feel it worthwhile to ream flash holes in their search for accuracy. The use of a flash hole reamer is discussed in the chapter dealing with reloading for accuracy.

13. Primer Pocket Reamer/Uniformer

Primer pocket reamers are required to remove the crimped material on military and some para-military style cases. Such cases have the primer firmly secured by the displacement of brass around the primer pocket. These cases often require a special heavy duty decapping rod assembly to remove the crimped in place primer. After removing the fired primer a new primer cannot be safely or correctly seated unless a reamer is used to remove the crimped material.

A primer pocket uniformer is used by many shooters to insure that the diameter and depth of every primer pocket is uniform and to make certain the bottom is flat. The benefits of this additional effort cannot be realized until groups are at or below 1/2 inch at 100-yards. It requires care to avoid overcutting the primer pocket as it can cause gas leaks at ignition. Such leaks can quickly ruin a bolt face and are potentially dangerous. Obviously, primer pocket uniforming is not for everyone. (For more info see chapter dealing with accuracy.)

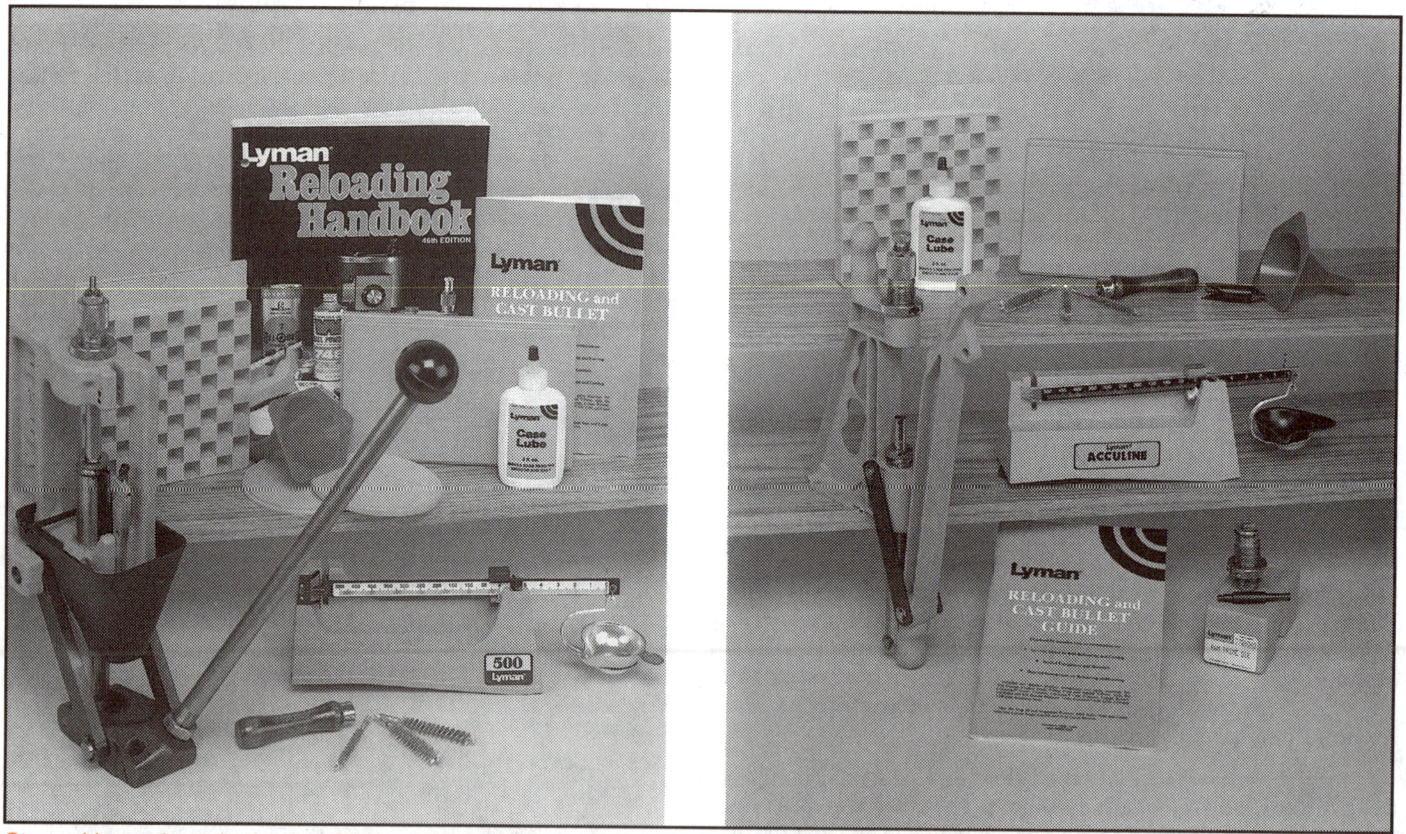

Starter kits such as the Lyman Orange Crusher Starter Kit (left) or Acculine Starter Kit (right) offer most of the essential tools needed to get started in reloading including a press, powder scale and funnel, case lube kit and loading block.

As time goes by, the reloader will find that the list of other available accessories appears endless. Let real needs or goals dictate which ones might fit your circumstances.

Getting started requires only the previously discussed nineteen items. Indeed one could load hundreds-of-thousands of rounds with the addition of a few of the listed accessories, a powder measure and primer pocket cleaner being the most likely candidates. After this, case cleaning equipment will add a lot of pride and satisfaction to reloading. From this point on, it should be obvious when you will actually benefit from an addition to your basic tools. Lacking an obvious need, chances are you will get along nicely without further additions.

The how-to of component selection will be covered in the appropriate chapters of this manual.

CHAPTER 3
HOW TO RELOAD RIFLE CARTRIDGES

The steps of reloading rifle cartridges are relatively simple operations. As the reloader becomes involved, it will be obvious that there are many considerations to be given to each step. Pertinent in-depth discussions will be undertaken in the appropriate component chapters. These are must reading. The following discussion is geared to the basic essential operations and to caution in areas of particularly frequent problems. Because of the many similarities between the reloading of rifle and handgun cartridges, the following chapters will have a great deal of similar material.

Reloading's mechanical steps essentially come down to removing a fired primer, resizing the case so as to insure it will chamber and hold the bullet firmly, replacing the primer, replacing the propellant, and seating a new bullet. The actual practice is more complex, but not overly so, and does require a few considerations. The proper procedure is outlined in the following twelve steps:

1. Selecting a load and components
2. Case inspection
3. Inside case neck brushing
4. Case cleaning and lubricating
5. Case resizing and fired primer decapping
6. Lubricant removal and secondary inspection
7. Case length measuring
8. Case trimming and/or deburring
9. Primer seating
10. Weighing and charging powder
11. Bullet seating (and crimping if required)
12. Final inspection

We will discuss each step as a basic function. Additional information will be found in many of the following chapters. Reloaders are cautioned to read all the pertinent material before actually beginning to reload ammunition.

Step One - Selecting A Load And Components

For the novice, choice of components is perhaps the most confusing aspect of getting started. It is no wonder many reloaders, with years and thousands of rounds of experience, still do not make appropriate selections; not because it is difficult to do so but simply because they have not been exposed to material which made the selection process easy.

To make the selection task as clear as possible, the data section of this handbook makes specific recommendations as to the propellant most likely to satisfy the needs of top accuracy.

Where no such suggestion is made (in the preamble to the data for each cartridge) there simply is no single best first choice. In these instances the reloader can select any of the propellants listed for the bullet weight and reasonably expect results to be on par with any other propellant choice. Often this situation is a result of the accuracy limitations of the firearms or limitations of the cartridge design. In rare calibers it might be a result of all listed propellants giving equally superb performance.

Uniform ballistics are normally part and parcel to the best accuracy, though at times the load providing the most uniform ballistics does not provide the very best accuracy. All accuracy loads do, however, fall well within the limits of acceptable ballistic uniformity.

The first efforts should always begin with a load listed in the "Suggested Starting Grains" column in our data. Heavier loads should not be used until the shooter has had some reloading experience and has understood all of the information in this handbook. Be absolutely certain that you use the exact bullet weight and the exact powder and charge weight shown in the data.

150 gr. Jacketed SP
3.200" OAL

Powder	Sugg. Starting Grains	Velocity fps	Pressure C.U.P.	Max. Load Grains	Velocity fps	Pressure C.U.P.
RX7	32.5	2372	40,300	37.0	2570	50,000
IMR-3031	42.0	2450	29,400	50.0	2932	48,000
748	45.0	2614	38,800	52.0	2902	50,000
H-335	40.4	2508	42,400	46.0	2755	49,300
BL-C(2)	46.0	2770	44,600	51.0	2932	48,000

The first loads assembled should always be taken from those listed in the Suggested Starting loads column of the data section. Do not use heavier loads until some experience has been gained.

Primer selection can also be kept quite simple. Naturally, a primer of the right size must be used, i.e. large rifle or small rifle sizes. The data lead-in for each cartridge clearly specifies the appropriate size.

Do not use "magnum" style primers unless the data specifically calls for these, as doing so can alter the ballistic uniformity, ballistic level, and safety of the load. Generally speaking, magnum primers are appropriate in calibers using the word magnum in their proper name, i.e. 357 Magnum, 44 Magnum, 7mm Remington Magnum, and 375 H&H Magnum. But this is not always so, for example the 224 Weatherby

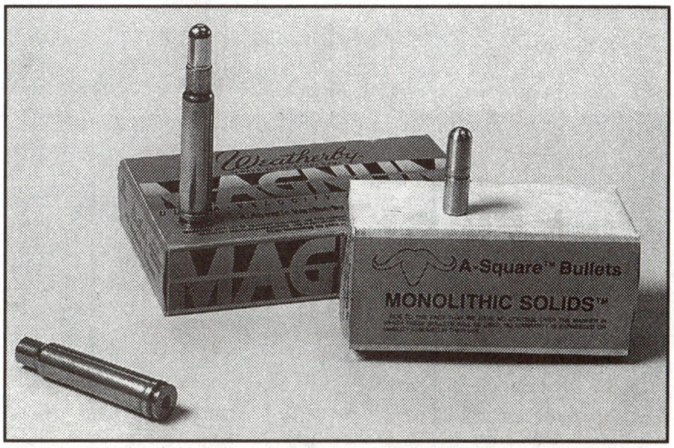

Often the reloader can purchase the same bullet as loaded in the factory ammunition that has been previously used successfully.

Magnum and 222 Remington Magnum are best loaded with non-magnum primers. And some cartridges without a magnum designation do require magnum primers for best results, i.e. the 416 Rigby. Always make primer selection based on the type of primer used to develop the data listed. This will insure that pressures are not adversely effected by primer choice.

The novice will do well to select a component bullet that most nearly duplicates the bullet that has been used as factory ammunition. For specific hunting applications the reloader may be wise to select nothing less than a premium grade hunting bullet. If doubt exists, refer to the chapter on jacketed bullets.

In some instances, the exact bullet that has been used in factory ammunition can be purchased for reloading. For example, Remington offers a large number of their factory bullets as components. Additionally, many of the bullets used in Federal Factory Premium ammunition are available from Nosler (Partition type) or Sierra.

The beginner will often have fired cases on hand as a result of saving them from the shooting of factory loads. Brass simply should be kept segregated by lots and loaded accordingly. Lots are homogeneous groups of cases from the same maker and bearing the same lot number on the box (more on this in a later chapter).

When purchasing brass cases many reloaders will buy new brass from bulk suppliers. This saves considerable money. Brass may also be purchased from the ammunition makers, packed in boxes of the same size as factory ammo. Never load fired cases from an unknown source. Reloading of brass that may have been fired in a gun having excessive headspace, or that may have been previously loaded to excessive pressure, can be dangerous. There is no way to be certain that the so called "once fired" cases being purchased are indeed so. If only one seriously weakened multifired case sneaks into a lot of brass it could be the basis of a serious accident.

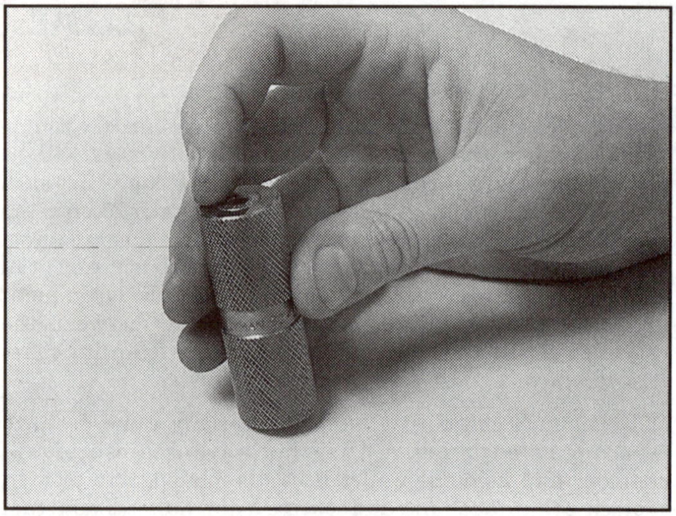

Carefully inspect each case for any defects. A case length/headspace gauge is recommended to check critical case dimensions.

After the appropriate components have been selected, set them aside in a remote area and bring them to the loading bench only when they are required.

Step Two - Case Inspection

Each case must be in appropriate condition for further use. Only your inspection can decide upon a case's suitability. After wiping each case with a clean cloth, look each one over for defects, discarding any cases which are not as they should be. To prevent accidental later use of a defective case, crimp shut the mouth using a pair of pliers.

Eliminate all cases that show any signs of deterioration, such as cracks and splits. Look carefully for any indication of incipient case separation - a bright ring partially or com-

Reloading Rifle Cartridges

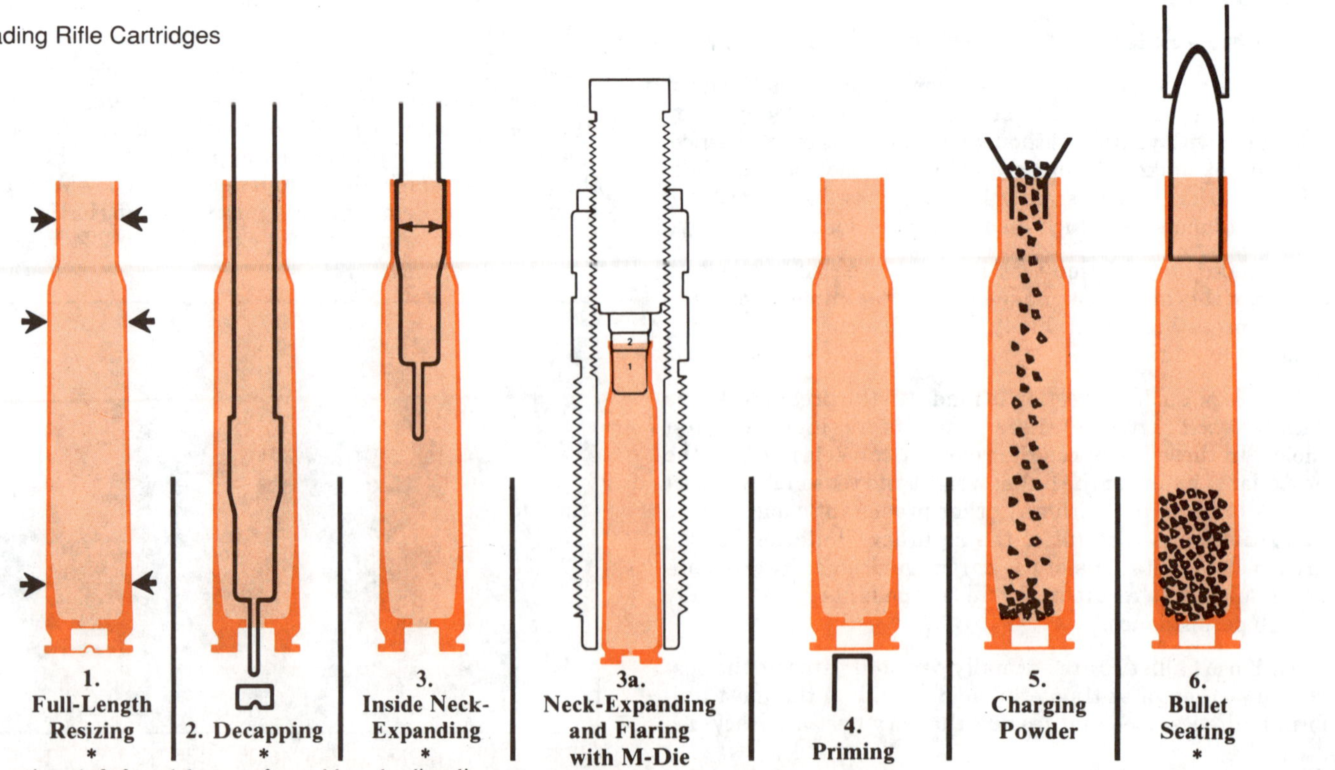

*Operations 1, 2, 3, and 6 are performed by reloading dies.

pletely around the case head . Any corrosion on the case or signs of gas leakage around the primer are other reasons to discard the case involved. For additional information on inspection see the chapter on cartridge cases.

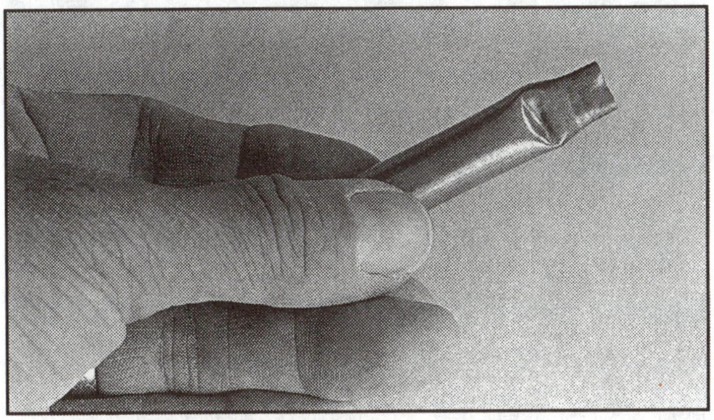

Crimp shut the mouth of any defective case using a pair of pliers. This will prevent inadvertent use of the case at a later date.

Be sure to turn the case over and allow anything that might have gotten into the case to fall free. As cases pass inspection place them into your first loading block which you should locate to one side of your work station. Put them into the block in a mouth up position.

Under typical circumstances cases may be reloaded and fired as few as two times or as much as six or more times. Be sure you understand the limitations of multiple firings as discussed in the chapter dealing with cases.

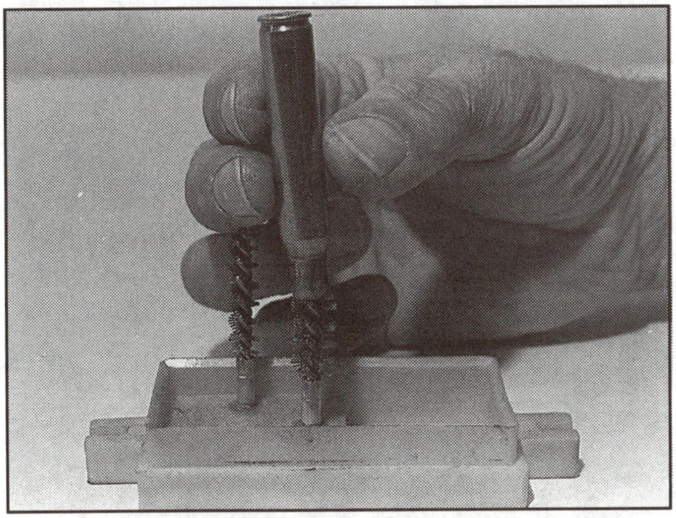

Brushing will rid the important neck portion of the case of firing residue, making it easier for the expanding button to slide in and out of the case and will also enhance bullet pull uniformity.

Step Three - Inside Case Neck Brushing

Take each case from the loading block and brush the inside of the case necks with a cleaning brush of appropriate caliber. This will loosen and remove excess carbon and other fouling. Doing this will help ease the passage of the expanding button when resizing and enhance accuracy by helping to maintain a uniform bullet pull. Transfer brushed cases to the second loading block located on the opposite side of your work station but place them mouth down so as to let the loosened crude fall from the cases.

To keep dirt from accumulating in the loading block, many reloaders prefer to tap the inverted case on the work bench. Even if you use this preferred method be certain to place the case upside down in the second loading block to offset any possibility of forgetting to do this.

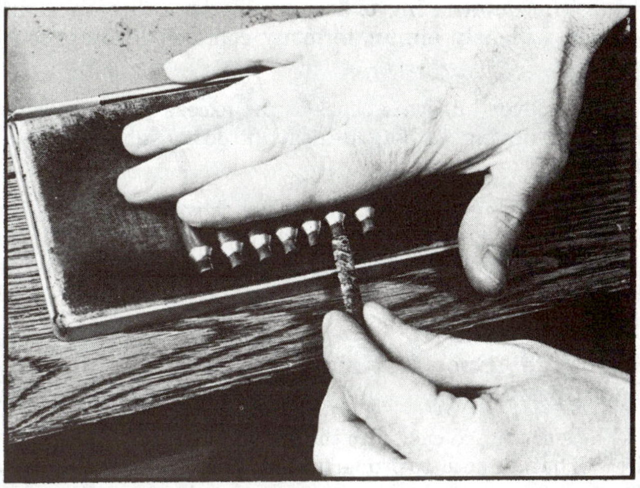

Lubricate several cases at a time by rolling them on a prelubed pad.

Step Four - Case Cleaning And Lubrication

If cases are to be cleaned in a tumbler or vibrator, such as one of the Lyman Turbo tumblers, now is the time to do it. Follow the manufacturer's directions on the use of a tumbler, being sure not to overload the cleaner and to allow sufficient time for the cases to clean properly. Cases that have been tumbled for a sufficient time with appropriate media will look almost as bright and shiny as new. When removing cases from the tumbling media be sure to tap all the media free from each case before placing it mouth down into the second loading block.

Lyman turbo tumblers will get used cases looking like new in less than two hours.

If cases are not to be cleaned in a tumbler, be certain they are free from any grime that might damage or scratch the sizing die.

Now lubricate each case by rolling it lightly across a prelubed pad. Several cases may be lubed at one time. After lubricating each case, place it mouth up in the first loading block. This alternating of loading blocks and case mouth position will help eliminate many common loading bench errors.

Do not over lubricate cases or the excess lube will cause case dents during resizing. Be neat and do not get lube into the case mouth or primer pocket.

Avoid getting lubricant on bottleneck case shoulders. Lubricant on the shoulder will cause dents to form when resizing. The neck portion of the case generally will require no lubrication when either full length resizing or when simply neck sizing.

Step Five - Case Resizing And Primer Decapping.

Place the lubed case into the shell holder and manipulate the tool handle so as to properly resize the case neck and body. Be sure to follow the die manufacturer's instructions for appropriate die adjustment. Generally, when full length resizing rifle cases, the shell holder needs to bottom firmly against the die while the ram cams past the top of the press stroke.

As the case is moved into the die the old primer will be removed by the decapping pin. The decapping rod should be adjusted downward only far enough to insure fired primer removal. If the rod is adjusted too low it and the case will be damaged when it strikes the case web.

If excessive force is needed to resize normal cases it is a sign that insufficient lube has been used.

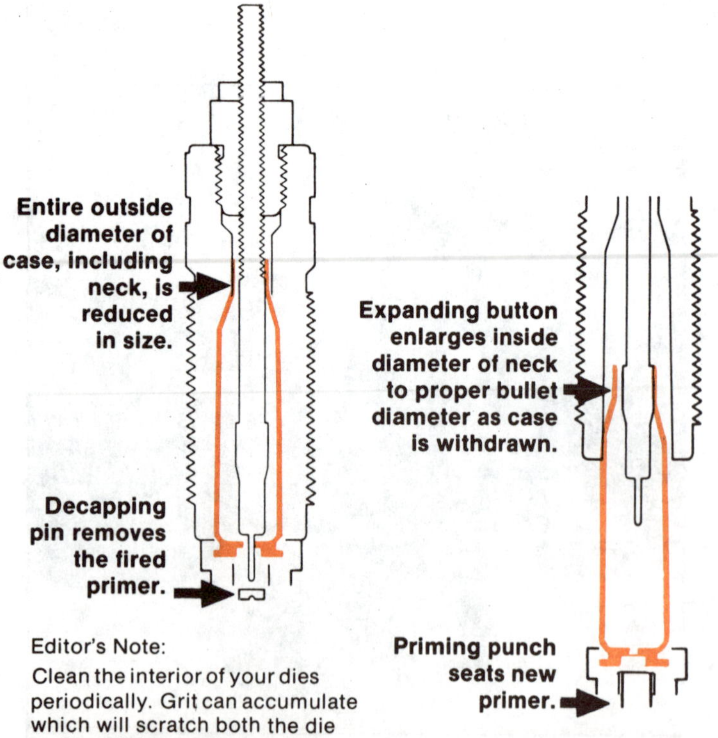

Editor's Note:
Clean the interior of your dies periodically. Grit can accumulate which will scratch both the die walls and the cartridge cases.

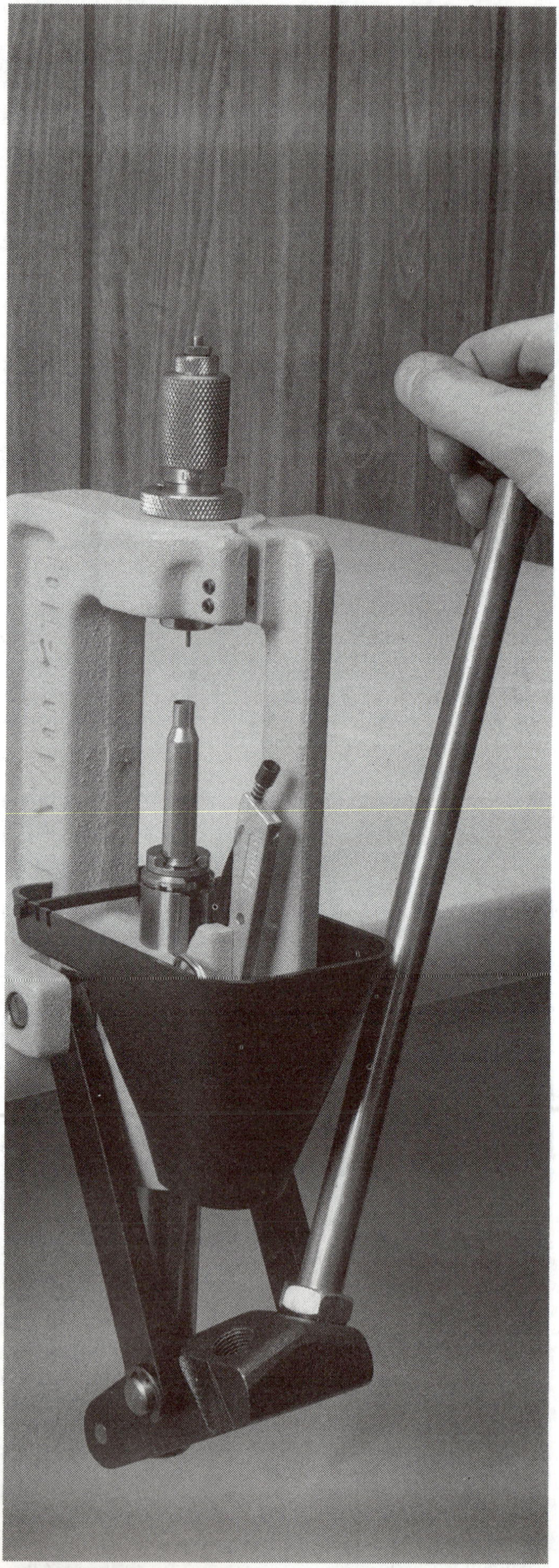

If dents form on the case, especially on the shoulder area, it means that too much lubricant is being used.

The inside diameter of the case neck will be returned to an ideal dimension to insure proper bullet retention in the finished round.

Place the resized cases in the second loading block in the mouth down position. Visually inspect each case as it is placed in the loading block to insure that the fired primer has been removed.

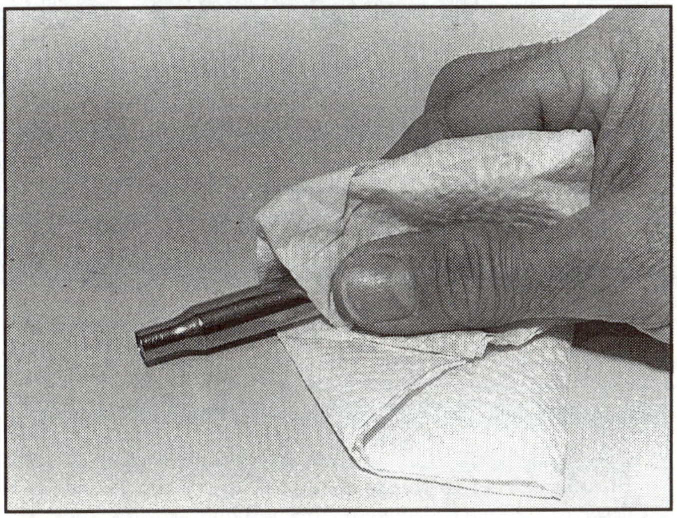

Wipe away all traces of case sizing lubricant and inspect each case for abnormalities.

Step Six - Lube Removal And Secondary Inspection

Carefully wipe away all traces of lubricant from each case using a clean cloth. As you do so inspect each case for defects. Case mouths and bodies sometimes split during resizing. It is possible that the first sign of incipient case separation will not be visible until after resizing. Also this is a backup inspection to your first one. Discard any case showing a defect, crimping its mouth closed before doing so.

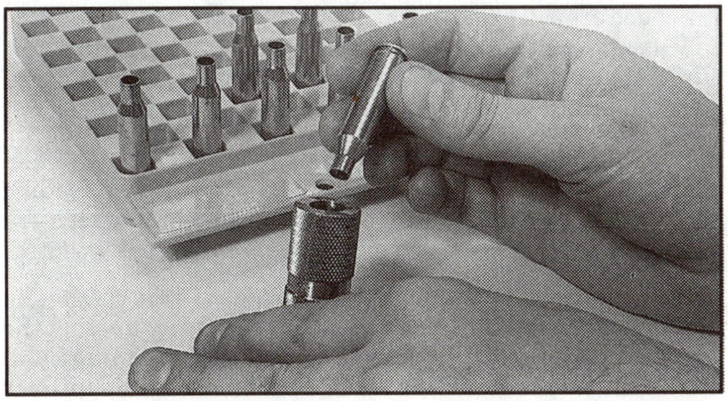

Each bottleneck cartridge case should be dropped into a cartridge case gauge after resizing and lubricant removal.

Next, drop the resized case into a cartridge headspace gauge. The case should not protrude above the maximum headspace step on the back of the die, nor should it be below the minimum headspace step at the same end of the die. If case headspace is too long, readjust the sizing die. If cases are too short they should not be used. A too short headspace dimension can be caused by a faulty resizing die and/or a faulty shell holder.

The gauging of resized bottleneck cases is a safety step overlooked by many reloader's. However, omission of this step means that the reloader is hoping everything is OK but has no assurance that it is so. This is not a very smart approach.

Remember that dirt or lube accumulated in the gauge can cause erroneous indications of a headspace measurement that is too long.

Many reloaders adjust case headspace (through sizing die adjustment) so as to best match individual chamber dimensions. More on this will appear in later pages.

After gauging, place the cases back into your first loading block in a mouth up position.

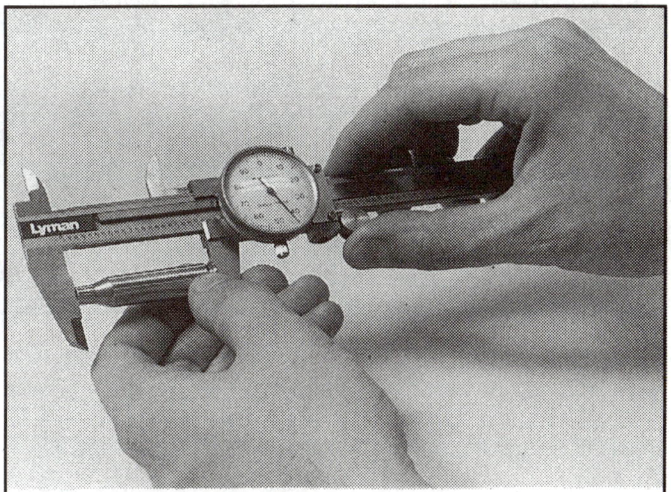

TEST COMPONENTS:

Cases	Winchester
Trim-to Length	2.484"
Primers	Winchester 8½-120
Primer Size	Large Rifle
Lyman Shell Holder	No. 2
Jacketed Bullets Used	Sierra HP #2110, 110 gr.
	Sierra SP #2120, 125 gr.
	Hornady SP #3020, 130 gr.

Carefully measure cartridge case length after resizing. Cases which are longer than the maximum allowable dimension must be shortened to the "Trim-to" length listed in the data tables.

Steps Seven And Eight - Case Length Measuring, Trimming And Deburring

Cases stretch when fired to conform to the length of the chamber. Resizing will bring the case body back to normal length but also tends to lengthen the case neck. The brass that flows forward into the neck as a result of firing and sizing will, sooner or later, make the case longer than acceptable. A too long case can dangerously increase chamber pressure and can be difficult or impossible to chamber.

Every case must be measured after every resizing. The data section of this manual shows a cartridge drawing for each cartridge. The length shown for the case is the maximum allowable length. Case length must never exceed this dimension. When one or more cases have stretched to the maximum, trim the entire lot of brass back to the listed trim-to length. This is indicated in the tabular information shown next to the cartridge drawing in the data section.

Trimmed cases must be deburred to remove rough edges left from cutting the case back to a correct length. A few quick twists with a deburring tool will remove outside burrs and another few quick twists will smooth out the inside of the case mouth. Do not over do this deburring operation.

After measuring (and trimming/deburring if required) return the cases to the second loading block in a mouth down position to be sure that no trimming residue is left inside the case.

Cases should never be trimmed more than 4 times. When a case needs its fifth trimming it must be discarded.

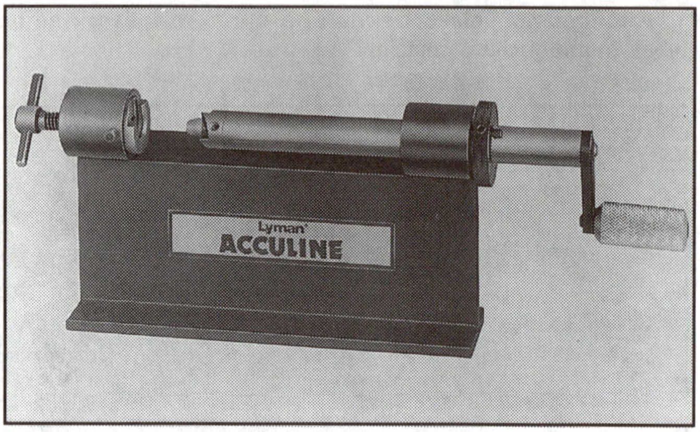

Cases are trimmed to acceptable lengths with a case trimmer such as the inexpensive Lyman Accutrimmer.

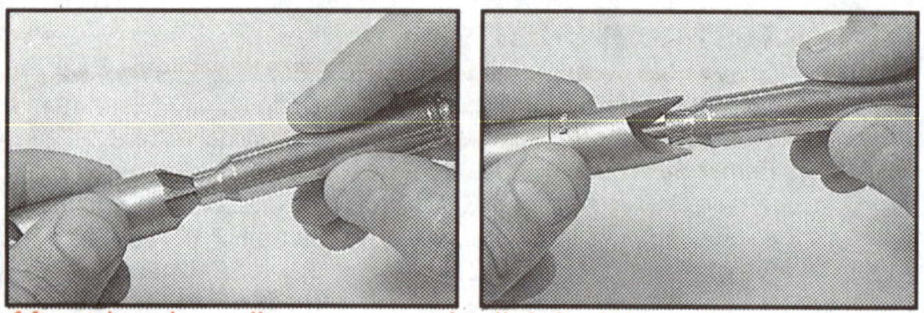

After trimming, all cases must be lightly deburred on the inside and outside of the case mouth.

Step Nine - Primer Seating

Seating primers requires great care. Primers and primer pockets must remain free from any contamination. Be sure the work area and primer tray are clean and free from any trace of solvent or lubricant. Then remove no more than one box of 100 primers from your remote storage area and place them on your bench. Carefully read the label with respect to primer size and type. Then refer to the data to insure that you have the right size primer on the bench. Now reread the label on the box. When you are certain that all is correct, assemble and adjust the Ram Prime unit in the die station according to the instructions supplied.

Now, wash and dry your hands to prevent contamination of the primer and later problems with your ammunition. This is an extremely important aspect of loading trouble-free ammunition. Do not simply wipe your hands on an old rag and proceed. Next, pour 100 or less primers onto the grooved half of your primer tray. Rotate the tray until all primers are cup side up.

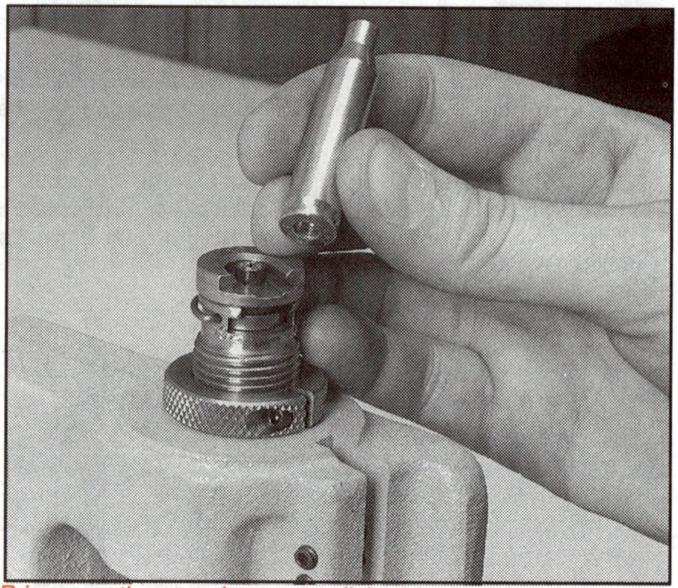

Primer seating may be accomplished in many ways and with differing tools. One of the best methods is to use a Ram Prime unit mounted in the reloading tool's die station.

Place a primer in the primer post, lower the primer post slightly, and place the case in the shell holder. Now lower the press handle to move the primer into the case's primer pocket. The Ram Prime unit should be adjusted so that the tool handle comes to a stop when the primer is seated to the correct depth.

Primers should normally be 0.003" to 0.007" below flush. Thus strive to adjust primer seating for a nominal of 0.005" below flush, which will allow for a plus or minus variation of 0.002 inch. The exception to this is when using CCI primers. With these the manufacturer suggests primers be seated to a depth of 0.002" to 0.004" below flush - a nominal of 0.003 inch.

Caution: primers seated higher or lower than recommended are a needless hazard. Check your primer operation by using your caliper to randomly measure seated primers from time to time. Naturally, the use of a caliper is essential for the initial adjustment of the priming set up.

After priming, return the primed cases to the first loading block but place them again in a mouth down position. This will allow you to see that all primers have been properly seated.

Proper priming suggestions are included in the chapter on priming. Be certain to read this information to insure that the job goes as intended. Primers require special storage and handling. This information is also contained in the primer chapter.

After priming all cases, return any unused primers to their original container. Once again read the label on the primer box to make a final verification that the correct primers were used. Finally, return the primers to their remote storage place.

Step Ten - Weighing And Charging Powder

The weighing of the powder charge is a critical aspect of reloading. Set up the scale and insure that it has been correctly leveled. It is a good practice to now verify the scales accuracy by using a weight check set.

Select the powder to be used from the data section and then remove a one pound can of it from its remote storage area. Bring the can to the bench, and again verify from the data the powder you wish to use. Read the can label once more. Double checking all components placed at the loading bench is a must for safe reloading.

Pour some powder into the powder trickler and place the trickler alongside of the powder scale pan so that powder can be fed from trickler to pan. Also place some powder in a small open container (or the powder measure if one is used).

Using a scoop, take a quantity of powder from the open container and place it on the scale pan. (Or meter a charge from a powder measure onto the scale pan and place the pan on the scale hanger.)

A handy scoop can be made from an appropriately large straight walled case such as a 30 M-1 Carbine, 357 Magnum, or 45-70. Cut the case to approximate length using a hacksaw. Final length can be accomplished with a case trimmer or careful hand filing. A handle can be fashioned from a stiff wire loop wound around the case's extractor groove.

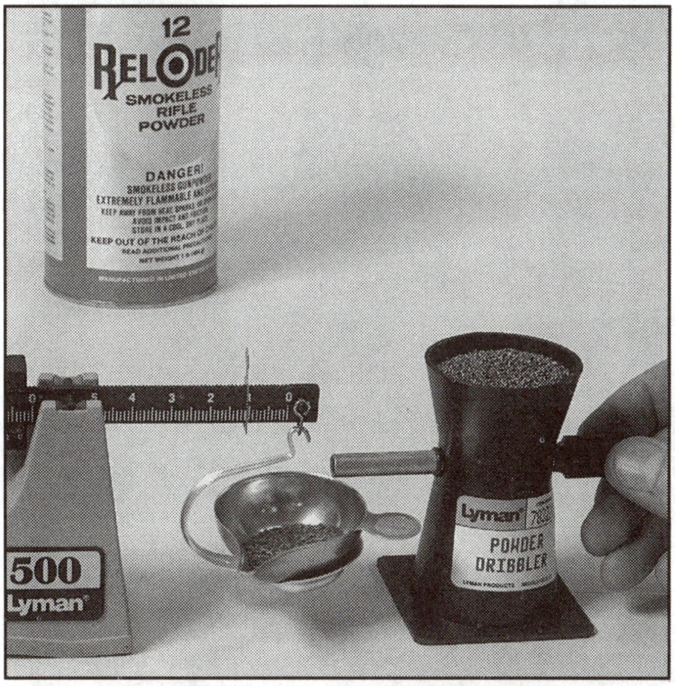

Each powder charge should be carefully weighed. The use of a powder dribbler will add speed and convenience.

Bring the scale into perfect balance by adding the necessary extra few granulations with the powder trickler. Then pour the weighed powder charge into a primed case with the aid of a powder funnel. Next, place the charged case into the second loading block in a mouth up position.

After charging all cases, carefully inspect each case for a powder charge of appropriate height. If any charges appear to be of greater or lesser amount than normal, reweigh the powder charge.

Make certain that the scale poises do not move inadvertently when removing and replacing the scale pan at each weighing. When the weighing is completed once again look up the required powder type and charge weight. Then reread the can label and again verify that the scale is adjusted properly to balance with the desired powder charge.

Return any unused propellant from the trickler, open container, or powder measure, to its original canister. Return the powder can to its remote storage area. Leave the scale on the bench.

Powder requires special handling and storage. The powder chapter contains information on this aspect of reloading.

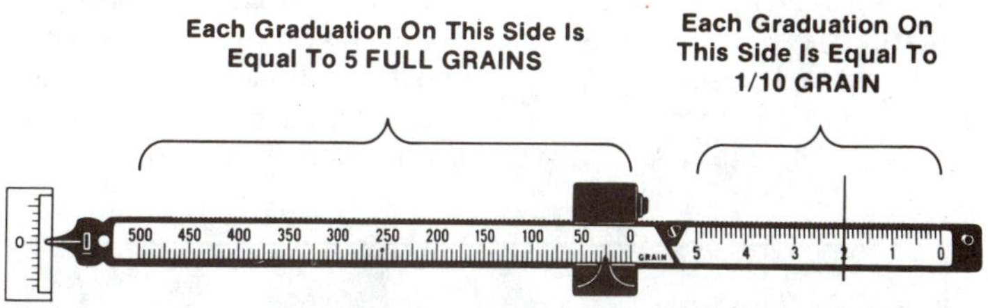

The illustration shows the beam of a modern reloading scale. Note how it is graduated on both sides of the pivot point. The scale is set by moving the two weights (poise) to the proper graduations. The large poise (on the left) is used to obtain multiples of FIVE GRAINS, while the small poise (on the right) is used for 1/10 FRACTIONS of a grain or SINGLE grains from one to five.

EXAMPLE: The illustration shows a setting of 27.0 grains. If you wanted to decrease this 1/10 grain, you would simply move the small poise one notch to the right.

Caution: Technicians in the Lyman lab have observed a potential serious phenomena involving powder scales and plastic loading blocks and/or Styrofoam packaging or other objects made of plastic material. These materials sometimes retain a static electricity charge, enough to create an electro-static field of varying radii.

This electro-static field has proven capable of causing radical deflection of uncharged and zeroed scales of all brands (available to them at the time). Powder in the scale pan tends to dampen the amount of deflection by varying degrees dependent upon circumstances.

It is strongly urged that you clear the surface of your loading bench before carefully setting up your scale. Then replace your equipment one piece at a time, paying particular attention to the scale pointer, especially as any plastic or Styrofoam item is replaced on the bench. Any item that causes a scale pointer deflection should be removed before powder weighing begins. Do this at each loading session.

The weighed powder charge is poured into the case carefully, using a powder funnel to prevent any spillage.

Charging with the Lyman 55 Powder Measure

The most efficient way to charge a large number of cases with powder is by the use of a precision powder measure such as the Lyman #55, used in conjunction with an accurate, sensitive reloading scale, like the Model 500.

First, establish the correct measure setting by adjusting the measure, throwing and weighing a charge. Please see the User's Guide that comes with the Powder Measures for complete set up instructions.

Next, when the measure is adjusted for the proper charge, tighten down all lock screws and begin metering powder into the cartridge cases.

To avoid accidentally "Double Charging" a cartridge, you should develop a foolproof system of loading. A suggested method is to place all the uncharged cases on your left. As you pick up each case for charging, turn it upside-down and shake it. This will ensure that the case is empty. Turn the case right-side-up, charge it and place it carefully on your right.

A powder scale is essential to verify the correct powder measure setting.

Important! Check every tenth charge on the scale to ensure that the proper weight of powder is being thrown. Minor variations of several tenths of a grain may well be due to inconsistent operation of the powder measure. Strive for a smooth, consistent operating technique of the measure and the consistency of your charge weight will improve.

Unless you are using a charge that is the absolute maximum for your rifle, a charge-to-charge weight variation of several tenths of a grain will have no practical effect on accuracy. If, however, your charge is a fraction of a grain from absolute maximum, there is a better, safer charging technique which goes as follows:

First, set up the measure to throw a charge approximately one grain lighter than the desired weight. Second, throw the charge from the measure into the scale pan—not the cartridge case.

Third, set the pan into the scale, ensure the poise is properly set for the desired charge, and steady the scale—it should read a bit too light.

Fourth, use a powder dribbler to add the last bit of powder to bring the scale up to the mark. Now, using a reloader's funnel, pour the powder into the cartridge case.

This technique is somewhat more elaborate, but much safer than use of the measure only when absolute maximum load must be used.

Step Eleven - Bullet Seating (And Crimping When Required)

Select the bullet required for the load being used and bring a single box of these to the loading bench. At the bench carefully double check the bullet box label for the correct type and weight. Then, using your caliper, verify that the bullets in the box are of the same diameter as the label indicates. Next, weigh several samples on your scale to verify that the weight is as specified. Packaging mistakes have occurred. Alert reloaders are not harmed by such errors. The scale can now be put away.

Install the bullet seating die in the press according to the manufacturers instructions. Be sure to adjust the seating die using a case without primer or powder. Note - when using compressed powder charges special seating die adjustments will be required. This is discussed in the powder chapter. Novice reloaders may wish to avoid compressed powder charges (so indicated in the data) for their first efforts.

Editor's Note:

After every 300-500 cast bullet rounds, remove the seating screw and clean accumulated lubricant from both the die interior and the seating screw.

Cast bullet lubricant build-up can cause misaligned bullets and a gradual increase in seating depth which often escapes detection.

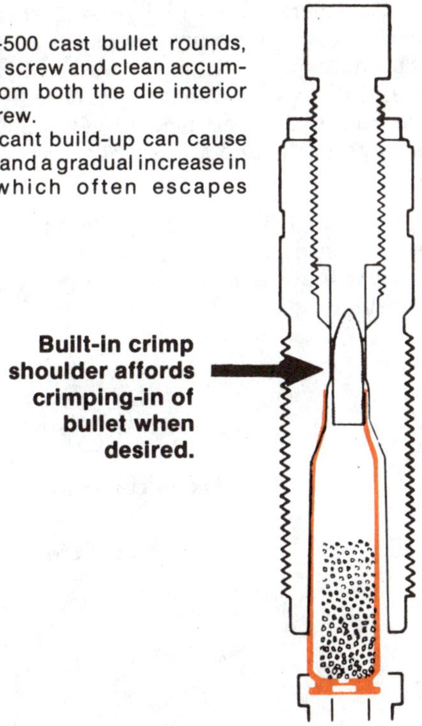

Bullets should generally be seated to the maximum overall length listed in the data section, or as close to this length as practical. Overly long or short loaded round lengths can create abnormally high pressure.

Place the charged case into the shell holder, hold a bullet in position on top of the case and run it smoothly into the seating die. Place the loaded round into the empty loading block.

CRIMPED

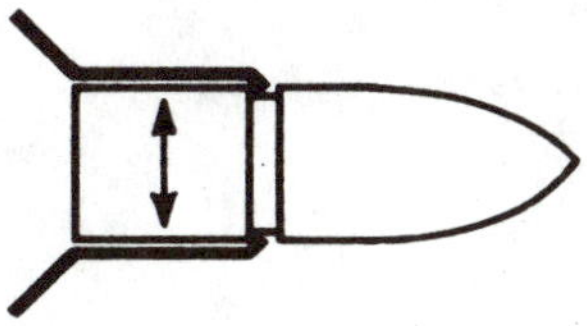

While bullet will not move forward or backward, it may be rocked sideways, out of alignment.

UNCRIMPED

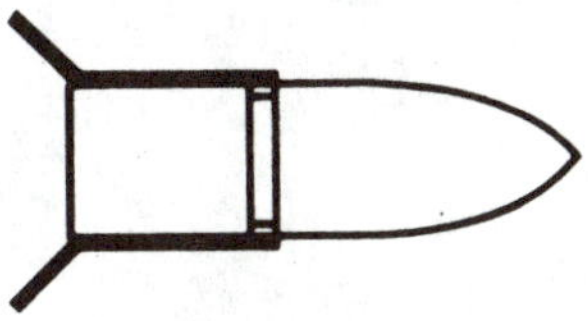

Most uniform bullet pull is afforded when the neck walls are tight against the bullet.

Some reloaders switch from batch loading to semi-progressive loading at the powder charging step. At this point they charge a case and then immediately seat the bullet. Doing so can avoid spilled powder should a loading block be accidentally jostled. I prefer this method, but it does eliminate the visual batch inspection of all charged cases for uniform powder levels. Therefore, when using this latter method, it is important to be doubly certain that the weighed charge is perfect, that it is carefully placed into the case and that it visually fills the case with an expected volume.

Crimping of the case to the bullet is done when ammunition is to be used in pump action or semiautomatic rifles. Crimping is sometimes done for cartridges with very heavy recoil. Crimping requires that cases be of a very uniform length or the case mouth will not correctly align with the required crimping groove in the bullet. Thus, reloaders often trim cases to be crimped at every reloading.

Step Twelve - Final Inspection

Examine each cartridge carefully, including the measurement of its overall length with your caliper. While looking over each round run a finger over the seated primer to insure that it has been seated to the proper depth. Should any round be found with an abnormality discard it in a safe manner. Be especially alert to split case necks, which can occur when a bullet is seated into a worn-out case.

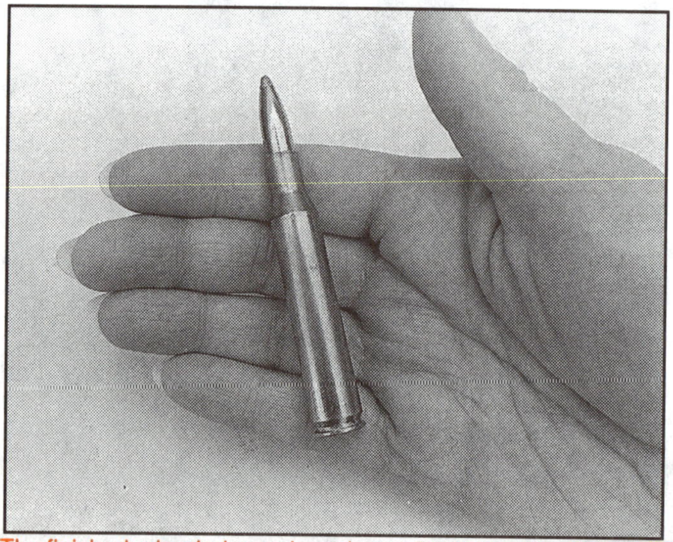

The finished reloaded round receives a final inspection before it is boxed.

Place the loaded ammo into a suitable container with a label indicating all the specifics of the load. Make your loading log entries and the job is completed.

The foregoing is appropriate for most rifle cartridges which will be loaded with jacketed bullets — that is, all bottleneck case cartridges. This material is equally appropriate to the loading of bottleneck handgun cartridges, i.e. 30 Luger and 30 Mauser. Rifle cartridges with a straight side wall, i.e. 32-40 Winchester, 45-70 Government, and 458 Winchester Magnum, require the use of an additional die to accomplish neck expansion. These are loaded in the same manner as most handgun cartridges. The following chapter dealing with the loading of handgun cartridges would be appropriate for the basic steps of reloading these style cases.

Reloader labels identify this group of fired reloads by components used and number of firings.

Reloaders who will be using cast lead bullets in rifle cartridges will require the use of an additional loading die. This die expands the case mouth so that it will accept the soft lead bullet without shaving its base. The reloading procedure for cast lead bullets is identical to the procedure for loading straight wall handgun cartridges and is described in the next chapter.

As stated earlier there is more to reloading than the basic steps just described. Reading the remaining portion of this book will acquaint you with a great many other important considerations.

CHAPTER 4
HOW TO RELOAD HANDGUN CARTRIDGES

The steps of reloading handgun cartridges are unto themselves rather simple operations, and they closely duplicate the procedures used to load bottleneck rifle cartridges. Therefore, much of the following copy is necessarily identical to that used to describe rifle cartridge reloading. There is, however, an additional die used in the reloading process for all straight walled handgun (and rifle) cartridges. (The loading of bottleneck handgun cartridges is identical to the loading of bottleneck rifle cases. The reader interested in these, i.e. 221 Fireball or 30 Luger, should refer to the previous chapter.)

As the reloader becomes involved it will be obvious that there are many considerations to be given to each step. Pertinent in-depth discussions are undertaken in the appropriate individual component chapters. These are must reading. The following are the basic essential operations and potential areas where caution is stressed.

When reloading handgun cartridges, the mechanical steps essentially are: resizing the case to insure it will chamber, removing the fired primer, expanding the case mouth so it can accept and hold the bullet firmly, replacing the primer, replacing the propellant, and seating (and crimping when required) a new bullet into place. The actual practice is a bit more complex, but not overly so, and does require a few considerations. The proper procedure is outlined in the following thirteen steps:

1. Selecting a load and components
2. Case inspection
3. Inside case neck brushing
4. Case cleaning and lubricating
5. Case resizing and decapping of fired primer
6. Lubricant removal and secondary inspection
7. Case length measuring
8. Case trimming and/or deburring
9. Case neck expansion
10. Primer seating
11. Weighing and charging powder
12. Bullet seating (and crimping if required)
13. Final inspection

We will discuss each step as a basic function. Additional information will be found in many of the following chapters. Reloaders are cautioned to read all the pertinent material before actually beginning to reload handgun ammunition.

Since experienced handgun reloaders often use a turret press to reload large quantities of ammunition, our illustrations show the use of the Lyman T-Mag turret tool. The rotating turret of this tool allows reloading to be conducted in a semi-progressive manner — that is, all operations on a single case are completed before the next case is loaded. However, it is suggested that the novice batch load his cases, performing the same operation on all cases before moving on to the next step, until some reloading experience has been gained.

Step One - Selecting A Load And Components

For the novice, choice of components is perhaps the most confusing aspect of getting started. It's no wonder that many reloaders with years and thousands of rounds of experience still do not make appropriate selections, not because it is difficult to do so but simply because they have not been exposed to the right information.

To make the selection task as clear as possible, the data section of this handbook makes specific recommendations as to the propellant most likely to satisfy the needs of top accuracy and uniform ballistics.

Where no such suggestion is made (in the preamble to the data for each cartridge) there simply is no single best first choice. In these instances the reloader can select any of the propellants listed for the bullet weight to be loaded and expect results to be on par with any other propellant choice. Often this situation is a result of the accuracy limitations of certain firearms or cartridge designs. In rare calibers it might be a result of all listed propellants giving superb performance.

Uniform ballistics are normally synonymous with accuracy, though at times the load providing the most uniform ballistics does not provide the very best accuracy. All accuracy loads do, however, fall well within the limits of acceptable ballistic uniformity.

170 gr. Jacketed HC
1.580" OAL

Powder	Sugg. Starting Grains	Velocity fps	Pressure C.U.P.	Max. Load Grains	Velocity fps	Pressure C.U.P.
SR-4756	10.6	1052	13,300	13.0	1360	36,700
IMR-4227	20.5	1131	20,000	23.5+	1340	33,000
2400	17.5	1040	15,000	22.0+	1311	28,600
Unique	9.8	1056	13,700	12.4	1378	36,100
H110	22.5	1217	15,200	26.3+	1507	33,700

The first loads assembled should always be taken from those listed in the Suggested Starting Loads column of the data section. Do not use heavier loads until some experience has been gained.

The first efforts should always begin with a load listed in the "Suggested Starting Grains" column in our data. Heavier loads should not be used until the shooter has had some reloading experience and has understood all of the information in this handbook. Be absolutely certain that you use the exact bullet weight and the exact powder and charge weight shown in the data.

Primer selection can also be kept quite simple. Naturally a primer of the right size must be used, i.e. large pistol or small pistol sizes. The data lead-in for each cartridge clearly specifies the appropriate size.

Do not use "magnum" style primers unless the data specifically calls for these. Doing so can alter the ballistic uniformity, ballistic level, and safety of the load. Generally speaking, magnum primers are appropriate in calibers using the word magnum in their proper name, i.e. 357 Magnum, 44 Magnum and 45 Winchester Magnum. Still, light target loads for these can often be best when assembled with standard non-magnum primers. Always make primer selection based on the type primer used to develop the data as listed. This will insure that pressures are not adversely effected by primer choice.

The novice will do well to select a component bullet that most nearly duplicates the bullet used in factory ammunition. For maximum economy, handgunners often choose lead bullets instead of jacketed ones. If doubt exists refer to the chapters on jacketed and cast bullets.

In some instances the exact bullet that has been used in factory ammunition can be purchased for reloading. As an example, Remington offers a large number of their factory bullets as a component. Many Lyman cast bullet designs very nearly duplicate popular factory lead bullets as well.

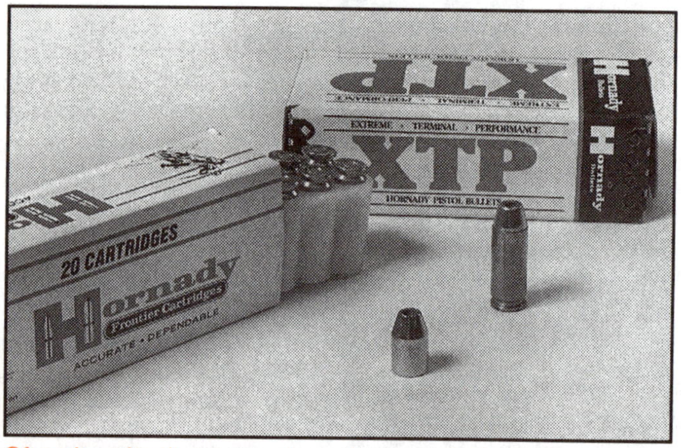

Often the reloader can purchase the same bullet as loaded in the factory ammunition that has been previously used successfully.

The beginner will often have fired cases on hand as a result of saving those from the shooting of factory loads. Brass should be kept segregated by lots and loaded accordingly. Lots are homogeneous groups of cases from the same maker and bearing the same lot number on the original ammo box.

When purchasing brass cases many reloaders will buy new brass from bulk suppliers. This saves considerable money. Brass may also be purchased new in conventional packaging as supplied by the major ammunition manufacturers. Never load fired cases from an unknown source. Reloading of brass that was fired in a gun with excessive headspace, or that may have been previously loaded to excessive pressure, can be dangerous. There is also no way to be certain that the so called "once fired" cases being purchased are indeed so. If only one seriously weakened multifired case sneaks into a lot of brass it could be the basis of a serious accident.

Handgunners often purchase reloaded ammunition before actually beginning to reload their own shells. Since it is impossible to tell how often such reloaded cases have been fired, or under what conditions, it is essential that none of the cases from such ammo be reloaded.

After the appropriate components have been selected, set them aside in a remote area and bring them to the loading bench only when they are required.

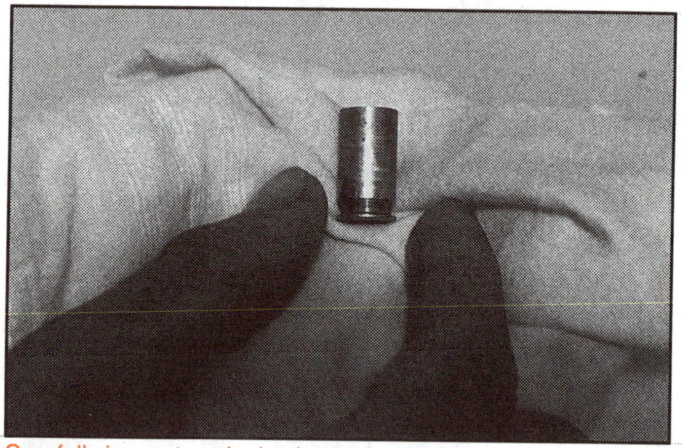

Carefully inspect each pistol case for defects.

Reloading Pistol Cartridges

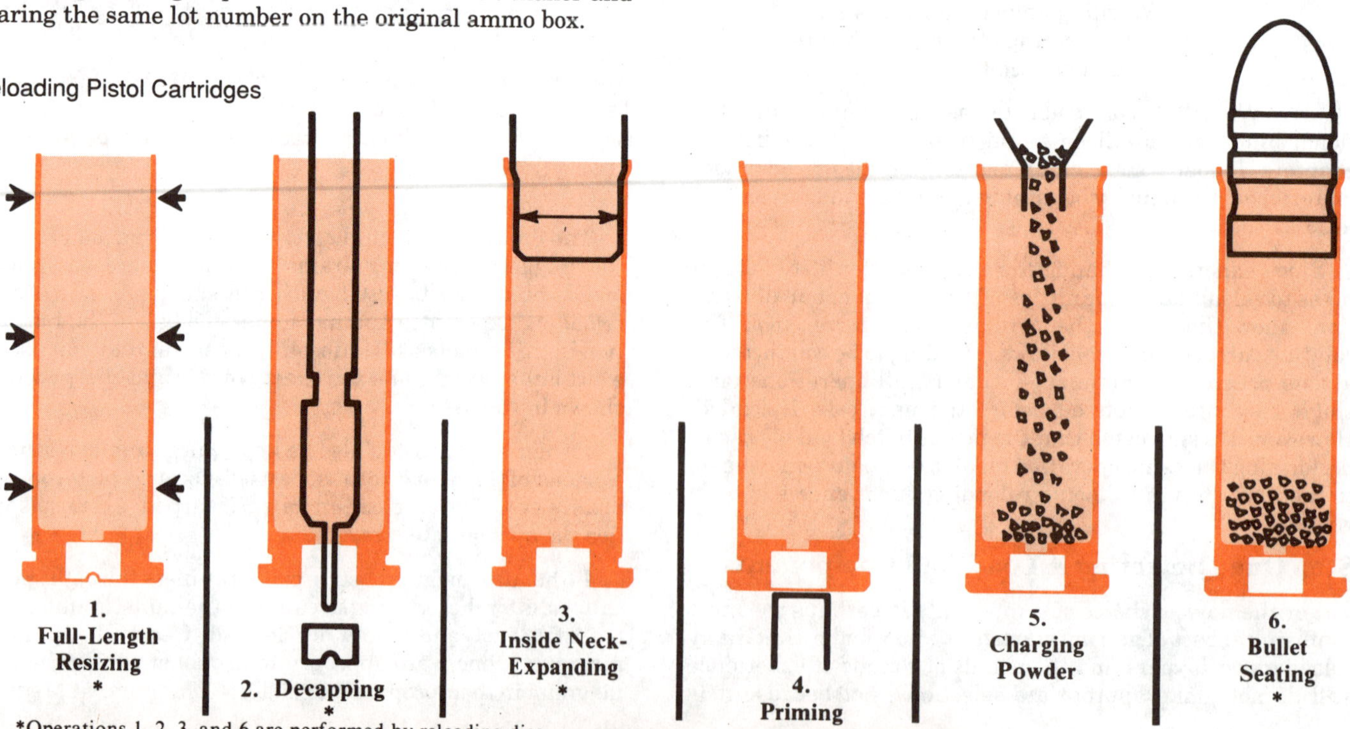

*Operations 1, 2, 3, and 6 are performed by reloading dies.

Step Two - Case Inspection

Each case must be in appropriate condition for further use. Only your inspection can decide upon a case's suitability. After wiping each case with a clean cloth, look each one over for defects, discarding any cases which are not as they should be. To prevent accidental use of a defective case, crimp the mouth shut using a pair of pliers.

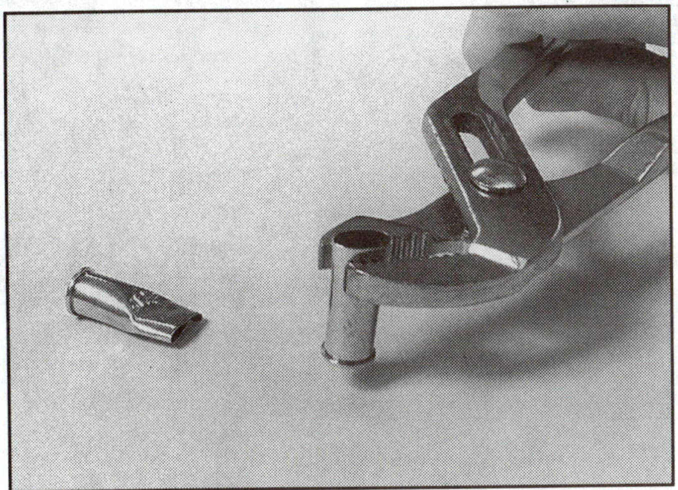

Crimp shut the mouth of any defective case using a pair of pliers. This will prevent inadvertent use of the case.

Eliminate all cases that show any signs of deterioration, such as cracks and splits. Look carefully for any indication of incipient case separation - a bright ring partially or completely around the case head . Any corrosion on the case or signs of gas leakage around the primer are other reasons to discard the case involved. Handgun cases sometimes have bulges at the base that suggest the cartridge was overstressed at the initial firing. These should be discarded also. For additional information on inspection see the chapter on cartridge cases.

Be sure to turn the case over and allow anything that might have gotten into the case to fall free. As cases pass inspection place them into your first loading block which you should locate to one side of your work station. Put them into the block in a mouth up position.

Under typical circumstances handgun cases may be reloaded and fired as few as four times or as much as eight or more times. Be sure you understand the limitations of multiple firings as discussed in the chapter dealing with cases.

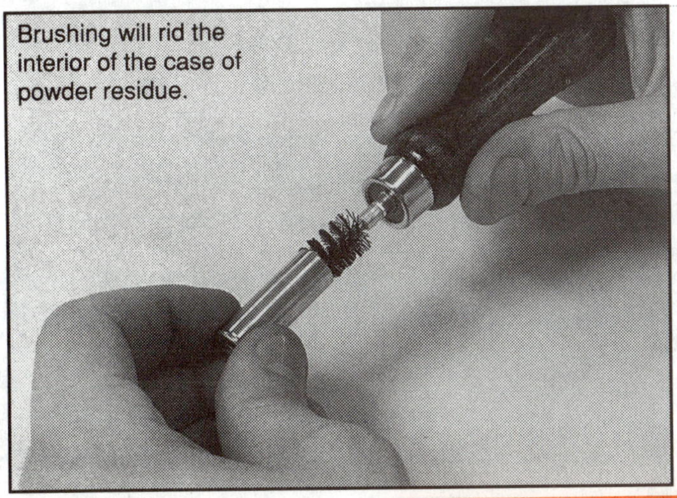

Brushing will rid the interior of the case of powder residue.

Step Three - Inside Case Brushing

Take each case from the loading block and brush the inside of the cases with a cleaning brush of appropriate caliber. This will loosen and remove excess carbon and other fouling. Doing this will help ease the passage of the expanding plug and enhance accuracy by helping to maintain a uniform bullet pull. Transfer brushed cases to the second loading block located on the opposite side of your work station but place them mouth down so as to let the loosened crud fall from the case.

Handloaders using semi-progressive or progressive loading methods will eliminate this step in exchange for speed.

To keep dirt from accumulating in the loading block, many reloaders prefer to tap the inverted case on the work bench. Even if you use this preferred method, be certain to place the case upside down in the second loading block to offset any possibility of forgetting to do this.

Tumblers will get used cases looking like new. Lyman Case Cleaner is an excellent precleaner for extremely fouled and dirty cases.

Step Four - Case Cleaning And Lubrication

If cases are to be cleaned in a tumbler or vibrator, it can be done at this point. Some reloaders will, however, clean all cases in a tumbler as their first step. Follow the manufacturers directions on the use of a tumbler. Be sure not to overload the cleaner and to allow sufficient time for the cases to clean properly. A case that has been tumbled for a sufficient time with appropriate media will look almost as bright and shiny as new. When removing cases from the tumbling media be sure to tap all the media free from each case before placing it mouth down into the second loading block.If cases are not to be cleaned in a tumbler, be certain they are free from any grime that might damage or scratch the sizing die.

Now lubricate each case by rolling it lightly across a prelubed pad. Several cases may be lubed at one time. After lubricating each case place it mouth up in the first loading block. This alternating of loading blocks and case mouth position will help eliminate many common loading bench errors.

Do not over lubricate cases or the excess lube will cause case dents during resizing. Be neat and do not get lube into the case mouth or primer pocket.

The use of a carbide sizing die does away with the need to lubricate cases and for the later removal of the lubricant. This convenience is highly recommended to all reloaders.

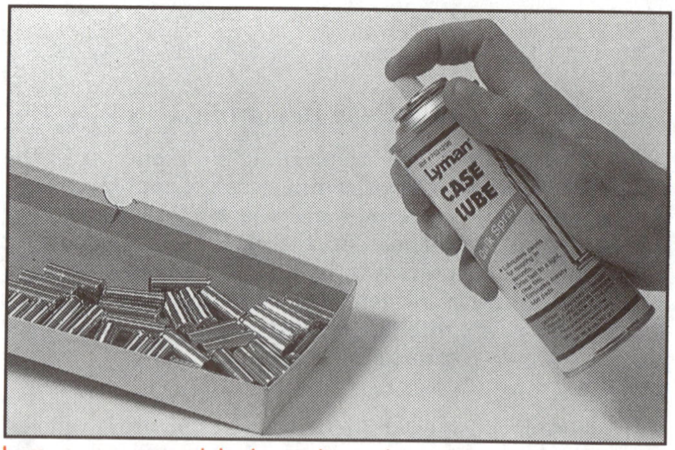

Lyman spray case lube is an alternative to lubing with a pad and is much faster. An entire loading block of cases can be lubed in seconds.

Step Five - Case Resizing And Primer Decapping.

Place the lubed case into the shell holder and manipulate the tool handle so as to properly resize the case. Be sure to follow the die manufacturer's instructions for appropriate die adjustment. Generally, when full length resizing handgun cases, the shell holder needs to bottom firmly against the die while the ram cams past the top of the press stroke. However this is NOT true when using carbide dies. In this instance the shell holder should never touch the loading die.

As the case is run into the die the old primer will be removed by the decapping pin. The decapping rod should be adjusted downward only far enough to insure fired primer removal. If the rod is adjusted too low it and the case will be damaged when it strikes the case web.

If excessive force is needed to resize normal cases it is a signal that insufficient lube has been used. If dents form on the case it means that too much lubricant is being used.

Place the resized cases in the second loading block in the mouth down position.

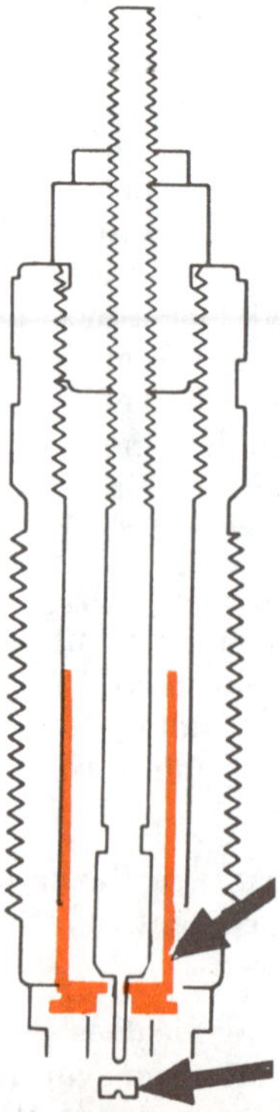

Step Six - Lube Removal And Secondary Inspection

Carefully wipe away all traces of lubricant from each case using a clean cloth. As you do so, inspect each case for defects. Case mouths and bodies sometimes split during resizing. It is possible that the first sign of incipient case separation may not be visible until after sizing and expanding. This is also a backup inspection to your first one. Discard any case showing a defect, crimping its mouth closed before so doing. If you use a carbide sizing die there will be no lubricant on the case to be removed, but inspection remains important.

Place the cases back into your second loading block, again in a mouth up position.

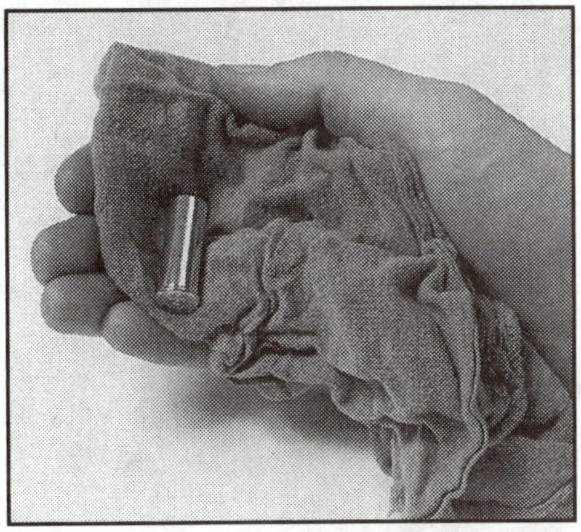

Wipe away all traces of case sizing lubricant and inspect each case, discarding any with defects.

Steps Seven And Eight - Case Length Measuring, Trimming And Deburring

Cases stretch when fired. Resizing also tends to lengthen the case. The brass that flows forward as a result of firing and sizing will sooner or later make the case longer than acceptable. A case that is too long can dangerously increase chamber pressure and can be difficult or impossible to chamber.

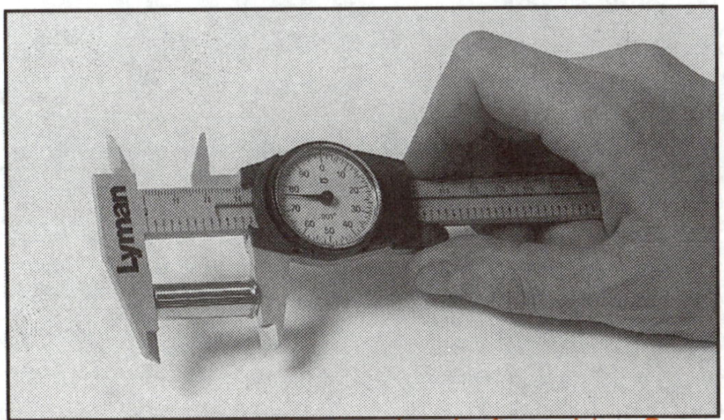

Carefully measure cartridge case length after resizing. Cases which are longer than the maximum allowable dimension must be shortened to the "trim-to" length listed in the data tables.

Every case must be measured after every resizing. The data section of this manual shows a cartridge drawing for each cartridge. The length shown for the case is the maximum allowable length. Case length must never exceed this dimension. When one or more cases have stretched beyond the maximum, trim the entire lot of brass back to the listed trim-to length as indicated in the tabular information shown next to the cartridge drawing in the data section.

When loading in a semi-progressive or progressive manner it is not practical to measure cases at this point. Cases should instead be measured before loading, at the first inspection, with allowance made for possible stretching in the sizing die. This will require some trial and error effort to be sure that all is correct.

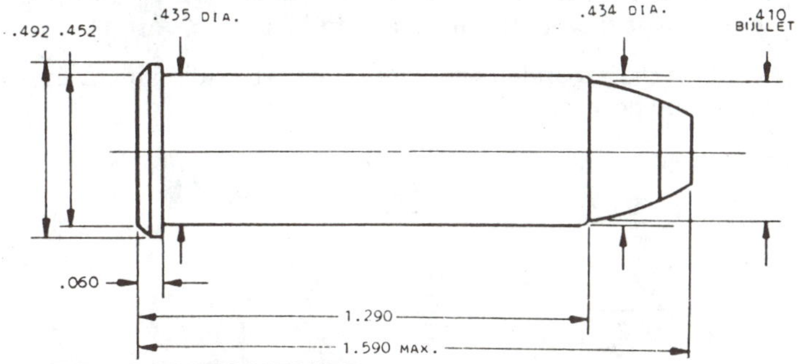

TEST COMPONENTS:

Cases	Winchester
Trim-to Length	1.285"
Primers	Remington 2½
Primer Size	Large Pistol
Lyman Shell Holder	No. 30
Jacketed Bullets Used	Sierra HC #8500, 170 gr.

Cases are trimmed to acceptable and uniform lengths using small hand or electrically powered lathe-like tools like the Lyman Power Case Trimmer shown above.

Trimmed cases must be deburred to remove rough edges left from cutting the case back to a correct length. A quick few twists with a deburring tool will remove outside burrs and another few quick twists will smooth out the inside of the case mouth. Do not overdo this deburring operation.

After measuring (and trimming/deburring if required) return the cases to the first loading block in a mouth down position to be sure that no trimmings are left inside the case.

Cases should never be trimmed more than 4 times. When a case needs its fifth trimming it must be discarded.

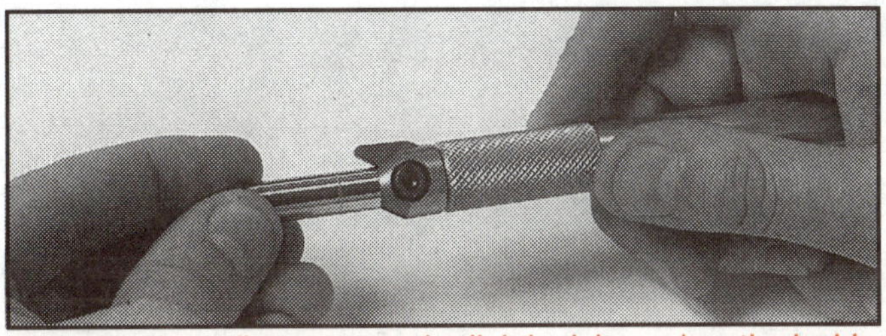

After trimming, all cases must be lightly deburred on the inside and outside of the case mouth.

Step Nine - Case Neck Expansion

Run the case into the expanding die. The expanding plug should be adjusted so that approximately 1/32" to 1/16" of the case mouth is expanded to a diameter of 0.001" over bullet diameter. This allows jacketed bullets and the relatively soft cast lead bullets used in handguns to start into the case without shaving material from their bases.

Place the expanded cases into the first loading block in a mouth up position.

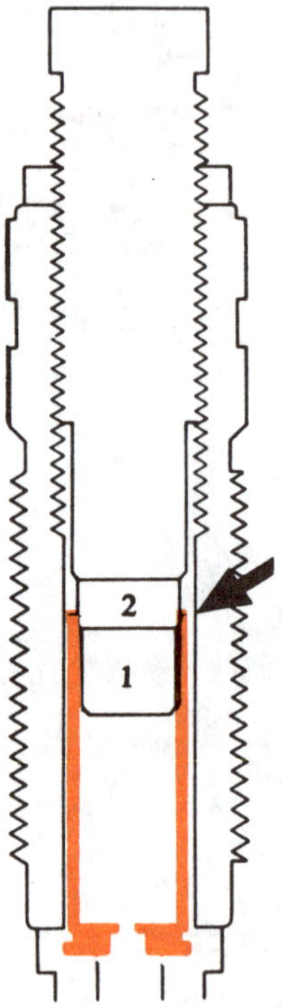

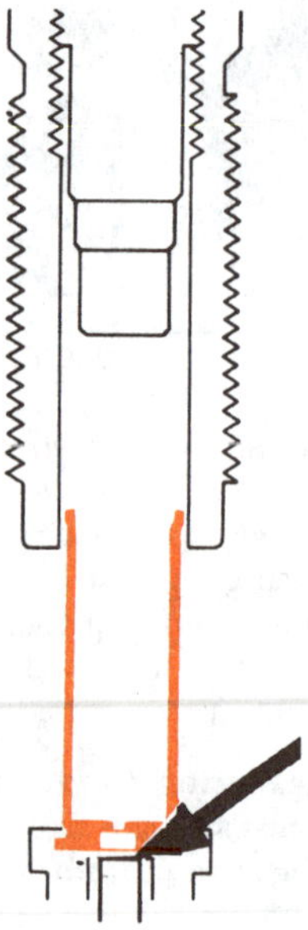

Step Ten - Primer Seating

Seating primers requires great care. Primers and primer pockets must remain free from any contamination. Be sure the work area, and primer tray are clean and free from any trace of solvent or lubricant. Then remove no more than one box of 100 primers from your remote storage area and place them on your bench. Carefully read the label with respect to primer size and type. Then refer to the data to insure that you have the right size primer on the bench. Now reread the label on the box. When certain that all is correct, assemble the Ram Prime unit to the press and adjust it correctly.

Now, wash and dry your hands. Failure to do so could cause contamination of the primer during handling and later problems with your ammunition. This point cannot be overstressed. Next, pour 100 or less primers onto the grooved half of your primer tray. Rotate the tray until all the primers are cup side up.

A favored way of priming is to use a Ram Prime unit mounted in a die station. This method, however, is not practical for semi-progressive reloading. The procedure described here is the use of the standard priming unit supplied with the tool. Place a primer in the primer post, lower the primer post slightly, and place the case in the shell holder.

Primer seating may be accomplished in many ways and with differing tools. One of the best methods is to use a Ram Prime unit mounted in one of the reloading tool's die stations.

Now move the press handle to bring the priming post up against the case until the handle comes to a stop. Primers should normally be 0.003" to 0.007" below flush. Thus, strive to adjust primer seating for a nominal of 0.005" below flush. The exception to this is CCI primers which the manufacturer suggests be seated to a depth of 0.003 to 0.005 inches below flush - a nominal of 0.004 inch.

> **Caution:** primers seated higher or lower than recommended are a needless hazard. Check your primer operation by using your caliper to randomly measure seated primers from time to time. Naturally the use of a caliper is essential for the initial adjustment of the priming set up.

After priming return each case to the second loading block but place it again in a mouth down position. This will allow you to visually see that all primers have been properly seated.

Proper priming suggestions are included in the chapter on priming. Be certain to read this information to insure that the job goes as intended. Primers require special storage and handling. This information is also in the primer chapter.

Return any unused primers to their original container. Once again read the label on the primer box to make a final verification that the correct primers were used. Finally return the primers to their remote storage place.

Step Eleven - Weighing And Charging Powder

The weighing of the powder charge is a critical aspect of reloading. Set up the scale and insure that it has been correctly leveled. It is a good practice to now verify the scale's accuracy using a weight check set.

Select the powder to be used from the data section and then remove a one pound can of it from its remote storage area. Bring the can to the bench, and again verify from the data the powder you wish to use. Read the can label once more. Double checking all components placed at the loading bench is a must for safe reloading.

Pour some powder into the powder trickler and place the trickler alongside of the powder scale pan so that powder can be fed from trickler to pan. Also place some powder in a small open container (or the powder measure if one is used). **See rifle section for instructions on powder charging with the Lyman 55 powder measure.**

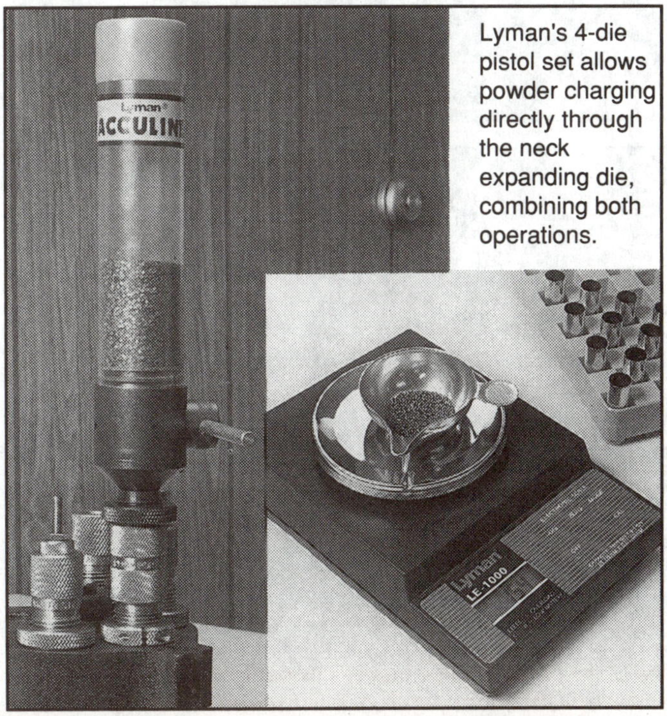

Lyman's 4-die pistol set allows powder charging directly through the neck expanding die, combining both operations.

Be sure to verify powder charges with an accurate scale.

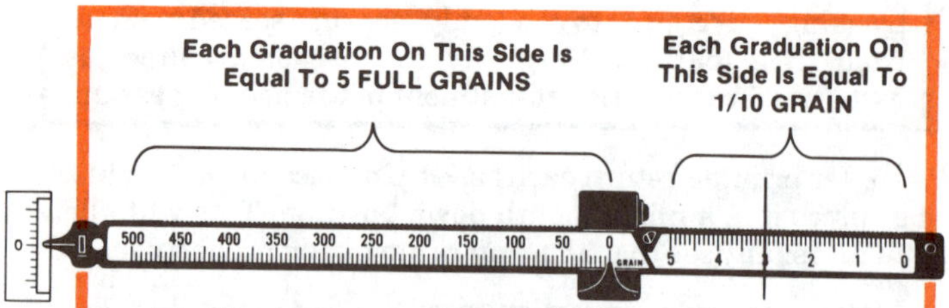

The illustration shows the beam of a modern reloading scale. Note how it is graduated on both sides of the pivot point. The scale is set by moving the two weights (poise) to the proper graduations. The large poise (on the left) is used to obtain multiples of FIVE GRAINS, while the small poise (on the right) is used for FRACTIONS of a grain or SINGLE grains from one to five.

EXAMPLE: The illustration shows a setting of 3.0 grains. If you wish to increase this to 8.0 grains, simply move the large poise one notch to the left.

Using a scoop take a measure of powder from the open container and place it on the scale pan. (Or meter a charge from a powder measure onto the scale pan and place the pan on the scale hanger.)

A handy scoop can be made from an appropriately large straight walled case such as a 380 Auto, 30 M-1 Carbine, or 357 Magnum. Cut the case to approximate length using a hacksaw. Final length can be accomplished with a case trimmer or careful hand filing. A handle can be fashioned from stiff wire loop wound around the case's extractor groove.

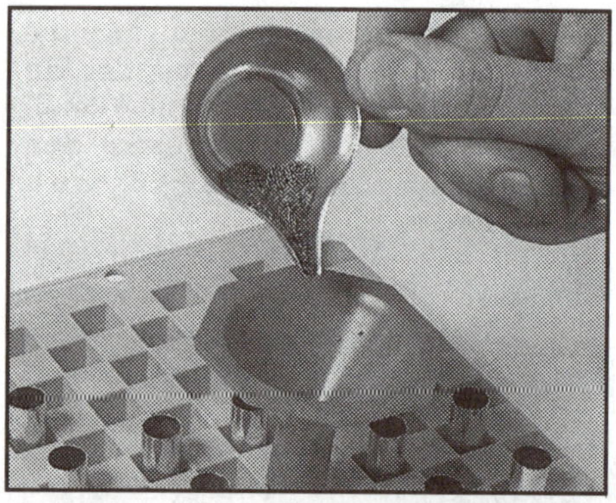

The weighed powder charge is poured into the case carefully, using a powder funnel to prevent any spillage.

The Lyman Auto Trickler makes powder trickling fast and easy. A simple push button operation dispenses powder at a consistent rate.

Bring the scale into perfect balance by adding the necessary few granulations with the powder trickler. Then pour the powder into a primed case with the aid of a powder funnel. Now place the charged case into the second loading block in a mouth up position.

After charging all cases carefully inspect each case for a powder charge of appropriate height. If any charges appear to be of greater or lesser amount than normal, reweigh the powder charge.

Make certain that the scale poises do not move inadvertently when removing and replacing the scale pan at each weighing. When the weighing is completed, once again look up the required powder type and charge weight. Then reread the can label and again verify that the scale is adjusted properly to balance with the desired powder charge.

Return any unused propellant from the trickler, open container, or powder measure to its original canister. Return the powder can to its remote storage area. Leave the scale on the bench.

Powder requires appropriate handling and storage. These requirements are discussed in the powder chapter.

Step Twelve - Bullet Seating (And Crimping When Required)

Select the bullet required for the load being used and bring a single box of these to the loading bench. At the bench carefully double check the bullet box label for the correct type and weight. Then, using your caliper, verify that the bullets in the box are of the same diameter as the label indicates. Next, weigh several samples on your scale to verify that the weight is as specified. Packaging mistakes have occurred. Alert reloaders are not harmed by such errors. The scale can now be put away.

Install the bullet seating die in the press according to the manufacturer's instructions. Be sure to adjust the seating die using a case without primer or powder. Note - when using compressed powder charges, special seating die adjustments will be required. This is discussed in the powder chapter. Novice reloaders may wish to avoid compressed powder charges (so indicated in the data) for their first efforts.

Bullets should generally be seated no shorter or longer than the lengths listed in the data panel directly beneath the bullet weight. Overly short or long loaded round lengths can create dangerously high pressure.

Place the charged case into the shell holder, hold a bullet in position on top of the case and run everything into the seating die. Place the loaded round into the empty loading block.

Some reloaders switch from batch loading to semi-progressive loading at the powder charging step. At this point they charge a case and then immediately seat the bullet. Doing so can avoid spilled powder should a loading block be accidentally jostled. I prefer this method, but it does eliminate the visual batch inspection of all charged cases for uniform powder levels. Therefore, it is important to be doubly certain that the weighed charge is perfect, that it is carefully placed into the case and visually fills the case to an expected volume when using this latter method.

Caution: After reloading and before firing, wipe your cases to remove all sizing lubricant. The presence of oil or grease on a cartridge may dangerously increase thrust on the bolt face.

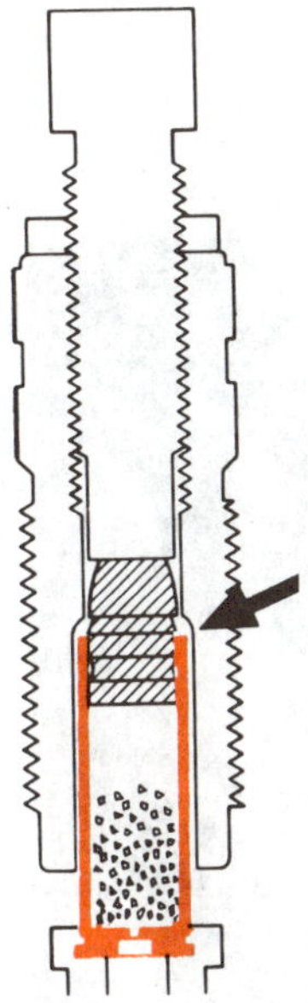

Editor's Note:

After every 300-500 cast bullet rounds, remove the seating screw and clean accumulated lubricant from both the die interior and the seating screw.

Cast bullet lubricant build-up can cause misaligned bullets and a gradual increase in seating depth which often escapes detection.

Note—The entire cartridge case does not enter this die.

Built-in crimp shoulder affords crimping-in of bullet when desired. Do not roll-crimp rimless autoloading cartridges which headspace on the case mouth; taper crimp only, using a fourth die.

Crimping of the case to the bullet is done on most ammunition which will be used in revolvers. It is not done, however, on semi-automatic pistol cases that headspace on the case mouth. Crimping requires that cases be of a very uniform length or the case mouth will not correctly align with the required crimping groove in the bullet. Thus reloaders often trim cases to be crimped at every reloading. VERY EXACT DIE ADJUSTMENTS ARE REQUIRED FOR CORRECT CRIMPING.

Caution: Technicians in the Lyman lab have observed a potential serious phenomena involving powder scales and loading blocks and/or Styrofoam packaging or other objects made of plastic materials. These materials sometimes contain a static electricity charge, enough to create an electro-static field of varying radii.

This electro-static field has proven capable of causing radical deflection of uncharged and zeroed scales of all brands (available to them at the time). Powder in the scale tends to dampen the amount of deflection by varying degrees dependent upon circumstances.

It is strongly urged that you clear the surface of your loading bench before carefully setting up your scale. Then replace your equipment one piece at a time, paying particular attention to the scale pointer, especially as any plastic or Styrofoam item is replaced on the bench. Any item that causes a scale pointer deflection should be removed before powder weighing begins. Do this at each loading session.

Step Thirteen - Final Inspection

Examine each cartridge carefully, including the measurement of its overall length with your caliper. While looking over each round run a finger over the seated primer to insure that it has been seated to the proper depth. Should any round be found to be abnormal discard it in a safe manner. Be especially alert to split case necks, which can occur when a bullet is seated into a worn-out case.

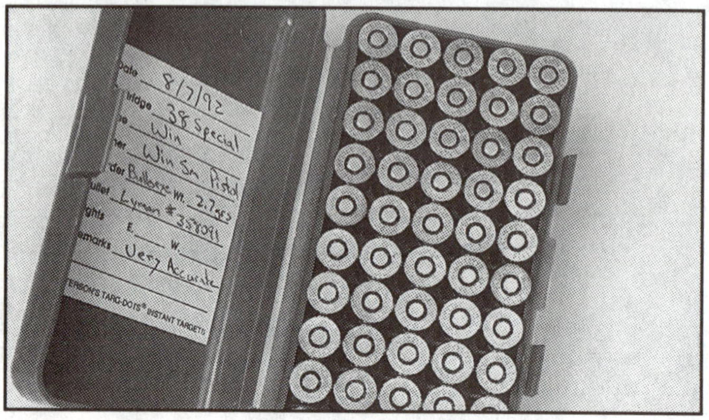

Each box of loaded cartridges should be clearly identified with all pertinent information before putting the ammo away for later use.

Place the loaded ammo into a suitable container with a label indicating all the specifics of the load. Make your loading log entries and the job is completed.

For Information
or
Technical Assistance
Call
1-800-22-LYMAN

As stated earlier there is more to reloading than the basic steps just described. Reading the remaining portion of this book will acquaint you with a great many other important aspects.

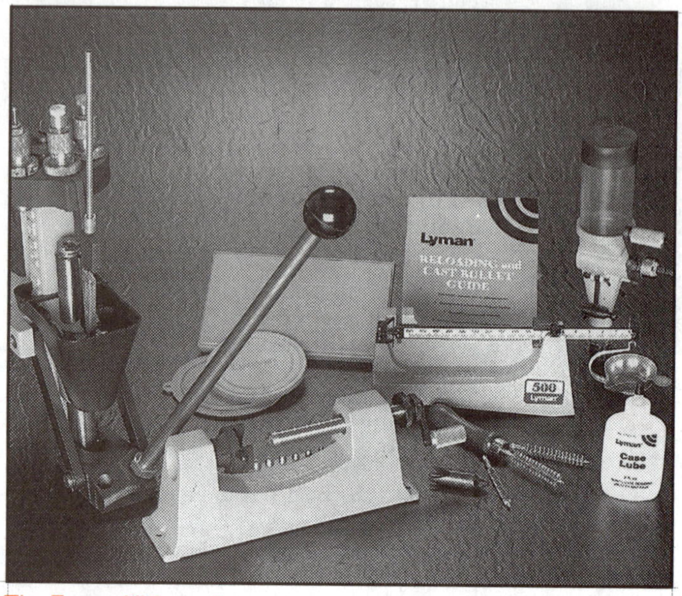

The Expert Kit has all the essential tools and accessories needed to start loading pistol cartridges.

2

THE COMPONENTS OF RELOADING

CHAPTER 5
CASES AND THEIR PREPARATION

Case Basics

Cartridge case base configurations are made in a number of styles or types. The basic nomenclature applied to the various types is straight forward. The proximate drawing shows case styles and nomenclature.

Rim Types

Rimmed cartridge cases were among the first types developed. The generous protrusion of the case rim allowed for ample extractor purchase in early firearm designs. This was important as some of the earliest metallic ammo and firearms resulted in fired cases were often difficult to extract. System components required sufficient extraction surface area to keep things going smoothly. Headspacing of both ammunition and firearms using rimmed cases is simple and easy to control. Today, rimmed cases still make sense in some firearm types, i.e. vintage type lever action models. Most revolver cartridges are of the rimmed type (such as the 38 Special, 357 Magnum and 44 Magnum) as is the popular 30-30 Winchester.

The rimless case, paradoxically, does have a rim. With this style case the rim is approximately the same diameter as the back end of the case body. Any included case body taper could result in a rimless case having a rim slightly larger in diameter than the case body. The "rim" of the rimless case is formed by an undercut made between it and the case body. It is this undercut which provides a purchase for the extractor. Rimless cases were first inspired by a need for cartridges which would reliably feed from a box magazine as opposed to a tubular type magazine. The lack of a protruding rim easily accomplished that goal.

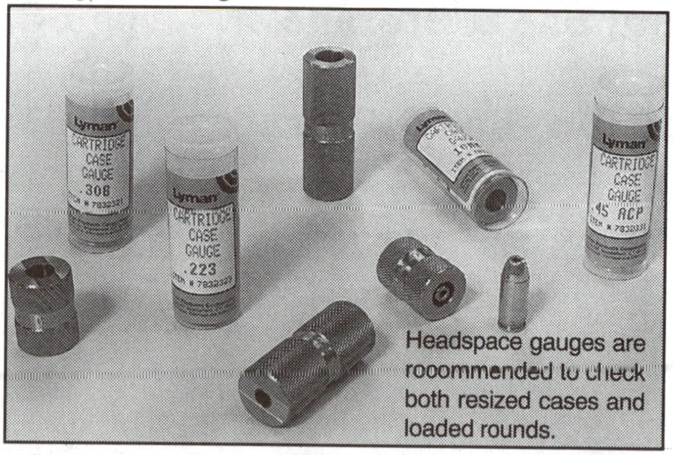

Headspace gauges are recommended to check both resized cases and loaded rounds.

Rimless cases are headspaced by one of two methods. The first uses the case shoulder to arrest forward movement of the case in the chamber. The second uses the end of the case, the case mouth, to halt forward movement when the case is chambered. Most semiauto handgun cartridges are of the rimless type (such as 45 ACP, 9mm Luger, and 380 Auto) as is the 30-06 and many other popular cartridges.

Rebated rim cases have a rim which is smaller in diameter than the case body. This style of case originated from the need to use large diameter cartridge cases with the same standard bolt face dimensions used in the manufacturing of many rifles. Some have claimed that rebated rim cases can cause feeding difficulties if the bolt rides over the narrow diameter rim and impacts with the large diameter case body. This is true only when firearm parts tolerances are not carefully controlled or when a magazine is not exactingly loaded. Nonetheless, it is this specific shortcoming that has prevented popular acceptance of rebated rim cases. An illustration of a rebated rim case is the 284 Winchester.

Semi-rimmed cases have a rim area slightly larger in diameter than the case body. This type case was originally designed to make use of existing bolt face sizes with a slightly smaller than normal case body diameters. The semi-rimmed case functions and headspaces the same as rimless cases. An example of a semi-rimmed case is the now defunct 225 Winchester.

The belted magnum case tries to incorporate the easy headspacing of a rimmed cartridge with the easy feeding from a box magazine of the rimless case. A belted case also allows easy interchangeability between the extractor systems of box magazine and double barreled rifles. It first saw light when the abilities of firearm and ammunition manufacturers to hold tight tolerances were still being perfected. Today, the belted case has become synonymous with "magnum" power cartridges. Nonetheless, there is no continuing technical justification for designing a case with a belt. Only the marketing aspects of giving a visual appearance of power causes the belt to be a continued part of cartridge case design. Typical of the belted type case is the 7mm Remington Magnum.

Case Styles

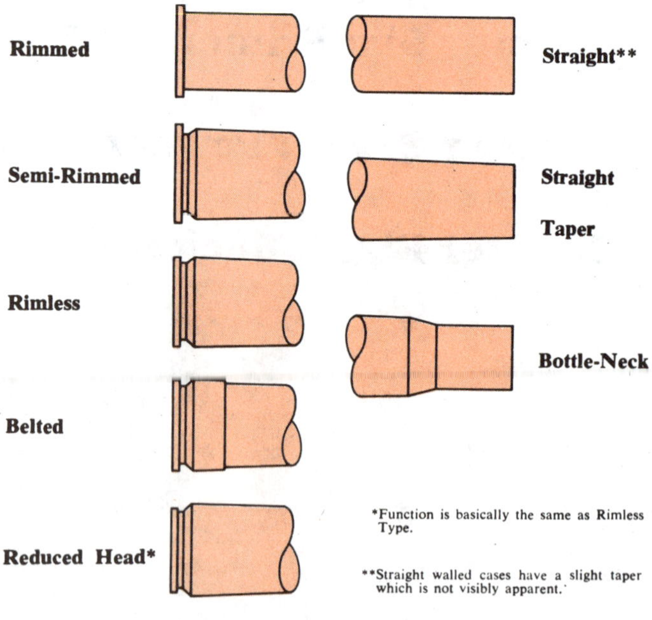

Body Types

The forward end of the cartridge case also varies with design requirements. The basically straight walled cases (these may have a very shallow included taper) worked great with black powder. Straight cases are also perfect for reliable feeding in semi-automatic handguns. A straight walled case

Cartridge Case Nomenclature

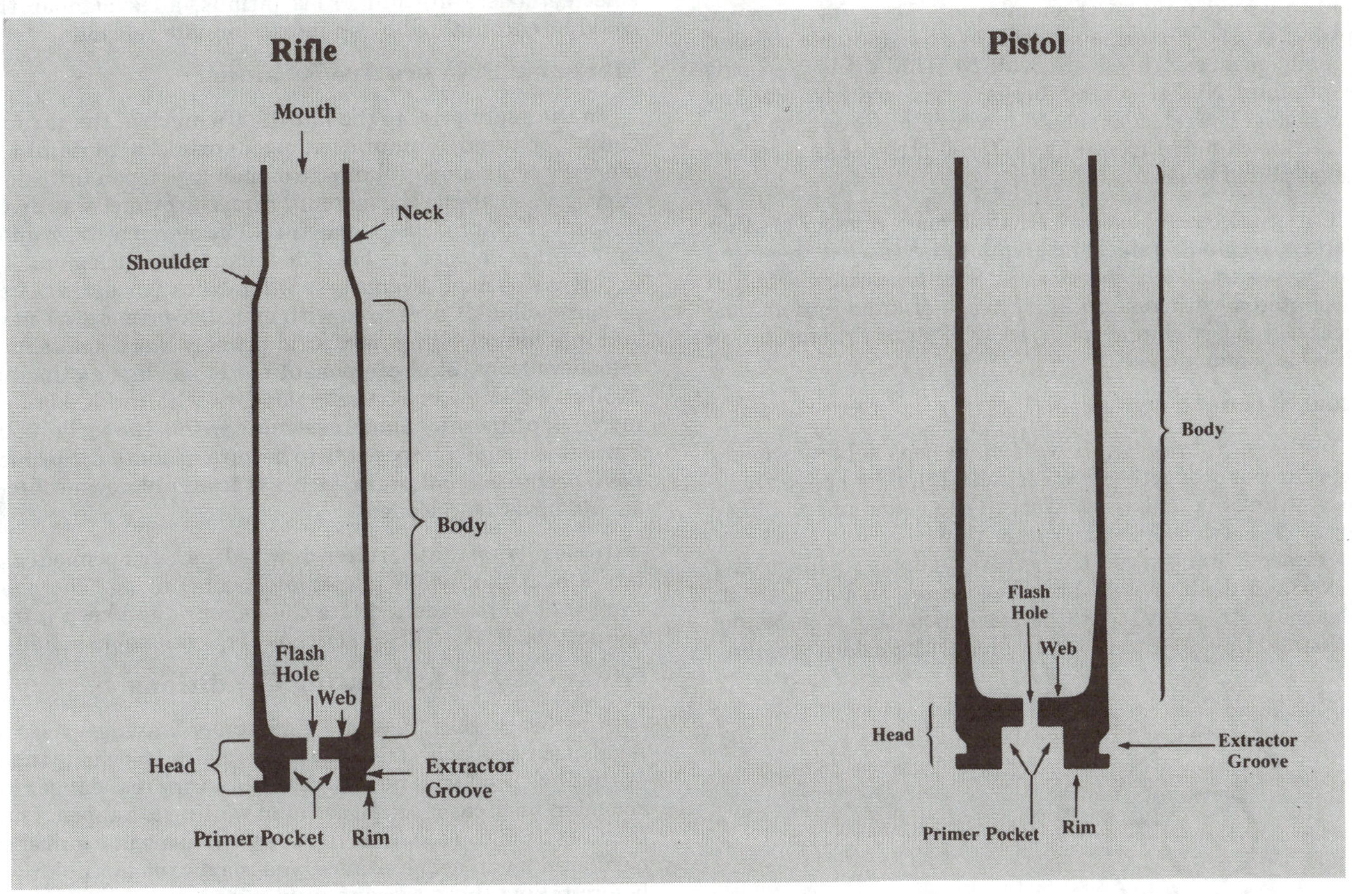

While cases vary in shape and size, the same basic nomenclature applies to all center fire cartridges.

may also be the answer to using faster burning powders when case space might otherwise interfere with the desired ballistics. The 45 ACP is an example of a straight walled case with a rimless head design. The 45-70 Government is a straight case with a rimmed head; while the 458 Winchester magnum is a straight case with a belted head.

Bottleneck cases are so called because the forward end of the case narrows from a large propellant capacity body to a smaller diameter bullet-holding neck. This case type usually allows for the maximum in ballistic potential. The 270 Winchester is a bottlenecked case with a rimless head. The 375 H&H is a bottleneck case with a belted head. The 218 Bee is a bottleneck case with a rimmed head. The 220 Swift is a bottleneck case of the semi-rimmed type.

Berdan And Boxer Primed Cases

Brass cases are currently constructed in two basic styles, having to do with the type primer that is used in loading. The American designed Berdan primer is generally now used only in some ammunition manufactured abroad. The European designed Boxer primer has become the standard for American manufactured cases and ammunition. It is also used by a number of foreign ammo manufacturers. The only notable exception is the American non-reloadable CCI Blazer ammo case which uses a Berdan primer.

Basically, the differences between the two are that Boxer primer type cases have a centrally located flash hole in the primer pocket while the Berdan types use two, non-central flash holes. Also, the Boxer method has the primer anvil as a small component of the primer while the Berdan method has the anvil as part of the case. The Boxer method has become prevalent because it makes case manufacturing less complicated. It results in a more sensitive priming system, and lends itself well to reloading.

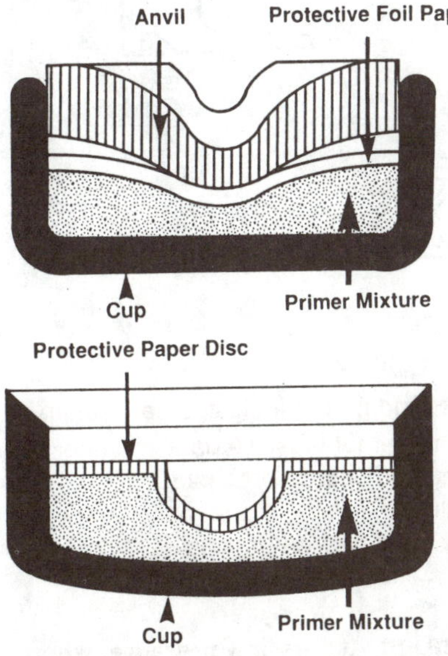

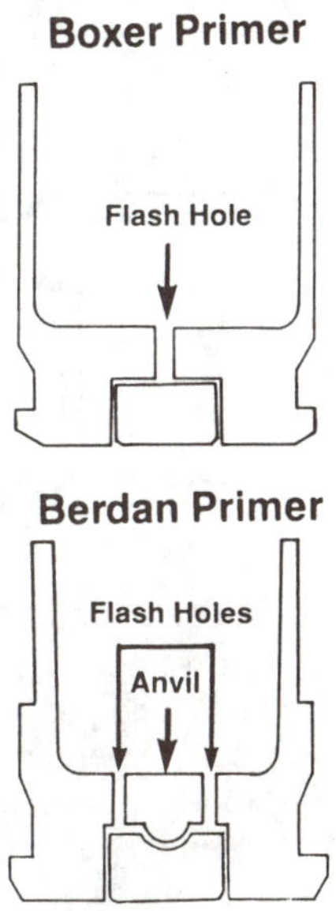

The European designed Boxer primer type case has become a standard. The American designed Berdan primer type case is almost never used in this country -the major exception being the non-reloadable CCI Blazer ammo cases.

Case Material

The cartridge case is generally the most expensive component in a loaded round of ammunition. Thus, its reuse can lower the cost of cartridge assembly by a considerable amount. Happily, most cartridges are made from brass and are easily reloadable. Nickel plated brass cases are also readily reloadable. The reloader should, however, make no attempt to reload cases made from any material other than brass or nickel-plated brass.

Cartridge cases have been made from a number of other materials including steel and aluminum. As stated, these and other types should not be reloaded. Such cases are most often encountered with rare military types of ammunition. The exception is the aluminum cased CCI "Blazer" ammunition which is commonplace.

Case Strength

In the past, brass cases were made considerably weaker than current production. For example, the older balloon head cases were of notably less strength than current solid head cases. This style case had a rather thin web, with a balloon-like internal hump to give the necessary support to the primer pocket and flash hole. Balloon head cases should never be reloaded with today's data because their thin web cannot withstand the pressures encountered with modern ammunition.

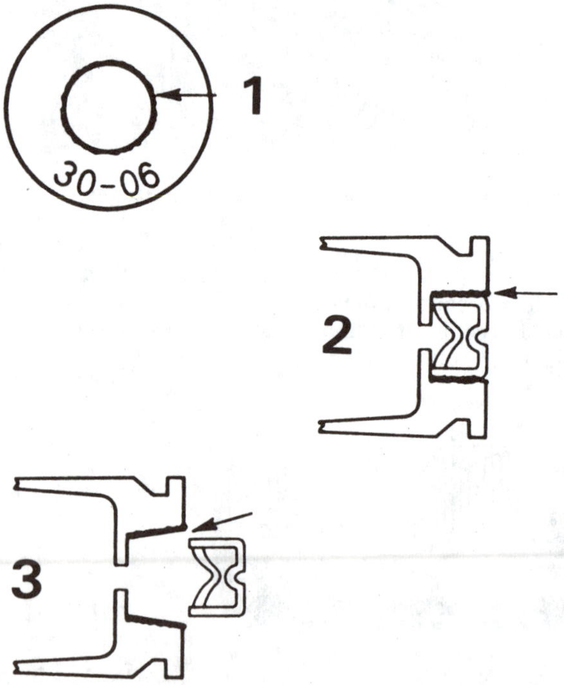

1. Dark smudge line around primer indicates gas leakage.
2. Pocket opens, leaving primer loose. A dark smudge, indicating gas leakage, is usually present.
3. Pocket opens completely, freeing the primer. A dark smudge, indicating heavy gas leakage, is usually present.

Visible Gas Leaks: At very low pressure, when the case walls are not expanding, gas may leak around the outside of the case. High pressure will usually be accompanied by other signs such as blown primers, stretched pockets, etc.

Strength is an important feature of any case to be reloaded. For this reason cases used for reloading must be carefully selected. There are numerous pitfalls, all avoidable, that could be encountered by an indiscriminate reloader.

Mercuric Primers Are Ruinous

In the years prior to the late 1940's much of the commercially produced ammunition used primers containing a mercury compound - fulminate of mercury. Upon firing mercuric primed ammo the mercury from the primer was driven deep into the brass case's structure. The mercury was ruinous to the brass causing its breakdown and embrittlement. The problem was made even worse when black powder was used as a propellant. Cases fired with mercuric primers are never suitable for reloading. Mercuric primers were available to reloaders from retail component outlets as late as the mid-1950's. We know of several reloaders who recklessly used mercuric primers for ammo assembly on into the early 1960's. Since it is usually impossible to be sure no mercuric primers have been used, old brass, or brass from unknown sources, should never be reloaded.

(Reloaders of the past often dropped mercuric primed cases into a bucket of water immediately after firing. The hopes were that water would dilute the mercury and keep it from ruining the brass. This practice is of questionable validity.)

Other Case Deteriorating Conditions

Any brass case fired in a chamber having excessive headspace, fired with an excessive load, fired too many times, or that has been trimmed too often, is a very real danger to a reloader. Such cases are an accident waiting to happen. Thus, the next time you see some fired cases for sale at a gun shop, or offered by an acquaintance, you may want to think twice before taking those cases to your reloading bench. Usually, there is absolutely no way you can be sure of their age, condition, number of firings, number of times trimmed, the headspace of the gun they were fired in, or the pressures generated at firing.

Even if the seller is extremely cautious to insure that all cases are from the first firing of factory ammo, it is impossible to be certain of the headspace. If only one old worn-out case was mistakenly picked up off the range with the once-fired cases, there may be an accident at your next loading session. (With modern case cleaning methods even old cases may look like new.)

Load only new unfired brass cases or cases that were once fired in the firearm for which you will be reloading ammunition. Any other course of action is less than prudent. Saving a few pennies on a case makes little sense when one considers the cost of a firearm and/or medical treatment, or worse. Sticking to new cases will keep the selection of brass worry free and help insure that your reloads will perform as hoped.

Case Life is Finite

Brass cases will not last forever. They do wear out and reach a point when they must be discarded. Reloaders must have a good working knowledge of the number of times a case can be reloaded. They should also understand what other specific problems need to be addressed during the mandatory case inspection done prior to the beginning of each reloading session.

While experienced benchrest shooters can sometimes obtain a seemingly incredible number of firings from each case, the average shooter using commercial dies and chambers should consider eight firings as a safety limit for brass case life. To go further will needlessly expose oneself to the danger of a sudden and catastrophic cartridge case failure. Even the most expensive brass is comparatively cheap when used over a case life of eight firings. Don't forget to count the first firing of factory ammo.

If maximum loads are routinely used, case life will be shorter - a good safety limit would then be about five firings.

Cases constructed with a belt, i.e. 7mm Rem. Mag., 300 H&H Mag., or 338 Win. Mag., etc., will not last as long as just suggested. Generally, belted cases should not be loaded more than two or three times. The exception is when a cartridge headspace gauge is used to set up the resizing die. When such a gauge is used, case life may be extended to 3 or 4 firings. Short case life for belted types is due to the overly long chamber dimensions that often occur in firearms chambered for this style case. Since headspace is controlled by the belt and not the case shoulder, many manufacturers use very lengthy bolt face to case shoulder dimensions. Top quality custom rifles from knowledgeable builders often do not have this problem.

Even on cases without a belt, reloading life may be notably shorter than suggested due to lengthy firearm headspace. When a cartridge is repeatedly shortened in a sizing die and fire formed lengthened in the chamber, it will quickly lose its ability to withstand additional firings. A form of case failure called incipient head separation will occur if such a case is loaded once too often. Again, a cartridge headspace gauge can be of use in preventing early case failure.

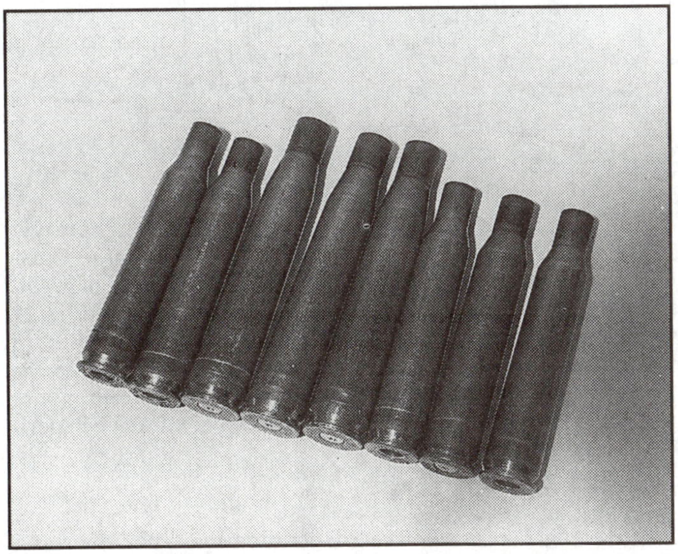

Incipient case separation can be catastrophic. Shown are some of the early warnings (bright partial rings on case near head) that show these cases have already been loaded too often.

Incipient Case Head Separation

The separation of the case head from the body, called incipient separation, is a dangerous condition caused by the weakening and thinning of the case at the junction of the web and sidewall. Incipient separation is often first preceded by the telltale sign of a bright ring, just in front of the case web, encircling part or all of the fired case's circumference. The next degree of failure could be a partial splitting of the case along the bright line. The final phase of failure is complete case separation. The actual separation failure of a case may not occur in steps as outlined. Incipient case separation can occur without early warning signs. Such failures can be potentially catastrophic. In a separation situation, if the rear section of the case has sufficient flexible material to expand and seal the chamber, no permanent damage may occur. However, if the case fails to seal the chamber, gases escaping rearward can damage the gun and/or shooter.

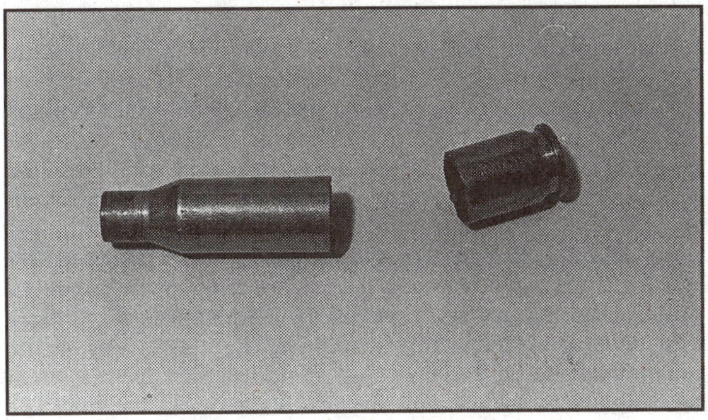

This case failed due to incipient separation. Such occurrences can be catastrophic to both firearm and shooter.

Every reloader should learn to detect the slightly lighter coloration of a ring-like area partially or fully encircling the case at the web/body junction that precedes the tell tale bright ring or actual failure from incipient case head separation. This is an important safety aspect of the reloading procedure.

Cartridge Case Gauges

The headspace of a cartridge/firearm relationship is very important. Headspace is a measurement of the distance between the headspacing portion of a cartridge and the headspacing surface of the chamber. Rimmed cartridges most often use the rim as the headspacing surface while rimless cartridges headspace most often from the front shoulder of the case. The belt of many magnum style cases serves as the headspacing surface. If the distance from the cartridge's headspacing surface to the matching chamber surface is too great, the cartridge/firearm system is said to have excessive headspace. When this condition is present the case is not adequately supported by the chamber and the cartridge may burst upon firing.

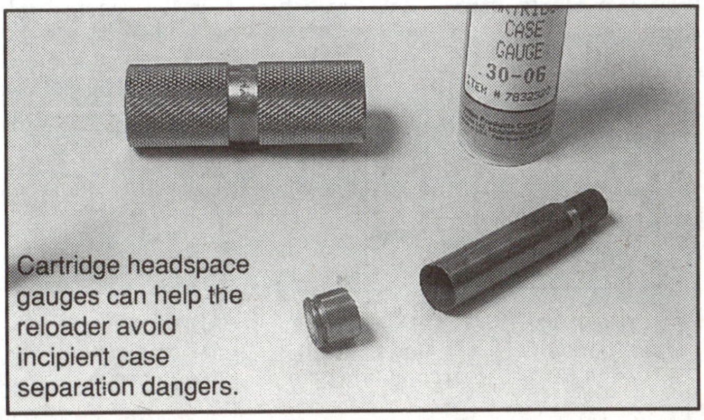

Cartridge headspace gauges can help the reloader avoid incipient case separation dangers.

The primers in the two cases on the left of this photo show signs of high pressure while the two on the right are normal. Primer condition is only one pressure indicator that you should monitor carefully when developing hunting handloads.

As stated, case headspace gauges will allow the reloader to set up full length resizing dies so that the fired case will not be shortened an excessive amount during the reloading process. This will help prevent incipient separation failures. However, if the chamber is excessively deep, no amount of die adjusting will prevent premature case failure as the case will be stretched beyond reasonable limits on its first firing. Gauges made for cartridges that headspace on the shoulder are of a fixed type while those that are built for belted cases are adjustable. Excellent case headspace gauges are available from several manufacturers including Lyman. Every ammo assembler should have a case headspace gauge for each cartridge reloaded.

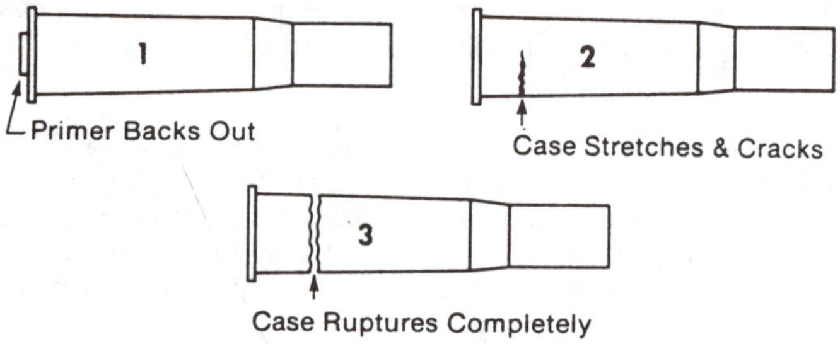

The above illustrations are indications of excessive headspace. Do not use any firearm which shows such signs of headspace problems.

Pressure Affects Case Life

Pressures generated in the cartridge at firing are also part of case life. For example, high pressures can enlarge primer pockets. If you feel a primer seat without the usual resistance, discard the case immediately. It is possible that a primer might seat without resistance because it is undersized. Either scenario allows for the potential dangerous leaking of hot gases from the case. Resist the temptation to squeeze one more firing from any case/primer combination that assembles without the usual feel of the primer being squeezed into the primer pocket. A few leaky pockets can ruin your bolt face. Worse, a blown primer can raise havoc with firearm and shooter.

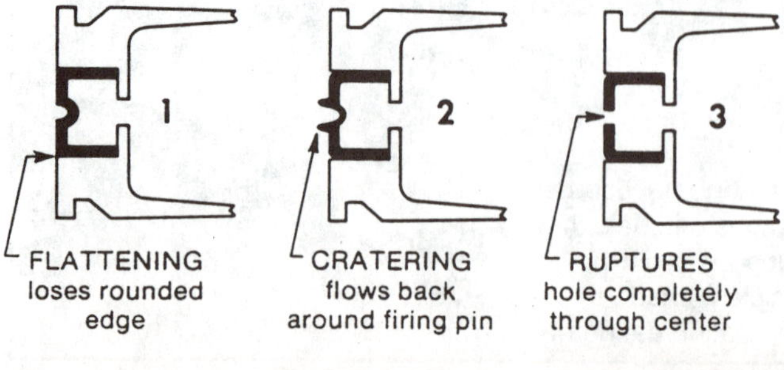

The degree of chamber pressure will cause notable changes in case usefulness. Some time back, we separated a lot of 270 brass into two portions. The first was used for a full-power hunting load with a chamber pressure of approximately 52,000 copper units of pressure (c.u.p.). These cases were fired five times before we found it necessary to discard the lot.

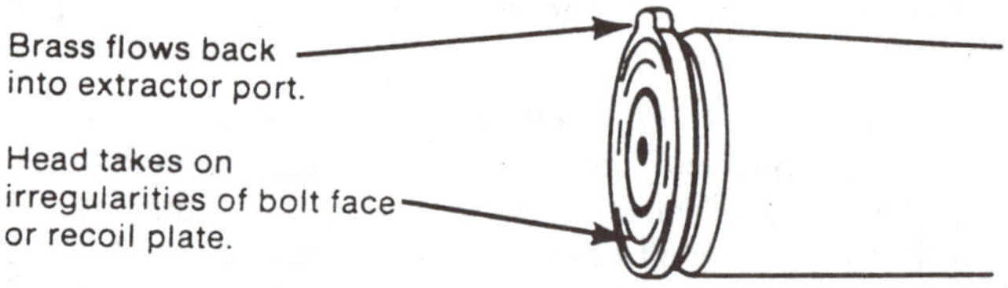

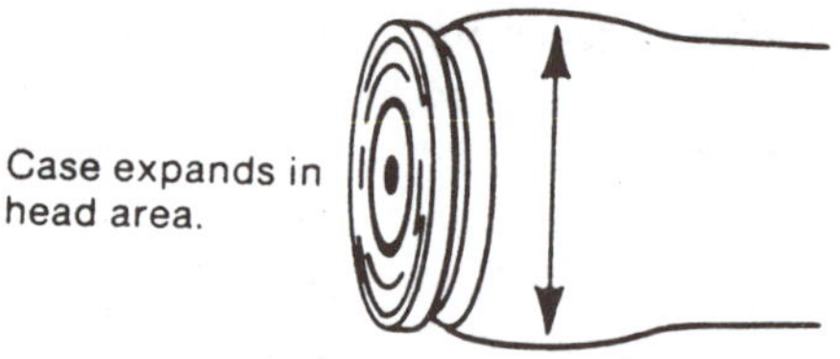

The cases in the second portion were loaded as light target loads with a chamber pressure of just over 40,000 c.u.p. These cases were fired eight times and showed no sign of wear. Indeed, they were only trimmed twice. We experienced 12 firings with these cases and loads before the need to discard the entire lot which were neck resized only. The novice reloader should not attempt so many loadings from a case until a full working knowledge, backed by extensive experience, has been gained.

Sizing And Case Life

Full-length resized cases, depending upon chamber and die dimensions, can also be worked excessively with respect to case diameters. Repetitive fired case diameter reduction, when resizing, can lead to premature longitudinal splits and cracks on the case. These frequently occur in the neck area. Sometimes such a problem can be eliminated by resizing only a short portion of the case neck. Neck sizing dies are available for most cartridges.

Sometimes reloaders opt for neck sizing based on expected increased case life. However, for hunting loads where smooth feeding, chambering, and extraction are important, Lyman suggests full length resizing.

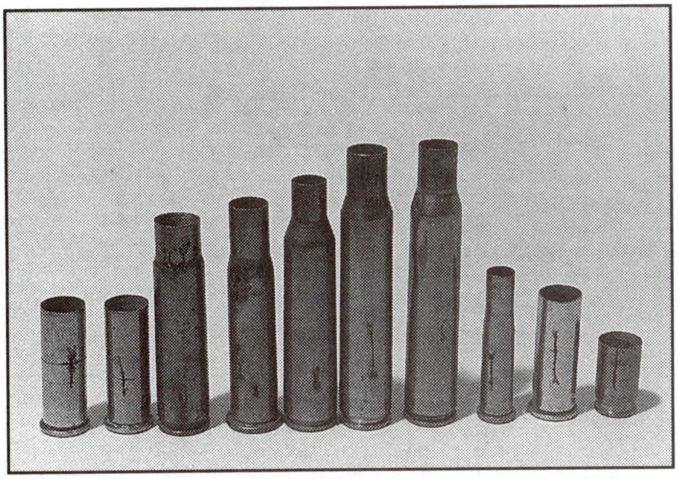

Split cases must be culled from the cases to be reloaded.

Case necks will eventually become work hardened and begin to split. This will occur after the same number of reloads whether the cases are neck sized only or are full-length sized. In the past it has been suggested that case life can be lengthened by annealing the brass. Lyman currently cautions against attempts to anneal case necks. By the time a case requires annealing at the neck, other factors will have usually brought about the end of the case's useful safe reloading life.

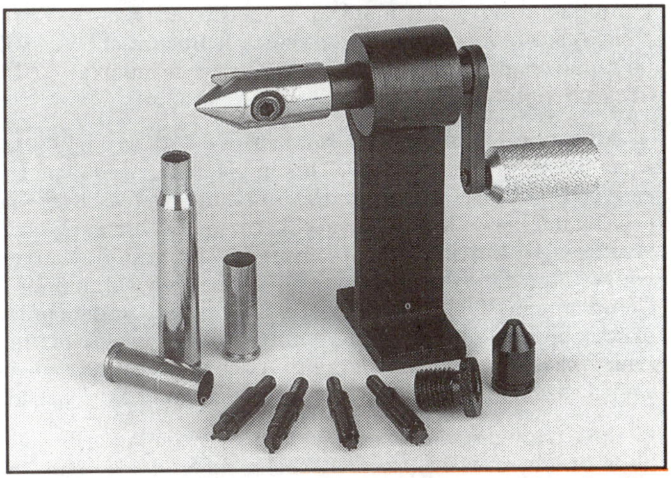

Careful case preparation is the secret to accuracy in reloading. The Lyman Case Care Kit has a bench mountable utility crank which speeds up small case prep jobs.

Additionally, it is almost impossible for the reloader to control and maintain the precise temperature required for proper annealing.

Cases Require Trimming

Cases stretch a bit with each firing and with each resizing. Sooner or later a case will exceed the maximum permissible length. Then it must be returned to a useful length by

Lyman offers three different types of case trimmers; the economical Accutrimmer, Universal Trimmer and Power Trimmer.

trimming away excess material from the case mouth. Every time a case requires trimming a certain amount of brass will be removed; perhaps as little as 0.005" or as much as 0.015" of case length.

After four trimmings, the removal of 0.020" to 0.060" case material will have occurred. This brass has flowed from the case body - from as far back as the junction of the case web and case wall. This area of the case will grow quite thin as the case stretches. By the time the case needs a fifth trimming the case wall often will be dangerously thin at the web/wall junction. Common sense tells us that it is impractical to determine exactly how safe or dangerous it may be to continue to further trim a case. So, good reloading practice dictates that whenever a case needs its fifth trimming, it should be discarded.

Case trimming may be accomplished with manually operated or electrically powered tools, dependent upon the reloader's preference and/or budget.

Case Necks Grow Thicker

While the walls of a case thin as it stretches from repeated firing and sizing, the neck section may thicken as a result of brass flowing forward in a manner that displaces it in several directions. This thickening can eventually produce a loaded

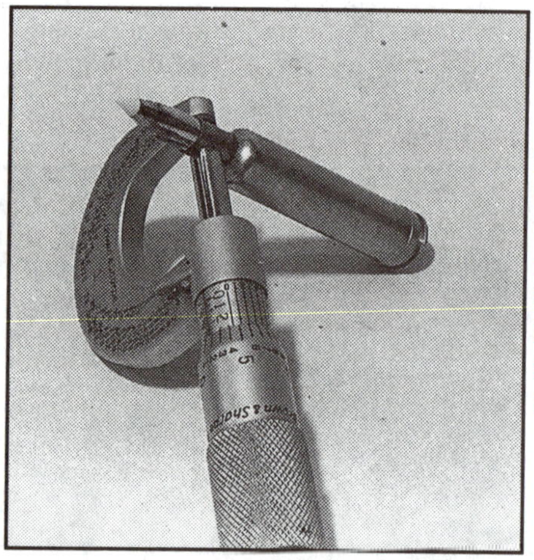

Sample measuring of the loaded round outside-neck diameter is an important safety step for every reloader.

round with an outside neck diameter that is incompatible with the firearm chamber. When a loaded round fits snugly into the neck area of the chamber, upon firing the brass will not be able to expand properly to insure an uninhibited bullet release. Thus, chamber pressure goes up higher than anticipated, perhaps becoming dangerously excessive.

While some reloaders do not pay attention to the outside neck diameter of loaded rounds, excessive case neck thickness is a definite problem and should be given the same amount of consideration as case length. Each problem can cause a serious hazard due to abnormally high chamber pressures. If pressures go high enough, depending on the degree of case lengthening, or neck thickening, a potentially dangerous accident could occur.

Case neck maximum dimensions are shown on the various cartridge drawings that appear at the beginning of the data for each specific cartridge. Take the time to measure a sampling of your cases after seating the bullet. Generally your case neck diameter will be well below the maximum. If cases should begin to get close to the maximum it is evidence of heavy brass flow from the case body. The wise reloader will discard the entire lot of brass when this condition begins to occur.

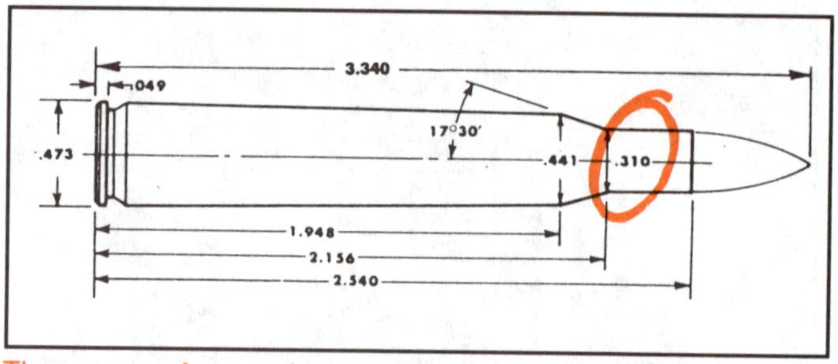

The proper reference for loaded round MAXIMUM outside-neck diameter is included in the cartridge drawing shown at the beginning of the cartridge's data listing.

Warning: Case neck turning equipment is designed to be used with new cases for the improvement of accuracy or the fitting of cases to the special dimension chambers that sometimes occur with benchrest rifles. It is also sometimes used when reforming new brass of a specific caliber to another caliber (not a undertaking for any but the most experienced persons). Case neck turning equipment is not intended for use in prolonging case life by the turning of overly thick necks.

Obviously, safe case handling demands that accurate records be kept for the number of firings, the likely pressure levels (max vs. less than max), and the number of trimmings. Don't forget to also record whether the cases were neck or full length resized. It is also recommended that you keep records on the number of cases discarded due to brass failures. When 10% of the cases in a lot have been discarded it is time to discard all the remaining cases in the lot. Doing so will help eliminate sudden and catastrophic case failure.

Case Inspection And Preparation
Segregate Cases By Lots

All new brass, or cases saved from the firing of factory ammunition, should be carefully segregated into homoge-

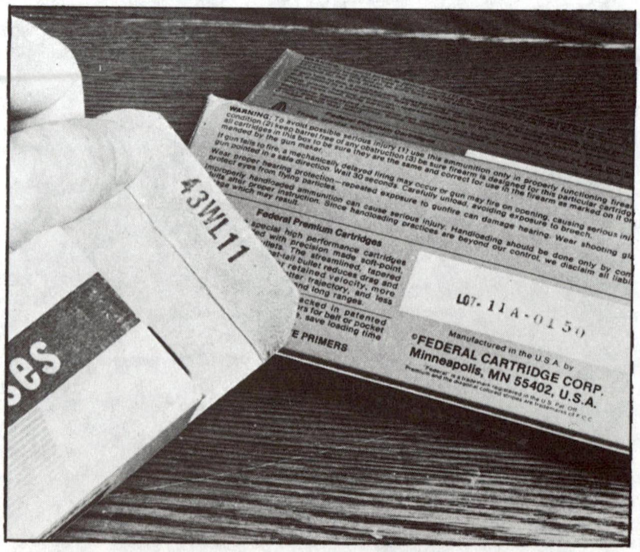

Lot numbers are important as the reloader needs to keep cases segregated to insure uniformity, accuracy and safety.

neous lots. Lot numbers are usually printed on the factory box of loaded ammo or brass - often on an inside flap or the rear panel. Brass purchased in bulk should be assigned a lot number for each box as originally packaged.

Such segregation is essential for uniformity of ballistics and accuracy, but more significantly, to ensure that brass cases are discsarded when they should no longer be used. Without segregation it may be necessary to discard perfectly good cases from another lot that may have been mixed in. More on this later in this manual.

Case Inspection

The reloading operation begins with case inspection. This is an easy but vital step to ensure the safety and reliability of reloaded ammunition. Thus, great care and sufficient time should be expended in the inspection process. Look carefully for any of the following:

- Cracks
- Splits
- Dents
- Perforations etc.
- Burn throughs
- Incipient partial head separation (or any tell tale early warning sign of eventual head separation)
- Corrosion
- Gas leakage around primer pocket
- Missing, backed out, or loose primer
- Abnormal case bulging or swelling
- Case Brass extrusion into ejector or extractor cuts of bolt face
- Foreign material in case.
- Any other condition that might suggest less than normal firing conditions

Case inspection must again be repeated after case sizing since cases can crack, split, incur damaged mouths, or have their rims bent during the resizing process. (Bent rims are an often overlooked cause for difficult chambering. Bent rims are usually caused by insufficient sizing lubricant. This causes undue force to be applied to the case rim when withdrawing the case from the sizing die.)

Case Cleaning

Cases can be cleaned sufficiently for reloading by simply wiping each one with an absorbent rag to remove any grit and grime that could destroy the fine micro finish polish of the resizing die.

Case tumblers come in various sizes to meet the need of each reloader.

However, most reloaders prefer to restore their cases to the bright shiny appearance of new factory cases. This can be easily accomplished by using one of Lyman's fine case polishing Turbo tumblers. Lyman offers a complete line of polishers from the economical Tubby Tumbler to the giant Mag Tumbler which is designed for high volume users. Each of the Lyman tumblers are vibratory cleaners that quickly restore cases to a look-like-new condition.

After using a tumbler it is important to insure that all polishing media has been removed from the inside of the case. Many reloaders choose Lyman's patented Auto-Flo tumblers which automatically separate cases from the polishing media when the tumbling is complete.

Case Resizing

Case sizing is generally accomplished in one of two types of resizing dies. Full length resizing, which resizes the case back to near factory dimensions is the preferred method for most reloaders. Neck sizing is a method used to lengthen case life when chamber dimensions are noticeably larger than full length resizing die dimensions. It is also a sizing method used to enhance accuracy for benchrest shooting. When neck sizing, the entire neck, or optionally only a portion of it may be reformed.

Cases can be damaged during resizing if care is not taken. For example, the use of too much resizing lubricant when full length resizing can cause dents to form on the case, especially on the case shoulder. In addition, the case mouth might be dented if the case does not enter the sizing die in correct alignment. Case necks, and to a lesser extent case bodies may also crack and split due to the stresses of resizing. If cases are neck expanded in a separate die, watch for any mouth splits that might form during the expanding operation.

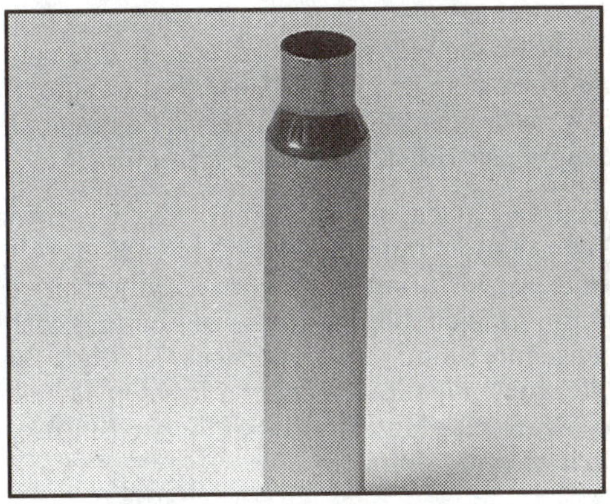
Cases can be dented in a full length or small base die as a result of using excess sizing lubricant (or a clogged air vent hole in the die).

After sizing the case, all traces of resizing lubricant should be removed by carefully wiping the case with a clean absorbent rag. Many reloaders accomplish this lubricant removal step by placing the decapped and resized cases into a case cleaning tumbler for a short period.

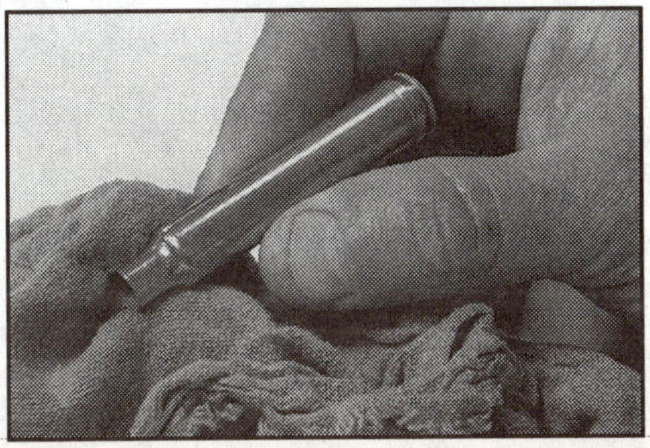
Carefully remove all traces of lubricant from the fired case after resizing.

Case Trimming

As mentioned earlier, cases stretch when fired and when resized. An overly long case will cause difficult chambering, or may actually prevent the chambering of a cartridge. Long cases will cause potentially dangerous increases in chamber pressure when the cartridge is fired. Obviously, the reloader must inspect each cartridge case's length after resizing.

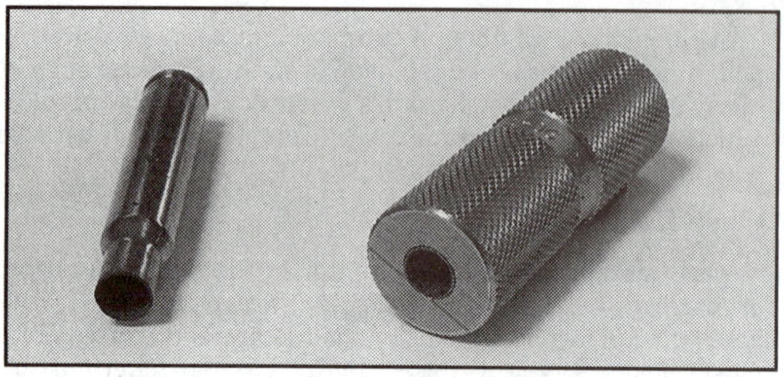

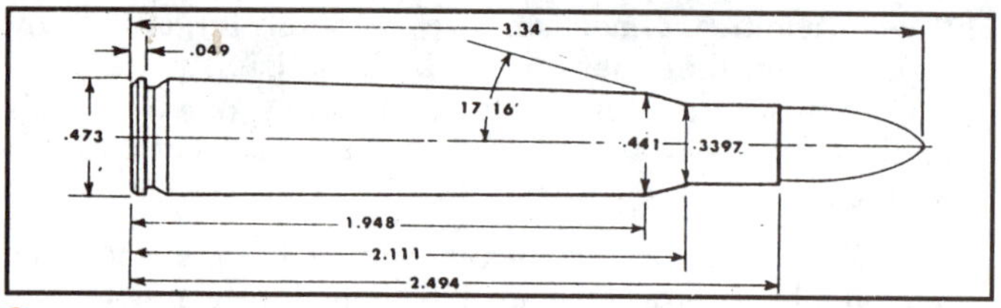

Case measurements are critical to ammo performance and safety. Checking the resized case for overall length and maximim headspace is essential during every reloading of the case.

Case measuring is usually accomplished by setting and locking a vernier or dial caliper at the maximum case length listed in the data section for each specific cartridge. Then slip each case between the measuring jaws. Those that will not fit require trimming. Case length gauges or cartridge gauges can also be used to isolate cartridges that need trimming.

As discussed earlier, it is important to keep careful records of the number of firings and trimmings of each case. Generally speaking, it is best to trim the entire lot of cases when one or more need trimming. This enables easy record keeping without the need to segregate small groups of cases. If only a few cases are trimmed at each loading session a lot of brass will quickly need to be segregated into small batches with differing numbers of firings and trimmings.

Cases are trimmed to the suggested trim-to length (as listed in the data section for each cartridge) using a small lathe-like case trimming tool. After trimming, the case mouth must be deburred. Use a deburring tool to quickly remove burrs from both the inside and outside of the case mouth. The concave end of the deburring tool is used to clean up the outside edge of the case while the convex end will remove the burrs from the inside edge.

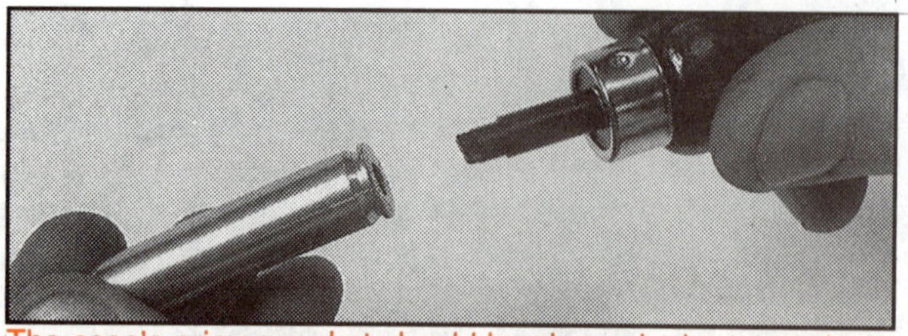

The case's primer pocket should be cleaned prior to seating a new primer. A few twists of an appropriate tool or brush will get the job done.

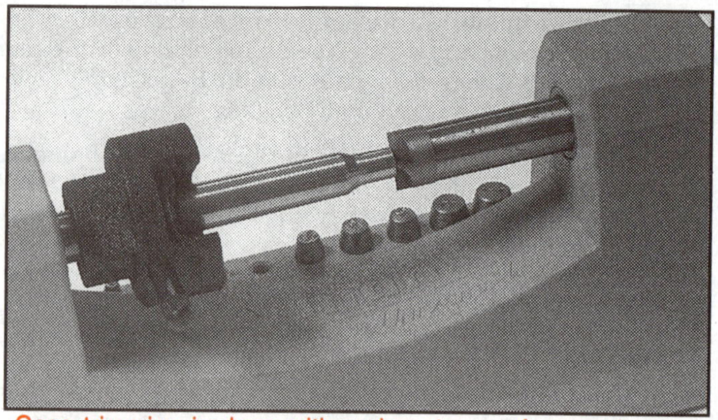

Case trimming is done with various types of case trimmers. Hand operated trimmers are the most common type used. Never trim any case more than 4 times.

Lyman offers the reloader several style trimmers. The most popular of these is the Universal Trimmer. This model has a patented chuck head which eliminates the need for various size collets to hold the case head. There is a power driven version of this model which appeals to high volume users. Reloaders on a budget might prefer the economical Accu Trimmer.

Again, never trim a case more than four times. Brass cases are severely weakened with the flow of notable amounts of brass into the neck.

Primer Pocket Cleaning

After decapping, the primer pockets will show traces of varying amounts of a carbon-like deposit. If the deposit is heavy enough it can interfere with the proper seating of the new primer. A few quick twists with a primer pocket cleaning tool or brush will rid the case of any such firing residue. Omitting this important step can lead to the dangers of high primers, or reduced accuracy.

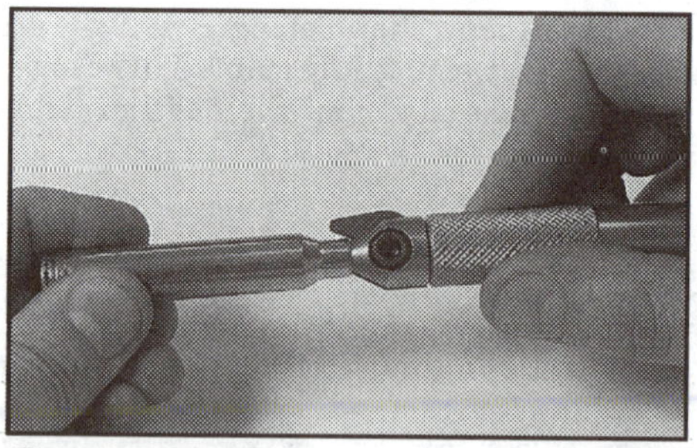

Deburring of both the inside and outside of the case neck must be done after each trimming.

Case Forming

Some reloaders choose to make brass of a given caliber from brass of a different caliber. This is often done due to less costly brass being available in a different caliber. For example, reforming military 308 Winchester (7.62 NATO) cases into 243 Winchester cases is often undertaken. This at first appears to be a simple matter of necking down the larger case by running it into the smaller die. However, such a simplistic approach could lead to serious problems.

When reforming cases to a smaller caliber, all of the material in the larger diameter neck must be moved into the smaller diameter neck. This will almost invariably result in an increase in the smaller caliber case neck wall thickness as well as neck length. The extra neck length is, of course, trimmed away. But the increase in case neck wall thickness is often overlooked.

Increased neck wall thickness can result in a loaded rounds outside neck diameter that noticeably exceeds the maximum allowable dimension. The safety of the load may thus be compromised as the case will fit the chamber so tightly that proper bullet release cannot occur at firing. When using reformed cases, careful measuring of the outside neck diameter of loaded rounds is essential. If a sample round exceeds the maximum then the case necks must be turned until they are of appropriate dimension.

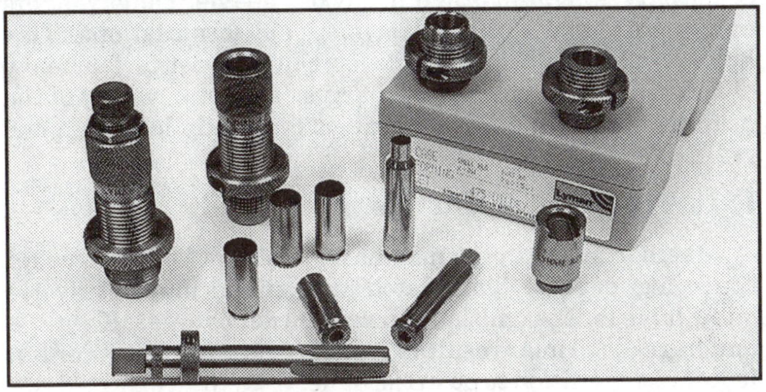

Case forming dies are used to make brass of one caliber into brass of another caliber.

Generally speaking, it is unwise to attempt case reforming unless there is a legitimate need. This operation should be undertaken only by experienced persons.

Case forming of a small caliber to a larger one is sometimes undertaken by reloaders. An example is necking up 30-06 cases to 35 Whelen cases. Necking up will noticeably thin the case neck. This can result in poor die performance, poor bullet retention and/or premature neck splitting.

Brass can be purchased in bulk at a a price low enough to make the time and expense of case alteration impractical. Even if time is considered valueless and thus the altering of cases should appear appealing, it is strongly suggested that the reloader consider the hazards involved with such operations. Additionally, the safety factor involved when using cases with the wrong headstamp, or perhaps with body annealing effected (as when resizing to a shorter cartridge) seems to be of far more concern than the purchase price of a hundred new cases of appropriate size.

Case Savvy

Cases are the weak link in the control of chamber pressure. Thus every effort must be made to insure that each case used is suitable for another firing with the load in use. Exacting case inspection cannot be overstressed. If there is any doubt with respect to case reusability - crimp it shut and throw it

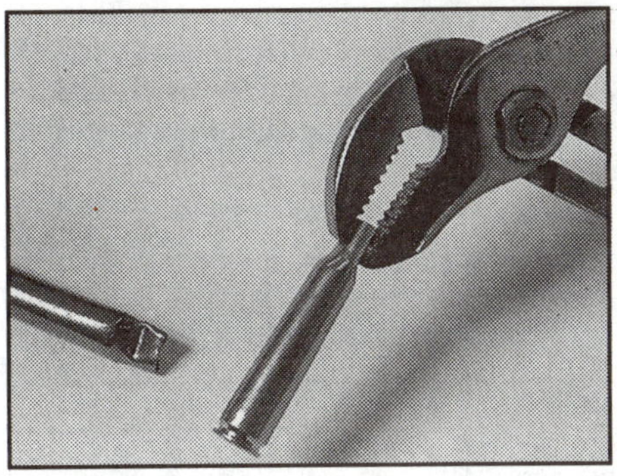

Always crimp the case mouth shut whenever discarding brass to prevent later unintentional use of it.

out. Remember that case inspection continues throughout the loading process. After the initial inspection, cases should be looked over after each operation. Did the case crack, split, or deform during sizing? When neck expanding is done separately did any cracks occur during the operation? Did the primer seat too easily? Or perhaps did the case mouth split during bullet seating?

The safety of the reloaded cartridge will always hinge heavily on the condition of the brass case. The reloader can assure the case's suitability with careful inspection. The time spent doing so is part of what makes a knowledgeable and conscientious reloader.

It is always best to load with clean cases. For the highest level of convenience, Lyman's Auto-Flo Tumblers automatically separate the cases from media after cleaning.

CHAPTER 6
PRIMERS - SELECTION, USE, AND STORAGE

Primers are taken for granted by some reloaders. They are tiny components that require just brief handling and they usually seem to work. Some give the primer and the priming operation no more than a passing thought.

Primers are made in basic sizes which include large and small rifle and handgun sizes. There are also magnum versions of each of the basic four sizes. There are substantial differences in both dimensional and hardness aspects of rifle and handgun primers. The differences in height are required for the differing depths of primer pockets. The hardness differences are required due to the strength of firing pin blows. Additionally, certain size cases and propellant charges require basic changes in the amount of ignition force.

The selection of a primer size for a new cartridge is sometimes based on specific ignition characteristics. Large magnum size cases and their heavy powder charges may demand the heaviest possible primer size, therefore large rifle magnum primers may be required. Sometimes a firearm design dictates a specific primer size. An example of this is the 40 S&W. If this cartridge had a large pistol primer, the live primer would be impacted by the ejector when a loaded round was ejected from the chamber. Hence, a small diameter primer is required for this cartridge. Primer manufacturers have carefully worked out the nuances of primers and the reloader should not be concerned with specific differences so long as the correct size and type primer is used. Proper primer choice is listed in the data section for each specific cartridge.

Despite this, serious thought should be given to the care in handling and storage of primers by reloaders. First and foremost, it must be noted that primers are designed to explode when the priming pellet is crushed between the anvil and the primer cup. This is accomplished by a percussion blow. The firing pin strikes the primer cup and crushes it against the primer anvil, with the priming pellet sandwiched between the two.

Primer Makeup

The compound used in primers is basically made by sulfonating resorcinol with sulfuric acid. Then nitric acid is used to nitrate the compound to form trinitro-resorcinol, commonly known as styphnic acid. This acid is then treated with lead nitrate under precisely controlled conditions to form the lead styphnate salt. Several forms of this salt are used in priming compounds the most common being "normal" structure. Another type used by one manufacturer is called "basic" structure.

Manufacturers add fuels such as tetracene and barium-nitrate and other materials such as small amounts of aluminum or glass to enhance specific desired characteristics and/or sensitivity. Magnum primers may use heavier pellet weights and/or somewhat different primer mix formulations. This helps to obtain desirable ignition characteristics for heavy difficult-to-ignite propellant charges. The total weight of the priming pellet used in each primer is usually less than one grain.

Primers Effect Ballistics

Match grade primers may have very exact tolerances for the primer pellet weight. Match type primers may not always be suitable for the ignition of large powder charges. Should a match grade primer result in a misfire (scorched powder but insufficient ignition to push the bullet from the case and down the barrel) or in a hangfire (delayed ignition with sometimes audible pause between firing pin strike and ignition), immediately discontinue (do not fire even one more round) the use of that primer with the cartridge being used. Such ignition characteristics can be hazardous.

All variations of different primers result in varying ballistics and uniformity, if otherwise identical components are used for loading. For example, one set of tests have been conducted wherein uniformity (the extreme spread of velocity and pressure) ran from as little as 38 fps. with one primer to 133 fps. with another primer. This for a string of 10 shots. Pressure comparisons among the same group of test loads showed extreme spreads of pressure at as little as 3,500 psi. to as much as 19,300 psi. The range of performance ran from very good to abysmal with the test load based solely on the primer used. In the same series the highest average velocity was 3000 fps. and the lowest average was 2960 fps. While there was no significant average velocity change, the ballistic worth of the loads using seven different primers was quite varied.

It is not possible to say that one specific primer does a better job than another. A given primer might perform poorly with a specific load in one caliber, satisfactorily with a second load in another caliber and outstanding with a third load in still another caliber. The point is that primer testing can be an important part of load tuning.

U.S. BOXER PRIMER CHART

	Large Rifle	Large Rifle Magnum	Small Rifle	Small Rifle Magnum	Large Pistol	Large Pistol Magnum	Small Pistol	Small Pistol Magnum
Remington	9½	9½M	6½	7½	2½		1½	5½
Winchester	WLR (8½-120)		WSR (6½-116)		‡WLP (7-111)		WSP (1½-108)	WSPM (1½M-108)
Federal	210 +210M	215	*200 **205 ++205M		150	155	100	
CCI	200 BR2	250 MAG	400 BR4	450 MAG	300	350 MAG	500	550 MAG

*Also may be used for magnum, small pistol loads.
**Thick cup design, especially for .17 Rem. and .22 centerfire loads.
+Match version of No. 210.
++Match version of No. 205.
‡Also may be used for Magnum, Large Pistol loads.

Primer Seating

Primers are assembled so the anvil rests with minimal contact on the priming pellet. In this condition primers are not fully sensitized for ideal ignition. Further compression of the priming pellet by the anvil, during proper seating of the primer into the cartridge case, completes the desired sensitizing of the primer.

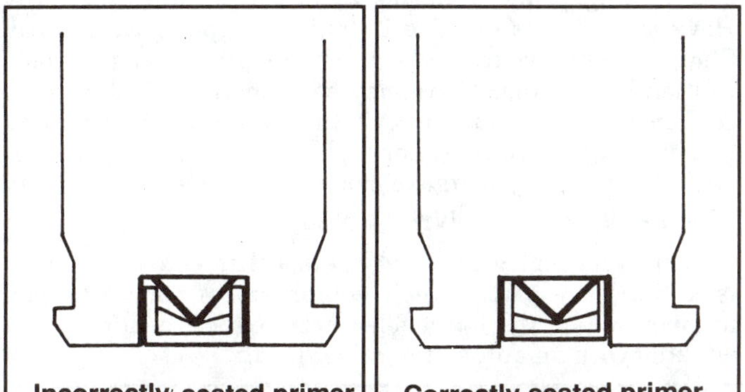

A correctly seated primer will insure positive ignition and the best ballistic uniformity. An improperly seated primer will result in unreliable ignition and poor ballistic uniformity.

Properly seated, the primer cup will be below the flush of the case head. Such positioning will insure that the anvil is lightly pushed down into the primer cup to properly stress the priming pellet. For all but CCI brand, primers should be seated 0.003" to 0.007" below flush. If the reloader tries to seat primers at a nominal depth of 0.005" below flush, normal variations found in primer heights and primer pockets will result in seating depths within the suggested tolerances.

CCI primers require a below flush seating of 0.003" to 0.005". There is a gauge available to easily measure the depth of CCI primers.

Seating primers flush, or somewhat above flush, will fail to cause the anvil to move down into the primer cup and the pellet will not be properly stressed. This can result in misfires or erratic ignition.

Warning: A primer that is above flush of the case head can be accidentally ignited by bolt or action parts during feeding. If ignition occurs before the action is closed and locked, the results can be catastrophic to the firearm, shooter and bystanders.

Excessively deep primer seating can also cause unwanted problems. The primer cup can collapse when seated overly deep. This condition can result in gas blow-by between cup and case. Serious bolt face erosion can then occur. In addition, excessively stressing the primer pellet can cause the primer to become too sensitive. This might lead to ignition as a result of rough handling of the loaded ammo. If a primer is seated too deeply, the priming pellet may crack due to excessive compression by the anvil. Usually a cracked pellet will fail to explode resulting in a misfire. On occasion, excessively deep seating causes ignition during the priming operation.

Deep primer seating can also cause ignition failure. This occurs when the firing pin cannot strike the primer with adequate force.

Priming on the T-Mag press may be accomplished with the use of a priming arm.

Many reloaders prime by feel, seating the primer until the cup is felt bottoming in the primer pocket. This method sometimes results in unnecessary variation in primer seating depths. It's hard to precisely apply pressure to the seating tool handle while attempting to "feel" the cup bottom through the chain of tiny anvil, primer cup, case, primer punch, tool linkage, tool handle, and the operator's hand. Seating by feel is also subject to variations in the primer cup fit in the primer pocket, and any crud build-up in the primer pocket. The "feel" method should be left to the very experienced person. For most reloaders the seating of primers by feel should be avoided.

Generally, the best method for seating primers is to use a mechanical stop to arrest the motion used to move primers into the primer pocket. Most presses are equipped with various types of mechanical stops for primer seating. The method in whch a priming station is assembled in the die position at the top of a press is recommended. This type of unit is adjusted so that the normal stop of the press linkage occurs when the primer is at the correct depth.

It is a wise practice to seat primers about halfway into the primer pocket and then rotate the case 180 degrees before finishing the seating operation. This insures that primers are squarely seated.

Properly calibrated, a dial indicating caliper makes a useful tool for checking primer depths. Start with the rod end of the vernier protruding about 0.015". Then place the rod

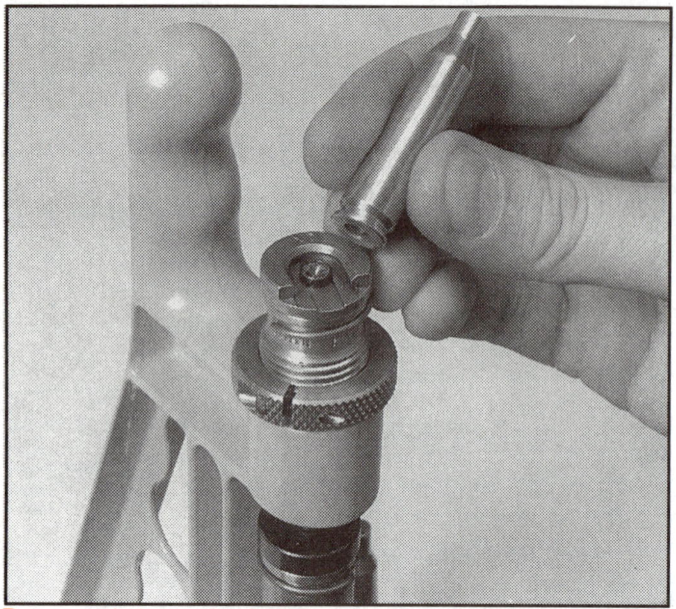

Remote priming tools of many types are available. Shown here is a RAM prime unit mounted on the Lyman Accupress.

against the primer while seating the base of the vernier against the case head. The primer seating depth (below flush) can then be read from the dial. Some practice is required to obtain accurate primer depth measurements.

Running a finger over the case head, after a primer has been seated to the correct depth, will give you a feel for a correctly seated primer. With practice, you will be able to "feel" when a primer has not been seated deep enough, when it is seated within an acceptable range, or if it is too deep in the primer pocket. At the beginning of each loading session, measure enough primers to insure they are correctly seated. Then "feel" each subsequently seated primer to insure none vary from the desired depth.

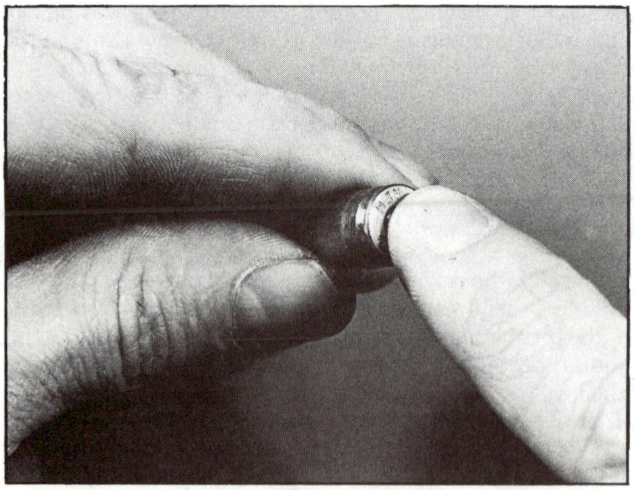

With practice, running a finger over the primed case will enable the reloader to detect high, properly seated, or excessively deep primers.

Important: Wash and dry your hands thoroughly immediately before priming. Before doing so, make certain all of the case sizing lubricant is removed from all cartridge cases. Develop a priming routine that will not allow primers to become contaminated with anything. One method is to manipulate the press handle and the cartridge cases with one hand and use the other hand solely to pick up primers from the primer plate and place them on the priming post. This method will give 100% assurance that your primers will never be contaminated.

CAUTION: Primer Feeds Require Extreme Care

Handling primers one at a time is a safe procedure. However, primers handled in bulk are a potential hazard. Should you choose to use an automatic primer feed extreme caution is required. There may be as many as 100 primers each in contact with the next in the automatic primer feed's magazine. Should one primer explode, all will do so spontaneously. Even a few primers exploding can cause serious property damage and personal injury.

Some very high volume reloaders feel an auto primer feed is a necessary part of their equipment. If you think you number among the latter folks, use extreme caution when working with an automatic primer feed.

The SAAMI pamphlet on "Sporting Ammunition Primers" is a good reference.

In the SAAMI pamphlet "Sporting Ammunition Primers", the major manufacturers of primers caution against the use of primer feeds unless adequate shielding separates the operator from the hazard of explosion. The actual use of an automatic primer feed unit also demands an uncompromising light hand to prevent the crushing of a primer should a jam or malfunction of any type occur.

Never, never push, bang, or otherwise apply any force to a primer or primer feed.

If you elect to use an automatic primer feed **stop** at the first sign of a primer jam. Then think about what must be done. Force nothing and do not be tempted to even jiggle the tool or any part of it. Rather carefully disassemble the primer feed, first removing the primer magazine and all primers contained therein.

Primers are explosive and if you abuse a jammed primer it might well show you its devastating force.

Caution: Extreme care must be used when filling a primer feed magazine tube with primers. Never force a primer into the pickup tube. A force of no more than five pounds should be required for the primer to go into place. Do not use any primer pickup tube requiring more than this amount of effort. Primers must be correctly oriented in the pickup tube. It is the users obligation to insure that this is so. Misoriented (upside-down or standing on end) primers are a serious potential hazard. Never store Primers in a primer feed tube. Return all unused primers to their original packaging. Dot not force any primer feed tube-follower down onto primers.

Caution: Primers Can Dust

Primers can and do "dust". Minute amounts of priming mix will accumulate in the loading area, and on primer handling equipment, especially when an auto primer feed is used. An accumulation of priming dust is a fire and/or explosion hazard. Occasionally wiping off priming trays, priming tools and the work area with a damp cloth will reduce the likelihood of an accident. Rinse the wiping cloth thoroughly in running water to remove any primer dust it picks up before storing it.

The entire loading area and all equipment must be kept clean and free of any contamination by primer dust (and powder accumulations). Fired primers, separated cups and anvils, or any trace of hard or abrasive material can be the catalyst that causes primers, primer dust or spilled powder to ignite.

Pick up spilled primers immediately and be sure they are all recovered. Stepping on a primer may cause it to explode.

Primers are Explosive

Primers by nature are explosive. As a result, they may explode if subjected to friction, percussion, crushing, or excess heat from any cause whether open flame or not. Static electricity and many other abuses may cause primers to explode.

The reloader should avoid such things as hammering, impacts or percussion blows of any kind, including bullet impact. Hot tobacco ash, sparks and a great many other abuses must be avoided. In short, treat primers as an explosive. Careful handling and storage is always a must.

Never transfer primers from their original factory container to any other container. Primers stored loose or in bulk, such as in a glass jar, will cause a violent explosion if one is ignited.

When priming, pour out no more primers than needed. Never place more than a single box of 100 primers into the primer flipper. After each priming session, immediately return any unused primers to their original packaging.

It is important to prevent the buildup of static electricity in any primer handling equipment or on the person handling or loading the primers. Electrically ground all loading equipment.

Never decap live primers. It is a needless hazard.

Primer Storage

Due to their explosive nature only an absolute minimum should be kept in storage. The National Fire Protection Association's NFPA 495 says that not more than 10,000 primers should be stored in a private residence. This recommendation is law in most communities. Even a deeply involved reloader should have no reason to store more than 1,000 each of large rifle, large rifle magnum, small rifle, large pistol, large pistol magnum, small pistol, small pistol magnum, and shotshell primers. That's 8,000 primers. With care in replacing exhausted supplies, it is not difficult to adhere to the 10,000 primer limit. Many, if not most, reloaders will be able to get along nicely with far fewer primers kept on hand.

Always store primers in a remote area away from any possible source of ignition including bullet impact. Be sure no heat, spark, electrical, percussion, or any other form of abuse can occur in the storage area.

Keep primers stored away from oxidizing agents, flammable liquids, and flammable solids (including handloading powders).

A storage cabinet is strongly recommended. A cabinet should be solidly constructed of 1" thick lumber to delay the transfer of heat to the contents in event of a fire or other mishap. Make sure the storage cabinet is remotely located with respect to trash, combustible materials, sources of heat (including the suns rays), open flames, electrical equipment, hot water heaters, mechanical equipment, furnaces, solvents, flammable gases, and so on. To avoid the potential of an unexpected or unusual accident, do not store anything else in the primer cabinet.

Naturally never smoke around primers. No smoking signs in the storage area and at the loading bench are suggested.

A bulletin is available from SAAMI entitled "Sporting Ammunition Primers: Properties, Handling, And Storage For Handloading". Every reloader should acquire a copy from:

SAAMI
P.O. Box 838
Branford, CT 06405

Color Coding

Some manufacturers color code the primer foil - the paper-like separator between the primer pellet and the anvil. This color coding combined with the primers physical size is a way of assuring that the appropriate primer is packaged in the correct box. The following table is a list of the color codes for those U.S. manufacturers who follow this commendable practice.

COLOR CODING OF PRIMER FOILS

PrimerType	Winchester	Remington	Federal
Small pistol	WSP red	#1-1/2 white* (green)	#100 green
Small pistol magnum	WSPM orange* (yellow)	#5-1/2 red	——
Large pistol	WLP orange* (yellow)	#2-1/2 white* (green)	#150 green
Large pistol magnum	——	——	#155 blue
Small rifle	WSR blue* (green)	#7-1/2 white* (green)	#200 red
Small rifle match	——	——	#205M purple
Small rifle magnum	——	——	#205 purple
Large rifle	WLR red	#9-1/2 blue	#210 red
Large rifle match	——	——	#210M red
Large rifle magnum	WLRM green	#9-1/2M white* (green)	#215 purple

*The lacquer color may appear to be the color in parenthesis due to color density.

With the foregoing in mind, it is obvious that giving a bit of thought to primers and the priming operation will result in more uniform ammo and perhaps prevent an accident. Both are worthwhile, even essential goals.

For Information
or
Technical Assistance
Call
1-800-22-LYMAN

CHAPTER 7
POWDERS - SELECTION, USE, AND STORAGE

This is just a small selection of the many propellants used in the data for this handbook. Even with more than five dozen powders to select from, the choice can be both quick and precise. When the reader understands why there may be more than a dozen listings for a specific bullet weight and follows our suggestions, testing of a load need not be difficult.

The smokeless powders used for the loading of modern rifle and handgun ammunition are all flammable solid propellants. As such they are designed and intended to burn rapidly and can do so without the presence of oxygen - they generate all required oxygen as they burn. The handloader must treat all propellants with respect. We will cover many specific safety factors later in this chapter.

Selection Of An Appropriate Propellant

There are more than 60 different propellants listed in the data section of this manual. To the beginner such wide choices can appear confusing, even intimidating. It is our purpose to help you make the most appropriate selection for your needs. We show why there is a need to list more than three or four of the best choices, and to generally acquaint you with the extensive data tables.

Many cartridges are best served with relatively few propellants. Best served means those propellants which will combine the best possible accuracy with a near maximum velocity and extremely uniform ballistics. Be assured that each powder listed in this data book meets stringent criteria for uniformity and peak pressure. Nonetheless only a few of the listed powders will optimize these criteria with the tightest groups and the highest velocities.

Why list all those other powders? There are many reasons to do so. Perhaps the first is the reloaders interest in safety. Many reloaders recognize the safety in stocking only a few types of powder. This minimizes the potential for using the wrong powder and keeps quantities stored at a minimum. Some reloaders will, therefore, select the propellant best suited to their most important or demanding application. Then they use this propellant in as many other cartridges as possible to meet their specific needs. Some even select additional cartridges based on their likelihood to perform with their favorite propellant.

Availability is another reason for not using one of the "best" selections. Not every retail outlet handles a complete line of powders. Those that are stocked may vary greatly in ballistic performance. Clearly, an appropriate powder must pass the test of availability.

There are also specific firearms which will not deliver the desired accuracy with any of the best powders. Often this is a direct or indirect result of some specific problem existing within the firearm, for example a barrel which has internal stress. Sometimes a diligent effort to try many powders will turn up one that will compensate for the firearms shortcoming and provide fully acceptable accuracy. There simply is no way to predict what propellant may be the likely candidate for this scenario.

Some reloaders have good reason not to want the highest possible velocity from a specific bullet weight. A desire for recoil reduction, or the use of a very powerful cartridge on relatively light game are just two reasons that a powder might be selected from those that are not the "best choices". What is best for one reloader may not be for another reloader. A reloader wishing to use a 375 H&H Magnum for deer hunting may deliberately select a fast burning propellant that offers noticeably lower velocity than other choices. Such selection will allow for a reduction in recoil and perhaps less meat damage.

When considering a ballistic reduction, some reloaders feel that simply loading a reduced charge of the ideal propellant makes sense rather than selecting another powder. This is not an accurate assessment of the problem. Loads that fall below the suggested starting charge often will not deliver acceptable ballistics. In some instances such loads can actually be unsafe to use. Pressure variations may be so extreme that they create individual pressure levels far beyond the limits of safety.

(Left) a 125-pound block of nitro cellulose prior to extrusion at the Hercules Plant, Kenville, N.J.

(Right) an extrusion press produces soft spaghetti-like strands of powder which is later cut into the appropriate lengths. (Hercules)

For the reasons just stated there is a limit to the low end ballistics to which each cartridge can be loaded; just as there is a limit to the reachable upper limits.

To help the average reloader, and the person who wishes to combine top accuracy, good ballistic uniformity and near maximum velocity, we clearly suggest specific propellants in the comments preceding the data for most cartridges. The suggestion may be limited to a single outstanding performer. Or, we may suggest two, three or even more propellants as being top choices. We fully expect that any of the suggested propellants will prove quite satisfactory.

In rare cartridges we will make no specific suggestion. In these instances, the performance level of one propellant versus another may be indistinguishable due to the cartridge or firearm chambered for it.

We believe that we have covered all of the popular applications, as well as a very wide range of ballistics, for each cartridge listed in the data section. Naturally, seldom used cartridges did not see the extensive testing of more popular rounds.

Propellant Versus Case Volume

Generally speaking, the powders that give the best ballistic uniformity and the best accuracy are often those that most nearly fill the case to the base of the bullet. Ideally the reloader should strive to fill the case from at least 90%, to perhaps a slight compression of the powder - about 110%. This insures that powder positioning in the case will not cause substantial changes in the ballistics - as a result of ignition characteristics.

An example of less than ideal ballistic uniformity would be a cartridge case that was only half filled with powder. When held horizontally the powder might be spread evenly over the entire bottom half of the case. Ignition of the primer would then result in the entire top surface of the powder, a relatively large area being ignited at once. If the same load were fired in a vertical position, the primer flash would work primarily on the much smaller surface of powder in direct contact with the case base. It is true that the strength of the primer flash would see ignition carried up into the powder column, but this characteristic would not equate to an ignition of the same magnitude as discussed in the first example. Filling the case, within plus or minus 10% is the ideal choice.

It is not always possible to fill a case based on ballistic considerations. But when it can be done, this is the best choice for the reloader.

Range testing of any propellant is always required. For the best combination of accuracy and ballistics the reloader should read the introductory comments at the beginning of the data for each cartridge.

Compressed Powder Loads

One consideration of the actual powder charge is how it fits into the case. Many of the loads in this handbook are noted as being compressed powder charges. This means that the listed powder charge fills the case beyond the point where the base of the bullet will be when it is seated to the depth used in our testing. Naturally, if you seat your bullets less or more deeply than we did, the amount of propellant compression may vary. The degree of variation you experience could range from no compression to excessive compression.

Compression of the powder will depend not only on the weight of the powder charge but on the size of the cartridge case. For example a 47 grain powder charge may be a lot to get into the case if you are loading a 243 Winchester. A charge of 65 grains might be questionable in a 7mm Remington Magnum, while it will be no special trick to get a 100 grain charge loaded into a 416 Rigby case.

Compressed powder loads (where the powder charge fills the case so much that the seated bullet must compress the powder down into a smaller volume) can cause the reloader difficulty. The text explains how to judge the degree of compression and the acceptability of a specific compressed powder charge.

Not all propellants should be compressed. Generally speaking we advise against the compression of powder in any rifle or handgun cartridge load that is not indicated by our data as being tested as a compressed powder load.

Heavily compressed powder charges (those that fill more than 110% of the available case volume with the used bullet seating depth) are to be avoided. The results of excessively compressed charges might include cartridge overall lengths that continue to grow after the loaded round is taken from the bullet seating die. This is caused by the elastic nature of the compressed charge trying to return to its original volume. This means when the internal upward force of the compressed powder exceeds the strength of the case neck to grip on the bullet the powder will slowly force the bullet out of the case. This can lead to jammed firearms and erratic ballistics, as well as very poor accuracy. None of these are desirable attributes.

Obviously, the reloader must be able to differentiate between acceptable light compression and excessively heavy compression of the powder charge. This is not a difficult evaluation.

The depth to which a bullet is seated will vary considerably with bullet weight, chamber dimensions and reloader preference. In a 338 Winchester Magnum, the base of a heavy bullet (i.e. 250 grains) may well need to be pushed past the shoulder/neck junction and into the case body. On the other hand, a light (200 grain) bullet being seated into the same 338 case may need to be seated so that it occupies only the neck area.

It is not necessary to know the actual percent of powder compression to come up with an answer to this question: "Is this load's powder charge compressed acceptably or excessively?"

To determine the degree of compression, seat the selected bullet in an empty case to the desired overall cartridge length. Carefully secure the seating screw to insure a uniform seating depth on succeeding rounds. Also be certain that the die body is locked in a secure position.

Typical round to round overall length variation might range from +/- 0.003" to +/- 0.010" in the case of hollow point bullets. The variation is caused by differences in individual bullet tips, the reloader's ability to measure repeatable lengths, tooling, and reloader techniques. You will, however, be able to establish an overall length for your components and tooling being employed by simply measuring a single cartridge. Make note of this overall length. Assume a potential variation of +/ -0.008" as the total range to be encountered. Experience has shown this to be a realistic approach.

168 gr. Jacketed HPBT
2.800" OAL

Powder	Sugg. Starting Grains	Velocity fps	Pressure C.U.P.	Max. Load Grains	Velocity fps	Pressure C.U.P.
IMR-3031	33.6	2290	36,500	38.2	2575	51,500
748	36.4	2280	36,200	41.4	2586	51,000
IMR-4064	35.6	2339	39,600	40.5+	2605	51,800
IMR-4895	34.2	2306	39,200	38.9	2555	51,300
IMR-4320	34.7	2251	37,300	39.5	2510	49,100
H-380	38.2	2256	37,000	43.5+	2555	51,600

Note: Loads shown in shaded panels are maximum.
+ Designates a compressed powder charge.

Then place the desired powder charge, for which you wish to determine of the degree of propellant compression, into a primed case. Let the powder charge pour into the case normally. No attempt should be made to tap the powder charge into a minimum volume. With the previously adjusted bullet seating die, seat a bullet into the powder charged case. Without delay measure and record the overall length of this round. If the charged cartridge's overall length is longer than the empty sample, plus the tolerance of +0.008", the loaded round has an excessively compressed powder charge.

If the loaded length is the same as the empty sample (again within the tolerance of +0.008"), remeasure it after waiting 24 hours. If it still remains within the anticipated overall length tolerance the load is probably compressed within an acceptable range.

Final verification of acceptable compression requires assembling two or three boxes of ammo, checking each round for overall length immediately after bullet seating and again after 24 hours. If none of these loaded rounds exceeds the range of earlier suggested empty case length, the degree of propellant compression is fully acceptable.

The amount of force applied to the bullet's base by compressed powder charges will vary. The method used when pouring the charge into the case and the propellant granulation size will both determine the extent of this force. The method described above allows a judgement as to whether or not the degree of compression is acceptable for a specific set of components and loading dimensions.

Note: Do not use any bullet crimp when using this method to judge propellant compression acceptability. Also, keep in mind that a neck sized case will produce less powder compression than a full length sized case. This is because the full length sized case will have a smaller internal volume.

If a heavily compressed load is acceptable, crimping the case to the bullet may provide sufficient bullet retention to prevent the problems that could arise with creeping bullets. Never use any compressed charge beyond the data's maximum load recommendations. A compressed charge should never be used unless the bullet is seated to the specified overall cartridge length listed in the data table.

There are steps that can be taken to get a specific recommended powder charge to occupy less case space. Using a powder funnel with an extra long drop tube section will allow powder to settle in the case at a higher level of density. Or you can try pouring the powder into the case at a very slow rate. Tapping the funnel or case during the powder pouring also helps settle the powder to maximum density.

The data in this handbook was not developed using any special methods to settle the powder into the case. Should any method beyond that already described be required to settle powder sufficiently to avoid excessive compression we suggest that the propellant in use be rejected in favor of another. This is because extraordinary powder settling might cause an unanticipated change in ballistics.

Sometimes a heavy powder charge can be made to fit simply by switching the brand of cases. Even a change in lot number of the same brand can sometimes show a change in case volume due to individual lot variations in case wall thickness.

Powder Characteristics

Propellants are made in various configurations. The most commonly encountered form is a tubular granulation with or with out longitudinal perforations. Another shape is commonly referred to as "Ball" or Spherical powder. This type is made under a process in which surface tension of the forming propellant causes it to take on a round configuration. Most of these powders are later flattened to varying degrees in order to control burning rates. They may lose much of their rounded shape in this process, sometimes being reduced to almost a flake type shape. Finally there are powders which are produced in flake or wafer type configuration. Often these are relatively fast burning handgun or shotgun type propellants.

Sporting rifle and handgun smokeless propellants are compounds made from certain chemicals, mainly nitrocellulose or a combination of nitrocellulose and nitroglycerine. These are designed to burn at predetermined pressure levels in the confines of a cartridge case.

Propellants which use only nitrocellulose as the main energy ingredient are called single base powders. Those that use a blend of nitrocellulose and nitroglycerine for their energy source are known as double base.

Winchester powders are spherical propellants that start out in a basically true ball shape; but in order to control burning rates the powder undergoes varying degrees of flattening. IMR powders are mostly single base powders which generally start out as very long spaghetti-like extrusions. These are then cut up into very short tubular grains. Alliant powders are generally double base types of an extruded tubular or flake shape. Hodgdon sells both spherical and tubular powders of both single and double base types. Hodgdon does offer one spherical powder with a true ball shape - H380.

POWDER CLASSIFICATION CHART

Powder Type	IMR Powders	Alliant	Hodgdon	Winchester	Accurate Arms
Extruded Tubular and Stick	IMR 4227, 4198, 3031, 4064, 4895, 4320, 4350, 4831, 7828, SR4759	Rx7, Rx12, Rx15 Rx19, Rx22	H4227, H4198, H322 H4895, H4350 H4831, H1000		2015BR, 4350, 3100 2495BR
Cut Round Flakes	700x, 800x, SR4756, SR7625, PB	Bullseye, Red Dot Green Dot, Unique Blue Dot, Herco 2400	Clays		Nitro 100
Ball – Round and Flattened			HP38, HS-6, HS-7, H110, H335, BL-C(2) H380, H414, H450 H870, Trap 100	296, 540, 571, 748, 760, 231 WSL, WSF, WST	No. 2, No. 5, No. 7, No. 9, 1680, 2230, 2460, 2520, 2700, 8700

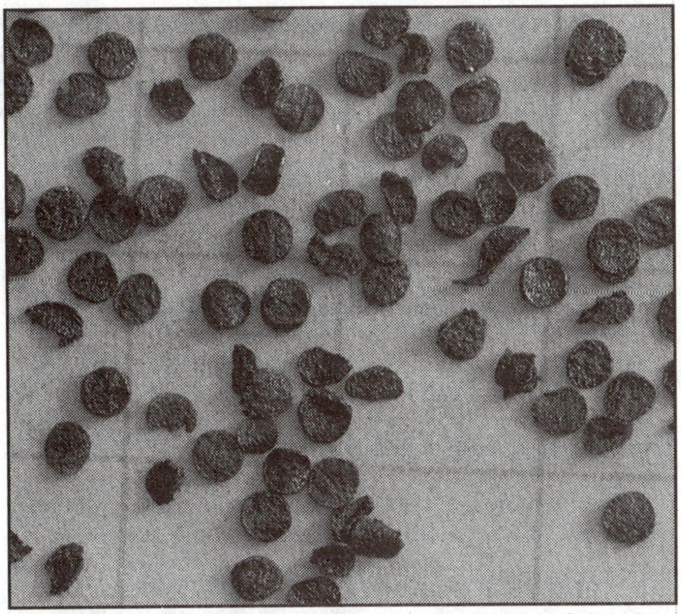

Alliants' Herco is a double base propellant with a flake shape granulation.

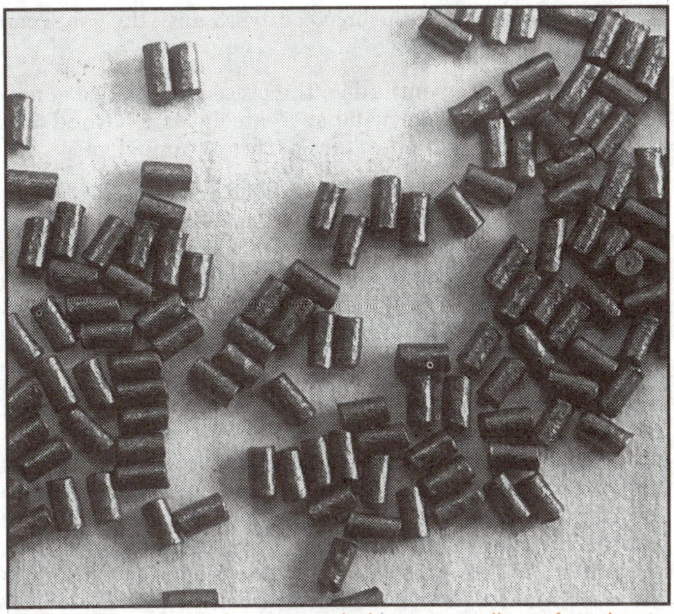

IMR 4895 is a single base extruded type propellant of moderate grain length.

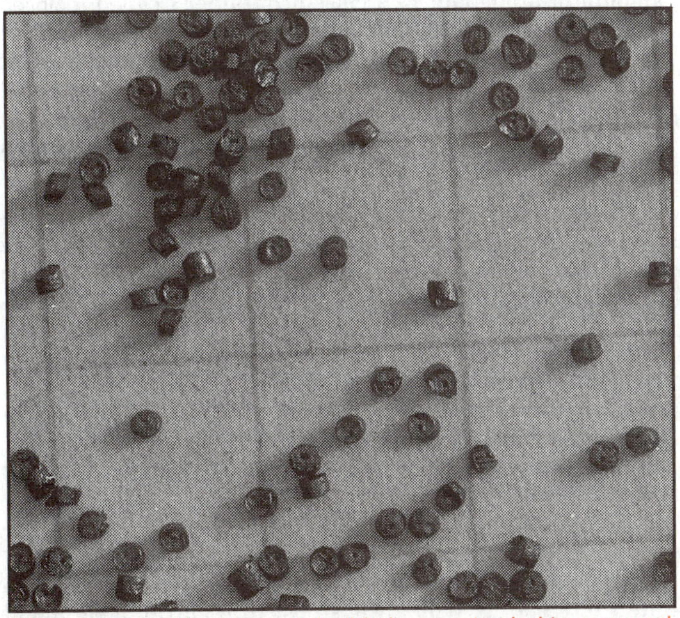

Alliant Reloder 7 is a typical double base extruded type propellant with a rather short grain length.

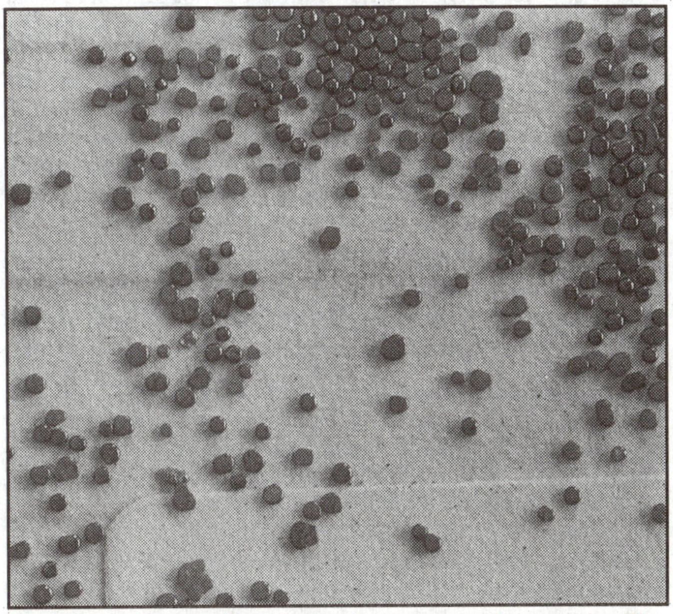

Winchester 296 is typical of a double base spherical powder. The true spherical shape is altered to obtain the desired burning characteristics.

Ignition of powder is normally caused by the heat generated by an exploding primer. However ignition can be caused under any circumstance which causes the granulations to be heated beyond ignition (kindling) temperature. Such causes can include sparks of any type or source, heat generated by any source, flame from any source, energy transfer which would raise the temperature to ignition level (this includes bullet impact), and other causes. Heat transfer to the powder need not be direct. For example, heat applied to the outside of the propellant container can result in ignition of the powder contained therein.

When powder burns it forms enormous quantities of gas with an extremely high temperature. Confined, this gas will cause very high pressure. In the confines of a cartridge case, a comparatively small amount of propellant can cause pressure as high as 65,000 pounds per square inch. The amount of pressure generated is a direct result of the amount of propellant burned and the degree of confinement.

If powder is burned in the open air, i.e. outdoors, there is adequate room for the hot gases to rapidly expand without creation of high pressure. Indeed, even a pound of powder so burned will not create a hazard due to pressure even to persons or things positioned more than several yards away.

If sufficient quantities of propellant burn within a confined area, such as a room, without the possibility of venting, the resulting gas, pressure in that area will rise. This pressure can push out windows and even closed doors. The results of such burning will not create an explosion but anyone in the area could be seriously burned or suffer injury. This is due to the force of expanding gases causing windows, doors or weak structural members to be forced outward until the gases can be vented.

Given adequate ventilation, the pressure increase generated by burning powder will not create damage beyond that caused by the high temperature of the expanding gases.

Generally, propellants used in sporting arms cartridges will not explode or detonate when ignited. Burning smokeless propellants do not generate the shock waves produced by an explosive. Unlike ignited smokeless propellants, the results of an ignited explosive defies the adequate venting of shock waves even when it occurs in the open.

Ignited stored powder can raise pressure within the storage area, thus there are specific requirements for keeping powder on hand. The first of these is never to transfer any propellant from its original container, which is designed to burst or partially open at a very low pressure. Some containers simply will have their cap pushed off if the contents are ignited. This will then allow for all the gases to be vented from the container. (Keeping powder in the original container is also essential for proper identification of the propellant.)

Ignited powder will continue to burn until it is consumed. Therefore, stored propellants should be kept away from any source of potential ignition, including but not limited to those earlier mentioned. Storage of propellants is best accomplished in a container that will protect the contents from outside heat sources. It is imperative that the storage container have one or more weak walls. These will open out at very low pressure to vent gases before they can cause damage in case of unintended ignition. Without this built-in protection, a strong enclosure could burst with considerable force possibly causing property damage or personal injury.

Other storage cabinet recommendations include construction with heat resistant fire-insulating materials. The cabinet should be sufficiently large to adequately vent the gases if a container were inadvertently ignited.

Storage cabinets are often made of 1" thick wood with one or more walls designed to open outward, or blow free at very low pressure. They should be many times larger than necessary to store the minimum quantities of powder.

Other Storage Considerations

Do not store propellants in the same area with solvents, flammable gases, primers, or any highly combustible material. Smoking should never be allowed in storage areas or while handling and using powder. Place conspicuous NO SMOKING signs in the storage area and at the loading bench. Remember to obey all legal requirements for storage as applicable to your community.

Many municipalities use the National Fire Protection Agency's publication NFPA 495 as law. It, in part, states that smokeless powders intended for individual use shall not exceed 20 pounds unless stored in a wooden cabinet or box with walls of at least 1" nominal thickness. Even then the store quantity shall never exceed a total of 50 pounds.

Every reloader should have a copy of the SAAMI pamphlet entitled "Properties And Storage Of Smokeless Propellants" which is available from:

SAAMI
P.O. Box 838
Branford, CT 06405

Always store powder in a cool dry place away from any potential source of ignition. Do not store powder where it will be subjected to the sun's rays or be influenced by any other source of heat. Be certain that children cannot gain access to the storage cabinet.

Spilled powder should never be picked up with a vacuum cleaner. Use only a brush and pan and dispose of the powder in a safe manner.

The nitroglycerin of a double base powder can be absorbed through the skin, by breathing vapors, or by ingestion. Symptoms can include headaches. The following precautions must be observed when handling double base powders. In case of accidental ingestion induce vomiting by placing fingers down throat and immediately call a physician. Prevent contact with food, chewing material and smoking material. Ensure adequate ventilation during handling and loading procedures. Do not carry powder in clothing. Wash hands thoroughly after handling. Follow specific manufacturers precautions.

Burning Rates

Some reloaders make assumptions on the burning rates of specific propellants. This is a dangerous practice. Manufacturers establish relative burning rates by placing a prescribed amount of propellant in a closed container - called a closed bomb (though it does not burst) - and then measure the pressure generated by the ignited propellant. Such a test offers the lab technician some vague guidelines to use when developing loads with a specific propellant. However, the pressures generated in cartridges do not necessarily relate in any fashion to those generated in a closed bomb test.

The pressure generated by a specific propellant will be the result of the weight of the bullet, the shape of the case, the pressure generated by burning and other factors. A specific powder often burns with greater or lesser quickness depending upon the specific cartridge and bullet weight being used. Information from a relative burning rate chart cannot be applied to reloading applications. Never assume any specific characteristics for a given propellant. Stick to the loading data contained in this manual, neither exceeding the maximum charge nor going below the minimum charge. And never substitute one propellant for another.

Use The Right Powder

Methods employed to prevent using the wrong powder are often interesting. It is always best to read the label aloud before and after each loading session and then compare the nomenclature with your data records or this manual. However, any safeguards that prove practical can be helpful. For example some reloaders have purchased their favorite powders in different size and shape cans. Their rifle powder might be purchased in a 8 pound container, handgun powder in a 1 pound container, and so on.

Powders Used In Our Data

The handloader must make absolute identification of a propellant before using it. There are many products with similar and even identical nomenclature. Pay specific attention to manufacturer identification. For example: Accurate Arms 4350 is different from Hodgdon 4350 and both of these are different from IMR 4350. NEVER use data developed for a specific brand powder with another brand powder containing similar or identical numerical identification. Each powder is different and requires specific data.

The remainder of this chapter deals with comments concerning each of the nearly sixty powders used in our data tables. This information is given as general interest material. Make no assumptions as a result of reading this material on potential applications not listed in the data table.

IMR Powders

IMR powders were at one time manufactured by DuPont in the U.S. Then manufacturing was moved to Canada. Finally the company was sold by DuPont and the powders are now distributed by IMR Powder Co. of Plattsburgh, New York 12901. Any questions regarding IMR propellants should be directed to this address.

"Hi-Skor" 700-X

This propellant was originally intended for application in shotshells. However it has practical use in many handgun cartridge applications. It is popular as a target load powder in many handgun cartridges. This is one of two double base powders marketed to reloaders by IMR. Sold in 1/2, 5, and 12 pound containers.

"Hi-Skor" 800-X

This propellant has similar design applications to "Hi-Skor" 700-X. Its burning rate is conducive to obtaining high velocities in certain handgun cartridge applications. The geometry of 800-X hinders uniform metering from powder measures and thus most users prefer to weigh every powder charge. This is the other double base powder sold in canisters to reloaders by IMR. Sold in 1/2, 5, and 12 pound containers.

PB

PB gets its name from the initials of its basic grain structure type - porous base. In this propellant the porosity of the powder is the method used to control burning rates. It was originally designed for shotshell applications but is now often used in handgun loads. Sold in 1/2, 5 and 12 pound containers.

SR 7625

This is a single base powder intended for shotshell and handgun loads. It is also used in many lead bullet rifle loads. This is the fastest burning propellant in the canister grade SR series. The SR in the nomenclature of this and other IMR propellants using the same letters stands for Small Rifle - a confusing name for seldom used powders. Sold in 1/2, 5, and 12 pound canisters.

SR 4756

This single base propellant is primarily designed for shotshell applications requiring a slow burning propellant. It is also used for many handgun loads and some lead bullet rifle loads. Sold in 1/2, 5, and 12 pound containers.

SR 4759

This single base powder is suitable for some small case rifle cartridges and a few heavy handgun cartridge loads. In the past its popularity waned so much that it was withdrawn from the market. After being reintroduced it became very popular with cast bullet shooters. SR 4759 is the slowest burning of the canister SR series of powders. Sold in 1/2, 5, and 12 pound containers.

IMR 4227

This is the fastest burning propellant in the IMR canister powders. The IMR stands for Improved Military Rifle - again a somewhat misleading title for today's canister powders. IMR 4227 is ideal for the 22 Hornet and other small case rifle cartridges. It is also used in many heavy handgun loads, cast bullet loads and the 410 bore shotgun. Sold in 1, 8, and 20 pound packaging.

IMR 4198

This propellant is relatively fast burning and at one time was the favorite of benchrest shooters using the 222 Remington. It is perhaps the premier selection for the 444 Marlin. It was at one time the factory powder for this cartridge. IMR 4198 has a fine reputation for accuracy in those applications for which it is well suited. Sold in 1, 8, and 20 pound containers.

IMR 3031

This is an extremely popular powder in many small to medium size rifle cases. It is a perennial favorite for the 30-30 Winchester. At one time this was used to load 250-3000 Savage Remington factory ammo. It is also highly recommended in the tiny 17 Remington and some shooters favor it in the giant 458 Winchester Magnum. It is extremely versatile and at its best in medium capacity cases, including the 358 Winchester. Sold in 1, 8, and 20 pound containers.

IMR 4895

This was at one time an extremely popular powder. Its fine granulation allows it to meter well through a powder measure. It gained an enviable reputation in the 30-06 for accuracy. Additionally many target shooters favor it for the 308 Winchester with 168 grain bullets. Sold in 1, 8, and 20 pound canisters.

IMR 4320

As with all IMR series powders this is a single base propellant. Factory 308 Winchester ammunition had been loaded with this powder, though it was long ago replaced. It is a versatile powder used in a wide range of rifle cartridges. Sold in 1, 5, and 20 pound packaging.

IMR 4064

This is an extremely versatile and generally accurate powder. It has been a favorite in the 308 Winchester, 30-06 Springfield with 150-grain or lighter bullets and in the powerful 375 H&H. At one time it was a factory choice for the latter cartridge. Its applications range from the 22-250 Remington to the 416 Remington, doing quite well at both extremes. Sold in 1, 5, and 20 pound containers.

IMR 4350

This is a relatively slow burning propellant that is used in cartridges from 22 to 45 caliber. IMR 4350 is perhaps the single best choice for such popular applications as the 243 Winchester for all bullet weights, the 257 Roberts and the 7x57mm Mauser with most bullet weights, and the 30-06 with bullets of 180 grains or heavier. It has been the choice of at least one factory for past loadings for the 243 Winchester, 270 Winchester, and the 300 H&H Magnum.

IMR 4831

This is a very slow burning propellant that is ideal for many magnum style cartridges including the 338 Winchester Magnum, the 375 H&H Magnum and the 416 Rigby. It is also likely to be the very best choice for the 270 Winchester and the 280 Remington. Accuracy in its best applications is usually outstanding and velocities are often at the highest practical levels. It works in a wide range of cartridges having a relatively large powder capacity. Sold in 1, 8, and 20 pound containers.

IMR 7828

The slowest burning of canister grade IMR powders, IMR 7828 is ideal for applications in large capacity cases with heavy bullets. It is the very best selection for the 378, 416, and 460 Weatherby Magnum cartridges. Many shooters feel it is also a fine choice for the 7mm Remington Magnum, 7mm, 300 and 340 Weatherby Magnums. Sold in 1, 8, and 20 pound canisters.

These Alliant powders are included in our data tables. Each is a double base type of either an extruded tube or flake shape.

Alliant (Formerly Hercules) Powders

Alliant canister powders are manufactured at Kenvil, New Jersey. All powder types remain the same whether manufactured under the Alliant or Hercules name. Questions regarding Alliant propellants should be addressed to Harry Thompson, Alliant Powder Co., Howard Blvd., Kenvil, New Jersey 07847.

Bullseye

Bullseye is a double base powder and the fastest burning of Alliant canister grades. This is perhaps the most popular handgun powder in the U.S. It has proven extremely accurate in a wide range of standard velocity handgun cartridges and often will deliver near maximum and extremely uniform ballistics. It is probably the powder used most often to load pistol and revolver cartridges. Bullseye's reputation for performance is well deserved and it has been used continuously for 90 years. It is ideal for the tiny 25 ACP as well as the 45 ACP. It is perhaps most often used for target loadings for the 38 Special. Sold in 1, 4, and 8 pound kegs.

Red Dot

This double base propellant was designed as a shot shell powder and may well be the most popular of its type. It is also a popular selection for many handgun and cast bullet rifle loads. It takes its name from the color coded granulations mixed into the propellant. In non-canister powders there are several different speeds of Red Dot, but the reloader deals with just one grade. Sold in 1, 4, and 8 pound cans.

Green Dot

Slightly slower burning than Red Dot this double base powder incorporates a small amount of green colored granulations, hence its name. It is popular for applications similar to Red Dot. Sold in 1, 4, and 8 pound containers.

Unique

Unique is a very versatile propellant suitable for a wide range of shotshells and handgun ammunition. It is also a fine choice for cast bullets in many rifle cartridges. It has long been considered the very best selection for the 45 Colt cartridge. It is a good choice for the handgunner wanting slightly higher velocities in a wide range of calibers. It works in the minuscule 25 ACP as well as the 44 Remington Magnum. Sold in 1, 4, and 8 pound containers.

Herco

A relatively slow burning shotshell and handgun powder, Herco is a double base type with coarse granulations. Its use is mostly for heavy or magnum type loadings. Sold in 1, 4, and 8 pound kegs.

Blue Dot

A very slow burning shotshell and magnum handgun cartridge propellant, this powder has somewhat limited applications. This propellant contains some blue colored identification granulations. Sold only in 1 or 5 pound canisters.

Alliant 2400

This double base powder has a fine granulation and is suitable for a limited number of small cased rifle cartridges such as the 22 Hornet. It is also popular as a magnum pistol powder. After 60 years of use its applications remain limited but effective. It is often used as a cast bullet rifle cartridge propellant. For the 357, 41 and 44 Magnums it may be the premier selection for full power loads. Sold in 1, 4, and 8 pound packaging.

Reloder 7

This is the fastest of the Reloder series and is a double base powder suitable for a wide range of cartridges from the 222 Remington to the 458 Winchester Magnum. It is perhaps the best choice for the latter with 500 grain bullets as it uses a minimum of case space and does not require heavily compressed powder loads. Accuracy in this application is outstanding and one inch groups or less are common if the shooter and rifle are up to the task. At one time Reloder 7 was the darling of benchrest shooters using the 222 Remington. This powder has a mix of red, yellow and blue identifying granulations. Sold in 1 or 5 pound containers. Reloder 7 is abbreviated as Rx7 in our data section.

Reloder 12

This is a moderately fast burning propellant used in a wide range of rifle cartridges and is a double base type. Applications range from 22 to 45 caliber. It is ideal for medium capacity cases. This propellant has proven very accurate in a wide range of calibers. Sold in 1 and 5 pound canisters. Reloder 12 is abbreviated as Rx12 in our data section.

Reloder 15

A double base powder of moderate burning speed, Reloder 15 fits a very wide range of applications such as the 223 Remington and the 416 Remington. In both extremes accuracy is superb. Performance of this propellant is like the, now unavailable, Norma 203 although not on a grain for grain basis. This propellant, as are the remaining Reloder speeds, is imported by Alliant. It is sold in 1 and 5 pound canisters. Reloder 15 is abbreviated as Rx 15 in our data section.

Reloder 19

A relatively slow burning double base powder that also has wide applications. It works in the 22-250 Remington as well as in the 340 Weatherby Magnum. While it should not be loaded on a grain for grain basis, performance will be similar to that of the discontinued Norma 204. Sold in 1 and 5 pound containers, Reloder 19 is abbreviated as Rx19 in our data section.

Reloder 22

Those who miss Norma MRP would do well to give serious consideration to double base Reloder 22. While it does not load grain for grain it is very similar to the now unavailable Norma propellant. It is basically a magnum rifle propellant, and is the same powder currently loaded in many factory Weatherby Magnum cartridges. Alliant suggests applications from the 220 Swift all the way to the powerful 416 Weatherby Magnum. Many finicky 7mm Remington Magnums have become top shooters when Reloder 22 was used. Reloder 22 is the slowest burning Alliant propellant. Sold in 1 and 5 pound containers. Reloder 22 is abbreviated as Rx22 in our data section.

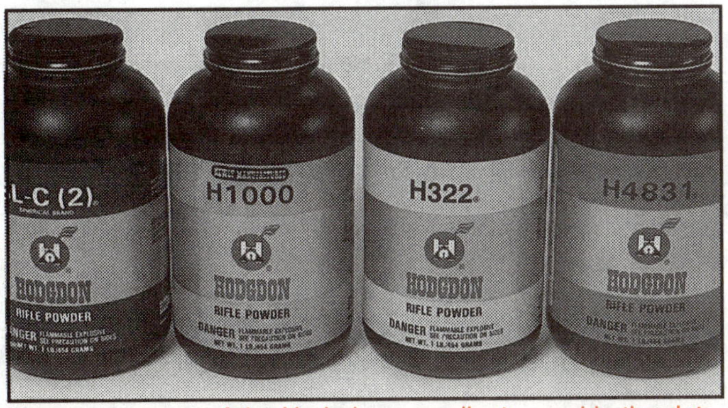

Shown are some of the Hodgdon propellants used in the data section of this handbook. They include spherical and extruded types.

Hodgdon Powders

Editors note: All Hodgdon propellants are currently available in 1 and 5 pound canisters. Some Hodgdon powders have been available both as newly manufactured types and as surplus types. Look for identifying statements on the powder can for those available in both types. If you develop a load for one type powder, and then switch to another, you must redevelop your load - starting at the lowest listed charge and slowly working up the load. If you have questions about Hodgdon propellants write to Hodgdon Powder Co., P.O. Box 2932, Shawnee Mission, Kansas 66201.

HP38

This double base spherical powder is manufactured by Olin and sold under the Hodgdon name. It will nearly duplicate the performance of Winchester 231 and is extremely fast burning. The name was derived from Handgun Propellant ideal for the 38 Special. Hodgdon suggests it as a perfect candidate not only for the 38 Special but also for 9mm Luger and the 45 ACP. It is suitable for use in almost all handgun cartridges and is an economical powder to use.

Clays

Clays is a new shotshell powder that has specific handgun cartridge applications. Because this is the newest propellant, current available data is somewhat limited. It is sold in 14 ounce containers.

HS6

This is another spherical Olin manufactured propellant not unlike the old Winchester 540MS. Nomenclature is short for Hodgdon Shotshell type number 6. It is basically a shotshell powder that has handgun and cast bullet applications. Hodgdon suggests it is well suited for many 9mm Luger and 45 ACP loads.

HS7

A relatively fast burning spherical Olin made propellant, HS7 is kin to Winchester 571. It is basically a shotshell powder that has handgun applications. This powder was named for being the 7th in the series of Hodgdon shotshell powders. Hodgdon suggests the 357 Magnum, 41 Magnum, 44 Magnum and 45 Winchester Magnum as well suited applications of this powder.

H322

This is an extruded type propellant that has been a universal favorite in the two most accurate cartridges ever designed - the 22 PPC and the 6mm PPC. In the latter application it has won more bench rest matches than any other propellant. It is also a perfect choice for the 222 Remington and the 6mm Remington BR cartridges. Its very short grains allow it to meter with a uniformity almost unrivaled by any other extruded propellant. It was named for its relative burning rate as compared to H335 - it is a bit faster burning.

H335

This is a very popular spherical propellant manufactured by Olin. H335 is an equivalent of Olin's WC 844. This powder enjoys an enviable reputation in cartridges like the PPC's,

222, and the 223 Remington. Nomenclature establishment has become unclear. Obviously it was some reference to relative burning speed. It is similar to the slightly slower burning BL-C(2). It currently can be found on dealer shelves as both newly manufactured or as an excess made to military specifications. Hodgdon suggests this powder as a strong candidate for the 308 Winchester and the 35 Remington.

BL-C (2)

This is a spherical Olin made propellant, spoken as Ball C 2. It started as a surplus supply of Olin WC 846. It currently can be found both as a newly manufactured type or as a surplus type made to U.S. military specifications. It is extremely popular in such diverse applications as the 223 to the 30-06 Springfield. Originally this powder was called Ball Type C. The C was a derivative of "cool" burning. Newly manufactured propellant appears to be a bit slower burning than the original surplus type.

H380

H380 was an early surplus type named for the fact that Bruce Hodgdon found 38.0 grains to be extremely well suited to the 22-250 with 55 grain bullets - using the early lots of GI powder. This powder is the only currently available spherical type that has a true ball configuration. It meters most uniformly as compared to all canister grade propellants because of its shape. It is a great favorite in the 7mm-08 and 308 cartridges. It has very wide applications for rifle cartridges.

H414

Spherical H414 is a ballistic kin to Winchester 760 and has wide applications in diversified cartridges. It is especially popular in the 30-06 and similar sized rounds. Use in cartridges as large as the 375 H&H is not uncommon.

H4831

An extruded propellant, H4831, is a well loved propellant that works superbly in many large capacity cases. It is the powder that helped make the 270's enviable reputation, and a favorite for use in most of the belted magnum calibers. Data for this propellant is not useable with IMR 4831.

H1000

H1000 was named for the 1000 yard target shooters who Hodgdon felt would like the performance characteristics of this propellant. It is a powder for very large capacity cases with heavy bullets. It is an extremely slow burning extruded type propellant.

H870

Spherical H870 is a very slow burning propellant with limited practical applications. It is a powder to be used in large capacity cases with heavy weight bullets.

Winchester Ball Powders

Winchester powders are manufactured by Olin in St. Marks, Florida. Questions regarding Winchester powders should be directed to Customer Service, Winchester, East Alton, Illinois 62024.

Winchester Ball Powders are an important part of the data tables in this manual.

231

This is the fastest burning Winchester canister powder. It has near universal application in typical handgun cartridges. It is highly favored for its well known accuracy potential. Sold in 1, 3, and 8 pound containers.

WSL

This is a relatively new Winchester double base spherical powder. It is among the first propellants to be made from Olin's revised manufacturing process (called IBP for Improved Ball Powder) which now uses virgin nitrocellulose. The nomenclature is short for Winchester Super Light, primarily a shotshell powder. It, however, has applications as a handgun propellant. Sold in 1, 3 and 8 pound canisters.

WST

Winchester Super Target is a shotshell powder by design and generally replaces the now discontinued 452AA. It is used by many handgun target shooters. Sold in 1, 3, and 8 pound containers.

WSF

Winchester Super Field is another of the new generation of Winchester shotshell powders having applications in handgun cartridges, especially for competition loads. Sold in 1, 3, and 8 pound packaging.

680

This is a very fast burning rifle powder suited to small cartridge cases such as the 22 Hornet or 30 Carbine. (Note: This powder has been discontinued by Winchester.)

748

A relatively fast spherical powder that is used in cases as small as the 222 Remington and as large as the 458 Winchester Magnum. Sold in 1 and 8 pound containers. This powder has been extremely popular with benchrest and target shooters. It is highly favored by some users of the 308 Winchester cartridge. As with all ball powders, it meters very uniformly through a powder measure.

760

The slowest of Winchester's current line of canister powder, 760 is suitable for a very wide range of applications as a rifle propellant. Sold in 1 and 8 pound canisters. It has given excellent accuracy with many cartridges including the standards such as the 270 Winchester, 280 Remington and 30-06 Springfield.

Many of the powders imported by Accurate Arms are included for the first time in a Lyman manual.

Accurate Arms Powders

Editors Note: Accurate Arms Powders are the newest line of powders to be included in Lyman's data. These are imported from Israel and are generally available in 1 pound cans.

No.2

This is a very fast burning double base spherical powder with a wide range of handgun cartridge applications. Our use is currently limited to the 9mm Luger and 38 Special cartridges. Identified as AA2 in our data section.

No.5

Our use of this fast burning double base spherical handgun cartridge propellant starts with the 32 H&R Magnum. We have also used it with the 9mm Luger, 357 Magnum, 38 Super Auto., 40 S&W (with superb accuracy), 10mm Auto., 44 Remington Magnum, and 45 ACP. Identified as AA5 in our data section.

No.7

Designed for NATO 9mm carbine ammo, we have used this double base spherical powder in the 9mm Luger, 357 Magnum, 38 Super Auto., 10mm Auto., and 44 Remington Magnum.Identified as AA7 in our data section.

No. 9

This is a relatively slow burning double base spherical handgun powder. We used it in the 357 Magnum and 44 Remington Magnum. Identified as AA9 in our data section.

1680

This is a double base ball powder designed specifically for the 7.62x39 Russian. It is very fast burning and has limited applications.

2015BR

This is a single base small grain extruded type powder. It is intended for small capacity cartridges and does well in the 22 and 6mm PPC cartridges, among others.

2230

This was previously called MR-223, and is a double base spherical type propellant. It was developed with the 5.56 NATO (223 Remington) in mind. It is a comparatively fast burning rifle powder suitable for many medium capacity cases.

2460

This is a double base spherical propellant slightly slower burning than 2230. It is useful from the 223 to the 30-06. Our applications begin with the 222 Remington.

2520

Designed for medium capacity rifle cartridge cases, this is also a double base spherical type. Our applications begin with the 22-250 Remington.

2700

At this writing this is a very new canister propellant, only months old. Thus data for it is somewhat limited at this time. It has a relatively moderate slow burning rate and is a double base spherical type.

3100

This is a relatively slow single base extruded type propellant designed to compete with IMR 4831 and Hodgdon H4831. Thus it is basically suitable as a magnum cartridge propellant.

8700

8700 is a very slow burning double base spherical propellant that was intended for use in large capacity cases.

CHAPTER 8
JACKETED BULLETS

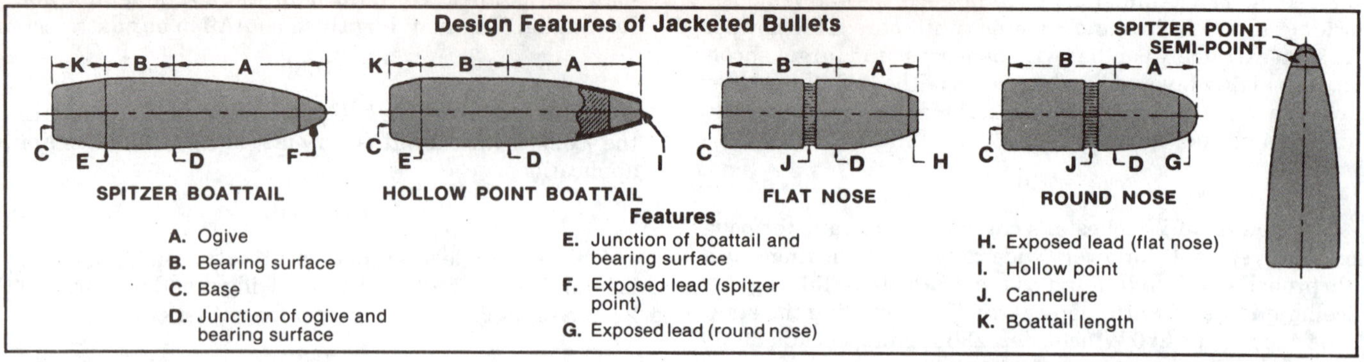

Jacketed bullets are made in many varieties to meet specific applications. For example, varmint hunters demand bullets that are highly accurate in order to meet the requirements of hitting tiny targets at long ranges. A woodchuck some 300 yards away is a very difficult target. Prairie dogs at the same range are even more challenging, requiring match grade bullets.

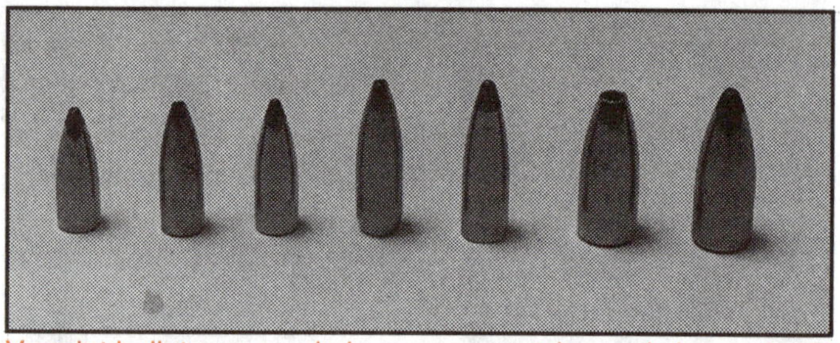

Varmint bullets are made in numerous styles and shapes in 22 to 30 caliber. Shown are: 22 caliber 50 grain spitzer, 53 grain hollow point, 55 grain semi-spitzer; 24 caliber 70 grain hollow point, 85 grain spitzer; and 30 caliber 110 grain hollow point, 125 grain spitzer.

Varmint Bullets

Varmint bullets must be highly frangible.This means they must expand violently and rapidly, almost completely disintegrating, when incurring the minimal resistance offered by most small game. If these bullets did not expand in this way, they would fail to transfer their energy to the quarry and simply pass through, creating a very small wound.

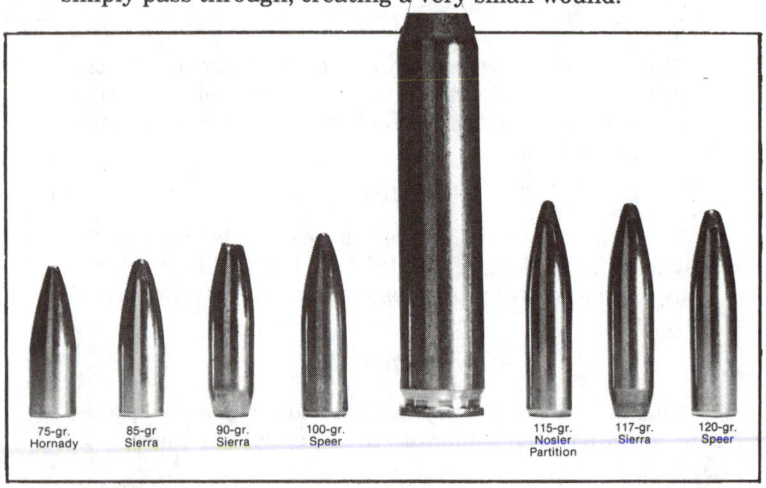

This photo illustrates the range of bullet weights that a handloader can choose for the .25-06. Those weighing from 75 to 100 grains are at their best for varmint shooting at 25-06 velocities while the heavier ones work well on big-game.

The frangibility of varmint bullets is also a safety asset. Fired at the low angles often associated with varmint hunting, strongly constructed bullets have a propensity to ricochet into the country side after striking the ground, often coming to rest at an unknown point. Frangible varmint bullets, however, tend to break up on impact with the ground and reduce the potential danger of ricochet.

Target Bullets

Serious target shooters demand very accurate bullets. Since they are punching holes in paper with safe backstops, they are not concerned with bullet expansion. Some target bullets are extremely frangible and are made with thin jackets. This method of assembly is known as soft swaging. Other target bullets are, in reality, non-expanding bullets.

Just because a bullet has lead exposed at the tip, or has a hollow point, is no guarantee of expansion. Examples of highly accurate target bullets that do not expand are some of the discontinued Winchester target bullets which incorporated a hollow point.

Bullets intended primarily for target applications often bear a trade name or generic nomenclature suggesting this type of application. However, it is the reloaders obligation to determine if a specific bullet is a target or hunting type.

Target bullets are made to exact specifications for weight and dimensions to insure a high degree of accuracy. Most commercial target grade bullets are capable of far better accuracy than the firearms they are fired in. It is not uncommon for bullet makers to get groups of ten shots measuring well less than half an inch when testing target bullets during production runs.

Big Game Bullets

Big game hunters place the most demands upon bullet design. They require good accuracy and controlled expansion. That is to say, a bullet must penetrate hide and fat before expansion begins. Then expansion must occur at a controlled rate so the bullet may penetrate deeply into vital organs before expansion stops the bullets forward progress. Deep penetration and controlled expansion are difficult design goals to combine.

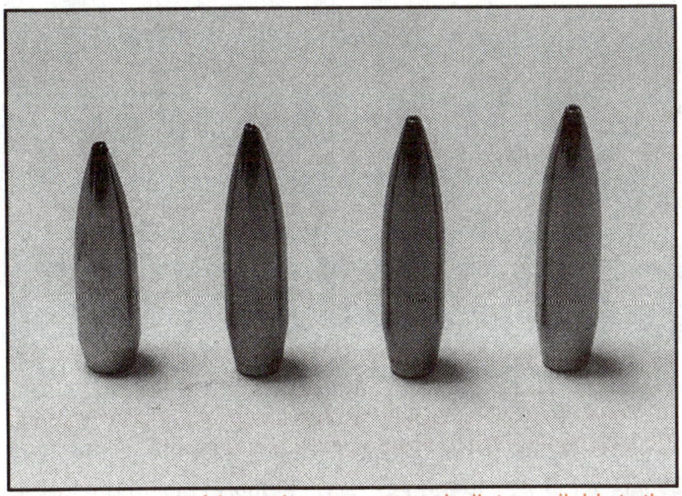

Shown are some of the various target type bullets available to the reloader. These are 30 caliber 168- 180- 190- and 200grain types.

Expansion often means that as the jacket peels back and the lead alloy core expands a portion of the jacket and core will be wiped away from the bullet mass. As this occurs, the loss of mass may be so great that the bullet's momentum will stop and penetration will cease. A specific amount of mass retention is a vital part of a hunting bullets design.

Big game bullets designed simply to retain most of their original weight may not expand properly. The combination of good expansion with high weight retention usually means a bullet design that is costly to produce with a premium price. The premium may be as little as a two fold price increase, or it may be considerably more when compared to popularly priced bullets. Still, for the comparatively few numbers of bullets required for sight-in and actual hunting, cost increases will go unnoticed. Practice, always an essential, can easily be accomplished with popularly priced bullets if the shooter sights in with the premium bullet right before hunting season.

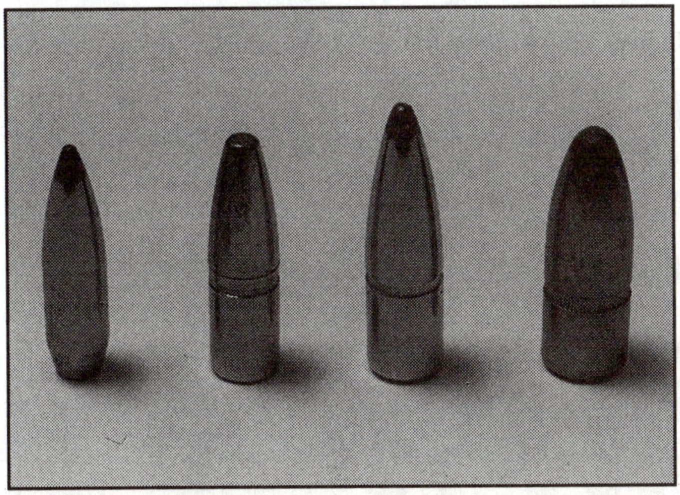

Big game bullet types include the following: 27 caliber 130 grain spitzer, 30 caliber 180 grain semi-spitzer, 33 caliber 225 grain premium grade Partition type and 37 caliber premium grade 285 grain bullet.

When hunting relatively lightweight thin-skinned game, bullet design criteria are different than that for hunting thick-skinned or very heavy game. In the former, expansion needs to be comparatively rapid. In the latter the bullet may have to penetrate up to three times the distance before the vitals are reached. Thus, for tough game, a heavily constructed bullet with considerably slower expansion is required.

Some normally priced bullets have proven perfectly adequate for the most popular big game - deer and antelope. Each shooter will have to carefully consider the size of the game to be hunted as well as the type of shots in order to determine if regular or premium grade bullets will be required.

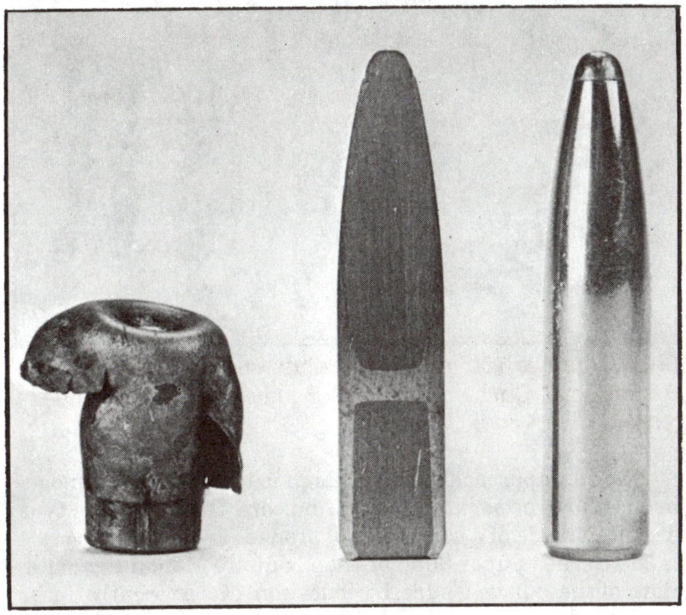

Unfired 175-grain Nosler Partition bullet, right, is shown with sectioned specimen and one recovered from a greater kudu in typical condition . . . front core gone, wings folded back close to shank, and about 65 percent of original weight intact.

Solids

For hunting very large dangerous game such as elephant, cape buffalo, or hippo, many hunters choose a "solid" bullet to insure deep penetration into thick hide, fat, tough muscle and huge bones. However, many solids fail when used in these applications.

In the past a "solid" was not really a solid. Rather it was a "full" metal cased bullet, having a lead core and a copper jacket which was closed on the nose end instead of the base end. As serious hunters have found, some bullets with this style construction sometimes bend, rivet at the nose, or even rupture, on truly heavy game. When this occurs, penetration often is not in a straight line and insufficient.

Improvements in solid designs include the use of a lead core with a copper plated steel jacket. While not infallible, these are a definite improvement over earlier designs, although short barrel life may be a side effect. We have seen numerous barrels which were worn to the point of being unable to properly stabilize a bullet after as few as 600 rounds were fired with steel jacketed bullets. The point is that such bullets should be used only for sighting in and actual hunting if long barrel life is important. Complaints of failures persist even with improved types of solids.

In recent years other "solid" type bullets have been designed. These include some with extremely heavy copper alloy or bronze jackets and hard alloy cores. The Trophy Bonded Sledgehammer solid is of the latter type. These have a fine reputation for getting the job done without failures. Each bullet costs several dollars but that is a small price considering the overall cost of a safari.

Shown are two 400 grain .416 caliber bullets. On the left is a Barnes solid. On the right is an A-Square soft point with heavy jacket.

Another approach to solid design is to use a true homogeneous alloy - no separate jacket and core. Bullets of this type are often made of a solid piece of bronze. Typical of this style is the Barnes Super Solid or the A-Square Monolithic Solid. Both of these have a superb reputation. More recently Speer has introduced a "solid" which uses a heavy bronze jacket and a hard tungsten core.

Appropriate Bullet Application

Each reloader needs to consider the kind of bullet performance required for their application and then make a wise selection. The deer and antelope hunter who takes only broadside, behind the shoulder, shots will place the least amount of demands on a soft point bullet. The hunter who wishes to take difficult angle shots, which require considerable penetration, may find need for a better bullet. As game gets larger, or ranges vary, a premium bullet may be the only type that will make sense to the reloading hunter.

Premium grade bullets are produced by two major market suppliers. These are the Nosler Partition and Speer Grand Slam types. Other premium bullets are available from semi-custom or custom manufacturers. The major ones of this type include bullets from Barnes, Trophy Bonded Bullets, Swift A-Frame and A-Square.

Velocity Versus Bullet Selection

The reloader needs to consider other points when selecting appropriate bullets. One of the criteria for selection can be the relative velocity level of the cartridge to be reloaded. Bullets driven at different velocities do require different construction. For a 22 caliber bullet to expand properly at 22 Hornet speeds, a very thin jacket with a good amount of exposed lead is generally required. Cartridges of approximately the velocity of a 222 Remington require the use of a slightly stronger bullet. To withstand the very high velocities of a 22-250 or 220 Swift, a 22 caliber bullet must be amply strong. The use of a

The lower velocities attained with handgun cartridges require thin jacketed bullets to assure proper expansion.

22 Hornet or 222 Remington style bullet in one of the latter could result in the bullet disintegrating, in flight, into a blue puff of smoke due to the high rotational and velocity forces.

Bullets of other calibers also need to be selected with concern for impact velocity. For example, the very low velocity of the 30 Carbine cartridge demands very "weak" bullet construction if there is to be any bullet expansion. A 30-30 demands considerably tougher bullets but not as tough as a 30-06 bullet. If you are using a 300 Weatherby Magnum only the toughest bullets will penetrate properly, while lesser bullets are prone to disintegrate at the high impact velocity of this very fast cartridge.

Appropriate Bullet Weight

Specific applications will generally dictate a weight, and range of applicable bullet weights. For example, 30 caliber bullets of 110 to 130 grains are most often thin jacketed types designed for varmint hunting. In this same diameter, bullets of 150 to 165 grains are usually intended for application on light big game - i.e. antelope, deer, small black bear, goat and sheep. The heavier bullets, from 180 to 220 grains, are generally meant for medium to heavy game such as elk or moose.

In most diameters, from 24 to 30 caliber, the lightest bullets are varmint types and the heavier bullets are for

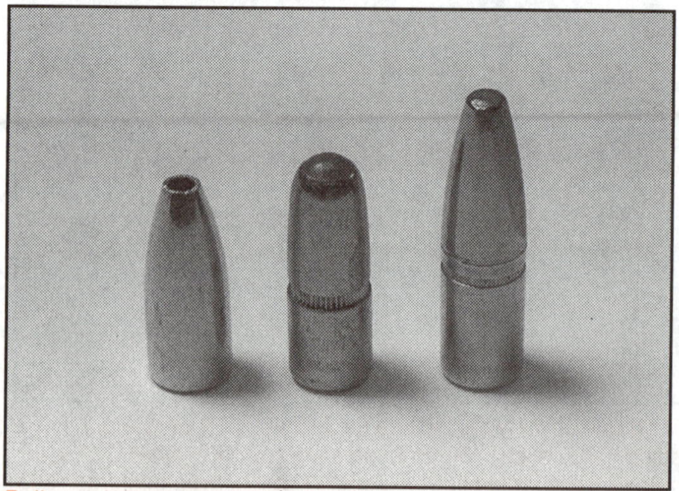

Bullet weights can sometimes give the handloader an idea of intended application. For example, in a given diameter, light weight bullets are often meant for varmints (30 caliber 110 grain on left), medium weights for light big game (30 caliber 150 grain in center) and heavier weights for heavy big game (30 caliber 180 grain on right).

progressively heavier big game animals. Bullets of 32 caliber, and larger seldom are made in frangible varmint types. Generally such bullets are all for big game, the lightest being for smaller big game animals.

Of course, bullets in 224 diameter are ordinarily never intended for any type of big game application. These are fundamentally varmint or target type bullets.

Bullet Nose Configuration

Sometimes it is essential to use flat or blunt nose bullets in specific rifles. For example, cartridges to be used in a tubular magazine should always have flat or very blunt nose configurations. The use of a spitzer in a tubular magazine, where the point of one bullet rests against the primer of the round immediately in front of it, could result in the igniting of that primer under the forces of recoil. Some fire arms were never designed to feed properly with any bullets other than those with a round nose. Some 458 Win. Mag. bolt action rifles are perfect examples of this.

At long range, it is possible for a 300 Savage cartridge, loaded with a spitzer bullet, to outperform a 30-06 loaded with a round nose bullet. A 30-06 using a blunt 180 grains bullet, with a muzzle velocity of 2700 feet per second, will have the bullet arrive at the 300 yard mark traveling at 1725 feet per second. A 300 Savage using a 180 grain spitzer bullet, with a muzzle velocity of 2350 feet per second will see the bullet arrive at 300 yards with a remaining velocity of 1750 feet per second. The 300 Savage with a spitzer bullet is now traveling 25 feet per second faster than the 30-06 with a round nose bullet. Thus, over 300 yards a spitzer bullet shape was able to overcome the 350 feet per second advantage the round nose possessed at the muzzle.

Looking at it another way, two 30 caliber 180 grain bullets starting at identical velocity will have drastically different down range ballistics depending upon bullet profile. A very sharp nosed bullet of this type, starting at 2700 feet per second, will have a 300 yard remaining velocity of 2075 feet per second versus the round nose remaining velocity of 1725 feet per second - a 350 feet per second advantage for the spitzer. At 400 yards the spitzer would have approximately a 450 feet per second superiority. At either range, the difference

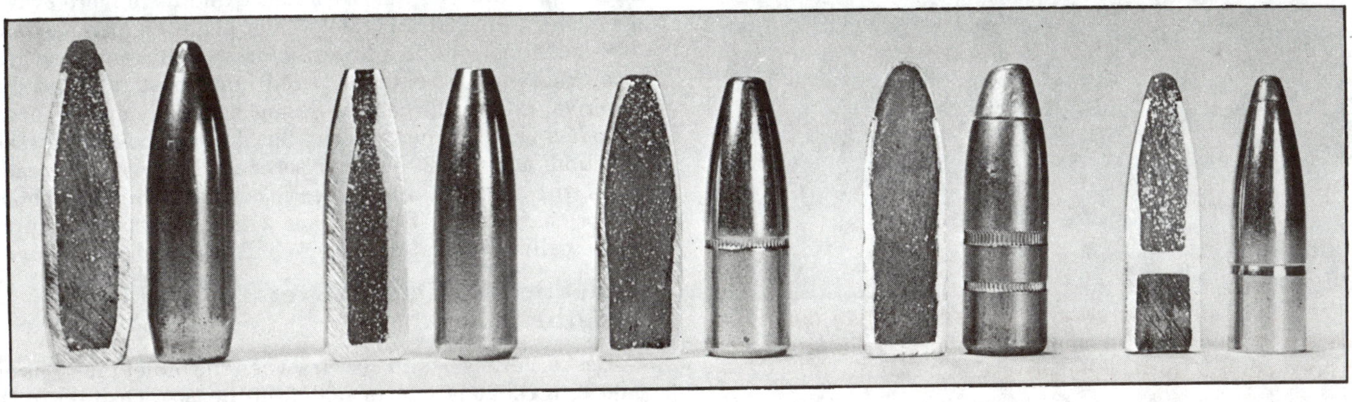

Five different ideas about controlling expansion in bullets for heavy game are shown here. At left, the Sierra 300-grain .375 with conventional core and thick jacket. Next, Bitterroot .358 250-grain slug with core soldered to the inside of the jacket. Third, Speer Grand Slam 285-grain .375 with dual cores of differing hardnesses (not apparent in this photo). Fourth, Winchester-Western Silvertip in 300-grain .375, showing aluminum cap over large exposure of lead at tip. At right, 210-grain Nosler Partition, with two cores separated by wall of gilding metal jacket material.

The selection of round nose bullets sometimes occurs because of misinformation. Many handloaders and factory ammo users alike still believe in the old wife's tale which states that only slow moving round nose bullets can plow through heavy brush and reach the intended quarry. Spitzer bullets, the same tale states, deflect considerably more when encountering brush. Finally, some believe that the greater the velocity, the more the bullet will be deflected.

This "theory" has been debunked many times. During the mid 1960's a Lyman employee, the late Lysle Kilbourn, and your editor ran extensive deflection tests. The findings of these tests, conclusively proved that all bullets deflect regardless of nose profile or velocity. Proven also, was that no one bullet form deflects more than any other. Those who use round nose or blunt bullets because they believe that these can better penetrate through brush are working with a misleading piece of "advice".

Others use blunt nose bullets because "they expand better than spitzers". About a hundred years ago, when smokeless powder and jacketed bullets were first coming into general use, this was quite factual. But, modern bullet design has created spitzer bullets that expand readily, outperforming or equalling the best round nose styles.

Sectional density has nothing to do with nose shape or size. These very dissimilar bullets all have very similar sectional densities. From left, a 250-grain .35 (SD .279), 270-grain .375 (SD .275), 400-grain .45 (SD .272), 154-grain 7mm (SD .273), and 180-grain .30 (SD .277).

is enough to create substantial inequality in trajectory and terminal ballistics.

Spitzers are ballistically superior because their sharp profile results in less air resistance to slow the bullet down. The reloader can judge the relative performance of any bullet by referring to its ballistic coefficient. This information can usually be found in most bullet manufacturers information. A round nose 180-grain 30 caliber bullet might have a ballistic coefficient of .248 while a spitzer of identical weight and caliber could have a rating of .438. The larger the number the better a bullet will resist deceleration.

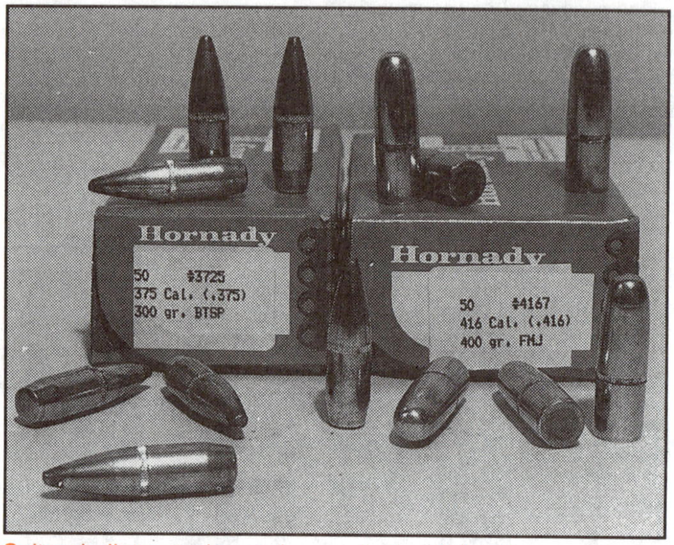

Spitzer bullets are almost always a better choice than round nose bullets as they deliver substantially higher velocities and game anchoring capability at long ranges.

Spitzers are not without weakness. Their sharp soft lead tips deform easily during recoil when the bullets impact with the front magazine-wall. This is most pronounced with magnum calibers. Loading to a shorter overall length can help reduce recoil damage to spitzer bullet noses.

Certain spitzers have been designed to preclude any flattened noses. The Remington Bronze point and the Nosler Ballistic Tip are good examples. Both use a tough insert placed into a hollow tip to form a very sharp and recoilproof bullet nose.

When bullet nose damage is a serious and continuing problem, it may become necessary to consider an alternative. Using a protected point bullet will eliminate the problem but select a bullet with as small a meplat (the blunt tip area) as possible in order to preserve down range ballistics. Bullets such as Speer's Grand Slam or Nosler's Protected Point Partitions will prevent tip damage while maintaining a fairly sharp profile.

It makes little sense to develop a load that squeezes the last foot per second out of your favorite cartridge but loses its down range performance because a blunt nosed bullet is used. Selecting sharp bullet profiles, when appropriate, pays big ballistic dividends. The individual who never shoots past 100 yards will never notice a difference, but as ranges increase variations in bullet performance will become obvious due to changes in nose shape.

Bullet Base Configuration

The bullet base shape can influence ballistics. Flat base bullets produce more air turbulence and have more drag to overcome than boattail bullets. This means an increased rate of velocity loss due to a lower ballistic coefficient. However, this difference is relatively small and most influential at lower velocities. The boattail shape is actually most effective in reducing drag when velocity has dropped below the speed of sound, about 1120 feet per second. The average reloader deals with substantially higher velocity and thus may choose flat or boattail bullet base types without concern for appreciable ballistic variations due to bullet base shape change.

Bullet Seating Depth (Overall Cartridge Length)

Best accuracy usually occurs when the bullet must "jump" 0.005" to O.020" from its seated-in-the-case position to the point at which its bearing surface first engages the rifling lands. The best seating depth for any specific load and firearm combination can be determined only by trial and error. Load 10 cartridges to one overall length and fire them for group size; then repeat the test with additional groups fired with cartridges seated to a different overall length. As little as 0.005" can make a noticeable difference in accuracy, thus, it is essential to test cartridges in increments of 0.005". More on this aspect is included in the chapter dealing with accuracy and load evaluation.

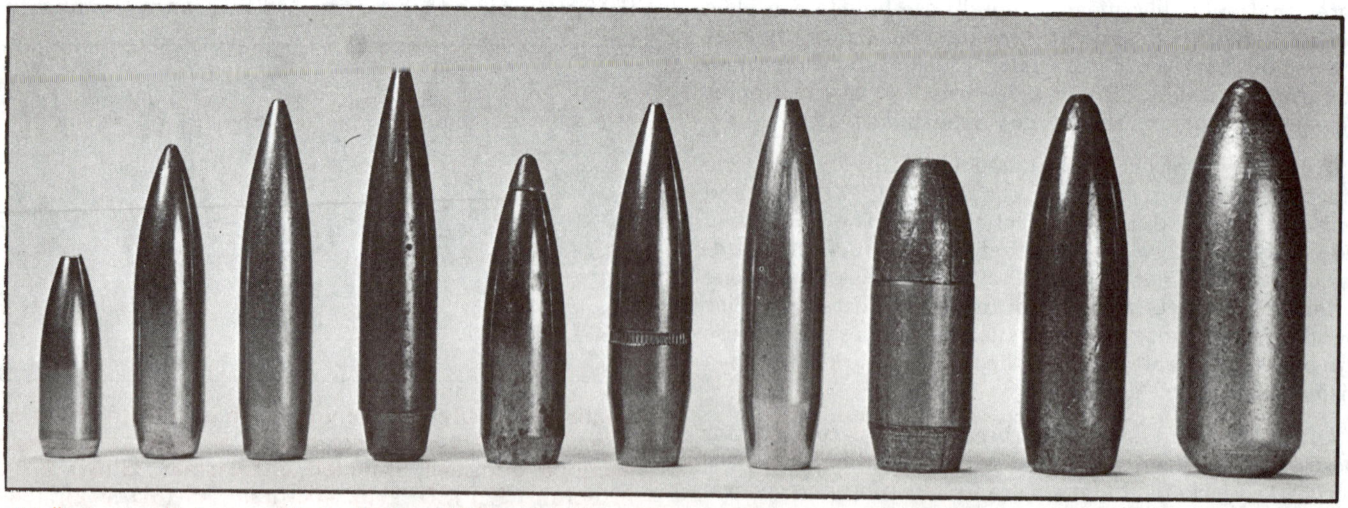

Boat-tails are used to improve the ballistic coefficients of both hunting and target bullets across the full range of calibers, as shown here. The third bullet from right is actually a boat-tailed cast bullet, poured into a copper-tubing jacket. From left, calibers are .224, .257, .264, .284, .30, .30, .30, .375, .375, and .458.

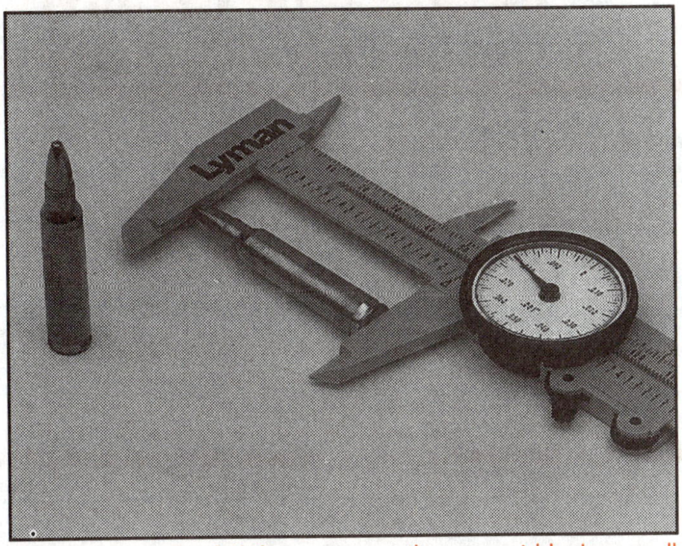

Bullet seating depth (often expressed as a cartridge's overall length) can effect accuracy and ballistics.

Handgun Bullets

Jacketed handgun bullets generally are not subjected to the extremes in velocity ranges or in game size as rifle bullets. Most hunters will restrict handgun shots to considerably less than 100 yards, therefore a relatively narrow range occurs with respect to impact velocity at the shortest or longest practical ranges. A minimum impact velocity of about 900 feet per second is usually required in order to obtain reliable expansion from jacketed handgun bullets. There are a few exceptions in either direction to this velocity requirement. Since handgun impact velocities seldom exceed 1600 feet per second, and are often considerably slower, even the differences from one caliber to another do not place extreme demands on a bullet.

Generally speaking handgun bullets with hollow point nose configurations expand more rapidly than soft point types. There are reasons to favor either type. The soft points slower expansion will allow the required deep penetration, when taking big game. Hollow points are often used on smaller game and varmints.

Some handgun bullets with lead exposed at the nose are not intended for expansion and are in fact "nonexpanding" target style bullets. Follow the manufacturers suggestions and guidelines.

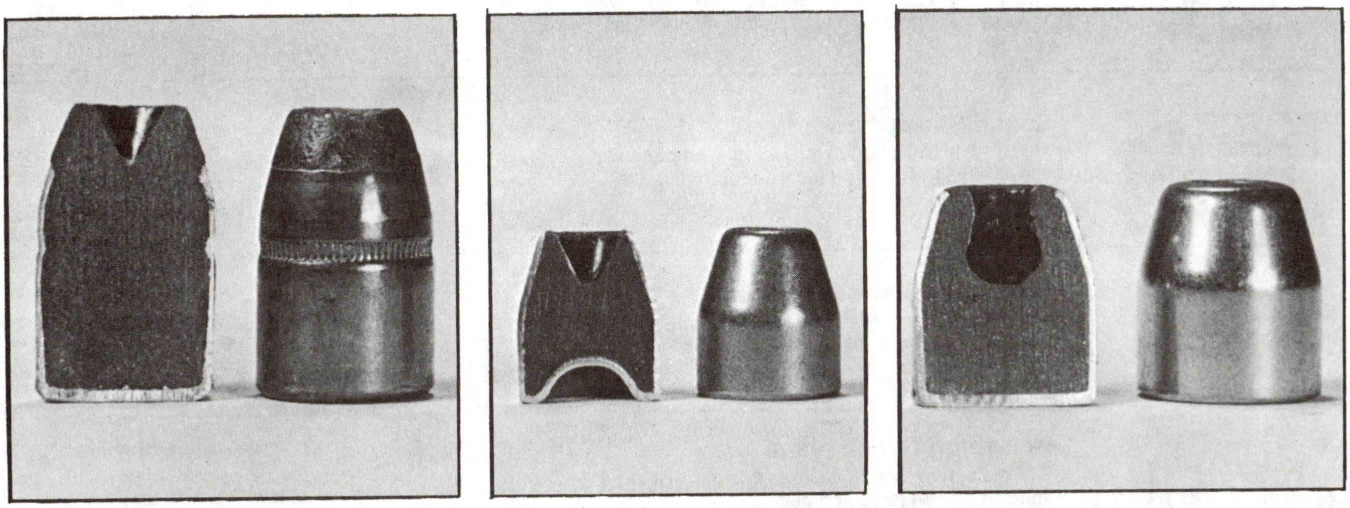

The lower the velocity, the bigger the hollow point must be to initiate reliable expansion, as shown by these pistol bullets. At left, a 240-grain .44, intended for 1,400 feet per second; center, a .380 90-grainer, intended for 1,050 fps; and, right a 185-grain .45, intended for about 1,000 fps.

Lyman is a member of and supports the National Reloaders Manufacturers Association (NRMA). Please feel free to contact them for more information on how to get started in reloading.

RELOADING
Order Form

The National Reloading Manufacturers Association makes several publications and audio-visual presentations available to shooters and organizations. Please use this form when ordering, and enclose your check or money order. Send to:

NATIONAL RELOADING MANUFACTURERS ASSOCIATION

NRMA
One Centerpointe Drive
Suite 300 Dept. LY
Lake Oswego, OR 97035

Ship to:

NAME ______________________

ADDRESS ______________________

CITY/STATE/ZIP ______________________

	ITEM	QUANTITY	PRICE EACH	TOTAL
	**NEW Set Yourself Up to Reload** A step-by-step guide to getting started in reloading rifle, pistol or shotgun ammo. Includes Reloading Bench Plans and Reloading Source Guide & Check List which are also sold separately. Includes list of equipment, accessories and components you need to start reloading.	_____	$ 5.00	_____
	**Reloading Video** A half-hour video on VHS cassette covers the basics of reloading, with detailed close-ups of all important operations in rifle, pistol and shotshell reloading.	_____	$19.95	_____
	**Sound/Slide Show** A colorful introduction to reloading. Complete show with 51 slides and recorded narrative on cassette tape.	_____	$19.95	_____
	**Reloading Course Outline** For introductory reloading classes. Developed for use by gun clubs, schools or stores as a guide in teaching reloading classes.	_____	$ 2.00	_____
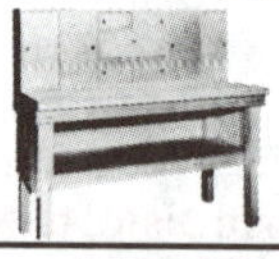	**Reloading Bench Plans** Complete step-by-step instructions for building a sturdy, wooden reloading bench.	_____	$ 3.00	_____
	**Reloading Source Guide & Checklist** A handy source of information for literature, books, films, etc., about reloading, PLUS a "Have/Need" list for reloading supplies.	_____	$ 1.50	_____
	**Reloading Catalog of Catalogs** For the first time, you can get catalogs for all leading reloading manufacturers...tools, presses, components, accessories...all at once without writing to each one individually. A wealth of reloading information at your fingertips.	_____	$ 4.00	_____
	**Reloader Patch** Embroidered jacket patch.	_____	$ 2.00	_____
	**Shooting Jacket Packet** A collection of brassards from various reloading manufacturers.	_____	$ 6.95	_____

My check or money order is enclosed in the amount of .... $_______

3

RELOADING THEORY

CHAPTER 9
RECORDS AND OTHER RELOADING SAFETY

Safety should be the most important concern of the reloader. In an effort to help reloaders, many publishers have developed arbitrary numbers of rules for reloading safety. These often are so strongly proclaimed to be the commandments of reloading that a novice might think such rules are all inclusive. However, it is not possible to come up with ten or twelve safety suggestions that will eliminate all potential hazards. Of course, there are basic safety rules that should never be broken. We will list some later in this chapter. The reloader must develop a frame of mind that will help prevent the "unforeseen" or "freak" accident.

Because safety is essential, some of the material in this section may be somewhat repetitive of information contained elsewhere in this handbook. Nonetheless, it is repeated here to insure that the message will not be overlooked. Remember, proper safety will include developing an attitude that does not allow the reloader to forget the nature of the components used or of the assembled ammunition.

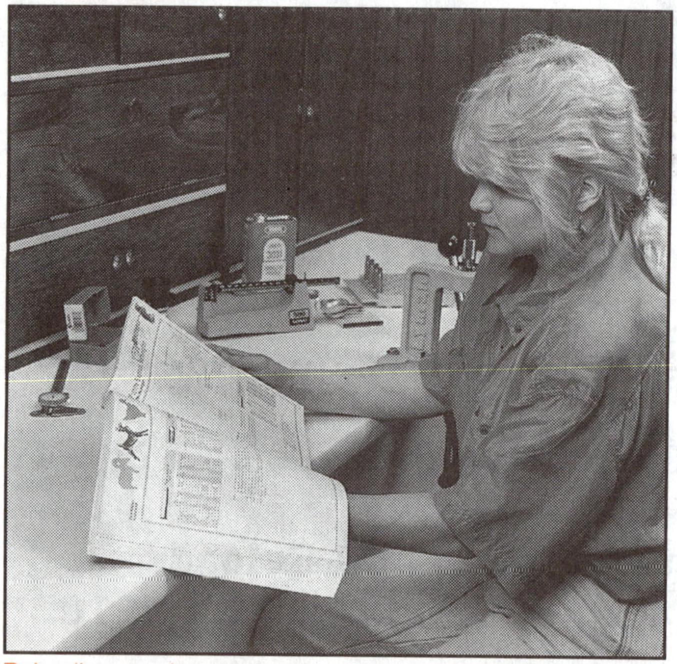

Reloading must be undertaken only by mature persons who seek the safest possible ammunition.

Of concern to many reloaders is the safety of propellants while stored or in use. Smokeless powder is a highly flammable substance designed to burn vigorously and to produce relatively enormous volumes of gas and pressure. Primers have somewhat different characteristics but are equally deserving of care. Always keep in mind that primers are by nature explosive. When these two components are assembled the same care given to them individually must be given to assembled cartridges even though characteristics of assembled ammo differ from those of bulk components.

In the chapters dealing with primers and propellants we discussed many of the safety aspects of their storage and handling. These chapters are vital to the reloaders understanding of safety. Other aspects are equally as important and will be covered in this chapter.

While components require care in handling, use, and storage, one of the most important steps a reloader can take to insure safety is to develop a thorough record keeping system. Good record keeping has many aspects, not the least of which is the ability to assemble the best possible ammo.

Need for Record Keeping

When a lot of reloaded ammo consistently groups under an inch at 100 yards, or is capable of hitting varmints at ranges up to 400 yards, a reloader will want to be able to duplicate the load as closely as possible. However, the scant information placed on a tiny reloading label may not be sufficient to allow precise duplication. Detailed records will allow precise duplication of a load.

In the event of an accident involving reloaded cartridges, most reloaders without complete records will be unable to adequately defend themselves against a charge of incompetence. A careful and detailed set of records can go a long way in proving that a reloader's ammo was as good as any made by a large factory. Records often can clearly show that reloads were not the reason why a gun blew up, or why the shooter or a bystander was injured.

Records and Recalls

Let's look at a hypothetical situation: A manufacturer wishes to recall a certain lot of components. The manufacturer places advertisements in all the usual trade magazines advising that the component is being recalled. The specific lot number of the faulty component is carefully indicated in the recall notice.

An alert reloader reads the recall notice and recognizes the component as being the same used to assemble much of his ammunition. He finds he has a quantity of 280 Remington, 30-06 Springfield, 35 Whelen, and even some 35 Remington ammo on hand that he has loaded with this type component.

Now, if the reloader has carefully segregated each lot of loaded ammo, and kept a clear record of the lot numbers of all the components used to load each of his lots, he will be able to identify which specific ammo had been assembled with the recalled component. If he has no such records all of his ammo must be suspect and discarded.

Records Help Maintain Accuracy

Good reloading records helped us solve a problem of lost accuracy not very long ago. The firearm involved was a favorite hunting rifle chambered for the 270 Winchester. It had previously fired consistent 1 minute-of-angle (moa) groups. However, a new lot of reloads would do no better than 2-1/4 moa. Consulting records proved that all was the same except for the bullet lot number. Careful comparison of a bullet from the current lot with one from the previous lot (always save a small sample of each lot of components) clearly showed a changed bullet ogive.

While both lots were loaded to an identical overall length there was a noticeably longer bullet "jump" occuring before bullets from the new lot engaged the rifling when compared to previous lots. Records of the bullet seating length based on the bullet ogive measured with a comparator clearly substantiated the difference. Adjusting the bullet seating depth to

LYMAN RELOADING RECORDS LOG (ONE)

Date Loaded	Lot No.	Caliber	Brass Brand Lot No.	Primer Brand Type Lot No.	Powder Brand Type Lot No. Charge	Times Fired	Times Trimmed	Bullet Brand Type Weight Lot No.
(sample entry) 11/7/91	76	300 H&H	Rem. S21K6601	Win LR SML200G	IMR 4350 P84FE08A 6405 48.0	3	2	Nosler Part 180 XC1501

LYMAN RELOADING RECORDS LOG (TWO)

Date	My Lot No.	Comments
11/7/91	76	Case Length - 2.840" OAL w/bullet - 3.600"

give an identical bullet jump as earlier lots resulted in the accuracy problem disappearing.

Sooner or later every reloader will find it necessary to know the exact dimensions to which previously reloaded ammo was assembled. Then only good records will supply the needed information.

Records are an important part of safe and sane reloading, with at least two separate logs being required.

Lot numbers Are Vital To Record Keeping

An indispensable aspect of record keeping is recording lot numbers. Almost all components are identified by the manufacturer with a lot number. These numbers usually appear on the label or some other portion of the package. For brass, the lot number often appears on an inside end flap or on the box back. This is the same for cases obtained by the firing of factory ammo. Primers may have the lot number on the backside side of the 100 unit box. Most bullets and powders are marked with the lot number some place on the label.

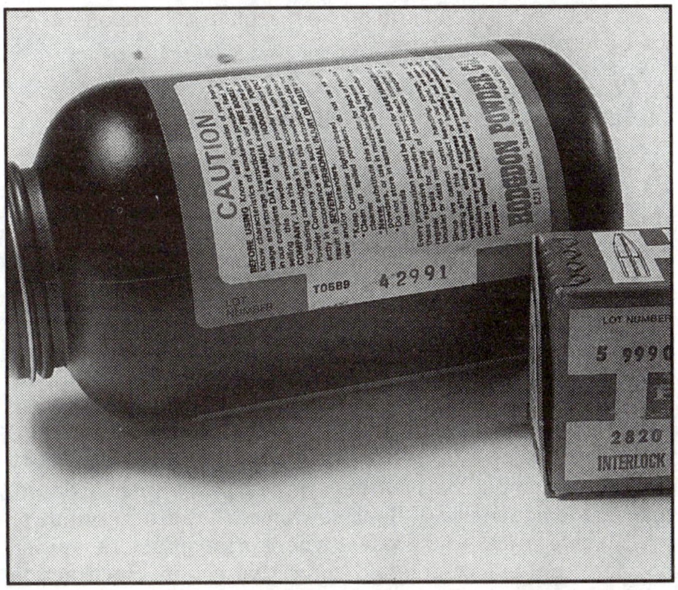

Lot numbers are important when segregating components.

Every batch of reloaded cartridges that the reloader puts together should be from a homogeneous lot of each specific component. This means that all bullets, or cases, or powder, used must have the same lot number. Should it become necessary to use components from a second lot, the reloader should separate the ammo loaded with the newer lot from that with the previous lot.

Such segregation is accomplished by the use of reloader established lot numbers for each separate group of assembled cartridges. Simply assign a specific lot number to each homogeneous group of reloads. Clearly label the boxes with this number. When a variation in a component lot number occurs due to the depletion of a previous lot, start a new personal lot number. Obviously the new lot of ammo will require segregation from the previous lot and must be clearly labeled with the new number.

Records Require Two Logs

There is a requirement for two individual sets of reloading records. The first chronicles lot numbers and other specifics. This record should list: a) date a homogeneous lot of ammo is assembled; b) lot number applied to that ammo; c) the brand, type, and lot number of each of the components used. Also include in this record the specific propellant charge used, the number of times the cases have been previously fired and the number of times the cases have been trimmed (there is a definite practical limit to both of these aspects). Finally, set down the total number of rounds loaded in the lot of ammo.

As discussed earlier in this manual, maintaining a record of the number of times cases have been fired is basic safety. Case life will vary widely with circumstances. The reloader must learn when a specific load in a specific firearm has reached maximum case life. Experience will dictate when it is time to throw away cases. The guidelines set down earlier will help keep things moving safely.

A record of the number of times the cases in a given lot has been trimmed is vital. Earlier it was shown that it is not possible to continue to remove brass from a case on an indefinite basis. Under normal circumstances and pressures when a case needs its fifth trimming it should be assumed that it is at the end of its useful life. Never trim a case a fifth time. The only exception to this rule is when excessive amounts of brass are being removed with each trimming. Depending on the amount of trimmed material it may be prudent not to trim a specific lot of cases more than two or three times. The first trimming to "true up" new cases or once fired factory ammo would not normally be included in these guidelines.

The format of the first set of records is an individual decision. It will, however, require about 19 columns to incorporate the previously mentioned aspects. A sample page and example is included for your convenience. Making copies and keeping them in a three ring binder can make the record keeping nearly effortless. Simply white out the examples before making copies.

Eventually you will come upon a component that either has an illegible or missing lot number. Avoid the temptation to not record a lot number; but apply a lot number of your own choosing, perhaps the date of purchase and the initials of the point of purchase. An example would be 11-11-91 RGS. Mark the component package; save it and two or three samples of the components (if powder, save perhaps 100 grains); and record your applied number in your records.

Common sense will dictate the length of time component samples need to be retained. Samples of components no longer available in loaded ammo or older lots may often be discarded. Store sample powder and primers appropriately.

The second set of records will complete the needs of safe and sane reloading. This log will show the date and lot number you applied to each homogeneous group of reloads. In it you record all observations made during the reloading process. It also records all of the physical dimensions used - for example, case length after sizing and after trimming, as well as overall cartridge length. It should make mention of any specific tools used if they are unique, and contain comments on any variations made from similar ammo and the reason for the change.

Records Should be Detailed

An example of a detailed record would be as follows: *"11/1/91, lot number 267, Loaded 86 rounds of 338 Win. Mag. Loaded this lot with Winchester Large Rifle Magnum primers instead of previously used Federal 215 primers in an effort to enhance cold weather performance. Cases were trimmed to 2.490. Inspection of cases after last firing showed no signs of case deterioration. Cases checked in cartridge headspace/length gauge, after sizing. Overall loaded length set at 3.340 inches."*

Keeping a record of all the details of your loading might result in some new habits like measuring bullet diameters; check weighing a few bullets; observing powder and primer-foil colors (not every manufacturer color codes primer-foils); and inspecting primers for anvils. Record all of these observations in your second log under the date and lot number you applied to the ammo. Are these extreme? Maybe; but it will take only one accident to change a reloader's thinking.

Reloading records are not only essential to show that you are a prudent reloader or to help in the duplication of a favorite load. Records can also be helpful in situations over which you have no control. For example, changes in components have been made which caused ballistic variations without the reloader being made aware by the component manufacturer. Some of these changes that we have run into were notable.

At one time a major manufacturer reduced the overall length of a specific bullet to accommodate changes in bullet ogive. The new bullet length resulted in enormous groups when used with preexisting data and specifications. When dies were carefully adjusted to eliminate the long bullet jump, the ballistics of the load were noticeably changed. Without a chronicle listing the characteristics and dimensions (and/or a sample of previous lots) of the components used would have been difficult.

We remember when shooting the data for a handbook published years ago, a specific primer suddenly changed ballistics compared to previous primers of the same designation. When discussing the problem with the manufacturer, it was only through the use of lot numbers that we were able to find out that they had changed the primer composition from previous lots. This speedy ending to the "mysterious" change in ammo ballistics was the result of good record keeping.

Even propellant burning rates have been changed and the consequences have led to some confusion. You may even want to keep a third set of records, by lot numbers, on the specific date and place of purchase.

It should by now be obvious that it is most unwise to assume that any reloading endeavor is a safe undertaking without the benefit of a detailed set of reloading records. Thankfully, good record keeping is neither difficult nor time consuming.

The maintenance of records is particularly convenient if the record books are kept right on the loading bench. If you are meticulous about records, it is unlikely there will ever be a problem that cannot be solved. Equally important, in the unlikely chance of an accident, records could possibly absolve the reloader of any responsibility for the mishap. Maintaining records will often cause the reloader to think about certain aspects of reloading that might otherwise have gone unnoticed. This will lead to greater knowledge and make for a more informed reloader.

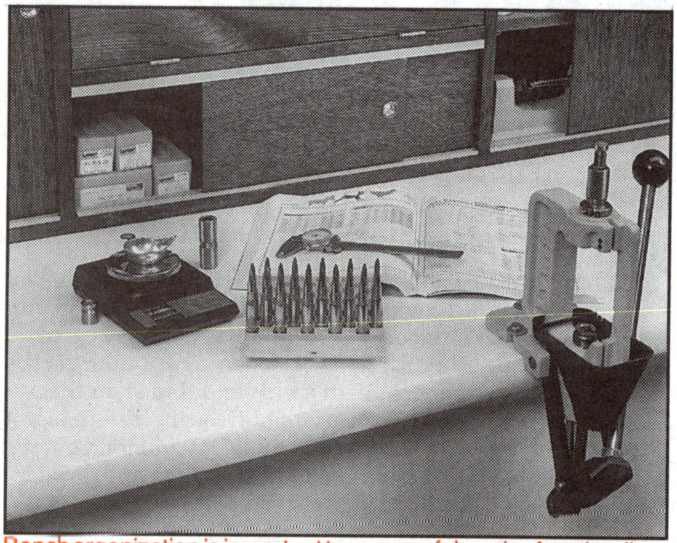

Bench organization is important to successful, and safe, reloading. Keep tools well spaced and only a minimum of components on the bench.

A Few Points Of Safe Reloading Bench Organization

The prime consideration in setting up a loading bench is the remote storage of powder and primers. When it is possible, the remote storage of all other components including dies, miscellaneous tools and accessories will help prevent problems caused by using the wrong component or tool. **(Always store powder and primers independent of each other and all other stored items.)**

The bench's top surface should be given consideration. As stated earlier, primer "dust"; (that is minute particles of priming pellet compound) will accumulate on and around areas and tools where primers are used. This dust is a safety hazard and must be routinely wiped away with a damp, disposable paper towel. Bench surfaces that are irregular or which have cracks and grooves where primer dust, (or spilled powder) can collect demand corrective action. Such action might mean resurfacing the benchtop with plywood, or

Formica, or perhaps some other smooth material. The solution could be as simple as using some photographers heavy background paper on the bench top, replacing it as needed. Benches with spaces between the boards making up the top are acceptable but the floor, or shelving, underneath must be cleaned when powder spills occur. The floor should also receive an occasional damp mopping with disposable paper toweling to remove primer dust accumulation.

Priming Equipment Cautions

Priming tools should be wiped with a clean dry cloth. Avoid using a damp cloth which eventually will cause rusting. Do not use an oily cloth as this could eventually result in priming pellet contamination. It is prudent to periodically remove automatic primer feeds from tools so equipped in order to flush them with water. Be sure the feeds are empty of primers. If boiling water is used the parts will become hot enough to cause rapid drying and help prevent rusting.

Keep in mind that any rusting could cause a jammed priming unit and create a potential hazard. Be sure the flushed priming units and feed parts are thoroughly dried. Replace any unit or part that shows even a trace of pitting, rust or any other deterioration.

Primer Seating Depths Are Critical to Safety

Safety demands a very careful check of primer seating depths. A primer that is seated slightly deep might create a sensitive primer condition. Excessively deep seating usually means the primer pellet is cracked and ignition becomes questionable. A primer that is not seated below flush can be ignited during the feeding cycle of the firearm causing the cartridge to ignite before it is fully chambered and the breach is locked. This can lead to personal injury or a seriously damaged firearm. It is amazing how often a high primer is detected when inspecting other shooter's reloads. More on primer safety is contained in the chapter dealing with primers.

Primer seating depth is an important aspect of reloading safety.

Powder Spills

All powder spills must be promptly cleaned up, but never use a vacuum to do so. A dust pan and broom are a safe way to pick up spilled propellant. The use of any electrical appliance to pick up spills must be avoided because of the potential of heat and sparks which could ignite the propellant. Be meticulous with the clean-up procedure.

Tool Positioning On The Loading Bench

The basic loading tool's location on the bench will usually work out best at a central position. A powder measurer will be convenient for most reloaders when located 1-1/2 to 2 feet to the left or right of the press. We find it handy if the powder measure is not permanently mounted to the bench. This frees up work space for other tasks.

If more than one tool is to be used on the bench it should be located at least three feet from all other permanently mounted tools. A lesser distance will find elbow room to be lacking and will result in a need to jam components, loading blocks, or other items in hard to reach or awkward positions.

The case trimmer should be mounted away from the loading bench. The accumulation of brass chips will make a mess of a bench unless the chips are cleared away after each trimming session.

If each powder charge is to be weighed, it will prove practical to place the scale in proximity to the powder measure. Consider placing the scale at eye level, perhaps on a shelf at the back edge of the bench. Having the scale at eye level will prevent any error in reading the scale due to the eye being positioned above (or below) the scale's pointer and index mark. NOTE: Electronic scales, such as the Lyman LE-1000, usually have the read-out window positioned so that the scale is best placed below eye level.

It is good reloading practice to seat the bullet immediately after charging the case with propellant. This is a safeguard against mixing up cases in the loading block and prevents a case from receiving a double powder charge, or no charge at all. It also prevents charged cases in a loading block from being accidentally upset resulting in wasted powder and lost time. Thus, the reloading press, powder measure and scale, should be located on the bench to be comfortably reached from a single seated position.

When setting up the bench, keep in mind that plenty of room should be available for one or more loading blocks, a primer tray and bullet box. The finished ammunition storage box also should be conveniently located.

Keep The Loading Bench Clean

One of the marks of a well organized loading bench is a general lack of clutter. The maximum amount of components that should ever be on the bench is one box of bullets, one can of powder, a single 100-pack of primers and up to 100 empty cases. Tooling on the bench should be limited to that needed for the specific loading operation. (Never use the reloading bench as a work station for firearm repair or cleaning when components or primer handling equipment are present.)

Those items that are properly stored below the bench, include loading data references, personal loading records, cartridge gauges, micrometer, extra dies, and other miscellaneous tools. Avoid storage over the bench to offset the possibility of heavy objects falling or being dropped on primers which may be on the bench. If storage shelves above the bench must be used, keep only very small, light objects on them.

Lighting Caution

Sufficient lighting is essential to bench set-up. A caution on bare bulbs of any type: When mounted directly over the work area, they can be a needless safety hazard. We are aware of several mishaps which occurred when bulbs where broken directly over a loading bench. Shields should be provided in the form of unbreakable translucent covers. This is especially important when long guns are cleaned or worked on at the bench and a low ceiling is involved. Hot sparks, falling onto exposed powder or primers, are potentially quite dangerous.

Careful inspection of each lot of new components is an essential part of maintaining safety.

Component Inspection

Every lot of components used should be carefully inspected and double checked against records of previous lots. For example: a few bullets from each box should be weighed and measured. Packaging errors have included putting the wrong weight and even wrong diameter bullets into a box. Pistol primers have been packaged as rifle primers and so on. It is only the reloaders vigilance and constant checking that can eliminate such errors and the possible catastrophic aftermath. While such errors are extremely rare it is always best to put caution first.

Case Records of Firings Are Vital

As stated elsewhere, case records and inspections are particularly important. Despite the number of firings that can occur when the cartridge case fits the chamber precisely (such as might occur with a custom bench rest chamber and cases), cartridge cases do wear out. As discussed in the case chapter, full length resizings of cartridges fired in commercial chambers usually result in an expected case life of less than 8 firings. Belted cases fired in most factory rifles and full length sized seldom allow for more than three safe firings.

Cartridge Headspace Gauges

Cartridge headspace gauges are a must to insure that headspace dimensions of fired cases versus reloaded ones are not drastically altered during full length resizings. The careful use of cartridge headspace gauges can increase useful case reloading life. In the instance of belted cases, such gauges can sometimes result in doubling the case life.

Excessive Pressure And Chamber Gauges

Excessive pressure indications are not always apparent when firing cartridges with a "too hot" load. This is especially true when pressures are only slightly over maximum. Nonetheless, such loads can slowly pound a bolt rearward until the firearm develops potentially dangerous excessive headspace. Chamber gauges that allow the reloader to check firearm headspace can lead to the discovery of slowly changing headspace and the discontinuance of the problem load before serious firearm damage has occurred.

Case Inspection Is Vital

Careful inspection of fired cases cannot be overstressed as an important safety step in reloading. Simply because a cartridge has fired in an apparently normal fashion, extracted, and ejected without undue difficulty, it is no assurance of a load's safety or of the cases ability to withstand another firing.

Each and every fired case should be checked for the condition of the fired primer and of the brass case itself. Inspect the fired primer for excessive flattening, cratering, rupture, piercing (blanking), or backout. All these are signals that must not be ignored. Any sign of gas leaking around the fired primer or of an enlarged primer pocket is a sure sign of a problem, from a simply worn out case, to a load that develops too much chamber pressure.

Proper interpretation of the specific problem is also essential. For example, a backed out primer might be a signal of excessive pressure, an indication of excessive headspace, or even a load that develops insufficient pressure.

Fired cases that show any sign of irregularity should be discarded immediately. However, first crimp the mouth of the case shut using a pair of pliers. This will prevent accidental use of the case at a later date.

Be especially alert for the early warning of a bright ring (partial or complete) appearing around a case's circumference at the head/body junction. This is the first sign of pending incipient case separation. However, separation of the case body from the head is not always preceded by early warning.

Examine the flat base portion of all fired cases for any bright areas which may be caused when excess pressure tries to extrude brass into the ejector hole of the bolt face. The raised metal will be wiped bright when the bolt is rotated to open the action.

Naturally, if a case shows signs of deterioration it is essential to eliminate the cause of the problem, except when deterioration is the result of normal case life due to multiple firings.

Case Trimming Cautions

Trimming cases an excessive number of times is a common safety violation. Generally, cases should be discarded after the fourth trimming. The brass that is removed with each trimming flows forward from the case body. Only so much brass can flow forward before the case is seriously weakened, usually at the junction of the head and body. Experience has shown that attempts to "feel" the step formed by such brass flow, with a bent paper clip or other device fail to reveal a hazardous condition.

Check Neck Diameters

Another safety requirement is the need to check the outside neck diameter of loaded cartridges. While not a common problem, neck diameters can become so thick as to enter the chamber with a near line-on-line fit. When fired, the case neck is prevented from expanding enough to properly release the bullet. Very high, possibly dangerous, chamber pressures will result. More on this, and other aspects of case safety are contained in the case chapter.

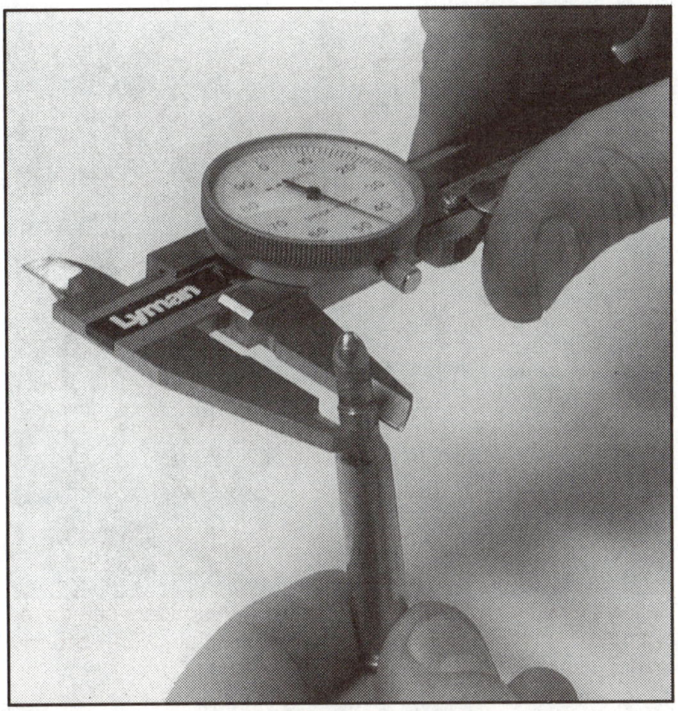

Carefully measure a sampling of the loaded ammo's neck diameters. The outside neck diameter must never exceed the maximum dimension listed in the data section of this handbook.

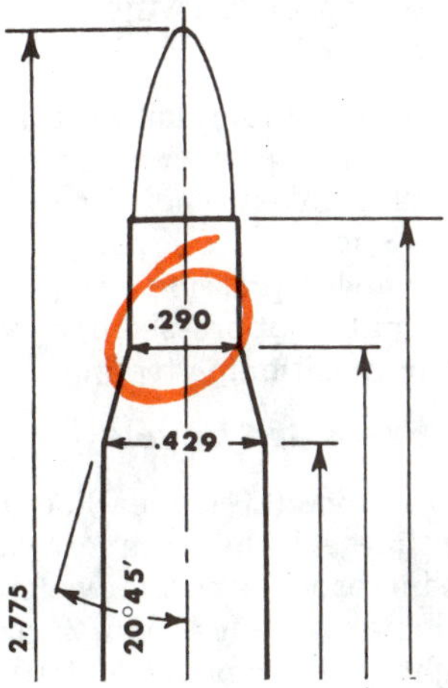

Shown is a section of a maximum cartridge drawing from the data section which indicates the maximum allowable neck diameter.

Keep Data Updated

All data publishers update their data recommendations from time to time. This is done in part to keep the listings current with new components and cartridges. But it is also an on-going effort to keep data safe. New listings may drop loads that have become problems due to changes in components, firearms or other reasons. Any data source should be considered obsolete as soon as the publisher releases a new edition. Such advice is not a ploy for repeat sales but rather an important factor in maintaining safety.

Stay Alert

Reloaders should remain constantly alert and learn all that is possible about reloading safety. No single publication can contain all the required knowledge. Additional reading is forever a part of good reloading.

Many accidents occur right at the loading bench. Adherence to safety practices and common sense can eliminate these occurrences and keep your reloading hazard free.

Ten Rules For Reloading Safety*

1. Set up your reloading bench and storage cabinets where they will not be exposed heat, sparks, sun rays, flame or any other source of ignition.
2. Establish a loading routine and do not vary from it.
3. Keep the loading bench, loading equipment, and surrounding area free of clutter.
4. Be absolutely certain of all component identification. Never use components from an unlabeled source. Never store components in anything but the factory container. Never purchase components that are not in a sealed factory container.
5. Understand what you are doing and why it must be done in a specific way. When in doubt stop. Then contact the manufacturer of the components and/or tools involved. Assume nothing!
6. Follow loading recommendations exactly. Never substitute components in a proven rifle or handgun load without backing down to the starting load and working up the load carefully. Never substitute any smokeless powder in any amount for any black powder or Pyrodex powder.
7. Remember to use proper protective equipment. Always wear safety glasses when loading and shooting. Always wear ear protection when shooting.
8. Stay alert. Do not load when potential distractions are near. No visitors, TV, radio, conversation with family or similar diversions when assembling ammo. Do not load when tired, when smoking or when under the influence of any substance or medication. Do not subject any component or loaded cartridge to any form of abuse. Be certain to read this entire handbook before beginning to reload.
9. Keep all components and tools out of the reach of children and incompetent persons.
10. Never shoot reloads assembled by another person.
* This is not a complete list of safety precautions.

CHAPTER 10
CONTROLLING PRESSURE

Reloaders, as a whole, tend to be meticulous people. It's no wonder then that one of their most often asked questions is which handbook shows the true maximum load for a specific caliber and bullet weight? The question is prompted by the fact that most reloaders tend to own several handbooks and that it is difficult to find any two that agree exactly. Indeed a specific handbook may vary the listed maximum loads from one issue to another. The differences in listed maximum loads, from one publisher or issue to another, can be substantial.

It is necessary to have at least a general understanding of those things which can cause a change in pressure and how these effect the data published in any specific handbook. With such an understanding the variations in published data will no longer cause confusion.

Test Conditions Often Vary

The actual barrel used for testing can and has caused velocity variations greater than 150 feet per second with or without a corresponding change in pressure. The same is true for pressure figures which can easily vary by 5,000 pounds per square inch from barrel to barrel, with or without any anticipated change in velocity. Since it is unlikely that a specific laboratory's barrel will match any other laboratory barrels internal dimensions, one should actually anticipate that published data will vary when supplied from different test equipment.

Pressure testing is a complex undertaking suitable only for ballistic labs. Still the reloader must learn to control pressure if satisfaction and safety are to be part of ammunition assembly.

To the previous cause of data variations we need to add the actual physical dimensions of the finished loaded round used for the testing. One lab may seat bullets to a different depth, or perhaps use a bullet that is 0.0005 inch larger or smaller in diameter than any other lab. This, and other dimension altering factors, suggest that data from different sources will show different maximum propellant charges.

The actual lot of components used for a test can provide substantially different ballistic results, even when compared to another lot of the same style and brand. Variations from one brand to another can cause changes in pressure that are noticeably at odds with other components that are of similar description.

Adding more fuel to the dilemma is the fact that individual loading methods can cause still other variations in test results. Thus, it should be recognized that when data closely agrees, from one manual to another, it speaks highly of the labs involved and suggests some unplanned and coincidental consistency in test components and/or procedures or equipment.

Ballistic labs cannot duplicate all of the handloaders criteria for ammo assembly. Thus the reloader must carefully work up to any load chosen.

Finally, it is unlikely that the reloaders barrel's internal dimensions, lots of components, or procedures will duplicate those of any specific lab.

Other Considerations For Establishing Maximum Loads

Keep in mind that pressure alone may not be the only criterion used to determine maximum powder charges. This can be especially true with data from extremely knowledgeable sources. For example: how much powder actually fits into the case without undue propellant compression is important, as are factors of ballistic uniformity, suppressed or elevated temperature testing and powder positioning testing.

Testing Loads For Safe Pressure Levels

Begin by firing at least 10 rounds (20 is ideal) of ammo loaded at the starting powder charge. If there are zero visible signs of pressure you may the increase the powder weight for the next string of cartridges. For charges less than 25 grains increase the propellant weight in increments of 1/10 of a grain. For charges over 25 grains, but under 50 grains, increments of 1/2 of a grain are appropriate. When using charges of 50 grains, or more, increases of 1 grain at a time are appropriate. Always fire at least 10 rounds at each test level.

Measuring crushers is but one part of safe load development. Tests are run at ambient, suppressed and elevated temperatures. The lab also considers such things as loading density, ballistic uniformity and other factors before accepting a load for publication.

Be certain that no visual signs of excess pressure occur. You may then proceed to the next heavier increment of propellant. Advance in this manner until you have reached the maximum charge. Never exceed our listed maximum charge.

Should any visual pressure sign(s) occur during test firing, even on a single round, immediately stop shooting and reduce the propellant charge by 4%. This should then be considered the maximum charge with the specific components and assembly procedures being used, unless the established charge results in a powder weight below the original starting weight. In this case discontinue any further use of the propellant and start over with another propellant.

Remember, in the future if any change in the specific components used (from the original testing) is made, it will be necessary to reestablish a new maximum load. This includes such changes as switching from one brand of component to another - even of a similar type. It also includes changing from one lot of components to another of the same type. Lot numbers are usually printed on the component box, but may be located on an inside surface.

Since a change in component lot means starting all over to verify the maximum load, most reloaders purchase enough components in a lot to last a year or so. This cuts down on load redevelopment time. However, do not be tempted to store excessive quantities of primers or propellants in order to avoid occasional re-establishing of a maximum load. It is an unnecessary hazard to store excessive quantities of powder and primers. Of course, any quantity of bullets or brass may be safely kept on hand.

Safe and accurate reloading begins with the establishment of a reasonable maximum load for each specific set of components. Anything less is fool hardy and will eventually lead the reloader to grief.

Controlling Pressure

Pressure, properly controlled, is essential to cartridge performance. Too little pressure will result in sub-standard ballistics. This could lead to firearm malfunction or unsatisfactory terminal ballistic performance. Too much pressure can slowly, or catastrophically, wreck a gun.

There is no such thing as just a little too much pressure. Perhaps an example is in order. A certain reloader used a particular 270 Winchester load for years. It was "just a half grain heavier than listed as maximum" in a respected loading manual. The reloader found the load to be accurate and effective and "knew" it was an OK load. He had used it for years and fired it with nothing but perfect satisfaction. There was never a single excessive pressure indication from the ammo.

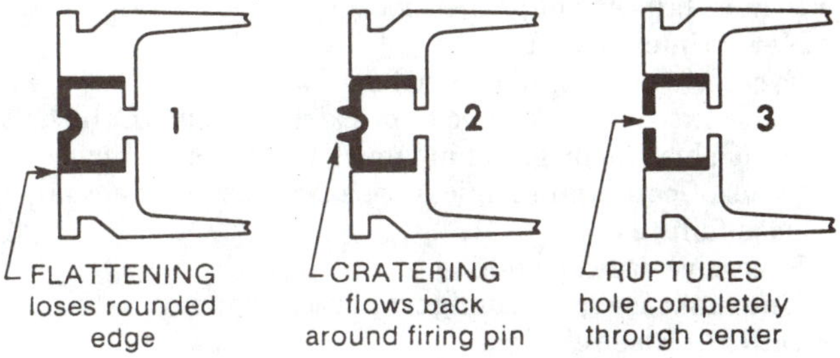

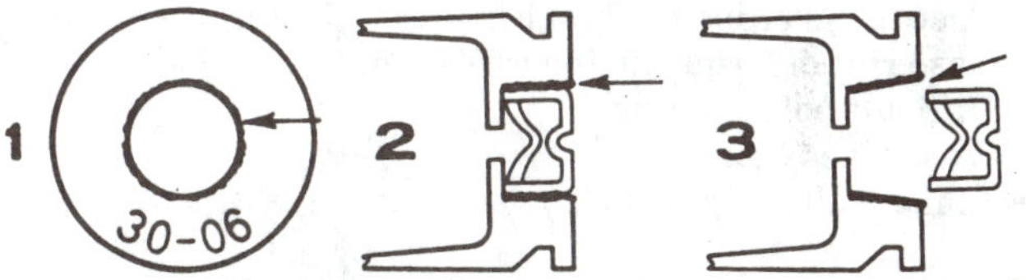

1. Dark smudge line around primer indicates gas leakage.
2. Pocket opens, leaving primer loose. A dark smudge, indicating gas leakage, is usually present.
3. Pocket opens completely, freeing the primer. A dark smudge, indicating heavy gas leakage, is usually present.

Then one day, coincidental with the start of the use of a new lot of brass, he suddenly had signs of incipient case separation. The reloader called the brass manufacture to complain of poor quality. After a brief conversation it was suggested that he have the rifle's headspace checked. Sure enough, his expensive custom rifle had slowly developed excess headspace due to continued use of a load that was "just slightly" hot.

Our reloader was lucky. His rifle was repairable and the problem caused no personal injury. Other instances of continued excessive pressure have caused irreparable property damage and even serious physical injuries.

Occurrences as described, and of other types, have proven time after time that ammo can be too hot and never give a single clue until the damage has been done. Too much pressure is not necessarily accompanied by early, visually detectable, excess pressure indication(s). Adhering to loading manual maximums is essential to avoid this type of problem.

Caution: Pressures can be too high with loads that do not exceed manual recommendations. Some of the causes can include variations in component lots, assembly procedure or finished dimensions. Savvy reloaders, as a whole, are aware that too much pressure is a hazard; something to be avoided at all cost. To this end, it is essential to watch for any telltale signs of excess pressure.

Excess Pressure Indications

Some of the generally accepted indications of excess pressure include:

- Abnormal primer appearance (flattened)
- Enlarged primer pockets
- Leaky primer pockets
- Loose primers
- Primer backout
- Cratered primers (partial extrusion around firing pin into firing pin hole)
- Blanked or pierced primers
- Primer falls out of primer pocket
- New primer seats loosely
- Excessive case head expansion
- Brass extrusion into ejector or extractor cuts in the bolt (may show as bright shiny marks on the case head).
- Signs of potential incipient case separation or actual case failure
- Excessive case stretching
- Unusual case head configuration after firing
- Difficult case resizing
- Short case reloading life
- Difficult case extraction from chamber
- Case stays stuck in chamber
- Case rim deforms during extraction
- Difficult bolt opening
- Excessive recoil
- Abnormally high velocity.

Signs of pending incipient case failure and actual case separation may be due to excess firearm headspace. Or, these may be indications of brass that has been overworked by repeated sizing and firing, or too much pressure.

These and other less often encountered clues should always be a tip that all is not well. However, it is possible to be in error when interpreting what any visible clue might mean. For example: if a primer backs out of the case upon firing, does it mean the load is developing more pressure than is acceptable? Or could it mean that the load was too light and insufficient rearward thrust occurred on the case to hold it tight to the bolt face? Comparison of the load being used to suggested loads in a handbook will give a good grasp on the correct interpretation; as will a lot of experience in examining cases fired at low and high pressure.

If any sign of higher than normal pressure is encountered it should always be interpreted as an indication that the load is excessive for the components used, for the assembly method, for the finished loaded round dimensions, for the specific firearm; or for some combination of all of these. Do not expect that all rounds fired will show similar pressure indications. A single round in a ten shot string may be the only one to give evidence of excessive pressure. But all rounds in the string may have developed too high a pressure level.

Pressure Varies From Shot To Shot

Keep in mind that in a ten shot test there could possibly be as much as a 10,000 psi difference between the highest and lowest pressure generated. This alone can cause a noticeable difference in the physical condition of the fired case. No two loads will usually give identical pressure even though the reloader takes great care to make the ammo as identical as possible. This simply is the nature of internal ballistics.

Take All Excess Pressure Indications Seriously

Any visible sign of excess pressure, even if it is displayed only on one cartridge in a box of fired cases, must be considered as a warning sign for the lot of components and load in use. It is true that one lot of brass, or primers, may show visible excess pressure signs at levels that may actually be substantially lower as compared with other lots of components. No indication of excess pressure should be ignored, even if it is found on only a single round.

A strong respect for any visual pressure indication(s) should be developed. Trust any warning of excess pressure. Do not be misguided and assume that a "soft" lot of components is causing the problem and that the load is still safe. The first half of such logic is valid; the second half is not!

While a specific lot of components may stand up well to excess pressure without giving a clue, gun damage may be occurring. This possibility obligates the reloader to use all cautions and never exceed the listed maximum loads.

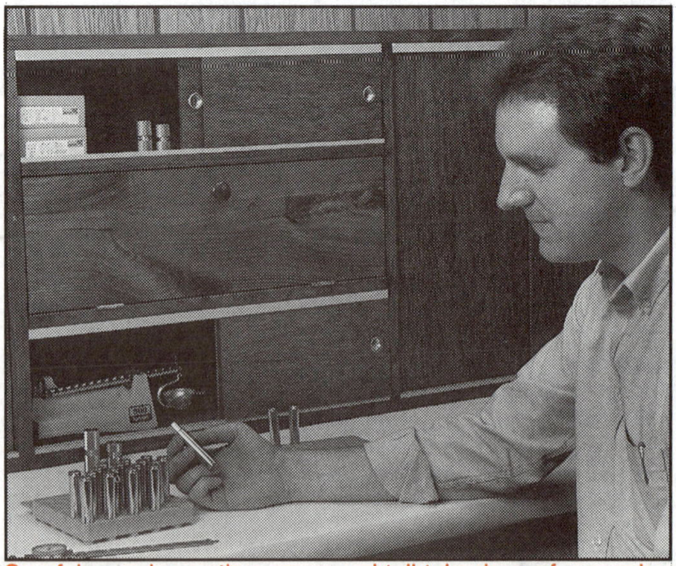

Careful case inspection can reveal tell tale signs of excessive pressure if the reloader understands what to look for.

Remember that every visible sign of excess pressure is a warning to be heeded. The absence of visible signs of excess pressure is also no guarantee of safety. Stay at or under the maximum listed loads and use no load that shows even a single sign of too much pressure.

Some Conditions That Can Raise Pressure

1. Actual case capacity. Thick walls and/or webs will reduce internal capacity and could increase pressure.

2. Case length. Cases exceeding maximum lengths could enter the chamber throat and cause pressure increases.

3. Case neck diameters. If over the maximum listed dimension, thick necks can be a source of increased pressure.

4. Crimping cases can alter chamber pressure. Never crimp a load unless the load is developed with a crimp.

5. Component variations. Never change component type, brand, or lot number without completely redoing the necessary load development.

6. Excessive, or improper storage of powder and primers. Always use fresh primers and propellants which have been properly stored. See the chapters on primers and propellants for more information on storage.

7. Excessively deep seated or overly long loaded cartridge lengths. Generally it is wise to load to the length listed in the data tables. When a specific overall length is not shown , it is wise to load ammo to the maximum length specified for the loaded cartridge. When it is not possible to load short bullets to maximum cartridge lengths, and there is no specified length, be certain to seat the bullet at least a full diameter deep. For example a 30 caliber bullet should be seated at least 0.308" into the case neck.

8. Firearm dimensional variations can cause substantial pressure changes.

9. Die dimensions, which vary from normal, can cause pressure changes.

Keep in mind that other factors can cause notable pressure changes.

No Two Shots Will Normally Duplicate In A 10 Shot Test

Round Number	Velocity fps.	Pressure CUP's	psi's	Case Head Expansion
1	2603	42,400	50,400 (low)	.004
2	2605	41,600 (low)	51,000	.006
3	2637	44,100	53,000	.006
4	2601 (low)	42,000	51,000	.006
5	2630	42,400	52,500	.006
6	2620	44,100	52,300	.006
7	2627	42,400	51,900	.006
8	2634	42,900	52,500	.006
9	2660 (high)	42,000	55,300 (high)	.006
10	2639	45,400 (high)	52,600	.006
Average	**2627**	**42,900**	**52,300**	**.006**
Extreme Spread	**60**	**3,800**	**4,900**	**.002**

CUP's Versus PSI's.

Perhaps the greatest area of misunderstanding of pressure terminology, has arisen from the early practice of referring to chamber pressure as some number of "pounds per square inch (psi)." Many early ballisticians believed that results of a firearm chamber pressure tests were accurately described as psi. The truth was their test results had little to do with actual pounds per square inch.

Their "crusher" testing and the testing still used in many ballistic labs, uses a barrel with a hole drilled into the chamber perpendicular to the bore's axis, usually 1" from the rear edge of the chamber. A slip fit piston is fitted into this hole with its end contoured to precisely fit the inside of the barrel chamber. In testing, a cartridge is loaded into the chamber and the piston slipped into place. A copper crusher is then stood on top of the piston and is securely held in position with an anvil. When fired, the cartridge case has a small disc rupture from it at the location of the piston hole. The hot and rapidly expanding gases in the chamber push equally on the bullet base and on the piston base. The piston in turn moves heavily against the copper crusher, which is forced to collapse to varying degrees depending on the total amount of pressure applied to it by the piston.

The amount of "crush" on a copper crusher can be converted to a pressure reading of so many copper units-of-pressure (CUP's).

The amount of "crush" of the copper cylinder is then measured carefully and this crush length is compared to a tarage table which lists a specific value for the amount of crush which occurs.

There is, of course, more to the crusher pressure testing system then the foregoing few words might suggest. The point to be made is that the copper (lead for shotguns and some handguns) crusher method is a valid and useful tool for ammunition evaluation, but it does not actually express pressures in true pounds per square inch. This did not escape the attention of newer generations of ballisticians. They set out to correct the misnomer of pounds per square inch and used instead the designation copper units of pressure (CUP's) or lead units of pressure (LUP's). However, the erroneous term PSI had become so accepted that it was frequently used interchangeably with CUP or LUP. While such use was technically wrong it created no major problems when everyone was talking about the same thing — the results of a copper or lead crusher pressure test.

Then it was discovered that an electric strain gauge could be mounted to a barrel to transmit firing stress signals that actually could interpret chamber pressure in true PSI. This newer Piezo system soon became commonplace with ammo manufacturers as it was a faster method than the crusher type testing. Nonetheless, the older crusher testing system remains valid and in general is used by many data labs, including Lyman.

Each system produces different numbers. Some confusion occurred with labs that never bothered to apply the correct CUP or LUP designation to crusher test results prior to switching to electronic testing. Now they are generating new and different pressure test results using the same psi nomenclature.

Almost from the beginning of publishing chamber pressures in true psi's, the reloader began to have difficulty interpreting psi test results. For years 52,000 psi (crusher method with erroneous designation) had been maximum pressure for the 270 Win. Then 52,000 CUP (crusher method with proper designation) was maximum. Now suddenly there is published data showing 65,000 psi as being an appropriate maximum pressure for the same cartridge. The resulting confusion is understandable for any reloader who was normally cautious and had not been aware of the industry testing changes.

Since so many factors enter into how much actual psi are generated in a cartridge there were some 'surprises' as true PSI data began to accumulate. For example, the size and shape of a cartridge was noted to influence pressure in varying directions. A specific 30-06 Springfield load that produces 50,000 CUP produces 60,000 psi. That's with all factors equal and only the test recording system being different. Equally, an 8mm Mauser load that results in 37,000 CUP will give only 35,000 psi. Some cartridges produce identical readings in both crusher and Piezo systems. For example, the 30 Carbine (40,000 CUP or PSI) or the 45-70 Gov't. (28,000 in either system).

The nearby table (developed by the Alliant Powder Co.), shows some comparisons between the older CUP and the newer psi method. It clearly points out that it would be an error to assume any correlation between the two test methods. Both systems have been used very successfully to develop data and manufacture ammunition when the system in use was thoroughly understood by the folks working with it. Both have merit and both are equally valid.

Lyman has chosen to stay with the crusher method in order to keep all test results in a common system. The crusher method remains as valid as it ever was, though it admittedly is slower and more costly to work with.

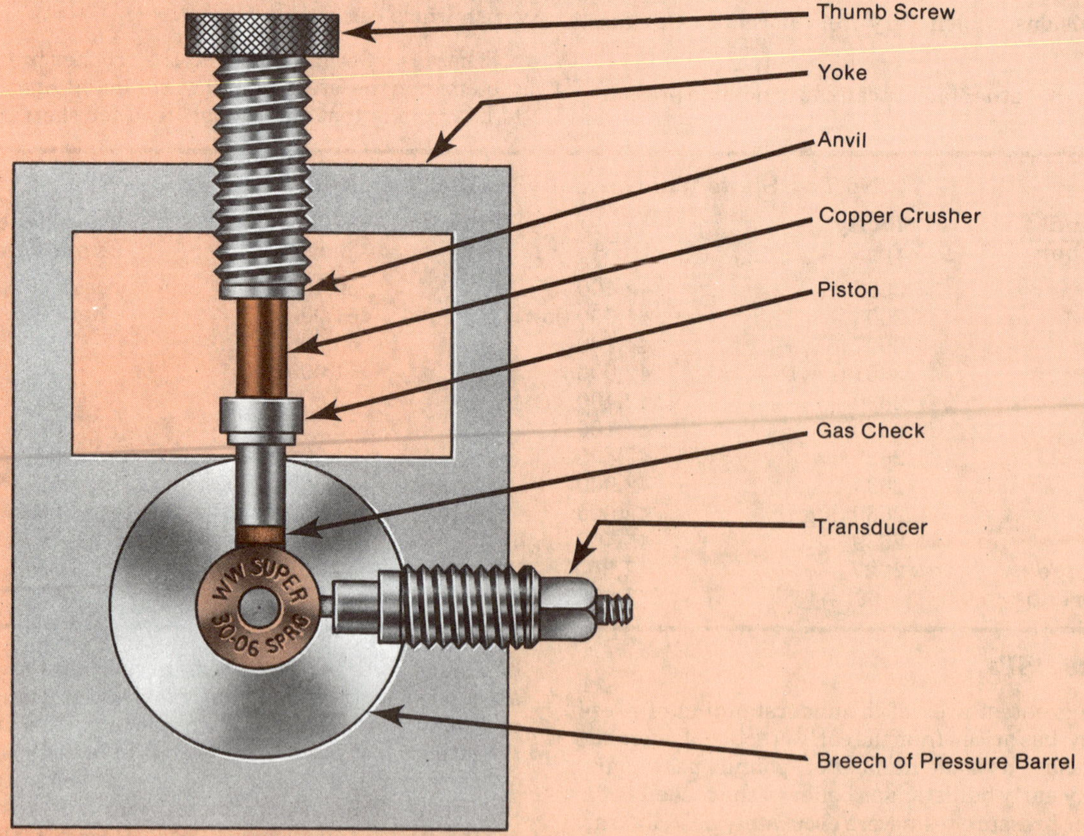

Installation for Simultaneous Pressure Measurements by Crusher and Transducer

CRUSHER/PIEZO PRESSURE TABULATION

Cartridge	CUP Value	PSI Value	Cartridge	CUP Value	PSI Value
22-250 Rem.	53,000	62,000	30 Carbine	40,000	40,000
222 Rem.	46,000	50,000	30-06 Spring.	50,000	60,000
223 Rem.	52,000	55,000	30-30 Win.	38,000	42,000
6mm Rem.	52,000	65,000	300 Savage	46,000	47,000
243 Win.	52,000	60,000	300 Win. Mag.	54,000	64,000
25-06 Rem.	53,000	63,000	303 British	45,000	49,000
257 Roberts	45,000	54,000	308 Win.	52,000	60,000
257 Roberts +P	50,000	58,000	8x57 Mauser	37,000	35,000
270 Win.	52,000	65,000	8mm Rem. Mag.	54,000	65,000
7mm-08 Rem.	52,000	57,500	338 Win. Mag.	54,000	64,000
7x57 Mauser	46,000	51,000	35 Rem.	35,000	33,500
7mm Rem. Mag.	52,000	61,000	45-70 Gov't.	28,000	28,000
280 Rem.	50,000	60,000			

NOTE: do not attempt to interpolate or extrapolate data for any other cartridge from this table. This table is intended to show the general type of variations that occur in the CUP and PSI systems but not intended to show all possibilities; nor should it be used as a loading guide.

CHAPTER 11
ACCURACY AND EVALUATING RELOADS

Handloading is sometimes undertaken solely to save money. Reloads, if we do not put value on the time used, can be had for a fraction of the cost of factory ammunition. But more often than not, reloading sooner or later becomes a quest for accuracy. That quest can bring about the use of tedious procedures and methods, not all of which will reward the shooter in proportion to the effort expended. Still, if one participates in benchrest shooting no step which will add an edge to accuracy is inconvenient. In this sport a group of 0.2" is not even going to get a second glance.

Accuracy of this type is often the legitimate quest of the reloader. It takes a few tricks and a bit of patience to make such precision a reality.

Accuracy And Resizing

Resizing of the cartridge case can be accomplished in a number of ways. Sizing only the neck portion of the case is one approach often used to enhance accuracy. Cases left at fired body dimensions allow the case to best center the bullet in the rifling. However, if the chamber is not perfectly round and concentric with the bore, the type of resizing will offer no specific advantage.

Most serious bench rest shooters will size only the portion of the case neck required to hold the bullet. With benchrest quality barrels and chambers this approach may result in increased accuracy.

The How To Of Special Neck Sizing

The accuracy-oriented shooter often will first consider changing from full-length resizing cases to neck sizing only. Over the course of many thousands of rounds fired during comparison testing, the usual neck sizing methods have been shown to provide a minimal but worthwhile increase in accuracy. Accuracy with neck-sized cases seems to be improved by about 1/4 minute-of-angle (M.O.A). That's a worthwhile gain, but the inevitable tight chambering of cartridges may not appeal to big game hunters who insist on smooth rifle actions. Neck sizing can sometimes lead to cartridges failing to properly chamber under hunting conditions, with actions other than bolt actions.

As mentioned earlier, the standard neck sizing operation, which sizes the case neck from mouth to shoulder, can be improved upon. Partial case neck sizing can offer a larger increase in accuracy for most rifle/cartridge combinations. Partial neck sizing done properly will result in the die being adjusted to size approximately the first 0.2 inch of the neck.

Neck Sizing And Bullet Pull

Uniform bullet pull (the amount of force required to pull a seated bullet from a case) is a part of accuracy. With a case neck sized fully (or at least to a length equal to the bullet seating depth) a specific load might have a bullet pull of 50 pounds, plus or minus the variable amount which will always be present. If a plus or minus 15% variation is encountered,

Neck sizing, properly done, can add a measurable degree of accuracy to reloads but is not ideal for every application.

bullet pulls might then range from 42.5 to 57.5 pounds. With a case loaded identically; except with only a short section of the neck sized, bullet pulls might be 20 pounds. Add the tolerance, and the range would be 6 pounds (17 to 23 pounds) as compared to a range of 15 pounds with full neck sizing. This is nearly a 2 1/2 times increase in the uniformity of bullet pull and will enhance overall accuracy.

Another advantage to partial neck sizing is that the unsized portion of the neck will fit the chamber more precisely and help align the bullet with the chamber's axis.

However, there is a disadvantage to low bullet pulls. Should a bullet touch or actually enter the firearms rifling, and the cartridge be withdrawn from the chamber without firing, a bullet might stick in the chamber throat and stay put. If the case is then withdrawn rearward, without firing the cartridge, it could result in powder spilling in the chamber and locking lug recesses. This certainly is not something one would look forward to in the field. If you elect to use partial neck sizing to improve accuracy, be certain that there is clearance between the bullet ogive and the chamber throat. Or better still, do not use ammo so assembled for field use, saving it instead for target applications.

Custom Sizing Dies May Be Needed

The worth of partial neck sizing can vary with the die's dimensions. Often, standard neck sizing die dimensions are made small to insure a snug bullet fit regardless of accumulated tolerance variations resulting from differences in case wall thickness or bullet diameter. This situation can be improved upon by using custom made neck sizing dies. Such dies use inserts which allow the reloader to tightly control the diameter of the sized case neck and also to eliminate the need for case neck expanding buttons. Such dies are made with the usual 7/8 X 14 threads for use in most reloading presses or without any threads for hand assembly methods. Dies of this type are produced by several custom makers.

Match individual lots of brass (case wall thickness) with a neck sizing insert which sizes the inside case neck diameter to 0.001 or 0.002 inches below bullet diameter. This results in minimal bullet pull and an almost negligible pull variation within a string of 10 shots. This technique has shrunk group sizes from 1 3/4 M.O.A. down to 1 M.O.A.

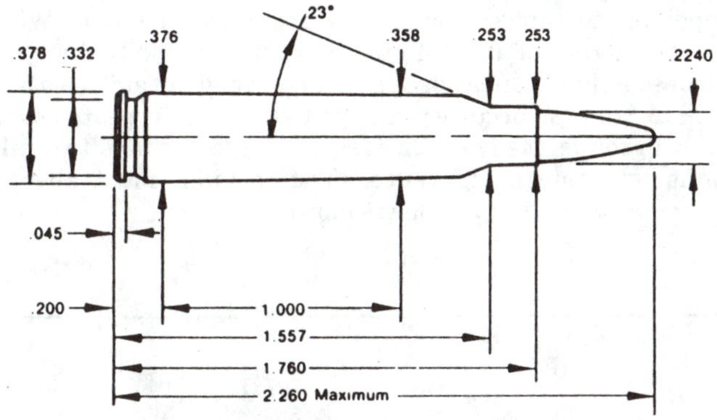

TEST COMPONENTS:

Cases	Remington
Trim-to Length	1.750"
Primers	Remington 7½ and CCI 450
Primer Size	Small Rifle, Magnum
Lyman Shell Holder	No. 26
Jacketed Bullets Used	Speer Spire Pt., 40 gr.
	Sierra Spitzer, 45 gr.
	Sierra SP Blitz #1340, 50 gr.
	Sierra HP #1410, 52 gr.
	Sierra Spire Pt. #1360, 55 gr.
	Sierra SMP #1370, 63 gr.

Uniform case lengths are basic and essential to getting the maximum in accuracy from your reloads. Always refer to the data section for the correct case dimensions.

Uniform Case Lengths

Another essential step to uniform reloads begins with trimming all new or fired cases to the same length. The trimming of once fired cases should be done after resizing, as this operation will cause some case stretching. Such case trimming is essential to the search for the best possible accuracy. In order to enhance shooting precision, case lengths will need to be uniform to a close tolerance - perhaps plus or minus 0.002 inch. Such uniformity is not too difficult to achieve with a good case trimmer and a bit of practice.

Flash Hole Uniforming And Deburring

The next step in case handling for optimum accuracy is optional, unless you demand the ultimate in precision. This is the making of the primer flash holes absolutely uniform and the elimination of any burrs which may have been left on the flash holes' termination on the inside of the case. These burrs occur when the manufacturers form the flash hole.

To make flash holes uniform in diameter, and to remove any burrs, a special hand held uniforming tool is used. This step should be done only after case trimming (and deburring) as the tool locates from the case mouth. If cases vary in length, the depth of the flash hole uniforming will vary accordingly. Tools to accomplish the uniforming of the flashhole are available from Lyman. This step is essential to most serious bench rest shooters.

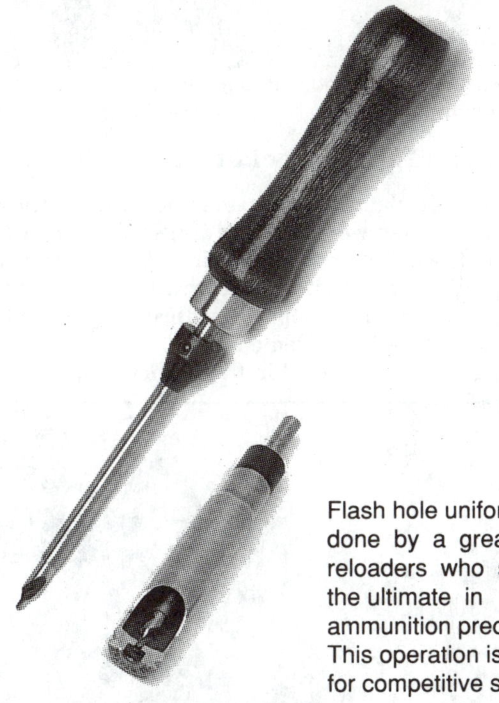

Flash hole uniforming is done by a great many reloaders who strive for the ultimate in ammunition precision. This operation is a "must" for competitive shooters.

Primer Pocket Uniforming

Some reloaders will perform other case handling steps, particularly benchrest or accuracy shooters. One of these is to ream the primer pocket to a uniform depth, while squaring off the bottom corner.

Experience shows that primer pocket reaming requires great care if the primer pocket is not to be inadvertently reamed to an oversize or out-of-round dimension. Premature case failure due to leaky primer pockets, caused by a less than expertly performed reaming, will be the clue that the undertaking has not been performed as it should have been. First attempts at primer pocket reaming should be limited to relatively few cases, perhaps ten. After these have been fired and reloaded several times the reloader will know if the undertaking was performed properly.

Case Neck Turning

Another step, useful to the accuracy shooter but of questionable value to others, is the turning of case necks to a uniform thickness. Though it remains an essential for competitive shooting, this step can bring on problems with bullet pull if done for the average chamber and sizing die. Difficulty is encountered when the case neck is turned to a dimension that prevents the sizing die from reducing it to a small enough diameter to properly hold the bullet in place. However, benchrest shooters usually use custom dimensioned chambers and neck sizing dies.

Primer pocket uniforming must be done with meticulous care to prevent damaging the case.

Always Clean Primer Pockets

Primer pocket cleaning does offer benefits to the field shooter or hunter, as well as to the target shooter. A badly fouled primer pocket can prevent a primer from seating to the desired depth and at extremes cause erratic ignition. It takes only a twist or two with an appropriate brush or tool to free a primer pocket of residue. Remember to follow the primer seating suggestions contained in the primer chapter.

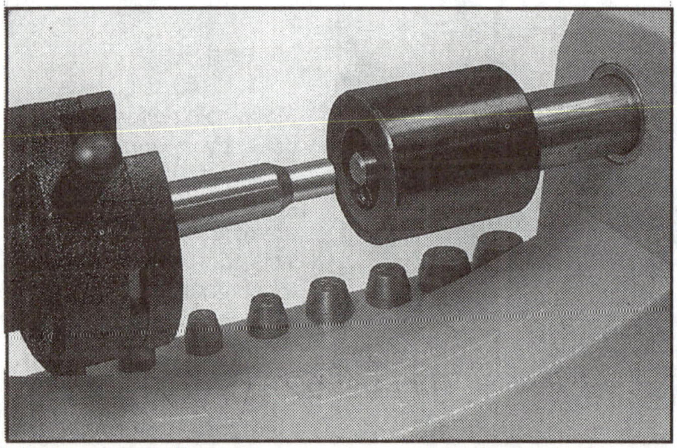

Neck turning is standard practice among benchrest shooters and many long range varmint shooters. It is easy to do, though special neck sizing dies may become essential after turning cases.

Match Grade Primers

Another good approach is to pay a few dollars more per 1000 primers and purchase match grade primers. Often going to a better grade primer can eliminate those occasional so-called flyers.

The Correct Propellant Is Essential

The selection of the correct powder for maximum uniformity of ballistics is essential to getting the most in accuracy. Most cartridge/bullet-weight combinations are ideally served by perhaps one, two, or maybe three propellants. Sure, IMR 4064 gives acceptable ballistics in a 270 Win; but IMR 4350, IMR 4831 or Hodgdon 4831 will all give outstanding uniformity with all weight bullets from 90 to 150 grains. The best powder (or several) can sometimes be selected by noting the propellants used by competitive target shooters or riflemen who always seem to have very fine shooting firearm/ammo combinations. We have tried to make specific recommendations in the comments prior to the data tables for most cartridges.

Most reloaders would do better to experiment with different brands and styles of bullets rather than using an endless variety of powders. If the powder choice is from among the most useful speeds, experimenting with bullets will usually show wider variations in obtainable accuracy as compared to experimentation with propellants.

Important in propellant selection is the need to fill the available case volume to at least 89%. Generally the less air space, the more uniform the velocity and pressure. Such uniformity will usually mean better groups. Heavily compressed powder charges are as self-defeating as inadequately filled cartridge cases.

Varying rifle cartridge powder charges by 1/10th of a grain is unnecessary and misleading in the search for accuracy. Any group fired which shows notable improvement over any group fired with a propellant charge of a few tenths of a grain difference is merely the result of chance. With powder charges of 20 to 35 grains expect that it will take at least 1/2 grain of propellant to cause a noticeable accuracy variation. With powder charges of about 40 to 60 grains it is likely that you will need a change in propellant charge weight of at least a full grain to cause an accuracy change that will be repeated over several groups. Heavier charges will take perhaps two full grains to create a difference. (The reloader must always remember to work up all loads slowly.)

The reloader in search of superb accuracy should confine shooting tests to the relatively few "best" propellants for the cartridge involved. It is more appropriate to try many bullets rather than many powders.

Never be mislead by the chance performance of a single group; rather base accuracy evaluations on the performance of multiple groups - five 5-shot groups is a good basis to form an opinion.

Bullets And Accuracy

Good bullets are essential to accuracy. The reloader should test a number of different types in the quest for precision shooting. It is essential that bullet seating depths be varied when testing for accuracy. A bullet jump of 0.005" to 0.020" will most often deliver the best accuracy.

When first testing a bullet use seating depths of 0.005" to 0.020" (the amount of "jump" of the bullet from its loaded length to when it first engages the rifling) in .005" increments. Once the best bullet seating depth is established for a specific bullet it will remain constant no matter what other load variations are tested. As the throat of the gun wears over the firing of a great many rounds it may become necessary to increase overall cartridge length by 0.005" to compensate for wear in the rifling lead.

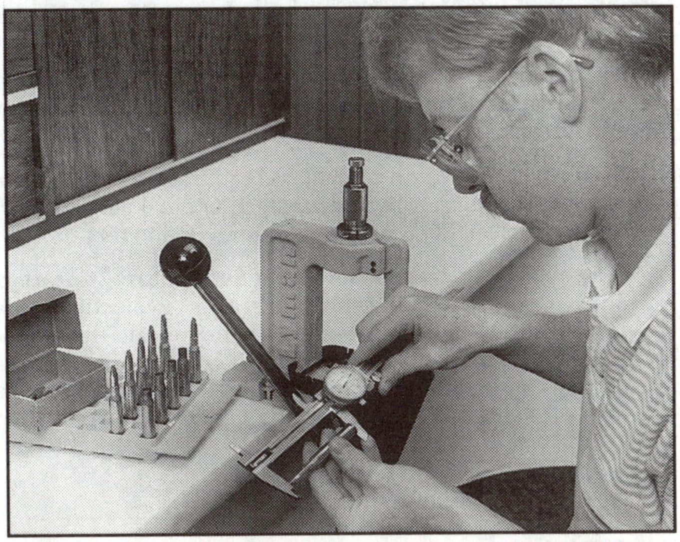

Bullet seating depth change in increments as little as 0.005" can cause substantial changes in accuracy.

Custom tools are available to help determine the overall length of a cartridge loaded with a specific bullet which will provide zero bullet jump. It is easy to load 10 cartridges at this length minus 0.005", minus 0.010", minus 0.015" and finally minus 0.020 inch.

There are also custom gauges that will allow the reloader to measure, and duplicate, the best bullet jump length. These gauges will measure from the cartridge case base to the beginning of the bullet ogive. This measurement will be the same no matter what the bullet brand, style or profile to be used. This can save a lot of testing when changing from one bullet to the next.

It is sometimes not possible to seat a bullet out far enough to obtain good accuracy and still have the loaded cartridge function through the magazine and action. When this is the case there are two options. The first is to have the barrel removed, cut a thread off the chamber end, and then reinstall it; rechambering and recutting the throat to more reasonable dimensions. The other option is to try round nose bullets. These, because of their short ogive, can sometimes reduce bullet jump sufficiently to deliver good accuracy. Of course, the use of a round nose bullet, with its lower ballistic coefficient will greatly reduce terminal performance at long ranges.

A Chronograph Can Be Important

Finally, consider the use of a good chronograph. Sometimes one can get good accuracy from a load with somewhat erratic velocity. Or inversely, poor accuracy from a load with very uniform velocity can occur. On a day-in-and-day-out basis, loads providing the most uniform velocities will prove superior in accuracy. Selecting an accurate load which provides a small velocity standard deviation always makes more sense than selecting an assumed accurate load with a large velocity standard deviation.

The use of a chronograph will allow the quick separation of an ideal powder choice from the not-so-ideal selections. It will also do it without having to depend upon rules of thumb or advice from someone who simply may never have loaded for a set of circumstances exactly like yours. A chronograph can go a long way to eliminating the guesswork sometimes involved when attempting to load for maximum accuracy. There will be more on using a chronograph later in this manual.

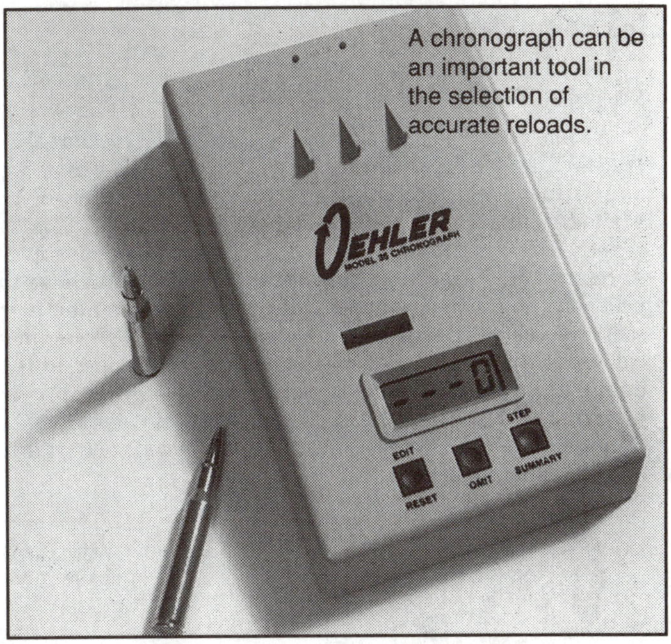

A chronograph can be an important tool in the selection of accurate reloads.

Weighing Powder Charges

Uniformity is essential to the best possible accuracy. Thus, it is natural to assume that weighing every powder charge is part of obtaining the maximum in ammo precision. Generally this is a fair assumption. However, competitive shooters have proven that metered powder charges can be very precise. One should not loose sight of the fact that most serious target shooters have learned how to meter very precise powder charges. Indeed their techniques are very refined and they often throw charges as accurate as those individually weighed on a scale.

The Basics Are Always Important

Of course, all the foregoing will be rewarding only if the basic skills of good reloading are mastered. If you don't segregate components by lot, if you're careless about case lengths or if your equipment allows for a lot of dimensional variations, most of the foregoing won't help to improve accuracy. Put it all together with a good rifle and the shooter can see that groups commonly ranging between 1/2 M.O.A. and 1 M.O.A. are achievable.

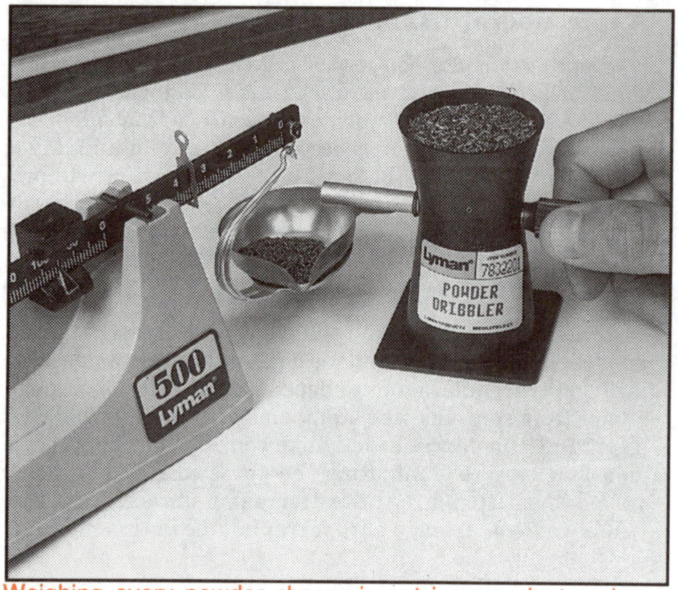

Weighing every powder charge is not inconvenient and can improve the accuracy of most loads.

Clean Barrels Are Important

A clean bore is essential to the best accuracy. If a barrel has traces of copper fouling or leading it will not perform as well as a clean barrel. A few passes of some general powder solvent soaked patches will not get a bore truly clean. It takes repeated over night soakings with a high grade copper solvent to remove all traces of jacketed bullet residue. Purchase the best solvent money can buy and use it per the manufacturers instructions. Remember, if any trace of "yellow" or brass colored fouling can be seen when looking into the muzzle at an angle, the barrel simply is not clean enough to provide maximum accuracy. The same is true if a wet cleaning patch comes out of the barrel with any blue or green stain on it after an overnight soaking.

Accuracy And Barrel Vibrations

When a firearm is discharged the barrel vibrates much as a tuning fork vibrates when it is struck. The barrel vibrations are caused by the stress imparted by cartridge ignition and by the bullet's passage through the bore. The barrel actually bulges ever so slightly around the bullet as the bullets moves through it.

The vibrations of the barrel cause the barrel to move with varying degrees of violence. It is this barrel movement that is largely responsible for the accuracy of a specific load. The vibrations take two forms, called fundamental and secondary.

Fundamental vibrations cause the muzzle of the rifle to trace an elliptical or circular path (see nearby illustration). The breech end of the barrel, anchored by the receiver, remains relatively motionless. The specific position of the muzzle as the bullet leaves the barrel can vary from shot to shot. If the barrel is in relatively the same position as each bullet exits, groups will be small, but, if the barrel is in relatively different positions as bullets exit, accuracy will be poor.

Secondary vibration takes the same form of the nodes of a snapped rope (see nearby illustration.) The uniformity of this type vibration can effect accuracy.

Heavy, or short and stiff barrels vibrate less than long slender ones. This is why benchrest shooters favor relatively short (most often 20") barrels with large diameters.

The amount of muzzle movement due to barrel vibrations is much larger than many shooters realize. Typically a 30-06 rifle will have angular movement at the muzzle that would equate to about four feet at 100 yards. It should be obvious that it is very desirable to have the bullet leave the barrel at nearly the same point in the muzzle's movement as possible.

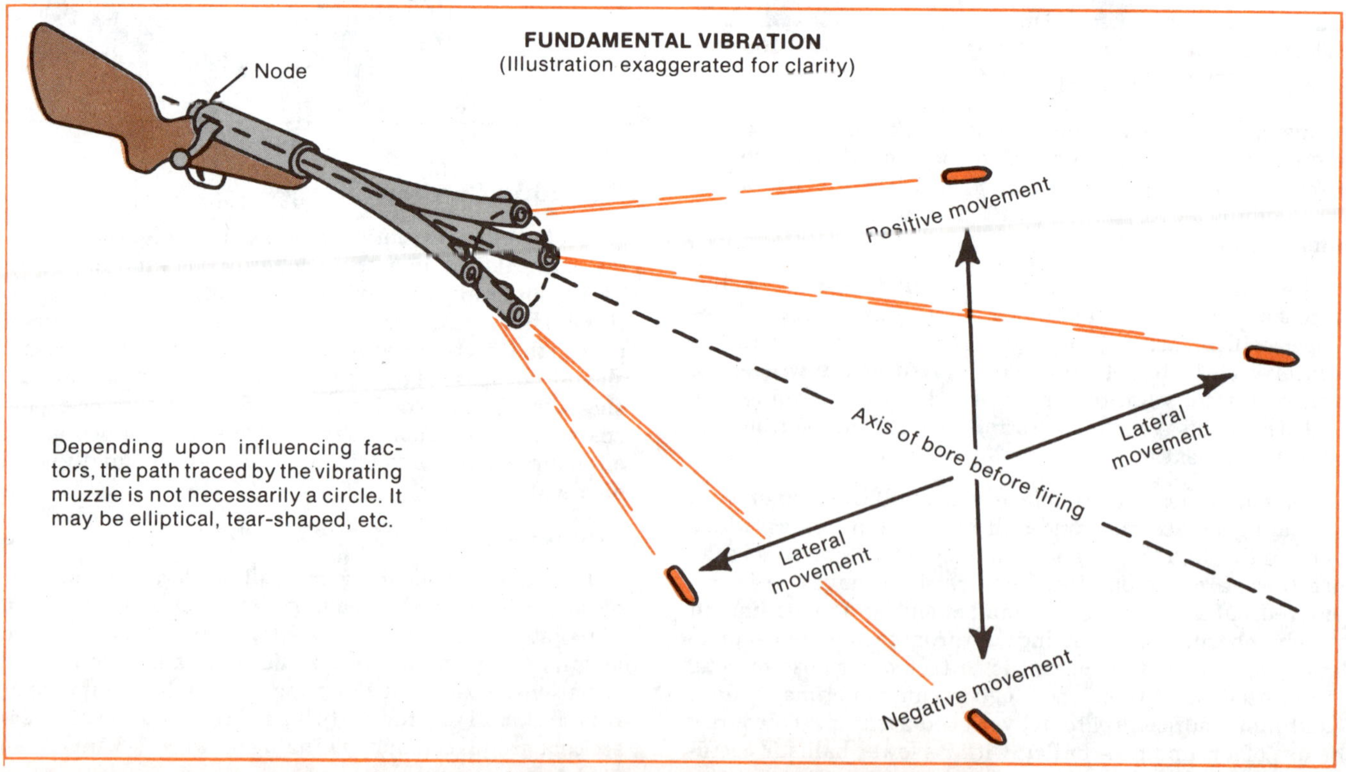

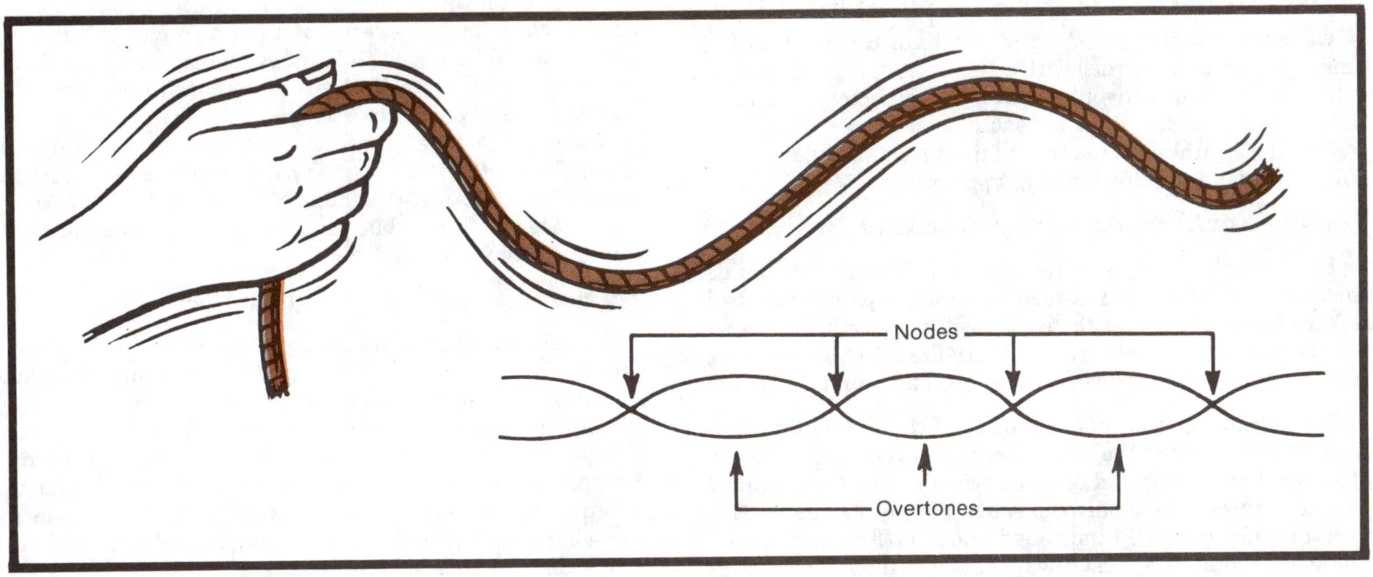

Muzzle Jump

The muzzle of a rifle jumps as a function of recoil as the bullet leaves the barrel. The amount of jump may vary considerably depending upon the weight of the firearm and the amount of recoil generated by the cartridge. Muzzle jump, in a rifle or handgun, is usually positive in direction (see illustration of A to B movement). That is to say the muzzle moves upward. However, it is possible to have negative movement (see illustration of A to C movement), or even horizontal movement, dependant upon the specifics of the firearm. The longer a bullet is in the barrel, the more it will be affected by the jump of the muzzle.

Any force or weight applied to the barrel will have considerable effect on the amount and direction of muzzle jump. For this reason the wise shooter will remember not to rest a barrel against anything when shooting.

Top Accuracy Is Obtainable

The number of steps taken to enhance accuracy should be geared in a practical way to the type of shooting and firearms involved. Assuming firearm and shooter are qualified, and that a 4X scope is used, groups ranging from 1 1/2 to perhaps 3/4 minute-of-angle are not overly difficult to obtain when a bit of extra thought and effort are put into load assembly. However, half inch groups will be very rare unless a superbly accurate firearm is used.

Special Care And Evaluation Required For Hunting Ammo

The only criterion (aside from safety) that target shooters must apply to the selection of their handloads is accuracy. Hunters, however, need to be concerned with a great many evaluation steps before they settle on a load. The firing of a few groups is not enough to accurately predict field performance.

Perhaps you have heard conversations that went like this: "I did everything right and still missed that buck". Or, "I hit that bull with a perfect shot and he still ran off. I never did find him."

Many hunters probably have no idea what went wrong in these cases. Just a few might be able to guess at the cause for such difficulties. It might be safe to assume that most unfortunate hunters based their load selection on limited testing.

MUZZLE JUMP DUE TO RECOIL
(Illustration exaggerated for clarity)

Firearm continues to recoil after bullet exits

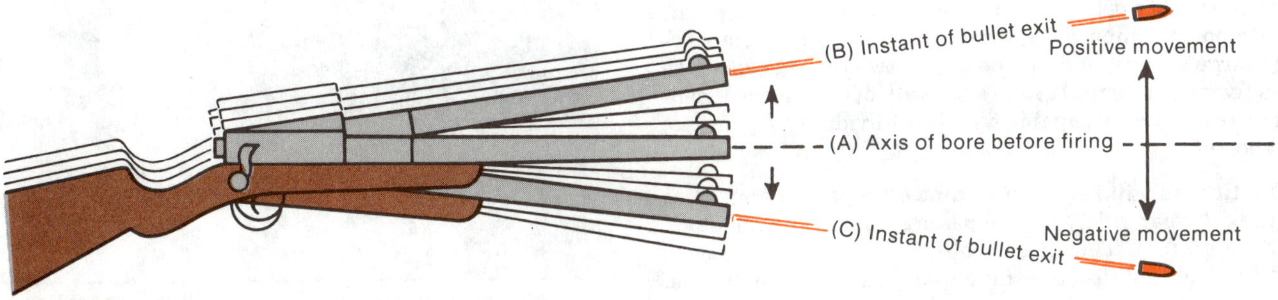

It is also reasonable to speculate that any load testing done was a comparison of group sizes fired from a bench type rest.

Of course, accuracy testing is required. But for hunting ammo it must be combined with other necessary evaluations. Whether the ammo involved is reloads or factory ammunition, it takes more than a casual look at a few groups containing a minimum number of fired rounds to determine suitability for a specific hunting application.

Testing For True Accuracy Potential

Knowing that every shot fired in the field counts demands that the hunter know his true accuracy capability and to refrain from taking any shots that will not result in a clean kill. To evaluate true accuracy potential requires a number of trips to the range and the firing of at least 25 rounds of ammo.

Begin hunting accuracy testing by carefully aligning two targets, one on top of the other. Hang the targets at a range of at least 100 yards and fire a five-shot group. Do not allow time for barrel cooling between shots. Simply shoot at a pace necessary for carefully controlled shots. (When in the field follow-up shots are almost always taken without barrel cooling time.)

Now remove the top target and set it aside for future reference. Place another target over the first one (which now has a five shot group on it), carefully align it and fire another five-shot group. However, before doing so, allow ample time for your barrel to completely cool. (One seldom has time for reloading in the field and subsequent shooting almost always begins with a completely cold barrel).

When the second group has been fired remove the top target, preserve it for later reference and carefully store the bottom target (which now has two 5 shot groups on it) for use on another day. Shooting for this test program is now over for the day. Return to the range on another day to fire another 5 shot group with a clean top target. In this case the bottom target is the one which already has the two 5 shot groups. Repeat the process on different days, firing only one 5 shot group until you have one bottom target with at least twenty-five rounds (five 5 shot groups) on it. Select the time of shooting so that you have one group fired in early morning light, another fired in mid day light and still another fired in late afternoon light. Try to have at least one group fired under bright sunshine and another under heavy overcast or rainy conditions. Mark each top target as to time of day and light condition.

When the shooting is done you will have five (minimum) targets, each with one 5-shot group on it. You will also have another target with at least 25 shots on it making a conglomerate group of all the test shots.

Inspection of the targets with the individual groups might disclose some comparatively small groups, some comparatively large groups, and maybe one or two targets with a so called flyer on it. Actual group sizes will, of course, be dependent on the potential capability of the individual rifle, ammo, and shooter.

The targets will reveal any point of impact change that occurs with your rifle with the passage of time due to stock moisture content. They should also reveal any tendency to "see" the sights differently under varying light conditions. This tendency, which can cause point of impact variations, will be most pronounced with open iron sights, and least noticeable with telescopic sights of good quality.

The "bottom" multiple group target will now show the true accuracy potential of your gun and load combination as it will occur in the field over a period of time and under differing light conditions. While each individual five-shot group might be relatively small, the collective group containing all shots will normally be several times larger than the smallest groups and somewhat bigger than the largest individual five-shot group. You will, in all likelihood, note that the few flyers from the individual targets "disappear" on the collective target, as they were not flukes, but part of the overall accuracy of the gun and ammo combination.

Do Some Testing At Field Ranges

To fairly evaluate the accuracy potential of the shooter/gun/ammo combination, set up a target at a range that would normally be considered maximum for shooting in the field. Shoot a few groups, not from a benchrest position, but from a position that would be employed in the field over a similar range. Group size most often will be quite a bit larger than when shooting from a bench. But, you will have a real basis for knowing what you and your outfit are capable of, rather than a misconception based on range testing.

You are now ready to evaluate the most important criterion of all, the value of the selected bullet as a hunting bullet versus its use as a target bullet.

Bullet Performance Is Vital To Big Game Hunting Ammo

Hunting bullet evaluation begins with an appraisal of personal experience and discussions with other people knowledgeable in this field. The lost game scenarios, described earlier in this dialogue, all could have been caused by a bullet that did not function as intended. Not all soft points expand uniformly from bullet to bullet. Indeed, some frequently fail to expand at all; or inversely they expand too rapidly, coming completely apart in failures known as fragmentation or core/jacket separations. Such failures can result in lost game.

Hunting bullets must expand in a classic mushroom and retain at least 60% of their original weight in order to consistently anchor game. The pictured bullets are all recovered Nosler Partition types.

If personal experience indicates that a particular bullet has occasionally resulted in a lost animal when a reasonable hit was made, that bullet should then be avoided when assembling future hunting loads. Bullet performance is akin to shots taken - every one counts in the evaluation process. Bullets which sometimes, or even most of the time, collect the game merely prove that on occasion it is possible to kill game with even poor bullets.

Big Game Bullet Performance Requires Careful Evaluation

Bullet evaluation is always an ongoing process. Trace, very carefully, the wound channel in every animal taken. Look for indications of proper expansion, neither too rapid or too slow. Large surface wounds always indicate a bullet that is too fragile. Excessively long wound channels of small diameter may be an indication of slow bullet expansion or no expansion at all.

Wound channel examination must take into consideration the spread of game weights that will be hunted with the same load and the angle of penetration. A bullet suitably matched to the game should always penetrate and exit the far side when broadside shots are taken, or at least come to rest, fully expanded, under the far side skin. If this amount of penetration is not achieved the bullet may fail to penetrate sufficiently when difficult angle shots are taken, for example, a shot at the south end of a northerly bound animal.

Naturally, range must be part of the evaluation process. At the longest ranges it might be reasonable to accept less than full penetration on broadside shots.

Game the size of deer or antelope put no heavy demands on bullet design, but the likes of a big moose will surely put a bullet design to the test, especially if it is used repeatedly. If all works well, a test of one bullet never proves a thing. Paradoxically a test of one bullet, in which a failure occurs, is evidence of a less than desirable bullet; remember every shot counts.

Firearm Functioning

The ammo evaluation process should also include a look at functioning in the firearm when the gun is used as it would be in the field under stressful conditions. This means cycling the ammo through the firearm as it would be fired as opposed to firing single rounds. Caution: this should be done only at a safe outdoor range facility. Ammo that results in feeding jams is never acceptable.

Bullet Impact Under Varying Conditions

You will need to consider a few more points. The first is: will the first shot from a cold, clean barrel have a point of impact similar to succeeding shots fired from a warm, fouled barrel? Your tests done to obtain a collective target will usually answer this question. Most often, shots fired from a cold barrel will evenly disperse themselves throughout the collective group. Rarely they will favor a smaller impact point, but still within the collective group. In cases where the "cold" shots occur outside of the collective group (this is extremely rare) there is little else to do but try rebedding the rifle in hopes of correcting the problem.

If the gun simply places shots from a clean bore outside of the collective group without regard to a barrel type, it is a simple matter to fire one or two fouling shots before taking the gun afield.

Hunting ammo evaluation is slow and tedious work, requiring frequent trips to the range and the consumption of expensive components. However, if it prevents one lost animal it will all prove worthwhile and rewarding.

CHAPTER 12
VELOCITIES: FACTORY AMMO AND RELOADS

Reloaders are often concerned with the duplication of factory ammunition ballistics that have served them well in the past. They may also wish to stay with a load that has a familiar trajectory. Whatever the reason, it often makes sense to reload a shell to closely parallel the performance of a factory load with the same weight bullet. Still, many reloaders are disappointed when comparing factory advertised ballistics to the velocities in a data table as discrepancies might seem to exist. Such disappointment is, however, generally unfounded.

It should be understood that factory advertised velocities are simply an industry advertised standard that suggest the highest likely average obtainable when using the ammunition in specially dimensioned velocity test barrels. These barrels are built to such minimum dimensions that they will generally produce higher pressure and velocity than the barrels used on standard commercial sporting firearms. Advertised velocities are based on arbitrary barrel lengths that often do not duplicate the barrel length used on most commercial firearms. For example, factory 30-30 Winchester ballistics are based on the use of a 24" test barrel whereas most shooters have 20" sporting barrels. Thus, advertised factory velocities may be somewhat higher than the results you obtain. Even the test barrel lengths used in our labs and reported in the data may differ from what you have.

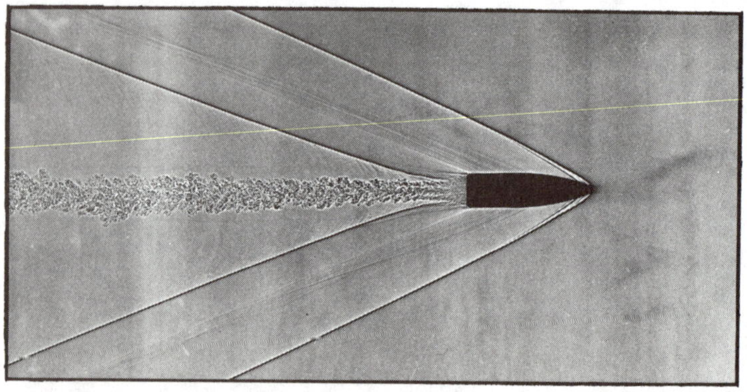

Bullets fired under controlled laboratory conditions will differ in performance from those fired in actual field conditions.

Remember also that factory advertised velocity is corrected upward from the actual velocity measured at perhaps 15 feet from the muzzle. The velocity listed in a reloading data table is, most often, an actual 15 foot velocity without any attempt to increase it to an estimated muzzle velocity.

For these and other reasons it makes sense to anticipate a different velocity from a specific firearm as opposed to a listed factory or data table velocity. As an example, the tables at the end of this chapter show some of the factory advertised loads and the actual velocities obtained from typical sporter barrels of the specified length. In most cases actual velocities are somewhat lower than listed but in a few cases velocities are equal or even faster.

If the reloader wants to know the exact velocity obtained with a specific set of components, finished cartridge dimensions, and specific barrel there is no recourse except to chronograph the load.

It is possible, however, to estimate velocity changes for different length barrels rather closely, usually within 100 feet per second or less. The following table will allow estimation of velocity gain or loss that might typically occur with a change from the 24" length used to develop the data contained in most of our ballistic tables. The table actually works well with barrels from 18 to 26 inches.

Barrel Length Change in inches	Approximate Velocity gain/loss in fps. for approximate advertised velocity of: 1000	1500	2000	2500	3000	3500	4000
1	5	10	15	20	25	30	35
2	10	20	30	40	50	60	70
3	15	30	45	60	75	90	105
4	20	40	60	80	100	120	140
5	25	50	75	100	125	150	175
6	30	60	90	120	150	180	210
7	35	70	105	140	175	210	245

The velocities obtained with handgun barrels often times differ substantially from that listed in factory of data book specifications. The reason for this is that the shorter barrels and wide chamber and barrel dimensions used to manufacture handguns create greater ballistic changes. Because these changes vary so much it is not practical to estimate velocities based on handgun barrel lengths. The shooter who must know the exact velocity of a specific load from a specific handgun will need to chronograph approximately ten rounds.

No two barrels are exactly alike with respect to duplication of velocity (or pressure) even when using ammunition from the same lot. Therefore it is not reasonable to expect duplication of velocity or trajectory with two different firearms even when they are of the same make and model. If long range trajectory is important to the shooter actual trajectory should be checked by range firing and/or chronograph testing.

Additional discussion of ballistics will be handled later in this manual.

ADVERTISED FACTORY CARTRIDGE VELOCITY AND ACTUAL VELOCITY

Rifle Cartridge	Bullet Weight In Grains	Advertised Velocity F.P.S. Muzzle	Actual Velocity At 3 Yards	From Bbl. Length In Inches
22 Hornet	45	2690	2688	22
222 Remington	50	3140	3012	18-1/2
22 PPC	52	3400	3399	24
223 Remington	55	3240	3014	18-1/2
223 Remington	64	3020	2687	20
222 Rem. Mag.	55	3240	3150	24
225 Winchester	55	3570	3425	24
224 Wea. Mag.	55	3650	3638	24
22-250 Rem.	53/55	3680	3591	24
6mm PPC	70	3140	3138	24
243 Winchester	75/80	3350	3175	22
243 Winchester	100	2960	2850	22
6 mm Remington	100	3100	2689	22
240 Wea. Mag.	87	3500	3500	24
240 Wea. Mag.	100	3395	3300	24
25-20 Win.	86	1460	1275	20
257 Roberts	100	2980	2710	24
257 Roberts	117	2780	2585	24
257 Wea. Mag.	87	3825	3675	24
257 Wea. Mag.	100	3555	3500	24
257 Wea. Mag.	120	3300	3275	24
264 Win. Mag.	140	3030	2950	24
270 Winchester	100	3430	3149	22
270 Winchester	130	3060	2883	22
270 Winchester	150	2850	2809	22
270 Wea. Mag.	100	3760	3686	24
7mm Mauser	175	2440	2433	24
7mm-08 Rem.	140	2860	2680	18-1/2
280 Remington	150	2890	2786	22
280 Remington	165	2820	2580	22
7mm Rem. Mag.	175	2860	2800	24
30 Carbine	110	1990	1950	20
30-30 Win.	150	2390	2350	20
30-30 Win.	170	2200	2150	20
300 Savage	150	2630	2700	24
300 Savage	180	2350	2350	24
30-40 Krag	180	2430	2275	22
308 Winchester	150	2820	2775	22
308 Winchester	165	2700	2487	18-1/2
308 Winchester	180	2620	2525	22
30-06 Spfd.	125	3140	3175	22
30-06 Spfd.	150	2910	2850	22
30-06 Spfd.	180	2700	2571	22
30-06 Spfd.	220	2410	2350	22
300 H&H Magnum	180	2880	2800	24
300 H&H Magnum	220	2550	2550	24
300 Win. Mag.	150	3290	3250	24
300 Win. Mag.	180	2960	3050	24
300 Win. Mag.	220	2680	2688	24

ADVERTISED FACTORY CARTRIDGE VELOCITY AND ACTUAL VELOCITY/cont.

Rifle Cartridge	Bullet Weight In Grains	Advertised Velocity F.P.S. Muzzle	Actual Velocity At 3 Yards	From Bbl. Length In Inches
300 Wea. Mag.	150	3600	3550	26
300 Wea. Mag.	180	3120	3075	24
300 Wea. Mag.	220	2850	2900	26
303 British	180	2460	2469	22
7.62x39mm Rus.	123/125	2300	2295	18-1/2
32 Win. Special	170	2250	2250	20
8mm Mauser	170	2360	2325	23
8mm Rem. Mag.	220	2830	2819	24
338 Win. Mag.	200	2960	2950	24
338 Win. Mag.	210	2830	2829	24
338 Win. Mag.	225	2780	2797	24
338 Win. Mag.	250	2660	2586	24
340 Wea. Mag.	250	3000	2775	26
348 Winchester	200	2520	2475	24
357 Magnum	158	1830	1741	22
35 Remington	150	2300	2375	20
35 Remington	200	2080	2000	20
358 Winchester	200	2490	2550	22
350 Rem. Mag.	200	2710	2700	20
38-55 Win.	255	1320	1322	26
375 H&H Magnum	270	2690	2669	24
375 H&H Magnum	300	2530	2478	23
378 Wea. Mag.	270	3180	3058	26
378 Wea. Mag.	300	2929	2824	26
416 Rigby	410	2370	2400	23
416 Rem. Mag.	350	2520	2495	22
416 Rem. Mag.	400	2400	2390	22
416 Wea. Mag.	400	2700	2715	24
44-40 Win.	200	1190	1212	24
44 Rem. Mag.	240	1760	1775	20
444 Marlin	240	2350	2400	24
45-70 Govt.	405	1330	1300	22
458 Win. Mag.	500	2040	2040	24
458 Win. Mag.	510	2040	2010	24
460 Wea. Mag.	500	2700	2583	26
Pistol Cartridge				
9mm Luger	115	1155	1123	4
9mm Luger	147	1010	954	4
38 Special + P	125	945	941	4
38 Special + P	158	890	852	4
357 Magnum	125	1450	1380	4
357 Magnum	158	1235	1215	4
40 S&W	155	1125	1086	4
40 S&W	180	990	911	4
44 Magnum	240	1180	1085	4
45 ACP	185	1000	983	5
45 ACP	230	835	825	5

4

CAST BULLETS

CHAPTER 13
INTRODUCTION TO CAST BULLETS

Caution: Making cast lead bullets exposes the reloader to the potential risk of burns and the accidental introduction of lead into the body. The reloader must take all the necessary steps to prevent either occurrence. It is important to read all of the precautions listed in this chapter. These and any other necessary safety steps are the responsibility of the reloader.

Cast bullets come in a wide variety and include handgun bullets, rifle bullets, hollow points, hollow bases, round ball and muzzleloading types.

Cast Bullets Are Very Economical

Cast bullets, made by the reloader, offer some unique aspects and challenges. The first advantage of cast bullets is their very low cost. Depending upon the specific alloy used, cast bullets can be prepared for just pennies apiece. In fact, it is common for a box of twenty 30-06 rounds assembled with cast lead bullets to cost less than $2.50; similar jacketed bullet ammunition could cost four or five times as much.

Cast Bullets Prevent Barrel Wear And Are Fun To Make

In addition to providing very inexpensive shooting, cast bullets allow for extensive shooting without barrel wear. Indeed, barrels often seem to last indefinitely when used exclusively with cast bullets.

The hobby aspect of using cast bullets cannot be ignored. Casting, sizing, and lubricating bullets creates many hours of reloading enjoyment.

Cast Bullet Applications

Cast bullets do have shortcomings. In rifle cartridges, often velocity and terminal performance of a cast bullet are well below that of jacketed bullets. It can also take a great deal of load development to equal the accuracy of jacketed bullet rifle ammo. This is especially true when maximum velocity cast bullet loads are assembled. These shortcomings are not usually present for most handgun cartridge applications. In fact, some of the most accurate handgun ammo assembled uses lead bullets.

Cast bullets are not well suited for use in barrels that are pitted or have a rough bore surface. Anything less than a smooth bore and leading will quickly build up, making accuracy with cast bullets almost impossible.

Cost Comparison of Factory vs. Handloaded Ammunition with Cast Bullets

		Handloaded Ammunition Cast Bullet — Rifle			Handloaded Ammunition Cast Bullet — Handgun	
	Jacketed Factory Ammunition	W/New Cases	W/Recycled Cases	Jacketed Factory Ammunition	W/New Cases	W/Recycled Cases
Cartridge	.30/06	.30/.06	.30/06	.357 Mag.	.357 Mag.	.357 Mag.
Bullet wt., grs.	180	170	170	158	158	158
Cartridges per box	20	20	20	50	50	50
Cost per box	$17.57	$7.53	$2.53	$18.55	$6.64	$3.12
Cost per cartridge	$.879	$.377	$.127	$.371	$.133	$.0624
Savings per box over Factory	—	$10.04	$15.04 (86%)	—	$11.91	$15.43 (83%)
Savings per cartridge over Factory	—	$0.50	$0.75	—	$0.24	$0.31

How many jacketed bullets do I have to shoot to make bulletcasting worthwhile?

Here's the answer to that very important question, referring to the .30/06 and .357 Magnum cartridges as detailed in the preceeding cost comparisons.

If you shoot 1032 jacketed rifle bullets (or 10.3 boxes of 100), you would equal the cost of Lyman's Starter Casting Kit, which lists (in '92) for $159.95.

Likewise, if you shoot 1163 jacketed pistol bullets (a little over 11.6 boxes of 100), you would equal the cost of the kit.

To pay off the cost of the Casting Kit (saving $12.90/box for rifle bullets and $11.15/box for pistol), you would need to cast 12½ boxes of rifle bullets or 14½ boxes of handgun bullets.

Once your equipment investment is paid off, you'll enjoy savings (at '92 prices) as follows:

		Net Savings — Per Ctg.	Net Savings — Per Box—100
Jacketed rifle bullets, 180 gr., .30 cal.	Unit Cost $0.1550	$.1290	$12.90
Cast rifle bullets, 180 gr., .30 cal.	Unit Cost $0.0296		
Jacketed pistol bullets, 158 gr., .38 cal.	Unit Cost $0.1375	$.1115	$11.15
Cast pistol bullets, 158 gr., .38 cal.	Unit Cost $0.0260		

Cast bullets are ideal when the cost of loading is very important. They are especially suitable in handgun loads. Cast bullets are a good choice for extensive target shooting and will almost insure a lifetime use with most barrels.

Casting Lead Alloy Bullets

Making cast lead alloy bullets requires the melting of a suitable metal alloy, pouring the melt into a mould, allowing the alloy to harden, dropping the bullet from the mould, and sizing the bullet to an appropriate diameter. The loading of cast bullets requires that the edge of the case mouth be slightly flared to allow the cast bullet to be started into the case without shaving metal from the bullet's base. All of these steps are straightforward. However, as in all aspects of reloading, specific cautions are important. These are as follows:

CASTING PRECAUTIONS

1. Cast only in a well ventilated area - the ideal place is outdoors. Never breath lead dust, lead fumes, or fluxing fumes. Never allow moisture to be present when casting. Even small amounts of moisture can cause violent steam eruptions which scatter molten lead in all directions.

2. If gas heat is used to melt the lead alloy it is important to take steps to prevent the inhalation of carbon monoxide.

3. Use only properly grounded electric melting pots. These are easily identified by the three prong plug. Never attempt to circumvent the proper grounding of the lead pot. Lack of grounding can introduce a risk of serious electric shock. Older lead furnaces with two prong plugs should not be used unless converted to a three prong plug and wiring by a qualified licensed electrician.

4. Wear heavy protective gloves, aprons, and goggles when casting. Splatters of hot molten lead can cause very serious burns. Be sure all parts of the body are covered with thick protective apparel that will prevent the lead melt from burning through clothing in case of an accidental spill or splatter.

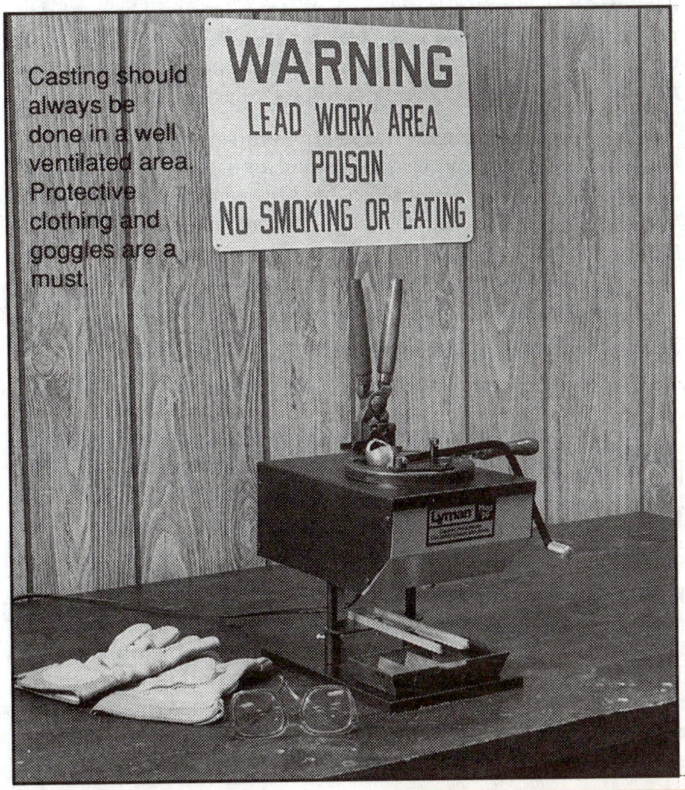

Casting should always be done in a well ventilated area. Protective clothing and goggles are a must.

5. Keep all persons away from the casting area. Many people do not realize the potential hazard represented by molten metal.

6. Be certain to place and use the melting pot where it cannot be tipped or knocked over. Consider anchoring the pot to the work surface. Never leave a pot unattended, not even for a few seconds.

7. Be cautious when using fluxing materials. Some are flammable. It is best to avoid this style of flux in favor of modern noncombustible types.

8. Be sure to place a suitable container under the pot spout to catch any leaking or spilled metal. This will prevent unintended damage to property or harm to people.

9. Never eat, drink, or smoke, and keep your hands away from your mouth when handling lead in any manner. Thoroughly wash hands when through with the lead handling operations.

10. Water, even in minute amounts, will cause a violent eruption of molten lead. Make sure all equipment is absolutely free of any moisture. Keep moisture of any type away from the casting area.

11. Primers, loaded ammunition and other substances can cause violent eruptions of the molten lead alloy if added to the hot melt. Take great care to keep all items out of the casting area which may cause accidents. Take meticulous care to sort all lead alloy to eliminate any chance of introducing anything but bullet alloy into the melting pot.

*Note: This is not a complete listing of all possible cautions.

Getting Started

Certain basic equipment is needed to cast bullets, in particular, protective apparel. Be certain that the apparel selected is strong enough to prevent molten lead from burning through and that the apparel chosen covers all parts of the body. Don't forget the importance of protecting ankles and feet from burns. There should be no area of exposed skin to be burned in an accident.

Lyman offers a complete selection of bullet moulds, furnaces and accessories for bullet casting.

The basic tools of bullet casting equipment include:

1. Bullet mould and handles.
2. Lead or lead alloy
3. Melting pot and dipper. (If a bottom pouring electric furnace is used, a dipper is not required.)
4. Alloy fluxing material.
5. Suitable mallet for tapping open the mould sprue cutter plate.
6. Soft pad of nonflammable material on which to drop bullets from mould.
7. Bullet sizing and lubricating equipment.
8. Bullet lubricant.
9. Casting thermometer.
10. Ingot mould (optional).

Lyman's Casting Starter Kit includes all the essential tools and accessories to get started in bullet casting.

Bullet Moulds

Bullet moulds come in an almost infinite variety. There are moulds for casting simple round balls (for muzzle loading firearms); shotgun slugs, rifle bullets and handgun bullets. These are often available in single, double, or four cavity types. There are hollow point or hollow base styles and almost anything else you might want. For the most part, when beginning to cast lead bullets it pays to select a bullet type that has a proven record of accuracy and a standard double cavity block.

The charts at the end of chapter 14 list Lyman's most popular moulds for rifle and handgun cartridges. Selection of a bullet style from these or from the data section will help insure a choice that will work well in the intended application.

Lead Alloys

Bullet metal can be a great many things. Some folks use any scrap they find. Sometimes such an approach works satisfactorily depending on the application of the finished bullet. However, a predetermined cast bullet diameter and exact bullet weight are desirable if the best accuracy is to be obtained.

Composition and Hardness (BHN) of Useful Bullet Alloys

Alloy	Lead	Percent Tin	Antimony	BHN
Monotype	72	9	19	28
Stereotype	80	6	14	23
Linotype	84	4	12	22
Lyman No. 2*	90	5	5	15
Taracorp Magnum*	92	2	6	15
1 to 1 Lead/Lino*	92	2	6	15
Electrotype	94.5	3	2.5	12
10 to 1	91	9	—	11.5
16 to 1	94	6	—	11
20 to 1	95	5	—	10
30 to 1	97	3	—	9
Wheelweights*	95.5	0.5	4	9
40 to 1	97.5	2.5	—	8.5
Pure Lead	100	—	—	5

*Responds well to heat-treatment.

The exact bullet alloy used should depend upon the intended application. Rifle bullets that will be loaded to near maximum cast bullet velocities should be made from a very hard alloy to prevent barrel leading which can harm accuracy. A good choice would be Lyman's #2 alloy. This is an alloy containing 5% antimony, 5% tin and 90% lead. This alloy has a Brinell Hardness Number (BHN) of 15. This, by the way, is the standard alloy used to determine the as-cast diameter and expected bullet weights for all Lyman moulds (except round ball and muzzleloading types). Pure lead has a Brinell Hardness Number (BHN) of 5, while Linotype has a BHN of 22.

Other alloys are of course suitable for bullet casting. The proximate table lists typical alloys commonly used for bullet making. Keep in mind that the hardest alloys will be useful in preventing barrel leading, but will make a bullet that is unlikely to expand. The softest alloys expand well but may result in barrel leading especially as velocities increase.

The bullet caster who wishes to make an alloy equivalent to Lyman's #2 can melt and mix 9 pounds of wheel weights and 1 pound of 50/50 bar solder. Another blend that results in a #2 alloy is 4 pounds of Linotype, 1 pound 50/50 solder and 5 pounds of pure lead.

The caster may purchase the required alloying material to make number 2 or Linotype metal from Brownells, Route 2 Box 1, Montezuma, Iowa 50171 (515 623-5401). They also sell several types of fluxing material.

Caution: Never attempt to salvage metal from any kind of battery. Doing so can be extremely hazardous.

Bullet Alloy Melting Pot And Dipper

Bullet alloy may be melted in several style pots. The most popular method is to use an electric furnace such as the Lyman Mag 20 or Magdipper. Both of these incorporate a thermostat for heat control and a wide anti- tip base for safety. The Mag 20 is a bottom pour furnace. With this style furnace the mould is placed beneath the pour spout and the handle is raised to release the bullet metal. The Magdipper model is similar except it does not have the bottom pour feature and is used with a ladle (dipper) to transfer molten bullet metal to the mould.

Lyman offers a smaller electric furnace called the Mini-Mag which has an 8-10 lb. capacity and requires the use of a dipper. A simple cast iron pot intended for use with an external heat source, such as an electric stove, is also available. This also requires the use of a casting dipper to transfer the melt to the bullet mould.

The lead pot, ingot mould and lead dipper are essential to casting your own bullets.

Fluxing Material

When the bullet alloy is melted a grey scum will form on the surface of the alloy. This is the dirt and other impurities that the alloy contained. Fluxing is done to remove the dirt and impurities from the alloy. These are skimmed off and removed using a lead dipper or large metal spoon type tool equipped with an insulated handle.

Mould Mallet

This need not be anything fancy. It can be as simple as a cut-off section of broom handle or as fancy as a wood mallet. It is used to strike a light blow on the sprue cutter plate to shear the sprue from the base of the bullet after the bullet has hardened in the mould.

Bullet Sizing And Lubricating Tools

As bullets fall from the mould they are slightly oversize. This is necessary since the bullet mould manufacturer has no way to control the exact alloy that will be used for casting bullets. Generally speaking the softest alloys will shrink the most when cooling. To reduce the bullet to the exact size desired and to ensure that it is perfectly round the bullet is pressed through a sizing die.

Cast bullets, depending on the alloy and velocity, will tend to leave deposits of the bullet alloy in the barrel. Such deposits will quickly destroy accuracy. To prevent the deposit of bullet alloy the bullet is lubricated. When properly done this will almost completely eliminate any barrel fouling, assuming an appropriate alloy and a proper velocity level are used.

Both sizing and lubricating are simultaneously accomplished with a tool such as the Lyman 450 Sizer/Lubricator.

Bullet Lubricant

Bullets are lubricated with one of several kinds of lubricants. Lyman Ideal bullet lubricant is a blend of sheep tallow, beeswax and lampblack. This is a traditional style lubricant that works well with most cast bullets at velocities up to a moderate level. Lyman's Alox formula uses a blend of a

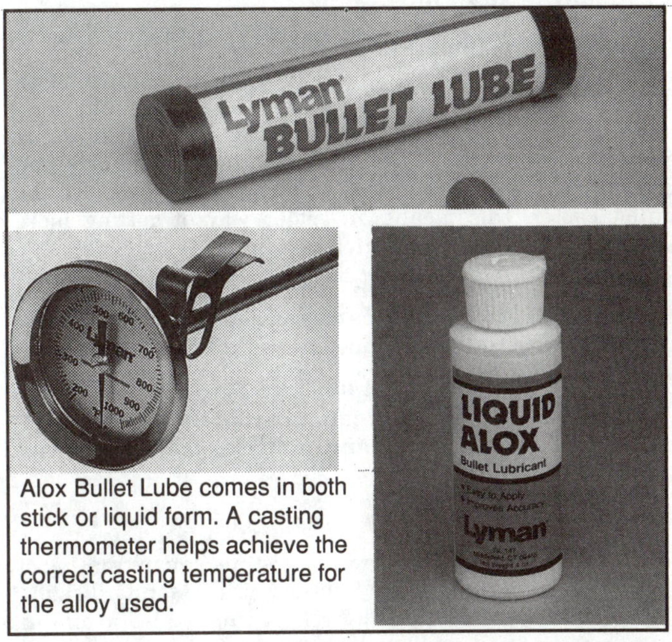

Alox Bullet Lube comes in both stick or liquid form. A casting thermometer helps achieve the correct casting temperature for the alloy used.

specific Alox compound and beeswax. Many cast bullet users favor this lubricant for higher velocity loads. There is a liquid version of the Lyman Alox bullet lubricant that allows bullets to be bulk lubricated easily. The bullets are simply placed in a container, the lube poured in and the container is then agitated.

Casting Thermometer

While many electric furnaces are equipped with a thermostat, the bullet caster who wishes to be able to duplicate casting conditions precisely will want to monitor bullet temperature with an accurate casting thermometer.

Ingot Mould

When preparing bullet alloys it is often convenient to mix the alloy desired and to cast it into ingots for later use. This will clearly identify the properly prepared metal from the base metals used to make the casting alloys. Lyman's ingot mould will make handy sized ingots for storage and later use.

Bullet Sizing and Lubricating Equipment

Installing the gas check seater on the Lyman 450 lube sizer.

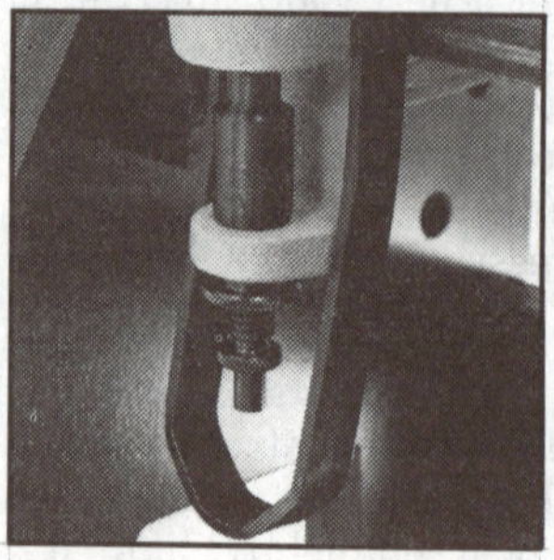

The gas check seater allows gas checks to be squarely seated on the base of cast bullets.

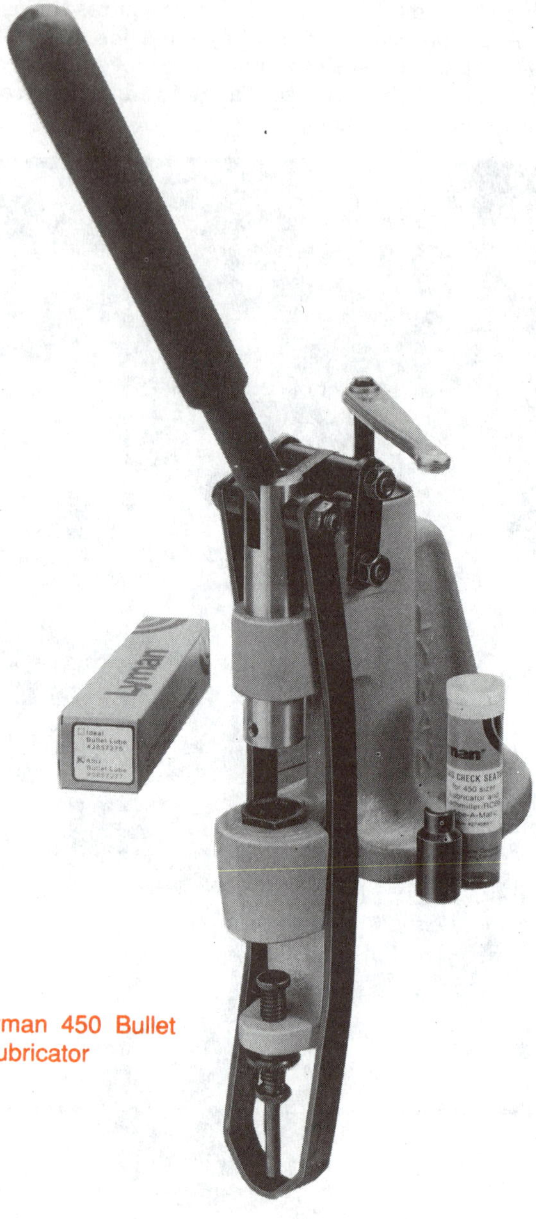

The Lyman 450 Bullet Sizer/Lubricator

The #450 Bullet Sizer/Lubricator

We've made the most popular Bullet Sizer/ Lubricator even better.

The new #450 Bullet Sizer/Lubricator is a perfect example of Lyman's continuing commitment to product improvement. The combined efforts of our ballistics laboratory and engineering department developed a better ram and leverage system for the #450. Strength, smoothness, stability and concentricity are all improved on our newest model. To the bullet caster, this means an easier way of getting more accurately sized and lubricated cast bullets.

Of course, the #450 Sizer/Lubricator still uses standard Lyman top punches and sizing dies as well as others of similar design. Changing bullet diameter requires only an inexpensive sizing die set.

Each #450 Sizer/Lubricator includes Lyman's patented gas check seater and a stick of alox/beeswax lubricant.

Gas Check Seater

This ingenious device allows consistent, square seating of gas checks using your Sizer/Lubricator and standard sizing dies. The new Lyman Gas Check Seater is also available separately and will work on all Lyman #450 Sizer/Lubricators as well as the Lachmiller/RCBS Lube-A-Matic.

Lyman Lubricants

Lyman offers two types: the generally popular Alox formula and Lyman's own original Ideal Lubricant. Both are essential to prevent barrel leading at higher velocities and may improve your shooting accuracy.

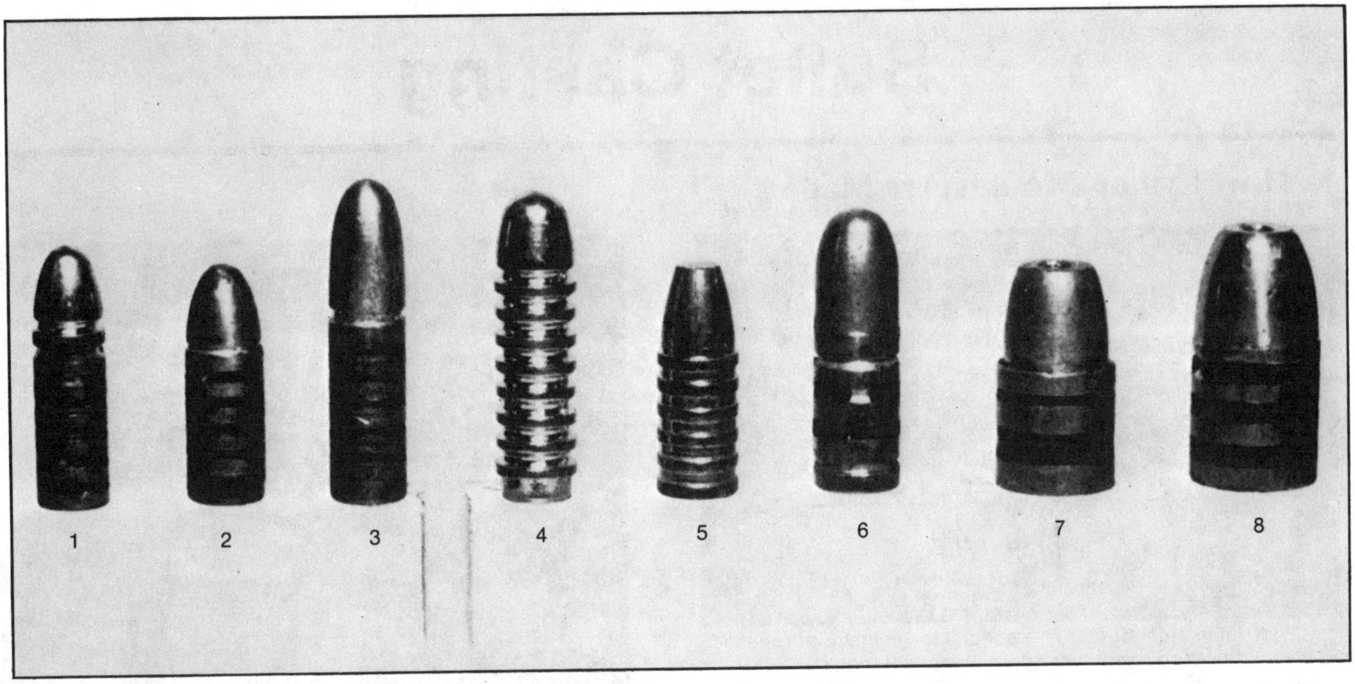

Shown above is a selection of the some of the past favorite Lyman cast bullet designs. (Several are still available).

(1) Loverin-designed #245497 for .243 and 6mm rifles;
(2) Loverin-designed #266455 for 6.5 mm Jap and other 6.5mm rifles;
(3) Loverin-designed #266469 for 6.5 mm Mannlicher rifles;
(4) Loverin-designed #287405 for 7mm rifles; (bullet is without lubricant or gas check).
(5) Ideal-Merrick designed #285221 (120-gr.) for .28-30 rifles (shown without lubricant).
(6) Lyman #311291 for .30-30 and other .30-caliber rifles.
(7) Keith-designed #429421 for .44 Special and .44 Magnum revolvers.
(8) Lyman-Gould #457122 for .45-70 and .45-90 Winchester rifles.

Predicted Physical Characteristics of Bullets Cast in Various Lead Alloys

		Predicted As-Cast Characteristics									Predicted Dimensions - Sized			
		Lead		Wheelweights		#2 Alloy		Linotype						
Caliber	Bullet Example	Wt. grs.	Dia.	Wt. grs.	Dia.	Wgt. grs.	Dis.	Wt. grs.	Dia.	H&I Sizing Die	Lead	Wheel-weights	#2 Alloy	Lino-type
Rifle														
.22	#225415	47	.2246″	46	.2246″	45	.2250″	44	.2252″	**.224**	.2237″	.2237″	.2240″	.2240″
.243, 6mm	#245496	87	.2437″	85	.2441″	83	.2445″	81	.2447″	**.243**	.2427″	.2427″	.2430″	.2430″
.25	#257464	92	.2576″	91	.2581″	88	.2585″	87	.2588″	**.257**	.2567″	.2567″	.2570″	.2570
.270	#280473	129	.2776″	127	.2781″	123	.2785″	121	.2788″	**.277**	.2765″	.2767″	.2770″	.2773″
.280, 7mm	#287308	170	.2846″	167	.2850″	162	.2855″	158	.2858″	**.284**	.2836″	.2837″	.2840″	.2844″
.30	#311291	176	.3095″	173	.3100″	168	.3105″	163	.3108″	**.308**	.3076″	.3077″	.3080″	.3081″
8mm	#323470	167	.3234″	165	.3240″	160	.3245″	155	.3248″	**.323**	.3226″	.3227″	.3230″	.3233″
.35	#358315	215	.3578″	212	.3584″	206	.3590″	200	.3594″	**.357**	.3566″	.3567″	.3570″	.3573″
.375	#375449	276	.3773″	272	.3779″	264	.3785″	256	.3789″	**.376**	.3756″	.3757″	.3760″	.3763″
.45/70, 458	#457193	439	.4575″	433	.4583″	420	.4590″	407	.4595″	**.457**	.4566″	.4567″	.4570″	.4574″
Pistol														
.25	#252435	53	.2522″	53	.2526″	51	.2530″	49	.2533″	**.251**	.2507″	.2507″	.2510″	.2510″
.32	#311252	80	.3125″	79	.3130″	77	.3135″	75	.3138″	**.309**	.3086″	.3087″	.3090″	.3100″
9mm	#358242	96	.3558″	95	.3564″	92	.3570″	89	.3574″	**.356**	.3556″	.3557″	.3560″	.3563″
.38/357	#358156	156	.3578″	153	.3580″	149	.3590″	144	.3594″	**.357**	.3566″	.3567″	.3570″	.3573″
.41	#410610	225	.4101″	221	.4108″	215	.4115″	209	.4119″	**.410**	.4096″	.4097″	.4100″	.4103″
.44	#429421	256	.4296″	252	.4303″	245	.4310″	238	.4314″	**.429**	.4286″	.4287″	.4290″	.4294″
.45	#452374	235	.4515″	232	.4523″	225	.4530″	218	.4535″	**.451**	.4506″	.4507″	.4510″	.4514″

Bullet Casting

1. How Lyman Moulds Are Made

1. Cold-rolled steel blocks of a special formula are purchased by Lyman Products for its complete line of moulds.

2. The rough blocks are milled so that the sides and faces are squared and correctly dimensioned.

3. The second cut is the groove which eventually will accommodate one of the pincer-type handles. At the same time the ends of the blocks are squared and milled to dimension.

4. The facing surfaces of the blocks are ground uniformly to a perfect match. They are then paired and numbered, and air-vent lines are milled on the faces.

5. Holes are drilled and tapped for dowel pins and setscrews.

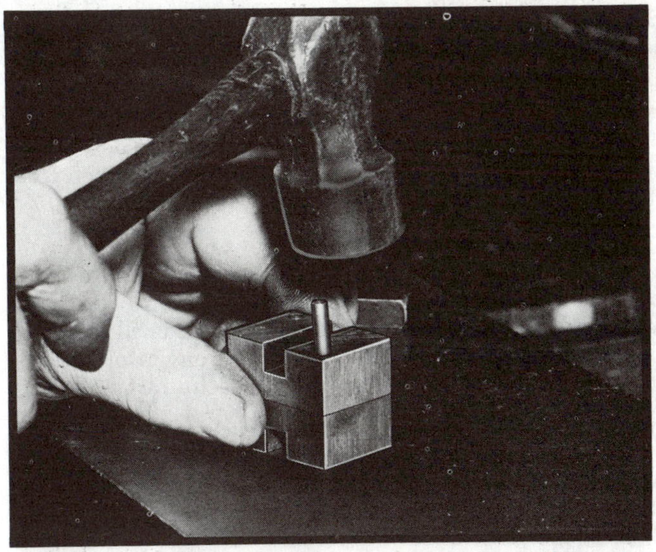

6. At this stage the dowel pins are driven home and the mould blocks checked for alignment.

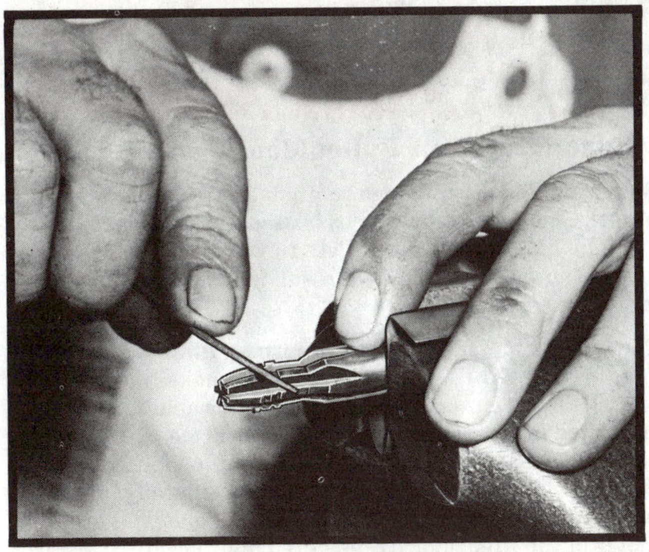

7. Each cutter, or "cherry", is carefully made by an experienced toolmaker following carefully detailed blueprints. However, machines can do only so much and the "final touches" are done by hand.

Once the cutter is brought into final form, it is sent out to be hardened by heat-treatment before being placed in the production cycle.

8. Cherrying drills, prepared by a toolmaker, cut the mould cavities. Hundreds of these drills, or "cherries", are needed to supply the variety of calibers and bullet designs within the Lyman line.

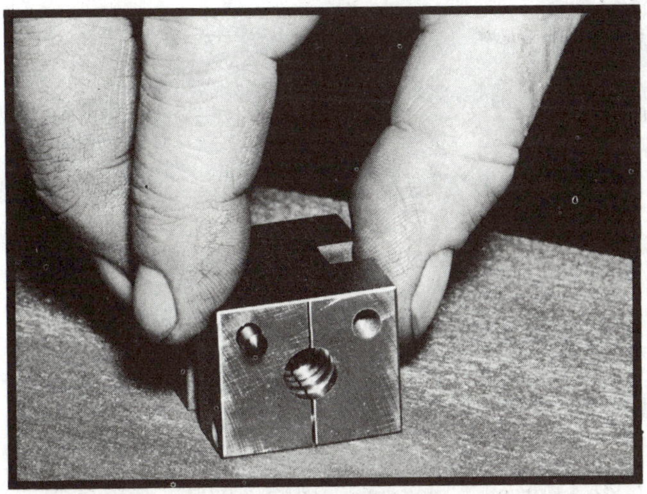

9. At several stages each block is strapped to remove any burrs that may have resulted from earlier handling and sharp edges are slightly broken.

10. The blocks are heat-cured to oxidize the metal. This "blues" the mould and minimizes the time required for the breaking-in process.

11. A screw fastens the sprue cutter in place and the blocks are ready for final inspection, packaging and shipment.

CHAPTER 14
CASTING AND USING LEAD BULLETS WITH ACCURACY

Understanding Your Bullet Mould

Bullet moulds are intended to produce a bullet of a given nominal diameter if an exact alloy is used; the same is true for the weight of the finished bullet. However, even a minor change in the specified alloy will result in a bullet of different diameter and weight.

The bore diameter of modern firearms can vary from gun to gun, even when they are of the same make, model and caliber. This is a natural result of the allowable tolerances incorporated in modern manufacturing. This means that any given firearm may need a cast bullet diameter slightly different from another rifle of the same caliber. Generally, as-cast bullets with a diameter a few thousandths over the groove diameter of the firearm are considered appropriate.

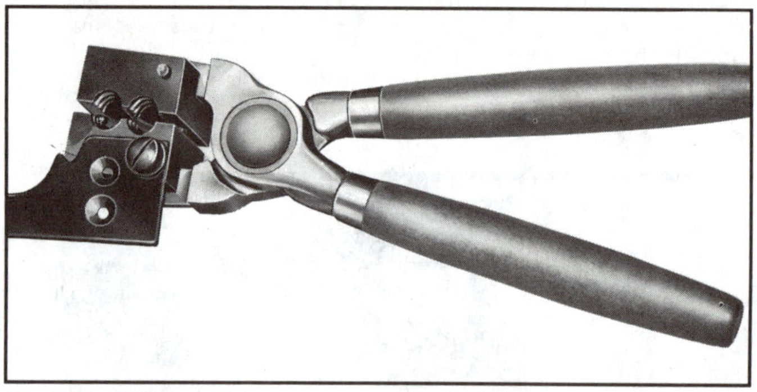

Moulds will cast a bullet close to nominal groove diameter. Sizing is usually required to match a particular firearm.

There is, of course a manufacturing tolerance for the cavity in the mould. Thus, the usual procedure is to have a finished bullet that comes from the mould slightly larger than ideally needed. The bullet is then sized to the exact diameter required. Sizing a small amount may improve cast bullet accuracy. However sizing an excessive amount may be detrimental to accuracy. A maximum sizing of 0.003" seems to be warranted for best results.

Tuning And Trouble Shooting your Mould

Mould blocks must be in good condition before the casting session starts. Inspection and repair as needed is a matter of common sense. Start by examining the blocks for proper fit.

With the sprue plate still removed, hold the blocks together, with a bright light source behind them. Look into the cavity and along the joining surface of the mould. There should be no light showing. Any light suggests that the mould halves are not coming into proper contact. This can be caused by burrs on the block surfaces or at the edge of the mould cavity. It can also be caused by excessive alignment-pin projection or foreign material in the pin positioning holes.

Burrs are removed by very careful stoning or a very light application of a fine jewelers file. Take extreme care to prevent damaging the edge of the bullet cavity.

Excessive pin protrusion can be remedied by lightly striking the engaging end of the pin with a ball peen hammer. Be certain to support the block on a hard flat surface when performing this operation.

Foreign material in the alignment pin hole can be flushed away with a pressurized degreasing solvent or wiped free with a dry cotton swab or pipe cleaner.

Next, check the top of the mould for any burrs that will prevent the sprue plate from lying in full contact with the blocks. This is also true for the sprue plate screw hole. If the sprue plate does not lay flat, the top of the mould will be galled when it is used or molten metal will build up on the top surface of the mould and/or the bottom surface of the sprue plate.

Inspect the vent lines on the face of each half of the block. These must be open from the cavity to the outer edge of the block. The vent lines are only 0.003" to 0.005" deep but may be carefully opened (if not already so) using a jewelers triangular file.

The sprue plate then must be checked for flatness and freedom from burrs. Remove any burrs from the edge of the sprue plate stamping with a fine jewelers file. If the plate is not flat, polish it using 400 grit wet or dry paper placed on a flat glass surface.

Assemble the blocks and try twisting them when they are correctly mated. Any twist should be removed by driving the locating pins inward so that they will fit a bit deeper into the engagement holes.

Inspect the handle mounting surfaces, especially around the screw holes. Remove any burrs that prevent the handles from pivoting freely.

Some Casting Tips

Avoid getting any lead on block interface surfaces when casting. Even minute spots of bullet metal will prevent the mould from closing properly. Wipe away such spots while the mould is hot. Do not use a synthetic rag for this purpose but rather one of a natural material such as cotton.

After dropping the bullet out of the mould, first close the blocks and then the sprue plate to avoid marring the mould.

Occasional hang-up of the bullet in one half of the mould is normal. Persistent hang ups in a specific half may be due to a tiny burr. Use a magnifying glass to locate the problem. A plug of steel wool rolled in the cavity may correct the problem.

Keep the sprue plate screw properly tensioned. The sprue plate should turn readily but not be loose. Excessive pressure of the plate on the mould may prevent proper venting. A sloppy fit will result in raised sprues on the bottom of the bullet which affect accuracy. These same raised sections can prevent proper seating when gas checks are required.

Never allow a mould to rust. However, rusted moulds can sometimes be cleaned up using steel wool. If you protect a mould with a lubricant between uses, it will be necessary to degrease the mould before each casting session. Some folks simply store cool moulds in air tight containers to prevent potential rusting. This eliminates the need for degreasing before each use.

Casting Alloys

All muzzleloading balls and bullets must be cast only from pure lead. Handgun bullets should be made with some hardening metal in the alloy. Depending upon the caliber and velocity some handgun bullets will need to be harder than others. Generally the higher the velocity, the harder the bullet needs to be. Barrels with rough interior surfaces will require the use of harder bullets. If you have a barrel leading problem due to velocity or bore surface roughness, make your bullets harder. Alloys made of a 20 to 1 mix of tin and lead are often used for low velocity handgun loads. As velocity is increased, harder alloys are a must. Sometimes more lubricant or a lubricant of another type can help leading and/or accuracy problems.

Rifle bullets are driven at relatively high velocity and therefore should be cast from relatively hard alloy - such as Lyman No. 2 alloy. For more on bullet alloys refer to the table in the preceding chapter.

Lyman has found that over the course of time specific alloys have varied in metals content and percentages. Since accuracy problems can be traced to changes in bullet alloy, do not assume that all wheel weights, or other alloyed metal, will always be the same. Wide discrepancies have, and undoubtedly, will continue to occur.

Many casters may do well to use a hard alloy such as Lyman No. 2 metal or its equivalent for all cast bullet applications (except muzzleloading). This will enable you to avoid almost all accuracy problems due to barrel leading.

Melting And Fluxing Bullet Metal

Caution: Be certain to wear protective clothing as previously described and heavy leather work gloves and goggles when working with molten bullet metal.

The melting pot must be capable of heating the bullet metal from 750 to 800 degrees (Fahrenheit). A Lyman cast iron pot placed on an electric or gas stove will prove satisfactory. However, the reloader will find an electric furnace made specifically for bullet casting to be faster and more convenient. Such units, properly used, are also cleaner and safer to work with. Their adjustable thermostats also make casting temperatures easier to maintain.

Prior to casting mould blocks should be pre-heated. Lyman's Mag 20 furnace has a convenient shelf for pre-heating.

Whether blending raw metals into bullet metal and forming ingots, or actually making bullets, the same melting and fluxing procedure is followed. It will take about 20 minutes for the solid metal to heat sufficiently to melt. When it is hot enough to flow freely the melt needs to be fluxed to remove all impurities which float to the top. If an alloy is not properly fluxed, the impurities will be poured into the bullet mould causing voids. The voids will in turn lead to poor accuracy.

To flux the molten alloy drop a small bit of fluxing material into the pot. Traditionally, tallow, beeswax, or Ideal bullet lubricant has been used to flux. These will cause a smokey gas to be given off from the pot. This smoke should immediately be ignited carefully with a match to eliminate it. Perhaps a better way is to use Marvelux (a dry fluxing material available from Brownells). There will be no smoke or greasy fumes when using Marvelux and a good flux will be obtained.

When fluxing, the melt must be stirred with a dipper. Stir deeply, introducing a dipper full of air into the melt frequently. The introduced air will help with the fluxing. The surface of a properly fluxed melt will appear almost mirror bright and will contain flecks of burned brown and black impurities. These must be skimmed off using a spoon like tool that has an insulated handle. Caution: the material skimmed off is very hot, perhaps 800 or more degrees. Dispose of it in a safe fire proof container that will not be accidentally turned over.

When you are casting, you may see impurities begin to rise to the surface. You must reflux as often as required. As soon as you see any grey scum forming on top of the melt, it is time to flux again.

Caution: Be extremely careful when fluxing to prevent tipping over the pot or splashing any of the molten alloy.
If the dipper is to be used for casting, allow it to remain in the melt when not in use in order to keep it hot.

Step By Step Bullet Casting

After the bullet alloy has been properly fluxed it is ready for casting. The melt temperature must be maintained or the cast bullets will not be as expected. The drawings on the next page show some common bullet problems related to temperature control. It is important to realize that proper temperature control also includes keeping the mould (and dipper if one is used) at the correct temperature. Thus, the casting pace may need to be increased or slowed, if you find the mould is too cold or hot.

Allow the mould to preheat by standing it on the edge of the electric furnace. If you are using a cast iron pot with an independent heat source, heat the mould to casting temperature by repeatedly filling it with lead until the bullets are properly formed.

Caution: Do not drop poorly formed bullets or sprue cut-offs back into the molten lead in the pot. Doing so can cause molten metal splashes which can result in serious burns or property damage.

You will need a small mallet, or a 10 to 12" piece of broom handle, to strike the sprue cutter plate open after the bullet has solidified. Also required is a soft pad to allow the bullets to fall on when opening the mould. This will prevent bullets

Cleaning the Bullet Alloy

Dirty bullet alloy will have gray impurities on the surface.

Add fluxing material to the bullet alloy and stir carefully.

After removing impurities alloy should have a clean, bright appearance.

from deforming. A piece of old ironing board cover is ideal as it is not highly flammable.

Each bullet caster tends to develop their own routine. The following suggested method works well for many casters. It uses a dipper as new casters may find this the easiest way to get good bullets quickly. Those who wish to start casting by using a bottom pour furnace can do so but the learning period for properly filling the mould may be somewhat longer.

Fill the dipper with melt and then, with the top of the bullet mould held vertically, place the dipper's pouring spout into the sprue plate hole on the mould. Next rotate the mould and dipper as a unit so that the mould is filled with lead. This is done by rotating the mould, with dipper in place against it, so that the top of the mould is horizontal and the dipper is above it.

While turning the mould back to the starting position slightly separate the mould and dipper to allow a small amount of alloy to fill the concave section of the sprue plate. If this is not done, as the melt solidifies, a hole will form in the base of the bullet.

Wait several seconds for the bullet to harden. Small light bullets, i.e. 22 caliber bullets, will harden faster than large caliber heavy bullets, i.e. 45-70 bullets. Pace yourself so that the bullet is hard enough that it will not be damaged as it falls from the mould. Additionally, speed up or slow down as required to keep the mould at an appropriate temperature for proper bullet forming. To keep mould temperature at an appropriate level, small diameter light bullets will require a faster pace than the larger bullets.

After the bullet cools in the mould, use a wood dowel or mallet to strike open the sprue plate. This will shear the sprue from the base of the bullet. Push the sprue plate far enough around to completely clear the opening in the mould. Do not use any more force on the spruce plate than necessary to shear the sprue from the bullet base.

Allow the sprue to drop into a fireproof container for later remelting. If bullet bases have a cavity or concave shape on the base, allow more of a sprue to form. The largest bullets will require the largest sprues.

If lead smears appear on the top of the mould, after rotating the sprue plate, allow the sprue to more completely solidify before cutting it off.

Do not expect the first cast to be perfect. Until the mould is at the proper temperature bullets will not fill out correctly. Generally, a half dozen or more casts may be required before well formed bullets are obtained. Imperfect bullets can be dropped into the sprue container and can be remelted at another time. Do not be tempted to heat the mould quickly by sticking it into the melt. Such action will bring more grief than may first come to mind.

If the mould overheats do not attempt to cool it in water or use any other artificial means to lower its temperature. The only safe method is to set the mould aside and allow a bit of time for it to cool.

Wrinkled bullets or bullets that are not fully filled out suggest that temperatures of the melt and/or mould are too cool. Frosted bullets, or bullets which break apart easily, suggest that the melt and/or the mould is too hot. Properly cast bullets will have crisp edges, and will appear clean and bright.

It is a good practice to weigh all cast bullets to eliminate any that have air voids. If you find that you have mastered casting techniques to the point that air voids are seldom encountered, you can then dispense with this procedure.

The actual weight of your bullets will vary with specific alloys. Bullet weights can easily vary from lot to lot due to variations (which are not at all uncommon) even when using the same type of alloy. Thus, it is a good practice to segregate one lot of bullets from those obtained from the next casting session.

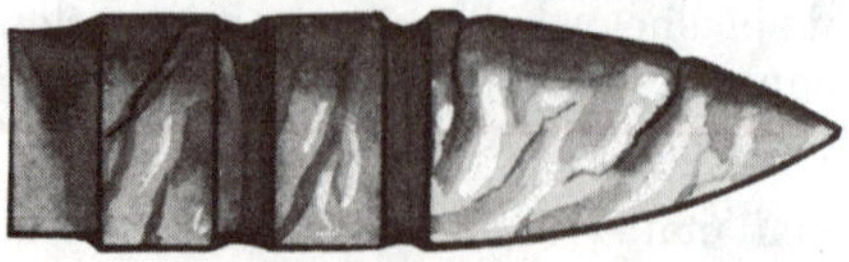

Wrinkled Bullet
Temperature Too Cool

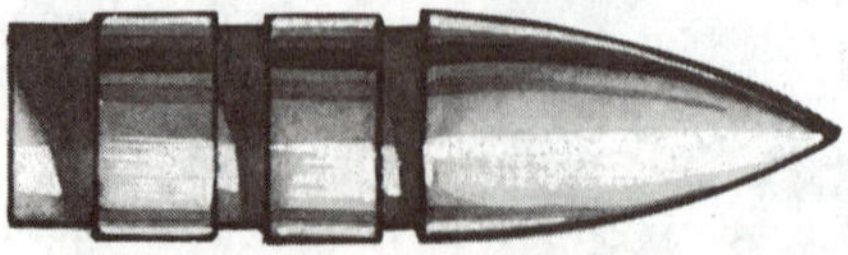

Good Bullet
Temperature Correct

Frosted Bullet
Temperature Too Hot

Basic Steps of Casting

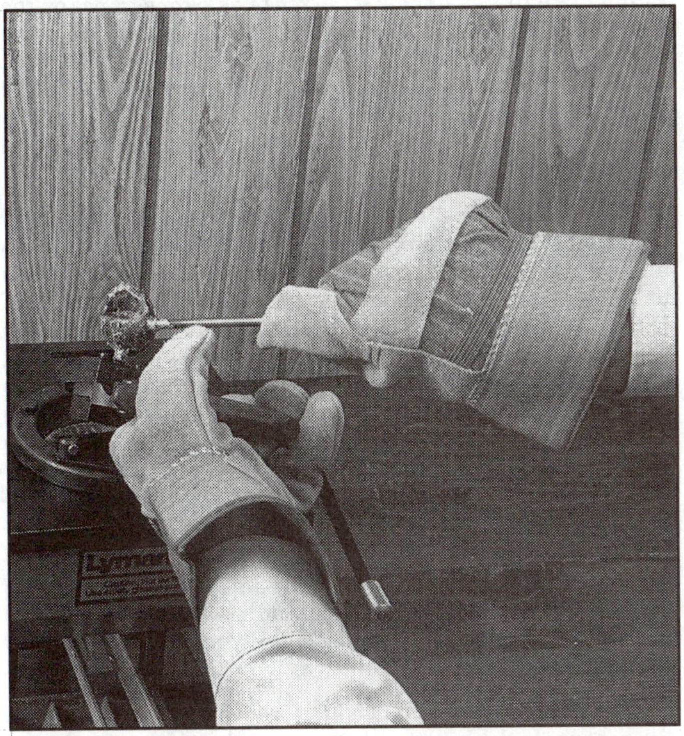

1. When casting with a dipper, pour lead into the mould by slowly rotating the mould and dipper together with the dipper on top.

2. When using a bottom pour feature, place the mould block directly under the pour spout resting on the mould guide. Lift the pouring handle to fill the mould.

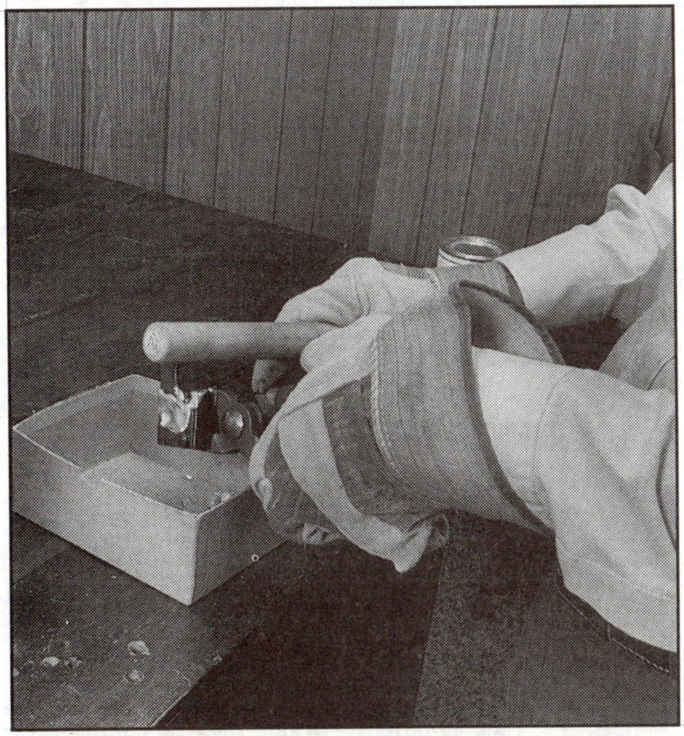

3. After waiting for the metal to harden, hit the sprue plate with a wooden mallet to break the sprue free from the bullet.

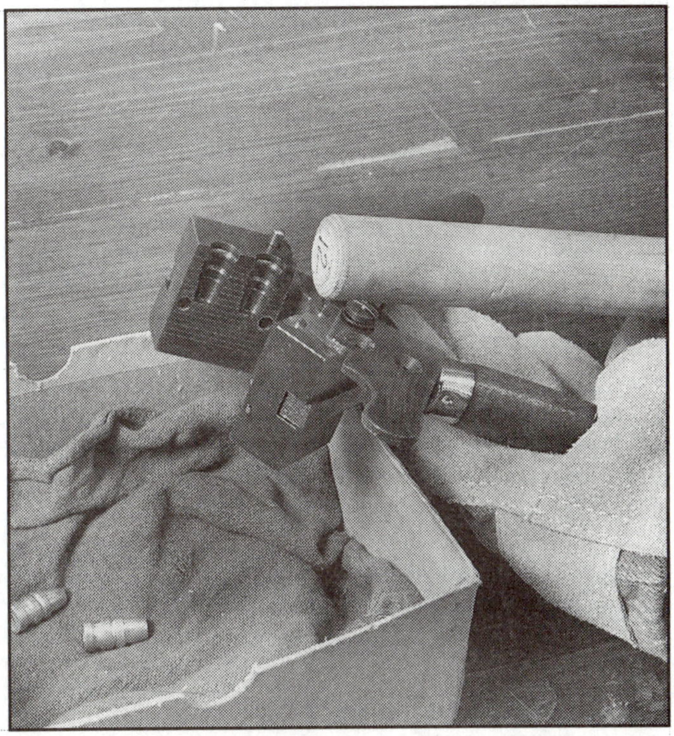

4. Open the mould handles and disengage the bullets from the mould by striking the mould handle hinge pin.

A good caster will find that most bullets will vary no more than a half grain from the average weight for the lot. Those that vary more than this amount should be culled and remelted at a later date.

Gas Checks Improve Cast Bullet Performance.

The metal caps often placed on a bullet's base are called gas checks. These protect the bullet base from being deformed by the hot gases of combustion. Gas checks are important whenever relatively high velocities will be used. By protecting the bullet base from having bits of lead brought to the melting point, the bore of the firearm is in turn protected from unnecessary leading.

When used with a high velocity bullet lubricant, such as Lyman Alox, bullets cast of a hard alloy and protected with a gas check can withstand velocities in excess of 2000 feet per second. Some cast bullet users find it possible to exceed even 2500 feet per second when everything is right and a smooth bore is present. Gas checks, attached when sizing the bullet, must be seated squarely or accuracy will be adversely affected.

There are two types of gas checks commonly used. There is the uniform wall-thickness type that fits to the bullet base by friction. The second style has a thicker lip section which "crimps" into the bullet. The friction seated types tend to fall off the bullet shortly after the bullet leaves the muzzle. Some have suggested this can affect accuracy if the separation of the gas check from the bullet is not uniform. The crimp-on gas check also can cause inaccuracy if it is not perfectly square with the bullet base. Trying both types may be in order if you have an accuracy problem. As with all bullet types, a square base is essential to accuracy. This applies equally to cast bullets used with or without a gas check.

Gas checks can be applied only to bullets specifically designed with a rebated base to accept them. Gas check style bullets may be used without the gas checks applied, if velocity is kept modest.

Press-on style gas checks should not be used when bullet bases go below the case neck. This will prevent a loose gas check from falling into the powder charge.

If you seat the bullet base below the case neck then it is suggested you use the crimp on style gas check. Gas checks and bullet bases must be compatible to insure proper gas check fit.

Sizing And Lubricating Cast Bullets

Sizing cast bullets allows the reloader to bring the bullet diameter to a nominal dimension. Usually bullets are sized to a dimension at, or just above, the actual groove diameter of the firearm in which they are to be used. To determine actually groove diameter of any firearm, the reloader "slugs" its bore.

If a slightly oversized lead slug is driven through the bore and then measured, the reloader will then know the smallest diameter in the barrel. (Some shooters prefer to size not to the smallest diameter but to the diameter of the barrel about two inches in front of the chamber.) When searching for the best accuracy it pays to first try sizing diameters equal to the bore dimension. When the accuracy of groove diameter bullets is known then a bullet diameter of 0.001" and 0.002" larger can be tried.

Great care must be used when slugging a bore so as not to damage the bore surface. The rod used to drive the slug through the barrel must be close fitting and have a slight radius on the outside edge. If the rod buckles or scrapes on the bore, irrepairable damage could occur. If in doubt it might be best to have a qualified and experienced gunsmith do the bore slugging. For these reasons, many shooters do not actually slug their bores, but rather start with a sizing diameter to match the diameter of jacketed bullets nominally used for a specific cartridge.

This is a practical method for the first series of accuracy tests using a specific diameter sizing die. Seldom are barrels undersized; rather when they vary from the nominal they are usually oversized. (It should be noted that some barrels of European manufacture have been found to be undersized.) Thus, if the first series of attempts do not provide the desired accuracy the second series of attempts can be with bullets sized 0.001" larger and the third sequences of attempts can be with bullets 0.002" larger. It is usually counter productive to use any larger sizing diameter.

Sizing And Lubricating Tips

Lubrication of bullets demands a proper flow of lubricant. This is usually not possible when the ambient temperature is less than 70 degrees Fahrenheit, and warmer is better. It is possible to damage a lubricator by applying too much force to the lubricant in the reservoir.

Use the exact top punch required for the bullet to be lubricated. If the bullet is forced into the lubricator with a poor fitting top punch, it may be sized without proper alignment. Such a condition is contrary to the goal of accuracy.

If bullet lube continually forms on the bullet base it may be due to excessive pressure on the lube in the reservoir. It could also be due to improper positioning of the bullet in the sizing die with respect to the dies lubricant orifices. Be sure to remove any lubricant from the bullet base in order to prevent contamination of the powder charge in a loaded round.

Do not store lubricated bullets as the lubricant may pick up contaminants that would be harmful to a bore. Size and lubricate only the number of bullets that will be loaded in a single session.

If a lubed bullet is dropped on the floor, or otherwise picks up contaminants, discard it or carefully remove all lubricant and contaminants from the bullet. It may then be relubed.

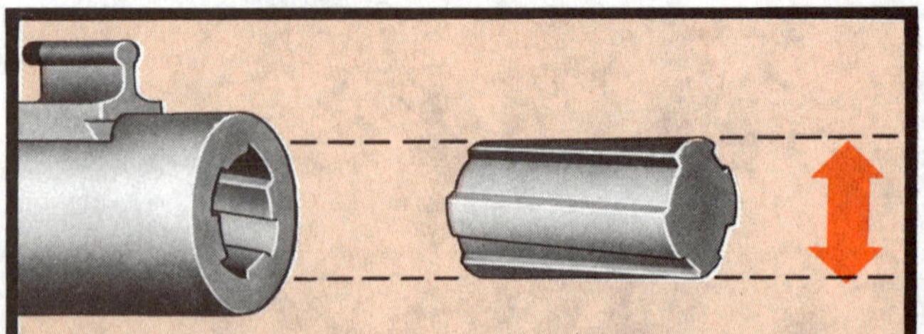

SLUGGING THE BARREL

Starting at the muzzle end, a soft lead slug (slightly larger than the groove diameter) is driven through the barrel. When the slug is retrieved on the breech end, it will be perfectly engraved and clearly marked by both the lands and the grooves. Measuring (with a micrometer) the diameter of this slug at its widest point will determine the groove diameter of the rifle. Bullets should conform to this dimension.

Keep in mind that undersized bullets can be a detriment to accuracy in most firearms. If sizing at nominal (and then increasing to 0.001" and 0.002" over nominal) does not give the desired results, slugging the bore becomes almost mandatory. If such slugging shows the bore to be as expected then it is reasonable to look for other causes of the inaccuracy. However, if the slugging shows a non-typical barrel then the appropriate bullet sizing diameters can be addressed. Keep in mind that it is usually not practical to reduce a bullets diameter by more than 0.003" without negatively affecting accuracy. Indeed, some shooters feel that no more than 0.002" in diameter reduction is practical before loss of accuracy is encountered.

As with all handloading steps, consistency is the secret to success. Every effort should be taken to insure that each bullet is the same as the last. When testing alloys, lubricants, or sizing diameters do so only by changing one aspect at a time. Keep good records of the results to be certain that things do not become jumbled in memory.

Important Measurements in a Rifle Barrel

When a soft lead slug is driven completely through a barrel, as in "A", it records only the smallest diameter of the bore.

Measuring "C" with a vernier caliper, using the inside jaws, will fairly well define bore and groove measurements at the muzzle only. Another measurement technique, used by some cast bullet riflemen, is to insert the nose of a bore-riding cast bullet design (like #311284) into the muzzle. If the bullet drops in without interference, the bore/bullet dimensional relationship is not as good as it could be for optimum accuracy.

This leaves us with "B"—the most important barrel dimension. Your cast bullet must be sized to match, or slightly exceed, the actual groove diameter found just ahead of your rifle's chamber.

Basic size-to diameters

Rifle Calibers	Sizing Diameters
All .22 cal. (except .22 Hi-power)	.224, .225†
.243, .244, 6 mm	.243, .244†
All .25 cals.	.257, .258†
.264 Win. Mag. & 6.5 mm.*	.264†, .266
.270 Win.	.277, .278†
7 mm, .280 Rem., .284 Win.	.284, .285†
.30 cals.	.308, .309†, .310
7.62 Russian* 7.62 x 39	.310, .311
.32/20 Win.	.311, .312†
7.65 Mauser*	.311, .312
.303 British*, 7.7 mm Jap.*	.313, .314†
8 mm Mauser (S-Bore)	.323, .325†
.32 Win. Spec. .32 S.L. & .32 Rem.	.321, .322†
.338 Win. & .33 Win.	.338†
9 x 56 mm & 9 x 57 mm	.354, .355†, .356
.35 cal.	.357, .358†, .359
.375 H & H Mag. & .375 Win.	.375, .377†, .378
.38/55*	.379
.38/40	.400, .401†
.44/40	.427, .428†
.44 Mag.	.429, .430, .431†
.444 Marlin	.430, .431†
.45/70 & .458 Win.	.457, .458†, .459
.50 Cal. Rifle	.509, .512†
Pistol Calibers	
.22 Jet & .221 Fireball	.224†, .225
.25 A.C.P.	.251†
.256 Win.	.257, .258†
.30 Luger	.310†
.32 A.C.P., .32/20, .32 S&W, 32 H&R	.311, .312, .313†, .314
9 mm Luger*, .38 A.C.P., .380 Auto	.354, .355†, .356
.38 Spec., .357 Mag., .38 S & W*	.357, .358†, .359, .360
.38/40, 40 S&W, 10 mm	.400†, .401
.41 S&W Mag., 41 AE	.410†
.44 S&W Spec. & .44 Mag.	.429†, .430, .431
.45 A.C.P., .45 Auto Rim, .45 Colt* (post-WW II), 45 Win. Mag.	.450, .451†, .452
.45 Colt (pre-WW II) & .445 Webley	.454†

*Wide variations in groove diameter.
†Indicates suggested useful diameter. This may vary from gun to gun.

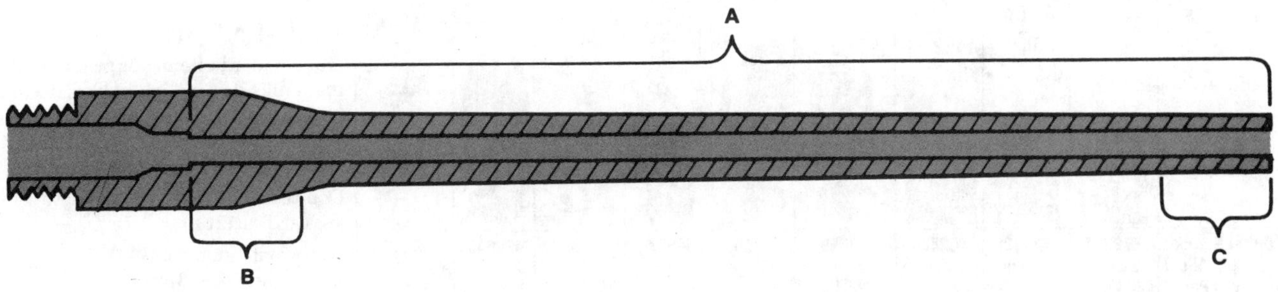

Lyman Bullet Moulds

WARNING:
Melting lead and casting lead objects will expose you and others in the area to lead, which is known to cause birth defects, other reproductive harm and cancer. See Instructions on Reducing Exposure supplied with product.

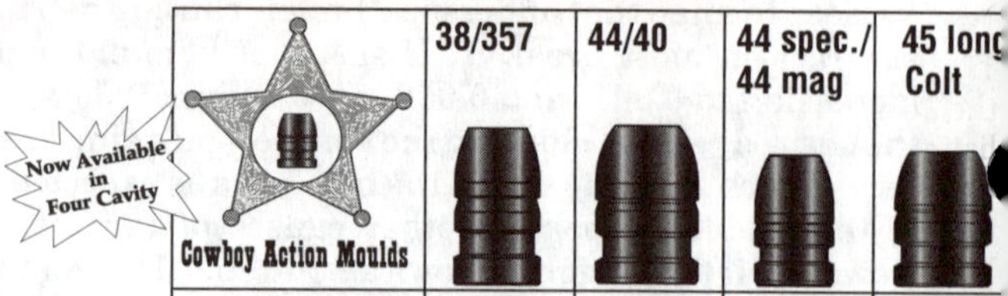

	38/357	44/40	44 spec./ 44 mag	45 long Colt
Bullet Number	**358665**	**427666**	**429667**	**452664**
Weight	158 gr.	200 gr.	240 gr.	250 gr.
Double Cavity Part#	2660665	2660666	2660667	2660664
Top Punch #	495	649	649	649
Top Punch Part #	2786769	2786778	2786778	2786778
Four Cavity Part#	2670665	2670666	2670667	2670664

Cowboy Action Moulds (New)

Like the first Ideal Moulds that were made over 100 years ago for the 44 and 45 caliber rifles and six guns that were used to tame the West and ensure that law and order prevailed, the Silver Star bullets have been designed to fill the needs of modern cowboy action shooters. All feature a nose that aligns perfectly in the cylinder throat to ensure greatest possible accuracy and a crimp groove that maintains proper overall length in both rifle and pistol use.

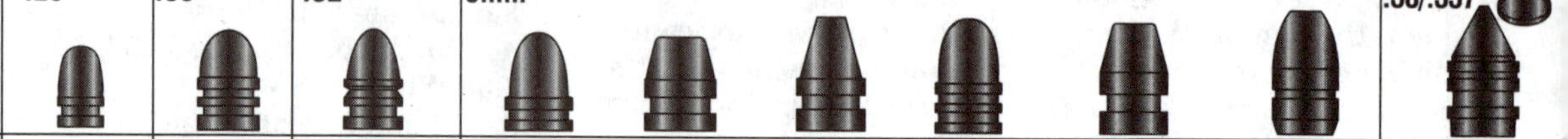

Pistol	.25	.30	.32	9mm						.38/.357
Bullet Number	**252435**	**311252**	**313249**	**356242**	**356632**	**356402**	**356242**	**356634**	**356637**	**358093**
Weight (#2 Alloy)	50 gr.	75 gr.	85 gr.	90 gr.	100 gr.	120 gr.	120 gr.	130 gr.	147 gr.	125 gr.
Double Cavity Part #	2660435	2660252	2660249	2661242	2660632	2660402	2660242	2660634	2660637	2660093
Top Punch Number	203	465	226	311	402	402	311	402	637	93
Top Punch Part #	2786748	2786742	2786703	2786710	2786723	2786723	2786710	2786723	2786749	2786754
Four Cavity Part #	—	—	—	—	—	2670402	—	2670634	2670637	—

Pistol cont'd.	.38/.357								40 S&W/10mm			.38/.40
Bullet Number	**358212**	**358091**	**358477**	**358156**	**358311**	**357446**	**358429**	**358430**	**401654**	**401638**	**401633**	**401043**
Weight (#2 Alloy)	145 gr.	150 gr.	150 gr.	155 gr.	160 gr.	165 gr.	170 gr.	195 gr.	150 gr.	175 gr.	200 gr.	175 gr.
Double Cavity Part #	2660212	2660091	2660477	2660156	2660311	2660446	2660429	2660430	2660654	2660638	2660633	2660043
Top Punch Number	311	495	429	429	311	429	429	430	43	43	43	43
Top Punch Part #	2786710	2786769	2786731	2786731	2786710	2786731	2786731	2786732	2786691	2786691	2786691	2786691
Four Cavity Part #	—	2670091	2670477	2670156	2670311	—	2670429	—	—	2670638	—	—

Pistol cont'd.	.41	.44/40	.44 & .44 mag											.45	
Bullet Number	**410610**	**427098**	**429348**	**429303**	**429215**	**429478**	**429360**	**429421**	**429383**	**429244**	**429640**	**429650**	**429649**	**452460**	**452630**
Weight (#2 Alloy)	215 gr.	205 gr.	180 gr.	200 gr.	210 gr.	210 gr.	235 gr.	245 gr.	245 gr.	255 gr.	275 gr.	300 gr.	325 gr.	200 gr.	200 gr.
Double Cavity Part #	2660610	2660098	2660348	2660303	2660215	2660478	2660360	2660421	2660383	2660244	2660640	2660650	2660649	2660460	2660630
Top Punch Number	43	43	348	303	421	251	360	421	251	421	649	421	649	460	460
Top Punch Part #	2786691	2786691	2786716	2786709	2786729	2786704	2786717	2786729	2786704	2786729	2786778	2786729	2786778	2786740	2786740
Four Cavity Part #	—	—	—	—	2670215	—	—	2670421	—	2670244	—	—	—	2670460	2670630

Pistol cont'd.	.45	.45 Colt .454 dia.	.45 Colt .452 dia.	.45 Colt .452 dia.	.45	Devastator Hollow Point Moulds 38 spec./ 9 mm ★	Devastator Hollow Point Moulds .40mm S&W ★	Devastator Hollow Point Moulds .44 ★	Devastator Hollow Point Moulds .45 ★
Bullet Number	**452374**	**454190**	**452424**	**452490**	**452651**	**356637**	**401638**	**429640**	**452374**
Weight (#2 Alloy)	225 gr.	250 gr.	255 gr.	255 gr.	325 gr.	125 gr.	155 gr.	250 gr.	180 gr.
Double Cavity Part #	2660374	2660190	2660422	2660490	2660651	2650637	2650638	2650640	2650374
Top Punch Number	374	190	424	424	649	637	43	649	374
Top Punch Part #	2786719	2786700	2786730	2786730	2786778	2786749	2786691	2786778	2786719
Four Cavity Part #	2670374	—	—	—	—				

Pistol Moulds

Our standard offering of pistol moulds covers popular bullet designs, in double cavity style. Four cavity blocks for selected designs are shown in the table below. Refer to the Hollow Point section for availability.

Double cavity and single cavity devastator mould blocks less handles (12 oz.)

Four-cavity mould block, less handles (2 lbs., 3 oz.)

★ Indicates single cavity mould

NOTE: All standard rifle and pistol moulds use handles #2735793, and four-cavity moulds use handle #2735794.
Bullet weights listed are nominal and may well vary from bullets dropped from a given mould. Variations between listed nominal weights and actual as-cast weights have absolutely no effect on accuracy.
These bullet illustrations are artists' renderings. They are for general reference only and are not intended to be precise representation.

Lyman Bullet Moulds

New

The **379670, 38-55, 335 grain** heavy weight bullet was designed for the shooters who requested a long heavy bullet for shooting the 38-55 at long range. The nose shape is the same design as the famous long range match winning Postell assuring the best ballistic coefficient and accuracy. The **446110, 43 caliber Mauser** has returned for all of those old Military Mauser rifles available today,

Rifle Moulds

Lyman rifle moulds are the standard by which all other rifle moulds are judged, regardless of price. Our standard offering of proven, popular designs is the widest available. The designs shown below picture gas checks where appropriate. All Lyman rifle moulds are available in double cavity only, except those moulds where the size of the bullet necessitates a single cavity (indicated below by a ★). All rifle moulds shown below require double cavity handles, regardless of the number of cavities indicated. Hollow point designs are noted above bullet.

(13 oz.) mould, less handles

Rifle	.22			6mm	.25	6.5mm		.270 Win.	7mm		.30 M1/ 7.62x39mm		.30	
Bullet Number	225438	225415	225646	245496	257420	266469	268645	280642	287346	287641	311359	311410	311466	311291
Weight (#2 Alloy)	44 gr.	55 gr.	55 gr.	84 gr.	65 gr.	140 gr.	150 gr.	150 gr.	135 gr.	160 gr.	115 gr.	130 gr.	152 gr.	170 gr.
Mould #	2660438	2660415	2660646	2660496	2660420	2660469	2660645	2660642	2660346	2660641	2660359	2660410	2660466	2660291
Top Punch Number	438	415	646	203	420	463	641	641	346	641	359	467	284	465
Top Punch Part #	2786733	2786726	2786759	2786748	2786728	2786741	2786753	2786753	2786715	2786753	2786772	2786743	2786707	2786742

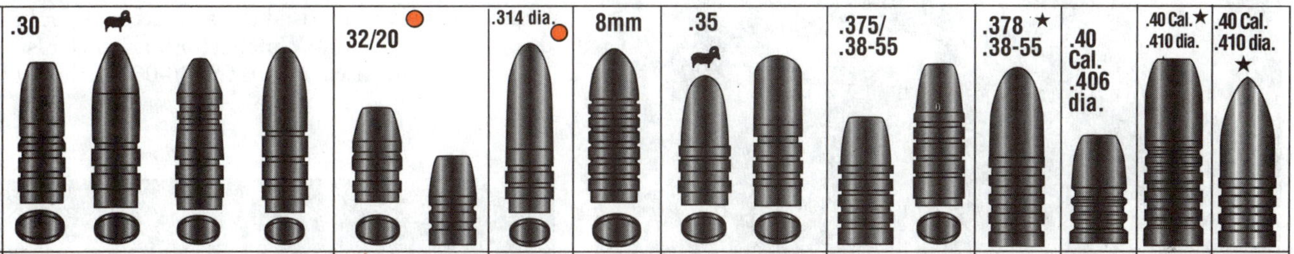

Rifle cont'd	.30	🐏			32/20 ●		.314 dia. ●	8mm	.35 🐏		.375/ .38-55		.378 ★ .38-55	.40 Cal. .406 dia.	.40 Cal.★ .410 dia.	.40 Cal. .410 dia. ★
Bullet Number	311041	311332	311644	311284	311316	311008	314299	323470	358315	358009	375248	375449	379670	403169	410655	410660
Weight (#2 Alloy)	173 gr.	180 gr.	190 gr.	210 gr.	112 gr.	115 gr.	200 gr.	165 gr.	204 gr.	280 gr.	249 gr.	264 gr.	335 gr.	240 gr.	400 gr.	385 gr.
	2660041	2660332	2660644	2660284	2660316	2660008	2660299	2660470	2660315	2660009	2660248	2660449	2660670	2660169	2640655	2640660
Top Punch Number	8	413	644	284	8	8	467	470	311	430	449	449	663	43	449	658
Top Punch Part #	2786690	2786725	2786758	2786707	2786690	2786690	2786743	2786745	2786710	2786732	2786736	2786736	2786779	2786691	2786736	2786755

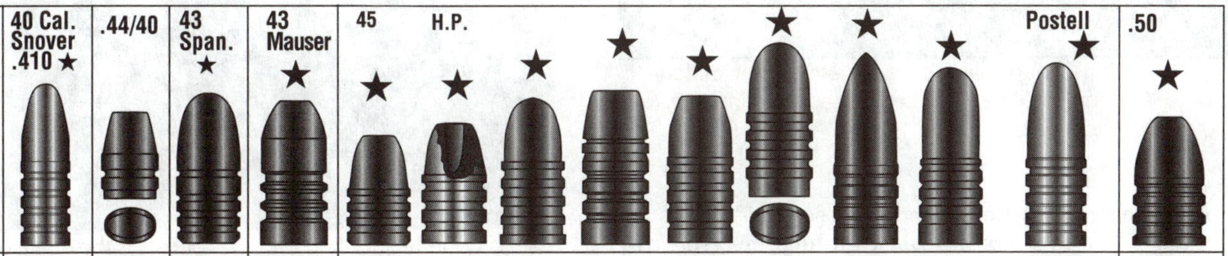

Rifle cont'd.	40 Cal. Snover .410 ★	.44/40	43 Span. ★	43 Mauser ★	45 ★	H.P. ★	★	★	★	★	★	★	Postell ★	.50 ★
Bullet Number	410663	429434	439186	446110	457191	457122	457124	457643	457193	457406	457658	457125	457132	515141
Weight (#2 Alloy)	400 gr.	215 gr.	370 gr.	340 gr.	292 gr.	330 gr.	385 gr.	400 gr.	405 gr.	475 gr.	480 gr.	500 gr.	535 gr	425 gr.
Mould #	2640663	2660434	2640186	2640110	2640191	2650122	2640124	2640643	2640193	2640406	2640658	2640125	2640132	2640141
Top Punch Number	663	43	251	251	191	191	374	191	191	374	658	374	132	141
Top Punch Part #	2786779	2786691	2786704	2786704	2786701	2786701	2786719	2786701	2786701	2786719	2786755	2786719	2786695	2786696

Silhouette Suggestions Look for the Ram

In both the "Pistol" and "Rifle" charts, we have indicated (🐏) bullets which should do well on silhouettes and have won many matches.

● Use 30 cal. gas check.

Mould Handles

Solid metal frame is easily attached to blocks with slotted screws. Hardwood handles provide a cool secure grip. Designed to provide uniform grip with different mould blocks.

Lyman mould handles are offered in three sizes. **Small**, for all older single cavity moulds. **Large**, for all current two cavity moulds, current single cavity round ball moulds, and all current Minie and Maxi moulds. **Four Cavity**, for all Lyman four cavity moulds.

Single-cavity handles (for old style small block, black powder and specialty moulds) (12 oz.) (#2735791)

Double-cavity handles (for all current two-cavity and single cavity moulds) (12 oz.) (#2735793)

Four-cavity handles (1 lb.) (#2735794)

Special Notice:

Lyman no longer manufactures small single cavity mould blocks. All current Lyman single and double cavity moulds are now made using our standard large mould blocks which require large mould handles.

CHAPTER 15
MAKING CAST BULLETS SHOOT ACCURATELY

By Ed Schmitt

More than a few years ago, while sighting in my first centerfire rifle (a 218 Bee Model 43 Winchester) my shooting was interrupted by a gentleman at the next bench. My limited experience made it seem as though his rifle was a thunder-maker that caused my fillings to vibrate under the tin firing line roof. In my youth I had not yet heard of ear protectors and if I had they would have been beyond my limited purchasing budget. My fingers were glued into my ears until he finished his group and indicated a desire to check his target.

On the way back from the targets he introduced himself and showed me his target. His 375 H&H had placed all ten shots into a ragged hole, perhaps an inch in diameter. His group made my five shot 2" group look less impressive than it had a few moments earlier. I wanted to know what brand of ammo he was using so I could make my groups as small as his. He explained he was using reloads and that more specifically he was using cast lead alloy bullets.

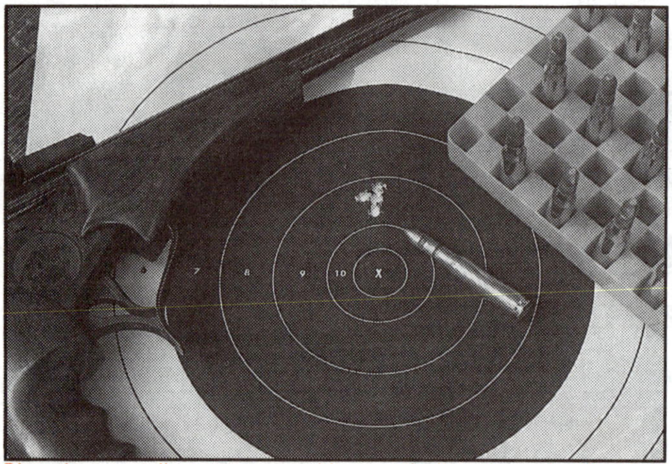

Shooting excellent groups with cast bullets is not difficult if shooters are willing to follow the basic steps outlined in this article.

In mock sincerity my new friend disclosed his "secret" for small groups. It was the bullet lubricant he preferred - two parts beeswax and one part chicken fat! Our talk concluded with him giving me a 30 caliber bullet mould so that I could also give cast bullets a try—if and when I got over the chicken fat part.

Since then I have fired thousands of rounds in practice and various competition disciplines, both in handgun and centerfire rifle, including benchrest. The placing of several bullet traps in the basement allowed for extensive shooting, using light loads in my 38 Special and 45 Auto. I found that cast bullets for handgun shooting quickly became essential if I were to avoid bankruptcy. Those traps and my reloading really paid for themselves over the years as I was able to reclaim the fired bullets, then remelt and recast them with very little alloy lost.

Over the years of shooting cast bullets in handguns I gained a modicum of cast bullet knowledge, though I never imagined how much more was required to shoot cast rifle bullets into small groups. When I got involved with club members who were shooting silhouette targets at 50, 100 and

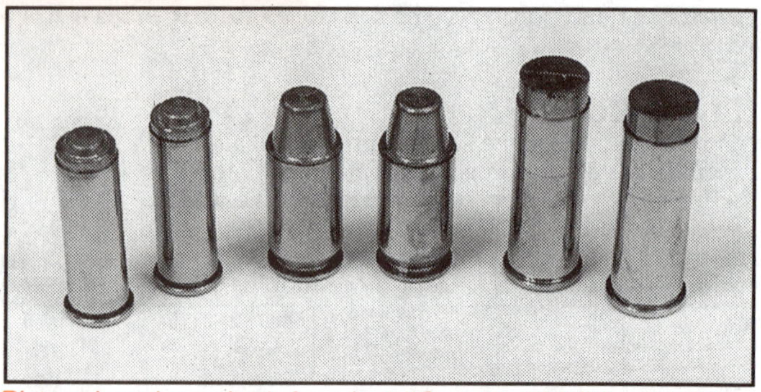

Pictured are the author's favorite 38 Special, 45 ACP and 44 Mag indoor practice rounds, all loaded with bullets he cast himself.

200 meters (all of the shooting was with cast lead rifle bullets), my cast bullet education really began.

I remembered and retrieved the gift 30 caliber bullet mould from among my unused treasures. It was not long before I discovered that it cast the well recognized Lyman #311291 bullet. This was supposed to be an accurate bullet originally designed for the 30-30 Winchester but which also had gained a reputation for accuracy in the 30-06 Springfield.

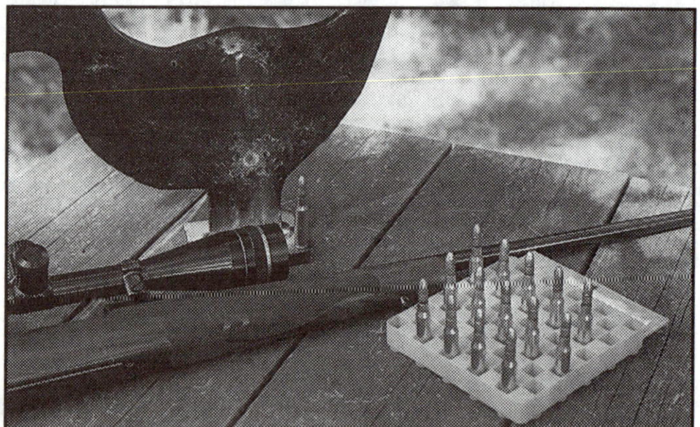

Steel silhouette shooting with cast bullets is the origin of Lyman's Super Silhouette line of bullet moulds.

I cast some bullets, sized them with a 0.308" H&I die, and loaded a supply of 30-06 ammo at about 2200 feet per second velocity. My sighting-in session proved to be a great disappointment. My 16X scope could not make up for any of the inaccuracy I suffered. Perhaps there was something to the chicken fat approach?

Seven inch groups are not what win matches. I scored a fat 5 in my first match with that ammo. Therefore, it was back to the drawing board and I began what turned into a four year effort to obtain top notch results with cast bullets. The fact that I was employed by Lyman made it a bit easier to test various approaches to cast bullet accuracy.

Some handloaders who were questioned suggested that 2" hundred yard groups were an accepted norm for "good" cast bullet accuracy. I felt if my target rifle would shoot 1" groups with jacketed bullets there was no need to settle for less with cast bullets. It was a trial and error effort to bring my idea to reality.

The first thing I worked on was lubricating the bullet, perhaps because it was one of the simplest areas to test. It became obvious that Guy Loverin's multi-grooved bullets worked well partially because the design allowed enough lubricant for higher velocities. Often the point that stands between success and failure of cast bullet accuracy at higher velocities is the leading that occurs in the last 8" or so of bullet travel. A modern bullet design obviously would require plenty of lubricant capacity.

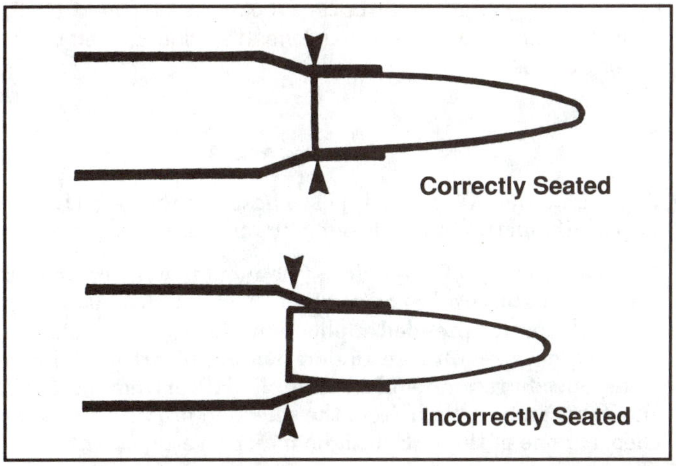

Shown above are examples of correctly and incorrectly seated cast rifle bullets. The bullet base should not extend below the juncture of the neck and shoulder.

Next it was learned, if not too soon, that top cast bullet accuracy occurs most often when the bullet is not seated below the neck of the case. Modern design would require that the length of the case neck be carefully considered when drawing the blueprint for the mould cutting cherry.

Additionally, it became painfully obvious that the three most critical points were fit, fit and fit of the bullet to the chamber and barrel of the rifle. When the bullet's nose was of a diameter that very closely matched the land diameter, accuracy was better than if it were smaller. The same applied to the bullets driving band diameter. If it closely matched the throat diameter (ball seat) things went better than when the diameter differed. This is not unlike the recommendation of many cast bullet experts.

However, it was also discovered that a third area was critical. This is the tapered area between the ball seat diameter and the lands of the rifling. The taper will vary depending on the caliber, but the cast bullet that matches this taper between the front driving band and the nose of the bullet will shoot the best.

Eventually, I had Lyman cut cherries that took the best of known designs and insured that they possessed the five major attributes of: **1. enough lubrication grooves; 2. proper nose bearing diameter; 3. proper length of the body of the bullet. 4. proper major bullet diameter (ball seat diameter) and 5. proper taper from ball seat diameter to nose riding diameter.**

1. Lubrication

To insure that a bullet would not run out of lubricant, each of the new bullet styles had shallow lubricating grooves on the bullet nose. This allowed for an excess of lubricant which is certain to prevent leading at most practical velocities.

2. Nose bearing diameter

Critical cartridge alignment is insured by making the cast bullet fit the known nominal rifling land dimension. Bullet nose diameter is relatively easy to establish as the tolerances of most land diameters are known. Ideally the nose diameter of the bullet should duplicate the land diameter of the rifle. A tolerance of 0.001" seems perfectly satisfactory and allows for the variations normally encountered in firearms. For example, in 30 caliber, bore diameters range from 0.300 to 0.302 inch. Thus, a cast bullet nose diameter of 0.301" is chosen. Bullets of this type are often called bore riders.

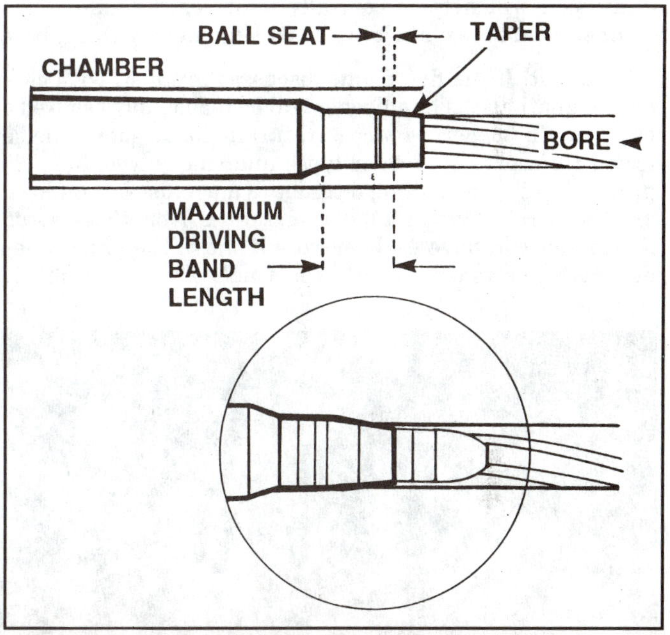

Shown above are the various parts that make up a rifle chamber. A correctly dimensioned cast bullet should match these parts precisely for proper bullet alignment.

3. Driving band length

The lengths of cartridge case necks are easily determined. In the case of 30 caliber, the bullet is designed to fit the shorter 308 Winchester rather than the longer neck of the 30-06 Springfield. For the 7mm diameter the bullet is made to fit the 7mm BR rather than the 7x57mm. When the new designs are to be used in longer neck cases they are simply seated to a lesser depth to insure best results.

4. Ball seat diameter

One approach is to use a bullet that is made to the exact groove diameter. This works well with jacketed bullets but only sometimes with cast bullets. The problem is the dimension of the throat is almost always larger than the groove diameter. At this ball seat diameter, the hot gases can get past a cast bullet base and melt some portion of it. This not only distorts the bullet but can also lead to unnecessary leading. Thus, the bullets major diameter, in my opinion, should be sized to the ball seat diameter rather than groove diameter. As an example, nominally this would be .309-.310" in most 30 caliber rifles as opposed to the sizing diameter of 0.308 inch.

As stated elsewhere in this manual, actual cast bullet diameters might vary from the nominal specification by perhaps 0.0015 inch. This makes it important that the cast bullet shooter measure cast bullets, and if necessary adjust bullet alloy to produce the desired diameters. The bullet sizing die is used to adjust slight major diameter concerns. But the nose diameter must be controlled with the mould and bullet alloy. Keep in mind that a "too large" nose diameter might result in the bullet being pushed back into the case when a cartridge is chambered. This can result in barrel leading, poor accuracy and/or high pressures.

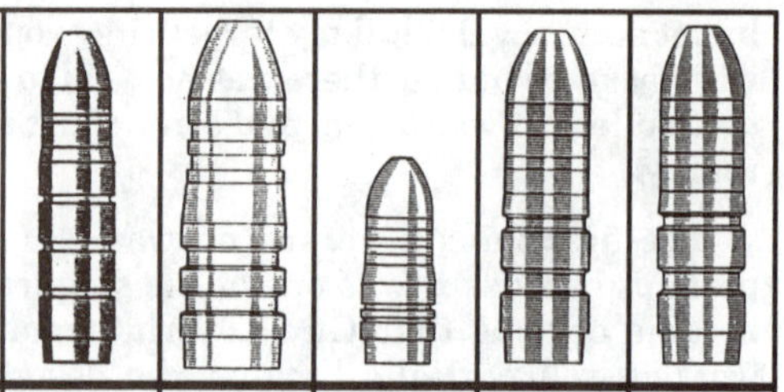

Bullet Number	268645	311644	225646	280642	287641
Weight (#2 Alloy)	150 gr.	190 gr.	55 gr.	150 gr.	160 gr.
Mould #	2660645	2660644	2660646	2660642	2660641
Top Punch Number	463	359	415	359	359
Top Punch Part #	2786741	2786772	2786726	2786772	2786772

Lyman's Super Silhouette bullet moulds as designed by Ed Schmitt, affectionately known as Lyman's "High Exalted Grand Bullet Goober."

5. Taper from major to nose diameter

This diameter is met by placing a taper on the bullet. This dimension will help to centrally locate each round in the chamber and thus place it in correct alignment with the bore.

That's it! If the five points discussed are met, with close approximation, the handloader will be reasonably assured of early top flight performance. Naturally loads should be selected from proven powder types and charge weights. The proper casting and culling techniques must also be employed and the use of a Lyman "M" die is essential. Nonetheless even a beginner can have early success if one of the proven new designs is selected. See nearby chart of the 22, 26 (6.5mm), 27, 28 (7mm), and 30 caliber types. These are the LSS (Lyman Super Silhouette) bullet designs, in my opinion, all winners.

Naturally, many of the older style moulds will continue to provide outstanding results when their dimensions are appropriate for the intended application. However, the selection of these does require an understanding of firearm dimensions. Specific recommendations in our data or from successful shooters can be a big help in the selection process. However, choosing one of the LSS designs may prove expedient.

Ed Schmitt, the author of this article has been a dedicated shooter and reloader for over 35 years. He is employed by Lyman and is actively involved in the sales/marketing and product management part of the business. He resides with his family in Guilford, CT.

5

GAME, TARGETS AND PERFORMANCE WITH RELOADS

CHAPTER 16
THE ULTIMATE TEST - AFRICAN GAME WITH RELOADS

By Edward A. Matunas

Using reloads with confidence is something that generally occurs with the use of many thousands of rounds of reloaded ammo over a period of many years. The hunting reloader has the most to gain from the use of his or her ammo with respect to the investment of total time and dollars. A trip to anywhere can be expensive and there are only so many weeks available each year for hunting. The reader may apply all the following comments and beliefs to a hunting trip from across the states to Alaska or even Africa.

Ammo For Africa

Many new reloaders use their ammo for practice sessions and less important hunting trips. When it comes to a major big game hunt they revert to the use of factory ammo. Perhaps this occurs due to the fact that a reloader may not have been loading long enough to have a strong feeling of confidence in loading highly accurate ammunition. The use of factory ammo may also be due to the recognized fact that the reloader has not yet learned enough about reloading and is occasionally experiencing difficulty with his or her reloads.

African game such as the tough cape buffalo requires the highest level of performance by reloads.

There is also the matter of choosing the right bullets. Much has been said about bullet selection. However, until the reloader gains experience, the ins and outs of bullet selection may be confusing.

I had been reloading for more than 30 years when I undertook my first safari. I would never have considered the use of factory ammo on that hunt. I wanted the very best that could be had in accuracy and performance - I wanted only my own ammo.

In truth, I did finally decide to take along a few boxes of factory premium ammo in 338 Winchester Magnum (as loaded with Nosler Partition bullets) on my first safari. This was because I had heard so much about the eternal African heat causing pressure problems with reloads. I was sure that my reloads would not give me a problem as they were modest in pressure and duplicated factory velocities. So it was - I gave the factory loads away before returning home. My dangerous game loads for that first safari were all handloads. Based on all the comments I had heard from heavily experienced African hunters, I could not get myself to use available factory bullets in heavy calibers.

The reloader heading for Africa should keep the 100 degree plus temperatures in mind. I have experienced days on end when the African daytime temperature never dropped below 110 degrees, peaking near 125 degrees. (I do remember one rare rainy 70 degree day that sent natives and professional hunters in search of sweaters and jackets.)

Ammo that is "hot" at home will give problems when subjected to tropical heat. Loads tested at home should show zero signs of high pressure if they are to go to the tropics. They should be tested on the hottest summer days after the ammo has been lying in the sun for an hour or more. Generally, keep pressures modest by using slow burning powders. Don't look for that last 50 feet per second - rather favor adequate velocity combined with pressure somewhat below maximum, and top accuracy.

Ammo Use Opportunities

I suspect the testing of ammo or bullets is something most do not want to do on a big game hunt - whether in the states or on safari. There simply is too much at stake to discover that a problem exists or the wrong bullets have been selected. Still, I had realized prior to my first safari, it takes a lot of bullets fired into game under a lot of varying conditions, to make an absolute determination on the worth of ammunition.

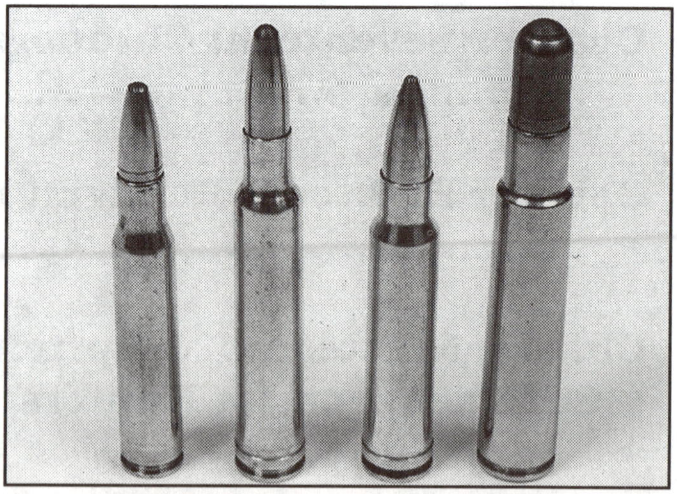

The four reloads shown accounted for several dozen big game animals, and varmints, during a single safari. They are: 30-06 Springfield, 180 grain/IMR 4350; 300 Weatherby Magnum, 200 grain/IMR 7828; 338 Winchester Magnum, 210 grain/IMR 4831; and 416 Rigby, 400 grain/IMR 4831.

I felt then, and still feel, that the ultimate test of reloads is the African game fields. Hunting and shooting day after day, for weeks on end, tends to bring out the worst, and the best, in assembled ammo.

Testing ammo on game is not something most folks get to do very often. Indeed, many big game hunters get to test but a single round each year. In fact, many never really get in enough shooting to say that their ammo has been fully field tested. I have been extremely fortunate as I have had the opportunity to test as many as fifty rounds of ammo on big game during some seasons. No, it's not generally possible to do that much hunting in the states in a given year. But combine a number of stateside hunts with an African safari and the number of game animals that can be taken in a year becomes enough to gain important insight into ammo performance.

I have also had the opportunity to harvest excess big game in large numbers when various entities saw the need to reduce the population of a specific species in given areas of Africa. Such shooting does not allow much in the way of time for recovering bullets or tracing wound channels. Nevertheless, it does allow the shooter to take notice of game reaction to various bullets and bullet placement. It also affords an excellent test of overall ammunition functioning.

With respect to bullet performance, more important than the number of animals that can be taken on a specific safari is the wide range of game size available. It is possible, during a two or three week safari to take "big" game weighing no more than 30 or 35 pounds. Included among this miniature sized game are such animals as klipspringer, duiker, grysbok and many other species. On the same hunt you might well shoot game the size of kudu and zebra, weighing from 500 to 600 pounds, using the same rifle and bullet. For the larger caliber rifle there will be dangerous game, the kind that enjoy being eager and belligerent antagonists. There are also some of the not-so-mean kinds that can weigh up to a ton (such as eland) or more (such as a giraffe).

Ammo testing in Africa is restrictive from the sense that usually only two cartridges and a total of three or four bullets can be used by a hunter during a given safari. The logistics of baggage weight restrictions during airline travel, as well as available space on a bush plane and in the safari vehicle, really limit the practical hunter to two rifles. And there is the international flight weight restriction of 5 kilos for ammunition that each traveler can check in with baggage. That's only eleven pounds - not enough to bring ammo for three or more rifles. Finally, there is another practical matter; the hunter must be able to take game as it comes - its not practical to attempt frequent rifle changes while various size game appear and disappear.

I have been fortunate enough to have one of my sons accompany me on several safaris. This has allowed a mutual agreement of no duplication of loads between hunters, thus doubling the number of loads and bullets that can be tested on a given trip.

Potent And Reliable Ammo Is Essential

African game animals are tough. They have survived a millennium of unrelenting hardships. They can survive years of seemingly endless drought and heat that would spell the end for less hearty creatures. They live with merciless predators on a daily basis. Thorns are present on almost everything that grows to tear at hides and toughen up a species after centuries of exposure. The outcome of such conditions is that most African game has a thick tough hide, heavy muscle, massive bone structure and a will to survive that at times defies description. Ammo for such game must deliver adequate punch. Africa is no place for puny cartridges or weak bullets.

Safari ammo must also be reliable as a hunter pays for all game wounded, whether recovered or not. Let one drop of blood be spilled and the trophy is yours to pay for. It won't do to let ammo that is less than reliable cost you a chance at a record book size animal, and such opportunities are not at all uncommon.

Samples of expanded bullets taken from African game which worked as expected.

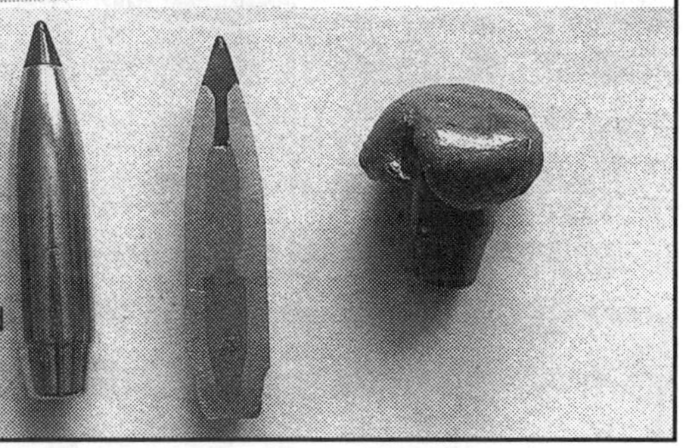

338 cal. Cascade Custom bullets shown sectioned and expanded.

Typical of an African hunt, and ammo test, is a recent safari my son, Eric, and I enjoyed in southeast Zimbabwe, on the Hunters Safaris Humani concession. This hunt was arranged by what I feel is one of the more qualified agencies in the states - Len Pivar's Safaris Africa (614-848-8449). Len is a dedicated and extensively experienced African hunter. He has taken, literally, more than a thousand African game animals. Many of these have been with reloads. Len's extensive experience allows him to supply a first class, no nonsense hunt. He also has proven, time after time, that he really understands what I want from a hunt, and he supplies it at affordable prices.

This hunt, on the famed Humani Ranch was under the direction of William (Willie) Finaughty (grandson of the author of the same name who wrote **The Recollections of an Elephant Hunter**). Hunting an area made famous by several notable African hunters of the last century, and having the grandson of one of these as our host, gave the hunt strong ties to the past. The hunt also proved to be memorable in that more than one dozen of the animals taken were of true record book stature.

In **My Life on the Ranch**, the general area of our hunt was then described as uninhabitable by Europeans because of intolerable heat, lack of water, and being infested with leopards. Contrary to the authors (father and daughter Sommerville's) early description, we found that Willie, and his charming wife Jen, were quite capable of making the region comfortable for a few stateside hunters. However, true to the earlier account of the area's game, we found the Humani concession thick with leopard (and a wide range of other game).

William Finaughty and the author with one of the many trophies Ed has taken through the years. This one was taken with a .338 Win Mag using a 250 grain A-Frame soft point.

Big tusks and plenty of meat to satisfy a hungry hunting party. This warthog was taken with a 30-06 with 200 grain Nosler Partition.

A Wide Variety Of Game With 30 And 33 Caliber Reloads

Ammo testing began in earnest after sight checks of all four rifles. Then no time was lost in collecting five impala for use as leopard baits. During the course of so doing we had noted a number of impala that appeared to be of good size with horns exceeding 24 inches. The need for promptly staking out leopard baits did not allow us time to carefully evaluate these animals.

The next morning's first order of business was to check baits. Two of the five had attracted hungry leopard, but the spoor at the bait suggested that neither were the really large animal we wanted. Two more baits were then dispatched and placed in new locations. We then set out to see if we could find those few impala that were better than average in size. Our efforts were interrupted by a good warthog which quickly disappeared into the high growth of a pan. After an hour of hunting we located the hog. His tusks were more than long enough to satisfy Eric and a difficult running shot was made to bring 250 pounds of warthog to a sudden stop.

The seven impala baits shot were taken using a 30-06 loaded with 180 grain Nosler Partition bullets and a charge of IMR 4350 sufficient to produce an actual 2800 fps. from a long 26" barrel. All the bullets passed completely through the animals, each of which weighed about 120 pounds. There was evidence of good bullet expansion for all shots, as anticipated performance occurred with every shot. The warthog was taken with a 200 grain 300 Weatherby Magnum load.

The remainder of the afternoon was spent looking for really big impala, without results. We had glassed more than a hundred males when it came time to once more check baits. We never made it to the baits as a fine bull kudu with horns easily exceeding 55" got our attention. When we finally caught up with him we could no longer agree that his horns exceeded 55" and we called it a day without firing another shot.

The next morning we sent two trackers to check the leopard baits while we went hunting for jackels. At noon we had accounted for about a dozen of these. These were deliberately taken with a 300 Weatherby Magnum using 200 grain bullets. We wanted to see the effects of this load on light game. Suffice it to say that all of these dropped immediately when the bullet struck. The 3000 feet per second velocity and the high ballistic coefficient of the bullet being used made long range hits on the relatively small quarry rather easy.

The trackers failed to show at noon and it was decided to take up a still hunt in an area loaded with a wide variation of game. A 4-1/2" duiker, another fine warthog, and a 25" impala were all cleanly killed by Eric and me with single shots, using a 338 Win Mag. loaded with 210 grain Partition bullets driven at 2800 feet per second. The duiker and impala are small animals for the cartridge used, but the selected bullet did a first class job on each.

The warthog was shot as it ran angling away. The 33 caliber Partition bullet did everything that could be asked and it was recovered under the skin of the neck. It weighed 140 grains - about 66 % of its original mass. That is about ideal for this style bullet when its energy has been fully expanded in the game. The bullet, with fully expanded nose section, measured 70 caliber. Most of the front core was gone and the rear portion was intact.

The trackers caught up with us at the evening meal to let us know that while several baits had been hit, none were eaten by a big cat. We planned to move our bait still further from our camp.

Three more baits were taken in the early morning light, and staked out in an area some local natives felt was frequented by a big leopard. The 300 Weatherby was used with the earlier mentioned load. Each bait was deliberately shot from north to south end, or in reverse. All three bullets exited and impacted an undetermined portion of the landscape. All three animals fell as though a ton of steel had fallen on them.

We were done with the baiting effort by mid-morning and began stalking, and sometimes chasing, kudu bulls. We hunted seventeen separate bulls before we located a good one.

It was just before dark when Eric placed the tape on its even 57" horns. A great finish to a day that saw four more animals fall to reloads from the Remington 700 chambered for Weatherby's classic 30 caliber magnum.

Another kudu and a fine zebra were added to the bag on the fifth afternoon to keep everything continuing on a high note, though none of the leopard baits were disturbed. The zebra, hit too far forward on the first shot, required a second shot when it was going straight away. On the second shot, the load being used (338 Winchester, 210 grain Nosler Partition) failed to penetrate far enough forward. A third shot was required to end it all. For the second shot, a 250 grain Partition bullet would have done the job in a better fashion.

A zebra will weight 500 to 600 pounds and is tough enough to test the mettle of most light rifles—such as the various 30 caliber magnums and on up to cartridges such as the 340 Weatherby Magnum. This stallion was taken with a 338 Win. Mag. with 210 grain bullets.

I have proven the superiority of the heavier bullet on larger game many times, but the versatility of the lighter one keeps bringing me back to it.

On the sixth morning we found two of the three recent baits had been half eaten. The spoor at one of these suggested a big leopard. We built a blind about 85 yards from the half eaten impala and planned to return in mid-afternoon. As it was past noon by the time we finished constructing the grass blind we decided simply to rest at a shaded area a mile and a half away. While so doing we were treated to the sight of perhaps a hundred or more impala along with a score of zebra coming and going to a bit of water in the middle of a pan. We also watched a young eland bull, perhaps 1000 to 1200 pounds, drink from the water hole. Then it was back to the blind and half eaten bait.

I am happy to say, the leopard came. However, the big cat got wise to us when circling the bait and simply melted away.

On the seventh day we decided to take a break and spend the day photographing game in another area. We "shot" nearly two hundred times during the day.

Day eight was time to again get serious about leopard hunting. Three baits were put in a dry river bed, about 3/4 of a mile apart, near spoor which was impressive enough to get our serious attention. The afternoon was spent hunting bushbuck. Two were taken, one with horns of 16-1/4 inches and the second with horns in excess of 17 inches.

The next morning, we found that one of the baits had been hit and nearly devoured. Spoor around the bait suggested that both a big leopard and a civet had dined, though it is safe to say they did so separately. We followed the bigger cat's tracks and decided that it was more likely to find the last bait in the line before it returned to this almost devoured kill. We gambled and built a blind near that bait. We were wrong. The cat returned to the original bait that afternoon.

We decided, once more, to stay on the uneaten third bait for the next afternoon vigil. To fill the day we goofed off on another photo session. As we settled into the blind the sun was still high. When we heard the cat coming, it was quickly becoming dark enough to make me question my ability to make a clean shot. The cat surprised us by coming along the top of the river bank. Outlined against the skyline the cat was an easy target, and was killed well before its attention was turned to chow. The 338 again performed as expected; complete penetration on a broadside shot with the wound channel indicating plenty of bullet expansion.

416 Proves Versatile With Reloads

With the major part of the hunt over we began an earnest effort to collect the remaining desired trophies. The eland bull, high on my list, proved to be very elusive. It was nearly three days and many miles of spooring before we caught up to the bull we wanted. However, my Ruger 416 Rigby, with its 400 grain Speer African Grand Slam bullets (loaded to 2425 fps.), was used along the way to take grysbok and klipspringer.

Eland have a reputation of soaking up a lot of hits from cartridges such as a 375 H&H, and calibers even bigger. I suspect that at least part of that reputation stems from using bullets that simply are not up to the task. The .416" Speer African Grand Slam, however, proved more than adequate. Hit solidly in the chest the eland buckled, ran and was down in about 10 seconds.

Large Kudu bulls weigh about 650 pounds and are tough to take. Only the best of ammo asembled with premium grade bullets should be used.

A young eland bull is quite massive—this one weighs somewhere between 900 and 1100 pounds. A fully mature bull will weigh about twice this.

CHAPTER 17
CHRONOGRAPHS AND BULLET VELOCITY

By Kenneth L. Oehler, Ph.D.

If The Bullet Hits The Target That's Good Enough!

If you use a chronograph, you have heard this comment before. The remark can be annoying, especially when you have spent your cash for a chronograph and have invested valuable time to measure velocities. To make matters worse, the statement may be incomplete, but it is not wrong. If the bullet misses the target, its velocity is irrelevant. A miss with a high velocity magnum puts no more meat on the table than a miss with light target load. If the bullet does not hit the target, the velocity is wasted, the bullet is wasted, the powder is wasted, and time used to assemble the round is wasted. Having the bullet hit the target is the most important consideration in shooting. If you have chosen a cartridge and bullet appropriate to your type of shooting and have chosen a "middle of the road" load, a proper hit on the target is all you have to worry about.

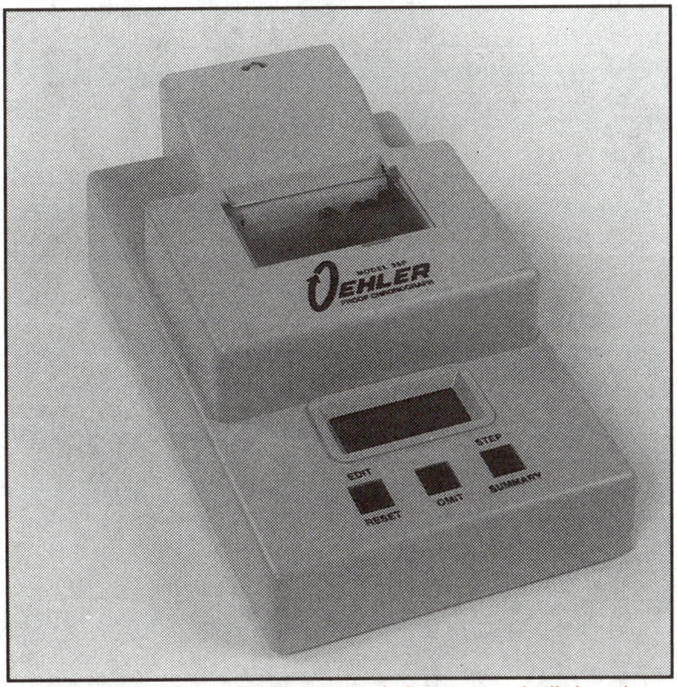

The Oehler Model 35P chronograph features a built-in printer and statistical summary function.

Now that we have agreed that only hits count, let us take a new look at velocity. Assume you have loaded up ten rounds of a new load gleaned from the loading tables (after carefully working it up from the suggested starting load). If you did not hope it to be better than your old load you would not have bothered with the work up procedure. How do you tell if the new load is better? You, of course, fire it at a target. If the new load makes a group that appears to be no larger than you are accustomed to seeing from your old load, you are encouraged. If the group looks small, you will load more ammo and measure several groups. If the groups average about the same as the original load, what else can you measure to distinguish the new load from the older one? They all go bang, they all recoil, and they all reached the target. It's hard to measure bang and recoil but we can measure velocity.

Assuming you have measured the velocities, what do you do with the numbers? The first thing most shooters want to know is the average velocity for a given load. In my opinion, doing a lot of shooting just to measure average velocity is a waste of time. It does not matter how often you measure velocities, you won't turn a 30-30 Winchester into a 308 Winchester, or a 30-06 Springfield into a 300 Magnum. There simply are no magic loads that give noticeably higher velocity without violating pressure restrictions.

How Much Velocity Makes a Difference?

Few shooters (or game animals) can tell the difference in even a 5% change in average velocity. For example, consider that you have a 308 Winchester and a 30-06 Springfield loaded with the same 150-grain bullet at typical respective velocities of 2800 and 2900 feet per second. The velocity of the 308 Winchester is a full 5% lower than the 30-06 Springfield. Zero both rifles at 200-yards and then fire groups at 100-yards to see the effect of the velocity difference. The 308 will be about 2.00" high and the 30-06 about 1.70" high. Even if both rifles are capable of minute of angle groups, it would be nearly impossible to reliably determine the difference of 0.30" in group centers between the two cartridges. The 30-06 still has a 10% advantage in energy - but you knew that before you started, based on existing information.

Consider again a 150 fps. velocity difference, but from a slightly different viewpoint. Imagine having five rounds of 30-06 ammunition with a velocity of 2950 fps. and five having a velocity of 2800 feet per second. Even if both loads group inside an inch, I'll bet that a combined group of both loads won't come close to an inch group. You will see a much larger group difference than just a 0.30" difference in drop. The larger difference is caused by minute changes in exactly where the muzzle was pointed at the instant the bullet left the barrel. You can not prevent the muzzle from whipping as the bullet goes through the bore, but you can try to make the muzzle follow the same whip for each shot. If you maintain the same velocity for each shot, you will have come a long way toward having the muzzle behave the same on each shot.

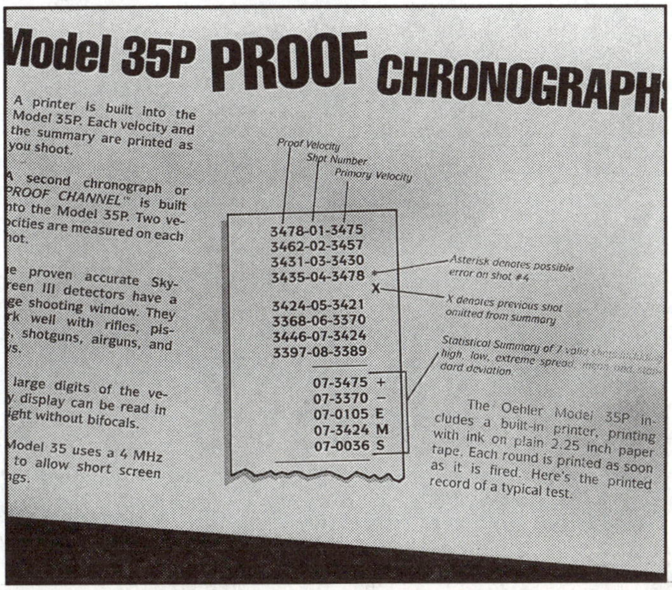

The chronograph printout feature gives the shooter a permanent record of his shot string.

Most shooters align the skyscreens with their target in order to record group size and velocities at the same time.

A chronograph measures velocity by timing the interval that a bullet takes to pass between skyscreens. Oehler's 3 screen system records 2 readings per shot, which enables the shooter to detect a faulty reading.

The chronograph can provide the shooter with such information as average velocity, high and low velocity in a shot string, velocity variation and in some cases standard deviation.

and the point at which the projectile is passing over the skyscreens. If you cannot find a cause for your difficulty and the unit's instructions fail to provide a clue, it is time to call the chronograph manufacturer.

If questionable readings are suddenly obtained with handloaded ammo, run a test with your reference ammo. If the chronograph gives unexpected velocities with the reference ammo, it often means that the light has begun to fail during your shooting session. If the chronograph gives the expected readings with the reference ammo, you know that the chronograph is operating properly. Either the reloads are not giving the expected velocity or there is something about the light that is fooling the chronograph. It is up to the shooter to determine what is fooling the machine.

Light Problems

Proper skyscreen operation is determined by three things light, light and light. There must be enough light, the light must be uniform, and the light must be from the right direction. When using a chronograph stay constantly alert to light conditions. If you don't pay attention to the light, chronograph errors will sneak up on you and you will have no clue as to the appropriate cure.

All skyscreens work the same way. They look at the total amount of light traveling from a long narrow strip of sky to an electric eye in the skyscreen. When a dark bullet passes between the electric eye and the strip of sky, the received light is slightly reduced. This light reduction is recognized by the skyscreens as a passing bullet. Various manufacturers use differing electronic techniques to assist the skyscreen in recognizing the slight reduction of light caused by the passing bullet. However, the shooter must make certain the skyscreens can do their job.

First, make sure there is enough light. Skyscreens are made to be used outdoors. They expect to "see" bright sky overhead. If your skyscreens are positioned under a roof overhang, tree branches, or other light barriers, problems may occur. Do not confuse the brightness of the day with the brightness of the sky. A dull day with the sky filled with rain clouds actually has a brighter sky than a crystal clear blue sky day. With solid cloud cover there is no direct light from the sun but there is a great deal of diffused light coming through the clouds. With a bright day there is a great deal of light coming from the sun but only a small amount from a clear blue sky.

Second, the light must be uniform across the strip of sky that the skyscreen views. If a skyscreen sees half bright white clouds and half dark blue sky, problems will likely occur. The light from the bright clouds will be dominant. The screen will then read a bullet passing the cloud but not one passing the darker sky. For this reason, as already stated, full cloud cover is the best possible light source. Recognizing that cloud cover cannot be controlled, most manufacturers have joined Oehler in providing a light diffuser mounted over the skyscreen window. As viewed by the electric eye the translucent diffuser forms an artificial cloud. When sunlight illuminates the top of the diffuser, the whole strip looks uniform and there is no danger of the sun shining directly in and blinding the skyscreen.

Third, as the bullet passes over the electric eye, the direction of the light becomes important. If the light strikes only the top half of the bullet, the bullet appears dark to the eye. The dark bullet contrasts with the brighter diffuser and is easy to recognize. Light can strike the bullet from below when it is reflected by snow or light colored sand. Light can come from the side if the sun is low in the sky as it would be in early morning or late afternoon. In either scenario the bottom of the bullet will be illuminated and may no longer appear dark to the skyscreen. In fact the bullet may fail to contrast at all with the light diffuser and the screen will fail to see it pass. It's like trying to read gray print on gray paper. To improve contrast, light must be kept from reflecting off the bottom half of the bullet. This can be done by placing the screens in the right place or in extreme cases you may want to paint the entire bullet ogive with a black felt-tip pen.

There is another light related occurrence that mostly plagues handgun shooters. You can't see it and therefore it is often overlooked. When a gun is fired, the crack of sound made as the bullet pops from the muzzle produces a shock wave. This shock wave is like a sphere expanding from the muzzle at the speed of sound (typically 1120 fps.). The shock wave behaves just like a big lens traveling through the air at the speed of sound. It is moving so fast that you can't see it, but it bends or changes light just as it would if it were standing still. If bullet velocity is slower than the speed of sound, the shock wave travels through the sky screen ahead of the bullet! When the sun is at the right angle, the passage of the shock wave will cause a change of light at the skyscreen. The electric eye then thinks the light change was caused by a bullet and the first screen falsely triggers. This typically results in an abnormally low velocity reading. The only positive cure for this is to place a blast screen midway between the muzzle and the first skyscreen. This sneaky problem is why you should use high speed 22 LR ammo as reference ammo. If you use standard velocity or match ammo, velocities will be subsonic and you may be caught by muzzle blast problems.

by [illegible]

The Southwest Pistol League was formed in Southern California during the [illegible] 1970s in an effort to establish action shooting as a legitimate competitive sport. From this beginning, The International Practical Shooting Confederation (I.P.S.C.) today has over 37 member nations and its U.S. affiliate, the United States Practical Shooting Association (U.S.P.S.A.) has about [illegible] active members from 300 clubs across the U.S. [illegible]

In addition to the growth of the Practical Shooting sport, there have been several new shooting games created along the same principles, but on a simplified basis. All of the I.P.S.C./U.S.P.S.A. matches are designed to simulate a situation wherein a firearm is used for self defense. There are no set courses of fire; every match is different. How each course of fire is shot is left up to the individual's discretion. The scoring system is based on hits and on how much time it takes to complete the course of fire. In addition, there are different scales of scoring depending upon the caliber (power factor) of the ammunition the shooter is using.

[illegible] The [illegible] system allows most all shooters to compete [illegible] handgun. A shooter who owns a handgun chambered for the 9mm Luger, along with a few extra [illegible], a set of shooting glasses and some ear protection, is ready to compete. The shooter will fire the same course as the Major competitors but will not compete directly against them and their $2000 handguns.

Matches Vary

Several specific shooting matches, which have fixed courses of fire, have evolved from all of this. The World Speed Shooting Championship (Steel Challenge), the NRA Action Shooting Championship (Bianchi Cup), and the Masters Tournament are examples. Each of these offer thousands of dollars in cash and merchandise to the winners. [illegible] competition.

The World Speed Shooting Championship, better known as the Steel Challenge [illegible]

Shooters on the line at the Bianchi Cup starting the rapid fire steel plate shooting. (Photo courtesy of Nyle Leatham)

Major and Minor

Each competitor must declare his ammunition as either Major or Minor power factor. The power factor is calculated by multiplying bullet weight (in grains) by the muzzle velocity (in feet per second) and dividing by 1000. For example a 45 ACP load with a 200 grain bullet with a velocity of 880 fps would be converted to power factor as follows: 200 times 880 divided by 1000 equals a power factor of 176. To make Major power a gun must have a factor of 175 minimum. To qualify for Minor power, the gun must have a [illegible] of at least 125. A 9mm Luger load using a 115 grain bullet at 1150 fps has a power factor of 132 and is considered a Minor load. The reason for the two power factors is to give credit to shooters who use more [illegible]. The scoring system favors the Major load and most of the top shooters use loads in this category.

drawing and firing until all targets are knocked down in as short a time as possible. Their time is their score and the lowest time wins. The pistols are the [illegible] "race" guns [illegible] developed. These are guns generally [illegible] and altered to shoot very light loads in order to give the shooter as much advantage as possible.

The Bianchi Cup has a Man-against-Man [illegible] top [illegible] competitors which is similar to the courses [illegible] the Steel Challenge match. The main difference is that the targets must be knocked down to score.

The [illegible] Event at the Masters Tournament also offers knock-down targets. For both the Bianchi Cup and Masters Tournament, the [illegible] is limited to six shots [illegible]. Therefore, many competitors [illegible] than used in The Steel Challenge.

Practical pistol shooting presents a wide variety of shooting challenges. Arnt Myhre takes aim through a cut away steel drum which limits left to right movement and range of vision. (Photo by Nyle Leatham)

Loads And Equipment

Typically, most Steel Challenge competitors use light bullets traveling at velocities between 900 to 1000 fps, i.e. a 9mm Luger with 115 grain bullets and fast burning powders such as 231 or WSL. Such a load produces minimal recoil and helps with the fast shooting required.

At the Bianchi Cup side events and top 20 Shoot-Off, the same handguns are usually used but with heavier loads. To insure the needed knock-down power the top shooters will increase velocity to perhaps 1150 fps., or shoot a heavier bullet at about 950 fps., i.e. a 9mm with a 147 grain bullet.

At the Steel Challenge, the most popular guns are those of the 1911 style, as made by Colt, Springfield Armory and Caspian Arms. The most popular cartridge appears to be the 38 Super. The top shooters favor optical sights which project an illuminated dot of approximately 10 minutes of angle (moa). That is, the dot would cover 10" at 100 yards. Optical sights have been part of the winners equipment at almost every major event for the past season.

Optical sights have been used for many years at the Bianchi Cup where the course of fire time limits are generous by speed shooting standards. Here, the most popular equipment includes a 38Special revolver and the mentioned optical sight. The power factor requirement for this match is anything at or exceeding 120 - typical of a standard 38 special with a 158 grain bullet. Shooters will decide what load to shoot based on the Moving-Target event. Many will shoot loads at higher velocities to reduce the lead required at the target which is traveling at 10 fps. The shooter will engage the moving target at 10, 15, 20 and 25 yards so your load is very important.

I develop loads not only for accuracy (capable of at least 4 moa), but also which allow an easy-to-hold-on lead point for the Moving Target. Light bullets, perhaps 110 grains, at approximately 1200 fps., will cut down on the amount of lead required. However, I prefer bullets of 140 to 150 grains at approximately 900 fps. because they allow me to hold on the outside edge of the target at 25 yards - a comparatively easy to maintain lead. I generally favor jacketed bullets for competition as they are easiest to obtain good accuracy and there is no smoke to obscure the target. Frequently, in still air, lead bullet loads will smoke badly enough that the target cannot be clearly seen after several shots. I do use lead bullets in my revolver for the Bianchi Cup match. I simply have not yet developed a jacketed bullet load that shoots as well in my

At the STEEL CHALLENGE the knock down steel silhouettes are set at various distances. It may be necessary to shoot them in a prescribed sequence; center—far right—near left, etc. (Photo courtesy of Nyle Leatham)

Shooting positions and shooting portals vary with the event, long and narrow for the standing or prone, limiting up and down or left to right vision and movement. Here Kipi Boykin, 1992 women's champion, fires in the prone position. (Photo by Nyle Leatham)

Frank Glen modified revolver. I use either a 142 grain truncated cone or 148 grain wadcutter bullet with this handgun.

The 45 ACP remains the most popular cartridge in IPSC/USPSA matches, though the top competitors seem to strongly favor the 38 Super Auto. However, there is a lot of controversy about the pressure levels of 38 Super loads that will provide Major power factors. The maximum pressure specified for the 38 Super Auto enables Major power factor with bullets weighing 145 grains or more. Many new shooters are confused by this as few of the top shooters shoot bullets this heavy. It must be kept in mind that the guns being used have very heavy barrels with extra support at the chamber end. **Please note: for your safety, and the preservation of your firearm, stick with the loads listed in this manual.**

Right hand shooters find it difficult shooting around the left side of a barrier. Here, Debbie James of Connecticut takes aim while the observer holds a timer. (Photo by Nyle Leatham)

In practical shooting there has been a tremendous increase in the popularity of the 9mm Luger cartridge combined with high capacity magazine handguns. In Europe, where safety standards are set differently, shooters have been shooting the 9mm Luger loaded to Major power factor for many years. When it was determined that such loads exceed the SAAMI specified maximum pressure levels, the 9mm Luger (Parabellum) was outlawed as a Major round. This U.S. ban has created a lot of controversy as it does not apply in any other country. As a result, fans of the 9mm diameter bullet have turned their attention to the slightly longer 9x21mm+P+ cartridge and the newer 356 Team Smith & Wesson (356 TSW).

Rob Leatham shoots in the prone position through a barrel at steel silhouettes set at various distances. Some are at the edge of his vision when shooting through this limited area. (Photo by Nyle Leatham)

Making Competition Ammo

Competition ammo requires the use of top quality components. I use only new cases for assembling match ammo. There is no reason to short change yourself with components that might result in inferior accuracy, or a gun malfunction. The cost of assembling good ammo is minor compared to the cost of equipment, air fare, hotel, rental car, and so on. Even for practice ammo, I will keep an eye on my components because if the gun does not work during practice, of what value will be the practice? I actually shot more in my native country, Norway, than I do here in my current home in Arizona. Thus, my practice sessions have become so important that I do everything I can to make each session successful. Top quality ammo is essential to this end.

I load my ammo using Dillon reloading machines. Working folks do not often have the time to assemble the necessary 500 rounds or more per week to meet practice requirements. My Dillon progressive loading tools allow for the rapid assembly of the needed ammo - an hour is all it takes to make 500 rounds. For an intense weekend of shooting I need devote only about two hours to reloading.

I make sure that all practice brass has been thoroughly cleaned before loading it. Even though I use carbide sizing dies, I lubricate all my brass with a spray-on lube. This makes loading effortless and I believe it improves uniformity. I prefer tight sizing dies that reduce case diameters to a

The Dillon XL650 progressive press is one of the types favored by shooters that do a lot of pistol cartridge reloading.

CHAPTER 19
THE OPTIMUM GAME WEIGHT FORMULA
A New Consideration For Hunters

By Edward A. Matunas

Hunters continually ponder the ideal (and elusive) caliber for use in an all around rifle. This, of course, is a futile effort. There simply is no cartridge ideally fitted for both a 13,000 pound elephant and a 30 pound duiker. Narrow the search to a single cartridge for North American hunting and there still is no individual selection suitable for Alaskan grizzly and Wyoming antelope. The more experienced hunter, however, may well wonder which cartridges are ideal for specific applications, or perhaps the broadest range.

The discussion of which cartridges are suitable for what game, or even which cartridges are suitable for a wide range of applications, has stemmed from the fact that all of the current and past methods for predetermination of cartridge application are quite subjective. An example of this is the often used ranking of cartridge potential by kinetic energy - those foot pounds (ft. lbs.) of energy listings contained in most ballistics tables. The justification for kinetic energy listings is that they are an absolute value derived at using well accepted laws of physics.

one twice that weight? And what about an animal standing broadside versus one at a difficult angle? Foot pounds begins to look like a weak argument.

Hunters with extensive experience have come to recognize that kinetic energy figures do no more than rank kinetic energy levels and have very little direct correlation to what is needed to kill an animal of a specific size. Still, the need for determining a cartridge's potential, at varying ranges, before the hunt is very real.

Hunters often do not have the opportunity to build adequate sophistication to know with certainty what cartridges and bullets weights are appropriate for what animals under what conditions. However, the need to know what to expect in the field is real enough that many attempts at using other ranking criteria have been made.

Formulas which included momentum figures and knock-out values have been proposed. Some of these systems have had merit, some did not. Not every hunter is content to simply take another more experienced hunters views on cartridge

Knowing what your chosen load is capable of before going into the field is an easy matter if you refer to the loads Optimum Game Weight rating.

Yet, it's apparent that kinetic energy will not accurately forecast the performance of a specific cartridge in the game fields. To illustrate: the 220 Swift has a 100 yard energy of 1415 ft.lbs. At the same range a 30-30 may deliver 1355 ft. lbs. of energy. Is the 220 Swift therefore a better deer cartridge than the 30-30? Not hardly. Or does the 100-yard kinetic energy of a 30-06 125 grain varmint bullet of 2145 ft. lbs. suggest it is an equivalent grizzly load as compared to the same 30-06's 220-grain bullet which has a kinetic energy of 2216 ft. lbs.? Don't bet your life!

Supporting the use of ft. lbs. as a measure of performance on game suggests that one might accept the often quoted advise that states it takes 1000 ft. lbs. to kill a deer or 1500 ft. lbs. to kill a moose or big bear. Is that 1000 ft. lbs. for a 140 pound deer or for a 250 pound deer? Does it take 1500 ft. lbs. to kill a 500 pound moose or is that level of energy suitable for

requirements. Such an approach is far too subjective for many.

It should be pointed out, that no evaluation system can be made to work if the wrong bullet is used for the wrong application. For example: a highly frangible 30 caliber varmint bullet cannot be used on 140 pound deer and deliver consistent results. Usually, such a bullet would penetrate only a few inches before disintegrating, or at best suffer a core/jacket separation. Such occurrence would result in a shallow surface wound that would allow the animal to escape, though it would eventually succumb to a slow and pain filled death. This is not a sporting scenario.

Regardless of cartridge choice, the hunter must always pick a bullet that will allow appropriate penetration to the animals vitals, at all of the angles at which he or she will take

According to the Optimum game weight formula this 45 pound Javelina could be taken with as light a cartridge as 257 Roberts.

Professionals like Len Pivar may know in advance what is required in the way of a special caliber. The rest of us use the Optimum Game Weight formula to be sure of our cartridge selection.

Even if you have never seen a moose you can make a good cartridge selection for the hunting of this animal by referring to the proximate animal weight table and the Optimum Game Weight table.

NOMINAL WEIGHTS (in pounds) OF MALE ADULT GAME
(weight of female*)

Game	Average	Very Large	Extreme
African elephant		13,000	
white rhino	5,000	8,000	
giraffe	2,600		
black rhino	2,100	2,860	
American bison	1,600 (930*)	2,000	3,000
African buffalo	1,000 (700*)	1,600	
polar bear	900	1,100	1,550 (900*)
muskox	775		
moose	600 (400*)	1,000	1,800+
grizzly (& brown) bear	700 (400*)	1,000	1,600+
elk	500	800 (600*)	1,100
zebra	500	600	
kudu	500	600	
waterbuck		600	
wildebeest	500	600	
lion	330	500	
black bear	300	500	650+
hartebeest		450	
caribou	240 (150*)	400 (250*)	700 (350*)
boar	200	350	450+
jaguar	200 (150*)	250	310
mule deer	200	300	400+
goat	190 (125*)	280 (190*)	310
sheep	175 (125*)	275 (175*)	325+
cougar	150 (90*)	225 (140*)	250
wart hog	140	175	
whitetail deer	125	225	350+
feral pig	125	250	
bushbuck	120		140
reedbuck	115		
fallow deer	110		440
axis deer	110		
impalla	100	120	140
antelope	90 (65*)	110	150
grey wolf	75	110	170
coyote	35	40	45
javelina	30	60	
duiker		30	
klipspringer		30	

AVERAGE FACTORY RIFLE CARTRIDGE [illegible]

Cartridge	Bullet Weight in grains	Muzzle Velocity (f.p.s.)	Muzzle	100	200	300	400
17 Remington	25	4040	[illegible]	[illegible]	[illegible]	[illegible]	[illegible]
[illegible] Hornet	45	2690	18	[illegible]	[illegible]	[illegible]	[illegible]
222 Remington	50	3140	[illegible]	[illegible]	[illegible]	[illegible]	[illegible]
222 Rem. Mag.	55	3020	[illegible]	[illegible]	[illegible]	[illegible]	[illegible]
223 Remington	55	3240	[illegible]	[illegible]	[illegible]	[illegible]	[illegible]
[illegible]	[illegible]	[illegible]	[illegible]	[illegible]	[illegible]	[illegible]	[illegible]
22-250 Remington	[illegible]	4000	[illegible]	[illegible]	[illegible]	[illegible]	[illegible]
22-250 Remington	55	3730	[illegible]	[illegible]	[illegible]	[illegible]	[illegible]
220 Swift	[illegible]	[illegible]	[illegible]	[illegible]	[illegible]	[illegible]	[illegible]
[illegible]	[illegible]	[illegible]	[illegible]	[illegible]	[illegible]	[illegible]	[illegible]
243 Winchester	100	[illegible]	[illegible]	[illegible]	[illegible]	[illegible]	[illegible]
[illegible]	[illegible]	[illegible]	[illegible]	[illegible]	[illegible]	[illegible]	[illegible]
[illegible] Remington	100	[illegible]	[illegible]	240	[illegible]	[illegible]	[illegible]
[illegible]	[illegible]	[illegible]	[illegible]	[illegible]	[illegible]	[illegible]	[illegible]
25-20 Win.	86	1460	[illegible]	[illegible]	[illegible]	[illegible]	[illegible]
250 Savage	100	2820	[illegible]	[illegible]	[illegible]	108	72
[illegible]	100	2980	[illegible]	[illegible]	[illegible]	[illegible]	[illegible]
[illegible]	[illegible]	[illegible]	[illegible]	[illegible]	[illegible]	[illegible]	[illegible]
[illegible]	[illegible]	[illegible]	[illegible]	[illegible]	[illegible]	[illegible]	[illegible]
[illegible]	[illegible]	3440	[illegible]	[illegible]	[illegible]	[illegible]	[illegible]
[illegible]	90	[illegible]	[illegible]	[illegible]	[illegible]	[illegible]	[illegible]
[illegible]	100	3220	[illegible]	[illegible]	[illegible]	[illegible]	[illegible]
[illegible]	[illegible]	2990	[illegible]	[illegible]	[illegible]	[illegible]	[illegible]
[illegible]	120	[illegible]	[illegible]	[illegible]	[illegible]	[illegible]	[illegible]
[illegible] Wea. Mag.	120	3530	[illegible]	[illegible]	[illegible]	[illegible]	[illegible]
[illegible]	[illegible]	[illegible]	[illegible]	[illegible]	[illegible]	[illegible]	[illegible]
[illegible] Swed.	[illegible]	[illegible]	[illegible]	[illegible]	[illegible]	[illegible]	[illegible]
[illegible] Swed.	156	2650	[illegible]	[illegible]	[illegible]	[illegible]	[illegible]
[illegible] Mag.	120	3210	[illegible]	[illegible]	[illegible]	[illegible]	[illegible]
[illegible] Mag.	[illegible]	3030	[illegible]	[illegible]	[illegible]	[illegible]	[illegible]
270 Winchester	100	3470	[illegible]	[illegible]	[illegible]	[illegible]	[illegible]
270 Winchester	130	[illegible]	[illegible]	[illegible]	[illegible]	[illegible]	210
270 Winchester	150	2960	[illegible]	[illegible]	[illegible]	[illegible]	[illegible]
270 Winchester	[illegible]	2850	[illegible]	[illegible]	[illegible]	[illegible]	[illegible]
270 Weatherby	[illegible]	[illegible]	970	769	601	[illegible]	[illegible]
[illegible]	150	[illegible]	1153	945	[illegible]	[illegible]	[illegible]
[illegible] Mauser	140	2660	[illegible]	[illegible]	322	[illegible]	[illegible]
[illegible]	175	2490	667	[illegible]	[illegible]	[illegible]	181
[illegible]	120	2700	[illegible]	263	[illegible]	[illegible]	[illegible]
[illegible]	120	3000	[illegible]	146	[illegible]	72	[illegible]
[illegible]	140	2860	[illegible]	[illegible]	[illegible]	[illegible]	[illegible]
[illegible] Remington	120	3150	225	170	[illegible]	[illegible]	68
280 Remington	[illegible]	3000	[illegible]	[illegible]	[illegible]	[illegible]	[illegible]
280 Remington	150	2890	[illegible]	[illegible]	[illegible]	[illegible]	[illegible]
280 Remington	[illegible]	[illegible]	[illegible]	[illegible]	[illegible]	[illegible]	[illegible]
[illegible] Rem. Mag.	[illegible]	3175	[illegible]	[illegible]	[illegible]	[illegible]	[illegible]
[illegible] Rem. Mag.	150	3110	[illegible]	[illegible]	[illegible]	[illegible]	[illegible]
[illegible] Rem. Mag.	165	2950	1048	[illegible]	[illegible]	[illegible]	[illegible]
[illegible] Rem. Mag.	175	2860	[illegible]	[illegible]	[illegible]	[illegible]	[illegible]
[illegible] Mag.	[illegible]	[illegible]	[illegible]	[illegible]	[illegible]	[illegible]	[illegible]
[illegible] Mag.	160	[illegible]	[illegible]	[illegible]	[illegible]	[illegible]	[illegible]
[illegible] Wea. Mag.	175	3070	[illegible]	[illegible]	[illegible]	[illegible]	[illegible]
[illegible]	125	2570	[illegible]	[illegible]	[illegible]	[illegible]	[illegible]
[illegible]	150	2390	[illegible]	[illegible]	[illegible]	[illegible]	[illegible]
[illegible]	170	2200	462	285	184	[illegible]	73
300 Savage	150	2630	614	440	310	215	145

AVERAGE FACTORY RIFLE CARTRIDGE OGW BALLISTICS/cont.

Cartridge	Bullet Weight in grains	Muzzle Velocity (f.p.s.)	Optimum Game Weight in lbs. at yards Muzzle	100	200	300	400
300 Savage	180	2350	630	474	352	258	188
308 Winchester	150	2820	757	549	391	274	188
308 Winchester	165	2700	804	593	431	309	218
308 Winchester	168	2680	815	656	525	417	328
308 Winchester	180	2620	874	666	502	374	275
30-06 Spfd.	125*	3140	242	168	115	76	50
30-06 Spfd.	150	2910	832	601	436	305	211
30-06 Spfd.	165	2800	896	665	485	350	248
30-06 Spfd.	180	2700	957	732	554	414	306
30-06 Spfd.	220	2410	1016	702	475	316	209
300 H&H Magnum	180	2880	1161	894	682	515	383
300 H&H Magnum	220	2550	1204	846	583	393	NA
300 Win. Mag.	150	3290	1202	867	618	434	299
300 Win. Mag.	180	2960	1260	1005	796	626	487
300 Win. Mag.	200	2830	1360	1155	972	809	675
300 Win. Mag.	220	2680	1397	1065	803	598	440
300 Wea. Mag.	150	3600	1575	1221	942	722	549
300 Wea. Mag.	180	3300	1747	1416	1143	918	732
300 Wea. Mag.	220	2905	1780	1132	698	414	240
32 Win. Special	170	2250	494	307	186	112	70
8mm Mauser	170	2360	570	331	185	103	61
8mm Mauser	220	2300	883	610	413	276	NA
8mm Rem. Mag.	185	3080	1500	1080	768	536	367
8mm Rem. Mag.	220	2830	1646	1248	937	695	508
338 Win. Mag.	210	2830	1499	1149	881	657	483
338 Win. Mag.	225	2780	1632	1292	1016	791	610
338 Win. Mag.	250	2660	1765	1389	1084	838	641
340 Wea. Mag.	210	3250	2271	1770	1370	1052	800
340 Wea. Mag.	250	3000	2531	2071	1688	1367	1100
35 Remington	150	2300	411	222	115	61	38
35 Remington	200	2080	540	294	156	89	60
358 Winchester	200	2490	926	614	396	250	157
350 Rem. Mag.	200	2710	1194	840	580	392	260
35 Whelen	200	2675	1148	807	556	374	249
35 Whelen	250	2400	1296	827	512	311	192
375 Winchester	200	2200	639	374	213	122	78
375 Winchester	250	1900	643	419	271	178	126
375 H&H Magnum	250	2750	1950	1500	1125	834	623
375 H&H Magnum	270	2690	2129	1550	1111	784	544
375 H&H Magnum	300	2530	2186	1381	845	504	301
378 Wea. Mag.	270	3180	3510	2882	2352	1909	1540
378 Wea. Mag.	300	2929	3378	2308	1542	1004	640
416 Rem. Mag.	350	2500	2871	2149	1546	1097	768
416 Rem. Mag.	400	2400	3318	2432	1755	1251	883
416 Rigby	400	2400	3318	2432	1755	1251	883
416 Wea. Mag.	400	2700	4723	3529	2598	1886	1351
44 Rem. Mag.	240	1760	471	227	119	79	59
444 Marlin	240	2350	1121	517	226	111	72
444 Marlin	265	2120	1004	548	292	164	109
45-70 Govt.	300	1810	801	453	260	168	123
45-70 Govt.	405	1330	579	392	289	230	190
458 Win. Mag.	465	2220	3549	2591	1863	1331	NA
458 Win. Mag.	500	2040	3184	2272	1603	1124	710
458 Win. Mag.	510	2040	3312	2164	1389	860	604

(*) = These are varmint style bullets and should never be used for big game. Optimum Game Weights shown for these bullets are for varmints—game that can be killed with highly frangible bullets, i.e. crow, prairie dog, woodchuck, coyote.

NA = The velocity of the bullet at this range is not available and hence no OGW can be computed.

A GUIDE TO OGW VALUES WITH SELECTED BULLET WEIGHTS AND VELOCITIES

Bullet Wgt. in grains	Velocity in f.p.s.	Bullet type	Optimum Weight in lbs.
100	3400	Big Game	590
	3200		492
	3000		405
	2800		329
	2600		265
	2400		207
	2200		160
	2000		120
120	3000	Big Game	585
	2800		474
	2600		380
	2400		299
	2200		230
	2000		173
	1800		125
130	3300	Big Game	910
	3100		755
	2900		620
	2700		499
	2500		395
	2300		308
	2100		235
	1900		174
	1700		125
140	3200	Big Game	965
	3000		794
	2800		645
	2600		517
	2400		405
	2200		313
	2000		235
	1800		171
	1600		120
145	3100	Big Game	940
	2900		769
	2700		620
	2500		493
	2300		385
	2100		292
	1900		215
	1700		155
150	3000	Big Game	911
	2800		740
	2600		593
	2400		465
	2200		359
	2000		270
	1800		197
	1600		140
165	2900	Big Game	996
	2700		805
	2500		638
	2300		495
	2100		378
	1900		280
	1700		200
	1500		140
170	3000	Big Game	1170
	2800		952
	2600		760
	2400		599
	2200		460
	2000		347
	1800		255
	1400		120
175	3000	Big Game	1240
	2800		1008
	2600		805
	2400		635
	2200		490
	2000		368
	1800		270
	1600		189
	1400		125
180	3000	Big Game	1310
	2800		1067
	2600		855
	2400		672
	2200		520
	2000		389
	1800		285
200	3000	Big Game	1620
	2800		1317
	2600		1055
	2400		829
	2200		640
	2000		480
	1800		350
	1600		246
210	3000	Big Game	1785
	2800		1452
	2600		1165
	2400		915
	2200		705
	2000		529
	1800		385
	1600		271

6

BALLISTICS

…326. The English records [illegible] …guns about as [illegible] …forms. [illegible] …nches in diameter [illegible] …ng, these large arrows were [illegible] …e. Large most [illegible] …ned at least [illegible] …se arrows and [illegible] …ving huge stones [illegible] …Ages.

…ws were not [illegible] …annon tubes [illegible] …two pounds [illegible] …The first of the [illegible] …c guns with [illegible] …iron balls weighing [illegible] …anges [illegible] …provements on [illegible] …s and were very effective [illegible] …stles.

…firearms [illegible] …guns [illegible] …nd appeared [illegible] …nd cannon [illegible] …meter. They were [illegible]

was able to give a reason why the trajectory is a curve. Galileo was a ballistics consultant to the arsenal at Venice for a number of years. The famous experiment in which he dropped two cannonballs from the Leaning Tower of Pisa was only one of a long series of experiments having to do with the effects of gravity. Galileo determined that the acceleration of a falling body by gravity is a constant. With this result he was able to show that, if the effects of air drag on a bullet are neglected, the trajectory is a type of curve called a parabola.

Galileo's work marks one of the great milestones in the history of ballistics. He published the first analytical descriptions of trajectories and is credited with laying the theoretical foundation for ballistics. However, it turned out that Galileo's parabolas were not very accurate representations of true bullet trajectories because air drag has a very important effect. Galileo was aware that air drag existed, but there was absolutely no way available in his time to measure any of its effects. He thought it was a small effect compared to gravity, and more than a hundred years went by before the effects of air drag even began to be understood.

Actually, Galileo's method was quite useful in his time, particularly for the low velocity, short range artillery of that day. It was necessary to know only the elevation angle of the muzzle and the horizontal range of the shot to calculate a parabolic trajectory for a test shot from any gun. These quantities could be measured with good accuracy in Galileo's time. Once the parabola was calculated to match the bore elevation angle and the range, a muzzle velocity could be derived from the curve. In a vacuum this derived muzzle velocity would have been almost exactly correct, but it was in error by a substantial amount due to air drag. In fact, it was always a good deal less than the actual muzzle velocity of the gun, because air drag always acts to slow a bullet. Nevertheless, the derived muzzle velocity could be used to analytically calculate firing tables (range versus elevation angle) for the gun. These calculated tables were in error by several tens of yards at ranges of a few hundred yards. but this accuracy was sufficient for using artillery against columns of foot soldiers or horse cavalry. For more than a hundred years Galileo's method was used to determine muzzle velocity, until a means was invented to measure velocity directly.

In 1740, in England, Benjamin Robins invented the ballistic pendulum and made it possible to measure velocity directly. The ballistic pendulum was a simple pendulum with a large, heavy wooden bob. To make a velocity measurement, the bullet was weighed and the bob was weighed. Then the pendulum was positioned with the bob hanging motionless and the bullet was fired into it. By measuring the height of the pendulum swing resulting from the bullet striking the bob, the velocity of the bullet could be determined. Robins made a series of measurements with .75 caliber (12 gauge) musket balls, including measurements of velocity near the muzzle and at several ranges from the muzzle. Robins reported muzzle velocities ranging from a little over 1400 to a little under 1700 feet per second. These numbers were astounding, so much so that they were widely disbelieved. When Galileo's method was used for the musket, the derived muzzle velocities were hundreds of feet per second less than the ones measured by Robins. Even more astounding, however, were the measurements of velocity decrease with range from the muzzle. In order to account for Robins' measured numbers, it was necessary to have an air drag force about 85 times stronger than the gravitational force.

Proposition VIII

To determine the Velocity, which any Ball moves with at any Distance from the Piece it is discharged from.

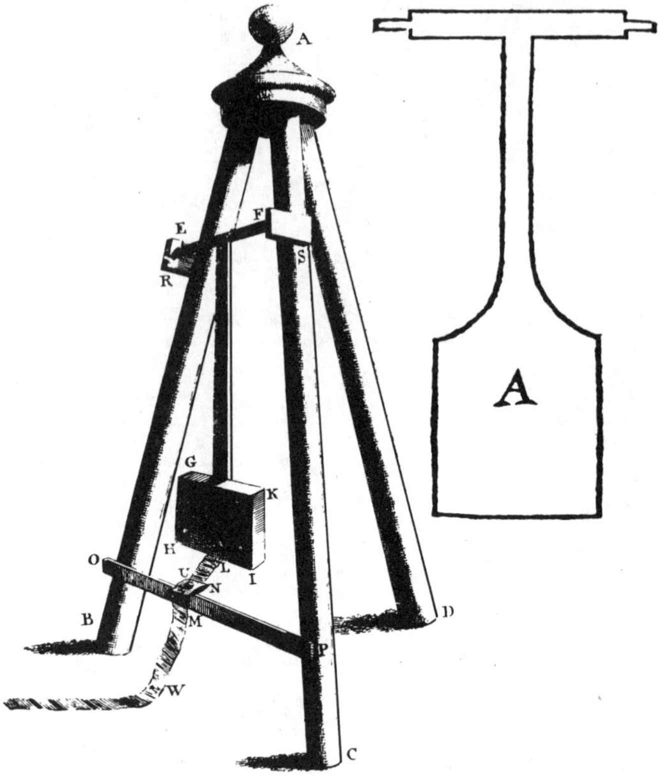

The simplest method of effecting this is by Means of an Instrument like that exhibited in the engraved Figure. **A B C D** *represent the Body of the Machine composed of the three poles* **B, C, D** *spreading at the bottom and joining together at the Top* **A**; *being the same with what is vulgarly used in the weighing and lifting of very heavy Bodies, and is called by Workmen the Triangles. On two of these Poles, towards their Tops, are screwed on the Sockets RS. On these Sockets the Pendulum* **E F G H I K** *is hung by means of its cross Piece* **E F** *which becomes its Axis of Suspension, and on which it must be made to vibrate with great Freedom. The Body of this Pendulum is made of Iron, having a broad Part at Bottom, which cannot be seen in this scheme; but its entire shape is represented in the annexed* **Figure A.**

The lower part of the Pendulum is covered with a thick piece of wood **G K I H**, *which is fastened to the Iron by Screws. Below the Bottom of the Pendulum there is a brace* **O P**, *joining the two Poles from which the Pendulum is suspended. To this Brace there is fastened a Contrivance* **M N U**, *made with two Edges of Steel, bearing on each other in the Line* **U N**, *something in the Manner of a Drawing-pen. The Strength with which these edges press on each other can be diminished or increased at Pleasure by Means of a Screw* **Z** *going through the upper Piece. There is fastened to the Bottom of the Pendulum a narrow Ribbon* **L N**, *which passes between these Steel Edges and which afterwards, by Means of an Opening cut in the Lower piece of Steel, hangs loosely down as at* **W.**

This Instrument thus fitted, if the Weight of the Pendulum be known, and likewise the respective Distances of its Center of Gravity and its Center of Oscillation from its Axis of Suspension, it will thence be known what Motion will be communicated to this Pendulum by the Percussion of a Body of a known Weight moving with a known Degree of Celerity and striking it in a given Point. That is, if the Pendulum be supposed at Rest before the Percussion, it will be known what Vibration it ought to make in Consequence of such a determined Blow. On the contrary, the velocity of the striking Body may be determined if the Pendulum, being at rest, is struck by a Body of a known Weight and the Vibration which the Pendulum makes after the blow is known.

From *New Principles of Gunnery*, by Benjamin Robins, 1742.

Newton's first solution of the ballistical problem, 1684.
From *Ballistics in the 17th Century*.

shapes are slightly dissimilar, to allow accurate ballistic computations for the actual bullet.

It also turned out that the ballistic coefficient **C** led to another advantage of major importance, which had to do with the method of computing bullet trajectories. It was mentioned earlier that calculus must be used to compute a bullet trajectory. The numerical computations necessary to calculate just a single trajectory are very lengthy and tedious when they must be done by hand, which was necessary until only a few years ago. The work involved in computing trajectories for many bullets and many muzzle velocities by hand is prohibitive. About 1880 an Italian ballistician named Siacci discovered a way to greatly simplify this problem. He showed that the ballistics of an actual bullet could be computed from the ballistics of the standard bullet in a simple way with the use of the ballistic coefficient, for the case of nearly horizontal fire. By this method a single trajectory was computed for the standard bullet by the tedious methods of calculus, and then a trajectory for any actual bullet with known C was computed from the standard trajectory with just simple algebra. Siacci's method shortcut a great amount of computation, and it has been used widely in ballistics ever since.

Between about 1865 and 1930, many firing tests were conducted worldwide to determine the drag characteristics of standard bullets adopted by different countries. Of particular note were tests made by Krupp in Germany in 1881 and by the Gavre Commission in France from 1873 to 1898, although many other tests were made. The Gavre Commission work was very comprehensive, including extensive firing tests going up to a velocity of 6000 fps, and a comprehensive survey of data available from tests in other countries. The Commission attempted to correlate all these data, and it published a composite drag characteristic for a certain standard bullet configuration. Unfortunately, the effects of varying atmospheric conditions during the many tests were not well understood at the time, and the Gavre drag function contained errors due to these effects. Just a few years after the Gavre Commission finished its work, standard atmospheric conditions were adopted by ballisticians for the expression of drag data, and these standard conditions are in use today.

The Krupp test data turned out to be the basis for ballistics tables for small arms right up to the present time. The standard bullet used by Krupp was a flat base design, 3 calibers long with a 2-caliber ogive head. **Figure 1** shows the shape of this bullet.

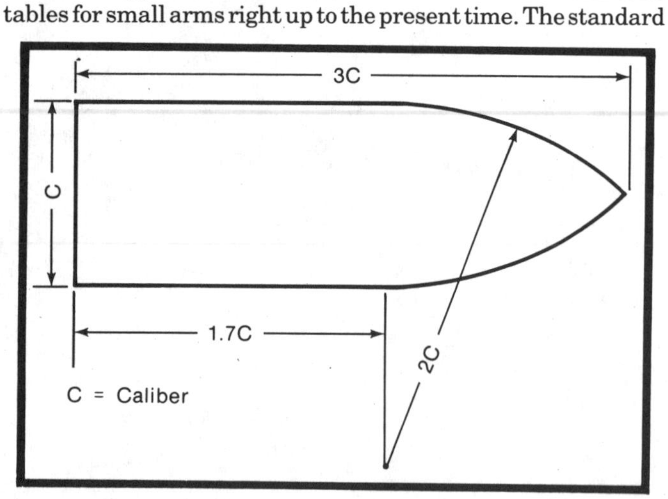

Figure 1

After the Krupp drag data were published, a Russian Army Colonel named Mayevski constructed a mathematical model for the standard drag deceleration for this bullet. Colonel James M. Ingalls, U.S. Army, used Mayevski's analytical model and computed and published his now-famous Ingalls Tables in 1884 (updated in 1917). The Ingalls Tables are the ballistics of a standard bullet of the Krupp shape. It turned out that this standard bullet has been a very fine model for use in computing the ballistics of most bullets for sporting use. This drag model has been used even for round balls, although three different values of **C** must be used in each of three velocity ranges to match the round ball drag to the standard bullet drag.

After World War II the Ballistics Research Laboratories of the Aberdeen Proving Ground undertook another long series of firing tests on military and sporting bullets, and they developed four standard drag models corresponding to four types of bullets. These four types are hollow point lead bullets, boattail bullets, very sharp pointed full patch bullets and all others except those in the above categories. It was found that the drag model for the last category, which includes the vast majority of hunting and target bullets in use today, is almost identical to Mayevski's model developed from the Krupp firing data. These models were published in 1965 by Winchester-Western. They are the most comprehensive known to the authors for sporting bullet types, and these models were adopted for use in the Lyman firing tests and ballistics computations for this manual.

Once the concept of the standard bullet was adopted, the one remaining problem was how to determine the ballistic coefficient **C** for an actual bullet. One method of determining **C** is firing tests. This method will be more fully described in a later chapter. The basic idea is to measure the ballistic performance of the actual bullet, compare it with the ballistic performance the standard bullet would have if fired at the same muzzle velocity and determine the right value of **C** to make them match. Tests are made at several muzzle velocities to see how well the standard drag model fits the actual bullet.

Another method of determining **C** by shape comparison was developed and published by Wallace Coxe and Edgar Beugless of the DuPont Company in the 1930's and has been widely used ever since. The method estimates the ballistic coefficient related to the drag model of the Krupp standard bullet. It consists of comparing the actual bullet point shape to a chart of point shapes for which ballistic characteristics are known. When a match is made as nearly as possible, the chart provides the form factor for the bullet, from which **C** can be easily calculated. This method is judgmental in nature, and therefore not as accurate as the firing test method. However, in the author's experience it gives a value of **C** which is accurate to ten or twenty percent, when it can be used. There are many bullets, particularly for blackpowder firearms, with shapes which are not included on the DuPont charts. Also, it does not provide for the deep lubricant grooves and the bands on cast lead bullets.

The main problems of ballistics are fairly well solved today, in the sense that bullet trajectories can be computed and actual firing test measurements will come very close to the computed values. The problem of drag is still under investigation, although in the last 25 years or so the emphasis has been placed on rocketry. Within this same period, high speed electronic digital computers have developed, and these machines have eliminated the tedium from ballistics computations. The ballistics tables in all the Lyman manuals were calculated on one of these machines. Special care was taken to use highly accurate computation methods to ensure that no appreciable computation errors would be contained in the tables.

The Development of the Bullet

As mentioned earlier, the first "hand-guns" (hand cannons) appeared before the year 1400. Records from the late 1300's show that hand-gunners had all necessary accessories, including bullet moulds for round lead balls, iron casting ladles, iron ramrods and match materials for lighting charges through touchholes in their hand-guns. From these records we can conclude that cast lead bullets have been used for at least six centuries, from the time of the very first guns that could be carried and fired by a single man.

Actually, it is not surprising that cast bullets for the first hand-guns were round balls made of lead. Lead projectiles were used in slings long before the age of firearms. Plentiful and inexpensive, with high density and low melting point, lead was ideal compared to other possible choices (iron, stone, precious metals) in those early times. The spherical projectile shape was also a heritage from antiquity. The sling had been used for thousands of years as a weapon for both hunting and warfare, and it was well known that nearly spherical stones were more accurate than stones with irregular shape. Today we know that the round ball projectile has a relatively poor ballistic shape. Nevertheless, this basic bullet shape is still with us for muzzleloaders, having survived more than six centuries of progress toward better ballistic shapes.

The round ball, of course, is the only practical projectile shape for smoothbores. If a nonspherical bullet is not spin-stabilized by rifling, it tumbles during flight, destroying accuracy and greatly reducing velocity and energy downrange. Round balls were used in rifles for the first two centuries of rifle development, but after that elongated pointed bullets began to prove better for use in rifled barrels. Improvements in accuracy and range were driving factors in early experiments that led to improved bullet shapes, but bullet design changes also occurred to solve some of the key problems in early rifles, like difficulty in loading, black powder fouling and sealing the bore against powder gases escaping past the bullet.

Wads between powder and ball to seal the bore, and over the ball to prevent it from rolling down the bore, were used in the very earliest firearms, and these measures were adequate for smoothbore cannons and hand-guns. When rifles appeared, though, different measures became necessary.

The first rifles came from Vienna around 1500. The idea of a spinning projectile came from the crossbow. Crossbow quarrels (short arrows) were stabilized in flight by feathered vanes which caused them to spin, and spinning quarrels were found to be more accurate than nonspinning ones. Consequently, the rifled bore was a natural idea for guns, and it was tried just about the same time that shoulder stocks were first attached to hand-guns. The rifle, then, is just about as old as the smoothbore gun fired from the shoulder.

It is an interesting sidenote that the first shoulder arms were given the name "arquebus" (sometimes spelled "harquebus"). At least two possible derivations of this name have been given in recent shooting literature. The one by G.W. P. Swenson in his book *Pictorial History of the Rifle* is particularly interesting. According to Mr. Swenson, "arquebus" derives from "arca bouza," meaning "bow with a hole." This explanation relates the first shoulder-fired guns to their ancestral crossbows.

Compared to the smoothbore arquebus, early rifles were very expensive and very hard to load. Rifling the barrel and holding a true bore diameter from end to end were very difficult mechanical operations required for the rifle but not for the smoothbore. In loading a rifle, the ball either matched the groove diameter or was a little larger, and it had to be driven into the barrel. Driving a ball down a fouled bore with an iron ramrod and a mallet was a slow process and the final bullet shape seldom resembled a round ball.

These circumstances slowed the development of the rifle for more than two centuries. The greased patch materialized in the early 1600's, and this eased the loading problem. However, people in those days did not understand how rifling spins a bullet. It was believed for some time that a patched ball fired in a rifle would not spin, because the rifling lands in the barrel did not engrave the ball surface. Consequently, the greased patch which was to become so successful in America in the next century, fell into disuse in Europe. Mainly, though, cost was the factor that kept rifles from being widely adopted for warfare in the seventeenth and eighteenth centuries. Rifles did see limited use in a very few European armies in the 1600's and 1700's, but the rifle mainly was an expensive hunting arm for prosperous gentlemen of those periods. The smoothbore musket was the standard shoulder arm of European armies, and armies here in the Americas as well, until about 1850. Gunmaking machinery developed during the Industrial Revolution of the mid-nineteenth century finally reduced the cost of rifles to the point where smoothbore muskets no longer had an overwhelming edge.

Also, the range and accuracy of the rifle had become important military advantages. The English learned this lesson brutally in the American War of Independence, where the Pennsylvania, New York and New England rifles with patched balls greatly outclassed the Brown Bess musket in more than a few skirmishes. Also by 1850, the hollow-base Minie' bullet for rifles had completely erased any loading speed advantage of the smoothbore musket and this bullet design gave even better accuracy than the patched ball.

Experiments with nonspherical bullets in rifles began in Europe and Russia about 1700. The flat base pointed bullet shape that we are familiar with today seems to have emerged in the early 1800's after a century of experiments. Looking back now, a hundred years seems a long time to accomplish so obvious a result. But little was known about interior or exterior ballistics in the 1700's. Bullets still had to be pounded down rifle bores, with attendant damage to point shapes. (The first breechloaders were invented specifically to work around this

Wilkinson and Greener Systems of Expanding Bullets. Left: The very deep cannelures in this soft lead Wilkinson bullet caused the two rear sections to be crushed and squeezed out into the rifling grooves upon discharge, even though the base of the bullet was entirely flat. Right: The Greener bullet had a tapered hollow in its base into which an expansion plug was driven by the force of the explosion. Both these systems are interesting mainly as variation on the more successful Minie system. *Lyman Archives.*

The Systems of Delvigne and Minie

In 1827, Captain Delvigne [illegible] a method of loading a spherical lead ball, either with or without a greased patch, down a barrel through which it would pass easily, and then expanding it against the lip of a chamber smaller than the bore proper. The bullet would [illegible] into the grooves of the rifling [illegible] of the ramrod. It was difficult, however, to expand it the same amount each time. Further, the bullet itself became wider than it was long and lost velocity quickly. Finally, the bullet tended to be pushed down into the chamber, crushing the powder and becoming T-shaped.

Captain Minie had been experimenting for years with a clean bore and tight-fitting elongated bullets. Sometimes these gave great range and accuracy. They weren't practical, however, because of loading troubles. Minie removed the stem from the breech of some experimental [illegible] rifles and placed an iron cup in the hollow base of undersized cylindro-conoidal bullets. Loading was easy; at the explosion of the propellant charge, the cup was driven into the bullet [illegible] and caused the bullet to expand and fill the grooves. Both accuracy and range were superior to anything yet achieved. Minie bullets were as accurate as the round bullets in any rifle at 100 and 200 yards. Man-sized targets could be hit at 400 yards consistently [illegible] accuracy and power were still apparent [illegible] yards. *Lyman Archives.*

problem. Proper rifling twist rates were being developed by experimental methods. And gunpowder was being improved and refined during this same period. Because [illegible] were changing at the same time, it wasn't until [illegible] that pointed bullets were proven clearly superior to round balls in both range and accuracy.

After the pointed bullet, the next major development in the projectile was the expanding bullet, which finally solved the loading problem for muzzleloading rifles because the early breechloaders just couldn't be sealed against escaping gases. The idea of the expanding bullet was simple. The bullet was cast slightly smaller than bore diameter. Then it could be pushed down even a fouled bore very easily, [illegible] so small that it would slip forward if the bore was tipped down before firing. When the gun was fired, the powder gases would expand the bullet into the rifling, sealing the bore against gases escaping around the bullet. Although the idea seems simple, more than 25 years were spent developing a practical expanding bullet. The first invention by W. Greener in England in 1836 was an elongated ball expanded by a tapered wooden plug in a tapered hole in the ball. Delvigne in France [illegible] an elongated hollow-base bullet in 1841. Minie added a tapered iron cup in the base of Delvigne's bullet in 1849. When the charge was ignited, this cup was [illegible] deeply into the base cavity, causing the sides of the bullet to expand outward. Before long it was discovered that the Minie bullet worked well without the tapered cup, provided that the center of gravity of the bullet design was properly placed.

Although Minie seems to have contributed least in an invention sense, he was successful in getting his expanded

bullet adopted [illegible] United States Army. The [illegible] was [illegible] developed by the time of the American Civil War. [illegible] in Europe and America [illegible] had a much higher velocity [illegible] and a much flatter trajectory [illegible] bullets, especially in .58 and .69 caliber, were used in the Civil War. It was a very important development, but it was to be important only for a very short period of time. When the Civil War ended in 1865, [illegible] center-fire cartridges and breechloading rifles and handguns to use them were about eight years away.

The period between 1825 and 1875 was a time of great [illegible] development in this nation and in Europe. The first metallic cartridges and the first successful breechloading arms using them were developed during this period. In France in 1835, Casimir Lefaucheux patented his pinfire cartridge system. Lefaucheux's invention was a fully self-contained metallic cartridge with bullet, powder, and a percussion cap inside. A pin extended from the side of the base of the cartridge, and when this pin was struck by the [illegible], the cap exploded [illegible] igniting the charge. The Lefaucheux system was popular in Europe and guns, especially shotguns and revolvers, were built [illegible] Lefaucheux [illegible] revolvers [illegible] in the American Civil War.

In America during 1848, W. Hunt patented a caseless cartridge which he called the Rocket Ball. Hunt used a hollow base lead bullet, filled the cavity with black powder and sealed it tightly so that the flash from a percussion cap would burn through the seal and ignite the charge. A few repeating rifles were built for Hunt's cartridges. His invention was modified by Horace Smith and Daniel Wesson in the early 1850's to include a primer within each cartridge. This became the Volcanic cartridge of the mid 1850's. The hollow bullet base contained powder and primer, and was sealed by cork, brass, or copper. The Volcanic Repeating Arms Company was founded in 1855 by Smith, Wesson, a shirt manufacturer named Oliver F. Winchester and some other investors, to manufacture rifles and pistols based on this cartridge invention. Volcanic Arms flopped, and the company went bankrupt in 1857. The main cause was weak bullet performance. The hollow bullets simply had insufficient capacity to hold a reasonable charge. It is doubtful that a Volcanic bullet ever achieved more than about 600 fps muzzle velocity, and very light bullets had to be used to achieve that much velocity. Also, the breeches of Volcanic guns could not be sealed against escaping powder gases.

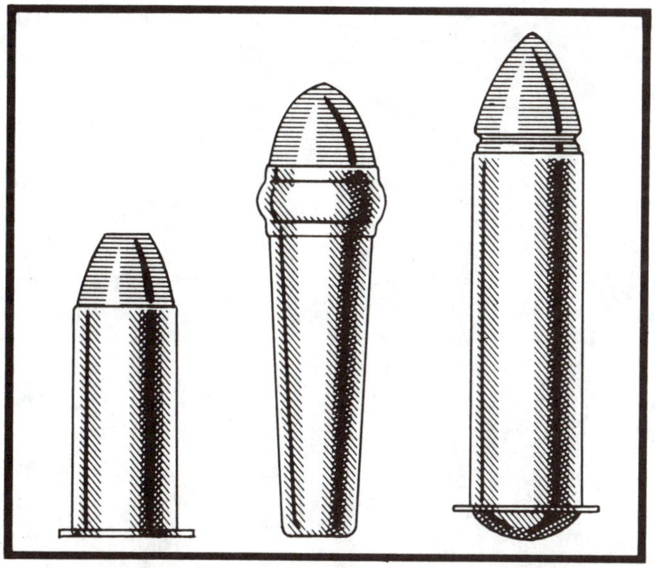

Early Metallic Unprimed Cases. These are typical of the early cases that did not have internal priming. Left: Cal. 50 Maynard, best known of the early Maynard cartridges, was ignited by the flash of a percussion cap. Middle: Cal. 54 Burnside, one of the earliest of American brass cases, was similarly fired. Right: Cal. 56 Billinghurst Requa was used in a volley gun with twenty-four barrels mounted in a row. One of the first of a quick-firing weapons, all barrels of this gun were fired by a single train of priming powder. *Lyman Archives.*

Despite these failures, the Volcanic experiment was important. The cartridge was the first caseless type invented in this country. The tubular magazine Volcanic repeater was basically a sound mechanical design. And Oliver F. Winchester was able to salvage the best ideas and go on to produce the Henry rifle and a long line of Winchester repeaters.

The first rimfire cartridge, the .22 BB cap, was invented by Flobert in Europe about 1845. The rimfire ignition system propagated to America well before the Civil War, and breechloading rimfire repeaters (the .56-.56 Spencer, .44 Henry, and .56-.50 Ball [Spencer cartridge]) saw action in that War. However, rimfire cartridges of that period were also weak, and ballistic performance was poor compared to the standard .58 caliber rifled musket. This muzzleloading rifle backed a 505

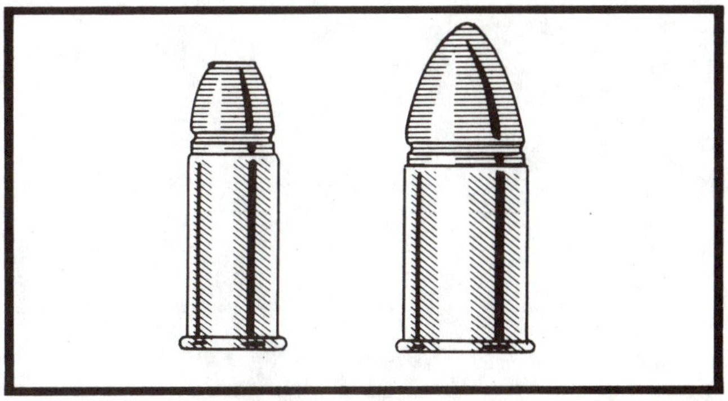

Early Rim-Fire Cartridges. These are noteworthy examples of early rim-fire cartridges. Left: Cal. 44 Henry, used in the famous rifles that led to the development of the Winchesters. Right: Cal. 56 Spencer carbine cartridge. Empty cartridge cases are still found where historic Indian fights and buffalo hunts took place. *Lyman Archives.*

grain Minie' bullet with 60 grains of black powder to produce 1000 fps muzzle velocity. The Spencer rifle with the .56-.56 Spencer cartridge propelled a 450 grain bullet with 40 grains of black powder, for a muzzle velocity of about 900 fps. The .44 Henry rifle delivered a 200 grain round nose or flat nose bullet at about 1100 fps, using 28 grains of black powder. The .56-.50 Spencer cartridge used a 350 grain bullet and 45 grains of black powder to produce about 1075 fps muzzle velocity. The mechanical strength of the foldedhead rimfire cases would not permit stronger loads.

Although the rimfire repeaters gave relatively poor ballistic performance, they had tremendous firepower compared to the muzzleloading arms, and this advantage was well understood at the end of the Civil War. It was also obvious from Civil War experience that the metallic cartridge had solved the gas-tight breech problem. It was clear in 1865 that the breechloading firearm with metallic cartridges was practical and manufacturable, but it was also clear that rimfire cartridges were not the right answer for either military or civilian needs at that time. Not only were rimfire loads weak, but rimfire cartridges could not be reloaded. This was a major drawback on the early frontier where guns meant survival and factory-loaded ammunition could be obtained only at widely separated locations.

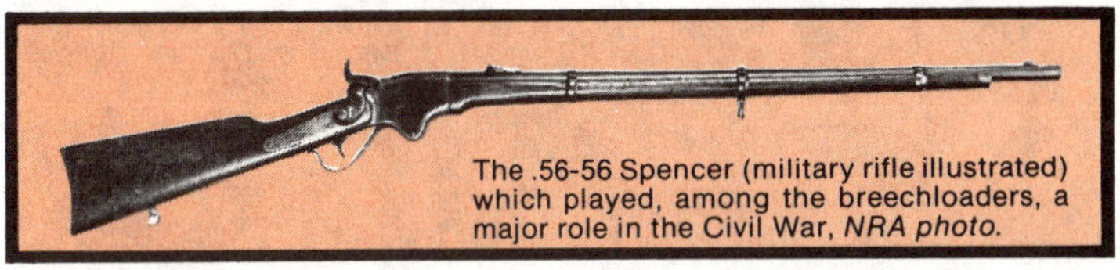

The .56-56 Spencer (military rifle illustrated) which played, among the breechloaders, a major role in the Civil War, *NRA photo.*

The right answer was the centerfire cartridge, which was developed in a seven or eight year period following the Civil War. The year 1873 marked a revolution in firearms development in this nation. That was the year the U.S. Army adopted the "Trapdoor" Springfield .45-70, Winchester introduced the Model 1873 .44-40 rifle and Colt brought out the Single Action Army .45 revolver—all using brand new centerfire black powder cartridges.

The centerfire era started with basic lead bullet shapes developed for muzzleloaders and rimfire cartridges. The round nose flat base shape was the first to be used, and inside lubrication was also used from the very beginning. The flat nose

Early Center-Fire Cartridges. Especially successful examples of early center-fire cartridges. Left: Cal. 577 Snider-Boxer was made considerably smaller than the breech diameter for easy loading, expanded upon ignition to seal the breech and prevent the escape of gas. Right: Cal. 450 Martini-Henry cartridge with thin coiled-brass case was easily damaged if roughly handled.
Lyman Archives.

Evolution of the Military Center-Fire Cartridge. Left: Cal. 50-70, issued between 1869 and 1873. Middle: Cal. 30-40 Krag, used in the U.S. Krag-Jorgensen rifle during the Spanish-American War. Right: The famous Cal. 30-'06 is the best known military cartridge and veteran of two World Wars and the Korean War.
Lyman Archives.

CHAPTER 21
TRAJECTORY VARIATIONS

By W.T. McDonald & T.C Almgren

Introduction

When a bullet leaves the muzzle of a gun, it flies a ballistic trajectory toward the target. This trajectory is always a curved path. The solid line in Figure 1 shows the curved trajectory for the particular case of sighting in a gun. The word "ballistic" means that the bullet's flight between muzzle and target is influenced only by the forces of gravity and aerodynamics; there is no further propulsion outside the barrel (the bullet is not a rocket), and the bullet cannot perform any aerodynamic maneuvers.

While gravitational and aerodynamic forces are the fundamental causes of the curved trajectory, several shooting conditions act together to control the exact shape of the trajectory. These are the following:

Muzzle velocity of the bullet - The higher the speed of the bullet, the faster it reaches the target, and the flatter it shoots.

Ballistic coefficient of the bullet - The higher the ballistic coefficient, the better the bullet retains its velocity, the flatter it shoots, and the better it resists the wind.

Altitude of the firing point - The higher the altitude above sea level, the less dense is the air. Therefore, air drag is lower at higher altitudes, and bullets shoot flatter than at lower altitudes. Moreover, the speed of sound is lower at higher altitudes, which also causes the air drag to be changed.

Weather conditions at the firing point - Temperature, barometric pressure, and relative humidity affect the air density and the speed of sound, and therefore the air drag on the bullet.

Shooting uphill or downhill - If a gun is sighted in under nearly level conditions, shooting at a target which is uphill or downhill from the firing point will cause the gun to shoot high.

Uphill shooting requires a knowledge of trajectory variations.

Wind conditions at the firing point - Crosswinds and vertical winds strongly affect bullet trajectories, causing deflections of the bullet to the right or left, or up or down. Headwinds or tailwinds acting on the bullet are not important usually.

A typical situation is that a hunter or target shooter uses a convenient shooting range, probably near home, to sight in a rifle or handgun, but then goes to a different locale for a hunt or a target match. There the shooting conditions are probably different from those at the time and place where the gun was zeroed in. An important question is whether the changes in shooting conditions will cause significant changes in the trajectory of the bullets. This is the question which this article attempts to answer. This question is especially important to hunters and silhouette competitors, and it may be important to competitors in paper target matches even though "sighters" may be allowed. For plinkers, changes in shooting conditions are a possible explanation for consistent hits in one location and consistent misses in another.

There are two types of trajectory variations. One type is random variations from shot to shot (see Figure 1). These variations cause bullet groups on a target. Even when the gun is in a machine rest so that human error cannot occur, groups happen. There are many causes, including random variations in some of the shooting conditions listed above. This article, though, is not concerned with these random trajectory variations, because group size depends mainly on the accuracy of the gun, uniformity of the ammunition, and skill of the shooter. We are concerned with the second type which are systematic variations which cause the groups to shift on the target, as shown in Figure 1. Changes in the shooting conditions listed above can cause systematic trajectory variations, and these are the subject of this article.

Not all changes in shooting conditions are independent of each other. For example, if after sighting in, a shooter goes to a higher altitude, air density and speed of sound are reduced, causing the gun to shoot higher. But at higher altitude the air temperature is usually lower, lowering the temperature of the powder in the cartridges. This in turn reduces muzzle velocity, causing the gun to shoot lower. Where the gun actually shoots is determined by both effects acting together. In any real world situation all changes in shooting conditions act together to result in a net change in bullet trajectory. But analyzing these changes one at a time will show us the importance of each one.

The effect of a change in any shooting condition on a bullet trajectory depends very much on the cartridge being fired and the bullet being used. If the ballistic coefficient of the bullet is low (0.1 to .2, for example) and the muzzle velocity is low (1000 to 2000 fps), the trajectory will be strongly affected by changes in most shooting conditions. On the other extreme, if the bullet ballistic coefficient is high (0.45 to 0.6) and the muzzle velocity is high (2500 to 3500 fps, for example), the trajectory will be much less sensitive to changes in most shooting conditions. Between these two extreme cases the effect of a change in any shooting condition will depend on the combination of bullet ballistic coefficient and muzzle velocity. Of course, the effect of any change depends also on the magnitude of the change.

To illustrate the effects of changes in shooting conditions we have selected ten cartridge and load combinations, four for

trajectory changes in the row beside each shooting condition change result from that change acting all by itself, and the trajectory changes always refer back to the baseline trajectory.

Variations in Ballistic Coefficient

Sometimes the ballistic coefficient (we will abbreviate it BC) of a bullet is not known accurately. This happens, for example, when the BC of an unfamiliar bullet is estimated by comparing it with a bullet of similar shape whose BC is known. These authors have found that an error of 10 percent or more in the estimated BC can result from using this method.

Also, it sometimes is necessary to substitute for the bullet used in a favorite handload, and the substitute bullet has a different BC.

The first shooting condition change considered is a change in bullet ballistic coefficient. Tables 2 through 11 show the effects of two changes in BC for each cartridge, first a decrease of 10 percent, and then a decrease of 20 percent. It is important to remember when examining the tables that muzzle velocity is the same for the substitute load as for the original load. Each table then shows the changes in remaining velocity and bullet path height that occur only because of the change in ballistic coefficient.

As an example, look at the 223 Remington cartridge in Table 2. A favorite load would send the Sierra 50 grain Blitz bullet from the muzzle at 2700 fps in a Thompson/Center Contender or Remington XP 100. The baseline trajectory data show remaining velocity and bullet path height at 50 yard intervals out to 250 yards from the muzzle. The first row of data for the 10 percent lower ballistic coefficient show that at 150 yards the substitute bullet would be traveling 55 fps slower than the 2160 fps for the favorite Blitz bullet. But, the new bullet would shoot only 0.12 inch lower than the Blitz at the 150 yard range point. If the substitute bullet had 20 percent lowered BC values than the Blitz, it would travel 123 fps slower and shoot 0.27 inch lower than the baseline trajectory at the 150 yard range point. If the BC changes happened to be increases instead of decreases, the trajectory changes would be nearly the same in magnitude, but the velocity of the new bullet would be higher and it would shoot flatter at the downrange points. That is, the higher BC would improve the bullet ballistics.

A similar interpretation applies to all the data in Tables 2 through 11. A minus sign (-) in front of any variation means that the bullet has less velocity, or shoots lower, than the baseline trajectory. A plus sign (+) means that the bullet has more velocity, or shoots higher, than the reference trajectory. In every case, the number following the + or - sign is the magnitude of the change in velocity or bullet path as measured from the baseline trajectory.

An interesting observation for all ten cartridges is that, while the BC reductions cause the bullet to have significantly lower velocities at the longer ranges, the changes in bullet path height are small. For example, the 1.52" change in bullet path height at 250 yards for the 223 Remington handgun cartridge (Table 2) with a 20 percent change in BC is a change of only 0.6 minute of angle (MOA). For the 223 Remington rifle cartridge (Table 6) a 20 percent change in BC causes a 219 fps change in velocity at 300 yards range, but the bullet path height change is only 1.7", which is approximately 0.6 MOA. For the 7mm-08 Remington and .30-06 cartridges (Tables 8 and 10) a 20 percent BC variation causes bullet path height changes of less than 1 MOA at ranges out to more than 400 yards, and less than 2 MOA at 550 yards. For the 300 Winchester Magnum cartridge (Table 11) a 20 percent BC variation causes a bullet path height variation of less than one MOA to ranges beyond 550 yards. Even the lowly 30-30 Winchester has less than one MOA bullet path height variation out to more than 200 yards.

This line of reasoning states that it is all right to load substitute bullets to the same muzzle velocity as a favorite load (Editors note: as long as the actual point of impact at 100 yards is verified as sometimes a bullet change even when of the same weight and basic shape can cause a noticeable point of impact shift due to actual velocities obtained and a change in barrel vibrations.) A more general conclusion is that for almost any gun reasonable changes in ballistic coefficient will not cause large changes in the bullet path at practical ranges. However, significant changes in remaining velocity can occur.

Variations in Muzzle Velocity

Shot to shot random variations in muzzle velocity are familiar to most handloaders. In fact, the standard deviation of muzzle velocities measured for a test group can be an indicator of the quality of the ammunition. But there also can be large, systematic changes in muzzle velocity in both handloaded and factory ammunition. For example, all data published in reloading manuals or by ammunition manufacturers were developed and tested in specific test guns. Other guns may produce muzzle velocities significantly different from the published values. This can occur because of differences in barrel length, chamber dimensions, or throating depth and other factors of different guns.

Even when a specific load is developed for a particular gun, systematic muzzle velocity variations can occur. Most handloaders know that a change in brands of primers or cases can affect velocity, and that they must be especially cautious if they change between brands of cases, and especially between military cases and commercial. A change in powder lots can cause a change in muzzle velocity. A change in almost any dimensional variable in the reloading process can cause a change in chamber pressure and hence in muzzle velocity. Examples are a change in bullet seating depth, neck resizing versus full length resizing, or using a crimp, and the list is by no means complete.

Cartridge temperature (specifically, powder temperature) at the time of firing has a very strong effect on chamber pressure and therefore on muzzle velocity. These authors have reported results of some tests on selected rifle and handgun cartridges in Reference 2. The tests showed that for the rifle cartridges tested, muzzle velocity changed about 2 fps for each degree of Fahrenheit change, rising as temperature increased and falling as temperature decreased. For the handgun cartridges tested, muzzle velocity changed about 1 fps for each degree of Fahrenheit change. The NRA Firearms Fact Book reports a temperature sensitivity of 1.7 fps per degree Fahrenheit for IMR powders. It is clear that a temperature change between sighting in and using the gun in a hunt or a match can cause a significant change in muzzle velocity. This effect can occur for both factory and handloaded ammunition.

It is not at all unusual to experience a temperature change of 50 degrees Fahrenheit between sighting in and hunting. For example, if you sight in on a balmy day with the temperature at 85 degrees and then hunt where the temperature is 35 degrees, the difference is 50 degrees, and that can cause your muzzle velocity to be about 100 fps lower if you are hunting with a high powered rifle.

This situation can be made even worse if when sighting in you happen to leave your cartridges exposed to the direct sunlight before you fire them. Brass absorbs heat from the sun very well, even when brightly polished. Cartridges left on the shooting bench exposed to the sunlight can reach a temperature much higher than the ambient temperature. This can increase the temperature difference between sighting in and hunting or shooting in a match, and it also can be dangerous if you are loading near maximum levels. A good practice is to keep your cartridges in the shade before firing them. (Editors note: the reverse might be true if you are heading for Africa. Leaving cartridges in the sun is what may happen there in 100+ degree heat. If cartridges left in the sun at the range show any abnormalities with respect to too much pressure, then the load will require reducing for positive African performance.)

The second pair of entries in Tables 2 through 11 show the effects of decreases in muzzle velocity. Two cases are considered for each cartridge. For the handgun cartridges decreases of 50 and 100 fps are used, and for the rifle cartridges decreases of 100 and 200 fps are used. Negative velocity changes have been adopted because the dominant influence should be temperature changes, which most likely are decreases. If instead the muzzle velocity change is an increase, then the variations in velocity and bullet path height versus range will be about the same in magnitude, but will be positive rather than negative.

in cold country, still sight in their rifles on balmy days, but carry their cartridges in ice chests and keep them there until they chamber and fire each round.

An interesting observation from the tables is that the variations in velocity grow smaller as range increases for every one of the example cartridges. For example, the first case for the 243 Winchester in Table 7 shows that the bullet starts out at the muzzle with a velocity 100 fps slower than the baseline (2900 versus 3000 fps). But at 400 yards the velocity difference is 82 fps (2054 fps versus the baseline 2136 fps). While this seems strange, there is a sound physical reason. As velocity decreases, air drag decreases even faster at all velocity levels. This means that the slower bullet retains its velocity a little better as it travels downrange. This in turn causes the velocity difference between the two bullets to grow smaller with increasing range. For those readers who are now wondering if the slower bullet catches up with the faster bullet, please be assured that this never happens!

Changes in Altitude

When a gun is sighted in at a low altitude and then is taken to a higher altitude for a hunt or a match, the gun will shoot higher and the bullet velocity downrange will be higher, compared to where the gun was sighted in. The main reason, of course, is that the air is thinner at higher altitude and therefore drag is lower. Very often, the changes in bullet ballistics with altitude are analyzed by changing the ballistic coefficient of the bullet to correspond with the change in air density. This is not quite correct, however, because air drag depends on the speed of sound as well as air density. Speed of sound also varies with altitude, but the changes in air drag caused by changes in speed of sound cannot be analytically modeled by only changing the bullet ballistic coefficient. The effects of the speed of sound changes on bullet ballistics are smaller (usually) than the effects of the air density changes, but they are not negligible.

Effects of Altitude Above Sea Level on Projectile Drop

Projectile	Muzzle Velocity (fps)	Ballistic Coefficient	Altitude (ft)	Drop (inches) at Range (yards)			
				50	100	150	200
Lyman #454613 (265 gr) Minie	1400	.156	Sea Level	−2.38	−10.44	−25.52	−49.00
			5000	−2.34	−10.19	−24.71	−47.12
			10000	−2.31	− 9.97	−23.98	−45.40
Lyman #311466 (151 gr)	1650	.250	Sea Level	−1.60	−6.99	−16.80	−31.78
			5000	−1.57	−6.83	−16.33	−30.64
			10000	−1.54	−6.69	−15.92	−29.69
Lyman #301620 (200 grs) Paper-Patched	1650	.370	Sea Level	−1.53	−6.63	−15.75	−29.30
			5000	−1.51	−6.50	−15.41	−28.55
			10000	−1.47	−6.39	−15.09	−27.88
Sierra 180 gr .308 Spitzer Boat Tail	2600	.540	Sea Level	−0.57	−2.51	−5.91	−10.86
			5000	−0.56	−2.47	−5.82	−10.67
			10000	−0.54	−2.42	−5.71	−10.46

Tables 2 through 11 show that a variation in muzzle velocity has an important effect on the trajectory of every one of the example cartridges. The velocity variations are substantial, and the bullet path height variations can exceed 1 MOA at medium ranges. The changes in bullet path are particularly severe for the cartridges with low muzzle velocities and bullets with low ballistic coefficients, i.e., the 357 Remington Maximum, 44 Magnum, and 30-30 Winchester. Shooters should be aware that relatively large changes in muzzle velocity can occur. Avoiding ballistics changes from this cause is important. This requires using the same reloading components and reloading procedures every time reloading is performed, and minimizing the temperature changes if possible. We know avid hunters who, when planning a hunt

In Tables 2 through 11 the third group of entries for each cartridge show the effects of altitude changes. The baseline trajectory for each cartridge is for sighting in at an altitude of 500 feet, changes in velocity and bullet path height are presented in the tables for shooting at 2500, 5000, and 7500 feet above sea level. Standard atmospheric conditions for all altitudes have been used to calculate these trajectory changes, and they are fully corrected for both air density and speed of sound changes with altitude.

For the high performance handgun cartridges, the 223 Remington and 7-30 Waters in Tables 2 and 3, the velocity increases are quite large for longer ranges at the higher altitudes, but the bullet path changes are less than a half MOA at 250 yards and at 7500 feet altitude. For any handgun cartridge in this class, altitude changes after sighting in should cause no problems unless the altitude at the shooting location is very high and the range to the target is very long. However, handgun silhouette shooters might consider a sight adjustment for the ram at 200 meters, even if they use a high performance cartridge similar to the 7-30 Waters.

The other two handgun cartridges, the 357 Remington Maximum and 44 Magnum with bullets with low ballistic coefficients at lower muzzle velocities, experience moderate velocity increases and moderate bullet path changes at longer ranges and at higher altitudes. Hunters using handguns in this class should make sight adjustments when they change altitudes by 4000 feet or more, for ranges beyond 150 yards. Silhouette shooters using cartridges in this class will find it necessary to adjust their sights or compensate their aim points for the turkeys and the rams, if they change altitudes by 3000 feet or more after sighting in.

All the rifle cartridges in Tables 6 through 11 (except for the 30-30 Winchester in Table 9) have large increases in velocity at the higher altitudes and longer ranges. However, all these cartridges have bullet path increases of less than one MOA for ranges up to 400 yards and altitudes beyond 7500 feet. The 30-30 cartridge, with low bullet ballistic coefficient and low muzzle velocity, is much more sensitive to altitude changes. However, for ranges out to 200 yards the bullet path changes are less than one MOA, even at 7500 feet, and no one recommends the 30-30 for ranges beyond 150 yards. Silhouette competitors using cartridges in the class of the 7mm-08 or 30-06 must consider sight adjustments or changing their aim points for both the turkey and the ram if they change altitudes by 2500 feet or more after sighting in.

The outstanding performer among our example cartridges is the 300 Winchester Magnum. This cartridge has bullet path changes of less than one MOA at ranges up to nearly 600 yards with an altitude change of 7000 feet. Our studies have shown that most cartridges in magnum calibers are very insensitive to changes in altitude. The data in Table 11 for the 300 Winchester Magnum are typical for the class. The velocity variations at longer ranges and higher altitudes are large, but the variations in bullet path height are quite small. In most practical hunting situations (400 yards extreme range) a hunter with a rifle in a magnum caliber need not worry about trajectory variations caused by changes in altitude after sighting in.

Variations in Atmospheric Conditions

Bullet trajectories are affected by variations in atmospheric temperature, barometric pressure, and relative humidity. This was discovered more than a century ago when ballisticians in Europe who tried to compare test data found differences in their results which could be traced to differences in their weather conditions. This led to the adoption of "standard" atmospheric conditions and to procedures for correcting test data acquired under any conditions to the standard conditions. This technique allows everyone's ballistic data to be compared on a common basis.

Standard Atmospheric Conditions at Altitude

Altitude (ft)	Temperature (°F)	Barometric Pressure mm Hg	Barometric Pressure inches Hg
Sea Level	59.0	750.0	29.53
1000	55.4	728.3	28.67
2000	51.9	707.1	27.84
3000	48.3	686.3	27.02
4000	44.7	660.1	26.22
5000	41.2	646.2	25.44
6000	37.6	626.9	24.68
7000	34.0	608.0	23.94
8000	30.5	589.4	23.21
9000	26.9	571.4	22.50
10000	23.3	553.8	21.80
11000	19.8	536.6	21.12
12000	16.2	519.8	20.47
13000	12.6	503.4	19.82
14000	9.1	487.4	19.19
15000	5.5	471.8	18.58

Early in this century the United States Army adopted a set of standard atmospheric conditions. These conditions are know as the Standard Metro Atmosphere, and they have been used by commercial ammunition manufacturers as well as the military. The Standard Metro conditions important for small arms ballistics are:

Altitude: sea level
Temperature: 59 degrees Fahrenheit
Barometric Pressure: 29.53 inches of Mercury
Relative Humidity: 78 percent

All ballistic coefficients and all ballistics tables published for commercial bullets are based on these conditions. While this makes it possible to compare everyone's ballistics, it leads to a problem. The problem is that shooting almost never takes place under standard conditions, and an important question is what happens to bullet trajectories when atmospheric conditions are nonstandard. Note that standard temperature and pressure change with altitude, that is, they are not defined only at sea level. This question then applies at any altitude.

Trajectory changes caused by changes in temperature, barometric pressure, and relative humidity at the shooting location are the subject of this section. Although wind conditions usually are included in weather reports, we will discuss effects of winds on bullet trajectories as a separate topic.

It is important to understand that barometric pressure obtained from a weather report is not the true pressure at the weather station. Rather, it is the weather station pressure "corrected" to sea level. That is, it is what a barometer would read at sea level, with the temperature at sea level being the same as the weather station temperature. For instance, on a nice day in Denver where the altitude is near 5000 feet, the official weather report would say that the barometer is around 30 inches of mercury, while the true atmospheric pressure at that altitude is nearer 25 inches. The reason why the barometer is corrected to sea level is so that the weather in Denver can be compared to the weather in New York or anywhere else. This is another case of standardizing for the purpose of comparison.

The pressure that must be used in ballistic calculations is the true atmospheric pressure. So, the barometer reading from a weather report must be recorrected to the true pressure at the weather station. Fortunately, modern computer programs for ballistics perform these calculations. Using one of these programs, two cases have been explored for our ten example cartridges. As the reader may recall, the altitude of the range where the guns are sighted in is 500 feet. In the first case we imagine that the guns are sighted in on a summery day when the temperature is 90 degrees, the barometric pressure reported by the local weather station is 30.00 inches, and the relative humidity is 80 percent. (As a matter of curiosity, the true pressure is 29.50 inches of mercury for the altitude and the temperature at the range. The Standard Metro pressure and temperature corrected to the 500 foot altitude are 28.98 inches and 57.2 degrees, respectively.) The first case then answers the question:

If we sight in under actual weather conditions at the range, what differences in bullet trajectories will result because the conditions are different from standard?

The first row (Case 1) in the fourth set of entries in Tables 2 through 11 answers this question for each cartridge.

In the second case we imagine that after sighting in under the true weather conditions described above, we go hunting or shoot in a silhouette match on a late fall day when the temperature is 35 degrees, the reported barometric pressure is 29.40 inches, and the relative humidity is 40 percent. We assume also that the hunt or match is in a location where the altitude is near 500 feet, not much different from the altitude where we sighted in. The second case then answers the question:

If we sight in under the summery conditions described above, what differences in bullet trajectories will occur when we shoot in the fall weather conditions?

The second row (Case 2) in the fourth set of entries in Tables 2 through 11 answers this question for each cartridge.

We believe that these cases are typical for some large regions of the continental United States, so that we can draw some meaningful conclusions from Tables 2 through 11. First, sighting in under warm weather conditions causes all the cartridges to shoot a little flatter and retain a little more downrange velocity than the trajectories predicted with standard atmospheric conditions only. However, the velocity differences at the longest range are no larger than 3 percent for any of the cartridges, and the difference in bullet path is a small fraction of a MOA for all but the .44 Magnum. For that cartridge the bullet path change approaches 1 MOA at 250 yards, which would cause a sight change for the ram at 200 meters in handgun silhouettes.

Also, the tables show for every cartridge that if the gun is sighted in under the balmy weather conditions and then is used under the cooler fall conditions, the bullet will have lower velocity downrange and shoot lower.

It is clear from the data in the tables that guns with cartridges firing bullets with low ballistic coefficients at low muzzle velocities are more sensitive to weather changes than guns having more efficient bullets at higher muzzle velocities. The outstanding performer among our ten example cartridges is again the 300 Winchester Magnum. But not far behind it are the 30-06, 7mm-08, and 243 Winchester. Handguns with cartridges like the 223 Remington and 7-30 Waters also are very insensitive to weather changes. If you are a handgun silhouette competitor and use a .44 Magnum, then changes from warm to cold weather can make sight changes of a MOA or so necessary for the farthest targets.

At practical ranges for hunting, variations in atmospheric temperature, pressure, and relative humidity will not cause important trajectory variations, as long as a ballistically efficient handgun or rifle is used. Note that this conclusion applies only to effects of changes in the atmosphere. We noted in a preceding section that the change in temperature of the powder in the cartridge (a 55 degree drop in our example) will cause a significant drop in muzzle velocity. The change in powder temperature is separate from the changes in atmospheric conditions, but it results from the change in temperature, and it in turn causes important trajectory changes also.

Variations Caused by Uphill or Downhill Shooting

Uphill or downhill shooting is a frequent experience when hunting in hilly or mountainous country, and such country constitutes some of the best hunting territory in North America. Uphill or downhill shooting is not a concern for silhouette games or other types of target matches, but it is a major concern for hunters.

Any gun which first is sighted in on a level range and then is fired either upward or downward relative to level will always shoot high compared to the baseline trajectory for level fire. Furthermore, the steeper the angle of fire, the higher the gun will shoot relative to its trajectory for level fire. Reference 2 contains a detailed explanation of this phenomenon.

The fifth group of entries in Tables 2 through 11 is trajectory variations for each cartridge caused by shooting either uphill or downhill at angles of 15, 30, and 45 degrees relative to level. This angle is called the elevation angle of fire, and a plus (+) sign indicates uphill shooting, while a minus (-) indicates downhill shooting. It is emphasized that either a positive or negative elevation angle causes a positive bullet path variation, that is, the bullet shoots high.

It is important to understand that the "range" values in Tables 2 through 11 are slant range distances for all entries in the Uphill/Downhill Shooting category. For example, for a plus or minus 15 degree angle, a 200 yard range figure means 200 yards measured along the line of sight, which is tilted upward or downward by 15 degrees relative to level.

For all the cartridges in the tables the variations in velocity are very small at all slant range distances and for all angles. Also, if the firing elevation angle is moderate (plus or minus 15 degrees or less), bullet path variations are less than one MOA at all practical ranges for hunting with any of the cartridges. Likewise, at steeper elevation angles bullet path changes are important with any cartridge. For example, if we are shooting downhill along a steep hillside at a rockchuck with a 223 Remington handgun and the load in Table 2, the bullet will be about one and a half inches high if that chuck is 200 yards away and the elevation angle is near 30 degrees, and about three and a half inches high at that distance if the angle is near 45 degrees.

A word about angles. Even a 15 degree hillside is uncomfortably steep for a human to climb or descend; 15 degrees corresponds to nearly a 30 percent slope! However, it is very reasonable for a hunter in hilly or mountainous terrain to have an opportunity for a shot down into a ravine or upward along a canyon wall at an angle in the 30 to 45 degree range. In such situations the hunter must understand the sensitivity of his or her rifle and load to uphill/downhill angles; be able to estimate the angle, range, and bullet path variation; and hold low by the right amount to compensate for the trajectory change. Shooting uphill or downhill is very much like shooting in the wind. Success depends on the skill of the hunter, which is a combination of understanding the shooting conditions, correctly estimating the trajectory change, and adjusting the aim point to compensate for the conditions.

Trajectory deflections caused by winds can be anything from negligible to very large, depending on the speed and direction of the wind. These authors have shown in Reference 2 that headwinds or tailwinds have negligible effects on bullet trajectories for wind speeds and ranges to the target which are reasonable for hunting and silhouette shooting. However, a crosswind, or a vertical wind, can drag the bullet left or right, or up or down, giving it a significant deflection at practical ranges. Deflections caused by winds often are called wind drifts, signifying that the bullet "drifts" in the direction of the wind. Both hunters and target shooters need to understand winds and how to compensate for wind effects on bullet trajectories. It is especially important for hunters to understand that vertical winds cause drifts just like crosswinds do.

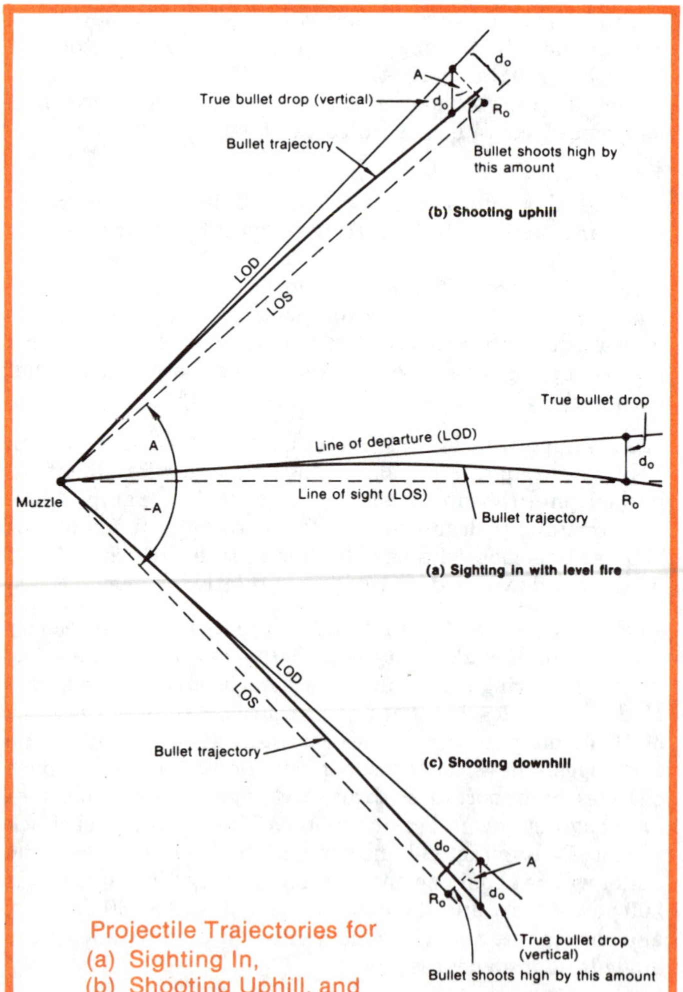

Projectile Trajectories for
(a) Sighting In,
(b) Shooting Uphill, and
(c) Shooting Downhill

For shooting purposes, the direction from which a wind blows is always referenced to the line of sight from the shooter to the target. Thus, a tailwind would blow from the shooter toward the target, while a headwind would blow in the opposite direction; a crosswind would blow from the shooter's left or right; and a vertical wind would blow either up or down relative to the line of sight. Of course, a wind can blow from any direction, and this complicates the situation because only those portions of the wind which blow across the line of sight either horizontally or vertically affect the trajectory. We can neglect the portion which blows directly toward the target or toward the shooter.

Any wind can be "resolved" into three "components" which are (1) along the line of sight to the target (a tailwind or headwind component), (2) horizontally across the line of sight (crosswind component), and (3) in the vertical direction perpendicular to the line of sight (vertical wind component). This is easiest to visualize when the shooting situation is level, that is, when no vertical component of wind exists. Then, the table which follows below gives the crosswind component of a wind with a total speed of 10 miles per hour and which blows from a direction related to the clock. We imagine the shooter at 6 o'clock and the target at 12 o'clock, and the wind blowing from a direction located by a time on the clock face. The table is as follows:

Wind Blowing from (hour angle)	Crosswind Speed Toward the Left (mph)
2:00 or 6:00	0.00
12:00 or 5:30	2:59
1:00 or 5:00	5:00
1:30 or 4:30	7:07
2:00 or 4:00	8:66
2:30 or 3:30	9:66
3:00	10:00

Wind Blowing from (hour angle)	Crosswind Speed Toward the Right (mph)
6:00 or 12:00	0.00
6:30 or 11:30	2:59
7:00 or 11:00	5:00
7:30 or 10:30	7:07
8:00 or 10:00	8:66
8:30 or 9:30	9:66
9:00	10:00

The speed of the crosswind component is proportional to the total wind speed. If, for example, the total wind speed were 5 mph, the crosswind component speeds in the table above would be divided by 2; if the total wind speed were 25 mph, then the crosswind component speeds in the table above would be multiplied by 2.5.

Table 2 lists the deflections (wind drifts) caused by either a crosswind or a vertical wind which has a speed of 10 mph. Note that 10 mph is the speed of either the crosswind component or the vertical wind component of the total wind. The wind drifts are listed for the ten cartridges and for two altitudes above sea level, the baseline 500 foot altitude and a 5000 foot altitude. It is apparent from Table 12 that a crosswind or vertical wind component of 10 mph can cause significant deflections of the bullet in either the horizontal or vertical direction at ranges reasonable for hunting or silhouette shooting. Again, the wind drifts are proportional to wind speed, so that a 20 mph crosswind, for example, would cause deflections twice as large as those shown in Table 12. Another observation is that guns chambered for cartridges firing bullets with higher ballistic coefficients at higher muzzle velocities are less sensitive to winds than guns with cartridges firing bullets with low ballistic coefficients at low muzzle velocities. However, wind drift is important for all guns, even those of magnum caliber.

It is apparent also from Table 12 that wind drifts are less at higher altitudes than at lower altitudes. This is because the bullet time of flight at higher altitudes is shorter as a result of reduced air drag.

Target shooters are concerned mainly with crosswinds, because nearly all target ranges are on flat ground. A hunter must be concerned with both crosswinds and vertical winds. If a wind is blowing in hilly or mountainous terrain, there must be a vertical wind component as well as a crosswind component. A hunter must be aware that all guns are just as sensitive to vertical winds as they are to horizontal crosswinds. This sensitivity to vertical winds explains many misses experienced by hunters shooting across canyons or along hillsides under windy conditions. Estimating vertical winds is very difficult. These authors do not know of any easy rules. Experience is the only teacher. If you are fortunate enough to have a guide, ask his advice about where to hold under windy conditions. If not, recall the wind sensitivities of your gun and use your best judgement.

Conclusions

In this article we have examined the changes in bullet trajectories caused by changes in shooting conditions. The shooting conditions considered include characteristics of the cartridge (specifically, changes in bullet ballistic coefficient and muzzle velocity) and characteristics of the environment at the shooting point (changes in altitude or weather, uphill or downhill shooting, and winds). We have chosen ten cartridges which in our judgment typify the range of effects which will be found with many handguns and rifles used for hunting and silhouette shooting.

The main conclusion of this study is that trajectory variations caused by changes in shooting conditions are important to both hunters and silhouette competitors, and perhaps to other target shooters as well. This is because significant trajectory changes can be caused by realistic variations in shooting conditions. Of course, the effects of changes in shooting conditions on a bullet trajectory depend on the cartridge and load used, and that is why we chose the ten cartridges to exemplify the effects. But we have found that the order of importance of the shooting conditions (from most important to least important) is the same for all cartridges:

1. Winds (horizontal crosswinds and vertical winds).

2. Uphill or downhill shooting.

3. Variations in muzzle velocity (caused by a change in temperature of the powder or by changes in cartridge components).

4. Changes in bullet ballistic coefficient (caused by uncertainty in the true value or by selecting a different bullet).

5. Changes in altitude.

6. Atmospheric temperature, pressure, or relative humidity different from the Standard Metro Atmosphere corrected for the altitude of the firing point.

Hunters must concern themselves with all these shooting conditions, although the last one becomes important only when the weather conditions of the hunt are severely different from the weather conditions when sighting in. Target shooters are not concerned with vertical winds or with uphill or downhill shooting, and suffering an unexpected change in bullet ballistic coefficient is not a realistic situation for serious target competitors. But other conditions in the list above can affect target shooters as well as hunters.

We found also that cartridges with high ballistic efficiency, that is, which fire bullets with high ballistic coefficients at high muzzle velocities, are the least sensitive to changes in all shooting conditions. However, variations in trajectories are still important, even for rifles in magnum calibers, because realistic changes in shooting conditions have strong effects.

We have evaluated the effects of changes in the shooting conditions listed above on a one by one basis. This was done in order to determine the effects of the individual changes and their relative importance. Of course, changes in shooting conditions usually occur in combinations, such as a change in altitude which results in a change in weather which causes a change in cartridge temperature which results in a change in muzzle velocity. A shooter needs to know the net effect of all changes acting together.

Having access to a personal computer with an exterior ballistics program is the best way to explore the sensitivities of your guns and loads to various combinations of changes in shooting conditions. In fact, it is the only way. It is not possible to determine these sensitivities by firing tests because you cannot control the shooting conditions very well, and even if you could, the tests would require an enormous number of rounds.

Several exterior ballistics programs are available from professional sources. We can personally recommend the exterior ballistics program available from Sierra Bullets, which we helped to develop. Regardless of which program you use, you will find that the computer is a powerful tool which accurately simulates bullet flight. It will have all the needed capabilities to help you evaluate the sensitivities of your loads to changes in shooting conditions. Understanding these sensitivities is one way to improve your marksmanship both on the target range and on the hunt.

References

1. These United States, The Reader's Digest Association, Inc., Pleasantville, New York, 1968.

2. Sierra Rifle (or Handgun) Reloading Manual, 3rd Edition, Sierra Bullets, L. P., Sedalia, Missouri, 1989.

TABLE 1
Cartridges Used as Examples

Handgun Cartridges	Bullet	Ballistic Coefficients	Muzzle Velocity (fps)	Zero Range (yards)	Sight Height (in)
.223 Remington	Sierra .224 dia. 50 grain Blitz (Spitzer)	.254 Above 2800 fps .233 Between .209 Below 1800 fps	2700	150	1.5
7-30 Waters	Sierra .284 dia. 120 grain Spitzer	.376 Above 2800 fps .380 Between .380 Below 1800 fps	2400	150	1.5
.357 Remington Maximum	Sierra .357 dia. 170 grain Full Metal Jacket Match	.175 Above 1300 fps .285 Between .222 Below 800 fps	1700	150	0.8
.44 Magnum	Hornady .430 dia. 240 grain Full Metal Jacket	.158 All velocities	1400	150	0.8
Rifle Cartridges	**Bullet**	**Ballistic Coefficients**	**Muzzle Velocity (fps)**	**Zero Range (Yards)**	**Sight Height (in)**
.223 Remington	Sierra .224 dia. 55 grain Spitzer	.279 Above 2800 fps .256 Between .230 Below 1800 fps	3100	250	1.5
.243 Winchester	Hornady .243 dia. 100 grain Boat Tail Spire Point	.400 All velocities	3000	250	1.5
7mm-08 Remington	Hornady .284 dia. 139 grain Spire Point	.396 All velocities	2800	250	1.5
.30-30 Winchester	Hornady .308 dia. 150 grain Round Nose (30-30) Interlock	.185 All velocities	2200	200	0.8
.30-06 Springfield	Sierra .308 dia. 180 grain Hollow Point Boat Tail Match	.514 Above 2200 fps .503 Between .478 Below 1800 fps	2700	300	1.5
.300 Winchester Magnum	Sierra .308 dia. 200 grain Spitzer Boat Tail	.597 Above 2500 fps .592 Between .555 Below 1800 fps	2900	300	1.5

TABLE 2

Cartridge: .223 Remington (Handgun) with Sierra .224 dia. 50 gr. Blitz at 2700 fps

Baseline (Reference) Trajectory		Velocity (fps) at Range (Yards)						Bullet Path Height (inches) at Range (Yards)					
		MZL	50	100	150	200	250	MZL	50	100	150	200	250
		2700	2513	2333	2160	1995	1838	-1.50	+0.45	+1.02	0.00	-2.88	-7.94
Changes from Reference Shooting Conditions		**Change in Velocity (fps) at Range (Yards)**						**Change in Bullet Path Height(inches) at Range (Yards)**					
Shooting Condition	**Change**	**MZL**	**50**	**100**	**150**	**200**	**250**	**MZL**	**50**	**100**	**150**	**200**	**250**
1. Ballistic Coefficient	-10%	0	-20	-38	-55	-70	-87	0.00	-0.01	-0.03	-0.12	-0.30	-0.65
	-20%	0	-45	-86	-123	-155	-197	0.00	-0.02	-0.08	-0.27	-0.70	-1.52
2. Muzzle Velocity (fps)	-50	-50	-48	-46	-44	-42	-40	0.00	-0.02	-0.10	-0.24	-0.46	-0.78
	-100	-100	-96	-92	-88	-84	-83	0.00	-0.05	-0.20	-0.50	-0.95	-1.61
3. Altitude (feet above sea level)	2500	0	+13	+24	+35	+45	+53	0.00	+0.01	+0.03	+0.08	+0.19	+0.40
	5000	0	+27	+53	+76	+98	+117	0.00	+0.01	+0.06	+0.17	+0.41	+0.85
	7500	0	+47	+79	+114	+147	+176	0.00	+0.02	+0.09	+0.25	+0.60	+1.24
4. Nonstandard Atmosphere	Case 1	0	+7	+14	+20	+25	+30	0.00	0.00	+0.01	+0.04	+0.10	+0.22
	Case 2	0	-14	-26	-37	-47	-57	0.00	0.00	-0.02	-0.07	-0.18	-0.41
5. Uphill/Downhill Shooting	± 15 deg	0	0	0	0	0	+1	0.00	+0.02	+0.09	+0.21	+0.40	+0.66
	± 30 deg	0	-1	-1	-1	0	+1	0.00	+0.08	+0.34	+0.82	+1.55	+2.58
	± 45 deg	0	-1	-1	-1	0	+1	0.00	+0.17	+0.75	+1.79	+3.38	+5.62

TABLE 3

Cartridge: 7-30 Waters (Handgun) with Sierra .284 dia. 120 gr. Spitzer at 2400 fps

Baseline (Reference) Trajectory		Velocity (fps) at Range (Yards)						Bullet Path Height (Inches) at Range (Yards)					
		MZL	50	100	150	200	250	MZL	50	100	150	200	250
		2400	2292	2186	2083	1983	1886	-1.50	+0.70	+1.26	0.00	-3.26	-8.72
Changes from Reference Shooting Conditions		**Change in Velocity (fps) at Range (Yards)**						**Change in Bullet Path Height (Inches) at Range (Yards)**					
Shooting Condition	**Change**	**MZL**	**50**	**100**	**150**	**200**	**250**	**MZL**	**50**	**100**	**150**	**200**	**250**
1. Ballistic Coefficient	-10%	0	-12	-23	-34	-44	-53	0.00	-0.01	-0.04	-0.11	-0.25	-0.48
	-20%	0	-27	-52	-75	-97	-117	0.00	-0.02	-0.08	-0.24	-0.54	-1.07
2. Muzzle Velocity (fps)	-50	-50	-49	-48	-46	-45	-43	0.00	-0.03	-0.14	-0.32	-0.59	-0.97
	-100	-100	-98	-95	-92	-90	-87	0.00	-0.06	-0.28	-0.66	-1.23	-2.02
3. Altitude (feet above sea level)	2500	0	+7	+15	+21	+28	+33	0.00	+0.01	+0.02	+0.06	+0.14	+0.28
	5000	0	+16	+32	+46	+60	+73	0.00	+0.02	+0.05	+0.14	+0.32	+0.62
	7500	0	+24	+47	+69	+90	+109	0.00	+0.03	+0.10	+0.25	+0.52	+0.98
4. Nonstandard Atmosphere	Case 1	0	+4	+8	+12	+16	+19	0.00	0.00	+0.01	+0.02	+0.07	+0.15
	Case 2	0	-8	-15	-22	-29	-35	0.00	0.00	-0.02	-0.05	-0.13	-0.29
5. Uphill/Downhill Shooting	± 15 deg	0	-1	-1	-1	-1	-1	0.00	+0.03	+0.10	+0.25	+0.46	+0.75
	± 30 deg	0	-1	-1	-2	-2	-2	0.00	+0.10	+0.41	+0.97	+1.81	+2.93
	± 45 deg	0	-1	-2	-3	-3	-2	0.00	+0.21	+0.90	+2.12	+3.94	+6.40

TABLE 4

Cartridge: .357 Remington Maximum with Sierra .357 dia 170 gr Full Metal Jacket at 1700 fps

Baseline (Reference) Trajectory		Velocity (fps) at Range (Yards)						Bullet Path Height (Inches) at Range (Yards)					
		MZL	50	100	150	200	250	MZL	50	100	150	200	250
		1700	1516	1354	1246	1172	1109	-0.80	+3.61	+4.16	0.00	-9.78	-25.87
Changes from Reference Shooting Conditions		**Change in Velocity (fps) at Range (Yards)**						**Change in Bullet Path Height (Inches) at Range (Yards)**					
Shooting Condition	**Change**	**MZL**	**50**	**100**	**150**	**200**	**250**	**MZL**	**50**	**100**	**150**	**200**	**250**
1. Ballistic Coefficient	-10%	0	-19	-32	-25	-28	-29	0.00	-0.01	-0.14	-0.46	-1.00	-1.87
	-20%	0	-43	-66	-56	-62	-62	0.00	-0.04	-0.32	-1.04	-2.28	-4.25
2. Muzzle Velocity (fps)	-50	-50	-45	-39	-19	-16	-13	0.00	-0.10	-0.44	-1.02	-1.77	-2.71
	-100	-100	-89	-70	-41	-34	-28	0.00	-0.22	-0.91	-2.10	-3.66	-5.61
3. Altitude (feet above sea level)	2500	0	+11	+18	+13	+13	+11	0.00	+0.01	+0.08	+0.25	+0.56	+0.99
	5000	0	+24	+40	+30	+32	+29	0.00	+0.03	+0.08	+0.57	+1.27	+2.27
	7500	0	+37	+62	+48	+52	+49	0.00	+0.05	+0.28	+0.87	+1.96	+3.53
4. Nonstandard Atmosphere	Case 1	0	+9	+17	+16	+20	+24	0.00	+0.01	+0.05	+0.22	+0.56	+1.10
	Case 2	0	-16	-30	-27	-36	-42	0.00	-0.03	-0.11	-0.40	-0.96	-1.89
5. Uphill/Downhill Shooting	± 15 deg	0	-1	-1	-2	-2	-3	0.00	+0.05	+0.24	+0.59	+1.13	+1.91
	± 30 deg	0	-1	-2	-3	-4	-5	0.00	+0.21	+0.93	+2.29	+4.42	+7.40
	± 45 deg	0	-2	-3	-4	-6	-7	0.00	+0.46	+2.04	+5.00	+9.62	+16.11

TABLE 5

Cartridge: .44 Magnum with Hornady .430 dia 240 gr Full Metal Jacket at 1400 fps

Baseline (Reference) Trajectory		Velocity (fps) at Range (Yards)						Bullet Path Height (Inches) at Range (Yards)					
		MZL	50	100	150	200	250	MZL	50	100	150	200	250
		1400	1241	1117	1030	966	914	-0.80	+5.55	+6.24	0.00	-14.41	-38.05
Changes from Reference Shooting Conditions		**Change in Velocity (fps) at Range (Yards)**						**Change in Bullet Path Height(Inches) at Range (Yards)**					
Shooting Condition	**Change**	**MZL**	**50**	**100**	**150**	**200**	**250**	**MZL**	**50**	**100**	**150**	**200**	**250**
1. Ballistic Coefficient	-10%	0	-16	-23	-24	-24	-25	0.00	-0.03	-0.19	-0.65	-1.48	-2.78
	-20%	0	-35	-48	-50	-52	-55	0.00	-0.06	-0.44	-1.46	-3.30	-6.24
2. Muzzle Velocity (fps)	-50	-50	-40	-29	-21	-16	-14	0.00	-0.18	-0.73	-1.65	-2.91	-4.46
	-100	-100	-79	-56	-41	-32	-27	0.00	-0.38	-1.52	-3.42	-5.98	-9.20
3. Altitude (feet above sea level)	2500	0	+6	+6	+3	+1	0	0.00	+0.02	+0.08	+0.21	+0.39	+0.58
	5000	0	+15	+16	+9	+3	+1	0.00	+0.03	+0.18	+0.51	+0.98	+1.49
	7500	0	+23	+28	+18	+7	+2	0.00	+0.05	+0.29	+0.84	+1.67	+2.64
4. Nonstandard Atmosphere	Case 1	0	+11	+20	+25	+28	+29	0.00	+0.02	+0.14	+0.49	+1.24	+2.45
	Case 2	0	-19	-35	-45	-49	-51	0.00	-0.03	-0.24	-0.88	-2.22	-4.41
5. Uphill/Downhill Shooting	± 15 deg	0	-1	-1	-2	-3	-3	0.00	+0.08	+0.36	+0.87	+1.69	+2.83
	± 30 deg	0	-1	-3	-4	-5	-6	0.00	+0.32	+1.40	+3.41	+6.55	+10.97
	± 45 deg	0	-2	-4	-5	-7	-8	0.00	+0.70	+3.05	+7.45	+14.27	+23.84

TABLE 6

Cartridge: .223 Remington with Sierra .224 dia 55 gr Spitzer at 3100 fps

Baseline (Reference) Trajectory		Velocity (fps) at Range (Yards)					Bullet Path Height (Inches) at Range (Yards)				
		MZL	100	200	300	400	MZL	100	200	300	400
		3100	2759	2420	2105	1815	-1.50	+2.71	+2.32	-4.02	-18.26
Changes from Reference Shooting Conditions		**Change in Velocity (fps) at Range (Yards)**					**Change in Bullet Path Height(Inches) at Range (Yards)**				
Shooting Condition	**Change**	**MZL**	**100**	**200**	**300**	**400**	**MZL**	**100**	**200**	**300**	**400**
1. Ballistic Coefficient	-10%	0	-41	-82	-117	-152	0.00	-0.02	-0.19	-0.81	-2.40
	-20%	0	-87	-160	-219	-285	0.00	-0.05	-0.42	-1.70	-5.01
2. Muzzle Velocity (fps)	-100	-100	-100	-92	-85	-84	0.00	-0.13	-0.60	-1.57	-3.21
	-200	-200	-200	-186	-172	-170	0.00	-0.28	-1.30	-3.37	-6.91
3. Altitude (feet above sea level)	2500	0	+24	+45	+63	+77	0.00	+0.01	+0.11	+0.43	+1.22
	5000	0	+50	+97	+136	+169	0.00	+0.04	+0.25	+0.91	+2.53
	7500	0	+72	+144	+204	+254	0.00	+0.05	+0.36	+1.33	+3.67
4. Nonstandard Atmosphere	Case 1	0	+14	+25	+35	+47	0.00	+0.01	+0.07	+0.28	+0.78
	Case 2	0	-26	-47	-65	-89	0.00	-0.01	-0.12	-0.51	-1.46
5. Uphill/Downhill Shooting	± 15 deg	0	0	0	+1	+3	0.00	+0.06	+0.29	+0.72	+1.46
	± 30 deg	0	-1	0	+3	+6	0.00	+0.25	+1.13	+2.81	+5.61
	± 45 deg	0	-1	0	+4	+9	0.00	+0.55	+2.46	+6.12	+12.16

TABLE 7

Cartridge: .243 Winchester with Hornady .243 dia 100 gr Boat Tail Spire Point at 3000 fps

Baseline (Reference) Trajectory		Velocity (fps) at Range (Yards)					Bullet Path Height (inches) at Range (Yards)				
		MZL	100	200	300	400	MZL	100	200	300	400
		3000	2767	2546	2336	2136	-1.50	+2.60	+2.17	-3.63	-15.79
Changes from Reference Shooting Conditions		**Change in Velocity (fps) at Range (Yards)**					**Change in Bullet Path Height(inches) at Range (Yards)**				
Shooting Condition	**Change**	**MZL**	**100**	**200**	**300**	**400**	**MZL**	**100**	**200**	**300**	**400**
1. Ballistic Coefficient	-10%	0	-25	-48	-68	-86	0.00	-0.02	-0.14	-0.44	-1.16
	-20%	0	-56	-106	-151	-190	0.00	-0.04	-0.30	-1.02	-2.69
2. Muzzle Velocity (fps)	-100	-100	-95	-90	-86	-82	0.00	-0.14	-0.61	-1.47	-2.84
	-200	-200	-190	-181	-172	-164	0.00	-0.29	-1.28	-3.12	-6.01
3. Altitude (feet above sea level)	2500	0	+15	+29	+42	+54	0.00	+0.01	+0.08	+0.27	+0.70
	5000	0	+33	+63	+91	+117	0.00	+0.04	+0.19	+0.60	+1.48
	7500	0	+48	+93	+136	+175	0.00	+0.06	+0.27	+0.88	+2.17
4. Nonstandard Atmosphere	Case 1	0	+9	+18	+25	+31	0.00	+0.01	+0.04	+0.16	+0.40
	Case 2	0	-18	-33	-47	-59	0.00	-0.01	-0.08	-0.30	-0.76
5. Uphill/Downhill Shooting	± 15 deg	0	-1	0	0	+1	0.00	+0.07	+0.29	+0.70	+1.34
	± 30 deg	0	-1	-1	0	+2	0.00	+0.26	+1.13	+2.73	+5.21
	± 45 deg	0	-2	-1	0	+3	0.00	+0.57	+2.46	+5.95	+11.32

TABLE 8

Cartridge: 7mm-08 Remington with Hornady .284 dia 139 gr Spire Point at 2800 fps

Baseline (Reference) Trajectory		Velocity (fps) at Range (Yards)						Bullet Path Height (inches) at Range (Yards)					
		MZL	100	200	300	400	550	MZL	100	200	300	400	550
		2800	2575	2362	2158	1965	1696	-1.50	+3.15	+2.56	-4.26	-18.54	-57.44
Changes from Reference Shooting Conditions		**Change in Velocity (fps) at Range (Yards)**						**Change in Bullet Path Height(inches) at Range (Yards)**					
Shooting Condition	**Change**	**MZL**	**100**	**200**	**300**	**400**	**550**	**MZL**	**100**	**200**	**300**	**400**	**550**
1. Ballistic Coefficient	-10%	0	-25	-47	-66	-83	-102	0.00	-0.03	-0.15	-0.54	-1.43	-4.44
	-20%	0	-54	-103	-146	-182	-219	0.00	-0.06	-0.35	-1.23	-3.26	-10.32
2. Muzzle Velocity (fps)	-100	-100	-95	-91	-86	-81	-73	0.00	-0.17	-0.75	-1.85	-3.57	-7.71
	-200	-200	-190	-181	-172	-162	-145	0.00	-0.37	-1.60	-3.92	-7.59	-16.40
3. Altitude (feet above sea level)	2500	0	+15	+29	+41	+52	+65	0.00	+0.02	+0.10	+0.34	+0.86	+2.57
	5000	0	+32	+62	+89	+114	+143	0.00	+0.05	+0.23	+0.72	+1.81	+5.41
	7500	0	+48	+92	+134	+171	+217	0.00	+0.07	+0.34	+1.08	+2.68	+7.92
4. Nonstandard Atmosphere	Case 1	0	+9	+17	+24	+30	+38	0.00	+0.01	+0.05	+0.18	+0.46	+1.45
	Case 2	0	-17	-31	-44	-56	-71	0.00	-0.02	-0.10	-0.34	-0.90	-2.79
5. Uphill/Downhill Shooting	± 15 deg	0	-1	-1	0	+1	+3	0.00	+0.08	+0.33	+0.81	+1.55	+3.36
	± 30 deg	0	-1	-1	0	+2	+5	0.00	+0.30	+1.30	+3.15	+6.03	+12.85
	± 45 deg	0	-2	-2	0	+2	+7	0.00	+0.65	+2.85	+6.87	+13.12	+27.79

TABLE 9

Cartridge: .30-30 Winchester with Hornady .308 dia 150 gr Round Nose Interlock at 2200 fps

Baseline (Reference) Trajectory	Velocity (fps) at Range (Yards)				Bullet Path Height (Inches) at Range (Yards)			
	MZL	100	200	300	MZL	100	200	300
	2200	1795	1447	1181	-0.80	+5.07	0.00	-21.92

Shooting Condition	Change	Change in Velocity (fps) at Range (Yards): MZL	100	200	300	Change in Bullet Path Height (Inches) at Range (Yards): MZL	100	200	300
1. Ballistic Coefficient	-10%	0	-41	-65	-63	0.00	-0.08	-0.71	-2.98
	-20%	0	-93	-145	-130	0.00	-0.17	-1.70	-7.14
2. Muzzle Velocity (fps)	-100	-100	-88	-71	-47	0.00	-0.40	-1.98	-5.30
	-200	-200	-175	-139	-90	0.00	-0.87	-4.26	-11.31
3. Altitude (feet above sea level)	2500	0	+26	+42	+40	0.00	+0.05	+0.44	+1.75
	5000	0	+57	+94	+95	0.00	+0.11	+0.93	+3.76
	7500	0	+86	+144	+155	0.00	+0.17	+1.38	+5.55
4. Nonstandard Atmosphere	Case 1	0	+16	+28	+37	0.00	+0.03	+0.26	+1.17
	Case 2	0	-29	-52	-66	0.00	-0.05	-0.49	-2.20
5. Uphill/Downhill Shooting	± 15 deg	0	-1	0	-1	0.00	+0.14	+0.66	+1.80
	± 30 deg	0	-1	-1	-1	0.00	+0.54	+2.57	+6.93
	± 45 deg	0	-2	-1	-2	0.00	+1.19	+5.59	+15.02

TABLE 10

Cartridge: .30-06 Springfield with Sierra .308 dia 180 gr Matchking at 2700 fps

Baseline (Reference) Trajectory	Velocity (fps) at Range (Yards)						Bullet Path Height (Inches) at Range (Yards)					
	MZL	100	200	300	400	550	MZL	100	200	300	400	550
	2700	2530	2366	2208	2053	1833	-1.50	+4.69	+5.45	-0.00	-12.57	-47.29

Shooting Condition	Change	Change in Velocity (fps) at Range (Yards): MZL	100	200	300	400	550	Change in Bullet Path Height (Inches) at Range (Yards): MZL	100	200	300	400	550
1. Ballistic Coefficient	-10%	0	-18	-35	-52	-66	-87	0.00	-0.04	-0.16	-0.45	-1.09	-3.21
	-20%	0	-42	-80	-117	-149	-196	0.00	-0.06	-0.32	-0.97	-2.53	-7.58
2. Muzzle Velocity (fps)	-100	-100	-90	-93	-91	-88	-83	0.00	-0.19	-0.82	-1.96	-3.74	-7.83
	-200	-200	-193	-186	-182	-174	-168	0.00	-0.41	-1.74	-4.18	-7.95	-16.71
3. Altitude (feet above sea level)	2500	0	+11	+22	+32	+43	+54	0.00	+0.01	+0.08	+0.28	+0.68	+1.96
	5000	0	+25	+48	+70	+92	+119	0.00	+0.04	+0.19	+0.60	+1.44	+4.12
	7500	0	+37	+72	+104	+138	+179	0.00	+0.07	+0.32	+0.94	+2.20	+6.12
4. Nonstandard Atmosphere	Case 1	0	+6	+13	+18	+24	+31	0.00	0.00	+0.03	+0.14	+0.35	+1.06
	Case 2	0	-12	-24	-34	-45	-58	0.00	-0.01	-0.08	-0.27	-0.69	-2.04
5. Uphill/Downhill Shooting	± 15 deg	0	-1	-1	-1	0	+1	0.00	+0.08	+0.34	+0.83	+1.57	+3.28
	± 30 deg	0	-1	-2	-2	0	+2	0.00	+0.31	+1.35	+3.24	+6.10	+12.62
	± 45 deg	0	-2	-3	-2	-1	+3	0.00	+0.68	+2.96	+7.07	+13.28	+27.37

TABLE 11

Cartridge: .300 Winchester Magnum with Sierra .308 dia 200 gr Spitzer Boat Tail at 2900 fps

Baseline (Reference) Trajectory	Velocity (fps) at Range (Yards)						Bullet Path Height (inches) at Range (Yards)					
	MZL	100	200	300	400	550	MZL	100	200	300	400	550
	2900	2746	2598	2454	2314	2112	-1.50	+3.78	+4.46	-0.00	-10.22	-37.96

Changes from Reference Shooting Conditions		Change in Velocity (fps) at Range (Yards)						Change in Bullet Path Height(inches) at Range (Yards)					
Shooting Condition	Change	MZL	100	200	300	400	550	MZL	100	200	300	400	550
1. Ballistic Coefficient	-10%	0	-17	-33	-48	-61	-79	0.00	-0.02	-0.10	-0.30	-0.74	-2.14
	-20%	0	-38	-72	-105	-135	-175	0.00	-0.05	-0.22	-0.71	-1.72	-4.93
2. Muzzle Velocity (fps)	-100	-100	-97	-93	-91	-88	-84	0.00	-0.15	-0.64	-1.52	-2.85	-5.86
	-200	-200	-193	-188	-183	-177	-168	0.00	-0.31	-1.35	-3.22	-6.05	-12.44
3. Altitude (feet above sea level)	2500	0	+10	+20	+29	+38	+50	0.00	+0.03	+0.09	+0.23	+0.50	+1.32
	5000	0	+22	+43	+63	+82	+108	0.00	+0.04	+0.17	+0.46	+1.04	+2.80
	7500	0	+32	+63	+93	+122	+162	0.00	+0.07	+0.25	+0.67	+1.52	+4.06
4. Nonstandard Atmosphere	Case 1	0	+6	+12	+17	+22	+29	0.00	0.00	+0.03	+0.10	+0.25	+0.72
	Case 2	0	-12	-22	-32	-41	-54	0.00	-0.01	-0.06	-0.20	-0.48	-1.38
5. Uphill/Downhill Shooting	± 15 deg	0	-1	-1	-1	0	+1	0.00	+0.07	+0.29	+0.70	+1.31	+2.68
	± 30 deg	0	-1	-2	-2	0	+2	0.00	+0.27	+1.15	+2.73	+5.10	+10.37
	± 45 deg	0	-2	-3	-2	-1	+3	0.00	+0.58	+2.52	+5.96	+11.11	+22.54

TABLE 12

Wind Drift (Horizontal or Vertical) Caused by a 10 mph Wind (Horizontal or Vertical)

	ALTITUDE 500 ft						ALTITUDE 5000 ft					
HANDGUN CARTRIDGES	Wind Drift (inches) vs. Range (yards)						Wind Drift (inches) vs. Range (yards)					
	MZL	50	100	150	200	250	MZL	50	100	150	200	250
.223 Remington	0.0	0.4	1.5	3.5	6.4	10.4	0.0	0.3	1.3	2.9	5.3	8.6
7-30 Waters	0.0	0.3	1.1	2.4	4.4	7.1	0.0	0.2	0.9	2.0	3.7	5.9
.357 Remington Magnum	0.0	0.9	3.8	8.7	15.1	22.7	0.0	0.8	3.3	7.6	13.4	20.4
.44 Magnum	0.0	1.2	4.8	10.6	18.3	27.5	0.0	1.1	4.4	9.9	17.4	26.6
RIFLE CARTRIDGES	Wind Drift (inches) vs. Range (yards)						Wind Drift (inches) vs. Range (yards)					
	MZL	100	200	300	400	550	MZL	100	200	300	400	550
.223 Remington	0.0	1.0	4.4	10.8	20.8		0.0	0.9	3.7	8.9	16.9	
.243 Winchester	0.0	0.7	3.0	7.1	13.1		0.0	0.6	2.6	5.9	10.9	
7mm-08 Remington	0.0	0.8	3.4	7.9	14.7	29.8	0.0	0.7	2.8	6.6	12.2	24.4
.30-30 Winchester	0.0	2.6	11.4	28.0			0.0	2.2	9.4	23.2		
.30-06 Springfield	0.0	0.6	2.7	6.2	11.5	23.0	0.0	0.6	2.3	5.2	9.6	18.9
.300 Winchester Magnum	0.0	0.5	2.1	4.8	8.7	17.2	0.0	0.4	1.8	4.0	7.3	14.4

CHAPTER 22
AIR DRAG EFFECTS AND THE BALLISTIC COEFFICIENT

by W.T. McDonald and T.C. Almgren

This section describes the effects of air drag on bullet flight, and the effects of bullet shape on air drag. Air drag is a most complex subject, and this discussion really will only scratch the surface. However, the important physical effects can be described simply; and, better yet, they can be seen in high speed photographs and shadowgraphs of bullets in flight.

Drag depends, first of all, on the velocity of the bullet *relative to the air through which the bullet moves.* At elevations near sea level the deceleration of a bullet by air drag can be mathematically expressed as

$$a = -\frac{1}{C}GV$$

where the symbol **a** stands for acceleration (deceleration is a negative acceleration), **C** is the ballistic coefficient of the bullet, **G** is called the drag function, and **v** is the velocity of the bullet relative to the air. If the air is moving (wind is blowing), then the velocity of the bullet relative to the air is a little different from the velocity of the bullet relative to the ground and this difference must be taken into account when computing the bullet trajectory. The negative sign in the equation above signifies that the air drag decelerates the bullet (slows it down).

The drag function **G** has been determined for the standard bullet by extensive firing test measurements. It has been found that **G** also depends very much on the velocity of the bullet relative to the air. **Figure 1** shows how **G** depends on velocity for the Krupp standard bullet. At low bullet velocities, **G** increases almost linearly with velocity. When the bullet velocity begins to approach the speed of sound (about 1120 fps at sea level), however, **G** grows very sharply, signifying that drag also increases very sharply at velocities near the speed of sound. **G** continues to grow larger as bullet velocity increases, but **G** grows more slowly when bullet velocity is well above the speed of sound than it does when bullet velocity is near the speed of sound.

This rapid change in drag when the bullet crosses the sonic barrier is the effect that Benjamin Robins revealed for the first time with his ballistic pendulum. In Galileo's method, air drag was considered negligibly small compared to gravity. Using the drag function in **Figure 1**, for Robins' .75 caliber musket ball at 1400 fps, the drag deceleration is about 70 times larger than the acceleration due to gravity.

When a bullet moves at a speed such that the air molecules which are pushed aside by the bullet move at speeds which exceed the speed of sound, a shock wave begins to form in the air. **Figure 1** shows that this begins to happen for the Krupp standard bullet at about 800 fps. At this velocity the bullet travels slower than the speed of sound by more than 300 fps, but the air molecules which must flow around the bullet point and out of its way must move at velocities higher than 1120 fps. When the bullet velocity exceeds 1120 fps, a shock wave forms around the entire bullet.

Figure 2 is a shadowgraph of the Sierra 130 grain, .270 caliber jacketed bullet in flight at a velocity of 2970 fps, and the shock wave around the bullet is dramatically illustrated. (This shadowgraph is shown with the permission of Sierra Bullets.) The shadowgraph shows two shock fronts, in fact, one starting at the bullet nose and a second emanating from the turbulent wake of the bullet in the air behind it. Both shock waves are funnel-shaped. The view in the shadowgraph is like slicing the funnel bowl through its center and then looking at the edge in a sideview. **Figure 2** shows that the turbulent wake extends far behind the bullet. This region of turbulence occurs because as the bullet moves, it pushes the air aside, leaving a void immediately behind it. The air then flows back into this void, causing the formation of the second shock wave emanating from this wake and following behind the bullet.

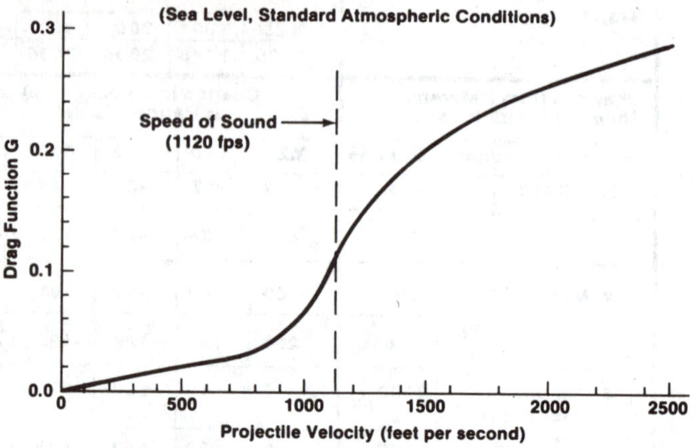

Figure 1

It is the buildup of these shock waves that causes the rapid change in drag near the speed of sound. As the bullet travels, it must push these shock waves, and it must provide the energy to move the air along the surfaces of both shock fronts. These shock waves thus produce a "drag" on the bullet, and the energy required to move the shock waves comes from the bullet motion, in turn slowing it down.

Lyman studied these shock wave effects for cast lead bullets a few years ago, and results of that study are described in the "Black Powder Ballistics" article in Lyman's *Black Powder Handbook.* There are some key differences between the shock

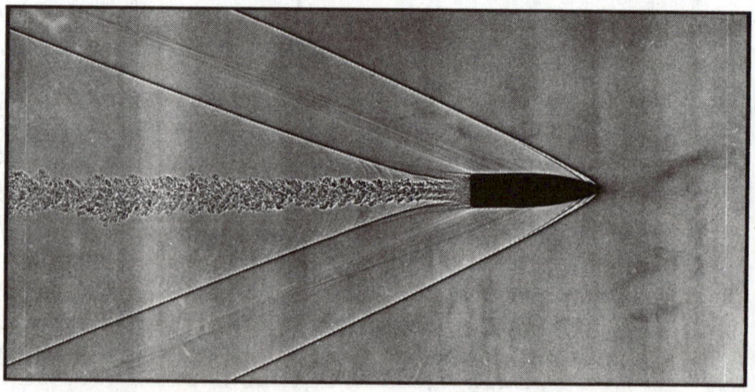

Figure 2 Shadowgraph of Sierra 130 Grain .270 Caliber Jacketed Bullet at 2970 fps Velocity.

wave patterns for cast bullets and jacketed bullets. First of all, at the lower velocities of cast bullets, the shock waves are more bowl-shaped than funnel-shaped. That is, the shock wave shape grows shallower as velocity decreases toward the speed of sound. Also, the shock front does not extend as far out into the air around the bullet at lower velocities (the "bowl" has a smaller diameter at its "rim" than the "funnel" does). This smaller extent of the shock front is consistent with a lesser drag on the bullet at lower velocities.

Almost any kind of surface irregularity on a bullet will cause a shock wave to form. Jacketed bullets are smooth, and so **Figure 2** shows just two shock fronts. Cast bullets, though, almost always have grooves to hold the lubricant (an exception is paper-patched bullets) and usually have a shoulder where the bullet point shape meets the cylindrical bearing surface. Lyman's studies showed that these surface irregularities can cause additional shock fronts to form. These secondary shock fronts caused by surface irregularities not only increase the drag on a bullet, but they also cause the drag function for the actual bullet to have a shape that is somewhat different from the standard bullet drag function shown in **Figure 1.**

There are other reasons why the drag characteristic of an actual bullet is not matched by the shape of the standard drag function. Point shapes of actual bullets are quite different from the Krupp standard bullet point shape for one thing. For another thing, the overall shape and diameter ratios of actual bullets do not match those of the standard bullet. These observations apply for jacketed bullets as well as cast lead bullets, but the drag characteristic mismatch is more pronounced for cast bullets because of the surface irregularities.

This situation immediately raises a question. If the drag function of the standard bullet does not fit the actual bullet, how can we use the standard bullet ballistics to calculate trajectories of the actual bullet? The answer is that we use more than one value of ballistic coefficient for an actual bullet, depending upon its velocity. The authors have found, through experience at both Lyman and Sierra Bullets that three values of ballistic coefficients allow very accurate trajectory computations for almost any kind of rifle or pistol bullet. One value is used for higher velocities, a second value for intermediate velocities and a third for lower velocities of each bullet.

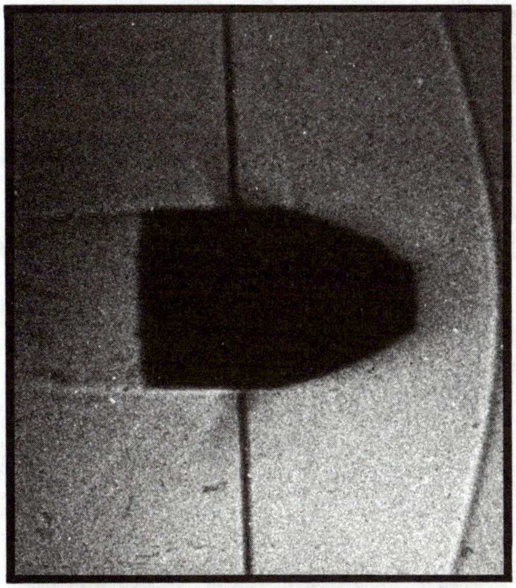

A

Shadowgraph of #575213 (Old Style) at 1135 fps velocity. Note shock wave in front of bullet, trip wire for camera (vertical straight line) and turbulent wake behind bullet.

B

Shadowgraph of #577611 at 1255 fps velocity.

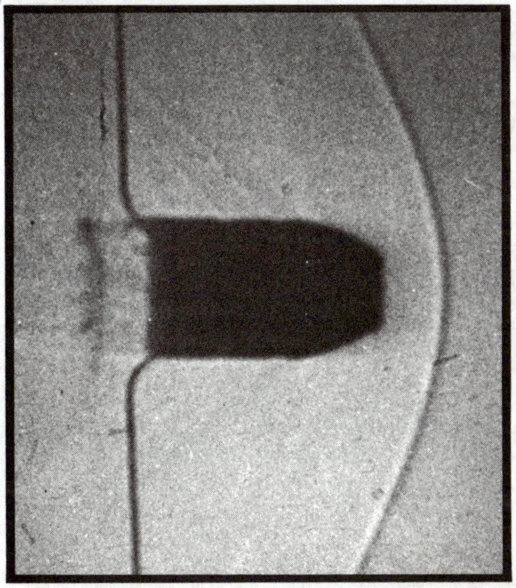

C

Shadowgraph of #454613 at 1604 fps velocity. Note shock waves emanating from lubricating grooves on cylindrical section of bullet.

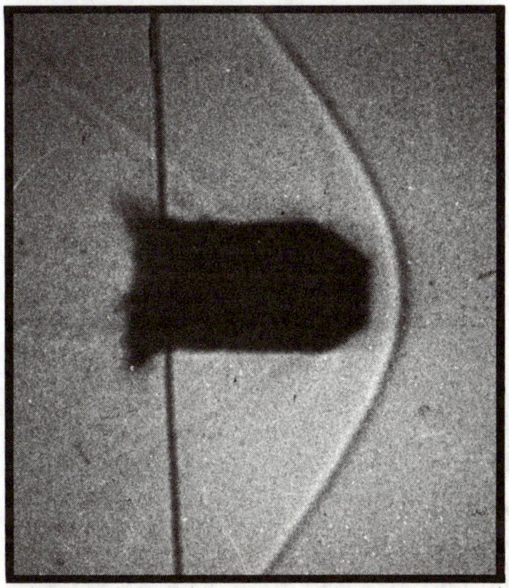

D

Shadowgraph of #445369 at 1636 fps velocity.

The bullet trajectory calculation starts at the muzzle. Muzzle velocity is usually in either the high or intermediate range, and the ballistic coefficient used to start the trajectory computation is the one that applies for whichever range the muzzle velocity lies in. The trajectory calculation then proceeds in a step-by-step manner. At each step the bullet velocity is decreased by a small amount (typically no more than 2 or 3 fps), and the incremental changes in range, drop and time of flight which the bullet undergoes for this small change in velocity are calculated. When bullet velocity crosses the velocity boundary for the range in which the initial ballistic coefficient applies, then the ballistic coefficient value is changed to that of the next range. The step-by-step trajectory calculations continue in this manner, changing ballistic coefficient value whenever a ballistic coefficient velocity range boundary is crossed, until the maximum range limit is reached.

Experiments have shown that this method will allow trajectories to be calculated with a high degree of accuracy. This method of changing ballistic coefficient with bullet velocity allows the drag function of the standard bullet to be matched to the drag characteristic of the actual bullet in a piecewise manner. To do this, the values of **C** for each bullet should be measured in each velocity range. It is reasonably easy to measure the ballistic coefficient of a bullet by means of firing tests, although expensive test equipment and a considerable amount of test time are required to do this. The only alternative approach would be to experimentally determine the drag function of every bullet. This would be extremely expensive and the accuracy of the calculated trajectories would not really be improved.

Both Lyman and Sierra Bullets have used the method described above for bullet trajectory calculations and have invested the resources necessary to measure the ballistic coefficients of their bullets. Sierra's experiments are reported in their reloading manuals. Lyman's experiments began with cast bullets for black powder guns and have been described in the Lyman *Black Powder Handbook, Centennial Journal* and *Cast Bullet Handbook*, 3rd Ed. Altogether Lyman has measured three ballistic coefficient values for Lyman cast bullets used in a centerfire rifle or handgun, and two values for Lyman conical cast bullets intended for black powder rifles or handguns. Three ballistic coefficient values have been determined for round ball projectiles.

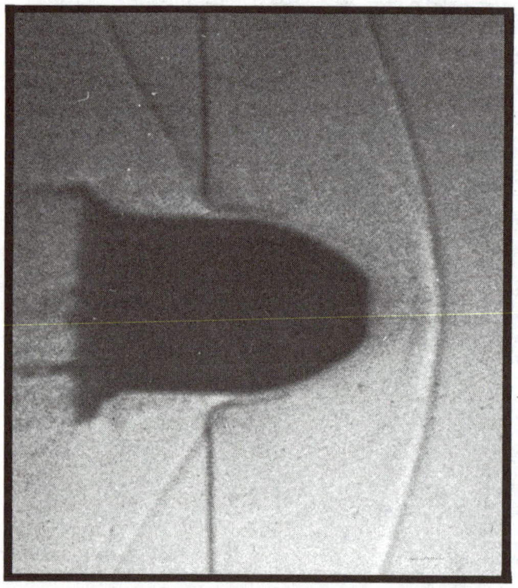

E

Shadowgraph of #575213 (New Style) at 1135 fps velocity. Note extended skirt on bullet and shock wave emanating from cylindrical section well in front of extended skirt.

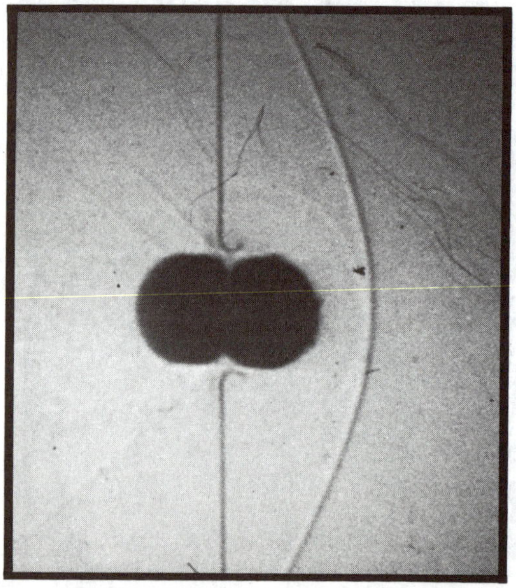

F

Shadowgraph of double .560" round ball load at 1247 fps velocity. The two balls have been pressed together by pressure and heat in the rifle barrel. Also note shock wave emanating from turbulent wake behind projectiles.

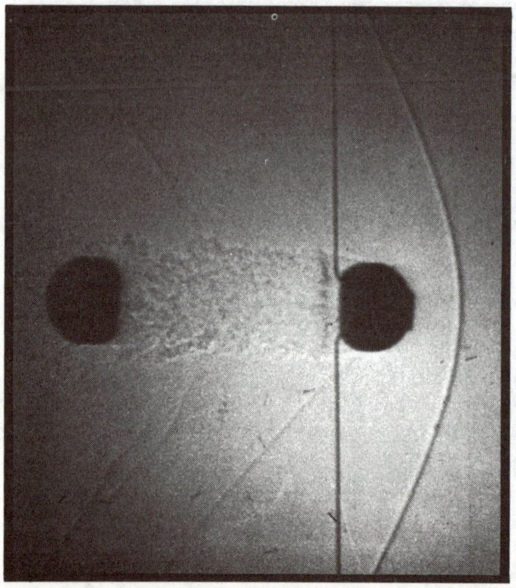

G

Shadowgraph of double .560" round ball load at 1150 fps velocity. The balls have been separated and each is pushing a shock wave.

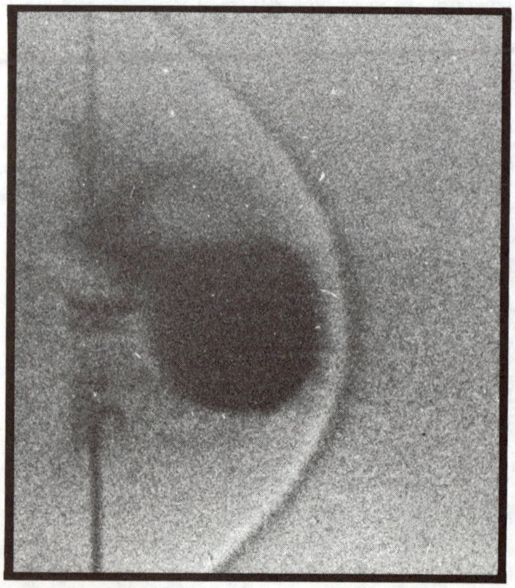

H

Shadowgraph of .445" round ball at 2220 fps velocity. Note the flattening of the ball by the pressure behind the shock wave.

The range instrumentation, firing test program and test method have been explained in detail for the various types of cast bullets in the handbooks cited above. The basic test approach is to measure both the muzzle velocity and time of flight over a precisely measured 50 yard range for each test round fired. Several test rounds (typically five) are fired for each of the three (or two) muzzle velocity levels chosen for the bullet type under test. The muzzle velocity and time of flight measured for each bullet are then fed to a digital computer which calculates the ballistic coefficient for that bullet. The computer first calculates a trajectory over the 50 yard range that the standard bullet would have if it started with the measured muzzle velocity of the test bullet. This trajectory calculation produces a time of flight for the standard bullet. The computer then determines a ballistic coefficient that makes the calculated time of flight equal the measured time of flight for the test bullet. This computed ballistic coefficient is then the value which the test bullet must have had to produce the measured time of flight.

There are two theories about determining ballistic coefficient values from such a collection of test data. One theory says that if all test bullets were perfect, all results for one bullet type at one velocity level would be identical. Then, since imperfections always reduce ballistic coefficient, the correct choice is the highest test value found.

Lyman feels, though, that a shooter really wants to know the *reasonable* performance he can expect from good bullets, and not performance based on bullet perfection. Lyman personnel used the kind of reasonable care in casting the test bullets that a shooter can be expected to use. Consequently, the test values of ballistic coefficient at each velocity level were averaged (since there were small shot-to-shot variations in muzzle velocity at each level).

After this analysis was completed, the trend in the average ballistic coefficient value as a function of velocity was examined for each bullet type. Appropriate velocity boundaries between high, intermediate and low velocity ranges for ballistic coefficient values were established for each type of bullet. A ballistic coefficient value for each velocity range was then determined by interpolating or extrapolating the test data.

The results of all of Lyman's ballistic coefficient investigations have been gathered together in **Tables 1-1, 1-2** and **1-3. Table 1-1** presents ballistic coefficients of cast bullets normally used in centerfire rifle cartridges; **Table 1-2** presents ballistic coefficients of cast bullets normally used in centerfire pistol cartridges; and **Table 1-3** presents ballistic coefficients of cast bullets for black powder shoulder arms and handguns, both rifled and smooth bore.

Tables 1-4 and **1-5** present ballistic coefficients of jacketed rifle and pistol bullets manufactured by Sierra Bullets. These tables cover many bullets in Sierra's catalog, and are published here with the permission of Sierra Bullets.

The accuracy of a calculated bullet trajectory depends on the accuracy of the ballistic coefficient. The ballistic coefficients in these tables are sufficiently accurate that the calculated bullet trajectories are accurate to better than one-half minute of angle. This means that if a shooter sights in his rifle at, say, 50 yards, then trajectory tables will tell him where his bullets will strike at other ranges with an accuracy better than one-half inch at 100 yards, one inch at 200 yards and so forth.

Table 1-1 Ballistic Coefficients of Cast Bullets for Centerfire Rifle Cartridges

Projectile	Weight (grains)	Ballistic Coefficient Velocity Boundaries (fps)		Ballistic Coefficient Values		
		Upper	Lower	High Velocities	Intermediate Velocities	Low Velocities
#225107	37	2400	1800	.091	.099	.108
#255438	41	2400	1800	.094	.107	.122
#225415	45	2400	1800	.116	.140	.168
#225462	54	2400	1800	.107	.131	.152
#245496	83	1600	1200	.202	.215	.230
#245497	90	1600	1200	.230	.240	.260
#245498	95	1600	1200	.210	.225	.245
#257420	68	2200	1800	.129	.152	.180
#257464	90	1800	1400	.204	.217	.234
#257312	88	1800	1400	.208	.236	.273
#257418	105	1800	1400	.221	.253	.296
#257325	112	1800	1400	.235	.263	.263
#280468	122	2200	1600	.193	.224	.254
#280473	124	1800	1400	.275	.290	.310
#280412	136	1800	1400	.245	.255	.270
#287448	119	2000	1600	.165	.205	.245
#287346	135	1800	1400	.235	.265	.305
#287405	150	1600	1200	.245	.265	.285
#287308	162	1600	1200	.325	.345	.365
#311359	113	2000	1600	.181	.182	.182
#311441	115	2200	1600	.143	.152	.163
#311576	120	1800	1400	.172	.196	.220
#311465	122	2200	1600	.163	.171	.195
#311410	130	1600	1200	.239	.250	.265
#311440	151	2200	1600	.134	.151	.171
#311466	151	2000	1600	.250	.250	.260
#301618	160	2400	1800	.300	.310	.305
#311291	169	2200	1600	.202	.231	.250
#31141	170	2200	1600	.220	.250	.280
#311407	173	1800	1400	.270	.300	.325
#311467	178	2200	1600	.320	.305	.300
#311334	187	1800	1400	.340	.312	.275
#301620	200	2200	1800	.379	.370	.362
#311299	200	1800	1400	.377	.358	.390
#311290	208	1800	1400	.305	.300	.275
#311284	210	1800	1400	.332	.331	.335
#323470	160	2200	1600	.187	.205	.225
#323471	214	1800	1400	.475	.450	.425
#323378	242	1600	1200	.410	.420	.430
#375248	248	1600	1200	.290	.310	.325
#375449	264	1600	1200	.315	.348	.375
#375167	264	1600	1200	.315	.290	.270
#457191	293	1800	1400	.201	.222	.240
#457122 H.P.	310	1800	1400	.270	.267	.265
#457122	322	1800	1400	.280	.274	.270
#457124	366	1800	1300	.300	.299	.275
#457193	420	1600	1200	.307	.335	.365
#457102	426	1600	1200	.359	.355	.350
#457121	433	1600	1200	.436	.430	.420
#457406	451	1600	1200	.387	.380	.370
#457125	464	1600	1200	.391	.380	.380
#457132	490	1500	1100	.384	.375	.370
#515141	422	1600	1200	.250	.250	.225

Table 1-2 Ballistic Coefficients of Cast Bullets for Centerfire Handgun Cartridges

Projectile	Weight (grains)	Ballistic Coefficient Velocity Boundaries (fps)		Ballistic Coefficient Values		
		Upper	Lower	High Velocities	Intermediate Velocities	Low Velocities
#225438	45	1300	800	.120	.150	.160
#225415	49	1300	800	.160	.208	.200
#257420	65	1300	800	.160	.200	.150
#257312	89	1300	800	.185	.200	.210
#358242	92	1300	800	.103	.103	.100
#358345	115	1300	800	.120	.120	.100
#358242	121	1300	800	.155	.210	.200
#356402	121	1300	800	.145	.110	.100
#358480	133	1300	800	.145	.125	.100
#358495	141	1300	800	.095	.065	.055
#358477	150	1300	800	.180	.150	.130
#358156	158	1300	800	.210	.160	.100
#358311	158	1300	800	.228	.220	.220
#358429	168	1300	800	.285	.265	.240
#41028	212	1300	800	.160	.170	.150
#410610	215	1300	800	.160	.170	.150
#410459	220	1300	800	.150	.195	.160
#410426	240	1300	800	.235	.300	.235
#429348	180	1300	800	.065	.045	.040
#429303	200	1300	800	.105	.095	.090
#429215	210	1300	800	.190	.215	.150
#429360	232	1300	800	.170	.195	.140
#429421	245	1300	800	.210	.250	.200
#429244	250	1300	800	.205	.260	.200
#45468	175	1300	800	—	.120	.120
#454190	250	1300	800	—	.270	.270
#454424	255	1300	800	—	.210	.160

Table 1-3 Ballistic Coefficients of Cast Bullets for Black Powder Guns

Projectile	Weight (grains)	Ballistic Coefficient Velocity Boundaries (fps)		Ballistic Coefficient Values		
		Upper	Lower	High Velocities	Intermediate Velocities	Low Velocities
#37583	145	—	1100	—	—	0.095
#445599	250	—	1100	—	.126	.110
#445369	291	—	1100	—	.151	.132
#450229	155	—	1100	—	—	.050
#45468	175	—	1100	—	—	.098
#45467	200	—	1100	—	—	.112
#454616	230	—	1100	—	.078	.067
#454613	265	—	1100	—	.156	.130
#454612	300	—	1100	—	.176	.140
#504617	370	—	1100	—	.095	.081
#533476	410	—	1100	—	.137	.123
#575494	315	—	1100	—	.067	.060
#575602	400	—	1100	—	.109	.097
#575213 O.S.	460	—	1100	—	.145	.112
#575213 N.S.	505	—	1100	—	.160	.145
#577611	530	—	1100	—	.180	.165
#57730	570	—	1100	—	.163	.144

(table continued)

Table 1-3 Ballistic Coefficients of Cast Bullets for Black Powder Guns (Continued)

Projectile	Weight (grains)	Ballistic Coefficient Velocity Boundaries (fps)		Ballistic Coefficient Values		
		Upper	Lower	High Velocities	Intermediate Velocities	Low Velocities
.319 RB	49	1300	1000	.045	.037	.031
.350 RB	64	1300	1000	.049	.041	.035
.375 RB	79	1300	1000	.053	.044	.037
.440 RB	128	1300	1000	.062	.051	.043
.445 RB	132	1300	1000	.063	.052	.044
.451 RB	138	1300	1000	.064	.052	.045
.457 RB	143	1300	1000	.064	.053	.045
.495 RB	182	1300	1000	.070	.058	.049
.535 RB	230	1300	1000	.075	.062	.053
.562 RB	267	1300	1000	.079	.065	.055
.735 RB	596	1300	1000	.104	.085	.072

Table 1-4 Ballistic Coefficients of Sierra Rifle Bullets

Projectile	Weight (grains)	Ballistic Coefficient Velocity Boundaries (fps)		Ballistic Coefficient Values		
		Upper	Lower	High Velocities	Intermediate Velocities	Low Velocities
.223 Hornet	40	2800	1800	.130	.132	.140
.223 Hornet	45	2800	1800	.150	.155	.160
.224 Hornet	40	2800	1800	.129	.133	.139
.224 Hornet	45	2800	1800	.149	.154	.159
.224 SMP	45	2800	1800	.213	.192	.170
.224 SPT	45	2800	1800	.226	.206	.195
.224 SMP	50	2800	1800	.235	.212	.188
.224 SPT	50	2800	1800	.250	.232	.210
.224 Blitz	50	2800	1800	.250	.232	.210
.224 SMP	55	2800	1800	.260	.235	.210
.224 SPT	55	2800	1800	.280	.260	.240
.224 Blitz	55	2800	1800	.280	.260	.240
.224 SPBT	55	2800	1800	.285	.260	.230
.224 FMJ	55	2800	1800	.280	.260	.235
.224 SMP	63	2800	1800	.290	.260	.235
.224 MKHPBT	52	2800	1800	.245	.220	.155
.224 MKHP	53	2800	1800	.235	.215	.170
.243 HP	60	2800	1800	.215	.190	.170
.243 HP	75	2800	1800	.265	.235	.210
.243 HPBT	85	2800	1800	.315	.280	.220
.243 SPT	85	2800	1800	.340	.320	.290
.243 SMT	100	2800	1800	.370	.335	.295
.243 SPT	100	2800	1800	.400	.388	.340
.243 SPBT	100	2800	1800	.465	.432	.375
.243 MKHPBT	70	2800	1800	.265	.240	.185
.257 HP	75	2800	1800	.240	.215	.190
.257 SPT	87	2800	1800	.303	.280	.255
.257 HPBT	90	2800	1800	.270	.243	.200
.257 SPT	100	2800	1800	.345	.317	.290
.257 SPT	117	2800	1800	.400	.370	.340
.257 SPBT	117	2800	1800	.435	.390	.350
.257 HPBT	120	2800	1800	.370	.335	.280

(table continued)

Table 1-4 Ballistic Coefficients of Sierra Rifle Bullets (Continued)

Projectile	Weight (grains)	Ballistic Coefficient Velocity Boundaries (fps)		Ballistic Coefficient Values		
		Upper	Lower	High Velocities	Intermediate Velocities	Low Velocities
.264 HP	85	2800	1800	.260	.235	.210
.264 HP	100	2800	1800	.305	.275	.245
.264 SPT	120	2800	1800	.430	.400	.365
.264 SPBT	140	2800	1800	.530	.480	.425
.264 MKHPBT	140	2800	1800	.545	.532	.515
.277 HP	90	2800	1800	.250	.225	.200
.277 SPT	110	2800	1800	.365	.337	.310
.277 SPT	130	2800	1800	.430	.410	.360
.277 SPBT	130	2800	1800	.450	.440	.380
.277 RN	150	2800	1800	.485	.475	.410
.277 SPBT	150	2800	1800	.250	.275	.310
.284 SPT	120	2800	1800	.420	.400	.355
.284 SPT	140	2800	1800	.490	.455	.415
.284 SPBT	140	2800	1800	.500	.460	.415
.284 SPBT	160	2800	1800	.560	.520	.475
.284 RN	170	2800	1800	.275	.300	.340
.284 SPBT	175	2800	1800	.645	.580	.520
.284 MKHPBT	150	2800	1800	.560	.545	.530
.284 MKHPBT	168	2800	1800	.630	.615	.595
.307 HP .30-30	125	2800	1800	.135	.135	.135
.307 FN .30-30	150	2800	1800	.160	.160	.160
.307 FN .30-30	170	2800	1800	.182	.182	.182
.308 RN Carbine	110	2800	1800	.145	.160	.185
.308 FMJ Carbine	110	2800	1800	.145	.160	.185
.308 HP	110	2800	1800	.190	.197	.225
.308 SPT	125	2800	1800	.340	.325	.290
.308 RN	150	2800	1800	.195	.215	.245
.308 SPT	150	2800	1800	.410	.390	.350
.308 SPBT	150	2800	1800	.449	.405	.360
.308 HPBT	165	2800	1800	.420	.380	.320
.308 SPBT	165	2800	1800	.490	.465	.420
.308 RN	180	2800	1800	.235	.260	.295
.308 SPT	180	2800	1800	.460	.460	.445
.308 SPBT	180	2800	1800	.540	.540	.520
.308 SPBT	200	2800	1800	.590	.580	.570
.308 RN	220	2800	1800	.270	.300	.330
.308 MKHPBT	150	2800	1800	.445	.425	.402
.308 MKHPBT	168	2800	1800	.500	.475	.450
.308 MKHPBT	180	2800	1800	.555	.542	.530
.308 MKHPBT	190	2800	1800	.575	.560	.545
.308 MKHPBT	200	2800	1800	.595	.582	.565
.308 MKHPBT	220	2800	1800	.655	.640	.625
.311 SPT	150	2800	1800	.400	.390	.350
.311 SPT	180	2800	1800	.460	.460	.430
.323 SPT	150	2800	1800	.375	.365	.335
.323 SPT	175	2800	1800	.470	.460	.440
.323 SPBT	220	2800	1800	.580	.575	.550
.338 SPBT	250	2800	1800	.600	.595	.575
.358 RN	200	2800	1800	.190	.210	.235
.375 SPBT	300	2800	1800	.590	.580	.560
.458 FN .45-70	300	2800	1800	.160	.160	.160
.458 HP .45-70	300	2800	1800	.160	.160	.160

Table 1-5 Ballistic Coefficients of Sierra Pistol Bullets

Projectile	Weight (grains)	Ballistic Coefficient Velocity Boundaries (fps)		Ballistic Coefficient Values		
		Upper	Lower	High Velocities	Intermediate Velocities	Low Velocities
.355 JHP	90	1300	900	.095	.115	.105
.355 JHP	115	1300	900	.127	.155	.130
.355 FMJ	125	1300	900	.138	.220	.175
.357 JHC	110	1300	900	.120	.135	.120
.357 JHC	125	1300	900	.133	.155	.140
.357 JSP	125	1300	900	.133	.155	.140
.357 JHC	140	1300	900	.145	.160	.135
.357 JHC	150	1300	900	.155	.170	.145
.357 JSP	158	1300	900	.165	.175	.155
.357 FMJ	170	1300	900	.175	.285	.222
.410 JHC	170	1300	900	.123	.122	.120
.410 JHC	210	1300	900	.165	.210	.160
.410 FMJ	220	1300	900	.182	.195	.175
.4295 JHC	180	1300	900	.130	.125	.140
.4295 JHC	210	1300	900	.167	.148	.155
.4295 FMJ	220	1300	900	.172	.185	.165
.4295 JHC	240	1300	900	.185	.172	.171
.4515 JHP	185	1300	900	.100	.110	.135
.4515 FMJ	185	1300	900	.130	.140	.135
.4515 FMJ	230	1300	900	.150	.215	.165
.4515 JHP	240	1300	900	.150	.160	.170

Notes

7

DATA SECTION

INTRODUCTION TO DATA

Important: Please Read

The data in this section cannot be used without a full understanding of what it represents or without complete reloading knowledge. Any attempts to use this data without the necessary cautions, procedures and background knowledge could be extremely dangerous causing personal and property injury, even death.

The Data listed in this section has been tested by Lyman technicians and found safe when loaded with their test components, tooling, procedures and finished cartridge dimensions and when fired under laboratory controlled conditions in the test firearms. The publisher and editors have no control over how reloading is conducted by the individual. Every change in equipment, procedure, component lot, finished cartridge dimensions and firearm used will effect ballistics and/or the safety and usefulness of a load. Therefore, no warranties are implied or expressed for the data and copy contained in this book. We specifically disclaim any warranties of fitness for any and all particular purposes and specifically disclaim any and all liability for consequential damages of any kind.

The individual assumes all risks for the safety of reloaded ammunition. Improperly loaded ammunition, or the failure to follow all necessary precautions, may result in serious personal injury and/or death to the shooter or bystanders.

There are many precautions to which the reloader must adhere. This introduction deals with many of these but cannot possibly foresee, or include, all possible cautions or procedures.

Maximum listed loads must always be worked up to carefully. Always start with the suggested starting load and develop the load slowly, carefully, firing at least ten shots at each incremental propellant charge weight.

Test Components

Keep in mind that the test components are not of Lyman manufacture. Therefore, it is possible that production changes which would effect ballistics or load safety could occur without Lyman's knowledge. If there is any doubt, created for any reason, including unexpected results from even a single shot, immediately stop. Do not proceed with reloading or fire another shot until you have contacted the component manufacturer and cleared up the difficulty.

Test Specifications

These are self explanatory and can sometimes help the shooter form some judgement as to performance in their firearm.

Cartridge Drawing

This information is based on the maximum allowable loaded cartridge dimensions according to accepted industry standards. With the exception of bullet diameters, it is unlikely that the reloader will ever encounter a cartridge with these dimensions unless a problem is present. The listed dimensions, when compared with the reloader's ammo dimensions can be a help in making certain maximum dimensions are not exceeded. For example, if your loaded cartridges have a neck diameter equal or greater than the listed dimension for your cartridge the ammo should not be fired.

Comments

The information given here is to help the reloader with component selection and/or to impart specific information about the cartridge. The comments should always be read for specific loading suggestions which may make the reloader's work go easier.

Bullets

Not every possible bullet has been tested. We have used bullets which we feel are of a popular weight and type. The listed data for each bullet is all of the data we can supply at this time. Keep in mind that seating depths can cause noticeable changes in ballistics. This is especially true for handgun ammunition. It is strongly suggested that the specific cartridge overall length be used whenever it is shown with the bullet weight. Otherwise load as close as possible to the maximum overall cartridge length shown in the cartridge drawing without ever exceeding it.

Powders

We have used powders that were generally popular at the time the specific testing was conducted. Not every cartridge was shot for this edition. Thus, some of the newer propellants may not be listed with some of the less popular loadings or bullets. Where a single or a few specific powders have proven the best possible choices with respect to ballistic uniformity combined with accuracy, we have so stated in the comments section. These comments have replaced the previously used call-outs for the most accurate loads. All such comments have the backing of years of actual shooting experience and the input from thousands of readers. Therefore, when we now suggest accuracy loads it means they have all been proven with actual shooting tests in many rifles.

Suggested Starting Grains

Loads shown in this column are the place to begin load development. Use the exact charge shown. Never go below the starting charge as to do so can sometimes cause dangerous conditions. Lighter charges can sometimes result in dangerously high pressure. Work up from the starting charge to the maximum charge slowly and in small increments. Be sure you have sufficient knowledge to interpret the results of your progressively heavier charges. Of course, you may choose not to use a maximum load and stay at the starting charge or some point between it and the maximum load.

As a beginning handloader you may want to confine all loads to the starting charges. When you are familiar with all the aspects of ammo assembly then you can begin to work up loads.

Maximum Load Grains

Maximum charges are never to be exceeded. THEY ARE NOT DELIBERATELY HELD TO CONSERVATIVE LEVELS. They are the exact charge that produced the maximum pressure level allowed or otherwise proved to be the maximum charge we could list. Never use any maximum charge without carefully working up to it from the Suggested Starting Grains.

It is quite possible that listed maximum loads may not be safe in your firearm, with your components, with your loading procedures or with your finished cartridge dimensions. Use caution when developing a load. If even a single cartridge shows evidence of excessive pressure stop using the load and propellant. Switch to another propellant and begin the process again.

Should trouble continue, stop all loading and contact the component manufacturer.

Maximum load listings are dependent on a number of criteria beyond average pressures of the loads. Ballistics must be uniform or individual high and low pressures could be dangerous. The amount of powder that will fit into the case without excess compression (this will vary with case lots, powder lots, and bullet seating depth), performance at high or low temperatures, muzzle flash and other important factors are considered when we develop maximum loads. Never assume listed data can be exceeded.

Compressed Loads

All compressed loads are indicated with a + sign. Depending upon the volume of the specific case lot being used the load may or may not fit into the case. Never attempt to force feed the powder into the case. If the bullet cannot be started into the case reduce the powder charge to allow for 1/10" of free space in the case neck. Cases which bulge after bullet seating should not be used. Never compress powder so that it causes case bulging. For more information on determining the suitability of a compressed load refer to the chapter on powder.

Filler Wads

A few cast bullet loads are listed which require the use of a Dacron filler wad in the form of 1/4" thick batting. Use these loads only as shown. This filler material can be purchased in most yard-goods stores. The material is cut into squares and placed into the case to seal the powder against the primer flash hole. Never use a wad where the load does not indicate it specifically. Never compress the wad so that a space is present between the wad and the bullet base. The wad should only fill the empty air space between the powder and the bullet base.

17 Remington

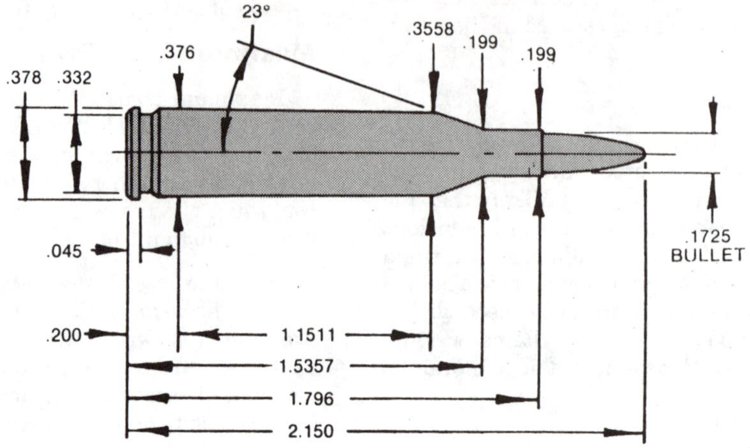

COMMENTS:

Introduced in 1971 by Remington this case, based on the 222 Remington has never been very popular. Most of the lack of popularity stems from practical reasons.

The 17 caliber cartridges tend to foul barrels quickly. With most rifles, if the bore is not thoroughly cleaned every 15 shots or so, accuracy quickly disappears. It is not every gun shop that stocks 17 caliber bullets or the skinny cleaning rods required.

Factory ammo achieves a velocity of about 4,000 fps. The reloader can duplicate this speed using IMR 4320.

TEST COMPONENTS:

Cases .... Remington
Trim-to Length .... 1.786"
Primers .... Remington 7½
Primer Size .... Small Rifle, Magnum
Lyman Shell Holder .... No. 26
Jacketed Bullets Used .... Hornady HP #1710, 25 gr.

TEST SPECIFICATIONS: (Velocity Only)

Firearm Used .... Remington Model 700
Barrel Length .... 24"
Twist .... 1-10"
Groove Dia. .... .172"

25 gr. Jacketed HP
2.150" OAL

Powder	Sugg. Starting Grains	Velocity fps	Pressure C.U.P.	Max. Load Grains	Velocity fps	Pressure C.U.P.
IMR-4198	17.0	3478	—	19.3	3830	—
RX7	17.0	3473	—	19.0	3759	—
IMR-3031	20.0	3452	—	22.0	3857	—
748	20.0	3418	—	22.8	3761	—
BL-C(2)	20.3	3450	—	21.9	3731	—
IMR-4064	21.0	3455	—	23.0	3819	—
IMR-4895	21.5	3616	—	23.3	3934	—
IMR-4320	22.5	3692	—	‡24.2	4023	—

Note: Loads shown in shaded panels are maximum.
‡ Designates a factory velocity duplication load.

22 Hornet
(5.6 x 35mm Rimmed)

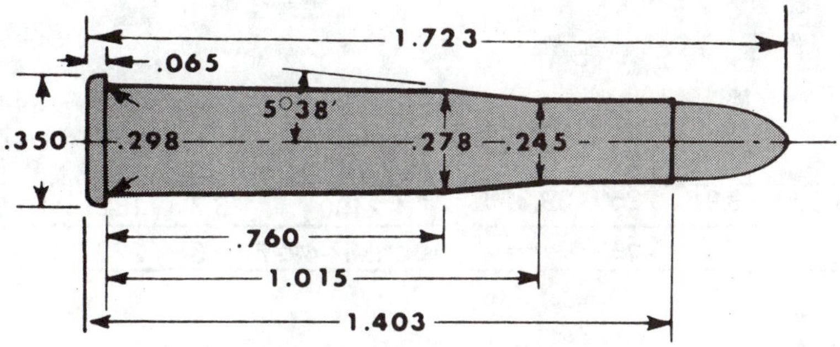

COMMENTS:

Two distinct bullet diameters are used to load this cartridge – .223" and .224 inch. The smaller bullet is used in older rifles which had this groove diameter. Current models have 0.224" groove diameters.

Only a few of the powders available are suitable to the very small case capacity of the 22 Hornet and Winchester has recently announced they are discontinuing one of these – 680. Perhaps the best powder to start with is IMR 4227, along with a 45 grain Hornet bullet.

Bullets with thin jackets, usually specifically designated for the Hornet cartridge are required if reliable expansion is to be obtained.

Hornet cases are not overly strong and the reloader should take great care during case inspection.

TEST COMPONENTS:

Cases .... Remington, Winchester
Trim-to Length .... 1.393"
Primers .... Remington 6½
Primer Size .... Small Rifle
Lyman Shell Holder .... No. 4
Jacketed Bullets Used .... Sierra JSP #1100, 40 gr.
Hornady JSP #2220, 45 gr.
Remington PSP #B22710, 50 gr.
Speer HP, 52 gr.
Remington PSP #B22924, 55 gr.
Cast Bullets Used .... (Sized to .223" dia.)
*Gas Check Bullets .... *#225415, 49 gr.

TEST SPECIFICATIONS:
(Velocity & Pressure)

Firearm Used .... Savage Model 219 and Universal Receiver
Barrel Length .... Savage 26"; Universal Receiver 24"
Twist .... 1-16"
Groove Dia. .... Savage, .224"; Universal Receiver .223"

40 gr. Jacketed SP
1.722" OAL

Powder	Sugg. Starting Grains	Velocity fps	Pressure C.U.P.	Max. Load Grains	Velocity fps	Pressure C.U.P.
H-110	9.5	2502	35,700	13.0+	2936	42,500
**2400	7.2	2124	24,200	8.8	2590	40,600
**IMR-4227	10.0	2427	—	11.7	2777	—
680	11.2	2506	28,300	13.8+	2918	42,500

45 gr. Jacketed SP
1.722" OAL

Powder	Sugg. Starting Grains	Velocity fps	Pressure C.U.P.	Max. Load Grains	Velocity fps	Pressure C.U.P.
H-110	10.0	2431	33,800	12.0+	2764	42,400
**2400	7.0	2065	24,600	8.3	2367	39,100
**IMR-4227	10.0	2364	—	11.3	2631	—
680	11.0	2466	34,500	‡12.5+	2696	42,400

Note: Loads shown in shaded panels are maximum.
\+ Designates a compressed powder charge.
** Designates use of Savage rifle and Remington cases.
‡ Designates a factory velocity duplication load.

22 Hornet
(5.6 x 35mm Rimmed)

50 gr. Jacketed Ptd. SP

Powder	Sugg. Starting Grains	Velocity fps	Pressure C.U.P.	Max. Load Grains	Velocity fps	Pressure C.U.P.
**2400	6.8	1905	24,800	8.2	2250	39,300
**IMR-4227	10.0	2375	—	11.0	2544	—

52 gr. Jacketed HP

Powder	Sugg. Starting Grains	Velocity fps	Pressure C.U.P.	Max. Load Grains	Velocity fps	Pressure C.U.P.
**2400	6.7	1821	21,900	8.0	2204	41,100
**IMR-4227	9.5	2237	—	10.5	2427	—

55 gr. Jacketed Ptd. SP

Powder	Sugg. Starting Grains	Velocity fps	Pressure C.U.P.	Max. Load Grains	Velocity fps	Pressure C.U.P.
2400	6.5	1750	23,000	7.8	2116	41,600
IMR-4227	8.5	2183	—	10.0+	2341	—
IMR-4198	9.5	2000	—	11.0+	2188	—

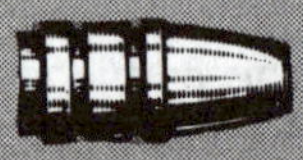

#225415
49 gr. (#2 Alloy) 1.694" OAL

Powder	Sugg. Starting Grains	Velocity fps	Pressure C.U.P.	Max. Load Grains	Velocity fps	Pressure C.U.P.
**Red Dot	2.0	1060	—	3.7	1645	—
**700X	2.0	1050	—	3.7	1665	—
**Green Dot	2.5	1230	—	4.2	1765	—
**PB	2.5	1185	—	4.2	1705	—
**Unique	3.0	1320	—	4.6	1805	—
**SR-7625	3.0	1350	—	4.4	1745	—
H-110	4.2	1124	—	6.5	1722	20,700
SR-4759	5.6	1266	13,500	7.5	1821	25,700
680	5.0	1199	—	7.5	1828	21,800
RX7	6.5	1159	—	9.5	1785	15,300

Note: Loads shown in shaded panels are maximum.
+ Designates a compressed powder charge.
** Designates use of Savage rifle and Remington cases.

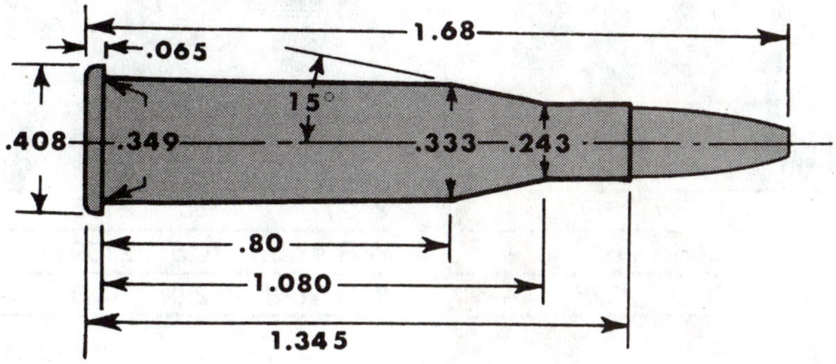

COMMENTS:

The reloader cannot generally duplicate advertised factory ballistics with canister powders currently available. When loading ammo for use in tubular magazines DO NOT USE ANY SHARP NOSED bullets. Cartridges for tubular magazines must be loaded only with bullets having blunt noses.

Hornet style bullets must be used if expansion is to be certain.

IMR 4227 seems the best powder to begin with, along with bullets of 45-grains.

Accuracy in lever action rifles is seldom impressive but when used in a bolt or single shot action the 218 Bee often gives accuracy approaching that of the 22 Hornet.

TEST COMPONENTS:

Cases .... Remington
Trim-to Length .... 1.335"
Primers .... Remington 6½
Primer Size .... Small Rifle
Lyman Shell Holder .... No. 10
Jacketed Bullets Used .... Speer Spire Pt., 40 gr.
Remington HP #B22702, 45 gr.
Remington Ptd. S.P. #B22710, 50 gr.
Speer HP, 52 gr.
Remington Ptd. S.P. #B22924, 55 gr.
Cast Bullets Used .... (Sized to .223" dia.)
*Gas Check Bullets .... *#225415, 45 gr.

TEST SPECIFICATIONS: (Velocity Only)

Firearm Used .... Winchester Low Wall; Marlin Over and Under
Barrel Length .... Winchester 20"; Marlin 26"
Twist .... 1-16"
Groove Dia. .... .223"

40 gr. Jacketed Spire Pt.

Powder	Sugg. Starting Grains	Velocity fps	Pressure C.U.P.	Max. Load Grains	Velocity fps	Pressure C.U.P.
2400	11.0	2645	—	12.5	2906	—
IMR-4227	11.0	2457	—	13.5	2932	—
IMR-4198	14.0	2564	—	15.0+	2762	—

45 gr. Jacketed HP

Powder	Sugg. Starting Grains	Velocity fps	Pressure C.U.P.	Max. Load Grains	Velocity fps	Pressure C.U.P.
2400	10.5	2463	—	12.0	2777	—
IMR-4227	11.5	2481	—	13.0	2754	—
IMR-4198	13.0	2309	—	14.0	2564	—

Note: Loads shown in shaded panels are maximum.
+ Designates a compressed powder charge.

218 Bee

50 gr. Jacketed Ptd. SP

Powder	Sugg. Starting Grains	Velocity fps	Pressure C.U.P.	Max. Load Grains	Velocity fps	Pressure C.U.P.
2400	10.0	2331	—	11.5	2617	—
IMR-4227	11.0	2331	—	12.5	2617	—
IMR-4198	12.0	2105	—	13.0	2325	—

52 gr. Jacketed HP

Powder	Sugg. Starting Grains	Velocity fps	Pressure C.U.P.	Max. Load Grains	Velocity fps	Pressure C.U.P.
2400	9.5	2178	—	10.5	2487	—
IMR-4227	10.5	2159	—	11.5	2506	—
IMR-4198	12.0	2083	—	13.5	2267	—

55 gr. Jacketed Ptd. SP

Powder	Sugg. Starting Grains	Velocity fps	Pressure C.U.P.	Max. Load Grains	Velocity fps	Pressure C.U.P.
2400	9.5	2183	—	10.5	2369	—
IMR-4227	10.0	2087	—	11.5	2380	—
IMR-4198	12.0	2145	—	13.2	2309	—

#225415
45 gr. (#2 Alloy) 1.595" OAL

Powder	Sugg. Starting Grains	Velocity fps	Pressure C.U.P.	Max. Load Grains	Velocity fps	Pressure C.U.P.
**Red Dot	2.8	1295	—	4.7	1805	—
**700X	2.8	1310	—	4.7	1805	—
**Green Dot	3.3	1410	—	5.2	1915	—
**PB	3.3	1365	—	5.2	1845	—
**Unique	3.7	1450	—	5.7	1960	—
**SR-7625	3.8	1480	—	6.0	2005	—

Note: Loads shown in shaded panels are maximum.
** Tested with a Winchester Low Wall

222 Remington
(5.7 x 43 mm)

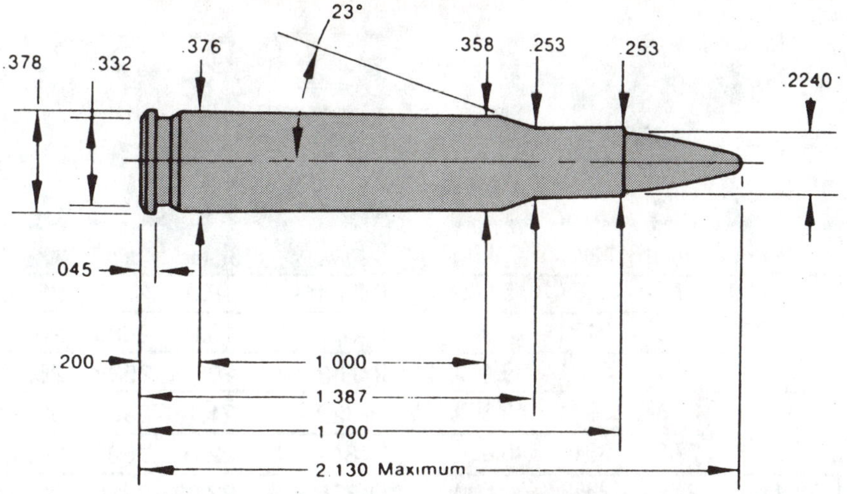

COMMENTS:

The best jacketed bullet weights are 50 to 53 grains and the first powder choice might be Hodgdon H322. But the triple deuce has shot spectacular groups with all of the listed powders. It is slightly more accurate than the 223 but it has been all but displaced by the latter.

The 222 Remington requires jacketed bullets designed for modest velocity expansion. These often carry names that suggest rapid expansion, i.e. Blitz. However do not use Hornet style bullets as these may suffer premature breakup.

Cast bullets when made hard and fired between 2,000 to 2,300 fps. work very well. Best results are with gas checks and Lyman Alox bullet lubricant.

TEST COMPONENTS:

Cases .... Remington
Trim-to Length .... 1.690"
Primers .... Remington 7½
Primer Size .... Small Rifle, Magnum
Lyman Shell Holder .... No. 26
Jacketed Bullets Used .... Speer Sp. Pt., 40 gr.
Sierra Sp., 45 gr.
Sierra Sp. #1340, 50 gr.
Sierra HP, #1400, 53 gr.
Sierra SPT, #1360, 55 gr.
Hornady HP, #2275, 60 gr.
Sierra SMP, #1370, 63 gr.
Cast Bullets Used .... (Sized to .224" dia.)
*Gas Check Bullets *#225415, 45 gr.
*#225462, 54 gr.
*#225646, 55 gr.

TEST SPECIFICATIONS:
(Velocity & Pressure)

Firearm Used .... Universal Receiver
Barrel Length .... 26"
Twist .... 1-14"
Groove Dia. .... .224"

40 gr. Jacketed Spire Pt.

Powder	Sugg. Starting Grains	Velocity fps	Pressure C.U.P.	Max. Load Grains	Velocity fps	Pressure C.U.P.
IMR-4198	18.0	2688	24,000	22.0+	3460	44,600
RX7	19.5	2994	29,400	23.5+	3584	44,600
IMR-3031	20.0	2283	19,200	23.5+	3076	33,500
H-335	24.0	2958	30,000	28.0+	3509	46,300
BL-C(2)	23.5	2898	30,000	27.5+	3484	44,600
IMR-4064	21.0	2481	25,200	23.6+	2849	30,500
IMR-4895	21.0	2590	26,400	25.0+	3125	36,400
IMR-4320	22.0	2583	27,600	26.0+	3067	36,900

Note: Loads shown in shaded panels are maximum.
\+ Designates a compressed powder charge.

222 Remington
(5.7 x 43 mm)

45 gr. Jacketed Spitzer

Powder	Sugg. Starting Grains	Velocity fps	Pressure C.U.P.	Max. Load Grains	Velocity fps	Pressure C.U.P.
IMR-4198	18.0	2672	24,600	21.5+	3333	45,400
RX7	19.0	2865	27,000	22.7	3472	45,900
IMR-3031	20.0	2538	27,600	23.5+	3134	39,400
H-335	23.5	2906	31,500	27.5+	3378	46,300
BL-C(2)	23.0	2857	30,500	27.0+	3345	45,400
IMR-4064	20.0	2331	24,600	23.6+	2857	32,000
IMR-4895	21.0	2583	27,000	25.0+	3115	39,400
IMR-4320	22.0	2570	27,000	26.0+	3039	37,700

50 gr. Jacketed SP
2.125" OAL

Powder	Sugg. Starting Grains	Velocity fps	Pressure C.U.P.	Max. Load Grains	Velocity fps	Pressure C.U.P.
IMR-4198	18.0	2717	28,200	20.5	3205	45,900
RX7	18.0	2680	25,200	21.5	3179	43,200
IMR-3031	20.0	2538	26,400	‡23.5+	3115	42,000
H-322	21.1	2907	33,500	24.0+	3235	45,500
748	22.0	2691	26,800	25.0+	3032	38,900
H-335	23.0	2816	32,000	26.8+	3333	43,400
BL-C(2)	23.0	2865	34,000	26.3+	3172	44,600
IMR-4064	20.0	2314	23,400	23.6+	2849	34,000
IMR-4895	22.0	2762	31,500	25.0+	3105	42,400
IMR-4320	23.0	2754	34,000	26.0+	3048	42,400
AA2015BR	20.7	2875	32,600	23.9	3267	46,200
AA2460	21.0	2749	32,400	23.5	3050	43,000

53 gr. Jacketed HP
2.130" OAL

Powder	Sugg. Starting Grains	Velocity fps	Pressure C.U.P.	Max. Load Grains	Velocity fps	Pressure C.U.P.
IMR-4198	17.1	2705	31,700	19.5	3025	45,500
RX7	18.0	2645	28,800	21.4	3205	46,300
IMR-3031	20.0	2544	27,600	23.5+	3039	41,100
H-322	19.8	2738	34,300	22.5+	3083	45,500
748	21.5	2651	29,000	24.5	3024	44,600
H-335	21.5	2781	31,800	24.5	3133	45,100
BL-C(2)	22.0	2638	30,000	25.0+	3025	45,300
IMR-4064	20.0	2341	25,800	23.6+	2840	36,900
IMR-4895	22.0	2747	34,000	25.0+	3058	45,900
IMR-4320	20.3	2409	27,800	23.0+	2807	39,500
AA2230	20.2	2601	29,200	23.5	3060	46,200
AA2015BR	19.2	2703	32,200	22.1	3084	45,500
AA2460	19.0	2410	23,900	23.6	3004	45,900
RX12	23.0	2699	26,800	25.6+	3022	35,300

55 gr. Jacketed Spitzer
2.130" OAL

Powder	Sugg. Starting Grains	Velocity fps	Pressure C.U.P.	Max. Load Grains	Velocity fps	Pressure C.U.P.
IMR-4198	17.0	2538	27,000	19.5	2980	45,100
RX7	18.0	2717	30,000	21.0	3174	46,300
IMR-3031	20.0	2531	28,800	23.5+	3048	45,400
H-322	19.3	2609	27,800	‡22.0	2990	45,500
748	22.0	2670	27,700	25.0+	3026	38,500
H-335	21.5	2544	29,400	26.3+	3225	45,600
BL-C(2)	22.0	2680	32,000	25.0	3086	45,900
IMR-4064	20.0	2347	25,800	23.6+	2849	37,700
IMR-4895	20.7	2454	28,000	23.5+	2837	38,800
IMR-4320	20.2	2394	26,300	23.0+	2744	39,000
AA2230	19.6	2550	27,700	23.0	2975	45,900
AA2015BR	19.0	2605	26,200	22.1	3098	46,700
AA2460	20.2	2555	30,200	23.2	2943	46,100
RX12	22.0	2507	25,800	25.7+	3053	40,700

Note: Loads shown in shaded panels are maximum.
+ Designates a compressed powder charge.
‡ Designates a factory velocity duplication load.

222 Remington
(5.7 x 43 mm)

60 gr. Jacketed HP
2.130" OAL

Powder	Sugg. Starting Grains	Velocity fps	Pressure C.U.P.	Max. Load Grains	Velocity fps	Pressure C.U.P.
IMR-4198	16.3	2473	30,000	18.5	2787	44,300
RX7	17.0	2590	31,000	20.6	2953	45,300
H-322	19.3	2579	30,800	22.0+	2906	42,700
748	21.5	2554	27,500	24.5+	2894	39,400
H-335	21.0	2487	28,800	25.0	3012	45,900
BL-C(2)	20.0	2433	28,200	24.1+	2808	43,100
IMR-4064	19.5	2267	28,200	23.6+	2785	41,600
IMR-4895	20.5	2469	32,000	23.0+	2748	38,700
IMR-4320	21.5	2450	34,000	23.0+	2680	37,200
AA2230	19.6	2462	31,000	23.0	2923	45,400
AA2015BR	18.5	2527	30,300	21.5	2907	46,000
AA2460	19.0	2366	30,000	22.1	2749	43,600

63 gr. Jacketed Ptd. SP
2.130" OAL

Powder	Sugg. Starting Grains	Velocity fps	Pressure C.U.P.	Max. Load Grains	Velocity fps	Pressure C.U.P.
IMR-4198	16.2	2314	27,000	19.3	2816	45,000
RX7	16.5	2398	28,800	20.0	2881	45,400
IMR-3031	19.0	2325	29,400	21.0+	2648	35,500
H-322	19.3	2586	32,500	22.0+	2926	45,500
H-335	21.0	2433	27,600	25.0	2949	43,700
BL-C(2)	20.0	2304	26,400	24.0	2915	45,900
IMR-4064	19.0	2141	25,200	23.6+	2785	40,700
IMR-4895	20.0	2364	30,000	23.0+	2747	40,400
AA2230	19.0	2384	28,800	22.3	2777	45,400
AA2015BR	17.8	2394	28,800	21.0	2836	45,200
AA2460	19.0	2358	30,600	22.4	2786	45,900

#225415
45 gr. (#2 Alloy) 2.107" OAL

Powder	Sugg. Starting Grains	Velocity fps	Pressure C.U.P.	Max. Load Grains	Velocity fps	Pressure C.U.P.
Red Dot	5.1	1785	25,200	7.4	2217	39,000
700X	4.7	1709	27,000	6.4	2053	39,900
Green Dot	5.3	1779	26,400	7.7	2232	40,700
PB	5.2	1718	25,800	7.0	2024	39,400
Unique	6.0	1972	25,800	8.0	2358	36,000
SR-7625	5.7	1785	24,000	7.0	2016	35,500
Herco	6.5	1956	27,000	8.8	2336	40,700
SR-4756	6.0	1769	21,600	8.7	2267	40,300
2400	11.0	2153	15,500	15.8	2835	31,300
SR-4759	12.2	2185	15,300	16.5	2895	36,700
IMR-4227	13.0	2314	19,700	15.9	2751	24,200
748	17.0	2019	12,100	25.0	2881	25,400

#225462
54 gr. (#2 Alloy) 2.031" OAL

Powder	Sugg. Starting Grains	Velocity fps	Pressure C.U.P.	Max. Load Grains	Velocity fps	Pressure C.U.P.
Red Dot	4.7	1590	27,000	7.0	2044	40,300
700X	4.3	1510	24,600	6.2	1855	39,400
Green Dot	5.0	1595	27,000	7.4	1980	40,700
PB	4.8	1510	25,800	6.5	1828	40,700
Unique	5.7	1795	25,800	8.0	2222	39,900
SR-7625	5.1	1565	27,600	6.5	1815	39,900
Herco	6.2	1808	27,000	8.6	2227	39,400
SR-4756	6.0	1666	26,400	8.1	2044	40,300
2400	11.0	2154	19,100	14.8	2820	43,700
SR-4759	11.8	2167	20,000	16.0	2805	43,500
748	19.2	2163	14,200	24.0	2836	30,700

Note: Loads shown in shaded panels are maximum.
+ Designates a compressed powder charge.

222 Remington
(5.7 x 43 mm)

#225646
55 gr. (#2 Alloy) 2.130" OAL

Powder	Sugg. Starting Grains	Velocity fps	Pressure C.U.P.	Max. Load Grains	Velocity fps	Pressure C.U.P.
Red Dot	5.5	1658	22,600	7.5	2033	38,300
700X	5.6	1728	23,700	7.6	2109	42,200
Green Dot	5.9	1728	21,800	7.7	2077	42,100
PB	5.7	1639	21,800	7.5	1958	41,100
Unique	6.2	1754	21,000	8.0	2115	36,400
SR-7625	5.6	1607	23,900	8.0	1988	41,000
Herco	6.2	1741	20,000	8.5	2172	39,700
SR-4756	6.0	1630	19,000	8.3	2045	41,800
2400	10.5	2170	21,800	12.1	2436	35,800
SR-4759	12.0	2261	26,200	14.0	2569	42,800
748	20.0	2261	17,100	26.0	3058	42,700

Note: Loads shown in shaded panels are maximum.

222 Remington Magnum

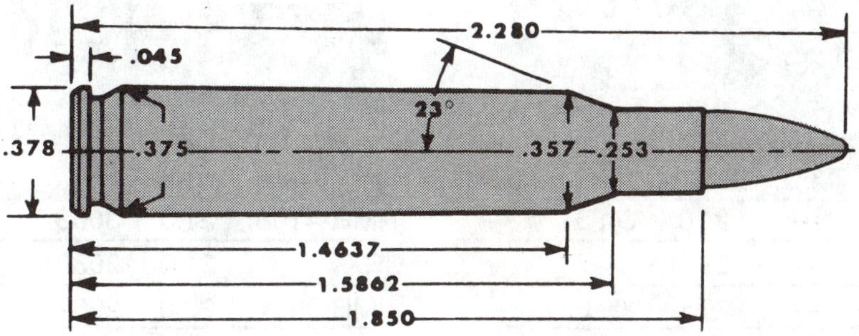

COMMENTS:

This cartridge is all but obsolete, having been the forerunner to the 223 Remington. Owners of rifles chambered for both the Remington 223 and 222 Magnum cartridges must use extreme care not to use the wrong ammo in a rifle.

Powders such as IMR 4198, IMR 4895, Reloder 7, and H335 are all good choices with jacketed bullets from 52 to 60 grains.

Cast bullets driven at 2,000 to 2,300 fps. perform best, especially when lubed with Lyman Alox and gas checked.

TEST COMPONENTS:

Cases .... Remington
Trim-to Length .... 1.840"
Primers .... Remington 7½
Primer Size .... Small Rifle, Magnum
Lyman Shell Holder .... No. 26
Jacketed Bullets Used .... Speer Spire Pt., 40 gr.
Sierra Spitzer, 45 gr.
Sierra Spitzer, 50 gr.
Speer HP, 52 gr.
Remington Ptd. SP #B22924, 55 gr.
Sierra SMPT, 63 gr.
Cast Bullets Used .... (Sized to .224" dia.)
*Gas Check Bullets *#225415, 45 gr.
*#225462, 54 gr.

TEST SPECIFICATIONS: (Velocity Only)

Firearm Used .... Remington 700, Jacketed Bullets; Remington 722, Cast Bullets
Barrel Length .... 24"
Twist .... 1-14"
Groove Dia. .... .224"

40 gr. Jacketed Spire Pt.

Powder	Sugg. Starting Grains	Velocity fps	Pressure C.U.P.	Max. Load Grains	Velocity fps	Pressure C.U.P.
IMR-4198	22.0	3322	—	24.5	3787	—
RX7	22.0	3333	—	24.0	3598	—
IMR-3031	23.0	3012	—	26.5+	3636	—
H-335	26.0	3205	—	29.0	3597	—
BL-C(2)	26.0	3125	—	29.0	3497	—
IMR-4064	24.0	2754	—	27.0+	3322	—
IMR-4895	25.0	3086	—	27.0+	3427	—
IMR-4320	26.0	3144	—	28.0+	3448	—

Note: Loads shown in shaded panels are maximum.
\+ Designates a compressed powder charge.

222 Remington Magnum

45 gr. Jacketed Spitzer

Powder	Sugg. Starting Grains	Velocity fps	Pressure C.U.P.	Max. Load Grains	Velocity fps	Pressure C.U.P.
IMR-4198	22.0	3278	—	24.0	3663	—
RX7	22.0	3236	—	23.5	3427	—
IMR-3031	23.0	3039	—	26.5+	3546	—
H-335	25.0	2906	—	28.5	3484	—
BL-C(2)	25.0	2958	—	28.5	3333	—
IMR-4064	24.0	2739	—	27.0+	3247	—
IMR-4895	24.0	2840	—	27.0+	3333	—
IMR-4320	25.0	2873	—	28.0+	3367	—

50 gr. Jacketed Spitzer

Powder	Sugg. Starting Grains	Velocity fps	Pressure C.U.P.	Max. Load Grains	Velocity fps	Pressure C.U.P.
IMR-4198	21.0	3003	—	23.5	3481	—
RX7	21.0	3030	—	23.0	3300	—
IMR-3031	23.0	2967	—	26.5+	3497	—
H-335	25.0	2976	—	28.0	3378	—
BL-C(2)	25.0	2906	—	28.0	3300	—
IMR-4064	24.0	2770	—	27.0+	3225	—
IMR-4895	24.0	2857	—	27.0+	3311	—
IMR-4320	25.0	2873	—	27.2+	3236	—

52 gr. Jacketed HP

Powder	Sugg. Starting Grains	Velocity fps	Pressure C.U.P.	Max. Load Grains	Velocity fps	Pressure C.U.P.
IMR-4198	21.0	3012	—	23.3	3436	—
RX7	20.0	2890	—	22.8	3268	—
IMR-3031	23.0	2941	—	25.5+	3367	—
H-335	24.0	2652	—	27.5	3236	—
BL-C(2)	24.0	2680	—	27.5	3125	—
IMR-4064	24.0	2564	—	27.0+	3194	—
IMR-4895	24.0	2808	—	27.0+	3279	—
IMR-4320	24.0	2754	—	26.2	3225	—

55 gr. Jacketed Ptd. SP

Powder	Sugg. Starting Grains	Velocity fps	Pressure C.U.P.	Max. Load Grains	Velocity fps	Pressure C.U.P.
IMR-4198	21.0	3039	—	23.0	3378	—
RX7	20.0	2890	—	22.5	3194	—
IMR-3031	23.0	2949	—	25.0+	3290	—
H-335	24.0	2793	—	27.2	3194	—
BL-C(2)	24.0	2724	—	27.0	3125	—
IMR-4064	24.0	2762	—	27.0+	3236	—
IMR-4895	24.0	2840	—	‡26.5	3247	—
IMR-4320	23.0	2617	—	25.5	3076	—

Note: Loads shown in shaded panels are maximum.
\+ Designates a compressed powder charge.
‡ Designates a factory velocity duplication load.

222 Remington Magnum

63 gr. Jacketed Ptd. SP

Powder	Sugg. Starting Grains	Velocity fps	Pressure C.U.P.	Max. Load Grains	Velocity fps	Pressure C.U.P.
IMR-4198	20.0	2840	—	22.5	3205	—
RX7	20.0	2816	—	22.0	3039	—
IMR-3031	21.0	2659	—	23.0	2915	—
H-335	23.0	2702	—	26.5	3086	—
BL-C(2)	23.0	2659	—	26.5	3021	—
IMR-4064	22.0	2506	—	25.5+	3012	—
IMR-4895	23.0	2672	—	25.5	3303	—
IMR-4320	23.0	2631	—	24.5	2865	—

#225415
45 gr. (#2 Alloy) 2.150" OAL

Powder	Sugg. Starting Grains	Velocity fps	Pressure C.U.P.	Max. Load Grains	Velocity fps	Pressure C.U.P.
Red Dot	5.6	1800	—	9.1	2335	—
700X	5.6	1800	—	8.0	2175	—
Green Dot	6.2	1870	—	9.2	2295	—
PB	6.3	1800	—	9.0	2190	—
Unique	6.8	1925	—	9.7	2325	—
SR-7625	6.7	1855	—	9.1	2205	—

#225462
54 gr. (#2 Alloy) 2.075" OAL

Powder	Sugg. Starting Grains	Velocity fps	Pressure C.U.P.	Max. Load Grains	Velocity fps	Pressure C.U.P.
Red Dot	5.4	1675	—	8.9	2175	—
700X	5.4	1660	—	7.8	2010	—
Green Dot	6.0	1750	—	9.0	2165	—
PB	6.1	1675	—	8.8	2045	—
Unique	6.6	1800	—	9.5	2195	—
SR-7625	6.5	1720	—	8.9	2030	—

Note: Loads shown in shaded panels are maximum.
+ Designates a compressed powder charge.

22 PPC

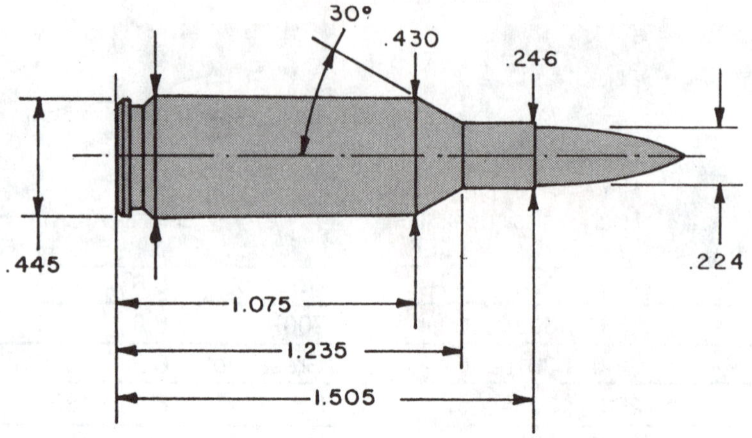

COMMENTS:

The 22 PPC is the necked down version of the 6 mm PPC. It is identical to the 6mm PPC except for the cartridge neck and bullet diameters. This is without doubt the second most accurate cartridge available, being bested only by its 6mm near twin.

The PPC nomenclature is taken from the initials of its designers — Pindell, Palmisano Cartridge. It has won many benchrest shoots, but the 6mm version is the real champ in competition. This round makes a good selection for varmint shooting. Hodgdon's H322 is the single best propellant selection for jacketed bullets, though all listed types will shoot very well indeed. Match grade bullets are required to get all that this cartridge can deliver.

If you use cases other than Sako PPC, USA, approach all loads with extreme caution.

TEST COMPONENTS:

Cases .... Sako PPC, USA
Trim-to Length .... 1.495"
Primers .... Remington 7½
Primer Size .... Small Rifle, Magnum
Lyman Shell Holder .... No. 3
Jacketed Bullets Used .... Speer HP, #1030, 50 gr.
Sierra HPBT, #1410, 52 gr.
Sierra SPT, #1360, 55 gr.
Hornady HP, #2275, 60 gr.

TEST SPECIFICATIONS:
(Velocity & Pressure)

Firearm Used .... Universal Receiver
Barrel Length .... 22"
Twist .... 1-14"
Groove Dia. .... .2245"

50 gr. Jacketed HP
2.025" OAL

Powder	Sugg. Starting Grains	Velocity fps	Pressure C.U.P.	Max. Load Grains	Velocity fps	Pressure C.U.P.
IMR-4198	21.0	2959	31,400	23.0	3295	44,200
H-322	23.0	3030	30,200	26.0	3359	41,200
AA2230	25.0	3039	29,700	28.0	3388	41,500
748	27.0	3028	29,400	30.0	3396	42,500
BL-C(2)	27.0	3051	32,600	30.0	3350	43,200
AA2015BR	22.0	2991	29,200	25.5	3391	41,300
IMR-4895	25.0	2979	30,300	28.0+	3353	42,400

Note: Loads shown in shaded panels are maximum.
\+ Designates a compressed powder charge.

22 PPC

52 gr. Jacketed HPBT
2.025" OAL

Powder	Sugg. Starting Grains	Velocity fps	Pressure C.U.P.	Max. Load Grains	Velocity fps	Pressure C.U.P.
IMR-4198	20.0	2995	33,000	23.0	3338	48,200
H-322	23.0	3038	34,400	26.0	3369	45,900
AA2230	24.5	3007	32,600	27.5	3340	44,100
748	27.0	3053	33,200	30.0	3405	45,900
BL-C(2)	27.0	3043	35,200	30.0	3335	46,800
AA2015BR	22.0	2957	31,200	25.5	3382	43,700
IMR-4895	25.0	2993	32,200	28.0+	3378	47,200

55 gr. Jacketed Spitzer
2.025" OAL

Powder	Sugg. Starting Grains	Velocity fps	Pressure C.U.P.	Max. Load Grains	Velocity fps	Pressure C.U.P.
IMR-4198	20.0	2979	35,400	23.0	3307	49,500
RX7	21.0	3003	40,800	23.0	3192	49,100
H-322	23.0	3040	38,100	25.5	3281	45,400
AA2230	24.0	2854	30,200	27.5	3311	45,700
748	26.5	2913	30,400	29.5	3262	41,700
BL-C(2)	27.5	3030	36,800	30.0	3320	48,600
AA2015BR	23.0	2914	31,600	26.0	3313	43,700
IMR-4895	25.5	2923	32,400	28.5+	3338	46,300

60 gr. Jacketed HP
2.115" OAL

Powder	Sugg. Starting Grains	Velocity fps	Pressure C.U.P.	Max. Load Grains	Velocity fps	Pressure C.U.P.
IMR-4198	19.0	2668	34,200	22.0	2998	46,400
RX7	19.0	2617	35,100	22.2	2920	45,800
H-322	22.5	2814	37,000	26.0	3176	49,200
AA2230	23.5	2760	34,200	27.2	3162	46,200
748	25.5	2827	32,800	29.4	3184	45,600
BL-C(2)	24.7	2739	33,400	29.5	3205	48,100
AA2015BR	21.5	2767	34,000	25.3	3194	48,000
IMR-4895	23.5	2681	29,200	27.3	3187	48,500

Note: Loads shown in shaded panels are maximum.
+ Designates a compressed powder charge.

223 Remington
(5.56 mm)

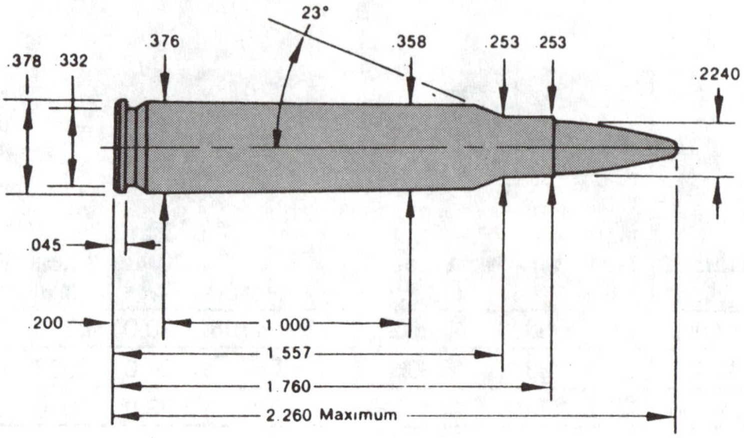

COMMENTS:

This is the current official cartridge of our armed forces. It is extremely accurate when fired in a suitable rifle and has a good range capability for varmint hunting.

Propellant choice can include numerous types which all deliver balanced accuracy and velocity. However if forced to suggest the most likely candidate we would pick Hodgdon's H335 with jacketed bullets. Match bullets will be needed to get the very best results.

Always use bullets with a cannelure and crimp the cases to them when assembling ammo to be used in a semi-automatic rifle.

Our favorite load for this cartridge is a 52 grain Sierra boat tail hollow point with 27.0 grains of H335.

Cast bullet accuracy is often best at velocities of 2,000 fps to 2,300 fps. Best of all was bullet #225646 at about 2,100 fps.

Rifles with fast twist rates (1-8") will do best with bullets of 60 grains and heavier while rifles with the standard twist (1-14") will do best with bullets of 60 grains or less.

TEST COMPONENTS:

Cases .... Remington
Trim-to Length .... 1.750"
Primers .... Remington 7½ and CCI 450
Primer Size .... Small Rifle, Magnum
Lyman Shell Holder .... No. 26
Jacketed Bullets Used .... Speer Spire Pt., 40 gr.
Sierra Spitzer, 45 gr.
Sierra SP Blitz #1340, 50 gr.
Sierra HPBT #1410, 52 gr.
Sierra Spire Pt. #1360, 55 gr.
Sierra SMP #1370, 63 gr.
Cast Bullets Used .... (Sized to .224" dia.)
*Gas Check Bullets
*#225415, 45 gr.
*#225462, 54 gr.
*#225646, 55 gr.

TEST SPECIFICATIONS:
(Velocity & Pressure)

Firearm Used .... Colt AR15; Universal Receiver
Barrel Length .... Colt 20"; Universal Receiver 24"
Twist .... 1-12"
Groove Dia. .... .224"

40 gr. Jacketed Spire Pt.

Powder	Sugg. Starting Grains	Velocity fps	Pressure C.U.P.	Max. Load Grains	Velocity fps	Pressure C.U.P.
IMR-4198	22.0	3215	—	24.0	3497	—
RX7	21.0	3215	—	23.5	3509	—
IMR-3031	22.0	2747	—	25.5+	3322	—
H-335	25.0	3125	—	28.5	3415	—
BL-C(2)	25.0	3012	—	28.5	3345	—
IMR-4064	23.0	2570	—	26.0+	3039	—
IMR-4895	23.0	2583	—	26.5+	3144	—
IMR-4320	24.0	2702	—	27.5+	3205	—

Note: Loads shown in shaded panels are maximum.
+ Designates a compressed powder charge.

45 gr. Jacketed Spitzer

Powder	Sugg. Starting Grains	Velocity fps	Pressure C.U.P.	Max. Load Grains	Velocity fps	Pressure C.U.P
IMR-4198	21.0	2967	—	23.0	3322	—
RX7	21.0	2985	—	23.0	3268	—
IMR-3031	22.0	2680	—	25.5+	3257	—
H-335	25.0	2801	—	28.0	3236	—
BL-C(2)	25.0	2717	—	28.0	3115	—
IMR-4064	23.0	2506	—	26.0+	2994	—
IMR-4895	23.0	2557	—	26.5+	3125	—
IMR-4320	24.0	2652	—	27.5+	3154	—

50 gr. Jacketed SP
2.235" OAL

Powder	Sugg. Starting Grains	Velocity fps	Pressure C.U.P.	Max. Load Grains	Velocity fps	Pressure C.U.P.
IMR-4198	20.0	2739	—	22.0	3115	—
RX7	20.0	2840	—	22.5	3154	—
IMR-3031	22.0	2688	—	25.5+	3257	—
748	26.4	2918	28,000	30.0+	3272	37,100
H-335	24.0	2808	—	27.5	3174	—
BL-C(2)	24.0	2666	—	27.5	3076	—
IMR-4064	23.0	2538	—	26.0+	2967	—
IMR-4895	23.0	2570	—	26.5+	3115	—
IMR-4320	24.0	2638	—	27.5+	3134	—
AA2230	23.0	2792	28,800	26.0	3527	49,100
AA2015BR	23.5	2976	30,200	26.5+	3397	46,600
AA2460	24.2	2833	31,900	28.2	3406	50,500

52 gr. Jacketed HPBT
2.250" OAL

Powder	Sugg. Starting Grains	Velocity fps	Pressure C.U.P.	Max. Load Grains	Velocity fps	Pressure C.U.P.
IMR-4198	19.0	2666	—	21.6	3039	—
RX7	20.0	2941	—	22.3	3076	—
IMR-3031	22.0	2645	—	25.0+	3125	—
748	26.4	2966	29,500	30.0+	3284	41,600
H-335	24.0	2583	—	27.0	3039	—
BL-C(2)	24.0	2564	—	27.0	2915	—
IMR-4064	23.0	2512	—	26.0+	2941	—
IMR-4895	23.0	2538	—	26.5+	3086	—
IMR-4320	24.0	2659	—	27.5+	3125	—
AA2230	23.0	2910	32,300	26.0	3472	49,100
AA2015BR	23.5	3012	32,000	26.8+	3405	47,900
AA2460	24.2	2916	36,200	28.3	3388	51,900

55 gr. Jacketed Spitzer
2.260" OAL

Powder	Sugg. Starting Grains	Velocity fps	Pressure C.U.P.	Max. Load Grains	Velocity fps	Pressure C.U.P.
IMR-4198	19.0	2645	—	21.7	3067	—
RX7	20.0	2770	—	22.0	3030	—
IMR-3031	21.0	2506	—	24.5+	3076	—
748	26.4	2894	28,500	30.0+	3281	43,400
H-335	23.0	2617	—	26.5	2967	—
BL-C(2)	23.0	2525	—	26.5	2949	—
IMR-4064	23.0	2531	—	26.0+	2949	—
IMR-4895	23.0	2564	—	26.0+	3030	—
IMR-4320	24.0	2672	—	27.5+	3144	—
AA2230	23.0	2756	31,100	25.5	3341	49,800
AA2015BR	23.2	2901	32,200	26.6+	3364	51,500
AA2460	24.0	2787	30,700	28.0	3305	51,400

Note: Loads shown in shaded panels are maximum.
+ Designates a compressed powder charge.

223 Remington
(5.56 mm)

63 gr. Jacketed Ptd. SP
2.260" OAL

Powder	Sugg. Starting Grains	Velocity fps	Pressure C.U.P.	Max. Load Grains	Velocity fps	Pressure C.U.P.
IMR-4198	18.0	2487	—	20.0	2544	—
RX7	19.0	2645	—	21.5	2890	—
IMR-3031	20.0	2403	—	22.5	2777	—
H-335	23.0	2695	—	26.0	2915	—
BL-C(2)	23.0	2544	—	26.0	2840	—
IMR-4064	22.0	2409	—	25.0+	2808	—
IMR-4895	22.0	2444	—	25.0	2865	—
IMR-4320	22.0	2421	—	25.5	2881	—
AA2230	22.0	2703	31,400	24.5	3154	52,000
AA2015BR	22.5	2739	34,400	26.0	3162	51,000
AA2460	23.5	2668	33,100	27.6	3147	50,000

#225415
45 gr. (#2 Alloy) 2.060" OAL

Powder	Sugg. Starting Grains	Velocity fps	Pressure C.U.P.	Max. Load Grains	Velocity fps	Pressure C.U.P.
Red Dot	5.6	1795	23,400	9.0	2280	40,700
700X	5.6	1790	23,400	7.7	2080	39,400
Green Dot	6.2	1855	23,400	9.2	2270	39,000
PB	6.3	1775	24,000	8.7	2120	39,400
Unique	6.8	1900	20,400	9.5	2300	35,000
SR-7625	6.6	1820	24,000	8.8	2120	39,900
**SR-4759	12.0	1853	18,200	16.0	2640	35,900

#225462
54 gr. (#2 Alloy) 2.090" OAL

Powder	Sugg. Starting Grains	Velocity fps	Pressure C.U.P.	Max. Load Grains	Velocity fps	Pressure C.U.P.
Red Dot	5.4	1675	22,800	8.6	2100	40,700
700X	5.4	1680	25,200	7.5	1970	39,000
Green Dot	6.0	1735	23,400	8.8	2110	38,100
PB	6.1	1675	24,000	8.5	1980	40,300
Unique	6.7	1805	22,000	9.0	2110	30,000
SR-7625	6.4	1715	25,800	8.3	1950	39,400
**H110	13.7	2120	23,700	18.0	2699	41,600
**SR-4759	11.3	1762	20,700	17.7	2611	43,600
**748	19.8	1902	21,600	26.0	2697	32,000

#225646
55 gr. (#2 Alloy) 2.260" OAL

Powder	Sugg. Starting Grains	Velocity fps	Pressure C.U.P.	Max. Load Grains	Velocity fps	Pressure C.U.P.
**Red Dot	6.6	1682	21,800	8.5	2022	38,300
**700X	6.4	1742	22,400	8.0	1995	36,600
**Green Dot	7.0	1752	21,500	8.8	2061	34,900
**PB	7.0	1719	25,600	8.3	1959	38,700
**Unique	7.6	1853	21,500	9.0	2081	33,500
**SR-7625	6.8	1640	23,300	8.6	1906	33,700
**SR-4759	13.0	2224	22,000	16.5	2657	44,100
**748	25.0	2563	22,400	29.4+	3037	38,000

Note: Loads shown in shaded panels are maximum.
\+ Designates a compressed powder charge.
** Signifies CCI 450 primers used.

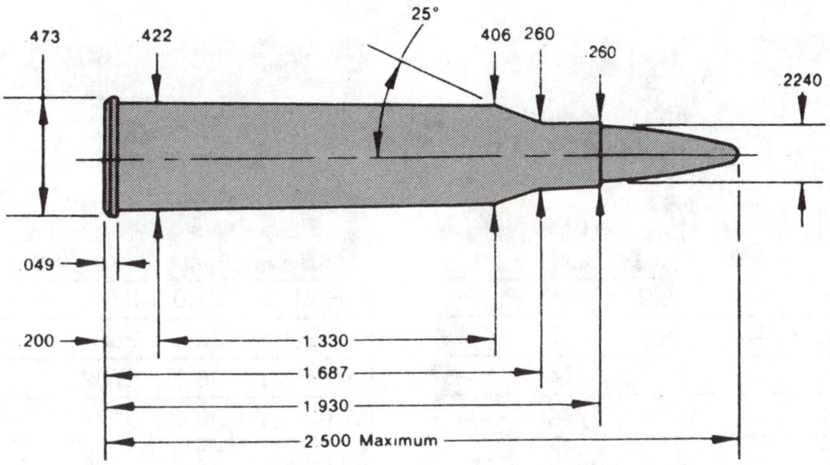

COMMENTS:

Introduced by Winchester in 1964, this case headspaces on the shoulder although it is technically a semi-rimmed case.

IMR 4320 proved to be the best powder for ballistic uniformity in our lab tests.

TEST COMPONENTS:

Cases .... Winchester
Trim-to Length .... 1.920"
Primers .... Winchester 8½ -120
Primer Size .... Large Rifle
Lyman Shell Holder .... No. 5
Jacketed Bullets Used .... Speer Spire Pt., 40 gr.
Sierra Spitzer, 45 gr.
Sierra Spitzer, 50 gr.
Speer Hollow Pt., 52 gr.
Remington Ptd. SP, 55 gr.
Sierra SMPT, 63 gr.
Cast Bullets Used .... (Sized to .224" dia.)
*Gas Check Bullets .... *#225415, 45 gr.
*#225462, 54 gr.

TEST SPECIFICATIONS: (Velocity Only)

Firearm Used .... Winchester Model 70
Barrel Length .... 22"
Twist .... 1-14"
Groove Dia. .... .224"

40 gr. Jacketed Spire Pt.

Powder	Sugg. Starting Grains	Velocity fps	Pressure C.U.P.	Max. Load Grains	Velocity fps	Pressure C.U.P.
IMR-4198	23.0	3086	—	26.0	3484	—
IMR-3031	26.0	3086	—	29.0	3497	—
IMR-4895	27.0	2849	—	30.0	3268	—
IMR-4064	31.0	3333	—	34.0	3846	—
IMR-4320	29.0	3067	—	31.0	3345	—
H-380	34.0	3356	—	38.0+	3786	—

Note: Loads shown in shaded panels are maximum.
+ Designates a compressed powder charge.

225 Winchester

45 gr. Jacketed Spitzer

Powder	Sugg. Starting Grains	Velocity fps	Pressure C.U.P.	Max. Load Grains	Velocity fps	Pressure C.U.P
IMR-4198	22.0	2865	—	25.0	3215	—
IMR-3031	25.0	2849	—	28.0	3257	—
IMR-4895	26.0	2710	—	29.0	3144	—
IMR-4064	30.0	3164	—	33.5	3745	—
IMR-4320	28.0	2857	—	31.0	3300	—
H-380	33.0	3184	—	37.5+	3571	—

50 gr. Jacketed Spitzer

Powder	Sugg. Starting Grains	Velocity fps	Pressure C.U.P.	Max. Load Grains	Velocity fps	Pressure C.U.P.
IMR-4198	21.0	2710	—	23.0	2941	—
IMR-3031	25.0	2865	—	28.0	3225	—
IMR-4895	26.0	2732	—	29.0	3125	—
IMR-4064	30.0	3174	—	33.0	3636	—
IMR-4320	27.0	2808	—	30.0	3194	—
H-380	33.0	3174	—	37.0+	3497	—

52 gr. Jacketed HP

Powder	Sugg. Starting Grains	Velocity fps	Pressure C.U.P.	Max. Load Grains	Velocity fps	Pressure C.U.P.
IMR-4198	20.0	2645	—	22.5	2906	—
IMR-3031	24.0	2793	—	27.0	3030	—
IMR-4895	25.0	2666	—	28.0	3012	—
IMR-4064	29.0	3067	—	32.5	3610	—
IMR-4320	26.0	2710	—	29.0	3048	—
H-380	32.0	3076	—	36.5+	3436	—

55 gr. Jacketed Ptd. SP

Powder	Sugg. Starting Grains	Velocity fps	Pressure C.U.P.	Max. Load Grains	Velocity fps	Pressure C.U.P.
IMR-4198	20.0	2638	—	22.0	2857	—
IMR-3031	23.0	2652	—	26.0	2976	—
IMR-4895	25.0	2659	—	28.0	2976	—
IMR-4064	29.0	3048	—	‡32.0	3534	—
IMR-4320	25.0	2610	—	28.0	2906	—
H-380	32.0	3039	—	36.0+	3401	—

Note: Loads shown in shaded panels are maximum.
+ Designates a compressed powder charge.
‡ Designates a factory velocity duplication load.

225 Winchester

63 gr. Jacketed Ptd. SP

Powder	Sugg. Starting Grains	Velocity fps	Pressure C.U.P.	Max. Load Grains	Velocity fps	Pressure C.U.P.
IMR-4198	19.0	2487	—	21.0	2695	—
IMR-3031	22.0	2544	—	25.0	2849	—
IMR-4895	24.0	2557	—	27.0	2832	—
IMR-4064	28.0	2923	—	31.5	3378	—
IMR-4320	24.0	2518	—	27.0	2816	—
H-380	31.0	2915	—	35.0	3215	—

#225415
45 gr. (#2 Alloy) 2.235" OAL

Powder	Sugg. Starting Grains	Velocity fps	Pressure C.U.P.	Max. Load Grains	Velocity fps	Pressure C.U.P.
Red Dot	6.0	1770	—	9.0	2235	—
700X	6.0	1775	—	9.0	2185	—
Green Dot	6.5	1845	—	9.5	2280	—
PB	6.5	1760	—	9.5	2170	—
Unique	7.0	1880	—	10.0	2315	—
SR-7625	7.0	1825	—	10.0	2195	—

#225462
54 gr. (#2 Alloy) 2.150" OAL

Powder	Sugg. Starting Grains	Velocity fps	Pressure C.U.P.	Max. Load Grains	Velocity fps	Pressure C.U.P.
Red Dot	5.5	1610	—	8.5	1965	—
700X	5.5	1590	—	8.5	1995	—
Green Dot	6.0	1660	—	9.0	2060	—
PB	6.0	1565	—	9.0	1970	—
Unique	6.5	1695	—	9.5	2045	—
SR-7625	6.5	1640	—	9.5	2025	—

Note: Loads shown in shaded panels are maximum.

224 Weatherby Magnum

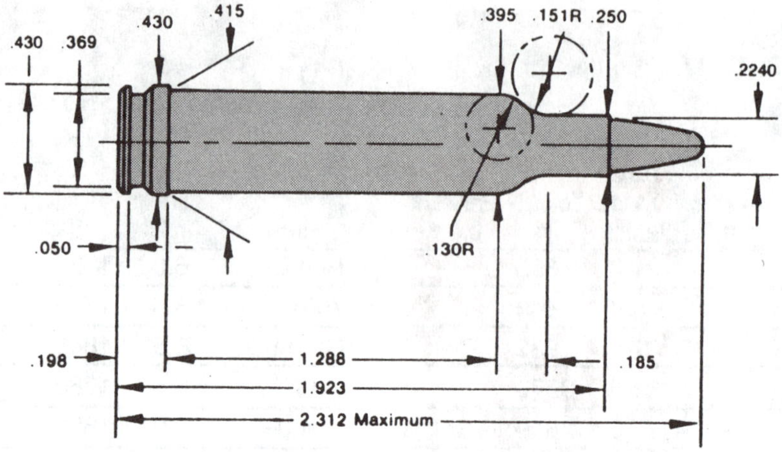

COMMENTS:

This is a very accurate cartridge which has never gained a great deal of popularity. Perhaps this is due to the fact that only Weatherby rifles have generally been chambered for the round and only Weatherby supplies ammo or brass. Do not use magnum style primers with this case as dangerous pressures will result. Jacketed bullets of 50 to 60 grains are the best choices. Use only bullets designed for high velocities to prevent premature bullet break up. Our recommended powder choice is IMR-4198.

TEST COMPONENTS:

Cases .... Weatherby
Trim-to Length .... 1.915"
Primers .... Remington 9½
Primer Size .... Large Rifle
Lyman Shell Holder .... No. 3
Jacketed Bullets Used .... Speer Spire Pt., 40 gr.
Sierra Spitzer, 45 gr.
Sierra Spitzer, 50 gr.
Speer HP, 52 gr.
Remington Ptd. SP, 55 gr.
Sierra SMPT, 63 gr.
Cast Bullets Used .... (Sized to .224" dia.)
*Gas Check Bullets *#225415, 45 gr.
*#225462, 54 gr.

TEST SPECIFICATIONS: (Velocity Only)

Firearm Used .... Weatherby Varmint Master
Barrel Length .... 26"
Twist .... 1-14"
Groove Dia. .... .224"

40 gr. Jacketed Spire Pt.

Powder	Sugg. Starting Grains	Velocity fps	Pressure C.U.P.	Max. Load Grains	Velocity fps	Pressure C.U.P.
IMR-4198	25.0	3521	—	28.5	4098	—
IMR-3031	29.0	3597	—	32.5+	4149	—
IMR-4895	30.0	3497	—	33.5+	4000	—
IMR-4064	30.0	3448	—	33.0+	3875	—
IMR-4320	32.0	3663	—	35.0+	4115	—

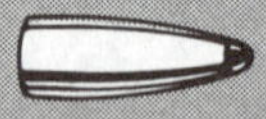

45 gr. Jacketed Spitzer

Powder	Sugg. Starting Grains	Velocity fps	Pressure C.U.P.	Max. Load Grains	Velocity fps	Pressure C.U.P
IMR-4198	25.0	3448	—	28.0	3891	—
IMR-3031	29.0	3584	—	32.0+	3984	—
IMR-4895	30.0	3546	—	33.0+	3875	—
IMR-4064	30.0	3460	—	33.0+	3816	—
IMR-4320	31.0	3509	—	34.5+	3891	—

Note: Loads shown in shaded panels are maximum.
+ Designates a compressed powder charge.

224 Weatherby Magnum

50 gr. Jacketed Spitzer

Powder	Sugg. Starting Grains	Velocity fps	Pressure C.U.P.	Max. Load Grains	Velocity fps	Pressure C.U.P.
IMR-4198	24.0	3247	—	27.5	3708	—
IMR-3031	28.0	3356	—	31.5+	3831	—
IMR-4895	29.0	3322	—	32.5+	3745	—
IMR-4064	30.0	3484	—	33.0+	3802	—
IMR-4320	31.0	3460	—	34.0+	3802	—

52 gr. Jacketed HP

Powder	Sugg. Starting Grains	Velocity fps	Pressure C.U.P.	Max. Load Grains	Velocity fps	Pressure C.U.P.
IMR-4198	24.0	3215	—	27.0	3623	—
IMR-3031	28.0	3356	—	31.0+	3773	—
IMR-4895	29.0	3247	—	32.0	3650	—
IMR-4064	30.0	3413	—	33.0+	3759	—
IMR-4320	30.0	3322	—	33.5+	3717	—

55 gr. Jacketed Ptd. SP

Powder	Sugg. Starting Grains	Velocity fps	Pressure C.U.P.	Max. Load Grains	Velocity fps	Pressure C.U.P.
IMR-4198	23.0	3095	—	26.5	3509	—
IMR-3031	27.0	3225	—	30.5+	3677	—
IMR-4895	28.0	3164	—	31.5	3597	—
IMR-4064	29.0	3279	—	32.8+	3717	—
IMR-4320	30.0	3311	—	33.0+	3650	—

63 gr. Jacketed Ptd. SP

Powder	Sugg. Starting Grains	Velocity fps	Pressure C.U.P.	Max. Load Grains	Velocity fps	Pressure C.U.P.
IMR-4198	23.0	2967	—	26.0	3356	—
IMR-3031	26.0	3039	—	29.5	3472	—
IMR-4895	27.0	3003	—	30.5	3413	—
IMR-4064	29.0	3247	—	32.0+	3559	—
IMR-4320	29.0	3144	—	32.0+	3484	—

#225415
45 gr. (#2 Alloy)

Powder	Sugg. Starting Grains	Velocity fps	Pressure C.U.P.	Max. Load Grains	Velocity fps	Pressure C.U.P.
Unique	6.0	1818	—	7.0	2008	—
2400	11.0	1964	—	13.0	2314	—

#225462
54 gr. (#2 Alloy)

Powder	Sugg. Starting Grains	Velocity fps	Pressure C.U.P.	Max. Load Grains	Velocity fps	Pressure C.U.P.
Unique	6.0	1727	—	8.0	2074	—
IMR-4227	13.0	2092	—	15.0	2386	—

Note: Loads shown in shaded panels are maximum.
\+ Designates a compressed powder charge.

22-250 Remington

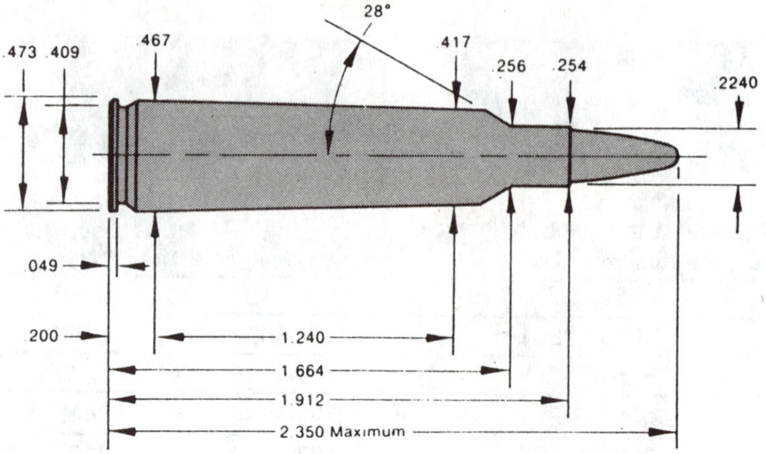

COMMENTS:

This is one of the best choices for a long range 22 caliber varmint rifle. Accuracy is generally excellent and its flat trajectory makes hitting small targets easier — even up to 300 yards.

Jacketed bullets of 50 grains or heavier are best. First recommendation on powder would be Hodgdon H380. In fact, a charge of 38.0 grains with a 55 grain bullet has become legendary. It began when Bruce Hodgdon first tried it and found it so accurate that he named the powder after the 38.0 grain charge — H380.

The data listed are intended only for standard chambers as manufactured to current industry specifications. Early chambers marked 22-250 or Varminter may differ from current standards by a substantial amount, to the point that this data may prove unsafe in such firearms.

Cast bullet #225646 at about 2,100 fps. is the best choice for lead bullet use in this cartridge, as it is in several other 22 caliber cartridges.

TEST COMPONENTS:

Cases .... Winchester
Trim-to Length .... 1.902"
Primers .... Winchester 8½-120
Primer Size .... Large Rifle
Lyman Shell Holder .... No. 2
Jacketed Bullets Used .... Speer Spire Pt., 40 gr.
Sierra Spitzer, 45 gr.
Sierra SP Blitz #1340, 50 gr.
Sierra HP #1400, 53 gr.
Sierra SPT #1360, 55 gr.
Hornady HP #2275, 60 gr.
Sierra SMP #1370, 63 gr.
Cast Bullets Used .... (Sized to .224" dia.)
*Gas Check Bullets
*#225415, 45 gr.
*#225462, 54 gr.
*#225646, 55 gr.

TEST SPECIFICATIONS: (Velocity & Pressure)

Firearm Used .... Universal Receiver
Barrel Length .... 26"
Twist .... 1-14"
Groove Dia. .... .224"

40 gr. Jacketed Spire Pt.

Powder	Sugg. Starting Grains	Velocity fps	Pressure C.U.P.	Max. Load Grains	Velocity fps	Pressure C.U.P.
RX7	29.0	3709	39,400	33.0	4130	52,800
IMR-3031	32.0	3606	39,400	36.5	4098	52,800
BL-C(2)	36.0	3778	39,900	40.0	4130	50,400
IMR-4064	34.0	3489	35,500	38.5+	4062	51,200
IMR-4895	33.5	3594	39,400	37.5	4022	52,000
IMR-4320	34.0	3566	41,600	38.5	4003	53,200
H-380	38.0	3861	33,000	42.0+	4195	51,600
IMR-4350	36.0	3270	34,000	40.0+	3648	38,600

Note: Loads shown in shaded panels are maximum.
+ Designates a compressed powder charge.

22-250 Remington

45 gr. Jacketed Spitzer

Powder	Sugg. Starting Grains	Velocity fps	Pressure C.U.P.	Max. Load Grains	Velocity fps	Pressure C.U.P
RX7	28.0	3525	38,600	32.0	3866	47,600
IMR-3031	31.0	3545	40,300	35.5	3965	50,800
BL-C(2)	35.0	3516	35,500	39.0	4000	50,000
IMR-4064	33.0	3373	34,500	37.5+	3945	52,400
IMR-4895	32.5	3353	36,400	36.5	3901	52,000
IMR-4320	33.0	3390	38,600	37.5	3836	50,000
H-380	37.0	3741	41,100	41.0	4055	51,600
IMR-4350	36.0	3239	33,500	40.0+	3613	43,300

50 gr. Jacketed SP
2.310" OAL

Powder	Sugg. Starting Grains	Velocity fps	Pressure C.U.P.	Max. Load Grains	Velocity fps	Pressure C.U.P.
RX7	28.0	3478	41,100	31.5	3779	52,400
IMR-3031	30.0	3181	33,000	35.3	3861	52,400
748	34.3	3391	36,800	39.0	3811	50,700
BL-C(2)	34.0	3480	36,900	38.5	3911	51,600
IMR-4064	32.0	3210	32,000	37.0+	3854	52,400
IMR-4895	31.0	3314	37,700	38.0+	3833	52,000
IMR-4320	33.0	3408	40,700	37.0	3742	51,200
H-380	36.0	3569	39,400	40.0	3912	52,000
760	37.4	3275	38,000	42.5+	3732	51,700
H-414	37.0	3294	37,500	42.0+	3776	50,200
IMR-4350	36.0	3182	34,000	40.0+	3591	45,000
AA2230	31.8	3290	40,500	35.4	3626	47,000
AA2015BR	29.7	3229	37,000	34.0	3716	50,600
AA2460	31.4	3154	37,300	36.2	3693	52,400
AA2520	32.5	3319	39,100	37.0	3724	51,200

53 gr. Jacketed HP
2.330" OAL

Powder	Sugg. Starting Grains	Velocity fps	Pressure C.U.P.	Max. Load Grains	Velocity fps	Pressure C.U.P.
RX7	27.0	3333	39,900	31.2	3561	52,200
IMR-3031	30.0	3181	33,000	34.5	3731	52,800
748	34.3	3395	40,800	39.0	3811	52,700
BL-C(2)	34.0	3426	39,000	38.2	3829	53,200
IMR-4064	31.0	3119	33,000	36.2	3743	52,800
IMR-4895	31.0	3219	35,500	36.0	3728	52,400
IMR-4320	32.0	3153	36,400	36.7	3645	51,200
H-380	35.0	3444	40,300	39.7	3830	53,200
760	37.0	3251	37,800	42.0	3716	52,300
H-414	37.0	3302	36,200	42.0+	3752	51,200
IMR-4350	36.0	3165	37,300	40.0+	3530	46,300
AA2230	31.4	3168	37,400	36.0	3624	52,100
AA2015BR	28.9	3151	36,900	33.8	3631	52,000
AA2460	31.1	3086	37,000	35.9	3626	52,400

55 gr. Jacketed Spitzer
2.350" OAL

Powder	Sugg. Starting Grains	Velocity fps	Pressure C.U.P.	Max. Load Grains	Velocity fps	Pressure C.U.P.
RX7	26.5	3295	38,600	30.5	3541	50,400
IMR-3031	30.0	3217	36,400	36.0+	3760	51,100
748	33.9	3291	36,500	‡38.5	3738	52,100
BL-C(2)	34.0	3478	42,400	38.0	3697	52,100
IMR-4064	31.0	3161	35,000	36.0	3715	52,400
IMR-4895	31.0	3197	36,900	35.5	3693	53,200
IMR-4320	32.0	3163	36,400	36.5	3630	52,000
H-380	35.0	3463	41,600	40.0+	3637	48,100
760	37.0	3248	39,800	42.0	3724	52,600
H-414	36.5	3197	36,000	41.5+	3598	47,700
IMR-4350	36.0	3195	37,700	40.0+	3561	49,600
AA2230	31.1	3096	37,400	35.8	3594	51,400
AA2015BR	29.0	3131	38,300	33.9	3602	52,500
AA2460	29.7	2988	34,900	35.8	3608	51,900

Note: Loads shown in shaded panels are maximum.
+ Designates a compressed powder charge.
‡ Designates a factory velocity duplication load.

22-250 Remington

60 gr. Jacketed HP
2.350" OAL

Powder	Sugg. Starting Grains	Velocity fps	Pressure C.U.P.	Max. Load Grains	Velocity fps	Pressure C.U.P.
RX7	25.0	3080	39,000	30.0	3434	51,900
IMR-3031	30.0	3193	37,700	34.0	3596	52,800
748	33.4	3217	37,700	38.0	3635	52,300
BL-C(2)	33.0	3346	45,000	36.7	3619	53,200
IMR-4064	31.0	3128	36,400	35.5	3589	52,000
IMR-4895	31.0	3173	38,100	38.0	3720	52,700
IMR-4320	31.0	3064	37,300	36.0	3503	51,600
H-380	34.5	3362	44,200	38.0	3622	52,400
760	36.1	3132	37,200	41.0+	3565	52,500
H-414	36.1	3138	35,800	41.0+	3538	48,900
IMR-4350	36.0	3136	38,600	40.0+	3497	50,400
AA2230	30.1	2973	37,900	34.8	3460	52,400
AA2015BR	27.2	2936	37,100	32.4	3460	53,000
AA2460	28.9	2859	34,600	35.0	3467	52,100
AA2520	29.8	2882	35,700	34.7	3392	52,300
RX15	31.0	2979	36,800	36.0	3524	52,400
RX19	36.4	2824	37,300	42.2+	3404	52,900

63 gr. Jacketed Ptd. SP
2.350" OAL

Powder	Sugg. Starting Grains	Velocity fps	Pressure C.U.P.	Max. Load Grains	Velocity fps	Pressure C.U.P.
RX7	25.9	2988	40,000	29.5	3315	52,600
IMR-3031	29.0	3045	36,000	35.5	3570	52,900
748	33.0	3140	39,000	37.5	3540	52,800
BL-C(2)	33.0	3223	42,400	37.0	3491	51,600
IMR-4064	31.0	3090	36,400	35.0	3490	50,800
IMR-4895	30.0	3056	36,900	34.5	3479	52,400
IMR-4320	31.0	3011	36,900	35.8	3480	52,800
H-380	34.0	3268	43,700	37.5	3554	52,800
760	36.1	3132	39,200	41.0	3528	53,000
H-414	35.6	3059	36,700	40.5+	3466	48,600
IMR-4350	36.0	3149	40,700	40.0+	3464	51,200
AA2015BR	26.8	2814	37,600	32.0	3311	52,000
AA2460	27.7	2729	35,300	33.8	3309	51,800
RX15	30.4	2945	38,300	35.4	3420	52,100
RX19	35.0	2772	37,700	40.8	3334	52,400

#225415
45 gr. (#2 Alloy) 2.325" OAL

Powder	Sugg. Starting Grains	Velocity fps	Pressure C.U.P.	Max. Load Grains	Velocity fps	Pressure C.U.P.
Red Dot	7.2	1981	22,800	9.5	2323	35,000
700X	7.0	1850	18,000	9.0	2268	33,500
Green Dot	7.5	1992	21,000	10.0	2356	34,500
PB	7.5	1920	18,000	9.5	2188	31,000
SR-7625	8.0	1981	20,400	10.0	2237	36,900
SR-4756	8.0	1915	18,000	11.0	2308	32,000
SR-4759	14.5	2176	13,000	19.0	2820	26,900
IMR-4198	16.9	2173	12,100	22.1	2867	23,900
748	22.3	2175	13,600	28.5	2814	21,500

Note: Loads shown in shaded panels are maximum.
+ Designates a compressed powder charge.

22-250 Remington

#225462
54 gr. (#2 Alloy) 2.347" OAL

Powder	Sugg. Starting Grains	Velocity fps	Pressure C.U.P.	Max. Load Grains	Velocity fps	Pressure C.U.P.
Red Dot	7.0	1828	21,000	9.5	2184	36,400
700X	7.5	1937	29,400	9.0	2143	36,900
Green Dot	7.0	1828	21,000	10.0	2236	36,900
PB	8.0	1880	27,000	9.5	2088	36,400
SR-7625	8.5	1955	30,500	10.0	2157	38,600
SR-4756	8.5	1935	18,600	10.5	2207	35,500
SR-4759	14.5	2154	15,700	20.0	2833	33,600
IMR-4227	15.6	2156	13,800	20.1	2858	35,300
IMR-4198	16.8	2167	14,500	23.0	2883	30,200
748	22.9	2127	12,900	28.6	2811	25,800

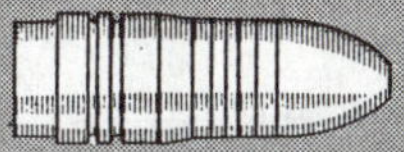

#225646
55 gr. (#2 Alloy) 2.350" OAL

Powder	Sugg. Starting Grains	Velocity fps	Pressure C.U.P.	Max. Load Grains	Velocity fps	Pressure C.U.P.
Red Dot	7.7	1797	22,100	10.4	2165	33,900
700X	8.0	1887	24,500	11.0	2286	40,200
Green Dot	8.2	1862	23,200	11.5	2292	39,700
PB	9.0	1892	27,300	11.0	2128	38,900
SR-7625	9.2	1905	26,700	11.7	2219	42,700
SR-4756	10.2	1986	26,700	12.6	2271	38,100
SR-4759	16.5	2218	22,000	18.5	2543	35,200
IMR-4227	16.5	2274	23,500	21.0	2754	36,200
IMR-4198	18.5	2172	18,400	24.6	2913	37,100
748	28.5	2380	19,000	34.0	3100	38,000

Note: Loads shown in shaded panels are maximum.

220 Swift

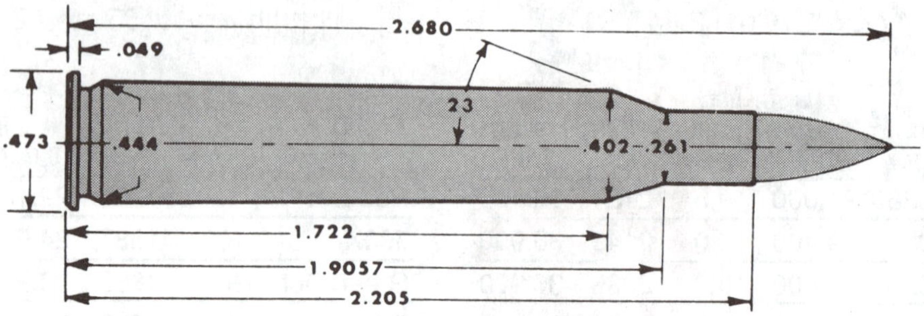

COMMENTS:

In its early years the Swift gained a mixed reputation. Ultimately at least one firearms company gave up the Swift due to the extent of problems with it. Many less than knowledgeable reloaders destroyed more than a few Swift rifles because they insisted on pushing past the limits of safety.

Currently Remington is loading the 220 Swift with a reasonable level of velocity as are a few other manufacturers. If the reloader forgets about visions of 4,000 + fps with 50 grain bullets, and confines efforts to the listed data, the Swift makes an excellent varmint cartridge. IMR 4064 with 55 grain jacketed bullets is about the right place to start in the search for accuracy.

TEST COMPONENTS:

Cases .... Winchester
Trim-to Length .... 2.195"
Primers .... Remington 9½, Winchester 8½ -120 and CCI 200
Primer Size .... Large Rifle
Lyman Shell Holder .... No. 5
Jacketed Bullets Used .... Speer Spire Pt., 40 gr.
Sierra Spitzer, 45 gr.
Sierra SP Blitz #1340, 50 gr.
Sierra SPT #1360, 55 gr.
Hornady HP #2275, 60 gr.
Sierra SMP #1370, 63 gr.
Speer SP #1053, 70 gr.
Cast Bullets Used .... (Sized to .224" dia.)
*Gas Check Bullets .... *#225415, 45 gr.
*#225462, 54 gr.

TEST SPECIFICATIONS: (Velocity & Pressure)

Firearm Used .... Winchester Model 70 and Universal Receiver
Barrel Length .... Winchester, 26"; Universal Receiver 24"
Twist .... 1-14"
Groove Dia. .... .224"

40 gr. Jacketed Spire Pt.**

Powder	Sugg. Starting Grains	Velocity fps	Pressure C.U.P.	Max. Load Grains	Velocity fps	Pressure C.U.P.
IMR-3031	35.0	3906	—	39.0	4385	—
IMR-4064	37.0	3891	—	41.0+	4347	—
IMR-4895	35.0	3663	—	39.0	4166	—
IMR-4320	37.0	3861	—	41.0	4291	—
H-380	40.0	3831	—	44.0+	4166	—
IMR-4350	39.0	3546	—	43.0+	3968	—

Note: Loads shown in shaded panels are maximum.
+ Designates a compressed powder charge.
** Tested in a Winchester Mod. 70 with Rem. 9½ primers.

45 gr. Jacketed Spitzer**

Powder	Sugg. Starting Grains	Velocity fps	Pressure C.U.P.	Max. Load Grains	Velocity fps	Pressure C.U.P
IMR-3031	34.0	3717	—	38.5	4219	—
IMR-4064	36.0	3626	—	40.5+	4184	—
IMR-4895	34.0	3559	—	38.5	4065	—
IMR-4320	36.0	3663	—	40.5	4166	—
H-380	39.0	3663	—	43.0	3984	—
IMR-4350	39.0	3521	—	43.0+	3861	—

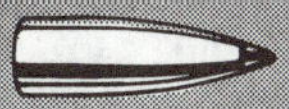

50 gr. Jacketed SP
2.670" OAL

Powder	Sugg. Starting Grains	Velocity fps	Pressure C.U.P.	Max. Load Grains	Velocity fps	Pressure C.U.P.
***RX7	26.5	3318	43,200	30.2	3597	52,900
**IMR-3031	34.0	3650	—	38.0	4065	—
***748	39.0	3425	41,000	44.0	3865	52,800
**IMR-4064	36.0	3650	—	40.0	4081	—
**IMR-4895	34.0	3559	—	38.0	3952	—
**IMR-4320	36.0	3597	—	40.0	4000	—
**H-380	38.0	3571	—	42.5	3906	—
***760	38.7	3107	33,900	44.0	3934	53,000
**IMR-4350	39.0	3509	—	43.0+	3921	—

55 gr. Jacketed Spitzer
2.675" OAL

Powder	Sugg. Starting Grains	Velocity fps	Pressure C.U.P.	Max. Load Grains	Velocity fps	Pressure C.U.P.
***RX7	24.6	3129	42,700	28.0	3395	53,000
**IMR-3031	33.0	3509	—	37.0	3921	—
***748	32.6	3332	43,500	37.0	3619	52,500
**IMR-4064	35.0	3472	—	39.0	3906	—
**IMR-4895	33.0	3427	—	37.0	3787	—
**IMR-4320	35.0	3460	—	39.0	3861	—
**H-380	38.0	3497	—	42.0	3773	—
***760	38.0	3382	42,800	‡42.0	3683	52,700
**IMR-4350	38.0	3356	—	42.5+	3787	—

60 gr. Jacketed HP***
2.675" OAL

Powder	Sugg. Starting Grains	Velocity fps	Pressure C.U.P.	Max. Load Grains	Velocity fps	Pressure C.U.P.
IMR-4198	26.0	3088	40,400	29.5	3382	53,000
748	31.7	3218	45,900	36.0	3479	52,900
IMR-4064	33.0	3304	43,900	‡37.5	3633	53,000
IMR-4895	33.0	3322	46,000	37.5	3655	52,900
IMR-4320	33.0	3227	43,800	37.5	3550	53,000
760	36.0	3191	41,300	40.9	3527	51,900
IMR-4350	37.0	3214	42,700	42.0+	3588	52,100

Note: Loads shown in shaded panels are maximum.
+ Designates a compressed powder charge.
‡ Designates a factory velocity duplication load.
** Tested in a Win. Mod. 70 with Rem. 9½ primers.
*** Tested with a Universal Receiver with Rem. 9½ primers.

220 Swift

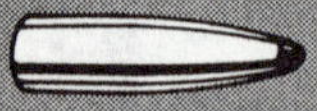

63 gr. Jacketed Ptd. SP
2.680" OAL

Powder	Sugg. Starting Grains	Velocity fps	Pressure C.U.P.	Max. Load Grains	Velocity fps	Pressure C.U.P.
+++IMR-4198	26.4	3092	44,400	30.0	3353	52,900
++IMR-3031	30.0	3215	—	33.5	3472	—
+++748	32.1	3169	44,400	36.5	3397	53,000
++IMR-4064	31.0	3144	—	35.0	3472	—
++IMR-4895	30.0	3067	—	33.5	3378	—
++IMR-4320	31.0	3105	—	35.0	3436	—
++H-380	37.0	3356	—	41.0	3623	—
+++760	36.0	3170	41,500	41.0	3488	52,300
++IMR-4350	37.0	3322	—	40.0	3584	—

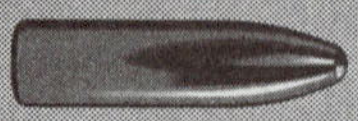

70 gr. Jacketed SP+++
2.600" OAL

Powder	Sugg. Starting Grains	Velocity fps	Pressure C.U.P.	Max. Load Grains	Velocity fps	Pressure C.U.P.
IMR-3031	28.6	2898	47,800	32.5	3241	53,000
748	29.0	2890	46,100	33.0	3173	52,900
IMR-4064	29.5	3025	47,700	33.5	3282	53,000
IMR-4895	29.5	2987	45,000	33.5	3284	53,000
IMR-4320	29.9	2937	43,800	34.0	3251	52,500
760	34.3	3002	44,300	39.0	3324	52,900
IMR-4350	34.3	3112	44,000	39.0	3399	53,000

#225415
45 gr. (#2 Alloy) 2.546" OAL

Powder	Sugg. Starting Grains	Velocity fps	Pressure C.U.P.	Max. Load Grains	Velocity fps	Pressure C.U.P.
***Red Dot	6.5	1730	—	10.0	2270	—
***700X	6.5	1710	—	10.0	2275	—
***Green Dot	7.0	1800	—	10.5	2305	—
***PB	7.0	1680	—	10.5	2210	—
***Unique	7.5	1805	—	11.0	2320	—
***SR-7625	7.5	1765	—	11.0	2280	—
**2400	10.5	1772	16,500	17.4	2723	31,500
**SR-4759	12.2	1748	15,500	18.5	2683	30,300

#225462
54 gr. (#2 Alloy) 2.525" OAL

Powder	Sugg. Starting Grains	Velocity fps	Pressure C.U.P.	Max. Load Grains	Velocity fps	Pressure C.U.P.
***Red Dot	6.0	1605	—	9.5	2035	—
***700X	6.0	1615	—	9.5	2040	—
***Green Dot	6.5	1660	—	10.0	2130	—
***PB	6.5	1610	—	10.0	2060	—
***Unique	7.0	1705	—	10.5	2165	—
***SR-7625	7.0	1635	—	10.5	2115	—
**2400	11.0	1724	18,000	18.5	2706	38,600
**SR-4759	12.4	1759	18,000	19.2	2685	40,000

Note: Loads shown in shaded panels are maximum.
** Tested with a Universal Receiver using CCI 200 primers.
*** Tested in a Win. Mod. 70 with Win. 8½-120 primers.
++ Tested in a Win. Mod. 70 with Rem. 9½ primers.
+++ Tested in a Universal Receiver with Rem. 9½ primers.

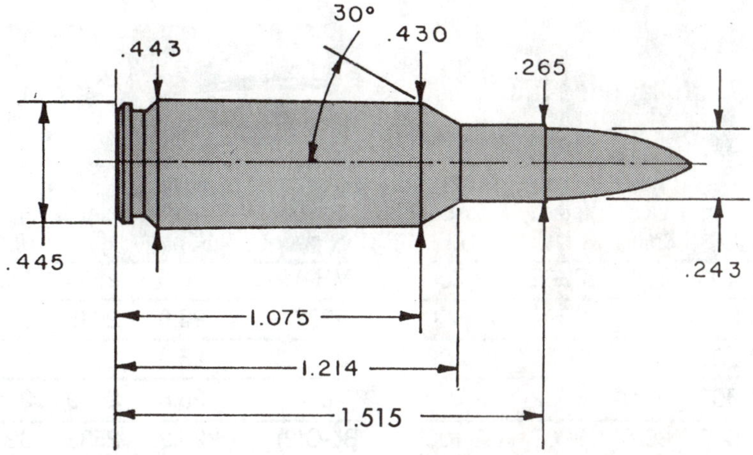

COMMENTS:

This cartridge has proven itself the most accurate round ever developed. Its originators, Ferris Pindell and Dr. Louis Palmisano were the first of many bench rest shooters to so prove. Today the "Six" holds more records than any other cartridge.

A group of 10 shots at 100 yards going into a 1/4" center to center group would not get the attention of a serious benchrest shooter with 6mm PPC experience. The cartridge will do better. Naturally such accuracy demands the very best in bullets, loading technique, and the appropriate charge of H322 or perhaps H335. Still, we have fired a production Sako single shot rifle with factory Sako ammunition and obtained frequent 1/4" to 1/2" groups. Nothing else has ever come close to this kind of out-of-the-box performance.

The twist rate of most 6mm PPC's works best with bullets of 68 to 75 grains. Heavier bullets will not always stabilize and poor results may occur with the latter.

TEST COMPONENTS:

Cases .... Sako PPC, USA
Trim-to Length .... 1.505"
Primers .... Remington 7½
Primer Size .... Small Rifle, Magnum
Lyman Shell Holder .... No. 3
Jacketed Bullets Used .... Sierra HP #1500, 60 gr.
Sierra HPBT #1505, 70 gr.
Hornady HP #2420, 75 gr.
Sierra HPBT #1530, 85 gr.

TEST SPECIFICATIONS: (Velocity & Pressure)

Firearm Used .... Universal Receiver
Barrel Length .... 22"
Twist .... 1-14"
Groove Dia. .... .2435"

60 gr. Jacketed HP
2.050" OAL

Powder	Sugg. Starting Grains	Velocity fps	Pressure C.U.P.	Max. Load Grains	Velocity fps	Pressure C.U.P.
IMR-4198	21.0	2734	32,200	23.5	3076	46,000
H-322	24.8	2846	33,000	27.5	3200	47,200
AA2230	26.2	2876	35,400	30.0+	3212	47,900
748	27.6	2819	32,800	31.7	3204	48,900
BL-C(2)	28.0	2803	34,400	31.2+	3092	46,400
AA2015BR	25.0	2961	34,400	27.5	3272	45,300
AA2460	26.5	2846	34,600	29.0	3133	46,600
H-335	26.0	2910	33,200	29.0	3237	46,500

Note: Loads shown in shaded panels are maximum.
+ Designates a compressed powder charge.

6mm PPC

70 gr. Jacketed HPBT
2.075" OAL

Powder	Sugg. Starting Grains	Velocity fps	Pressure C.U.P.	Max. Load Grains	Velocity fps	Pressure C.U.P.
IMR-4198	21.0	2611	32,400	23.5+	2962	48,700
H-322	24.5	2742	34,200	26.5	2999	45,500
AA2230	26.0	2732	34,800	29.0	3050	48,200
748	27.5	2697	34,000	31.0+	3032	47,700
BL-C(2)	27.5	2675	35,000	30.5	2962	47,400
AA2015BR	24.5	2801	35,000	27.0	3138	48,900
AA2460	26.0	2705	35,200	28.5	2980	46,300
H-335	25.7	2770	34,000	28.5	3073	47,200

75 gr. Jacketed HP
2.120" OAL

Powder	Sugg. Starting Grains	Velocity fps	Pressure C.U.P.	Max. Load Grains	Velocity fps	Pressure C.U.P.
IMR-4198	20.5	2555	34,000	23.0+	2835	46,800
H-322	23.8	2645	35,400	26.0	2930	47,200
AA2230	25.0	2620	33,800	28.0	2932	47,100
748	26.8	2649	33,600	30.2+	2976	48,100
BL-C(2)	26.2	2533	32,400	29.5	2880	47,400
AA2015BR	23.2	2629	33,800	26.0	2990	48,300
AA2460	25.3	2570	33,200	28.5	2930	48,100
H-335	25.4	2700	35,000	28.2	3007	47,700

85 gr. Jacketed HPBT
2.035" OAL

Powder	Sugg. Starting Grains	Velocity fps	Pressure C.U.P.	Max. Load Grains	Velocity fps	Pressure C.U.P.
IMR-4198	19.5	2366	35,000	21.5+	2622	47,300
H-322	22.5	2465	35,500	24.5	2702	45,800
AA2230	24.3	2473	34,000	27.0	2739	46,700
748	25.8	2500	34,400	29.0+	2790	48,000
BL-C(2)	25.2	2429	34,400	28.5	2729	47,300
AA2015BR	22.0	2483	34,400	24.5	2776	46,400
AA2460	24.0	2440	34,600	27.2	2757	48,900
H-335	24.3	2532	35,000	27.0	2823	48,100

Note: Loads shown in shaded panels are maximum.
\+ Designates a compressed powder charge.

243 Winchester
(6.2 x 52mm)

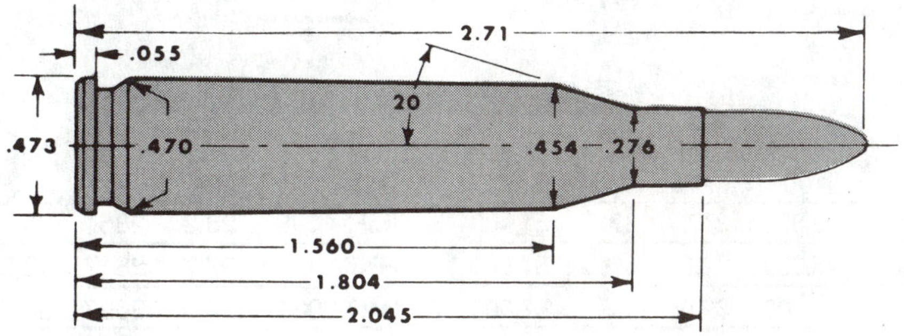

COMMENTS:

This is the most popular of all the 6mm diameter cartridges. While many would disagree, the late Les Bowman felt it the ideal deer or antelope cartridge for the shooter who really was not a frequent and comfortable user of larger cartridges.

The 243 is a fine varmint cartridge or an effective light big game round. Jacketed bullets up to 90 grains should generally be considered varmint types while those of 95 grains or more are light big game styles. Regardless of the bullet weight IMR 4350 is the universal first choice propellant. Premium grade bullets are a big asset to this small bore cartridge when it is used for light big game.

Cast bullets work best with velocities below 2,000 fps. Bullet #245496 is a prime choice for accuracy.

TEST COMPONENTS:

Cases .... Remington
Trim-to Length .... 2.035"
Primers .... Remington 9½
Primer Size .... Large Rifle
Lyman Shell Holder .... No. 2
Jacketed Bullets Used .... Sierra HP, 60 gr.
Sierra HPBT #1505, 70 gr.
Speer HP #1205, 75 gr.
Speer SP #1211, 80 gr.
Sierra HPBT #1530, 85 gr.
Remington Ptd. SP, 90 gr.
Hornady BTSP #2453, 100 gr.
Cast Bullets Used .... (Sized to .243" dia.)
*Gas Check Bullets .... *#245496, 83 gr.

TEST SPECIFICATIONS: (Velocity & Pressure)

Firearm Used .... Universal Receiver
Barrel Length .... 26"
Twist .... 1-10"
Groove Dia. .... .243"

60 gr. Jacketed HP

Powder	Sugg. Starting Grains	Velocity fps	Pressure C.U.P.	Max. Load Grains	Velocity fps	Pressure C.U.P.
IMR-3031	36.0	3365	36,400	41.0	3773	49,600
BL-C(2)	37.0	3323	37,300	41.5	3731	50,000
IMR-4064	38.0	3409	38,100	42.5	3774	49,200
IMR-4895	38.0	3488	43,300	42.0	3772	52,000
IMR-4320	39.0	3385	41,100	43.5	3707	50,000
H-380	40.0	3448	42,000	45.0	3778	51,200
IMR-4350	42.0	3291	37,300	47.0+	3726	51,600
H-4831	43.0	3057	34,000	49.0+	3596	48,400

Note: Loads shown in shaded panels are maximum.
+ Designates a compressed powder charge.

243 Winchester
(6.2 x 52mm)

70 gr. Jacketed HPBT
2.625" OAL

Powder	Sugg. Starting Grains	Velocity fps	Pressure C.U.P.	Max. Load Grains	Velocity fps	Pressure C.U.P.
IMR-4198	28.0	2946	38,800	31.8	3214	51,400
RX7	28.2	2969	45,900	32.0	3195	51,900
IMR-3031	35.0	3281	41,600	39.5	3482	52,000
H-335	32.6	3011	42,300	37.0	3291	51,900
BL-C(2)	35.0	3142	41,100	39.0	3419	50,800
IMR-4064	37.0	3316	42,900	41.0	3610	52,000
IMR-4895	36.0	3257	41,100	40.0	3539	51,200
IMR-4320	37.0	3177	38,600	42.0	3526	49,600
H-380	37.0	3136	39,900	41.0	3440	50,800
760	41.6	3146	39,300	47.3	3492	52,000
H-414	43.1	3263	43,100	49.0+	3592	52,000
IMR-4350	40.0	3193	39,900	45.5	3578	52,400
H-4831	43.0	3142	38,100	49.0+	3552	52,000
AA2015BR	31.0	3081	36,800	37.0	3448	51,900
AA2520	34.0	3031	38,600	39.0	3345	50,500

75 gr. Jacketed HP
2.600" OAL

Powder	Sugg. Starting Grains	Velocity fps	Pressure C.U.P.	Max. Load Grains	Velocity fps	Pressure C.U.P.
IMR-4198	28.2	2851	39,500	32.0	3125	52,000
RX7	27.7	2817	43,200	31.5	3058	51,200
IMR-3031	34.0	3097	38,600	38.0	3407	49,000
H-335	32.6	2930	46,800	37.0	3193	52,000
BL-C(2)	34.0	2974	41,100	38.0	3264	50,400
IMR-4064	36.0	3127	39,000	40.5	3460	50,400
IMR-4895	34.0	3041	37,700	39.0	3407	50,800
IMR-4320	37.0	3155	42,000	41.0	3437	50,000
H-380	36.0	3020	42,000	40.0	3294	52,000
760	40.5	3027	38,000	46.0	3375	51,200
H-414	41.4	3077	41,300	47.0	3435	51,900
IMR-4350	40.0	3160	44,200	45.0	3425	51,200
H-4831	43.0	3099	41,100	48.0+	3486	51,600
AA2015BR	30.0	2937	40,600	34.5	3224	51,900
RX12	34.0	2957	39,500	39.0	3273	51,200

80 gr. Jacketed SP
2.655" OAL

Powder	Sugg. Starting Grains	Velocity fps	Pressure C.U.P.	Max. Load Grains	Velocity fps	Pressure C.U.P.
IMR-4198	27.7	2792	45,100	31.5	3017	51,900
IMR-3031	33.0	3020	38,100	37.5	3309	49,200
BL-C(2)	33.0	2931	42,000	37.0	3209	51,600
IMR-4064	34.0	2958	36,400	‡39.5	3378	49,600
IMR-4895	32.0	2873	35.000	38.5	3308	50,400
IMR-4320	36.0	3057	40,700	40.5	3329	49,200
H-380	35.0	2956	41,600	39.0	3210	51,600
760	39.2	2982	43,100	44.5	3271	51,700
H-414	40.0	3058	46,300	45.5	3314	52,000
IMR-4350	39.0	3035	39,400	41.5	3194	52,000
H-4831	42.0	3059	41,600	47.0+	3406	51,600
AA2015BR	26.0	2697	38,900	32.0	3046	52,100
RX15	33.0	2951	43,100	37.0	3151	51,000
RX19	43.0	3087	42,600	48.0	3373	51,400

85 gr. Jacketed HPBT
2.615" OAL

Powder	Sugg. Starting Grains	Velocity fps	Pressure C.U.P.	Max. Load Grains	Velocity fps	Pressure C.U.P.
IMR-4198	27.3	2683	41,500	31.0	2922	51,900
IMR-3031	32.0	2836	36,900	37.5	3190	49,200
BL-C(2)	31.0	2620	34,000	36.5	3035	49,200
IMR-4064	34.0	2824	33,500	39.0	3236	49,200
IMR-4895	33.0	2883	38,600	38.0	3211	49,600
IMR-4320	35.0	2920	38,600	40.0	3255	50,000
H-380	33.0	2686	36,400	38.5	3094	50,400
760	39.2	2907	41,000	44.5	3176	50,800
H-414	40.0	2924	41,600	45.5	3234	51,200
IMR-4350	38.0	2844	36,900	44.0	3283	51,200
H-4831	41.0	2832	36,400	46.5+	3266	49,600
AA2015BR	26.5	2632	37,800	32.0	2992	51,200
RX15	32.0	2787	36,800	38.0	3147	50,000
RX19	42.0	2935	38,400	48.0	3326	50,900

Note: Loads shown in shaded panels are maximum.
+ Designates a compressed powder charge.

‡ Designates a factory velocity duplication load.

243 Winchester
(6.2 x 52mm)

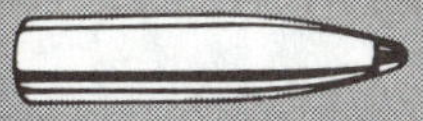

90 gr. Jacketed Ptd. SP

Powder	Sugg. Starting Grains	Velocity fps	Pressure C.U.P.	Max. Load Grains	Velocity fps	Pressure C.U.P.
IMR-3031	32.0	2778	37,700	37.0	3084	49,200
BL-C(2)	31.0	2560	34,000	36.0	2902	48,800
IMR-4064	35.0	2899	40,300	38.5	3097	49,200
IMR-4895	33.0	2758	38,100	37.5	3106	49,200
IMR-4320	35.0	2856	38,600	39.5	3146	49,200
H-380	33.0	2598	33,500	38.0	2922	48,800
IMR-4350	37.0	2703	34,000	43.5	3161	49,200
H-4831	41.0	2734	33,500	46.0+	3080	49,600

100 gr. Jacketed BTSP
2.650" OAL

Powder	Sugg. Starting Grains	Velocity fps	Pressure C.U.P.	Max. Load Grains	Velocity fps	Pressure C.U.P.
IMR-3031	30.5	2594	42,000	35.0	2869	50,800
BL-C(2)	30.5	2463	40,700	34.5	2723	50,800
IMR-4064	32.0	2658	43,300	36.5	2912	52,400
IMR-4895	31.2	2637	45,300	35.5	2868	52,000
IMR-4320	31.7	2593	44,000	36.0	2853	52,000
H-380	32.0	2549	41,600	36.0	2781	52,400
760	36.1	2678	44,300	41.0	2919	50,900
H-414	36.1	2666	44,700	41.0	2900	52,000
RX22	25.0	2229	32,600	42.5	3041	51,900
IMR-4350	35.0	2506	41,500	38.0	2863	51,800

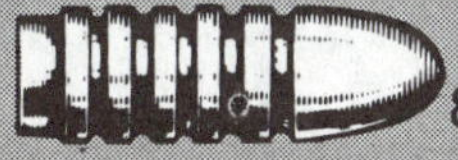

#245496
83 Gr. (#2 Alloy) 2.480" OAL

Powder	Sugg. Starting Grains	Velocity fps	Pressure C.U.P.	Max. Load Grains	Velocity fps	Pressure C.U.P.
Red Dot	10.0	1845	33,000	12.0	2032	40,300
700X	8.5	1684	30,500	10.5	1904	39,000
PB	9.0	1653	29,400	11.0	1836	36,400
SR-7625	9.5	1703	30,500	11.5	1909	39,400
SR-4756	11.0	1846	31,500	13.0	2013	38,600
SR-4759	14.9	1668	12,800	20.0	2156	21,400
IMR-4227	14.5	1623	11,900	19.7	2156	23,200
IMR-4198	16.0	1638	10,800	22.7	2175	19,200
RX7	16.0	1676	12,400	21.0	2092	20,200
748	26.1	2114	18,200	39.0	2906	43,800
H-335	25.8	2191	21,200	37.8	2926	46,000
H-4895	25.0	2126	17,300	37.2	2970	47,400
IMR-4320	28.2	2178	20,800	40.2	2996	45,700

Note: Loads shown in shaded panels are maximum.
+ Designates a compressed powder charge.

6mm Remington
(244 Remington)

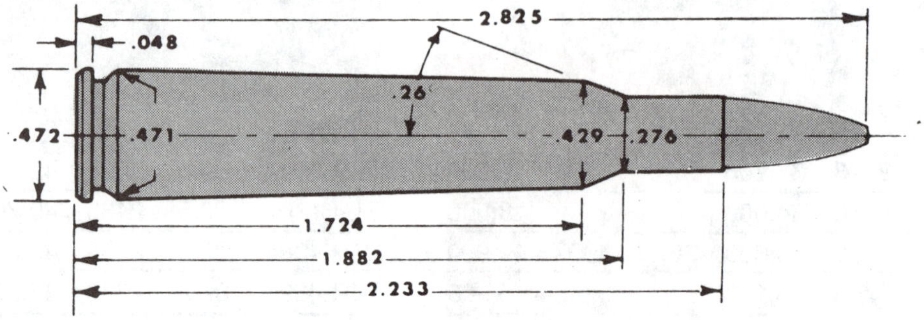

COMMENTS:

The 6mm Remington has a slightly larger case capacity than the 243 suggesting a bit more velocity. However since most shooters have found it slightly less accurate than the Winchester round it has never been extremely popular.

Early rifles marked 244 Remington had a 1 in 12" twist and shot best with bullets of 90 grains or less. Rifles with the 6mm Remington markings usually have a 1 in 9" twist and shoot well with bullets up to 100 grains.

IMR 4350 is the ideal first propellant selection for all jacketed bullet weights.

Cast bullet #245496 at 1,700 fps., or less, is the best choice for lead bullet shooting.

TEST COMPONENTS:

Cases .... Remington
Trim-to Length .... 2.225"
Primers .... Remington 9½
Primer Size .... Large Rifle
Lyman Shell Holder .... No. 2
Jacketed Bullets Used .... Sierra HP, 60 gr.
Sierra HPBT #1505, 70 gr.
Speer HP #1205, 75 gr.
Speer SP #1211, 80 gr.
Sierra HPBT #1530, 85 gr.
Speer Spitzer, 90 gr.
Hornady BTSP #2453, 100 gr.
Cast Bullets Used .... (Sized to .243" dia.)
*Gas Check Bullets .... *#245496, 83 gr.

TEST SPECIFICATIONS:
(Velocity & Pressure)

Firearm Used .... Universal Receiver
Barrel Length .... 24"
Twist .... 1-9"
Groove Dia. .... .243"

60 gr. Jacketed HP

Powder	Sugg. Starting Grains	Velocity fps	Pressure C.U.P.	Max. Load Grains	Velocity fps	Pressure C.U.P.
IMR-3031	37.0	3290	39,400	41.5	3677	50,400
BL-C(2)	38.0	3322	39,000	45.0	3773	52,000
IMR-4064	38.0	3164	36,900	44.0	3759	52,800
IMR-4895	38.0	3279	42,400	42.5	3623	51,200
IMR-4320	39.0	3194	39,900	44.5	3663	52,400
H-380	40.0	3279	37,700	46.5	3759	52,000
IMR-4350	42.0	3021	36,400	49.0+	3623	52,000
H-4831	46.0	3154	39,000	50.5+	3497	48,800

Note: Loads shown in shaded panels are maximum.
\+ Designates a compressed powder charge.

6mm Remington

(244 Remington)

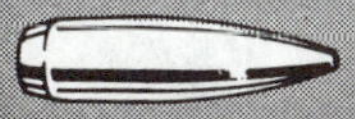

70 gr. Jacketed HPBT
2.825" OAL

Powder	Sugg. Starting Grains	Velocity fps	Pressure C.U.P.	Max. Load Grains	Velocity fps	Pressure C.U.P.
IMR-4198	28.6	3071	38,500	32.5	3374	51,900
RX7	26.9	3070	44,800	30.5	3301	51,900
IMR-3031	36.0	3105	36,400	41.0	3509	49,600
H-335	35.0	3308	45,900	40.0	3598	52,000
BL-C(2)	37.0	3067	36,400	44.0	3559	52,000
IMR-4064	37.5	3067	35,000	43.5	3584	51,600
IMR-4895	37.0	3067	38,600	42.0	3448	49,600
IMR-4320	38.5	3105	40,300	44.0	3509	52,000
H-380	39.0	3095	36,400	45.0	3588	52,000
760	40.0	3222	42,800	46.0	3611	50,700
H-414	40.0	3299	40,300	46.0	3643	51,400
IMR-4350	42.0	2976	38,600	48.5+	3484	51,200
H-4831	45.0	2967	36,400	50.5+	3413	48,800
AA2015BR	29.0	2821	38,600	34.0	3212	48,300
RX12	31.5	2777	36,600	39.5	3282	48,400
AA2520	33.0	2875	37,700	41.0	3423	49,300

80 gr. Jacketed SP
2.825" OAL

Powder	Sugg. Starting Grains	Velocity fps	Pressure C.U.P.	Max. Load Grains	Velocity fps	Pressure C.U.P.
IMR-4198	28.1	2900	37,000	32.0	3170	48,900
RX7	25.0	2719	39,800	28.5	2978	52,000
IMR-3031	35.0	2994	37,300	39.5	3412	50,600
H-335	29.4	2757	39,000	33.5	3049	50,300
BL-C(2)	36.0	2923	38,600	40.0	3415	52,000
IMR-4064	36.5	2915	36,000	42.0	3390	50,400
IMR-4895	36.0	2976	38,100	41.0	3333	50,400
IMR-4320	37.0	2906	36,900	43.0	3345	51,200
H-380	35.0	2785	34,500	43.0	3290	50,800
760	41.3	3152	40,700	‡47.0	3467	51,800
H-414	40.4	3209	42,600	46.0	3447	48,900
IMR-4350	41.0	2873	36,000	47.5+	3367	52,000
H-4831	44.0	2890	36,400	50.5+	3322	48,400
AA2015BR	25.0	2410	32,500	33.0	3018	48,200
AA3100	38.0	2616	35,700	49.5	3243	50,700
RX22	40.0	2922	40,600	49.0	3294	49,600

75 gr. Jacketed HP
2.825" OAL

Powder	Sugg. Starting Grains	Velocity fps	Pressure C.U.P.	Max. Load Grains	Velocity fps	Pressure C.U.P.
IMR-3031	36.0	3105	40,300	40.5	3485	50,700
BL-C(2)	37.0	3058	40,300	41.5	3447	51,500
IMR-4064	37.0	3048	37,300	42.5	3448	51,600
IMR-4895	37.0	3105	40,300	39.0	3385	51,400
IMR-4320	37.5	2994	37,700	41.0	3400	52,000
H-380	36.0	2865	34,500	43.5	3423	48,100
IMR-4350	41.5	2976	37,300	48.0+	3448	52,000
H-4831	44.0	2941	37,700	48.0	3288	45,000
AA2015BR	30.0	2817	42,000	35.5	3183	51,600
RX12	34.0	2800	38,800	42.0	3325	51,100
AA2520	32.0	2790	40,000	40.5	3333	52,000
AA3100	42.0	2761	37,700	48.5+	3249	50,700

85 gr. Jacketed HPBT
2.825" OAL

Powder	Sugg. Starting Grains	Velocity fps	Pressure C.U.P.	Max. Load Grains	Velocity fps	Pressure C.U.P.
IMR-4198	26.9	2785	39,200	30.5	3036	50,200
RX7	25.9	2755	41,200	29.5	2994	51,900
IMR-3031	34.0	2906	38,100	39.0	3354	52,000
BL-C(2)	35.0	2857	40,300	39.0	3134	52,000
IMR-4064	36.0	2865	37,300	41.0	3279	51,200
IMR-4895	35.0	2915	39,900	40.0	3236	51,600
IMR-4320	36.0	2849	36,900	42.0	3268	51,200
H-380	35.0	2770	37,300	41.0	3362	52,000
760	39.0	3055	41,100	45.0	3417	51,700
H-414	38.0	3029	41,800	45.0	3436	52,000
IMR-4350	40.0	2808	35,500	46.5+	3300	52,000
H-4831	44.0	2958	38,600	47.5+	3394	51,600
AA2015BR	29.0	2669	36,900	33.5	2994	50,100
RX19	42.0	2836	39,100	46.5	3211	50,800
AA3100	42.5	2715	39,500	47.5	3130	49,900
RX22	44.0	2879	38,000	49.4	3301	51,900

Note: Loads shown in shaded panels are maximum.
+ Designates a compressed powder charge.
‡ Designates a factory velocity duplication load.

6mm Remington
(244 Remington)

90 gr. Jacketed Spitzer

Powder	Sugg. Starting Grains	Velocity fps	Pressure C.U.P.	Max. Load Grains	Velocity fps	Pressure C.U.P.
IMR-3031	33.5	2793	38,100	38.5	3134	51,600
BL-C(2)	34.0	2739	40,700	38.5	3039	52,800
IMR-4064	35.5	2816	39,000	40.5	3164	50,800
IMR-4895	34.5	2785	37,700	39.5	3125	50,400
IMR-4320	35.5	2785	38,600	‡42.0	3184	52,000
H-380	34.0	2645	35,500	40.5	3067	52,800
IMR-4350	39.5	2754	36,400	46.0+	3194	51,600
H-4831	44.0	2857	38,600	48.5+	3225	50,000

#245496
83 gr. (#2 Alloy) 2.690" OAL

Powder	Sugg. Starting Grains	Velocity fps	Pressure C.U.P.	Max. Load Grains	Velocity fps	Pressure C.U.P.
Red Dot	9.0	1726	30,500	11.5	1973	40,700
700X	8.5	1635	29,400	10.7	1879	40,700
Green Dot	9.0	1703	27,600	11.5	1947	39,900
PB	9.0	1642	27,600	11.0	1829	38,600
SR-7625	9.0	1628	29,400	11.0	1818	38,600
SR-4756	10.0	1725	29,400	12.5	1974	40,700
SR-4759	13.0	1661	16,100	19.5	2202	34,000
IMR-4227	13.7	1711	17,300	19.7	2181	28,200
IMR-4198	16.8	1828	17,000	21.0	2144	22,500
RX7	14.5	1651	17,100	20.0	2117	25,800
748	20.8	1772	14,700	26.0	2173	23,400
H-4895	19.6	1760	16,300	24.5	2132	23,300

100 gr. Jacketed BTSP
2.820" OAL

Powder	Sugg. Starting Grains	Velocity fps	Pressure C.U.P.	Max. Load Grains	Velocity fps	Pressure C.U.P.
IMR-3031	32.0	2652	40,300	37.0	3049	51,500
BL-C(2)	32.5	2590	41,100	36.5	2849	52,000
IMR-4064	33.8	2811	38,500	38.5	3089	50,100
IMR-4895	34.0	2680	39,400	39.0	3003	51,600
IMR-4320	34.5	2631	38,100	41.0	3030	52,000
H-380	34.0	2570	39,000	38.5	2865	51,600
760	37.8	2800	39,600	43.0	3093	48,000
H-414	36.9	2769	36,300	‡43.5	3122	50,500
IMR-4350	37.4	2811	39,500	42.5	3143	51,100
H-4831	43.0	2739	41,600	48.0+	3086	52,000
AA2015BR	28.0	2485	43,500	32.4	2758	51,200
AA2520	31.0	2457	39,500	37.2	2869	51,000
RX19	38.0	2559	38,100	46.0	3058	51,700
AA3100	40.0	2576	39,400	45.0	2955	51,000
RX22	41.5	2723	38,300	46.6	3084	50,300

Note: Loads shown in shaded panels are maximum.
+ Designates a compressed powder charge.
‡ Designates a factory velocity duplication load.

240 Weatherby Magnum

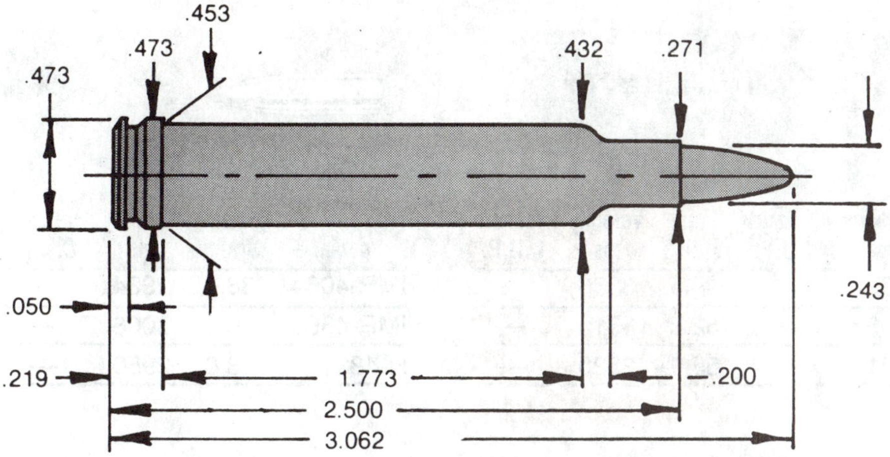

COMMENTS:

The 240 Weatherby is very fast, very accurate, but not very popular. This is largely due to limited availability of cartridges and reloading components. Therefore our data is limited to only the best powder choices. Accuracy with jacketed bullets tends to be excellent however no cast bullet testing has been done.

TEST COMPONENTS:

Cases .... Weatherby
Trim-to Length .... 2.490"
Primers .... Federal 215
Primer Size .... Large Rifle
Lyman Shell Holder .... No. 2
Jacketed Bullets Used .... Hornady Spire Pt., 70 gr.
Hornady HP, 75 gr.
Remington PL Match, 80 gr.
Sierra Spitzer, 85 gr.
Remington Ptd. SP, 90 gr.
Hornady Spire Pt., 100 gr.

TEST SPECIFICATIONS:
(Velocity Only)

Firearm Used .... Weatherby Mark V
Barrel Length .... 26"
Twist .... 1-10"
Groove Dia. .... .243"

70 gr. Jacketed Spire Pt.

Powder	Sugg. Starting Grains	Velocity fps	Pressure C.U.P.	Max. Load Grains	Velocity fps	Pressure C.U.P.
IMR-4064	41.0	3192	—	46.5	3564	—
IMR-4350	48.0	3184	—	54.0	3700	—
H-4831	50.0	3066	—	57.0+	3683	—

75 gr. Jacketed HP

Powder	Sugg. Starting Grains	Velocity fps	Pressure C.U.P.	Max. Load Grains	Velocity fps	Pressure C.U.P.
IMR-4064	40.0	3093	—	45.5	3417	—
IMR-4350	48.0	3164	—	53.0	3634	—
H-4831	49.0	3044	—	56.5+	3648	—

Note: Loads shown in shaded panels are maximum.
+ Designates a compressed powder charge.

240 Weatherby Magnum

80 gr. Jacketed HP

Powder	Sugg. Starting Grains	Velocity fps	Pressure C.U.P.	Max. Load Grains	Velocity fps	Pressure C.U.P.
IMR-4064	39.0	2992	—	44.5	3336	—
IMR-4350	47.0	3063	—	52.0	3515	—
H-4831	49.0	3014	—	55.0+	3520	—

85 gr. Jacketed Spitzer

Powder	Sugg. Starting Grains	Velocity fps	Pressure C.U.P.	Max. Load Grains	Velocity fps	Pressure C.U.P.
IMR-4064	38.0	2934	—	43.5	3262	—
IMR-4350	46.0	3008	—	51.0	3395	—
H-4831	48.0	2950	—	54.5	3426	—

90 gr. Jacketed Ptd. SP

Powder	Sugg. Starting Grains	Velocity fps	Pressure C.U.P.	Max. Load Grains	Velocity fps	Pressure C.U.P.
IMR-4064	37.0	2808	—	42.5	3129	—
IMR-4350	45.0	2917	—	50.0	3290	—
H-4831	48.0	2857	—	54.0	3367	—

100 gr. Jacketed Spire Pt.

Powder	Sugg. Starting Grains	Velocity fps	Pressure C.U.P.	Max. Load Grains	Velocity fps	Pressure C.U.P.
IMR-4064	37.0	2702	—	41.5	3020	—
IMR-4350	43.0	2772	—	48.0	3115	—
H-4831	47.0	2858	—	53.0	3210	—

Note: Loads shown in shaded panels are maximum.
+ Designates a compressed powder charge.

25-20 Winchester
(25 WCF)

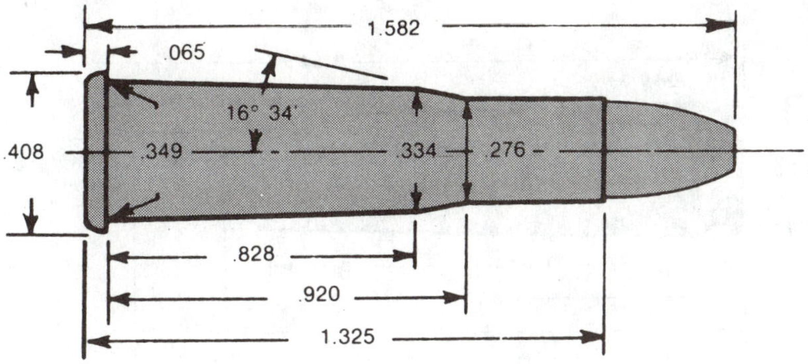

COMMENTS:

This data is not for use in old firearms designed for black powder. It is intended only for use in firearms designed for smokeless powder loads and in good condition.

When loading for tubular magazines you must use flat, blunt or round nose bullets. It is suggested that a light crimp be employed in ammo assembly for cartridges that will be used in a tubular magazine.

IMR 4198 is a good first powder choice. Data is brief due to the limited popularity of the cartridge.

TEST COMPONENTS:

Cases .... Remington
Trim-to Length .... 1.320"
Primers .... Remington 6½
Primer Size .... Small Rifle
Lyman Shell Holder .... No. 10
Jacketed Bullets Used .... Remington SP, 86 gr.

TEST SPECIFICATIONS:
(Velocity Only)

Firearm Used .... Winchester Model 92
Barrel Length .... 20"
Twist .... 1-14"
Groove Dia. .... .256"

86 gr. Jacketed SP

Powder	Sugg. Starting Grains	Velocity fps	Pressure C.U.P.	Max. Load Grains	Velocity fps	Pressure C.U.P.
Unique	4.5	1219	—	5.5	1508	—
2400	7.0	1150	—	9.0	1610	—
IMR-4227	8.0	1213	—	10.7	1713	—
IMR-4198	10.5	1283	—	13.0+	1751	—
H-110	7.0	1060	—	8.0	1294	—

Note: Loads shown in shaded panels are maximum.
\+ Designates a compressed powder charge.

250 Savage
(250/3000 Savage)

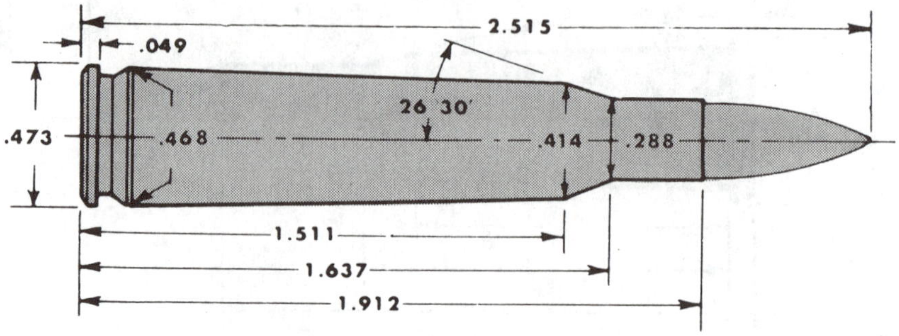

COMMENTS:

This cartridge is extremely accurate, but the proper length (weight) bullets must be selected. Early firearms had a 1 in 14" twist because the only factory bullet weight was 87 grains (at 3000 fps.) for the cartridges original name (250/3000). In these barrels 100 grain bullets seldom perform well. When the 100 grain bullet was introduced a twist rate of 1-10" was adopted. This twist often works with bullets as light as 75 grains and as heavy as 117 grains.

Due to the cases small volume the 100 grain bullet is the best for light big game. Be sure to select a big game style bullet as some 100 grain bullets are of the varmint type. Due to the limited ballistics of the cartridge it would be wise to select a premium grade bullet to optimize performance on light big game.

IMR 3031 is the best choice for accuracy in most rifles with all bullet weights.

When loading for lever action rifles use extra caution in developing a maximum load. Many lever actions will not allow the listed maximum load to be used safely.

TEST COMPONENTS:

Cases .... Remington
Trim-to Length .... 1.902"
Primers .... Remington 9½
Primer Size .... Large Rifle
Lyman Shell Holder .... No. 2
Jacketed Bullets Used .... Sierra HP, 75 gr.
Remington SP, 87 gr.
Remington SPCL, 100 gr.
Remington SPCL, 117 gr.

TEST SPECIFICATIONS:
(Velocity Only)

Firearm Used .... Custom Mauser 98
Barrel Length .... 24"
Twist .... 1-14"
Groove Dia. .... .257"

75 gr. Jacketed HP

Powder	Sugg. Starting Grains	Velocity fps	Pressure C.U.P.	Max. Load Grains	Velocity fps	Pressure C.U.P.
IMR-3031	32.0	3003	—	36.0	3472	—
IMR-4895	33.0	2881	—	37.0	3300	—
IMR-4064	34.0	2906	—	38.0+	3322	—
IMR-4320	35.0	2976	—	39.0	3427	—
BL-C(2)	35.0	2985	—	39.0	3300	—
H-335	33.0	2793	—	37.0	3058	—
H-380	38.0	2967	—	42.0+	3164	—

Note: Loads shown in shaded panels are maximum.
+ Designates a compressed powder charge.

250 Savage
(250/3000 Savage)

87 gr. Jacketed SP

Powder	Sugg. Starting Grains	Velocity fps	Pressure C.U.P.	Max. Load Grains	Velocity fps	Pressure C.U.P.
IMR-3031	31.0	2840	—	35.0	3247	—
IMR-4895	32.0	2747	—	36.5	3194	—
IMR-4064	33.0	2816	—	37.0+	3205	—
IMR-4320	34.0	2873	—	38.0	3236	—
BL-C(2)	34.0	2906	—	38.0	3154	—
H-335	33.0	2645	—	36.0	2923	—
H-380	37.0	2824	—	‡41.0+	3012	—

100 gr. Jacketed SP

Powder	Sugg. Starting Grains	Velocity fps	Pressure C.U.P.	Max. Load Grains	Velocity fps	Pressure C.U.P.
IMR-3031	30.0	2695	—	33.0	3012	—
IMR-4895	31.0	2624	—	35.5	3003	—
IMR-4064	31.0	2577	—	35.5	3021	—
IMR-4320	33.0	2754	—	37.0	3115	—
BL-C(2)	31.0	2463	—	35.0	2801	—
H-335	32.0	2544	—	34.0	2702	—
H-380	35.0	2645	—	‡39.0	2865	—

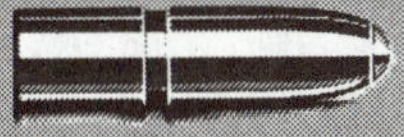

117 gr. Jacketed SP

Powder	Sugg. Starting Grains	Velocity fps	Pressure C.U.P.	Max. Load Grains	Velocity fps	Pressure C.U.P.
IMR-3031	28.0	2444	—	31.0	2672	—
IMR-4895	31.0	2570	—	33.5	2747	—
IMR-4064	30.0	2463	—	33.5	2710	—
IMR-4320	32.0	2557	—	35.0	2793	—
BL-C(2)	29.0	2283	—	32.0	2481	—
H-335	30.0	2369	—	32.0	2506	—
H-380	32.0	2427	—	36.0	2702	—

Note: Loads shown in shaded panels are maximum.
\+ Designates a compressed powder charge.
‡ Designates a factory velocity duplication load.

257 Roberts

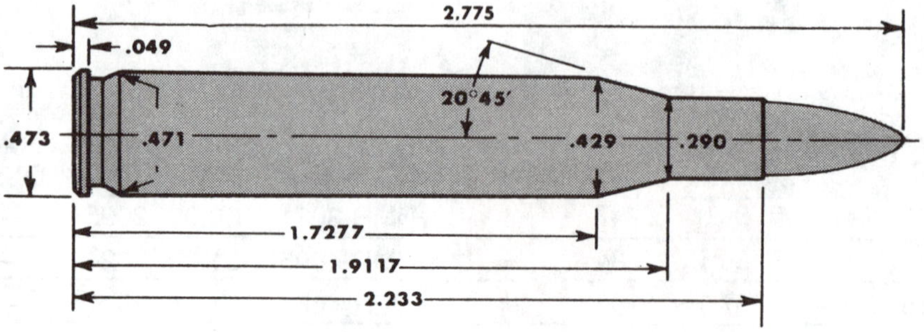

COMMENTS:

This cartridge could offer a worthwhile ballistic advantage over the 250 Savage. However, it is difficult to get top accuracy and performance from many older rifles as they use a short magazine while chamber throats are of a length that can accommodate a cartridge which is longer than the magazine will allow. Most new rifles have corrected this condition.

The reloader may encounter brass cases that have two different rim/extractor cut dimensions. Therefore it may be necessary to use two different shell holders.

The slower burning powders are best for accuracy and ballistic uniformity with IMR 4064 being the best single choice among the listed propellants with all bullet weights.

TEST COMPONENTS:

Cases .... Remington
Trim-to Length .... 2.223"
Primers .... Remington 9½
Primer Size .... Large Rifle
Lyman Shell Holder .... No. 2 or 8
Jacketed Bullets Used .... Sierra HP #1600, 75 gr.
Hornady SP #2530, 87 gr.
Sierra SPT #1620, 100 gr.
Nosler SPT #16319, 117 gr.
Cast Bullets Used .... (Sized to .257" dia.)
*Gas Check Bullets .... *#257464, 89 gr.

TEST SPECIFICATIONS: (Velocity & Pressure)

Firearm Used .... Universal Receiver and Custom Mauser 98
Barrel Length .... 24"
Twist .... 1-12"
Groove Dia. .... .257"

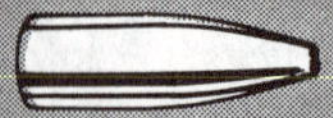

75 gr. Jacketed HP
2.735" OAL

Powder	Sugg. Starting Grains	Velocity fps	Pressure C.U.P.	Max. Load Grains	Velocity fps	Pressure C.U.P.
IMR-3031	36.0	2976	—	40.0	3311	—
BL-C(2)	36.0	2801	—	42.0	3311	—
IMR-4064	40.0	3012	—	44.0	3509	—
IMR-4895	38.0	2890	—	42.0	3268	—
IMR-4320	38.0	2808	—	42.0	3144	—
H-380	41.0	2906	—	45.0	3236	—
760	40.2	2849	33,500	45.7	3153	41,200
H-414	42.5	3016	37,800	47.0	3330	44,900

Note: Loads shown in shaded panels are maximum.

257 Roberts

87 gr. Jacketed SP
2.750" OAL

Powder	Sugg. Starting Grains	Velocity fps	Pressure C.U.P.	Max. Load Grains	Velocity fps	Pressure C.U.P.
IMR-3031	35.0	2840	—	39.0	3134	—
BL-C(2)	37.0	2824	—	41.0	3030	—
IMR-4064	39.0	2932	—	43.0	3311	—
IMR-4895	36.0	2724	—	40.0	3030	—
IMR-4320	36.0	2659	—	40.0	2932	—
H-380	40.0	2849	—	44.0	3086	—
760	41.1	2879	38,700	46.7	3092	44,200
H-414	40.5	2896	38,500	‡46.0	3140	44,200

100 gr. Jacketed Spitzer
2.775" OAL

Powder	Sugg. Starting Grains	Velocity fps	Pressure C.U.P.	Max. Load Grains	Velocity fps	Pressure C.U.P.
IMR-3031	34.0	2717	—	38.0	2985	—
BL-C(2)	36.0	2747	—	40.0	2898	—
IMR-4064	36.0	2631	—	40.0	2958	—
IMR-4895	35.0	2577	—	39.0	2881	—
IMR-4320	34.0	2463	—	38.5	2793	—
H-380	38.0	2564	—	42.0	2857	—
760	38.5	2591	35,700	44.7	2928	43,200
H-414	38.3	2683	38,000	43.5	2973	45,200

117 gr. Jacketed Spitzer
2.775" OAL

Powder	Sugg. Starting Grains	Velocity fps	Pressure C.U.P.	Max. Load Grains	Velocity fps	Pressure C.U.P.
IMR-3031	30.0	2341	—	33.0	2551	—
BL-C(2)	32.0	2380	—	35.0	2564	—
IMR-4064	32.0	2347	—	36.0	2610	—
IMR-4895	31.0	2288	—	34.0	2493	—
IMR-4320	33.0	2369	—	36.0	2595	—
H-380	34.0	2325	—	38.0	2544	—
760	36.3	2319	34,300	41.2	2688	44,800
H-414	35.0	2394	36,500	39.7	2706	45,100

#257464
89 gr. (#2 Alloy) 2.610" OAL

Powder	Sugg. Starting Grains	Velocity fps	Pressure C.U.P.	Max. Load Grains	Velocity fps	Pressure C.U.P.
SR-4759	14.0	1584	11,500	17.5	1933	16,700
IMR-4227	14.5	1584	11,100	19.0	2034	19,200
IMR-4198	16.0	1646	10,100	21.5	2063	16,500
RX7	14.5	1497	9,100	20.0	1981	15,200
748	18.0	1468	8,000	24.5	1959	13,600
H-335	17.3	1544	9,200	23.0	1954	13,600

Note: Loads shown in shaded panels are maximum.
‡ Designates a factory velocity duplication load.

25-06 Remington

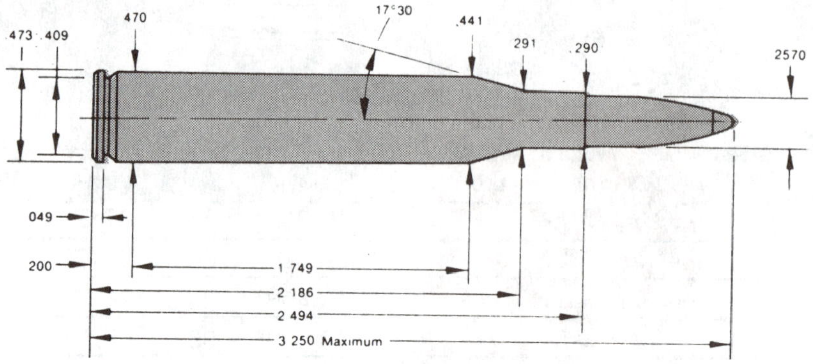

COMMENTS:

This was for years a wildcat cartridge until Remington decided to legitimize it in 1969. However, due to the great variations in chamber dimensions the listed data is not intended for use in custom chambers. The data is intended only for use in standard commercial rifles originally chambered for this round.

IMR 4350 and Hodgdon H4831 are the best choices for accuracy. Ballistic uniformity generally improves noticeably with bullets of 100 grains or heavier.

If ever there was a perfect antelope cartridge the 25-06 Remington is it. It shoots very, very, flat and has more than sufficient punch to make long shots on light game animals possible.

Best cast bullet results were with #257464 at speeds below 2200 fps. Beyond 2200 fps. accuracy became erratic.

TEST COMPONENTS:

Cases .... Remington
Trim-to Length .... 2.484"
Primers .... Winchester 8½-120 and CCI 200
Primer Size .... Large Rifle
Lyman Shell Holder .... No. 2
Jacketed Bullets Used .... Sierra HP #1600, 75 gr.
Hornady SP #2530, 87 gr.
Sierra SPT #1620, 100 gr.
Sierra SPT #1640, 117 gr.
Hornady HP #2560, 120 gr.
Cast Bullets Used .... (Sized to .257" dia.)
*Gas Check Bullets .... *#257464, 90 gr.

TEST SPECIFICATIONS: (Velocity & Pressure)

Firearm Used .... Universal Receiver
Barrel Length .... 26"
Twist .... 1-10"
Groove Dia. .... .257"

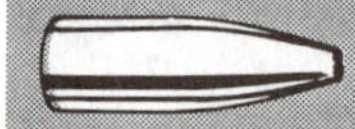

75 gr. Jacketed HP
3.019" OAL

Powder	Sugg. Starting Grains	Velocity fps	Pressure C.U.P.	Max. Load Grains	Velocity fps	Pressure C.U.P.
IMR-4198	33.3	3009	37,800	37.8	3315	50,700
IMR-3031	43.0	3304	42,400	48.0	3632	49,600
IMR-4064	45.0	3304	42,000	51.0	3740	52,800
IMR-4895	43.5	3170	42,000	49.0	3595	50,800
IMR-4320	45.0	3199	40,700	51.5	3671	52,400
H-380	47.0	3373	42,900	52.0	3663	51,200
760	50.6	3160	39,300	57.5	3608	51,800
H-414	50.5	3134	33,200	57.4	3663	49,300
IMR-4350	50.0	3128	41,600	56.5	3575	52,000
H-4831	53.0	3105	41,100	59.0+	3559	52,400

Note: Loads shown in shaded panels are maximum.
\+ Designates a compressed powder charge.

25-06 Remington

87 gr. Jacketed SP
3.096" OAL

Powder	Sugg. Starting Grains	Velocity fps	Pressure C.U.P.	Max. Load Grains	Velocity fps	Pressure C.U.P.
IMR-4198	31.6	2775	39,600	36.0	3059	49,500
RX7	32.7	2806	41,000	37.5	3074	51,600
IMR-3031	42.0	3113	41,600	47.0	3452	50,800
IMR-4064	44.0	3153	42,900	49.0	3485	50,800
IMR-4895	42.5	3120	42,900	48.5	3434	51,200
IMR-4320	44.0	3080	41,100	49.5	3420	51,200
H-380	45.0	3111	41,600	50.0	3436	52,400
H-414	49.5	3115	38,900	‡56.0	3444	51,200
IMR-4350	49.0	3078	41,100	55.0	3492	52,800
H-4831	52.0	3015	40,300	58.0+	3407	50,400

100 gr. Jacketed Spitzer
3.155" OAL

Powder	Sugg. Starting Grains	Velocity fps	Pressure C.U.P.	Max. Load Grains	Velocity fps	Pressure C.U.P.
RX7	27.2	2409	43,300	31.1	2633	47,900
IMR-3031	36.2	2726	38,800	40.0	2994	50,200
IMR-4064	40.0	2856	40,300	44.5	3158	51,200
IMR-4320	41.0	2822	41,600	46.5	3139	50,800
H-380	43.0	2951	45,900	47.5	3141	51,600
H-414	44.5	2803	38,700	‡51.0	3209	49,500
IMR-4350	46.5	2622	37,300	52.0	3202	50,400
H-4831	52.0	2978	43,700	56.0	3285	52,800

117 gr. Jacketed Spitzer
3.155" OAL

Powder	Sugg. Starting Grains	Velocity fps	Pressure C.U.P.	Max. Load Grains	Velocity fps	Pressure C.U.P.
RX7	28.4	2316	40,900	32.3	2525	48,500
IMR-3031	36.4	2590	37,500	41.0	2842	48,900
IMR-4064	37.1	2566	40,300	42.4	2866	49,600
IMR-4320	40.0	2631	41,600	46.0	2969	52,000
H-380	41.0	2681	43,300	44.5	2843	51,200
IMR-4350	46.0	2735	41,600	51.0	3013	50,400
H-4831	50.0	2803	42,000	54.0	3062	50,400

120 gr. Jacketed HP
3.250" OAL

Powder	Sugg. Starting Grains	Velocity fps	Pressure C.U.P.	Max. Load Grains	Velocity fps	Pressure C.U.P.
RX7	30.6	2375	44,100	34.8	2598	51,700
IMR-3031	37.8	2542	39,200	42.0	2789	49,900
IMR-4064	39.3	2629	40,200	44.7	2861	50,700
IMR-4350	44.4	2631	39,200	50.5	2963	50,600
H-4831	48.4	2569	40,300	55.0+	2981	49,500

#257464
90 gr. (#2 Alloy) 2.911" OAL

Powder	Sugg. Starting Grains	Velocity fps	Pressure C.U.P.	Max. Load Grains	Velocity fps	Pressure C.U.P.
**2400	20.0	2162	22,000	29.0	2781	47,800
**SR-4759	22.2	2193	22,100	32.8	2859	47,900
IMR-4227	22.0	2225	23,600	30.0	2753	46,300
**IMR-4198	24.5	2182	19,000	33.0	2770	35,900
**748	30.0	2127	16,800	46.3	2960	39,700

Note: Loads shown in shaded panels are maximum.
+ Designates a compressed powder charge.
‡ Designates a factory velocity duplication load.
** Signifies use of CCI 200 primers.

257 Weatherby Magnum

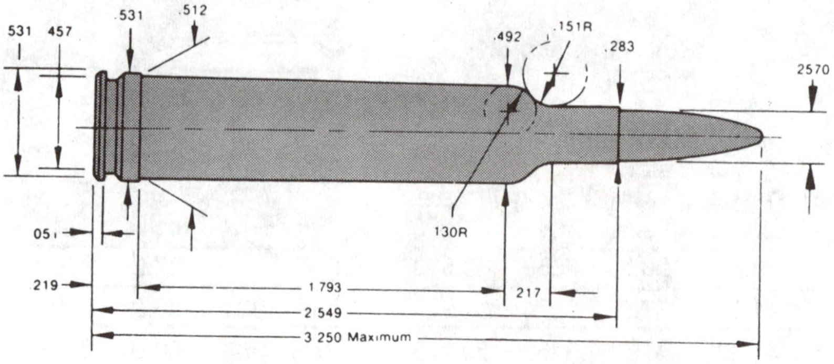

COMMENTS:

This cartridge never has gained a great deal of popularity, mostly because it has been a proprietary round available from but a single source — Weatherby. Thus data for this round has been limited.

The data is intended only for use in Weatherby rifles and not in custom chambers where dimensions may vary considerably from nominal.

Bullets of strong construction and weighing 117 grains are suggested for all light big game applications because of the very high bullet velocities. Lesser bullets tend to explode or disintegrate instead of penetrating as needed.

Bullets of 87 and 100 grains are suggested as best suited for varmint hunting.

TEST COMPONENTS:

Cases .... Weatherby
Trim-to Length .... 2.540"
Primers .... Federal 215
Primer Size .... Large Rifle
Lyman Shell Holder .... No. 13
Jacketed Bullets Used .... Sierra HP, 75 gr.
Remington SP, 87 gr.
Remington SPCL, 100 gr.
Remington SPCL, 117 gr.

TEST SPECIFICATIONS: (Velocity Only)

Firearm Used .... Weatherby Mark V
Barrel Length .... 26"
Twist .... 1-12"
Groove Dia. .... .257"

75 gr. Jacketed HP

Powder	Sugg. Starting Grains	Velocity fps	Pressure C.U.P.	Max. Load Grains	Velocity fps	Pressure C.U.P.
IMR-4064	55.0	3367	—	61.0	3745	—
IMR-4320	57.0	3460	—	63.0	3875	—
IMR-4350	64.0	3390	—	71.0	4048	—
H-380	62.0	3472	—	68.0	3952	—
H-4831	68.0	3367	—	75.0	3891	—

Note: Loads shown in shaded panels are maximum.

257 Weatherby Magnum

87 gr. Jacketed SP

Powder	Sugg. Starting Grains	Velocity fps	Pressure C.U.P.	Max. Load Grains	Velocity fps	Pressure C.U.P.
IMR-4064	52.0	3125	—	58.0	3521	—
IMR-4320	54.0	3290	—	60.0	3597	—
IMR-4350	62.0	3279	—	69.0	3759	—
H-380	60.0	3401	—	66.0	3690	—
H-4831	66.0	3184	—	73.0	3731	—

100 gr. Jacketed SP

Powder	Sugg. Starting Grains	Velocity fps	Pressure C.U.P.	Max. Load Grains	Velocity fps	Pressure C.U.P.
IMR-4064	50.0	2898	—	56.0	3279	—
IMR-4320	52.0	3021	—	58.0	3390	—
IMR-4350	59.0	3048	—	66.0	3472	—
H-380	58.0	3086	—	63.0	3367	—
H-4831	63.0	2923	—	70.0	3436	—

117 gr. Jacketed SP

Powder	Sugg. Starting Grains	Velocity fps	Pressure C.U.P.	Max. Load Grains	Velocity fps	Pressure C.U.P.
IMR-4064	48.0	2739	—	53.0	3012	—
IMR-4320	50.0	2801	—	55.0	3105	—
IMR-4350	58.0	2976	—	64.0	3322	—
H-380	56.0	2906	—	61.0	3164	—
H-4831	60.0	2824	—	67.0	3215	—

Note: Loads shown in shaded panels are maximum.

6.5x55mm Swedish Mauser

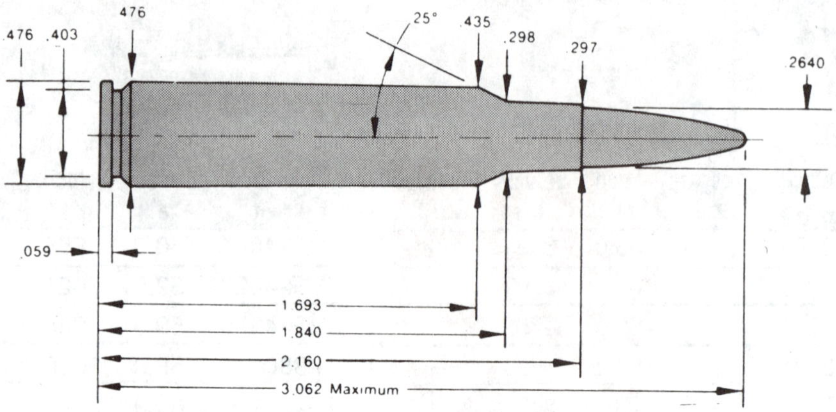

COMMENTS:

As with many foreign rifles, groove diameters tend to vary considerably from the nominal 0.264" diameter. We suggest you slug your bore before beginning to reload. Bores that are oversized will seldom provide accuracy and undersized bores should not be used with reloads.

When used with appropriate bullet styles this is a fine deer cartridge with ballistics far exceeding the legendary 6.5x54 Mannlicher Schoenauer cartridge. Several U.S. manufactured rifles have been made available for this cartridge on a limited basis. H380 with 140 grain bullets is a good place to start for an accurate light big game load.

TEST COMPONENTS:

Cases .... Norma
Trim-to Length .... 2.150"
Primers .... Remington 9½
Primer Size .... Large Rifle
Lyman Shell Holder .... No. 27
Jacketed Bullets Used .... Remington Ptd. SPCL, 100 gr.
Sierra Spitzer, 120 gr.
Hornady RN, 129 gr.
Remington SPCL, 140 gr.
Hornady RN, 160 gr.
Cast Bullets Used .... (Sized to .266" dia.)
*Gas Check Bullets .... *#266469

TEST SPECIFICATIONS:
(Velocity Only)

Firearm Used .... Swedish Military Carbine
HVA Bolt Action (cast bullets)
Barrel Length .... Swed. 18", HVA 23½"
Twist .... 1-7½"
Groove Dia. .... .266"

100 gr. Jacketed Ptd.SP

Powder	Sugg. Starting Grains	Velocity fps	Pressure C.U.P.	Max. Load Grains	Velocity fps	Pressure C.U.P.
IMR-3031	39.0	2739	—	43.0	3021	—
IMR-4895	41.0	2702	—	46.0	3058	—
IMR-4064	41.0	2666	—	46.0	3058	—
IMR-4320	42.0	2717	—	47.0	3067	—
H-380	46.0	2747	—	51.0	3076	—

Note: Loads shown in shaded panels are maximum.

6.5x55mm Swedish Mauser

120 gr. Jacketed Spitzer

Powder	Sugg. Starting Grains	Velocity fps	Pressure C.U.P.	Max. Load Grains	Velocity fps	Pressure C.U.P.
IMR-3031	37.0	2500	—	41.0	2770	—
IMR-4895	39.0	2551	—	43.0	2793	—
IMR-4064	39.0	2500	—	43.0	2770	—
IMR-4320	41.0	2617	—	45.0	2873	—
H-380	44.0	2597	—	49.0	2890	—

129 gr. Jacketed RN

Powder	Sugg. Starting Grains	Velocity fps	Pressure C.U.P.	Max. Load Grains	Velocity fps	Pressure C.U.P.
IMR-3031	35.0	2375	—	39.0	2583	—
IMR-4895	38.0	2463	—	42.0	2688	—
IMR-4064	38.0	2427	—	42.0	2638	—
IMR-4320	40.0	2506	—	44.0	2754	—
H-380	42.0	2444	—	47.0	2688	—

140 gr. Jacketed SP

Powder	Sugg. Starting Grains	Velocity fps	Pressure C.U.P.	Max. Load Grains	Velocity fps	Pressure C.U.P.
IMR-3031	33.0	2242	—	37.0	2415	—
IMR-4895	36.0	2304	—	40.0	2512	—
IMR-4064	36.0	2267	—	40.0	2506	—
IMR-4320	39.0	2421	—	43.0	2645	—
H-380	41.0	2325	—	46.0	2615	—

160 gr. Jacketed RN

Powder	Sugg. Starting Grains	Velocity fps	Pressure C.U.P.	Max. Load Grains	Velocity fps	Pressure C.U.P.
IMR-3031	31.0	2032	—	34.0	2207	—
IMR-4895	33.0	2040	—	37.0	2288	—
IMR-4064	33.0	2036	—	37.0	2237	—
IMR-4320	35.0	2132	—	39.0	2331	—
H-380	39.0	2178	—	43.0	2392	—

#266469

143 gr. (#2 Alloy) 2.825" OAL

Powder	Sugg. Starting Grains	Velocity fps	Pressure C.U.P.	Max. Load Grains	Velocity fps	Pressure C.U.P.
Red Dot	7.0	1135	—	9.5	1335	—
700X	7.0	1135	—	9.5	1330	—
Green Dot	8.0	1205	—	10.5	1465	—
PB	8.0	1145	—	11.5	1465	—
Unique	9.0	1275	—	11.5	1525	—
SR-7625	9.0	1220	—	12.5	1525	—
IMR-4198	15.0	1443	—	26.0	2091	—
RX7	15.0	1425	—	27.0	2182	—
IMR-4895	17.0	1316	—	30.0	2177	—
SR-4759	13.0	1375	—	25.0	2199	—

Note: Loads shown in shaded panels are maximum.

6.5mm Remington Magnum

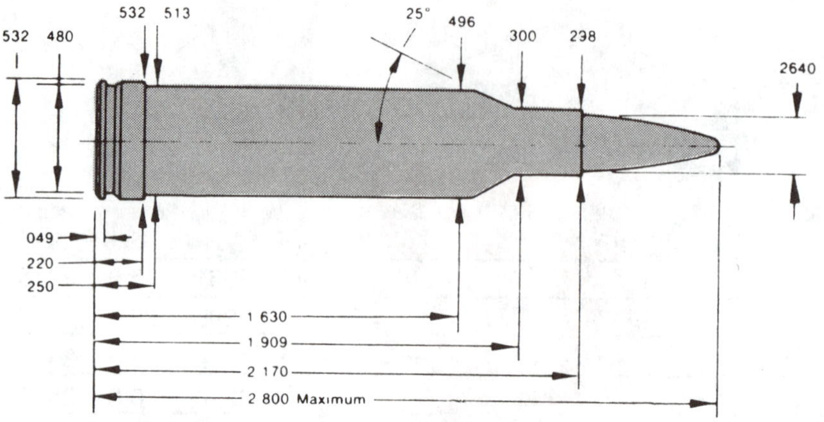

COMMENTS:

This is a cartridge that never had much general acceptance. It was available primarily in Remington carbines with short barrels and actions. Those folks who wanted short light woods rifles usually found other alternatives.

Still it makes a suitable light big game round when loaded with 129 grain bullets. We used magnum primers for the development of the limited data shown.

Because of limited popularity we have not conducted any cast bullet tests with this cartridge.

TEST COMPONENTS:

Cases .... Remington
Trim-to Length .... 2.160"
Primers .... Remington 9½M
Primer Size .... Large Rifle, Magnum
Lyman Shell Holder .... No. 13
Jacketed Bullets Used .... Remington Ptd. SPCL, 100 gr.
Sierra Spitzer, 120 gr.
Hornady RN, 129 gr.

TEST SPECIFICATIONS: (Velocity Only)

Firearm Used .... Remington Model 600
Barrel Length .... 18½"
Twist .... 1-9"
Groove Dia. .... .264"

100 gr. Jacketed Ptd.SP

Powder	Sugg. Starting Grains	Velocity fps	Pressure C.U.P.	Max. Load Grains	Velocity fps	Pressure C.U.P.
IMR-4064	43.0	2597	—	47.0	2801	—
IMR-4320	45.0	2702	—	49.0	2906	—
IMR-4350	52.0	2762	—	56.0	3012	—
H-380	47.0	2732	—	51.0	2881	—

120 gr. Jacketed Spitzer

Powder	Sugg. Starting Grains	Velocity fps	Pressure C.U.P.	Max. Load Grains	Velocity fps	Pressure C.U.P.
IMR-4064	39.0	2398	—	45.0	2638	—
IMR-4320	42.0	2481	—	47.0	2747	—
IMR-4350	49.0	2570	—	53.0	2808	—
H-380	45.0	2518	—	49.0	2695	—

129 gr. Jacketed RN

Powder	Sugg. Starting Grains	Velocity fps	Pressure C.U.P.	Max. Load Grains	Velocity fps	Pressure C.U.P.
IMR-4320	41.0	2398	—	46.0	2638	—
IMR-4350	48.0	2531	—	52.0	2747	—
H-380	44.0	2518	—	48.0	2624	—

Note: Loads shown in shaded panels are maximum.

264 Winchester Magnum

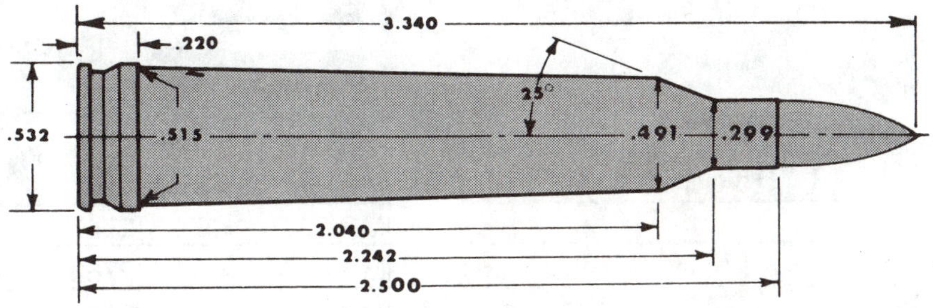

COMMENTS:

This is the first of the true general purpose big game cartridges. It has been used successfully on game as small as deer and antelope and as big as elk. For light big game the 140 grain bullet weight works well. For heavier game the 160 grain weight is to be preferred. With all bullet weights Hodgdon H4831 is extremely popular and supplies good ballistic uniformity and superb accuracy.

Magnum primers are suggested. Approach maximum loads slowly and with caution. Bullets less than 129 grains may be best left to varmint hunting.

Due to short barrel life few hunters would select this round over the 270 Winchester which offers similar performance .

TEST COMPONENTS:

Cases .... Winchester
Trim-to Length .... 2.490"
Primers .... Remington 9½M
Primer Size .... Large Rifle, Magnum
Lyman Shell Holder .... No. 13
Jacketed Bullets Used .... Remington Ptd. SPCL, 100 gr.
Sierra Spitzer, 120 gr.
Hornady RN, 129 gr.
Remington Ptd. SPCL, 140 gr.
Hornady RN, 160 gr.

TEST SPECIFICATIONS: (Velocity Only)

Firearm Used .... Winchester Model 70
Barrel Length .... 24"
Twist .... 1-9"
Groove Dia. .... .263"

100 gr. Jacketed Ptd.SP

Powder	Sugg. Starting Grains	Velocity fps	Pressure C.U.P.	Max. Load Grains	Velocity fps	Pressure C.U.P.
IMR-3031	48.0	3154	—	‡52.0	3356	—
IMR-4064	50.0	3144	—	55.5	3378	—
IMR-4895	52.0	3257	—	56.0	3436	—
IMR-4320	52.0	3300	—	56.0	3484	—
IMR-4350	58.0	3333	—	62.0	3509	—
H-380	54.0	3215	—	60.0	3390	—
H-4831	64.0	3322	—	71.0	3650	—

Note: Loads shown in shaded panels are maximum.
‡ Designates a factory velocity duplication load.

264 Winchester Magnum

120 gr. Jacketed Spitzer

Powder	Sugg. Starting Grains	Velocity fps	Pressure C.U.P.	Max. Load Grains	Velocity fps	Pressure C.U.P.
IMR-3031	46.0	2865	—	50.0	3058	—
IMR-4064	49.0	2923	—	52.5	3076	—
IMR-4895	50.0	2985	—	54.0	3144	—
IMR-4320	50.0	2985	—	54.0	3154	—
IMR-4350	56.0	3174	—	60.0	3290	—
H-380	52.0	2923	—	58.0	3205	—
H-4831	62.0	3164	—	66.5	3311	—

129 gr. Jacketed RN

Powder	Sugg. Starting Grains	Velocity fps	Pressure C.U.P.	Max. Load Grains	Velocity fps	Pressure C.U.P.
IMR-3031	45.0	2747	—	49.0	2923	—
IMR-4064	47.0	2762	—	51.0	2949	—
IMR-4895	49.0	2865	—	53.0	3039	—
IMR-4320	49.0	2881	—	53.0	3048	—
IMR-4350	55.0	2994	—	59.0	3164	—
H-380	50.0	2762	—	56.0	3039	—
H-4831	59.0	2958	—	64.5	3174	—

140 gr. Jacketed Ptd. SP

Powder	Sugg. Starting Grains	Velocity fps	Pressure C.U.P.	Max. Load Grains	Velocity fps	Pressure C.U.P.
IMR-3031	44.0	2638	—	48.0	2801	—
IMR-4064	46.0	2645	—	50.0	2801	—
IMR-4895	48.0	2724	—	52.0	2898	—
IMR-4320	48.0	2717	—	52.0	2906	—
IMR-4350	53.0	2840	—	58.0	2958	—
H-380	49.0	2617	—	54.0	2832	—
H-4831	57.0	2793	—	63.0	2976	—

160 gr. Jacketed RN

Powder	Sugg. Starting Grains	Velocity fps	Pressure C.U.P.	Max. Load Grains	Velocity fps	Pressure C.U.P.
IMR-3031	42.0	2409	—	46.0	2597	—
IMR-4064	44.0	2244	—	48.0	2597	—
IMR-4895	46.0	2538	—	50.0	2688	—
IMR-4320	46.0	2493	—	50.0	2659	—
IMR-4350	51.0	2645	—	56.0	2816	—
H-380	46.0	2398	—	52.0	2659	—
H-4831	52.0	2551	—	59.0	2793	—

Note: Loads shown in shaded panels are maximum.

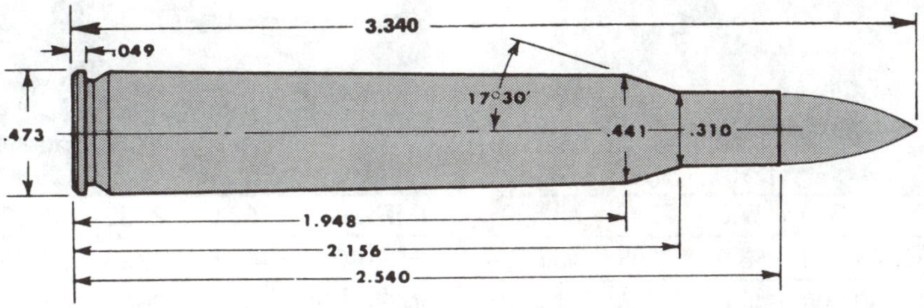

COMMENTS:

This is the smallest of what is generally considered an all around big game cartridge in the U.S. and is our number two big game cartridge with respect to popularity. The late Jack O'Connor, famed outdoor writer, did more to make its reputation than any other person.

If the 270 has a specific advantage over the number one popular hunting cartridge — the 30-06 Springfield — it is that it is milder in the recoil department. This feature makes it more suitable for the shooter who does not get a chance to practice a lot. The 270 does shoot a bit flatter with the 130 grain bullet than the comparable 150 grain 30-06 load.This makes hitting distant targets a bit easier. Accuracy with the 270 tends to be superb.

IMR 4350 will supply outstanding accuracy with all bullet weights up to 150 grains. However, either IMR 4831 or Hodgdon H4831 will duplicate this level of accuracy with 130 grain or heavier bullets and are the most popular choices with big game hunters.

Cast bullet #280642 can provide excellent accuracy.

TEST COMPONENTS:

Cases .... Winchester
Trim-to Length .... 2.530"
Primers .... Winchester 8½-120
Primer Size .... Large Rifle
Lyman Shell Holder .... No. 2
Jacketed Bullets Used .... Winchester SP, 100 gr.
Sierra SPT #1810, 110 gr.
Sierra SBT #1820, 130 gr.
Hornady BTSP #2735, 140 gr.
Hornady SP #2740, 150 gr.
Speer RN, 170 gr.
Cast Bullets Used .... (Sized to .277" dia.)
*Gas Check Bullets .... *#280473, 124 gr.
*#280642, 150 gr.

TEST SPECIFICATIONS:
(Velocity & Pressure)

Firearm Used .... Universal Receiver
Barrel Length .... 26"
Twist .... 1-10"
Groove Dia. .... .277"

100 gr. Jacketed SP

Powder	Sugg. Starting Grains	Velocity fps	Pressure C.U.P.	Max. Load Grains	Velocity fps	Pressure C.U.P.
IMR-3031	45.0	3154	43,700	50.0	3427	52,800
IMR-4064	46.0	3067	40,300	51.5	3413	53,600
IMR-4895	46.5	3134	44,600	50.0	3356	53,600
IMR-4320	46.0	2976	39,900	51.5	3333	53,200
H-380	48.0	3086	39,900	54.0	3436	54,000
IMR-4350	50.0	2949	41,100	57.0+	3345	54,400
H-4831	55.0	2702	40,300	62.0+	3356	51,600

Note: Loads shown in shaded panels are maximum.
\+ Designates a compressed powder charge.

270 Winchester

(6.9x64mm)

110 gr. Jacketed Spitzer
3.285" OAL

Powder	Sugg. Starting Grains	Velocity fps	Pressure C.U.P.	Max. Load Grains	Velocity fps	Pressure C.U.P.
IMR-4198	33.2	2627	33,300	37.8	2947	48,600
RX7	35.2	2736	44,000	43.0	3098	53,500
IMR-3031	44.0	2923	40,300	48.0	3236	52,400
IMR-4064	45.0	2881	40,300	50.5	3257	52,400
IMR-4895	45.0	2915	42,000	49.5	3225	53,200
IMR-4320	45.0	2793	38,100	51.0	3225	54,000
H-380	47.0	2932	40,300	53.0	3300	53,600
H-414	49.2	2833	35,500	56.0	3266	50,400
IMR-4350	49.2	2872	38,700	56.0	3274	49,600
H-4831	54.0	2824	38,600	61.0+	3174	49,200
RX12	43.0	2786	38,200	52.0	3142	51,600
RX15	45.0	2861	38,000	52.7	3296	51,400
AA3100	52.0	2655	36,100	61.5	3230	51,800
IMR-4831	54.0	2846	37,500	61.0+	3376	51,900

130 gr. Jacketed SBT
3.250" OAL

Powder	Sugg. Starting Grains	Velocity fps	Pressure C.U.P.	Max. Load Grains	Velocity fps	Pressure C.U.P.
IMR-3031	42.0	2739	41,600	47.0	3030	53,200
IMR-4064	43.5	2724	39,000	49.0	3039	52,400
IMR-4895	42.0	2680	41,100	47.5	3032	51,800
IMR-4320	44.0	2695	39,400	50.5	3058	52,600
H-380	44.0	2680	41,100	50.0	2994	54,000
760	48.8	2796	43,400	55.5	3080	51,600
H-414	47.9	2775	42,700	54.5	3062	52,800
IMR-4350	49.0	2739	39,400	54.5	3138	52,900
H-4831	53.0	2762	40,700	60.0+	3125	53,600
RX12	43.0	2650	38,600	50.2	2948	51,900
RX19	48.4	2486	35,000	57.0	3066	51,900
AA3100	50.1	2536	36,300	59.0	2939	51,900
RX22	51.0	2564	35,600	60.0	3173	51,800
IMR-4831	53.0	2823	40,800	57.0+	3127	52,100

140 gr. Jacketed BTSP
3.335" OAL

Powder	Sugg. Starting Grains	Velocity fps	Pressure C.U.P.	Max. Load Grains	Velocity fps	Pressure C.U.P.
IMR-3031	40.0	2636	41,800	45.5	2864	50,600
IMR-4064	41.3	2612	39,000	47.0	2929	52,000
IMR-4895	40.9	2666	44,400	46.5	2960	53,200
IMR-4320	40.7	2501	35,200	46.3	2879	53,600
H-380	43.8	2585	41,300	49.8	2878	53,800
760	47.7	2656	40,500	54.2	2975	52,300
H-414	46.2	2684	44,200	52.5	2914	51,700
IMR-4350	47.0	2649	38,500	53.5	3028	52,800
H-4831	51.0	2603	35,000	58.0+	3004	51,000
RX12	39.0	2483	39,500	47.8	2773	51,700
AA3100	48.8	2453	36,000	57.5	2979	51,900
IMR-4831	52.0	2703	39,900	56.3	3010	51,500

150 gr. Jacketed SP
3.285" OAL

Powder	Sugg. Starting Grains	Velocity fps	Pressure C.U.P.	Max. Load Grains	Velocity fps	Pressure C.U.P.
IMR-3031	38.7	2424	40,300	44.0	2755	53,600
IMR-4064	42.0	2564	42,000	47.0	2824	52,800
IMR-4895	39.0	2457	39,400	46.0	2801	54,000
IMR-4320	40.9	2505	37,700	46.5	2778	53,000
H-380	40.4	2434	41,300	46.0	2655	51,500
760	46.6	2496	41,000	53.0	2831	52,500
H-414	45.3	2493	43,200	51.5	2788	52,700
IMR-4350	47.0	2610	40,700	‡52.5	2906	54,000
H-4831	50.0	2557	39,000	57.0+	2941	54,000
RX19	46.9	2386	36,300	55.2	2857	51,600
AA3100	48.1	2354	36,300	56.6	2852	51,900
RX22	48.8	2466	36,300	57.5	2852	51,000
IMR-4831	50.3	2614	42,000	54.3	2860	52,000

Note: Loads shown in shaded panels are maximum.
+ Designates a compressed powder charge.
‡ Designates a factory velocity duplication load.

170 gr. Jacketed RN

Powder	Sugg. Starting Grains	Velocity fps	Pressure C.U.P.	Max. Load Grains	Velocity fps	Pressure C.U.P.
IMR-3031	37.5	2293	38,100	44.5	2631	53,600
IMR-4064	40.0	2331	38,600	46.5	2659	52,000
IMR-4895	38.0	2277	39,000	45.5	2631	52,800
IMR-4320	41.0	2358	37,300	48.0	2695	54,000
H-380	39.0	2252	39,000	46.0	2564	54,400
IMR-4350	46.0	2433	39,400	52.0	2739	52,000
H-4831	49.0	2398	36,000	56.0+	2801	53,600
IMR-4831	49.5	2459	39,700	52.9	2672	48,700

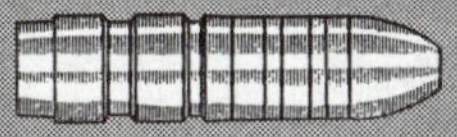

#280642
150 gr. (#2 Alloy) 3.073" OAL

Powder	Sugg. Starting Grains	Velocity fps	Pressure C.U.P.	Max. Load Grains	Velocity fps	Pressure C.U.P.
SR-4759	23.0	1926	34,200	31.0	2381	51,400
AA1680	24.0	1896	32,100	38.5	2534	51,800
RX7	26.0	1973	27,900	39.0	2660	52,000
IMR-3031	33.0	2155	28,100	39.0	2460	39,100
AA2015BR	30.0	2058	27,300	34.0	2254	33,500
H-335	31.0	2017	27,100	39.0	2422	39,700
IMR-4895	33.0	2007	24,900	44.0	2666	46,400
Unique	12.0	1519	26,300	19.0	2025	48,200
SR-7625	13.0	1533	29,800	18.0	1876	50,100
SR-4756	13.0	1500	26,100	19.0	1911	47,800

#280473
124 gr. (#2 Alloy) 3.150" OAL

Powder	Sugg. Starting Grains	Velocity fps	Pressure C.U.P.	Max. Load Grains	Velocity fps	Pressure C.U.P.
Red Dot	11.0	1587	29,400	13.5	1780	39,400
700X	10.0	1489	25,200	13.0	1729	39,900
Green Dot	11.5	1607	31,500	14.5	1839	40,300
PB	11.0	1499	27,000	13.5	1695	39,400
Unique	12.0	1698	28,200	15.3	1962	40,300
SR-7625	11.0	1505	25,800	13.5	1683	39,000
Herco	12.5	1647	30,500	15.5	1894	39,400
SR-4756	12.0	1554	24,600	15.5	1833	39,900
SR-4759	24.0	2134	27,000	32.0	2598	49,200
RX7	25.0	2065	19,900	39.0	2778	49,200
IMR-3031	31.0	2185	19,900	42.8	2914	48,900
748	34.0	2237	20,200	49.5	2955	45,600

Note: Loads shown in shaded panels are maximum.
\+ Designates a compressed powder charge.

270 Weatherby Magnum

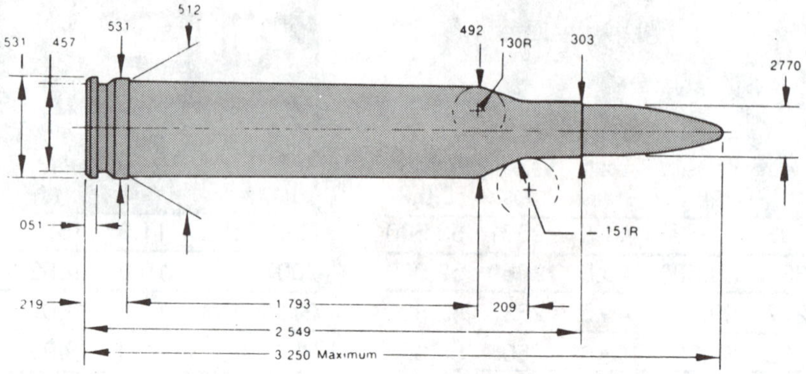

COMMENTS:

Despite the enormous popularity of the 270 Winchester, this is the only other commercial cartridge to use the 0.277" diameter bullet. Since it has been a proprietary cartridge it has never been as popular. Nonetheless, it does offer a sufficient increase in velocity over the standard Winchester cartridge and deserves the attention of serious long range shooters. If one wishes to use a 27 caliber cartridge on game as big as elk this is the better choice.

The data is intended for use only in commercial chambered rifles, not for use in custom chambered firearms. The exclusive use of a magnum primer is suggested for all jacketed bullet loads. Do not use magnum primers with cast bullet data.

Hodgdon H4831 is the best of the listed propellants with respect to combining accuracy and ballistic uniformity with jacketed bullets. Due to the high velocities obtained, premium bullets are a wise choice for big game applications.

TEST COMPONENTS:

Cases .... Weatherby
Trim-to Length .... 2.540"
Primers .... Federal 215 and Remington 9½
Primer Size .... Large Rifle
Lyman Shell Holder .... No. 13
Jacketed Bullets Used .... Sierra HP, 90 gr.
Remington Ptd. SP, 100 gr.
Remington Ptd. SP, 130 gr.
Remington SPCL, 150 gr.
Speer RN, 170 gr.
Cast Bullets Used .... (Sized to .277" dia.)
*Gas Check Bullets .... *#280473, 124 gr.

TEST SPECIFICATIONS: (Velocity Only)

Firearm Used .... Weatherby Mark V
Barrel Length .... 26"
Twist .... 1-12"
Groove Dia. .... .277"

90 gr. Jacketed HP

Powder	Sugg. Starting Grains	Velocity fps	Pressure C.U.P.	Max. Load Grains	Velocity fps	Pressure C.U.P.
IMR-4064	56.0	3086	—	62.0	3559	—
IMR-4320	59.0	3268	—	65.0	3677	—
IMR-4350	64.0	3125	—	73.0	3773	—
H-380	60.0	3067	—	68.0	3623	—
H-4831	71.0	3257	—	79.0+	3802	—

Note: Loads shown in shaded panels are maximum.
+ Designates a compressed powder charge.

270 Weatherby Magnum

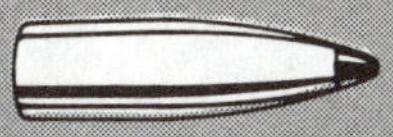

100 gr. Jacketed Ptd.SP

Powder	Sugg. Starting Grains	Velocity fps	Pressure C.U.P.	Max. Load Grains	Velocity fps	Pressure C.U.P.
IMR-4064	55.0	2994	—	62.0	3413	—
IMR-4320	58.0	3125	—	64.0	3497	—
IMR-4350	63.0	3003	—	72.0	3597	—
H-380	61.0	3003	—	67.0	3448	—
H-4831	70.0	3154	—	78.0+	3663	—

130 gr. Jacketed Ptd. SP

Powder	Sugg. Starting Grains	Velocity fps	Pressure C.U.P.	Max. Load Grains	Velocity fps	Pressure C.U.P.
IMR-4064	54.0	2890	—	61.0	3236	—
IMR-4320	57.0	3039	—	63.0	3345	—
IMR-4350	60.0	2923	—	67.0	3247	—
H-380	60.0	3039	—	66.0	3322	—
H-4831	66.0	2890	—	73.0	3345	—

150 gr. Jacketed SP

Powder	Sugg. Starting Grains	Velocity fps	Pressure C.U.P.	Max. Load Grains	Velocity fps	Pressure C.U.P.
IMR-4064	52.0	2732	—	59.0	3048	—
IMR-4320	56.0	2949	—	61.0	3157	—
IMR-4350	59.0	2801	—	65.0	3115	—
H-380	58.0	2898	—	64.0	3164	—
H-4831	63.0	2777	—	70.0	3184	—

170 gr. Jacketed RN

Powder	Sugg. Starting Grains	Velocity fps	Pressure C.U.P.	Max. Load Grains	Velocity fps	Pressure C.U.P.
IMR-4064	50.0	2531	—	56.0	2793	—
IMR-4320	52.0	2624	—	58.0	2890	—
IMR-4350	55.0	2518	—	61.0	2793	—
H-380	55.0	2610	—	60.0	2824	—
H-4831	59.0	2518	—	65.0	2816	—

#280473**
124 gr. (#2 Alloy)

Powder	Sugg. Starting Grains	Velocity fps	Pressure C.U.P.	Max. Load Grains	Velocity fps	Pressure C.U.P.
Unique	14.0	1756	—	16.0	1882	—
2400	20.0	1785	—	26.0	2118	—
IMR-4227	25.0	1964	—	27.0	2053	—

Note: Loads shown in shaded panels are maximum.
+ Designates a compressed powder charge.
** Designates use of Rem. 9½ primers.

7-30 Waters

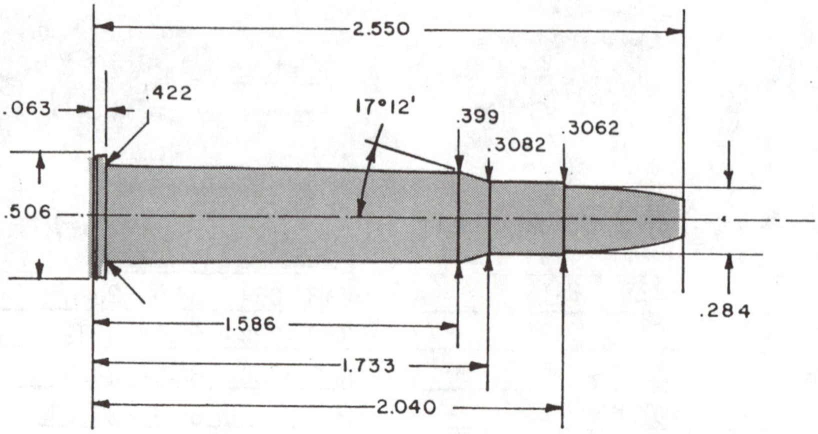

COMMENTS:

This is a modified version of the 30-30 that has gained some popularity, perhaps because of the current popularity of the 0.284" diameter bullets.

Caution: Never use pointed bullets in tubular magazines. Round or blunt nose bullets are required for safety in all such applications. Pointed bullets for this cartridge are intended only for single shot rifles.

This cartridge is suitable for light big game at modest ranges. Best results may be achieved with Winchester 748 and IMR-3031. When loading with cast bullet #287346 we recommend starting with IMR-3031.

TEST COMPONENTS:

Cases .... Federal
Trim-to Length .... 2.030"
Primers .... CCI 200
Primer Size .... Large Rifle
Lyman Shell Holder .... No. 6
Jacketed Bullets Used .... Sierra SP #1900, 120 gr.
Hornady FP #2822, 139 gr.
Cast Bullets Used .... (Sized to .285" dia.)
*Gas Check Bullets *#287346, 135 gr.

TEST SPECIFICATIONS: (Velocity & Pressure)

Firearm Used .... Universal Receiver
Barrel Length .... 20"
Twist .... 1-9.9"
Groove Dia. .... .284"

120 gr. Jacketed SP
2.550" OAL

Powder	Sugg. Starting Grains	Velocity fps	Pressure C.U.P.	Max. Load Grains	Velocity fps	Pressure C.U.P.
RX7	27.5	2166	26,000	31.0	2454	35,800
IMR-3031	30.5	2168	25,800	34.5+	2557	38,300
748	34.0	2264	28,500	39.5+	2621	37,100
BL-C(2)	32.3	2117	27,800	36.8	2466	39,400
AA2460	29.5	2084	25,000	33.8	2424	38,400
H-335	31.0	2232	26,800	34.8	2539	37,800
IMR-4895	31.8	2205	29,200	35.8+	2534	39,000

Note: Loads shown in shaded panels are maximum.
\+ Designates a compressed powder charge.

7-30 Waters

139 gr. Jacketed FP
2.545" OAL

Powder	Sugg. Starting Grains	Velocity fps	Pressure C.U.P.	Max. Load Grains	Velocity fps	Pressure C.U.P.
RX7	24.0	2002	31,000	27.9	2182	37,400
IMR-3031	28.2	1986	24,300	31.9+	2334	37,700
748	32.0	2132	27,500	36.0	2415	39,600
BL-C(2)	31.4	2019	26,000	35.7	2327	38,000
AA2460	29.0	1965	25,300	32.8	2277	37,400
H-335	29.5	2093	28,900	33.7	2352	38,600
IMR-4895	29.3	2005	25,000	34.0+	2370	39,700

#287346
135 gr. (#2 Alloy) 2.475" OAL

Powder	Sugg. Starting Grains	Velocity fps	Pressure C.U.P.	Max. Load Grains	Velocity fps	Pressure C.U.P.
SR-4759	16.5	1746	18,900	20.0	2017	34,600
RX7	23.0	1859	19,200	26.9	2145	31,000
IMR-3031	25.5	1816	16,800	31.0+	2277	34,900
748	29.5	1975	20,000	33.5+	2264	32,800
AA2460	26.0	1882	19,500	32.0	2273	34,600
H-335	26.2	1904	18,500	32.0	2300	34,500

Note: Loads shown in shaded panels are maximum.
+ Designates a compressed powder charge.

7mm-08 Remington

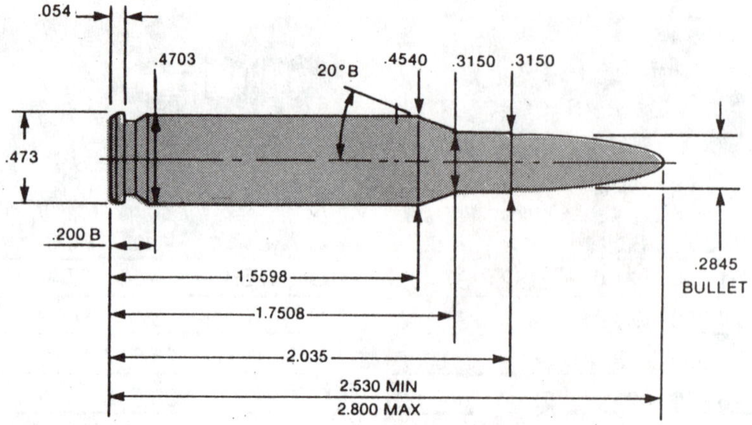

COMMENTS:

This cartridge was designed to fill a need for silhouette shooting. It is a necked down version of the 308 and has the advantage of less recoil as compared to its parent round. The 7mm-08 ranks right near the top among deer hunting cartridges. It has a flat trajectory and great accuracy. It is somewhat handicapped with the heavier bullets due to the short magazines used in rifles chambered for this round.

The factory ammo we have looked at was loaded with a Ball Powder that was just a tad faster burning than Winchester 760. Thus, 760 is as close as possible to duplicating factory specs. Nonetheless best accuracy often occurs with all weight bullets when using Hodgdon H380. Second choice would be IMR 4350 but this will mean the inconvenience of compressed powder loads.

TEST COMPONENTS:

Cases .... Remington
Trim-to Length .... 2.025"
Primers .... Remington 9½
Primer Size .... Large Rifle
Lyman Shell Holder .... No. 2
Jacketed Bullets Used .... Sierra SPT #1900, 120 gr.
Hornady SP #2820, 139 gr.
Hornady SP #2830, 154 gr.
Hornady BTSP #2845, 162 gr.
Sierra HPBT #1930, 168 gr.
Speer GSSP #1643, 175 gr.
Cast Bullets Used .... (Sized to .284" dia.)
*Gas Check Bullets *#287346, 135 gr.
*#287405, 150 gr.
*#287641, 160 gr.

TEST SPECIFICATIONS: (Velocity & Pressure)

Firearm Used .... Universal Receiver
Barrel Length .... 24"
Twist .... 1-9½"
Groove Dia. .... .284"

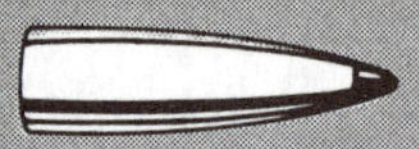

120 gr. Jacketed Spitzer
2.790" OAL

Powder	Sugg. Starting Grains	Velocity fps	Pressure C.U.P.	Max. Load Grains	Velocity fps	Pressure C.U.P.
RX7	33.1	2655	38,300	37.7	2911	50,500
IMR-3031	37.4	2706	34,800	42.6	3012	47,000
748	41.1	2704	35,700	46.7	3077	49,900
IMR-4064	39.1	2657	37,000	44.5+	3023	48,100
IMR-4895	37.8	2705	38,700	43.0	2999	48,100
H-380	43.5	2765	36,500	49.5+	3100	50,000
760	44.8	2652	34,200	51.0+	3062	50,200
H-414	44.0	2680	36,500	50.0+	3007	48,400

Note: Loads shown in shaded panels are maximum.
+ Designates a compressed powder charge.

7mm-08 Remington

139 gr. Jacketed SP
2.750" OAL

Powder	Sugg. Starting Grains	Velocity fps	Pressure C.U.P.	Max. Load Grains	Velocity fps	Pressure C.U.P.
RX7	30.0	2325	38,500	35.0	2593	47,200
IMR-3031	37.5	2564	39,000	40.0	2765	47,200
748	40.0	2533	37,500	45.0	2822	50,100
IMR-4064	38.0	2456	34,000	43.2	2854	50,200
IMR-4895	39.0	2523	35,000	43.3	2839	49,800
IMR-4320	37.5	2381	33,500	42.6	2727	49,300
H-380	41.7	2513	36,200	47.0	2815	50,500
760	43.5	2517	33,700	49.0	2852	49,200
H-414	42.3	2516	33,500	47.6	2840	46,100
IMR-4350	43.0	2508	36,200	48.3+	2822	47,000
AA2460	36.9	2433	36,700	41.4	2766	50,900
RX12	38.3	2485	40,500	43.6	2742	50,400
RX15	37.3	2467	35,400	43.0	2824	50,200
RX19	46.0	2512	37,500	52.0+	2857	49,300

154 gr. Jacketed SP
2.800" OAL

Powder	Sugg. Starting Grains	Velocity fps	Pressure C.U.P.	Max. Load Grains	Velocity fps	Pressure C.U.P.
RX7	29.9	2268	41,600	34.0	2481	52,000
IMR-3031	34.3	2335	34,000	39.0	2632	47,200
748	37.5	2376	37,700	42.7	2678	50,000
IMR-4064	36.7	2388	36,300	41.8+	2702	51,600
IMR-4895	35.4	2358	38,000	40.3	2664	51,200
IMR-4320	36.6	2315	37,000	41.6+	2641	50,100
H-380	38.5	2388	38,300	43.8	2668	51,000
760	40.4	2315	33,300	46.0	2702	50,800
H-414	40.4	2357	34,800	46.0	2731	51,700
IMR-4350	39.6	2324	35,000	45.0+	2622	46,900
AA2460	35.3	2262	38,500	40.5	2598	51,100
RX19	44.0	2394	37,800	49.5+	2676	47,400

162 gr. Jacketed BTSP
2.800" OAL

Powder	Sugg. Starting Grains	Velocity fps	Pressure C.U.P.	Max. Load Grains	Velocity fps	Pressure C.U.P.
IMR-3031	34.3	2339	35,700	39.0	2631	51,300
748	36.7	2312	35,000	41.7	2609	50,300
IMR-4064	35.6	2317	35,000	40.5	2626	49,700
IMR-4895	35.4	2349	38,300	40.3	2632	51,700
IMR-4320	35.9	2301	36,200	40.9	2589	51,200
H-380	38.7	2325	35,800	44.0+	2627	50,900
760	40.0	2269	33,300	45.5+	2644	49,100
H-414	39.9	2312	34,500	45.4	2677	51,300
IMR-4350	39.6	2343	36,000	45.0+	2622	47,300
AA2460	33.3	2181	37,000	39.0	2533	50,700
RX12	28.0	2010	33,500	39.7	2487	50,200
RX19	42.0	2264	34,800	49.0+	2702	50,200

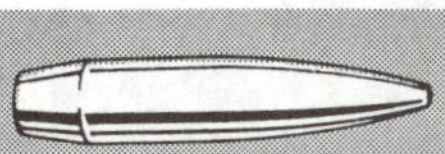

168 gr. Jacketed HPBT
2.800" OAL

Powder	Sugg. Starting Grains	Velocity fps	Pressure C.U.P.	Max. Load Grains	Velocity fps	Pressure C.U.P.
IMR-3031	33.6	2290	36,500	38.2	2575	51,500
748	36.4	2280	36,200	41.4	2586	51,000
IMR-4064	35.6	2339	39,600	40.5+	2605	51,800
IMR-4895	34.2	2306	39,200	38.9	2555	51,300
IMR-4320	34.7	2251	37,300	39.5	2510	49,100
H-380	38.2	2256	37,000	43.5+	2555	51,600
760	39.6	2280	36,200	45.0+	2632	51,800
H-414	39.1	2287	36,800	44.5	2628	51,900
IMR-4350	39.6	2357	38,300	45.0+	2605	50,700

Note: Loads shown in shaded panels are maximum.
+ Designates a compressed powder charge.

7mm-08 Remington

175 gr. Jacketed SP
2.735" OAL

Powder	Sugg. Starting Grains	Velocity fps	Pressure C.U.P.	Max. Load Grains	Velocity fps	Pressure C.U.P.
IMR-3031	32.3	2112	37,200	36.8	2408	50,400
748	34.3	2128	38,000	39.0	2396	51,500
IMR-4064	35.5	2260	41,900	39.2	2487	51,400
IMR-4895	34.3	2207	40,300	39.0	2473	50,600
IMR-4320	34.3	2171	40,200	39.0	2425	51,400
H-380	35.8	2103	38,000	40.7	2361	50,200
760	38.5	2208	37,500	43.8	2485	48,500
H-414	38.2	2252	39,000	43.5	2549	51,900
IMR-4350	38.7	2251	39,000	44.0+	2529	51,300
AA2460	31.0	1997	39,800	37.3	2323	51,300
RX12	25.0	1729	32,200	34.0	2205	50,900
RX19	40.3	2191	38,700	47.0	2539	51,400

#287346
135 gr. (#2 Alloy) 2.600" OAL

Powder	Sugg. Starting Grains	Velocity fps	Pressure C.U.P.	Max. Load Grains	Velocity fps	Pressure C.U.P.
SR-4759	15.3	1529	13,300	23.5	2189	36,900
IMR-4227	15.8	1584	14,600	24.0	2183	34,400
IMR-4198	16.8	1546	11,500	25.5	2161	32,300
RX7	17.4	1559	12,200	26.5	2164	31,600
IMR-3031	19.8	1495	10,900	30.0	2171	27,300
748	21.1	1509	9,600	32.0	2179	28,000

#287405
150 gr. (#2 Alloy) 2.520" OAL

Powder	Sugg. Starting Grains	Velocity fps	Pressure C.U.P.	Max. Load Grains	Velocity fps	Pressure C.U.P.
SR-4759	21.0	1969	32,200	24.0	2155	39,100
IMR-4227	21.0	1933	30,700	24.0	2104	35,200
IMR-4198	23.3	1962	29,200	26.5	2153	32,300
RX7	24.0	1972	28,200	28.0	2183	32,100
IMR-3031	26.4	1906	25,300	30.0	2121	29,600
748	28.4	1861	20,900	32.3	2115	26,500
H-335	27.5	1946	25,500	31.5	2187	28,100
IMR-4895	28.0	1915	19,600	32.0	2139	28,100

#287641
160 gr. (#2 Alloy) 2.705" OAL

Powder	Sugg. Starting Grains	Velocity fps	Pressure C.U.P.	Max. Load Grains	Velocity fps	Pressure C.U.P.
SR-4759	20.7	1761	21,500	26.0	2154	45,100
RX7	22.0	1716	18,200	28.0	2080	33,300
IMR-3031	28.7	1906	18,800	33.0	2238	32,000
748	34.5	1949	20,000	39.0	2299	33,500
AA2015BR	30.2	2060	26,900	34.0	2322	41,400
H-335	31.5	1999	23,800	36.0	2266	34,400
IMR-4895	29.5	1871	19,200	35.3	2296	36,000
Unique	12.0	1497	24,000	15.5	1769	42,100
IMR-4227	21.0	1748	20,800	27.0	2105	38,500
IMR-4198	23.0	1805	19,500	29.5	2181	36,100

Note: Loads shown in shaded panels are maximum.
+ Designates a compressed powder charge.

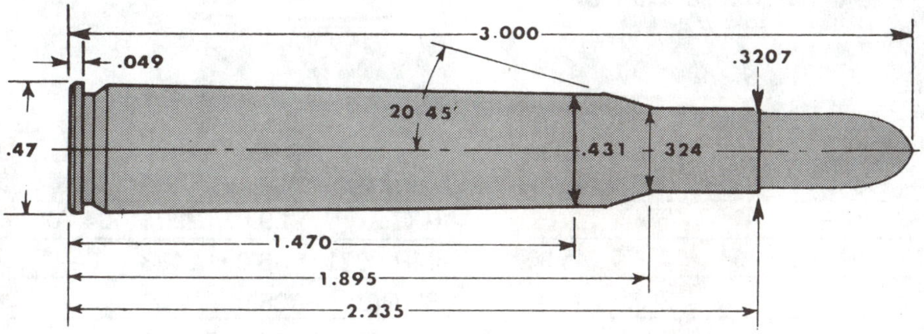

COMMENTS:

Land and groove diameters vary widely with this cartridge. As always, oversized bores seldom shoot well and undersized ones should not be used with reloads. The nominal groove diameter for this cartridge is 0.284". It is suggested that the bore of any foreign barrel be measured before reloading begins.

The 7mm Mauser is a very efficient light to medium game cartridge, when loaded to the industry pressure levels which should not exceed 46,000 C.U.P.

The most universal powder to combine accuracy and uniform ballistics seems to be IMR 4350. This applies to all bullet weights.

The popular trend today is to load this round with bullets of 139 to 150 grains for use in big game hunting. This trend ignores the fact that the reputation of this cartridge as an efficient big game killer was developed with the use of 175 grain bullets. In testing some extremely accurate results were attained with cast bullet #287641.

TEST COMPONENTS:

Cases ....Winchester (jacketed bullets), Federal (cast bullets)
Trim-to Length ....2.225"
Primers ......Winchester 8½-120 (jacketed bullets), Federal 210 and CCI 200 (cast bullets)
Primer Size ....Large Rifle
Lyman Shell Holder No.2
Jacketed Bullets Used ....Sierra SPT #1900, 120 gr.
Hornady SP #2820,139 gr.
Sierra HPBT #1915, 150 gr.
Hornady BTSP #2845, 162 gr.
Sierra HPBT #1930, 168 gr.
Speer GSSP #1643, 175 gr.
Cast Bullets Used ....(Sized to .284" dia.)
*Gas Check Bullets *#287405, 150 gr.
*#287641, 160 gr.

TEST SPECIFICATIONS:
(Velocity & Pressure)

Firearm Used ....Mauser 95 and Universal Receiver
Barrel Length ....Mauser 29"; Universal Receiver 24"
Twist..........................Mauser 1-10"; Universal Receiver 1-8¾"
Groove Dia. ....Mauser .2865"; Universal Receiver .284"

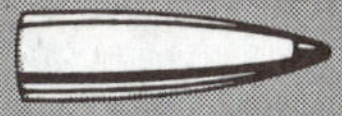

120 gr. Jacketed Spitzer
2.945" OAL

Powder	Sugg. Starting Grains	Velocity fps	Pressure C.U.P.	Max. Load Grains	Velocity fps	Pressure C.U.P.
RX7	30.0	2353	32.200	33.0	2630	41,600
IMR-3031	39.0	2717	—	43.0	2958	—
H-335	37.3	2693	39,100	41.6	2901	45,700
BL-C(2)	41.0	2688	—	45.0	2958	—
IMR-4064	41.0	2770	—	45.0	3048	—
IMR-4895	41.0	2824	—	45.0	3067	—
IMR-4320	42.0	2816	—	47.0	3067	—
H-380	43.0	2652	—	48.0	2949	—
760	46.7	2669	34.700	52.0	3001	44.500
IMR-4350	45.0	2652	—	50.0+	2932	—
H-4831	50.0	2666	—	53.5+	2865	—

Note: Loads shown in shaded panels are maximum.
+ Designates a compressed powder charge.

7mm Mauser

(7 x 57mm)

139 gr. Jacketed SP
2.970" OAL

Powder	Sugg. Starting Grains	Velocity fps	Pressure C.U.P.	Max. Load Grains	Velocity fps	Pressure C.U.P.
IMR-3031	37.0	2493	—	41.0	2762	—
H-335	36.0	2457	37,200	40.5	2687	45,300
BL-C(2)	39.0	2427	—	43.0	2695	—
IMR-4064	39.0	2557	—	43.0	2824	—
IMR-4895	38.0	2564	—	43.0	2881	—
IMR-4320	41.0	2666	—	45.0	2898	—
H-380	41.0	2525	—	46.0	2785	—
760	44.0	2430	32,700	49.0	2721	39,700
IMR-4350	43.0	2475	—	48.0+	2785	—
H-4831	49.0	2583	—	53.5+	2816	—
RX12	31.6	2059	32,400	40.5	2600	44,700
AA2520	31.4	2050	32,700	42.0	2605	44,500
RX19	40.9	2040	31,000	50.8	2681	45,400
RX22	44.8	2180	32,700	54.8+	2796	45,700

150 gr. Jacketed HPBT
3.025" OAL

Powder	Sugg. Starting Grains	Velocity fps	Pressure C.U.P.	Max. Load Grains	Velocity fps	Pressure C.U.P.
RX7	29.0	2224	34,700	33.0	2441	45,900
IMR-3031	36.0	2433	—	40.0	2672	—
BL-C(2)	38.0	2409	—	42.0	2645	—
IMR-4064	38.0	2487	—	42.0	2732	—
IMR-4895	38.0	2380	—	42.0	2785	—
IMR-4320	40.0	2570	—	44.0	2824	—
H-380	41.0	2506	—	45.0	2710	—
760	41.3	2390	34,300	47.0	2664	43,900
H-414	41.3	2398	31,200	47.0	2701	43,800
IMR-4350	42.0	2427	—	47.0	2739	—
H-4831	48.0	2538	—	53.0+	2816	—
RX12	31.2	2016	32,500	41.0	2501	44,800
AA2520	28.9	1894	31,200	38.4	2418	44,300

162 gr. Jacketed BTSP
3.037" OAL

Powder	Sugg. Starting Grains	Velocity fps	Pressure C.U.P.	Max. Load Grains	Velocity fps	Pressure C.U.P.
IMR-3031	34.9	2317	35,200	39.7	2548	44,700
BL-C(2)	38.3	2323	32,500	43.5	2624	46,200
IMR-4064	36.3	2358	35,000	41.3	2597	44,200
IMR-4320	35.6	2181	32,800	40.5	2498	42,800
H-380	42.0	2255	34,800	47.7	2487	45,000
760	41.9	2288	31,300	47.6	2623	44,800
H-414	41.6	2299	30,500	47.3	2635	43,600
IMR-4350	39.4	2188	29,200	44.8	2564	44,700
H-4831	44.0	1905	24,600	50.0+	2610	44,900

168 gr. Jacketed HPBT
3.050" OAL

Powder	Sugg. Starting Grains	Velocity fps	Pressure C.U.P.	Max. Load Grains	Velocity fps	Pressure C.U.P.
IMR-3031	34.0	2229	30,800	38.4	2496	43,600
BL-C(2)	36.5	2300	37,000	43.0	2531	44,500
IMR-4064	35.9	2327	24,600	40.8	2531	45,000
IMR-4320	35.0	2178	32,300	39.8	2475	45,400
760	40.9	2225	33,800	46.5	2605	46,100
H-414	40.8	2237	30,200	46.4	2588	43,800
IMR-4350	39.0	2199	31,300	44.4	2537	44,900
H-4831	42.2	2165	29,800	48.0+	2494	39,400
RX19	38.1	1940	32,700	47.0	2460	45,100
AA3100	38.2	1859	32,000	47.8	2427	45,300
RX22	40.2	2017	32,500	49.2	2537	45,400

Note: Loads shown in shaded panels are maximum.
+ Designates a compressed powder charge.

7mm Mauser
(7 x 57mm)

175 gr. Jacketed SP
2.935" OAL

Powder	Sugg. Starting Grains	Velocity fps	Pressure C.U.P.	Max. Load Grains	Velocity fps	Pressure C.U.P.
IMR-3031	34.0	2293	—	38.0	2463	—
BL-C(2)	35.0	2096	—	39.0	2347	—
IMR-4064	35.0	2178	—	39.0	2439	—
IMR-4895	35.0	2237	—	39.0	2512	—
IMR-4320	38.0	2341	—	42.0	2583	—
H-380	39.0	2320	—	43.0	2500	—
760	39.1	2169	33,000	44.5	2478	45,700
H-414	39.0	2165	32,500	44.4	2465	44,300
IMR-4350	41.0	2079	—	45.0	2531	—
H-4831	45.0	2272	—	50.0+	2570	—
AA3100	38.0	1854	31,200	47.6	2365	45,000
RX22	38.6	1948	33,000	48.6	2473	45,800

#287405
150 gr. (#2 Alloy) 2.800" OAL

Powder	Sugg. Starting Grains	Velocity fps	Pressure C.U.P.	Max. Load Grains	Velocity fps	Pressure C.U.P.
Red Dot	8.5	1245	22,800	11.5	1525	38,100
700X	8.5	1240	24,600	11.0	1485	39,000
Green Dot	9.0	1315	25,200	12.0	1540	37,700
PB	9.5	1245	25,200	12.0	1475	38,600
Unique	10.0	1405	27,000	14.5	1685	39,900
SR-7625	9.5	1230	24,000	12.0	1465	38,600
SR-4759	20.0	1778	23,300	24.3	2039	37,700
RX7	23.0	1755	18,700	27.0	2006	27,700
748	27.2	1685	13,700	34.0	2066	24,100
H-335	26.4	1734	16,100	33.0	2153	31,200
H-4895	25.6	1772	18,300	32.0	2156	31,300
IMR-4350	31.0	1771	17,400	38.0	2178	28,900

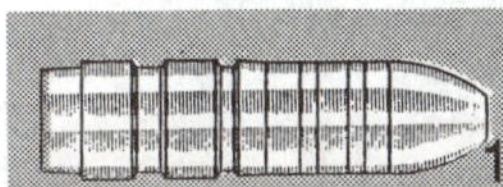

#287641
160 gr. (#2 Alloy) 2.900" OAL

Powder	Sugg. Starting Grains	Velocity fps	Pressure C.U.P.	Max. Load Grains	Velocity fps	Pressure C.U.P.
SR-4759	19.7	1581	20,500	24.0	1919	35,200
AA1680	23.2	1686	23,100	30.0	2094	38,200
RX7	22.4	1647	21,900	26.5	1869	26,000
IMR-3031	27.0	1687	20,500	34.8	2195	32,600
AA2230	29.4	1832	23,600	35.0	2155	33,100
748	33.5	1869	22,800	37.0	2104	32,700
Unique	9.7	1258	20,500	14.2	1625	33,400
SR-7625	10.0	1153	20,200	11.7	1351	28,100
IMR-4227	22.0	1723	22,700	27.0	2066	38,600

Note: Loads shown in shaded panels are maximum.
+ Designates a compressed powder charge.

284 Winchester

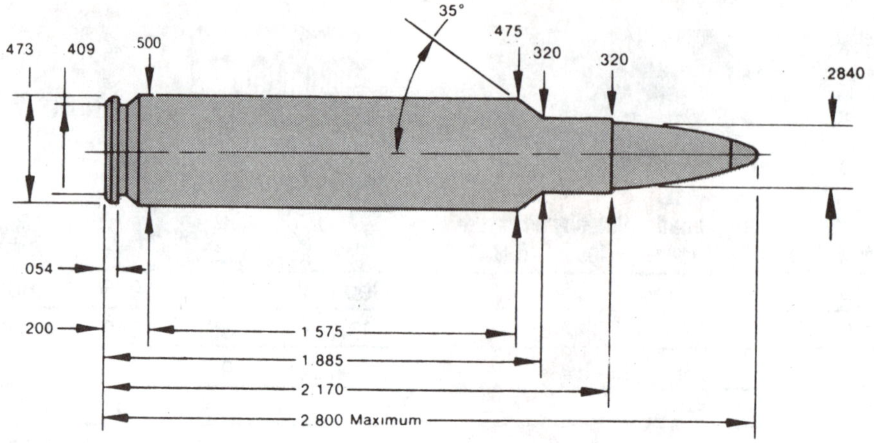

COMMENTS:

This is one of the very few rebated rim cartridges in existence (the rim is noticeably smaller in diameter than the case body). Many claim that this type of cartridge design lends itself to feeding jams when the magazine (or clip) rails are not perfectly dimensioned.

The purpose of this cartridge was to try to obtain full length cartridge ballistics in a case that would feed through a short action. The cartridge never saw much interest and now it is very close to extinction.

IMR 4350 seems as good a selection as any other propellant, and bullets of 150 grain weight seem to be the most accurate.

TEST COMPONENTS:

Cases .... Winchester
Trim-to Length .... 2.160"
Primers .... Winchester 8½-120
Primer Size .... Large Rifle
Lyman Shell Holder .... No. 2
Jacketed Bullets Used .... Sierra Spitzer, 120 gr.
Hornady Spire Pt., 139 gr.
Remington Ptd. SPCL, 150 gr.
Remington SP, 175 gr.
Cast Bullets Used .... (Sized to .284" dia.)
*Gas Check Bullets .... *#287405, 150 gr.

TEST SPECIFICATIONS: (Velocity Only)

Firearm Used .... Winchester Model 88 and Ruger Model 77
Barrel Length .... 22"
Twist .... 1-10"
Groove Dia. .... .284"

120 gr. Jacketed Spitzer

Powder	Sugg. Starting Grains	Velocity fps	Pressure C.U.P.	Max. Load Grains	Velocity fps	Pressure C.U.P.
IMR-3031	41.0	2645	—	45.0	2932	—
IMR-4895	42.0	2544	—	47.0	2890	—
IMR-4064	44.0	2660	—	49.0	3003	—
IMR-4320	44.0	2610	—	49.0	2967	—
IMR-4350	51.0	2710	—	57.0+	3076	—
H-380	47.0	2590	—	53.0	2932	—
H-4831	53.0	2557	—	59.0+	2949	—

Note: Loads shown in shaded panels are maximum.
+ Designates a compressed powder charge.

284 Winchester

139 gr. Jacketed Spire Pt.

Powder	Sugg. Starting Grains	Velocity fps	Pressure C.U.P.	Max. Load Grains	Velocity fps	Pressure C.U.P.
IMR-3031	40.0	2525	—	44.5	2793	—
IMR-4895	41.0	2427	—	46.0	2785	—
IMR-4064	42.0	2506	—	47.5	2832	—
IMR-4320	42.0	2444	—	47.5	2777	—
IMR-4350	50.0	2617	—	55.0+	2906	—
H-380	47.0	2551	—	52.0	2792	—
H-4831	52.0	2444	—	58.0+	2793	—

150 gr. Jacketed Ptd.SP

Powder	Sugg. Starting Grains	Velocity fps	Pressure C.U.P.	Max. Load Grains	Velocity fps	Pressure C.U.P.
IMR-3031	37.0	2298	—	41.0	2518	—
IMR-4895	40.0	2403	—	45.0	2672	—
IMR-4064	41.0	2487	—	46.0	2762	—
IMR-4320	41.0	2386	—	46.0	2666	—
IMR-4350	49.0	2610	—	54.0+	2816	—
H-380	46.0	2518	—	51.0	2739	—
H-4831	51.0	2439	—	57.0+	2724	—

175 gr. Jacketed SP

Powder	Sugg. Starting Grains	Velocity fps	Pressure C.U.P.	Max. Load Grains	Velocity fps	Pressure C.U.P.
IMR-3031	34.0	2053	—	38.0	2298	—
IMR-4895	38.0	2136	—	42.0	2380	—
IMR-4064	39.0	2207	—	43.0	2427	—
IMR-4320	39.0	2173	—	43.0	2403	—
IMR-4350	46.0	2341	—	51.0	2604	—
H-380	44.0	2309	—	48.0	2512	—
H-4831	48.0	2192	—	53.0+	2427	—

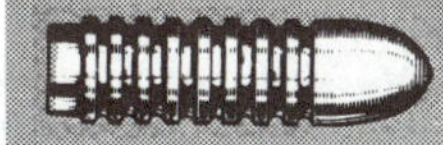

#287405
150 gr. (#2 Alloy) 2.794" OAL

Powder	Sugg. Starting Grains	Velocity fps	Pressure C.U.P.	Max. Load Grains	Velocity fps	Pressure C.U.P.
Red Dot	10.0	1370	—	15.5	1760	—
700X	10.0	1380	—	15.0	1735	—
Green Dot	11.0	1435	—	16.0	1780	—
PB	11.0	1375	—	16.0	1715	—
Unique	12.0	1500	—	17.0	1840	—
SR-7625	11.5	1410	—	16.5	1740	—
IMR-4227	17.5	1535	—	21.0	1745	—

Note: Loads shown in shaded panels are maximum.
+ Designates a compressed powder charge.

280 Remington

(7mm Express Remington)

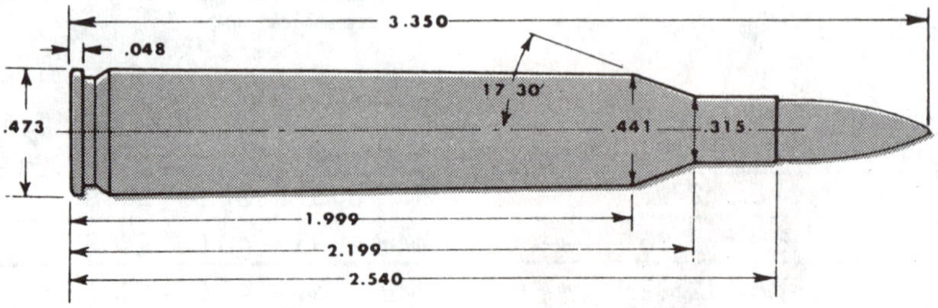

COMMENTS:

This cartridge started life as the 280 Remington. When it proved less than popular it was subjected to a name change — 7mm Express Remington. With the name change came some modifications of the factory loads. Eventually, Remington went back to the original designation.

Ballistically the 280 fills a very narrow gap between the 30-06 Springfield and the 270 Winchester. It seems best with bullets of about 160 to 162 grains, though it does well with a wide range of weights.

Hodgdon H4831 is particularly well suited to balancing uniform ballistics and accuracy in this cartridge. IMR 4350 is also a fine performer with this round.

TEST COMPONENTS:

Cases .... Remington
Trim-to Length .... 2.530"
Primers .... Remington 9½
Primer Size .... Large Rifle
Lyman Shell Holder .... No. 2
Jacketed Bullets Used .... Sierra SPT #1900, 120 gr.
Hornady SP #2820, 139 gr.
Nosler SP #16326, 150 gr.
Hornady BTSP #2845, 162 gr.
Sierra HPBT #1930, 168 gr.
Speer GSSP #1643, 175 gr.
Cast Bullets Used .... (Sized to .284" dia.)
*Gas Check Bullets .... *#287405, 150 gr.
*# 287641, 160 gr.

TEST SPECIFICATIONS: (Velocity & Pressure)

Firearm Used .... Remington Model 700, 742 and Universal Receiver
Barrel Length .... Remington Rifles 22" Universal Receiver 24"
Twist .... 1-9½"
Groove Dia. .... .284"

120 gr. Jacketed Spitzer
3.237" OAL

Powder	Sugg. Starting Grains	Velocity fps	Pressure C.U.P.	Max. Load Grains	Velocity fps	Pressure C.U.P.
RX7	34.5	2555	37,900	40.0	2852	45,800
IMR-3031	41.0	2717	—	45.0	2906	—
BL-C(2)	42.0	2645	—	48.0	2949	—
IMR-4064	44.0	2680	—	49.0	2976	—
IMR-4895	42.0	2631	—	47.0	2915	—
IMR-4320	44.0	2680	—	49.0	2976	—
H-380	49.0	2754	—	54.0	3039	—
H-414	46.0	2693	35,700	52.0	3003	44,200
IMR-4350	51.0	2747	—	57.0+	3144	—
785	52.8	2652	35,600	60.0	3006	45,000
H-4831	53.0	2597	—	59.0+	3003	—

Note: Loads shown in shaded panels are maximum.
\+ Designates a compressed powder charge.

280 Remington

(7mm Express Remington)

139 gr. Jacketed SP
3.270" OAL

Powder	Sugg. Starting Grains	Velocity fps	Pressure C.U.P.	Max. Load Grains	Velocity fps	Pressure C.U.P.
RX7	33.9	2378	37,800	38.5	2630	47,800
IMR-3031	41.0	2577	—	45.0	2801	—
BL-C(2)	42.0	2518	—	47.0	2724	—
IMR-4064	43.0	2531	—	48.5	2857	—
IMR-4895	42.0	2531	—	47.0	2808	—
IMR-4320	43.0	2531	—	48.5	2832	—
H-380	48.0	2645	—	53.0	2890	—
H-414	47.8	2662	39,500	53.5	2955	49,100
IMR-4350	48.5	2616	39,100	54.0	2912	49,000
H-4831	52.0	2518	—	58.0+	2873	—
RX12	41.4	2546	40,800	46.4	2767	49,800
RX15	43.1	2581	37,800	48.5	2866	48,800
RX19	52.4	2631	37,800	58.2	2989	49,500

150 gr. Jacketed SP
3.280" OAL

Powder	Sugg. Starting Grains	Velocity fps	Pressure C.U.P.	Max. Load Grains	Velocity fps	Pressure C.U.P.
IMR-3031	40.0	2577	—	45.0	2785	—
BL-C(2)	41.0	2444	—	46.0	2680	—
IMR-4064	43.0	2551	—	48.0	2840	—
IMR-4895	42.0	2554	—	47.0	2793	—
IMR-4320	43.0	2551	—	48.0	2793	—
H-380	47.0	2551	—	52.0	2808	—
760	46.6	2546	39,600	53.0	2829	48,800
H-414	46.5	2555	38,900	52.5	2871	49,000
IMR-4350	46.2	2480	38,500	52.5	2787	48,800
H-4831	51.0	2550	—	57.0+	2816	—
RX12	37.6	2360	38,800	42.6	2571	47,700
RX15	39.0	2413	38,700	46.0	2698	48,500
AA3100	50.5	2473	38,700	56.3	2761	49,700

162 gr. Jacketed BTSP
3.320" OAL

Powder	Sugg. Starting Grains	Velocity fps	Pressure C.U.P.	Max. Load Grains	Velocity fps	Pressure C.U.P.
IMR-3031	38.0	2343	37,200	44.0	2656	48,800
IMR-4064	41.2	2426	38,000	47.0	2707	48,800
IMR-4895	40.5	2400	37,500	46.2	2683	47,200
IMR-4320	40.3	2339	40,700	45.8	2634	49,600
H-380	41.5	2316	34,800	48.5	2597	48,400
760	45.5	2441	38,300	51.7	2745	48,300
H-414	44.0	2417	37,500	50.3	2729	49,900
IMR-4350	45.0	2379	34,700	51.5	2718	48,400
H-4831	51.0	2493	37,700	56.5+	2784	46,900
RX19	48.1	2425	37,300	53.8	2751	48,800
AA3100	49.3	2446	38,700	54.4	2722	49,100
RX22	49.2	2465	37,300	54.3	2747	47,900

168 gr. Jacketed HPBT
3.290" OAL

Powder	Sugg. Starting Grains	Velocity fps	Pressure C.U.P.	Max. Load Grains	Velocity fps	Pressure C.U.P.
IMR-3031	38.5	2334	37,300	44.0	2601	48,900
IMR-4064	40.5	2345	37,800	46.3	2632	49,300
IMR-4895	41.0	2412	38,700	47.0	2683	49,900
IMR-4320	41.0	2332	39,500	47.0	2634	50,000
H-380	41.0	2288	39,300	47.7	2544	49,200
760	44.0	2365	39,200	50.0	2653	49,400
H-414	43.5	2377	37,700	49.5	2667	49,100
IMR-4350	46.2	2451	38,700	52.2	2735	50,000
H-4831	47.5	2338	36,000	54.0	2658	49,500
RX19	45.7	2375	38,000	52.6	2675	49,800
AA3100	46.3	2298	37,300	52.0	2575	49,400
RX22	47.0	2402	38,300	53.0	2692	49,800

Note: Loads shown in shaded panels are maximum.
+ Designates a compressed powder charge.

280 Remington

(7mm Express Remington)

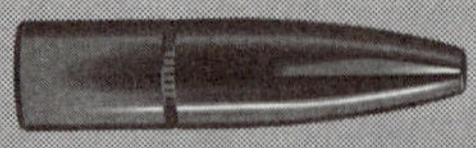

175 gr. Jacketed SP
3.220" OAL

Powder	Sugg. Starting Grains	Velocity fps	Pressure C.U.P.	Max. Load Grains	Velocity fps	Pressure C.U.P.
IMR-3031	35.0	2145	—	39.0	2331	—
BL-C(2)	39.0	2232	—	43.0	2398	—
IMR-4064	40.0	2252	—	44.0	2463	—
IMR-4895	39.0	2232	—	43.0	2444	—
IMR-4320	40.0	2257	—	44.0	2457	—
H-380	44.0	2283	—	49.0	2564	—
760	41.3	2219	38,200	47.2	2493	49,100
H-414	40.5	2227	39,000	46.3	2502	49,800
IMR-4350	41.5	2247	37,500	47.5	2520	47,700
H-4831	48.0	2212	—	53.0	2518	—

#287405
150 gr. (#2 Alloy) 2.794" OAL

Powder	Sugg. Starting Grains	Velocity fps	Pressure C.U.P.	Max. Load Grains	Velocity fps	Pressure C.U.P.
Red Dot	10.5	1405	—	15.0	1710	—
700X	10.0	1365	—	14.0	1645	—
Green Dot	11.0	1425	—	15.5	1735	—
PB	10.5	1330	—	15.0	1640	—
Unique	11.5	1440	—	16.5	1785	—
SR-7625	11.0	1370	—	15.5	1670	—

#287641
160 gr. (#2 Alloy) 3.205" OAL

Powder	Sugg. Starting Grains	Velocity fps	Pressure C.U.P.	Max. Load Grains	Velocity fps	Pressure C.U.P.
SR-4759	24.6	1872	26,000	29.6	2145	39,800
AA1680	28.4	1864	21,300	34.4	2155	34,300
RX7	25.8	1792	18,800	30.4	2040	27,900
IMR-3031	33.5	1984	20,000	38.6	2301	32,600
AA2230	34.0	2017	24,800	39.5	2298	35,800
748	38.2	1930	20,000	44.0	2362	35,000
AA2015BR	31.0	1991	22,500	36.0	2206	31,600
IMR-4895	32.3	1908	21,000	35.3	2020	22,700
Unique	15.5	1596	26,800	19.0	1815	40,400
SR-7625	13.0	1349	21,500	17.0	1637	41,200
SR-4756	15.0	1483	23,500	19.0	1773	41,700

Note: Loads shown in shaded panels are maximum.

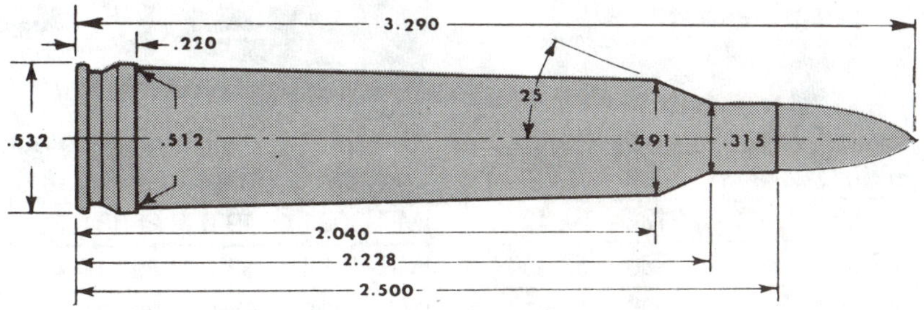

COMMENTS:

This is easily the third most popular big game cartridge in this country. It started with the need for a cartridge with more punch than the 30-06 but not more recoil.

The 7mm Remington magnum has been used to take every big game animal that walks on this continent. It really peaks out however with game as large as elk, although some would disagree. Bullets of 160 to 175 grains are best when the game gets tough.

Magnum primers and slow burning propellants such as Hodgdon H4831 and IMR 7828 are naturals for this application. Hercules Reloder 22 would also be a good choice.

Cast bullets need to be kept under 1800 fps for acceptable accuracy.

TEST COMPONENTS:

Cases .... Remington and Winchester
Trim-to Length .... 2.490"
Primers .... Remington 9½ M
Primer Size .... Large Rifle, Magnum
Lyman Shell Holder .... No. 13
Jacketed Bullets Used .... Sierra SPT #1900, 120 gr.
Winchester PPSP, 125 gr.
Speer Spitzer, 130 gr.
Nosler SP #16325, 140 gr.
Barnes "x" #28425, 140 gr.
Speer Spitzer, 145 gr.
Remington Ptd. SPCL, 150 gr.
Hornady SP #2830, 154 gr.
Hornady BTSP #2845, 162 gr.
Sierra HPBT #1930, 168 gr.
Speer GSSP #1643, 175 gr.
Cast Bullets Used .... (Sized to .284" dia.)
*Gas Check Bullets .... *#287405, 150 gr.
*#287641, 160 gr.

TEST SPECIFICATIONS: (Velocity & Pressure)

Firearm Used .... Universal Receiver
Barrel Length .... 26"
Twist .... 1-9½"
Groove Dia. .... .284"

120 gr. Jacketed Spitzer**
3.260" OAL

Powder	Sugg. Starting Grains	Velocity fps	Pressure C.U.P.	Max. Load Grains	Velocity fps	Pressure C.U.P.
RX7	39.6	2689	39,300	45.0	2960	49,400
IMR-3031	49.0	2935	40,300	55.0	3224	51,600
IMR-4064	55.0	3181	44,600	61.0	3389	51,600
IMR-4895	54.0	3083	41,600	60.0	3368	50,800
IMR-4320	56.0	3072	40,300	62.0	3381	50,800
H-380	53.0	2978	41,600	59.0	3233	51,200
H-414	56.8	3009	38,800	64.5	3333	50,800
IMR-4350	61.0	3012	36,900	68.5	3429	51,600
H-4831	67.0	3056	39,900	74.0+	3478	52,400

Note: Loads shown in shaded panels are maximum.
+ Designates a compressed powder charge.
** Remington cases used.

7mm Remington Magnum

125 gr. Jacketed SP**

Powder	Sugg. Starting Grains	Velocity fps	Pressure C.U.P.	Max. Load Grains	Velocity fps	Pressure C.U.P.
IMR-3031	49.0	2932	41,600	54.0	3138	52,400
IMR-4064	54.0	3031	42,000	60.0	3303	52,000
IMR-4895	52.0	2968	41,100	‡59.0	3306	52,800
IMR-4320	55.0	2994	41,100	61.5	3325	51,600
H-380	51.0	2858	41,100	58.0	3180	52,800
IMR-4350	61.0	3026	41,600	67.5	3367	52,400
H-4831	64.0	2951	38,600	71.5	3384	52,000

130 gr. Jacketed Spitzer**

Powder	Sugg. Starting Grains	Velocity fps	Pressure C.U.P.	Max. Load Grains	Velocity fps	Pressure C.U.P.
IMR-3031	49.0	2833	39,900	53.5	3071	52,000
IMR-4064	53.0	2914	42,400	59.5	3185	52,000
IMR-4895	51.0	2836	40,700	58.0	3199	52,400
IMR-4320	54.5	2912	42,400	61.0	3179	51,200
H-380	50.5	2733	39,400	57.5	3036	51,200
IMR-4350	60.5	2931	41,100	67.0	3247	52,000
H-4831	64.0	2887	38,600	71.0	3310	52,400

140 gr. Jacketed SP**
3.290" OAL

Powder	Sugg. Starting Grains	Velocity fps	Pressure C.U.P.	Max. Load Grains	Velocity fps	Pressure C.U.P.
IMR-3031	46.5	2718	40,300	53.0	2987	52,000
IMR-4064	46.3	2674	37,700	54.5	3038	51,700
IMR-4895	50.0	2744	41,100	57.0	3054	52,000
IMR-4320	47.6	2668	37,700	56.0	3070	51,900
H-380	47.5	2611	39,400	54.5	2860	43,000
H-414	49.5	2702	39,100	58.0	3017	51,500
IMR-4350	58.0	2842	42,400	66.0	3133	52,000
H-4831	62.0	2791	39,000	67.5	3146	50,600
RX12	48.0	2712	39,300	56.0	3036	51,800
RX19	61.7	2902	38,500	70.5	3247	49,900
AA3100	65.0	2828	38,200	70.5	3156	51,400
RX22	64.5	3037	42,900	72.5	3305	51,500
IMR-7828	67.0	2860	37,500	74.5	3258	52,000

140 gr. Barnes "X" ***
3.290" OAL

Powder	Sugg. Starting Grains	Velocity fps	Pressure C.U.P.	Max. Load Grains	Velocity fps	Pressure C.U.P.
IMR-3031	44.5	2562	39,700	50.6	2839	49,400
RX12	48.0	2615	40,000	55.5	2912	51,600
IMR-4895	48.0	2663	40,500	55.4	2947	49,500
IMR-4064	48.0	2631	40,000	58.5	3008	51,800
IMR-4320	50.0	2638	39,200	58.0	3010	51,600
IMR-4350	58.0	2798	40,000	66.0	3129	50,100
RX19	61.0	2835	40,000	69.3	3198	51,900
AA3100	61.5	2742	39,200	68.3	3036	49,900
H-4831	53.0	2713	42,100	69.5	3127	50,800
RX22	58.0	2739	37,500	69.5	3198	51,100
AA8700	77.0	2877	41,000	85.5+	3192	50,400
IMR-7828	68.0	2801	38,000	72.0	3159	50,800

Note: Loads shown in shaded panels are maximum.
‡ Designates a factory velocity duplication load.
** Remington cases used.
+ Designates a compressed powder charge.
*** Winchester cases used.

7mm Remington Magnum

145 gr. Jacketed Spitzer**

Powder	Sugg. Starting Grains	Velocity fps	Pressure C.U.P.	Max. Load Grains	Velocity fps	Pressure C.U.P.
IMR-3031	45.5	2618	39,000	52.0	2932	52,400
IMR-4064	48.5	2653	39,000	55.5	2970	52,400
IMR-4895	47.0	2617	39,000	54.0	2954	52,000
IMR-4320	50.0	2680	38,600	57.0	2987	52,400
H-380	46.5	2535	38,600	53.5	2831	52,000
IMR-4350	54.0	2677	38,600	63.0	3039	52,400
H-4831	59.0	2747	39,900	66.0	3049	52,400
RX12	46.0	2608	40,800	52.7	2881	51,400
RX19	59.0	2803	39,000	68.0	3158	51,600
AA3100	64.0	2824	42,200	69.5	3073	51,300
RX22	61.0	2901	41,500	70.0	3219	52,000
IMR-7828	64.5	2870	40,000	72.0	3172	51,800

150 gr. Jacketed Ptd. SP**

Powder	Sugg. Starting Grains	Velocity fps	Pressure C.U.P.	Max. Load Grains	Velocity fps	Pressure C.U.P.
IMR-3031	45.0	2590	36,000	52.0	2880	52,400
IMR-4064	50.0	2729	39,000	55.0	2971	50,000
IMR-4895	48.5	2657	38,100	53.5	2963	51,200
IMR-4320	49.5	2685	38,600	56.5	3044	52,000
H-380	47.5	2579	37,300	53.5	2814	48,400
IMR-4350	54.0	2611	33,000	63.0	3040	49,200
H-4831	59.0	2643	33,500	66.0	3008	48,400

154 gr. Jacketed SP**
3.290" OAL

Powder	Sugg. Starting Grains	Velocity fps	Pressure C.U.P.	Max. Load Grains	Velocity fps	Pressure C.U.P.
IMR-3031	45.0	2589	38,600	52.0	2891	52,000
IMR-4064	50.0	2754	41,100	55.0	2950	50,000
IMR-4895	48.5	2712	42,900	53.5	2945	51,600
IMR-4320	49.5	2692	41,100	56.0	2912	49,200
IMR-4350	55.0	2723	40,300	63.0	3012	51,400
H-4831	59.0	2710	37,700	66.0	3051	51,200
RX19	59.0	2717	37,500	68.0	3067	49,300
AA3100	63.0	2664	37,300	69.5	3017	51,200
RX22	61.0	2790	39,700	70.0	3121	48,900
IMR-7828	64.0	2666	35,600	71.5	3089	51,600

162 gr. Jacketed BTSP**
3.290" OAL

Powder	Sugg. Starting Grains	Velocity fps	Pressure C.U.P.	Max. Load Grains	Velocity fps	Pressure C.U.P.
IMR-3031	42.9	2472	38,000	50.5	2805	52,000
IMR-4064	43.4	2465	37,000	51.0	2781	51,700
IMR-4895	42.9	2465	38,500	50.5	2789	51,800
IMR-4320	45.4	2519	39,000	53.5	2863	51,800
IMR-4350	50.1	2555	36,700	59.0	2900	51,700
H-4831	56.1	2637	38,500	66.0	3010	52,000

Note: Loads shown in shaded panels are maximum.
** Remington cases used.

7mm Remington Magnum

168 gr. Jacketed HPBT**
3.290" OAL

Powder	Sugg. Starting Grains	Velocity fps	Pressure C.U.P.	Max. Load Grains	Velocity fps	Pressure C.U.P.
IMR-3031	44.5	2522	38,600	50.5	2776	50,000
IMR-4064	44.4	2469	38,800	50.5	2725	51,500
IMR-4895	47.0	2619	42,400	53.0	2834	52,000
IMR-4320	48.0	2543	38,600	53.5	2823	51,600
H-380	46.0	2479	40,700	52.0	2723	52,000
IMR-4350	53.0	2571	37,700	61.0	2925	52,000
H-4831	59.0	2685	38,600	65.0	2971	52,000
RX22	57.0	2594	38,500	66.0	2978	50,400
AA8700	70.0	2629	37,800	78.5	2948	49,900
IMR-7828	61.5	2622	37,500	69.0	2946	52,100
H-1000	65.5	2645	38,500	73.0	2971	51,100

175 gr. Jacketed SP**
3.290" OAL

Powder	Sugg. Starting Grains	Velocity fps	Pressure C.U.P.	Max. Load Grains	Velocity fps	Pressure C.U.P.
IMR-3031	45.0	2492	39,900	50.5	2720	49,600
IMR-4064	47.0	2516	38,600	53.0	2772	51,600
IMR-4895	45.0	2468	38,600	53.0	2755	50,400
IMR-4320	48.0	2542	41,100	53.5	2786	52,000
H-380	46.0	2401	36,400	52.0	2680	52,000
IMR-4350	53.0	2515	36,400	‡61.0	2876	52,000
H-4831	53.7	2521	40,200	60.0	2748	46,400
H-870	59.0	2269	28,700	72.0+	2674	40,900
RX22	56.0	2592	40,500	64.8	2888	50,800
AA8700	71.3	2610	38,800	78.5	2907	50,900
IMR-7828	61.3	2620	40,200	68.5	2910	51,100
H-1000	61.0	2559	40,200	71.0	2897	51,400

#287405***
150 gr. (#2 Alloy) 3.052" OAL

Powder	Sugg. Starting Grains	Velocity fps	Pressure C.U.P.	Max. Load Grains	Velocity fps	Pressure C.U.P.
Red Dot	16.0	1782	30,000	19.0	1943	38,100
700X	13.5	1601	25,800	16.5	1798	34,000
Green Dot	15.0	1675	25,800	18.0	1849	31,000
PB	15.5	1652	26,400	18.5	1820	36,000
SR-7625	15.5	1635	25,800	19.0	1851	37,700
SR-4756	19.0	1813	25,800	22.5	2004	36,900
SR-4759	28.4	2164	33,500	35.5	2396	47,100
IMR-4198	33.3	2186	27,500	42.0	2574	46,900
RX7	34.0	2157	29,000	42.5	2560	47,800
748	41.4	2174	27,400	51.8	2676	46,100
H-4895	40.0	2171	28,600	50.0	2739	49,500

#287641***
160 gr. (#2 Alloy) 3.170" OAL

Powder	Sugg. Starting Grains	Velocity fps	Pressure C.U.P.	Max. Load Grains	Velocity fps	Pressure C.U.P.
SR-4759	29.0	2023	25,500	38.0	2440	48,100
AA1680	36.0	2095	25,000	46.5	2555	46,000
RX7	35.3	2151	26,000	46.0	2613	45,000
IMR-3031	40.0	2221	24,800	50.5	2656	43,500
748	48.0	2246	26,300	53.5	2603	39,900
AA2015BR	37.0	2163	24,300	47.0	2592	42,200
H-335	40.5	2127	23,500	47.0	2458	36,000
IMR-4895	41.0	2180	23,800	50.0	2570	37.600
Unique	17.0	1680	24,300	21.5	1917	36,200
SR-4756	18.5	1694	26,000	22.0	1880	36,700
IMR-4227	31.0	2067	25,800	36.0	2282	35,000
IMR-4198	32.3	2082	24,000	41.0	2462	39,300

Note: Loads shown in shaded panels are maximum.
\+ Designates a compressed powder charge.
‡ Designates a factory velocity duplication load.
** Remington cases used.
*** Winchester cases used.

7mm Weatherby Magnum

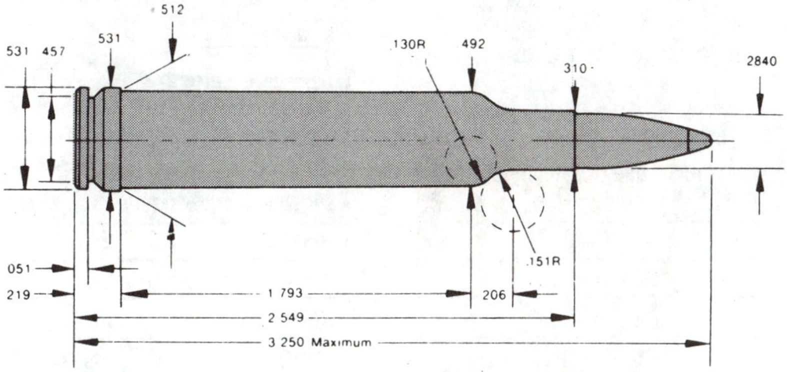

COMMENTS:

There is not much ballistic difference between this round and the far more popular 7mm Remington Magnum.

Due to limited popularity the data for this cartridge is brief. The data is intended only for commercially produced and chambered rifles. It is not for use in custom chamberings.

The best performing propellant of those listed is Hodgdon's H4831.

TEST COMPONENTS:

Cases .... Weatherby
Trim-to Length .... 2.540"
Primers .... Federal 215 and Remington 9½
Primer Size .... Large Rifle
Lyman Shell Holder .... No. 13
Jacketed Bullets Used .... Sierra Spitzer, 120 gr.
Hornady Spire Pt., 139 gr.
Remington Ptd. SPCL, 150 gr.
Remington Ptd. SPCL, 175 gr.
Cast Bullets Used .... (Sized to .284" dia.)
*Gas Check Bullets .... *#287405, 150 gr.

TEST SPECIFICATIONS: (Velocity Only)

Firearm Used .... Weatherby Mark V
Barrel Length .... 26"
Twist .... 1-12"
Groove Dia. .... .2825"

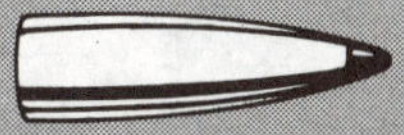

120 gr. Jacketed Spitzer

Powder	Sugg. Starting Grains	Velocity fps	Pressure C.U.P.	Max. Load Grains	Velocity fps	Pressure C.U.P.
IMR-4064	56.0	2949	—	62.0	3311	—
IMR-4320	57.0	3115	—	63.0	3345	—
IMR-4350	64.0	2985	—	71.0	3448	—
H-380	58.0	2932	—	66.0	3290	—
H-4831	70.0	3144	—	77.0	3546	—

Note: Loads shown in shaded panels are maximum.

7mm Weatherby Magnum

139 gr. Jacketed Spire Pt.

Powder	Sugg. Starting Grains	Velocity fps	Pressure C.U.P.	Max. Load Grains	Velocity fps	Pressure C.U.P.
IMR-4064	55.0	2816	—	61.0	3134	—
IMR-4320	56.0	2881	—	62.0	3194	—
IMR-4350	63.0	2915	—	69.0	3260	—
H-380	59.0	2832	—	65.0	3154	—
H-4831	69.0	3003	—	75.0	3340	—

150 gr. Jacketed Ptd. SP

Powder	Sugg. Starting Grains	Velocity fps	Pressure C.U.P.	Max. Load Grains	Velocity fps	Pressure C.U.P.
IMR-4064	54.0	2762	—	60.0	3048	—
IMR-4320	55.0	2857	—	61.0	3105	—
IMR-4350	61.0	2849	—	68.0	3225	—
H-380	58.0	2824	—	64.0	3115	—
H-4831	67.0	2949	—	74.0+	3300	—

175 gr. Jacketed Ptd. SP

Powder	Sugg. Starting Grains	Velocity fps	Pressure C.U.P.	Max. Load Grains	Velocity fps	Pressure C.U.P.
IMR-4064	51.0	2531	—	56.0	2732	—
IMR-4320	53.0	2652	—	59.0	2873	—
IMR-4350	59.0	2659	—	65.0	2949	—
H-380	56.0	2672	—	62.0	2801	—
H-4831	64.0	2695	—	70.0	2975	—

#287405**
150 gr. (#2 Alloy)

Powder	Sugg. Starting Grains	Velocity fps	Pressure C.U.P.	Max. Load Grains	Velocity fps	Pressure C.U.P.
Unique	15.0	1683	—	18.0	1872	—
2400	18.0	1559	—	24.0	1930	—
IMR-4227	23.0	1760	—	26.0	1930	—

Note: Loads shown in shaded panels are maximum.
+ Designates a compressed powder charge.
** Designates the use of Remington 9½ primers.

30 M-1 Carbine

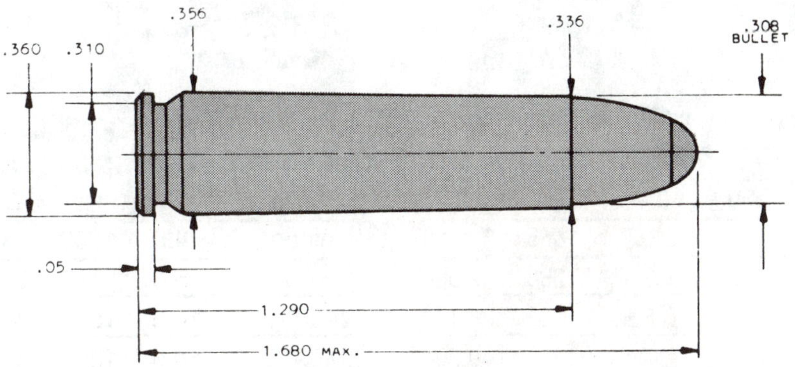

COMMENTS:

This cartridge headspaces on the mouth and therefore case trimming must be uniform and accurate. Do not reduce cases below the trim-to length. Additionally do not crimp bullets as this will prevent the case from properly headspacing on its mouth.

Best performance is with IMR 4227 with both jacketed and cast bullets.

Ballistics with this cartridge are too light for any big game applications. M1 carbines are best suited for plinking or casual shooting.

TEST COMPONENTS:

Cases .... Remington
Trim-to Length .... 1.286"
Primers .... Remington 6½
Primer Size .... Small Rifle
Lyman Shell Holder .... No. 19
Jacketed Bullets Used .... Sierra RN, 110 gr.
Cast Bullets Used .... (Sized to .308" dia.)
*Gas Check Bullets .... *#311359, 113 gr.
#311410, 130 gr.

TEST SPECIFICATIONS: (Velocity & Pressure)

Firearm Used .... Universal Receiver
Barrel Length .... 20"
Twist .... 1-20"
Groove Dia. .... .308"

110 gr. Jacketed RN

Powder	Sugg. Starting Grains	Velocity fps	Pressure C.U.P.	Max. Load Grains	Velocity fps	Pressure C.U.P.
SR-4756	6.5	1355	27,000	7.3	1509	38,600
IMR-4227	13.5	1642	28,200	15.0+	1845	39,000
2400	11.0	1602	29,400	14.0+	1945	39,400
AA1680	15.5	1642	22,300	17.3+	1897	34,400

Note: Loads shown in shaded panels are maximum.
+ Designates a compressed powder charge.

30 M-1 Carbine

#311359
113 gr. (#2 Alloy)

Powder	Sugg. Starting Grains	Velocity fps	Pressure C.U.P.	Max. Load Grains	Velocity fps	Pressure C.U.P.
Unique	5.5	1388	27,600	6.5	1555	38,600
SR-7625	4.6	1194	22,800	5.6	1368	39,000
SR-4756	5.5	1298	23,400	6.5	1463	38,100
2400	11.0	1633	31,500	12.5+	1828	40,300
IMR-4227	11.5	1497	27,000	13.3+	1733	39,000

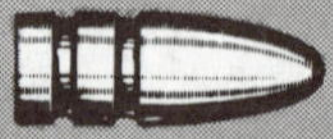

#311410
130 gr. (#2 Alloy)

Powder	Sugg. Starting Grains	Velocity fps	Pressure C.U.P.	Max. Load Grains	Velocity fps	Pressure C.U.P.
Unique	4.9	1260	23,400	6.1	1468	35,000
SR-4756	5.0	1180	22,000	6.0	1371	37,300
2400	10.5	1564	28,800	12.0+	1733	38,600
IMR-4227	10.6	1366	24,600	12.7+	1647	38,600

Note: Loads shown in shaded panels are maximum.
+ Designates a compressed powder charge.

30-30 Winchester

(30 WCF)
(7.62 x 51Rmm)

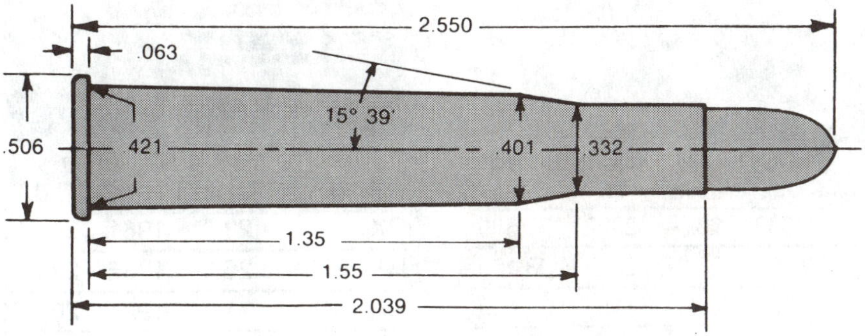

COMMENTS:

This cartridge works very well with a number of powders including IMR 3031, IMR 4064, Reloder 7, Reloder 12, and Winchester 748. Reloder 12 might be the best first choice. It has an advantage in velocity and is just as accurate as the others.

All ammo loaded for use in tubular magazines must be assembled with blunt or flat nosed bullets for safety. In addition ammo used in tubular magazines should also have the case crimped lightly to the bullet. Naturally proper crimping will demand very uniform case lengths. The 110 grain bullet is for varmints and plinking.

When used in bolt action rifles this is an extremely accurate cartridge. (Ditto for single shot pistols). It is also very accurate with cast bullets with velocities from 1,600 to 2,100 fps.

TEST COMPONENTS:

Cases .... Winchester
Trim-to Length .... 2.028"
Primers .... Winchester 8 1/2-120 and CCI 200
Primer Size .... Large Rifle
Lyman Shell Holder .... No. 6
Jacketed Bullets Used .... Hornady RN #3015, 110 gr.
Hornady RN #3035, 150 gr.
Hornady FP #3060, 170 gr.
Cast Bullets Used .... (Sized to .308" dia.)
*Gas Check Bullets .... *#311291, 169 gr.
*#311041, 170 gr.

TEST SPECIFICATIONS:
(Velocity & Pressure)

Firearm Used .... Universal Receiver
Barrel Length .... 26"
Twist .... 1-12"
Groove Dia. .... .308"

110 gr. Jacketed RN
2.530" OAL

Powder	Sugg. Starting Grains	Velocity fps	Pressure C.U.P.	Max. Load Grains	Velocity fps	Pressure C.U.P.
IMR-4198	27.0	2358	28,200	29.5	2617	38,100
IMR-3031	26.5	2085	28,400	33.5	2492	37,900
748	36.0	2280	25,600	41.0+	2631	31,400
H-335	37.8	2547	28,900	43.0+	2842	34,800
BL-C(2)	36.9	2308	27,300	42.0+	2645	33,800
IMR-4064	31.0	2168	27,000	35.0+	2568	37,300
IMR-4895	31.0	2258	28,200	34.0	2545	37,700
IMR-4320	28.0	2044	27,700	33.0	2418	37,100
IMR-4350	35.0	2181	29,400	39.0+	2419	35,000
RX12	31.0	2437	23,500	35.0	2729	29,200

Note: Loads shown in shaded panels are maximum.
+ Designates a compressed powder charge.

30-30 Winchester
(30 WCF)
(7.62 x 51Rmm)

150 gr. Jacketed RN
2.540" OAL

Powder	Sugg. Starting Grains	Velocity fps	Pressure C.U.P.	Max. Load Grains	Velocity fps	Pressure C.U.P.
IMR-4198	20.6	1884	30,000	23.5	2092	37,600
RX7	25.0	2154	30,500	28.0	2350	38,100
IMR-3031	26.5	1876	25,200	28.5	2145	38,000
748	33.8	2164	29,700	‡38.5+	2426	37,100
H-335	34.3	2233	29,200	39.0+	2473	35,800
BL-C(2)	32.5	2083	27,400	37.0	2304	37,000
IMR-4064	29.0	2055	32,000	31.5	2199	37,800
IMR-4895	26.4	1944	30,500	30.0	2145	37,400
IMR-4320	29.0	1956	31,000	31.5	2171	37,800
760	33.8	1889	26,800	38.5+	2154	31,400
IMR-4350	32.0	1980	31,000	36.0+	2218	38,100
AA2230	29.9	2108	27,800	31.7	2244	36,400
RX12	30.0	2311	29,300	33.5	2559	38,000

170 gr. Jacketed FP
2.540" OAL

Powder	Sugg. Starting Grains	Velocity fps	Pressure C.U.P.	Max. Load Grains	Velocity fps	Pressure C.U.P.
RX7	23.0	1961	32,500	26.0	2127	37,700
IMR-3031	25.0	1786	27,600	28.5	2054	36,900
748	31.7	1921	26,100	35.0	2116	30,100
H-335	32.1	2044	29,400	36.5	2282	37,700
BL-C(2)	27.0	1869	27,000	30.5	2128	37,300
IMR-4064	27.0	1845	28,200	30.5	2150	38,100
IMR-4895	24.6	1807	31,200	28.0	1973	36,800
IMR-4320	24.6	1708	30,200	28.0	1948	37,100
IMR-4350	30.0	1816	28,200	34.5+	2104	38,100
RX12	28.5	2108	28,300	32.0	2349	37,900
AA2520	29.0	1974	27,500	32.4	2170	36,000

#311291
169 gr. (#2 Alloy) 2.521" OAL

Powder	Sugg. Starting Grains	Velocity fps	Pressure C.U.P.	Max. Load Grains	Velocity fps	Pressure C.U.P.
Red Dot	6.5	1145	24,600	8.5	1363	34,500
700X	6.0	1101	26,400	7.5	1253	34,000
Green Dot	7.0	1126	23,400	9.0	1348	33,500
PB	6.5	1093	24,000	8.7	1311	36,900
Unique	7.0	1211	19,800	10.6	1568	36,000
SR-7625	6.5	1107	22,800	9.0	1348	36,400
Herco	9.0	1328	28,200	11.5	1538	38,100
SR-4756	7.0	1123	19,800	10.5	1431	35,000
**SR-4759	14.8	1530	19,900	18.5	1862	36,100
**RX7	19.0	1635	20,200	28.0	2152	34,400
**IMR-3031	22.5	1599	18,300	28.5	2095	32,500
**748	23.8	1604	16,900	37.3+	2355	35,100
**H-335	20.3	1638	18,500	32.5	2230	32,800

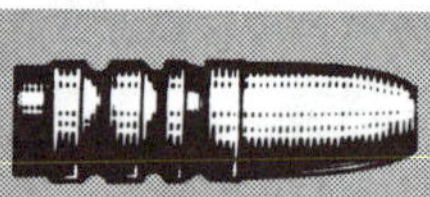

#311041
170 gr. (#2 Alloy) 2.510" OAL

Powder	Sugg. Starting Grains	Velocity fps	Pressure C.U.P.	Max. Load Grains	Velocity fps	Pressure C.U.P.
Red Dot	6.5	1159	24,600	8.5	1348	34,500
700X	6.0	1124	24,600	7.5	1269	34,000
Green Dot	7.0	1192	25,200	9.0	1373	33,500
PB	6.5	1100	23,400	8.7	1286	36,900
Unique	7.0	1240	21,600	10.6	1555	35,500
SR-7625	6.5	1118	22,800	9.0	1337	36,900
Herco	8.5	1281	25,800	11.0	1505	33,000
SR-4756	7.0	1102	19,800	10.5	1423	36,400
**SR-4759	15.5	1613	23,700	17.7	1803	32,600
**RX7	20.0	1632	19,800	28.6	2165	34,500
**IMR-3031	21.6	1555	20,100	27.0	2016	31,300
**748	24.8	1613	22,500	35.0+	2270	32,900
**H-335	22.0	1596	23,200	30.0	2254	33,800

Note: Loads shown in shaded panels are maximum.
+ Designates a compressed powder charge.
‡ Designates a factory velocity duplication load.
** Signifies CCI 200 primers used.

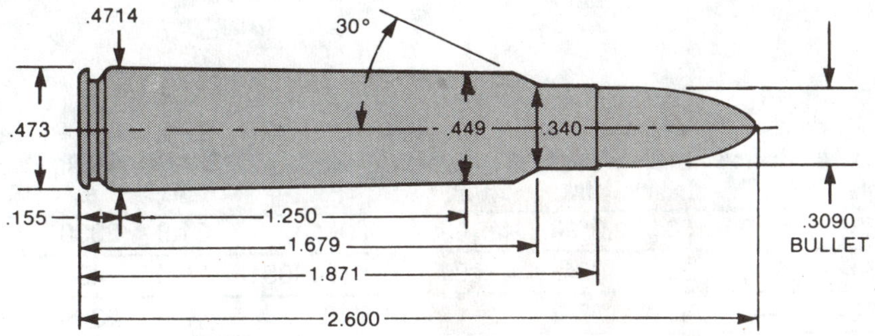

COMMENTS:

While no longer very popular this cartridge is a fine light big game round with only modest recoil.

It is an accurate round but its small case capacity makes most loads compressed powder ones, a bit inconvenient. The 110 and 130 grain bullets are of course intended for varmints.

Winchester 748 provides the highest velocity and combines it with good accuracy. Because of its dense nature it does not result in compressed powder loads.

The best cast bullet performance was with bullet #311291.

TEST COMPONENTS:

Cases .... Remington and Federal
Trim-to Length .... 1.865"
Primers .... Remington 9½ and Federal 210
Primer Size .... Large Rifle
Lyman Shell Holder .... No. 2
Jacketed Bullets Used .... Remington Ptd. SP, 110 gr.
Hornady SP #3020, 130 gr.
Hornady SP #3031, 150 gr.
Sierra HPBT #2220, 180 gr.
Cast Bullets Used .... (Sized to .309" dia.)
*Gas Check Bullets .... *#311359, 113 gr.
*#311291, 169 gr.
*#311467, 178 gr.

TEST SPECIFICATIONS: (Velocity & Pressure)

Firearm Used .... Savage Model 99 and Universal Receiver
Barrel Length .... Savage 22", Universal Receiver 25"
Twist .... 1-12"
Groove Dia. .... Savage .3075"; Universal Receiver .309"

110 gr. Jacketed Ptd. SP

Powder	Sugg. Starting Grains	Velocity fps	Pressure C.U.P.	Max. Load Grains	Velocity fps	Pressure C.U.P.
IMR-3031	38.0	2672	—	42.0	2967	—
BL-C(2)	41.0	2695	—	45.0	2857	—
IMR-4064	41.0	2624	—	45.0+	2890	—
IMR-4895	41.0	2659	—	45.0+	2967	—
IMR-4320	42.0	2702	—	47.0+	3021	—
H-380	41.0	2409	—	46.5+	2645	—

Note: Loads shown in shaded panels are maximum.
+ Designates a compressed powder charge.

300 Savage

130 gr. Jacketed SP
2.565" OAL

Powder	Sugg. Starting Grains	Velocity fps	Pressure C.U.P.	Max. Load Grains	Velocity fps	Pressure C.U.P.
H-4198	29.0	2336	29,100	33.0	2596	42,100
RX7	34.7	2602	36,300	39.5	2843	46,000
IMR-3031	37.0	2525	—	41.0	2808	—
748	42.2	2508	28,100	48.0+	2820	37,100
BL-C(2)	40.0	2512	—	44.0	2672	—
IMR-4064	40.0	2493	—	44.0+	2754	—
IMR-4895	40.0	2564	—	44.0+	2801	—
IMR-4320	41.0	2544	—	46.0+	2857	—
H-380	41.0	2320	—	46.5+	2538	—
760	43.1	2311	26,800	49.0+	2629	33,600

150 gr. Jacketed SP
2.577" OAL

Powder	Sugg. Starting Grains	Velocity fps	Pressure C.U.P.	Max. Load Grains	Velocity fps	Pressure C.U.P.
RX7	31.8	2330	32,900	36.2	2599	46,300
IMR-3031	36.0	2433	—	40.0+	2695	—
748	40.5	2369	28,900	‡45.0	2656	40,100
BL-C(2)	39.0	2358	—	43.0	2577	—
IMR-4064	39.0	2439	—	43.0+	2666	—
IMR-4895	37.0	2325	—	41.5+	2652	—
IMR-4320	40.0	2481	—	43.5+	2672	—
H-380	41.0	2242	—	46.5+	2487	—
760	41.0	2156	26,400	46.5+	2501	38,700

180 gr. Jacketed HPBT
2.597" OAL

Powder	Sugg. Starting Grains	Velocity fps	Pressure C.U.P.	Max. Load Grains	Velocity fps	Pressure C.U.P.
IMR-3031	33.0	2145	—	‡37.0+	2369	—
748	38.0	2209	29,100	43.0	2489	43,300
BL-C(2)	36.0	2105	—	40.0	2325	—
IMR-4064	36.0	2159	—	40.0+	2375	—
IMR-4895	36.0	2197	—	40.0+	2427	—
IMR-4320	37.0	2192	—	41.0+	2439	—
H-380	40.0	2132	—	45.0+	2336	—
760	39.0	2074	28,900	44.8+	2380	40,200

#311359***
113 gr. (#2 Alloy) 2.280" OAL

Powder	Sugg. Starting Grains	Velocity fps	Pressure C.U.P.	Max. Load Grains	Velocity fps	Pressure C.U.P.
Red Dot	8.0	1490	25,200	11.0	1845	39,900
700X	8.0	1475	25,200	11.0	1775	42,000
Green Dot	8.5	1525	22,800	12.0	1850	39,900
PB	9.0	1490	25,800	12.0	1800	41,600
Unique	9.0	1530	—	13.0	1960	36,400
SR-7625	9.5	1550	28,200	12.5	1815	42,900
SR-4756	10.0	1555	23,400	14.0	1920	41,600

Note: Loads shown in shaded panels are maximum.
+ Designates a compressed powder charge.
‡ Designates a factory velocity duplication load.
*** Tested in a Universal Receiver with Fed. 210 primers.

#311291***
169 gr. (#2 Alloy) 2.358" OAL

Powder	Sugg. Starting Grains	Velocity fps	Pressure C.U.P.	Max. Load Grains	Velocity fps	Pressure C.U.P.
**Red Dot	7.5	1185	24,000	10.5	1460	37,700
**700X	8.0	1245	26,400	10.0	1420	39,000
**Green Dot	8.0	1205	23,400	11.0	1485	36,400
**PB	9.0	1250	30,500	11.0	1410	42,000
**Unique	8.5	1235	22,200	13.0	1625	39,400
**SR-7625	9.0	1240	30,500	11.0	1415	41,600
**SR-4756	10.0	1310	25,800	12.5	1520	39,900
SR-4759	20.0	1888	28,800	22.5	2047	40,200
RX7	22.3	1812	19,300	32.4	2339	39,500
IMR-3031	26.7	1823	18,700	35.0	2388	38,400
748	29.4	1817	17,000	41.4	2456	36,600

#311467***
178 gr. (#2 Alloy) 2.525" OAL

Powder	Sugg. Starting Grains	Velocity fps	Pressure C.U.P.	Max. Load Grains	Velocity fps	Pressure C.U.P.
**Red Dot	7.5	1185	24,600	10.5	1425	38,100
**700X	8.0	1230	28,800	10.0	1400	39,900
**Green Dot	8.0	1210	25,200	11.0	1470	38,100
**PB	8.5	1205	27,000	10.5	1370	38,100
**Unique	8.5	1240	22,800	13.0	1600	38,100
**SR-7625	8.5	1195	25,800	10.5	1360	36,400
**SR-4756	10.0	1290	24,600	12.0	1450	36,000
SR-4759	18.0	1685	22,400	20.8	1920	36,900
RX7	26.0	1746	15,800	32.0	2198	33,900
IMR-3031	26.0	1689	15,100	30.0	2103	29,700
748	28.0	1737	15,700	36.0	2113	23,200

Note: Loads shown in shaded panels are maximum.
** Federal cases used.
*** Tested in a Universal Receiver with Fed. 210 primers.

30-40 Krag

For Ruger No. 3 Carbine Only

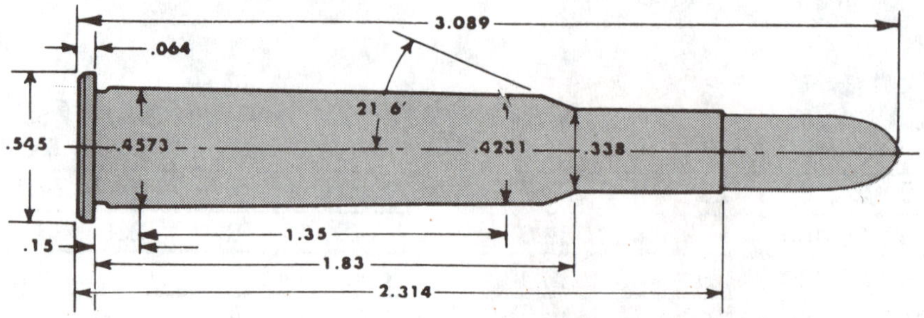

COMMENTS:

This round was dead for years but had a brief resurgence in the limited production Ruger No. 3 Carbine.

Many of the original Krag rifles chambered for this cartridge developed hair line receiver and bolt cracks. Therefore, we do not suggest this data for use in these firearms. The data is intended for use in the Ruger single shot only.

IMR 4350 provides the best ballistic uniformity and the highest potential for accuracy.

TEST COMPONENTS:

Cases .... Remington
Trim-to Length .... 2.304"
Primers .... Remington 9½
Primer Size .... Large Rifle
Lyman Shell Holder .... No. 7
Jacketed Bullets Used .... Remington Ptd. SP, 110 gr.
Hornady Spire Point, 130 gr.
Remington SPCL, 150 gr.
Remington SPCL, 180 gr.
Speer RN, 200 gr.
Remington SPCL, 220 gr.
Cast Bullets Used .... (Sized to .308" dia.)
*Gas Check Bullets
*#311359, 113 gr.
*#311466, 151 gr.
*#311291, 169 gr.

TEST SPECIFICATIONS: (Velocity Only)

Firearm Used .... Ruger No. 3
Barrel Length .... 22"
Twist .... 1-10"
Groove Dia. .... .3095"

110 gr. Jacketed Ptd. SP

Powder	Sugg. Starting Grains	Velocity fps	Pressure C.U.P.	Max. Load Grains	Velocity fps	Pressure C.U.P.
IMR-3031	41.0	2604	—	45.0	2849	—
IMR-4895	42.0	2433	—	47.0	2793	—
IMR-4064	42.0	2375	—	47.0	2695	—
IMR-4320	43.0	2347	—	48.5	2762	—
IMR-4350	48.0	2293	—	55.0+	2732	—
H-380	47.0	2403	—	52.0	2710	—
H-4831	51.0	2192	—	56.0+	2433	—

Note: Loads shown in shaded panels are maximum.
+ Designates a compressed powder charge.

30-40 Krag

For Ruger No. 3 Carbine Only

130 gr. Jacketed Spire Pt.

Powder	Sugg. Starting Grains	Velocity fps	Pressure C.U.P.	Max. Load Grains	Velocity fps	Pressure C.U.P.
IMR-3031	39.0	2386	—	43.0	2680	—
IMR-4895	42.0	2444	—	46.0	2659	—
IMR-4064	41.0	2242	—	46.5	2631	—
IMR-4320	42.0	2336	—	47.0	2638	—
IMR-4350	47.0	2242	—	54.0+	2617	—
H-380	45.0	2304	—	50.0	2570	—
H-4831	51.0	2173	—	56.0+	2403	—

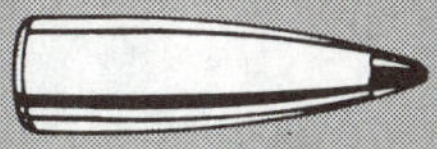

150 gr. Jacketed SP

Powder	Sugg. Starting Grains	Velocity fps	Pressure C.U.P.	Max. Load Grains	Velocity fps	Pressure C.U.P.
IMR-3031	38.0	2331	—	42.0	2557	—
IMR-4895	41.0	2415	—	45.0	2590	—
IMR-4064	41.0	2314	—	46.0	2564	—
IMR-4320	41.0	2283	—	46.0	2570	—
IMR-4350	47.0	2267	—	52.0	2538	—
H-380	44.0	2257	—	49.0	2531	—
H-4831	51.0	2207	—	56.0+	2415	—

180 gr. Jacketed SP

Powder	Sugg. Starting Grains	Velocity fps	Pressure C.U.P.	Max. Load Grains	Velocity fps	Pressure C.U.P.
IMR-3031	34.0	2004	—	38.0	2257	—
IMR-4895	38.0	2132	—	42.0	2375	—
IMR-4064	38.0	2114	—	42.0	2304	—
IMR-4320	39.0	2109	—	43.0	2331	—
IMR-4350	44.0	2109	—	49.0	2386	—
H-380	41.0	2096	—	46.0	2336	—
H-4831	49.0	2159	—	54.0+	2347	—

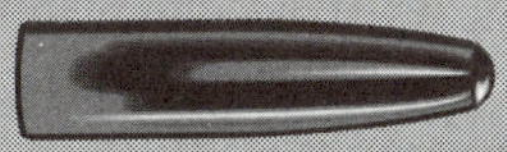

200 gr. Jacketed RN

Powder	Sugg. Starting Grains	Velocity fps	Pressure C.U.P.	Max. Load Grains	Velocity fps	Pressure C.U.P.
IMR-3031	33.0	1897	—	37.0	2141	—
IMR-4895	37.0	2032	—	41.0	2247	—
IMR-4064	37.0	1937	—	41.0	2207	—
IMR-4320	36.0	1934	—	40.0	2136	—
IMR-4350	42.0	1968	—	47.0	2222	—
H-380	39.0	1976	—	43.0	2145	—
H-4831	47.0	1996	—	52.0+	2217	—

Note: Loads shown in shaded panels are maximum.
\+ Designates a compressed powder charge.

30-40 Krag

For Ruger No. 3 Carbine Only

220 gr. Jacketed SP

Powder	Sugg. Starting Grains	Velocity fps	Pressure C.U.P.	Max. Load Grains	Velocity fps	Pressure C.U.P.
IMR-3031	32.0	1831	—	36.0	2074	—
IMR-4895	36.0	1992	—	40.0	2183	—
IMR-4064	36.0	1897	—	40.0	2127	—
IMR-4320	35.0	1882	—	39.0	2079	—
IMR-4350	41.0	1934	—	45.0	2127	—
H-380	38.0	1879	—	42.0	2079	—
H-4831	45.0	1879	—	50.0+	2096	—

#311359
113 gr. (#2 Alloy)

Powder	Sugg. Starting Grains	Velocity fps	Pressure C.U.P.	Max. Load Grains	Velocity fps	Pressure C.U.P.
Unique	9.0	1492	—	13.0	1872	—
2400	19.0	1782	—	23.0	2061	—
IMR-4227	20.0	1709	—	25.0	2061	—

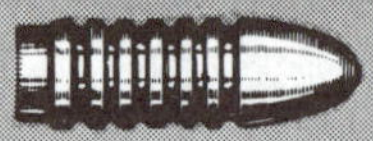

#311466
151 gr. (#2 Alloy)

Powder	Sugg. Starting Grains	Velocity fps	Pressure C.U.P.	Max. Load Grains	Velocity fps	Pressure C.U.P.
Unique	9.0	1381	—	13.0	1694	—
2400	19.0	1712	—	24.0	2036	—
IMR-4227	19.0	1628	—	24.0	1934	—
IMR-4198	24.0	1818	—	28.0	2032	—

#311291
169 gr. (#2 Alloy)

Powder	Sugg. Starting Grains	Velocity fps	Pressure C.U.P.	Max. Load Grains	Velocity fps	Pressure C.U.P.
Unique	8.0	1220	—	12.0	1562	—
2400	18.0	1647	—	23.0	1964	—
IMR-4227	19.0	1633	—	24.0	1923	—
IMR-4198	23.0	1739	—	27.0	1980	—

Note: Loads shown in shaded panels are maximum.
\+ Designates a compressed powder charge.

308 Winchester
(7.62 NATO)

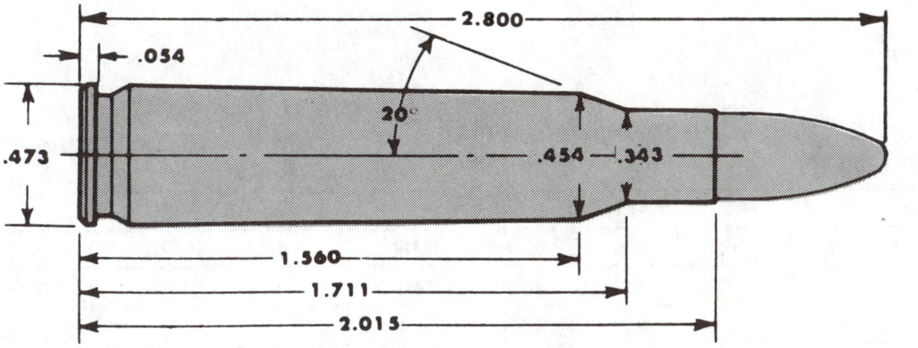

COMMENTS:

There are a lot of shooters who insist on reloading military brass. If you do so reduce maximum charges by a full two grains. This is perhaps the most accurate of the 30 caliber cartridges, and it works well with a wide variety of propellants and bullet weights. It is perhaps the most popular 30 caliber cartridge for cast bullet shooting.

The 110, 125, and 130 grain bullets are, of course, varmint weights. While it is difficult to single out a best propellant for these bullets, we suggest Winchester 748 as a good starting point. Hodgdon H380 is also a fine selection.

For heavier bullets all of Hodgdon H380, IMR 4064 and Winchester 760 are good first choices.

Best cast bullet accuracy usually occurs between 1,600 to 2,000 fps. Bullet #311644 is an excellent choice for long range target shooting or silhouettes.

Those who favor short action rifles will find the 308 among the best possible choices of available cartridges. It will come close to 30-06 performance with all but the heaviest bullets.

TEST COMPONENTS:

Cases .... Remington and Winchester
Trim-to Length .... 2.005"
Primers .... Winchester 8½-120 and Remington 9½
Primer Size .... Large Rifle
Lyman Shell Holder .... No. 2
Jacketed Bullets Used .... Sierra HP #2110, 110 gr.
Sierra SP #2120, 125 gr.
Hornady SP #3020, 130 gr.
Hornady SP #3031, 150 gr.
Nosler SP #16330, 165 gr.
Sierra HPBT #2200, 168 gr.
Sierra HPBT #2220, 180 gr.
Sierra HPBT #2210, 190 gr.
Sierra HPBT #2230, 200 gr.
Cast Bullets Used .... (Sized to .308" dia.)
*Gas Check Bullets
*#311359, 113 gr.
*#311466, 151 gr.
*#311291, 169 gr.
*#311041, 170 gr.
*#311467, 178 gr.
*#311332, 180 gr.
*#311644, 190 gr.

TEST SPECIFICATIONS:
(Velocity & Pressure)

Firearm Used .... Universal Receiver
Barrel Length .... 26"
Twist .... 1-12"
Groove Dia. .... .308"

110 gr. Jacketed HP***
2.603" OAL

Powder	Sugg. Starting Grains	Velocity fps	Pressure C.U.P.	Max. Load Grains	Velocity fps	Pressure C.U.P.
RX7	37.5	2898	36,900	42.0	3194	49,200
IMR-3031	39.6	2604	30,300	45.0+	3017	42,200
748	44.0	2656	27,200	50.0+	3023	36,200
H-335	42.0	2823	31,400	49.0	3282	49,400
BL-C(2)	44.9	2765	31,000	51.0+	3115	42,000
IMR-4064	45.0	2849	36,900	‡50.0+	3184	49,200
IMR-4895	44.0	2890	39,000	50.0+	3225	51,600
IMR-4320	44.0	2702	34,500	51.5+	3154	51,600
H-380	49.0	2994	41,600	55.5+	3268	50,800
IMR-4350	46.0	2481	28,800	52.0+	2849	39,400

Note: Loads shown in shaded panels are maximum.
+ Designates a compressed powder charge.
‡ Designates a factory velocity duplication load.
*** Used Remington cases and 9½ primers.

308 Winchester

(7.62 NATO)

125 gr. Jacketed SP***
2.622" OAL

Powder	Sugg. Starting Grains	Velocity fps	Pressure C.U.P.	Max. Load Grains	Velocity fps	Pressure C.U.P.
RX7	35.0	2624	31,000	40.2	3021	51,200
IMR-3031	38.7	2656	41,200	44.0+	2996	52,100
748	45.4	2687	29,500	‡51.2+	3034	41,600
H-335	41.0	2685	31,600	47.0	3108	49,400
BL-C(2)	45.0	2793	36,400	50.0	3086	50,000
IMR-4064	42.5	2702	36,300	48.0+	3017	50,500
IMR-4895	43.0	2747	38,600	47.0	3003	49,600
IMR-4320	45.0	2724	38,600	49.5+	2994	51,200
H-380	48.0	2890	42,400	54.0+	3115	49,600
H-414	45.3	2479	29,800	51.0+	2779	39,000
IMR-4350	45.0	2433	28,800	51.6+	2785	40,300

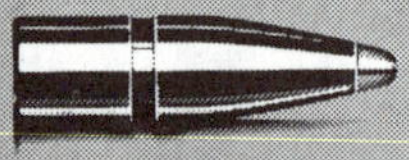

150 gr. Jacketed SP***
2.685" OAL

Powder	Sugg. Starting Grains	Velocity fps	Pressure C.U.P.	Max. Load Grains	Velocity fps	Pressure C.U.P.
RX7	35.0	2475	35,000	39.0	2739	48,800
IMR-3031	40.0	2544	36,000	43.5+	2942	49,100
748	44.0	2690	32,700	50.0+	2996	43,200
H-335	38.0	2457	32,300	45.0	2923	52,000
BL-C(2)	45.0	2717	40,700	49.0	2915	52,400
IMR-4064	43.0	2645	41,100	48.0+	2890	52,000
IMR-4895	40.0	2403	32,000	45.0	2777	50,000
IMR-4320	43.0	2493	37,300	‡48.5+	2840	52,800
H-380	47.0	2695	42,900	53.0+	2949	51,200
760	45.4	2415	28,500	52.0+	2759	37,100
H-414	44.0	2419	28,500	50.0+	2733	37,500
IMR-4350	46.0	2380	33,500	51.6+	2659	41,600
AA2230	40.5	2538	39,300	44.5	2803	51,100
AA2015BR	38.0	2588	39,300	42.0	2846	49,800
AA2460	40.0	2533	40,500	44.5	2786	51,800
RX12	42.0	2588	37,700	48.0	2917	52,000
RX15	41.0	2521	36,300	46.2	2847	50,000

130 gr. Jacketed SP***
2.690" OAL

Powder	Sugg. Starting Grains	Velocity fps	Pressure C.U.P.	Max. Load Grains	Velocity fps	Pressure C.U.P.
RX7	36.0	2652	36,000	40.0	2941	48,800
IMR-3031	40.0	2624	33,000	45.0+	2949	49,200
748	44.0	2645	30,200	50.6+	2992	44,200
H-335	41.0	2664	33,400	47.0	3085	48,800
BL-C(2)	45.0	2770	36,400	50.0	3058	50,800
IMR-4064	44.0	2770	39,000	47.0+	2983	49,000
IMR-4895	43.0	2754	40,700	47.0	2958	50,000
IMR-4320	45.0	2702	37,300	49.5	2941	50,000
H-380	48.0	2873	43,300	54.0+	3125	50,400
H-414	44.8	2465	29,700	51.0+	2750	39,500
IMR-4350	47.0	2531	34,500	51.6+	2770	40,700
AA2230	41.5	2665	35,400	47.5	3056	51,600
AA2015BR	40.8	2830	38,800	45.0	3123	51,200
RX12	45.0	2836	38,200	51.0+	3160	51,000

Note: Loads shown in shaded panels are maximum.
\+ Designates a compressed powder charge.
‡ Designates a factory velocity duplication load.
*** Used Remington cases and $9^{1/2}$ primers.

308 Winchester
(7.62 NATO)

165 gr. Jacketed SP***
2.780" OAL

Powder	Sugg. Starting Grains	Velocity fps	Pressure C.U.P.	Max. Load Grains	Velocity fps	Pressure C.U.P.
RX7	31.0	2277	37,300	36.5	2577	52,400
IMR-3031	37.0	2331	35,000	42.0	2652	49,600
748	42.3	2449	33,800	47.5	2756	49,800
H-335	36.0	2334	32,300	42.0	2674	47,000
BL-C(2)	41.0	2398	35,000	46.0	2732	52,000
IMR-4064	39.1	2406	35,300	44.5+	2698	50,300
IMR-4895	38.0	2426	39,300	43.2	2704	51,800
IMR-4320	42.0	2463	42,400	45.0	2632	51,200
H-380	45.0	2551	41,700	51.0+	2793	50,000
760	45.3	2328	31,800	51.5+	2646	43,500
H-414	44.0	2350	34,000	50.0+	2635	43,100
IMR-4350	46.0	2409	38,100	50.0+	2631	45,900
AA2015BR	36.0	2384	37,800	40.5	2647	50,400
AA2460	39.0	2377	38,700	44.0	2645	50,600
RX12	41.0	2423	37,000	45.0	2715	50,600
AA2520	40.0	2466	39,800	46.0	2724	51,200
RX15	41.0	2424	34,800	44.5	2684	47,900

168 gr. Jacketed HPBT***
2.775" OAL

Powder	Sugg. Starting Grains	Velocity fps	Pressure C.U.P.	Max. Load Grains	Velocity fps	Pressure C.U.P.
RX7	31.0	2267	36,400	36.5	2522	48,400
IMR-3031	37.0	2336	35,000	42.0	2645	49,600
748	40.3	2371	32,800	45.7+	2714	50,200
H-335	36.0	2308	32,700	42.0	2656	47,300
BL-C(2)	41.0	2444	37,700	45.0	2695	48,800
IMR-4064	40.0	2415	35,500	45.0+	2717	52,000
IMR-4895	38.0	2331	35,500	42.5	2624	51,200
IMR-4320	42.0	2475	40,700	45.0	2659	50,800
H-380	45.0	2557	40,700	51.0+	2824	51,200
760	43.8	2311	33,300	49.0+	2624	48,200
H-414	42.0	2274	33,500	48.0+	2574	45,000
IMR-4350	46.0	2398	36,000	50.0+	2617	44,200
AA2015BR	36.0	2420	39,000	41.0	2717	52,000
AA2460	38.0	2336	36,900	42.0	2608	51,000
RX12	40.0	2418	35,600	43.5	2687	49,500
AA2520	40.5	2510	39,200	46.5+	2773	51,100
RX15	40.8	2495	39,200	45.0	2731	51,400

Note: Loads shown in shaded panels are maximum.
\+ Designates a compressed powder charge.
*** Used Remington cases and 9½ primers.

308 Winchester
(7.62 NATO)

180 gr. Jacketed HPBT***
2.780" OAL

Powder	Sugg. Starting Grains	Velocity fps	Pressure C.U.P.	Max. Load Grains	Velocity fps	Pressure C.U.P.
RX7	30.6	2178	38,500	35.0	2442	51,700
IMR-3031	35.0	2183	33,500	40.5	2538	51,200
748	39.1	2320	34,500	44.0	2634	51,700
H-335	35.0	2258	32,800	41.0	2604	49,900
BL-C(2)	39.0	2293	36,900	44.5	2583	51,200
IMR-4064	40.0	2392	40,700	45.5+	2604	51,600
IMR-4895	38.0	2267	36,000	42.5	2564	52,400
IMR-4320	40.0	2293	38,600	44.5	2557	52,400
H-380	40.8	2320	34,300	‡47.0+	2620	45,300
760	41.1	2220	36,300	46.3	2535	48,500
H-414	41.7	2199	31,500	47.0+	2502	44,700
IMR-4350	45.0	2277	36,000	51.0+	2624	51,600
H-4831	45.0	2096	33,000	50.5+	2398	43,300
AA2460	38.0	2350	40,200	42.0	2560	50,500
AA2520	38.5	2355	38,800	45.5+	2663	51,400
RX15	39.0	2372	38,200	43.5	2638	51,600

190 gr. Jacketed HPBT***
2.775" OAL

Powder	Sugg. Starting Grains	Velocity fps	Pressure C.U.P.	Max. Load Grains	Velocity fps	Pressure C.U.P.
RX7	30.2	2100	35,700	34.6	2368	51,400
IMR-3031	35.0	2150	33,000	39.5	2469	50,000
748	38.8	2281	35,500	44.1	2573	51,100
H-335	34.0	2156	33,000	40.0	2503	47,500
BL-C(2)	38.0	2207	35,000	43.5	2557	51,200
IMR-4064	35.6	2192	37,100	41.0+	2470	48,100
IMR-4895	37.0	2217	35,500	41.5	2500	51,200
IMR-4320	39.0	2247	39,000	43.0	2481	50,800
H-380	43.0	2409	42,400	48.0+	2624	51,200
760	42.3	2225	36,300	47.2+	2522	50,600
H-414	40.5	2166	32,400	46.0+	2426	44,300
IMR-4350	38.5	2101	35,000	44.0+	2366	45,800
H-4831	45.0	2127	33,500	50.0+	2398	43,300
AA2460	36.5	2227	38,800	41.5	2496	51,200
AA2520	37.0	2274	39,800	44.5+	2577	51,400
RX15	38.5	2332	38,500	42.5	2549	50,500

Note: Loads shown in shaded panels are maximum.
\+ Designates a compressed powder charge.
‡ Designates a factory velocity duplication load.
*** Used Remington cases and 9½ primers.

308 Winchester
(7.62 NATO)

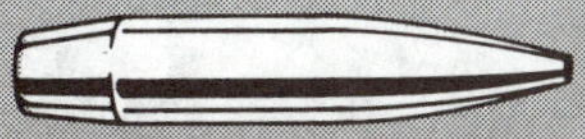

200 gr. Jacketed HPBT***
2.780" OAL

Powder	Sugg. Starting Grains	Velocity fps	Pressure C.U.P.	Max. Load Grains	Velocity fps	Pressure C.U.P.
RX7	29.2	2025	36,500	33.6	2282	51,700
IMR-3031	35.0	2109	34,500	39.5	2380	50,800
748	38.6	2209	34,700	43.9+	2517	50,400
H-335	33.0	2054	31,600	39.0	2411	47,400
BL-C(2)	38.0	2123	36,000	42.5	2392	50,800
IMR-4064	38.0	2217	37,300	42.5	2444	52,000
IMR-4895	37.0	2150	36,900	40.5	2392	51,600
IMR-4320	39.0	2178	37,300	42.5	2409	52,400
H-380	38.8	2178	33,200	45.0+	2447	43,800
760	40.8	2124	35,000	46.2+	2423	47,500
H-414	39.4	2075	32,500	45.0+	2347	43,200
IMR-4350	43.0	2145	36,000	48.0+	2457	51,200
H-4831	45.0	2061	32,500	49.5+	2309	42,000
AA2520	34.5	2105	36,300	40.5	2396	51,200
RX15	37.8	2257	38,500	42.2	2507	51,800
AA2700	42.0	2211	40,700	47.5+	2458	51,300

#311466**
151 gr. (#2 Alloy) 2.525" OAL

Powder	Sugg. Starting Grains	Velocity fps	Pressure C.U.P.	Max. Load Grains	Velocity fps	Pressure C.U.P.
Red Dot	10.0	1466	27,600	13.0	1730	40,700
700X	9.0	1407	28,200	11.5	1625	39,900
Green Dot	10.5	1461	25,200	13.5	1724	38,600
PB	10.0	1405	29,400	12.5	1613	39,400
Unique	11.0	1586	26,400	15.0	1933	41,000
SR-7625	11.0	1490	30,500	13.0	1650	37,700
Herco	13.0	1623	27,600	15.5	1824	39,000
SR-4756	12.0	1512	24,600	16.0	1826	39,900
SR-4759	17.5	1760	22,600	27.0	2401	47,800
IMR-4227	20.0	1771	18,500	31.0	2503	48,300
IMR-4198	21.8	1781	15,200	36.0+	2727	49,900
RX7	23.0	1779	14,800	39.5	2748	47,100
IMR-3031	27.0	1767	15,300	42.0+	2810	49,700
748	38.4	1767	16,500	48.0+	2920	48,000

#311359**
113 gr. (#2 Alloy) 2.430" OAL

Powder	Sugg. Starting Grains	Velocity fps	Pressure C.U.P.	Max. Load Grains	Velocity fps	Pressure C.U.P.
Red Dot	10.0	1623	24,000	14.0	2000	39,400
700X	9.0	1540	24,600	12.5	1879	40,300
Green Dot	11.0	1692	25,200	15.0	2049	40,300
PB	10.5	1601	24,600	13.5	1879	38,600
Unique	12.0	1831	24,600	17.0	2272	40,700
SR-7625	12.0	1704	26,400	15.0	1953	39,400
Herco	14.0	1886	25,800	17.0	2159	38,600

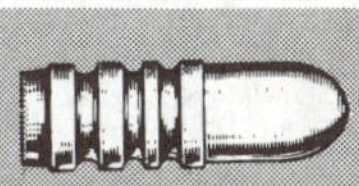

311291**
169 gr. (#2 Alloy) 2.510" OAL

Powder	Sugg. Starting Grains	Velocity fps	Pressure C.U.P.	Max. Load Grains	Velocity fps	Pressure C.U.P.
Red Dot	10.0	1367	28,800	12.5	1576	39,900
700X	8.0	1209	23,400	11.5	1547	39,900
Green Dot	11.0	1434	30,000	13.0	1582	37,700
PB	9.0	1220	22,200	12.5	1553	38,100
Unique	10.0	1368	23,400	15.0	1816	37,700
SR-7625	10.0	1290	28,800	13.0	1560	39,900
Herco	12.0	1479	26,400	15.5	1718	37,700
SR-4756	12.0	1416	30,000	14.5	1607	37,700
SR-4759	19.5	1847	22,300	26.0	2255	47,800
IMR-4227	21.6	1867	22,400	27.8	2270	49,500
IMR-4198	24.0	1897	22,700	30.5	2341	48,200
RX7	23.0	1870	21,200	41.5	2602	49,700
IMR-3031	28.5	1868	20,200	39.5+	2653	49,800
748	31.7	1853	21,100	46.0+	2764	49,300

Note: Loads shown in shaded panels are maximum.
+ Designates a compressed powder charge.
** Used Winchester cases and 8½ -120 primers.
*** Used Remington cases and 9½ primers.

308 Winchester
(7.62 NATO)

#311041** 170 gr. (#2 Alloy) 2.617" OAL

Powder	Sugg. Starting Grains	Velocity fps	Pressure C.U.P.	Max. Load Grains	Velocity fps	Pressure C.U.P.
Red Dot	10.0	1355	27,000	12.5	1565	39,000
700X	8.0	1206	24,000	11.5	1505	37,700
Green Dot	11.0	1434	28,800	13.0	1587	35,500
PB	9.0	1201	23,400	12.5	1479	40,300
Unique	10.0	1377	21,600	15.0	1802	37,700
SR-7625	10.0	1272	24,600	13.0	1499	38,600
Herco	12.0	1470	25,800	15.5	1733	37,700
SR-4756	12.0	1392	27,000	14.5	1587	37,700
SR-4759	21.5	1897	26,100	27.7	2345	50,300
IMR-4198	25.0	1898	22,700	34.0	2460	38,700
RX7	27.5	1924	24,100	39.0	2627	48,900
IMR-3031	29.0	1858	23,500	41.8	2709	51,200
748	32.0	1856	20,300	47.0	2790	50,400
H-335	30.0	1847	19,800	44.2	2787	50,200

#311467** 178 gr. (#2 Alloy) 2.750" OAL

Powder	Sugg. Starting Grains	Velocity fps	Pressure C.U.P.	Max. Load Grains	Velocity fps	Pressure C.U.P.
Red Dot	10.0	1352	27,600	12.5	1555	39,900
700X	8.0	1200	22,200	11.5	1513	39,900
Green Dot	11.0	1424	30,000	13.0	1570	37,300
PB	9.5	1240	22,200	12.5	1508	39,400
Unique	10.0	1373	23,400	14.5	1730	38,100
SR-7625	10.5	1320	27,000	13.0	1508	36,900
Herco	12.0	1457	25,800	15.5	1712	37,300
SR-4756	12.0	1379	24,600	14.5	1600	40,700
SR-4759	19.0	1813	28,600	26.3	2243	47,600
IMR-4198	22.5	1782	24,400	32.0	2376	48,100
RX7	23.0	1783	25,000	34.5	2409	41,100
748	30.0	1851	23,000	44.5	2269	47,800
H-335	26.5	2007	29,500	40.0	2575	46,400

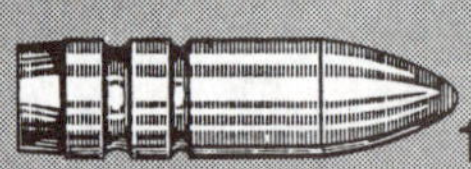

#311332** 180 gr. (#2 Alloy) 2.705" OAL

Powder	Sugg. Starting Grains	Velocity fps	Pressure C.U.P.	Max. Load Grains	Velocity fps	Pressure C.U.P.
SR-4759	23.5	1827	25,000	26.3	2035	38,600
RX7	28.0	1827	20,800	32.0	2172	35,800
IMR-3031	33.0	2001	20,500	37.5	2361	37,800
AA2015BR	31.0	1891	19,500	36.0	2324	37,900
H-335	32.0	1902	21,800	39.0	2316	36,800
IMR-4895	33.5	1909	20,000	39.0	2372	39,100
Unique	12.2	1425	23,000	15.5	1692	38,000
SR-7625	11.5	1270	21,500	14.0	1492	36,400
SR-4756	13.0	1378	22,700	15.5	1587	36,400
IMR-4227	25.0	1867	24,300	29.0	2126	36,900
IMR-4198	26.5	1856	22,000	32.0	2229	40,200

#311644*** 190 gr. (#2 Alloy) 2.675" OAL

Powder	Sugg. Starting Grains	Velocity fps	Pressure C.U.P.	Max. Load Grains	Velocity fps	Pressure C.U.P.
SR-4759	20.0	1682	21,600	26.0	2029	47,100
RX7	25.0	1738	18,500	33.0	2222	41,700
IMR-3031	32.0	1943	22,700	40.0+	2435	48,100
AA2015BR	28.0	1833	21,200	38.0	2340	47,700
H-335	31.0	1875	22,300	40.0	2409	46,800
IMR-4895	33.0	1923	22,800	42.0+	2447	49,200
Unique	10.5	1289	21,000	14.5	1614	40,700
SR-7625	10.0	1174	21,300	14.0	1501	42,200
SR-4756	11.5	1292	22,000	14.5	1531	37,900
IMR-4198	23.5	1760	21,900	31.0	2192	45,400

Note: Loads shown in shaded panels are maximum.
+ Designates a compressed powder charge.
** Used Winchester cases and 8½ -120 primers.
*** Used Remington cases and 9½ primers.

30-06 Springfield
(7.62 x 63mm)

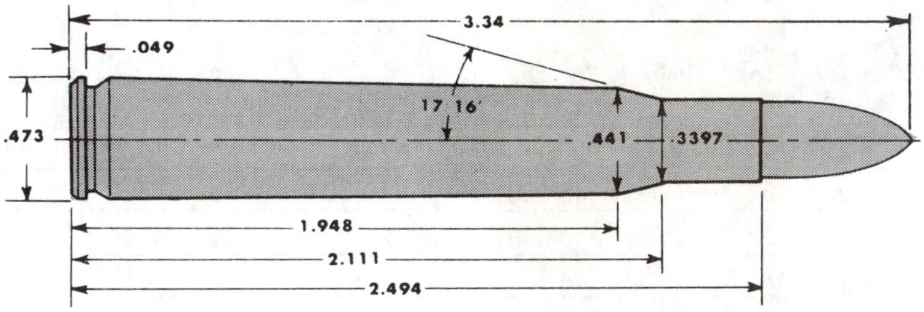

COMMENTS:

This is the most popular hunting cartridge in North America. As a result there are a great many bullets available. Weights from 110 to 220 grains are commonly loaded, though the most popular is the 180 grain.

Bullets of 110, 125, and 130 grains are for varmints and informal target shooting. The 150 to 168 grain weights are ideal for light big game, especially if ranges tend to be long. The 180 grain and heavier bullets are for game as large as elk, moose and so on.

It is almost impossible to select a best propellant as one of the 30-06's virtues is that it works very well with a large number of powders. Still if forced to suggest a starting powder we would pick IMR 4064 for bullets of 110 to 150 grains. From 165 to 180 grains we would first try IMR 4350. For heavier bullets it would be hard to choose from among IMR 4350, IMR 4831, Hodgdon H4831 and Hercules Reloder 22.

The 30-06 is also a top performer with cast bullets. Bullet #311284 is a fine choice for hunting and #311644 is excellent for target shooting.

TEST COMPONENTS:

Cases .... Winchester
Trim-to Length .... 2.484"
Primers .... Winchester 8½-120
Primer Size .... Large Rifle
Lyman Shell Holder .... No. 2
Jacketed Bullets Used .... Sierra HP #2110, 110 gr.
Sierra SP #2120, 125 gr.
Hornady SP #3020, 130 gr.
Hornady SP #3031, 150 gr.
Nosler SPBT #27585, 165 gr.
Sierra HPBT #2200, 168 gr.
Sierra HPBT #2220, 180 gr.
Barnes "X" #30835, 180 gr.
Sierra HPBT #2210, 190 gr.
Sierra HPBT #2230, 200 gr.
Hornady RN #3090, 220 gr.
Cast Bullets Used .... (Sized to .308" dia.)
*Gas Check Bullets
*#311359, 113 gr.
*#311466, 151 gr.
*#311291, 169 gr.
*#311041, 170 gr.
*#311644, 190 gr.
*#311284, 210 gr.

TEST SPECIFICATIONS:
(Velocity & Pressure)

Firearm Used .... Universal Receiver
Barrel Length .... 24"
Twist .... 1-10"
Groove Dia. .... .308"

110 gr. Jacketed HP
3.000" OAL

Powder	Sugg. Starting Grains	Velocity fps	Pressure C.U.P.	Max. Load Grains	Velocity fps	Pressure C.U.P.
RX7	37.0	2736	35,500	42.0	3073	50,000
IMR-3031	47.0	2930	35,000	52.0	3257	46,700
748	51.0	3060	39,000	‡60.0	3387	49,000
H-335	44.8	2923	40,400	51.0	3203	48,900
BL-C(2)	51.0	3144	41,100	55.0	3378	50,400
IMR-4064	50.0	3030	39,000	55.5	3356	48,800
IMR-4895	49.0	3012	39,400	53.5	3266	50,000
IMR-4320	50.0	2906	38,600	55.5	3184	49,600
760	55.4	2922	33,800	63.0+	3206	42,900
IMR-4350	56.0	2898	37,300	61.0+	3194	45,400
H-4831	55.0	2557	28,800	61.5+	2985	39,400

Note: Loads shown in shaded panels are maximum.
\+ Designates a compressed powder charge.
‡ Designates a factory velocity duplication load.

30-06 Springfield

(7.62 x 63mm)

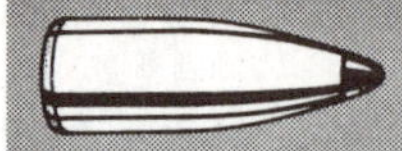

125 gr. Jacketed SP
3.080" OAL

Powder	Sugg. Starting Grains	Velocity fps	Pressure C.U.P.	Max. Load Grains	Velocity fps	Pressure C.U.P.
RX7	38.5	2710	38,500	43.0	2932	47,600
IMR-3031	45.0	2732	33,000	50.0	3095	45,000
748	51.5	2890	35,300	58.0	3241	48,300
H-335	45.3	2815	37,400	‡51.5	3144	48,500
BL-C(2)	49.0	2890	35,000	54.0	3215	48,000
IMR-4064	48.0	2762	31,500	55.0	3215	47,200
IMR-4895	47.8	2815	36,500	53.5	3096	47,300
IMR-4320	49.0	2770	36,000	55.0	3105	46,700
H-380	51.0	2967	39,000	56.5	3194	49,600
760	56.3	2801	34,200	63.0+	3215	47,600
IMR-4350	55.0	2754	34,000	61.0+	3105	46,000
H-4831	55.0	2544	30,000	61.5+	2906	40,000
IMR-4831	57.0	2700	33,600	63.0+	3017	43,200

130 gr. Jacketed SP
3.090" OAL

Powder	Sugg. Starting Grains	Velocity fps	Pressure C.U.P.	Max. Load Grains	Velocity fps	Pressure C.U.P.
RX7	37.8	2675	40,300	43.0	2937	49,800
IMR-3031	45.0	2793	36,400	50.0	3048	45,000
748	50.1	2844	38,300	59.0	3169	48,500
H-335	45.7	2813	40,800	52.0	3081	50,000
BL-C(2)	49.0	2932	38,100	53.5	3174	48,400
IMR-4064	48.0	2808	36,000	54.0	3184	49,600
IMR-4895	48.0	2890	42,000	51.0	3036	49,900
IMR-4320	49.0	2793	39,000	54.0	3076	48,400
H-380	51.0	2967	42,000	54.5	3125	50,000
760	55.4	2822	35,000	63.0+	3111	44,200
IMR-4350	51.0	2623	34,200	58.0+	2949	44,200
H-4831	55.0	2570	32,500	61.5+	2923	41,600
IMR-4831	57.0	2686	32,900	62.0+	2984	44,400

150 gr. Jacketed SP
3.200" OAL

Powder	Sugg. Starting Grains	Velocity fps	Pressure C.U.P.	Max. Load Grains	Velocity fps	Pressure C.U.P.
RX7	32.5	2372	40,300	37.0	2570	50,000
IMR-3031	42.0	2450	29,400	50.0	2932	48,000
748	45.0	2614	38,800	52.0	2902	50,000
H-335	40.4	2508	42,400	46.0	2755	49,300
BL-C(2)	46.0	2770	44,600	51.0	2932	48,000
IMR-4064	48.0	2695	36,000	53.0	3012	49,600
IMR-4895	46.0	2680	36,900	51.5	2958	49,200
IMR-4320	47.0	2624	37,300	‡53.0	2923	48,400
H-380	48.0	2724	40,300	53.0	2976	50,000
760	48.0	2535	34,200	56.0	2878	49,600
IMR-4350	49.0	2545	37,300	57.0+	2887	47,800
H-4831	55.0	2551	33,000	61.5+	2890	43,300
AA2230	46.0	2597	39,800	51.0	2870	49,300
AA2460	46.5	2597	38,500	51.7	2869	48,500
RX12	48.0	2617	37,300	55.7	2948	49,100
RX15	50.5	2777	40,000	54.0	2987	49,300
RX19	58.5	2690	40,800	64.3+	2925	49,600
IMR-4831	55.0	2571	33,800	61.0+	2903	47,100

165 gr. Jacketed SPBT
3.285" OAL

Powder	Sugg. Starting Grains	Velocity fps	Pressure C.U.P.	Max. Load Grains	Velocity fps	Pressure C.U.P.
RX7	34.0	2254	34,700	39.5	2542	46,500
IMR-3031	42.0	2481	35,500	‡48.0	2801	48,000
748	45.8	2527	38,000	52.0	2823	48,800
H-335	39.5	2336	38,700	46.0	2641	48,000
BL-C(2)	43.0	2444	34,000	48.5	2754	48,000
IMR-4064	45.0	2531	34,500	50.0	2824	48,000
IMR-4895	43.0	2469	36,400	48.5	2732	46,000
IMR-4320	44.6	2418	35,000	50.7	2742	48,600
H-380	45.0	2475	36,000	51.5	2824	50,000
760	50.2	2479	33,000	57.0	2782	47,600
IMR-4350	52.0	2583	38,100	57.0+	2824	48,000
H-4831	55.0	2557	35,500	61.5+	2881	45,900
AA2460	44.2	2485	39,800	49.9	2727	48,300
RX15	47.0	2546	41,500	51.0	2807	48,100
RX19	54.2	2463	35,500	60.4+	2785	49,000
IMR-4831	55.0	2562	36,800	61.0+	2876	49,400

Note: Loads shown in shaded panels are maximum.
\+ Designates a compressed powder charge.
‡ Designates a factory velocity duplication load.

30-06 Springfield
(7.62 x 63mm)

168 gr. Jacketed HPBT
3.300" OAL

Powder	Sugg. Starting Grains	Velocity fps	Pressure C.U.P.	Max. Load Grains	Velocity fps	Pressure C.U.P.
IMR-3031	42.0	2463	34,000	48.0	2808	49,600
748	45.7	2547	39,000	52.0	2822	50,000
H-335	43.1	2510	39,700	49.0	2753	50,000
BL-C(2)	43.0	2538	42,000	48.5	2777	48,400
IMR-4064	45.0	2551	34,000	50.0	2816	46,700
IMR-4895	43.0	2444	35,500	48.5	2762	47,600
IMR-4320	45.0	2475	36,400	51.0	2816	50,000
H-380	48.4	2564	40,500	51.5	2710	45,400
760	50.1	2541	37,300	57.0	2839	49,700
IMR-4350	52.0	2590	38,000	57.0+	2873	50,400
H-4831	55.0	2564	37,300	61.5+	2873	47,200
AA2015BR	40.0	2429	36,100	45.5	2689	48,000
AA2520	42.7	2383	36,200	48.0	2653	48,800
RX15	47.3	2592	38,800	51.8	2832	50,000
RX19	54.8	2489	37,800	60.9+	2810	50,000
IMR-4831	54.5	2544	35,700	60.6+	2847	47,800

180 gr. Barnes "X"
3.275" OAL

Powder	Sugg. Starting Grains	Velocity fps	Pressure C.U.P.	Max. Load Grains	Velocity fps	Pressure C.U.P.
IMR-3031	39.5	2274	35,700	44.0	2539	47,500
748	42.5	2235	34,200	48.0	2533	47,200
H-335	38.7	2132	37,700	45.0	2454	48,500
IMR-4895	41.5	2307	37,500	47.0	2578	48,100
IMR-4064	43.2	2447	34,000	48.5	2614	48,300
AA2520	40.0	2209	38,700	45.5	2467	48,500
IMR-4320	41.6	2178	35,700	47.5	2497	45,700
760	47.0	2325	37,100	53.3	2557	48,000
IMR-4350	51.0	2399	37,500	55.7+	2688	49,200
RX19	52.0	2336	37,200	57.0+	2627	49,700
AA3100	54.5	2419	40,500	59.8+	2625	48,400
H-4831	52.0	2390	38,200	57.4+	2649	48,100
RX22	54.5	2441	38,200	61.5	2715	48,200
IMR-4831	52.0	2374	34,100	57.6+	2717	48,700

180 gr. Jacketed HPBT
3.280" OAL

Powder	Sugg. Starting Grains	Velocity fps	Pressure C.U.P.	Max. Load Grains	Velocity fps	Pressure C.U.P.
IMR-3031	41.0	2380	35,500	46.0	2688	48,000
748	42.2	2365	37,000	48.0	2632	49,600
H-335	39.6	2370	41,900	45.0	2583	50,000
IMR-4064	43.0	2403	35,000	49.0	2710	47,200
IMR-4895	42.0	2386	36,000	47.5	2659	48,000
IMR-4320	40.7	2281	35,800	46.3	2584	49,900
760	45.0	2360	35,500	52.0	2666	49,900
IMR-4350	50.0	2469	38,300	56.0+	2801	50,000
H-4831	55.0	2604	40,300	60.0+	2840	50,000
AA2520	41.3	2275	35,400	47.3	2572	49,700
RX19	52.3	2497	41,200	‡58.3+	2698	48,800
AA3100	53.5	2425	38,300	59.0+	2648	48,300
RX22	55.4	2536	39,300	60.8+	2780	48,400
IMR-4831	52.5	2525	39,000	58.4+	2788	48,700

190 gr. Jacketed HPBT
3.280" OAL

Powder	Sugg. Starting Grains	Velocity fps	Pressure C.U.P.	Max. Load Grains	Velocity fps	Pressure C.U.P.
IMR-3031	40.0	2309	34,500	45.0	2590	47,000
BL-C(2)	41.0	2247	34,000	46.5	2518	47,000
IMR-4064	42.0	2309	32,500	47.5	2624	48,400
IMR-4895	41.0	2288	35,000	46.0	2569	50,000
IMR-4320	44.0	2341	34,500	49.0	2631	48,800
H-380	42.0	2309	37,300	47.5	2577	48,000
760	46.0	2350	36,200	53.0	2662	50,000
H-414	47.0	2368	36,200	54.0	2694	49,800
IMR-4350	44.8	2205	31,500	51.0	2586	50,000
H-4831	53.0	2450	36,900	59.0+	2724	47,200
AA2520	41.0	2233	36,800	46.0	2469	48,500
RX19	51.5	2341	37,300	56.0+	2606	48,300
AA3100	53.3	2368	38,800	59.0+	2614	48,600
RX22	55.0	2461	37,800	60.5+	2722	48,700
IMR-4831	51.0	2395	37,300	56.8+	2686	49,300

30-06 Springfield

(7.62 x 63mm)

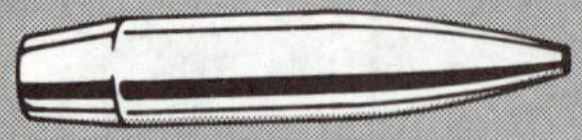

200 gr. Jacketed HPBT
3.280" OAL

Powder	Sugg. Starting Grains	Velocity fps	Pressure C.U.P.	Max. Load Grains	Velocity fps	Pressure C.U.P.
IMR-3031	40.0	2276	38,100	44.0	2457	48,400
IMR-4064	42.0	2309	38,100	47.0	2552	50,000
IMR-4895	41.0	2262	37,300	46.5	2512	48,400
IMR-4320	43.0	2294	38,600	47.5	2502	47,600
H-380	41.0	2155	37,300	46.5	2441	50,000
760	45.0	2224	36,300	52.0	2565	50,000
H-414	46.0	2258	35,000	53.0	2606	49,800
IMR-4350	49.0	2380	39,900	54.0+	2638	48,400
H-4831	52.0	2343	36,400	58.0+	2620	46,300
RX19	51.3	2302	36,500	55.6+	2526	45,800
AA3100	53.1	2336	39,000	58.8	2573	48,600
RX22	54.5	2441	40,000	60.0+	2680	50,100
IMR-4831	50.0	2323	35,800	55.5+	2587	47,400

220 gr. Jacketed RN
3.220" OAL

Powder	Sugg. Starting Grains	Velocity fps	Pressure C.U.P.	Max. Load Grains	Velocity fps	Pressure C.U.P.
IMR-3031	38.0	2087	35,000	43.0	2336	48,000
IMR-4064	41.0	2183	37,300	46.5	2439	49,200
IMR-4895	40.0	2197	41,100	44.5	2375	48,400
IMR-4320	41.0	2118	36,400	47.5	2421	49,200
H-380	41.0	2145	39,400	45.5	2341	50,000
760	44.4	2106	37,200	50.5	2381	49,400
H-414	44.9	2153	40,200	51.0	2379	49,200
IMR-4350	49.0	2309	40,700	53.0+	2500	49,200
H-4831	51.0	2283	38,600	57.5+	2583	50,000
RX19	50.0	2127	34,600	55.8	2451	48,800
AA3100	51.6	2172	37,800	57.5	2427	49,000
RX22	53.0	2247	37,700	58.8+	2507	48,500
IMR-4831	49.5	2192	35,700	54.6+	2458	47,100

#311359
113 gr. (#2 Alloy) 2.909" OAL

Powder	Sugg. Starting Grains	Velocity fps	Pressure C.U.P.	Max. Load Grains	Velocity fps	Pressure C.U.P.
Red Dot	11.0	1683	23,400	15.0	2000	36,900
700X	10.0	1610	22,000	14.0	1953	38,100
Green Dot	13.0	1792	28,800	16.0	2020	38,600
PB	12.0	1689	27,000	15.5	1941	39,000
Unique	14.0	1960	27,000	19.0	2331	42,000
SR-7625	12.5	1677	21,600	16.5	2008	42,400
Herco	14.5	1876	28,200	18.0	2132	39,000

#311466
151 gr. (#2 Alloy) 3.000" OAL

Powder	Sugg. Starting Grains	Velocity fps	Pressure C.U.P.	Max. Load Grains	Velocity fps	Pressure C.U.P.
Red Dot	11.0	1479	24,600	14.0	1715	35,500
700X	10.0	1408	22,800	13.5	1694	37,300
Green Dot	12.0	1547	29,400	15.0	1748	38,600
PB	11.5	1449	24,000	15.0	1697	38,600
Unique	12.5	1636	25,200	17.0	1941	37,300
SR-7625	12.5	1512	27,000	15.5	1733	39,400
Herco	13.5	1618	26,400	17.0	1858	39,400
SR-4756	13.5	1572	26,400	17.0	1805	37,300
2400	20.0	1723	16,400	30.5	2446	42,000
SR-4759	19.5	1652	13,700	32.0	2477	48,100
IMR-4227	21.5	1706	14,900	34.0	2495	44,500
IMR-4198	23.0	1670	12,100	37.0	2515	38,000
RX7	23.0	1704	13,700	36.0	2476	38,500
IMR-3031	29.0	1682	12,000	42.5	2522	34,100

Note: Loads shown in shaded panels are maximum.
+ Designates a compressed powder charge.

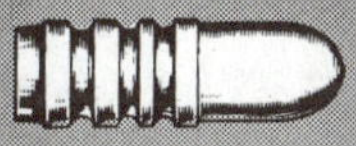

#311291
169 gr. (#2 Alloy) 3.013" OAL

Powder	Sugg. Starting Grains	Velocity fps	Pressure C.U.P.	Max. Load Grains	Velocity fps	Pressure C.U.P.
Red Dot	10.5	1390	24,000	13.5	1625	35,000
700X	9.5	1307	21,000	13.0	1580	36,400
Green Dot	11.5	1436	26,400	14.0	1597	35,500
PB	11.0	1340	21,000	14.5	1595	36,900
Unique	12.0	1508	24,000	16.5	1842	36,400
SR-7625	11.5	1363	23,400	14.5	1585	36,000
Herco	12.5	1488	26,400	16.0	1697	35,500
SR-4756	12.5	1418	22,200	16.5	1692	35,500
2400	18.6	1635	18,200	29.5	2296	43,100
SR-4759	20.0	1666	16,700	31.0	2325	46,900
IMR-4198	25.0	1729	15,100	38.5	2501	45,100
RX7	25.0	1733	15,100	38.3	2480	47,600
IMR-3031	29.0	1710	13,900	42.0	2469	38,200
748	30.0	1721	14,300	46.5	2531	38,700
H-4895	28.0	1705	14,400	44.0	2526	43,100

#311041
170 gr. (#2 Alloy) 2.968" OAL

Powder	Sugg. Starting Grains	Velocity fps	Pressure C.U.P.	Max. Load Grains	Velocity fps	Pressure C.U.P.
Red Dot	10.5	1355	22,800	13.5	1585	32,500
700X	9.5	1288	22,800	13.0	1545	36,000
Green Dot	11.5	1416	25,800	14.0	1582	34,000
PB	11.0	1326	23,400	14.5	1557	37,700
Unique	12.0	1504	24,000	16.5	1811	35,500
SR-7625	11.5	1346	24,000	14.5	1570	36,900
Herco	13.0	1497	27,000	16.0	1689	36,000
SR-4756	12.5	1406	22,800	16.5	1653	36,400
2400	20.0	1751	22,200	27.0	2141	34,500
SR-4759	20.0	1642	13,900	31.5	2329	47,400
IMR-4198	25.0	1717	13,100	38.5	2463	43,100
RX7	25.0	1661	12,800	40.5	2515	46,900
IMR-3031	30.0	1636	11,100	44.0	2522	40,800
748	31.0	1695	12,800	47.0	2503	37,500
H-4895	28.5	1706	14,100	45.0	2496	40,700

#311644
190 gr. (#2 Alloy) 3.075" OAL

Powder	Sugg. Starting Grains	Velocity fps	Pressure C.U.P.	Max. Load Grains	Velocity fps	Pressure C.U.P.
Red Dot	10.5	1245	18,000	14.5	1532	33,100
Green Dot	12.5	1360	24,800	15.5	1578	34,700
PB	11.5	1227	20,700	15.0	1492	34,500
Unique	13.0	1404	21,800	17.2	1677	36,800
SR-7625	12.2	1295	23,400	15.2	1514	38,000
Herco	14.5	1437	20,500	18.0	1664	31,100
SR-4756	13.5	1378	23,600	17.0	1587	34,100
SR-4759	23.0	1716	21,000	28.2	1975	37,700
H-4895	32.0	1875	21,800	44.0	2419	45,200

#311284
210 gr. (#2 Alloy) 3.063" OAL

Powder	Sugg. Starting Grains	Velocity fps	Pressure C.U.P.	Max. Load Grains	Velocity fps	Pressure C.U.P.
Red Dot	10.5	1244	27,600	13.5	1418	39,400
Green Dot	11.0	1248	26,400	13.5	1402	36,400
PB	11.0	1194	27,000	14.0	1384	37,000
Unique	11.5	1326	26,400	15.5	1577	36,400
SR-7625	11.5	1242	30,000	13.0	1320	36,400
Herco	12.5	1335	30,500	15.5	1510	39,000
SR-4756	12.5	1264	26,400	15.5	1445	35,000
2400	19.0	1592	27,000	24.0	1827	32,000
SR-4759	22.0	1679	27,800	27.8	2004	45,300
RX7	25.0	1685	22,200	38.2	2254	47,000
H-4895	29.0	1730	20,900	43.0	2335	45,600
IMR-4064	32.0	1769	22,100	44.2	2364	46,000
IMR-4350	35.8	1774	19,200	50.0	2422	47,800

Note: Loads shown in shaded panels are maximum

300 H&H Magnum

(300 Holland and Holland)
(7.63 x 72mm)

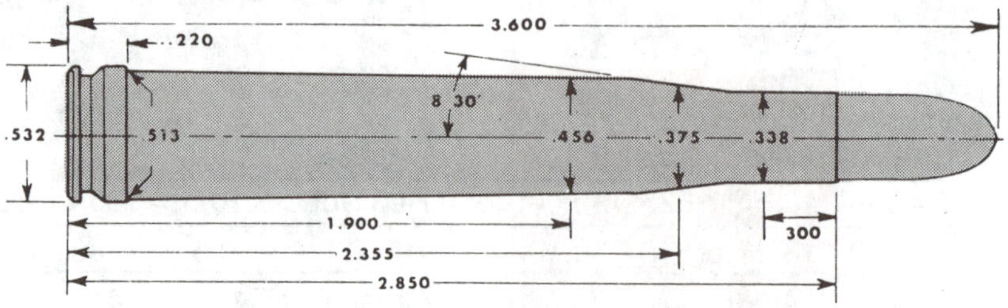

COMMENTS:

This is one of the most accurate of the belted magnum cartridges. It performs best with bullets of 150 grains and heavier (lighter bullets are varmint and target types). IMR 4350 and Hodgdon H4831 are particularly accurate in this cartridge. Many rifles will show an accuracy preference for 180 grain bullets.

Due to the relatively high velocities achieved, the use of premium grade bullets for hunting might be a wise decision.

TEST COMPONENTS:

Cases .... Winchester
Trim-to Length .... 2.840"
Primers .... Winchester 8½-120
Primer Size .... Large Rifle
Lyman Shell Holder .... No. 13
Jacketed Bullets Used .... Remington Ptd. SP, 110 gr.
Hornady Spire Pt., 130 gr.
Remington Ptd. SPCL, 150 gr.
Speer Spitzer, 165 gr.
Remington Ptd. SPCL, 180 gr.
Speer RN, 200 gr.
Remington SPCL, 220 gr.
Cast Bullets Used .... (Sized to .308" dia.)
*Gas Check Bullets *#311466, 151 gr.
*#311291, 169 gr.

TEST SPECIFICATIONS:
(Velocity Only)

Firearm Used .... Remington Model 721
Barrel Length .... 26"
Twist .... 1-10"
Groove Dia. .... .308"

110 gr. Jacketed Ptd. SP

Powder	Sugg. Starting Grains	Velocity fps	Pressure C.U.P.	Max. Load Grains	Velocity fps	Pressure C.U.P.
IMR-3031	58.0	3427	—	64.0	3708	—
IMR-4895	61.0	3390	—	68.0	3731	—
IMR-4064	61.0	3322	—	68.0	3708	—
IMR-4320	63.0	3427	—	70.0	3773	—
IMR-4350	72.0	3460	—	80.0+	3773	—
H-380	68.0	3390	—	76.0	3731	—
H-4831	73.0	3134	—	80.0+	3401	—

Note: Loads shown in shaded panels are maximum.
+ Designates a compressed powder charge.

300 H&H Magnum

(300 Holland and Holland)
(7.63 x 72mm)

130 gr. Jacketed Spire Pt.

Powder	Sugg. Starting Grains	Velocity fps	Pressure C.U.P.	Max. Load Grains	Velocity fps	Pressure C.U.P.
IMR-3031	57.0	3205	—	63.0	3448	—
IMR-4895	60.0	3205	—	67.0	3521	—
IMR-4064	60.0	3164	—	67.0	3497	—
IMR-4320	61.0	3225	—	68.0	3509	—
IMR-4350	71.0	3322	—	77.0	3623	—
H-380	66.0	3164	—	73.0	3460	—
H-4831	73.0	3058	—	80.0+	3333	—

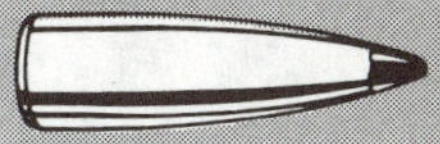

150 gr. Jacketed Ptd. SP

Powder	Sugg. Starting Grains	Velocity fps	Pressure C.U.P.	Max. Load Grains	Velocity fps	Pressure C.U.P.
IMR-3031	55.0	3012	—	61.0	3215	—
IMR-4064	59.0	3012	—	65.0	3268	—
IMR-4320	59.0	3039	—	66.5	3322	—
IMR-4895	60.0	3115	—	66.5	3390	—
IMR-4350	67.0	3086	—	75.0+	3427	—
H-380	64.0	3003	—	71.0	3268	—
H-4831	72.0	2923	—	80.0+	3236	—

165 gr. Jacketed Spitzer

Powder	Sugg. Starting Grains	Velocity fps	Pressure C.U.P.	Max. Load Grains	Velocity fps	Pressure C.U.P.
IMR-3031	52.0	2777	—	58.0	2994	—
IMR-4895	56.0	2865	—	62.0	3095	—
IMR-4064	56.0	2816	—	62.0	3076	—
IMR-4320	57.0	2840	—	63.0	3095	—
IMR-4350	63.0	2816	—	70.0	3144	—
H-380	61.0	2808	—	68.0	3105	—
H-4831	70.0	2793	—	77.0+	3095	—

180 gr. Jacketed Ptd. SP

Powder	Sugg. Starting Grains	Velocity fps	Pressure C.U.P.	Max. Load Grains	Velocity fps	Pressure C.U.P.
IMR-3031	50.0	2624	—	55.0	2808	—
IMR-4895	53.0	2672	—	59.5	2923	—
IMR-4064	54.0	2672	—	60.0	2941	—
IMR-4320	55.0	2739	—	61.0	2949	—
IMR-4350	60.0	2672	—	67.0	2985	—
H-380	59.0	2695	—	65.0	2923	—
H-4831	66.0	2624	—	73.0+	2949	—

Note: Loads shown in shaded panels are maximum.
\+ Designates a compressed powder charge.

300 H&H Magnum
(300 Holland and Holland)
(7.63 x 72mm)

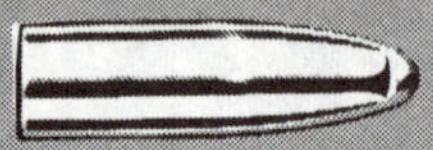

200 gr. Jacketed RN

Powder	Sugg. Starting Grains	Velocity fps	Pressure C.U.P.	Max. Load Grains	Velocity fps	Pressure C.U.P.
IMR-3031	48.0	2433	—	53.5	2652	—
IMR-4064	52.0	2481	—	58.0	2724	—
IMR-4895	52.0	2506	—	58.0	2754	—
IMR-4320	53.0	2531	—	59.5	2793	—
IMR-4350	59.0	2538	—	65.0	2808	—
H-380	57.0	2525	—	63.0	2777	—
H-4831	64.0	2506	—	71.0	2801	—

220 gr. Jacketed SP

Powder	Sugg. Starting Grains	Velocity fps	Pressure C.U.P.	Max. Load Grains	Velocity fps	Pressure C.U.P.
IMR-3031	47.0	2341	—	52.5	2557	—
IMR-4064	50.0	2369	—	56.5	2631	—
IMR-4895	51.0	2427	—	57.5	2710	—
IMR-4320	52.0	2450	—	58.5	2724	—
IMR-4350	57.0	2450	—	64.0	2732	—
H-380	56.0	2450	—	62.0	2698	—
H-4831	62.0	2398	—	69.0	2710	—

#311466
151 gr. (#2 Alloy) 3.345" OAL

Powder	Sugg. Starting Grains	Velocity fps	Pressure C.U.P.	Max. Load Grains	Velocity fps	Pressure C.U.P.
Red Dot	12.0	1490	—	15.0	1695	—
700X	12.0	1495	—	15.0	1700	—
Green Dot	13.0	1545	—	16.0	1735	—
PB	13.0	1485	—	16.5	1710	—
Unique	15.0	1665	—	18.0	1850	—
SR-7625	13.5	1500	—	17.0	1720	—

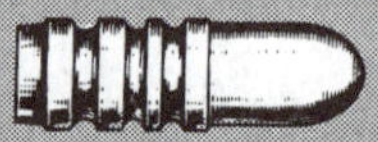

#311291
169 gr. (#2 Alloy) 3.335" OAL

Powder	Sugg. Starting Grains	Velocity fps	Pressure C.U.P.	Max. Load Grains	Velocity fps	Pressure C.U.P.
Red Dot	12.0	1400	—	15.0	1615	—
700X	12.0	1385	—	15.0	1590	—
Green Dot	13.0	1445	—	16.0	1640	—
PB	13.0	1380	—	16.5	1595	—
Unique	15.0	1570	—	18.0	1750	—
SR-7625	13.5	1395	—	17.0	1620	—

Note: Loads shown in shaded panels are maximum.

300 Winchester Magnum

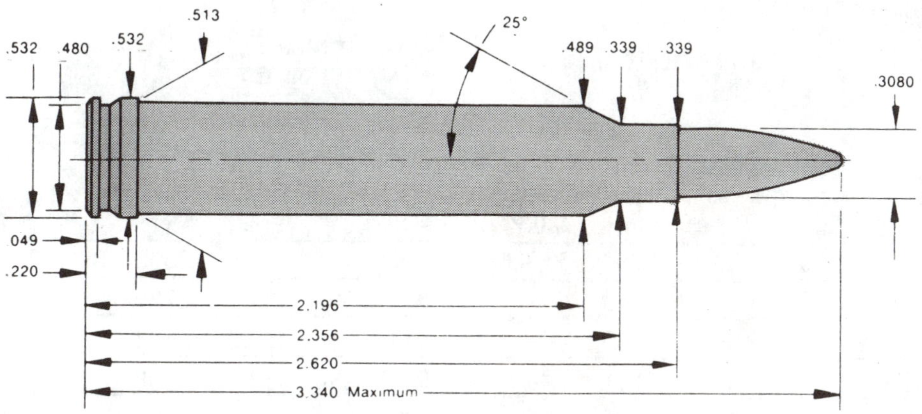

COMMENTS:

This cartridge will duplicate the ballistics of the 300 H&H Magnum. Its main advantage is that it will work in a standard length action whereas the H&H round requires a magnum length action.

This cartridge tends to give the best results with Winchester primers. Our data was developed using the standard Winchester primer and the use of a magnum primer will cause a pressure increase. When using magnum type primers work up loads cautiously and stay at least one grain below the maximum listed loads.

Favored powders should be Hodgdon H4831 and IMR 7828. Strongly constructed bullets of 180 to 220 grains usually give the best balance of internal ballistic uniformity and accuracy.

TEST COMPONENTS:

Cases .... Winchester
Trim-to Length .... 2.610"
Primers .... Winchester 8½-120
Primer Size .... Large Rifle
Lyman Shell Holder .... No. 13
Jacketed Bullets Used .... Remington Ptd. SP, 110 gr.
Hornady SP #3020, 130 gr.
Hornady SP #3031, 150 gr.
Nosler SP #16330, 165 gr.
Sierra HPBT #2220, 180 gr.
Barnes "X" #30835, 180 gr.
Sierra HPBT #2210, 190 gr.
Sierra HPBT #2230, 200 gr.
Hornady RN #3090, 220 gr.
Cast Bullets Used .... (Sized to .308" dia.)
*Gas Check Bullets *#311466, 151 gr.
*#311291, 169 gr.

TEST SPECIFICATIONS: (Velocity & Pressure)

Firearm Used .... Winchester Model 70 and Universal Receiver
Barrel Length .... 24"
Twist .... 1-10"
Groove Dia. .... .308"

110 gr. Jacketed Ptd. SP

Powder	Sugg. Starting Grains	Velocity fps	Pressure C.U.P.	Max. Load Grains	Velocity fps	Pressure C.U.P.
IMR-3031	59.0	3194	—	66.0	3546	—
IMR-4064	63.0	3174	—	70.0	3546	—
IMR-4895	62.0	3236	—	69.0	3521	—
IMR-4320	60.0	3067	—	68.0	3413	—
H-380	67.0	3144	—	74.0	3460	—
IMR-4350	73.0	3236	—	81.0+	3677	—
H-4831	75.0	3021	—	83.5+	3367	—

Note: Loads shown in shaded panels are maximum.
\+ Designates a compressed powder charge.

300 Winchester Magnum

130 gr. Jacketed SP
3.215" OAL

Powder	Sugg. Starting Grains	Velocity fps	Pressure C.U.P.	Max. Load Grains	Velocity fps	Pressure C.U.P.
RX7	47.1	2829	44,300	53.5	3096	53,900
IMR-3031	56.0	2976	—	63.0	3268	—
IMR-4064	61.0	2985	—	68.0	3300	—
IMR-4895	60.0	2976	—	67.0	3311	—
IMR-4320	60.0	3012	—	65.0	3215	—
H-380	65.0	2985	—	72.0	3215	—
H-414	66.3	3036	43,500	78.0	3401	53,900
IMR-4350	71.0	3086	—	79.0+	3497	—
H-4831	75.0	2932	—	83.0+	3290	—

150 gr. Jacketed SP
3.340" OAL

Powder	Sugg. Starting Grains	Velocity fps	Pressure C.U.P.	Max. Load Grains	Velocity fps	Pressure C.U.P.
RX7	42.5	2562	39,800	49.5	2840	53,700
IMR-3031	54.0	2785	—	60.0	3048	—
IMR-4064	59.0	2824	—	65.0	3086	—
IMR-4895	58.0	2808	—	65.0	3115	—
IMR-4320	56.0	2724	—	62.0	2967	—
H-380	63.0	2840	—	70.0	3105	—
H-414	59.8	2800	43,000	68.5	3111	52,200
IMR-4350	68.0	2890	—	‡76.0	3290	—
H-4831	73.0	2793	—	81.0+	3194	—
IMR-7828	75.5	2944	38,600	81.0+	3268	52,500

165 gr. Jacketed SP
3.340" OAL

Powder	Sugg. Starting Grains	Velocity fps	Pressure C.U.P.	Max. Load Grains	Velocity fps	Pressure C.U.P.
RX7	40.0	2401	35,600	47.0	2664	52,700
IMR-3031	52.0	2631	—	58.0	2852	—
IMR-4064	58.0	2724	—	64.0	2967	—
IMR-4895	56.0	2672	—	62.0	2906	—
IMR-4320	55.0	2631	—	61.0	2832	—
H-380	62.0	2747	—	69.0	2976	—
H-414	58.0	2700	42,700	69.2	3008	53,000
IMR-4350	67.0	2865	—	74.0	3194	—
H-4831	71.0	2724	—	79.0+	3086	—
RX19	70.0	2746	36,700	78.0	3144	50,100
AA3100	70.0	2669	37,500	77.5	3032	51,400
RX22	72.5	2786	37,700	79.5	3119	50,300
IMR-7828	74.0	2892	41,600	79.5+	3149	51,200

180 gr. Jacketed HPBT
3.340" OAL

Powder	Sugg. Starting Grains	Velocity fps	Pressure C.U.P.	Max. Load Grains	Velocity fps	Pressure C.U.P.
IMR-3031	50.0	2493	—	56.0	2717	—
IMR-4064	56.0	2590	—	62.5	2840	—
IMR-4895	54.0	2577	—	60.5	2793	—
IMR-4320	54.0	2518	—	60.5	2777	—
H-380	60.0	2610	—	67.0	2857	—
H-414	57.0	2572	39,000	66.0	2876	52,400
IMR-4350	65.0	2724	—	72.5	3048	—
H-4831	70.0	2666	—	77.0+	2958	—
RX19	68.0	2647	38,200	‡74.0	2962	49,700
AA3100	68.0	2562	37,200	75.5	2936	51,400
RX22	70.5	2696	37,800	77.0	3017	50,300
IMR-7828	70.7	2740	41,500	78.2+	3082	53,600

Note: Loads shown in shaded panels are maximum.
\+ Designates a compressed powder charge.
‡ Designates a factory velocity duplication load.

300 Winchester Magnum

180 gr. Barnes "X"
3.340" OAL

Powder	Sugg. Starting Grains	Velocity fps	Pressure C.U.P.	Max. Load Grains	Velocity fps	Pressure C.U.P.
IMR-3031	49.0	2417	38,800	57.0	2716	49,900
IMR-4895	50.0	2400	37,300	59.5	2756	50,400
IMR-4064	53.0	2435	37,800	62.0	2782	50,800
IMR-4320	52.5	2364	36,000	62.0	2771	51,300
H-380	54.5	2392	38,000	63.0	2679	50,300
IMR-4350	61.0	2509	36,700	68.0	2788	48,200
RX19	64.5	2587	37,000	72.5	2894	50,200
AA3100	64.0	2428	36,000	73.0	2794	50,600
H-4831	64.0	2567	39,700	74.0	2832	50,100
RX22	64.0	2575	38,000	73.5	2928	51,300
IMR-7828	67.0	2452	35,600	76.0	2864	50,200
H-414	56.5	2423	36,600	66.0	2736	48,900

190 gr. Jacketed HPBT
3.340" OAL

Powder	Sugg. Starting Grains	Velocity fps	Pressure C.U.P.	Max. Load Grains	Velocity fps	Pressure C.U.P.
RX7	41.3	2336	44,400	47.0	2514	53,600
IMR-3031	46.0	2347	36,400	51.5	2546	44,700
IMR-4064	49.6	2451	38,700	59.5	2774	52,600
IMR-4320	46.6	2343	36,100	53.0	2583	47,000
H-380	62.0	2550	35,400	70.0	2804	45,000
IMR-4350	59.4	2583	42,800	68.0	2880	53,700
H-4831	64.2	2522	37,600	73.0	2897	52,800
IMR-7828	69.5	2682	41,400	77.2+	3003	53,800

200 gr. Jacketed HPBT
3.340" OAL

Powder	Sugg. Starting Grains	Velocity fps	Pressure C.U.P.	Max. Load Grains	Velocity fps	Pressure C.U.P.
RX7	35.0	2064	35,500	45.5	2420	52,900
IMR-3031	49.0	2380	—	54.0	2564	—
IMR-4064	53.0	2439	—	59.0	2624	—
IMR-4895	52.0	2403	—	58.0	2631	—
IMR-4320	51.0	2369	—	57.0	2544	—
H-380	58.0	2469	—	64.0	2659	—
IMR-4350	62.0	2597	—	69.0	2849	—
H-4831	67.0	2531	—	74.0+	2816	—
RX19	64.5	2480	35,600	73.0	2877	51,800
AA3100	64.0	2392	36,000	72.5	2751	51,000
RX22	65.0	2489	35,800	74.5	2878	51,500
IMR-7828	67.0	2476	37,700	74.5	2828	51,800
H-1000	70.0	2442	36,000	80.0+	2798	49,900

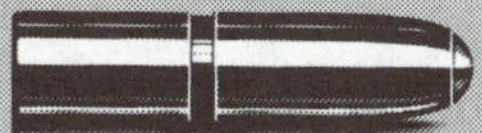

220 gr. Jacketed RN
3.340" OAL

Powder	Sugg. Starting Grains	Velocity fps	Pressure C.U.P.	Max. Load Grains	Velocity fps	Pressure C.U.P.
RX7	39.1	2113	43,600	44.5	2269	52,200
IMR-3031	47.0	2237	—	52.5	2439	—
IMR-4064	50.0	2298	—	56.5	2487	—
IMR-4895	50.0	2283	—	55.5	2475	—
IMR-4320	49.0	2227	—	54.0	2403	—
H-380	55.0	2331	—	61.0	2500	—
IMR-4350	60.0	2463	—	67.5	2747	—
H-4831	65.0	2415	—	‡72.0+	2688	—
RX19	64.0	2415	37,000	72.0	2730	50,900
AA3100	63.5	2312	36,500	71.5	2630	51,800
RX22	63.5	2374	36,700	72.5	2717	51,400
IMR-7828	66.0	2360	36,700	73.5	2664	51,000
H-1000	69.0	2337	36,800	78.0+	2653	49,400

Note: Loads shown in shaded panels are maximum.
\+ Designates a compressed powder charge.
‡ Designates a factory velocity duplication load.

300 Winchester Magnum

#311466
151 gr. (#2 Alloy) 3.185" OAL

Powder	Sugg. Starting Grains	Velocity fps	Pressure C.U.P.	Max. Load Grains	Velocity fps	Pressure C.U.P.
Red Dot	16.0	1725	29,400	19.5	1910	39,000
700X	15.0	1660	29,400	18.0	1840	38,600
Green Dot	16.0	1700	27,600	19.5	1890	35,500
PB	15.5	1605	26,400	18.5	1765	35,500
Unique	18.0	1790	26,400	22.0	2000	35,500
SR-7625	16.0	1620	28,200	19.0	1785	39,000

#311291
169 gr. (#2 Alloy) 3.135" OAL

Powder	Sugg. Starting Grains	Velocity fps	Pressure C.U.P.	Max. Load Grains	Velocity fps	Pressure C.U.P.
Red Dot	16.0	1640	31,000	19.0	1790	40,700
700X	14.0	1515	28,800	17.0	1690	39,400
Green Dot	16.0	1620	30,000	19.0	1775	38,600
PB	15.0	1495	28,200	18.5	1660	36,900
Unique	18.0	1705	29,400	22.0	1910	38,600
SR-7625	15.0	1480	27,600	18.0	1650	36,000
SR-4759	30.5	2075	28,400	40.0	2472	50,900
RX7	37.0	2184	27,200	48.0	2575	46,100
748	45.0	2176	25,400	60.0	2729	46,400
H-4895	40.5	2157	25,900	56.0	2748	51,200
IMR-4064	42.0	2073	21,100	60.0	2810	45,900

Note: Loads shown in shaded panels are maximum.

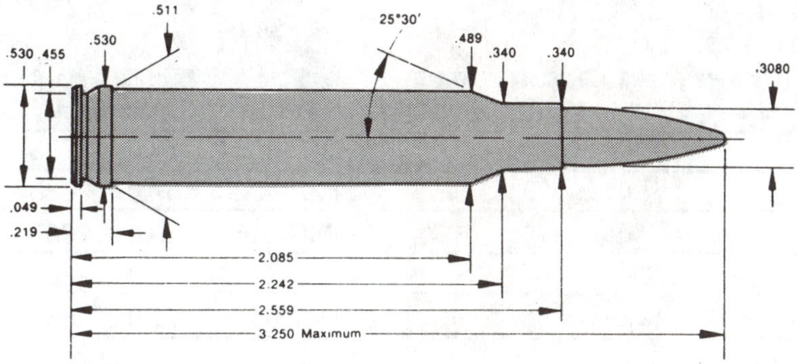

COMMENTS:

The test rifle had a considerable amount of free bore. Therefore this data may not be suitable in a custom chamber which lacks any equivalent amount of free bore (some custom chambers have no free bore at all).

We suggest that only standard large rifle primers be used with cast bullets.

Of the listed propellants Hodgdon H4831 is the single best choice.

TEST COMPONENTS:

Cases .... Norma
Trim-to Length .... 2.550"
Primers .... Federal 215 and Remington 9½
Primer Size .... Large Rifle
Lyman Shell Holder .... No. 13
Jacketed Bullets Used .... Remington Ptd. SP, 110 gr.
Hornady Spire Pt., 130 gr.
Remington Ptd. SPCL, 150 gr.
Speer Spire Pt., 165 gr.
Remington Ptd. SPCL, 180 gr.
Speer RN, 200 gr.
Remington SPCL, 220 gr.
Cast Bullets Used .... (Sized to .308" dia.)
*Gas Check Bullets .... *#311466, 151 gr.
*#311291, 169 gr.

TEST SPECIFICATIONS: (Velocity Only)

Firearm Used .... Schultz & Larsen 65 DL
Barrel Length .... 24"
Twist .... 1-10"
Groove Dia. .... .308"

110 gr. Jacketed Ptd. SP

Powder	Sugg. Starting Grains	Velocity fps	Pressure C.U.P.	Max. Load Grains	Velocity fps	Pressure C.U.P.
IMR-4064	59.0	2890	—	66.0	3413	—
IMR-4320	61.0	2958	—	68.0	3448	—
IMR-4350	71.0	3164	—	79.0+	3584	—
H-380	67.0	3067	—	74.0	3460	—
H-4831	72.0	2840	—	80.5+	3225	—

Note: Loads shown in shaded panels are maximum.
\+ Designates a compressed powder charge.

308 Norma Magnum

130 gr. Jacketed Spire Pt.

Powder	Sugg. Starting Grains	Velocity fps	Pressure C.U.P.	Max. Load Grains	Velocity fps	Pressure C.U.P.
IMR-4064	59.0	2865	—	65.0	3268	—
IMR-4320	60.0	2873	—	67.0	3300	—
IMR-4350	70.0	3095	—	78.0+	3484	—
H-380	66.0	3030	—	73.0	3345	—
H-4831	72.0	2859	—	80.5+	3184	—

150 gr. Jacketed Ptd. SP

Powder	Sugg. Starting Grains	Velocity fps	Pressure C.U.P.	Max. Load Grains	Velocity fps	Pressure C.U.P.
IMR-4064	58.0	2793	—	64.0	3164	—
IMR-4320	57.0	2688	—	65.0	3134	—
IMR-4350	68.0	2967	—	75.0	3345	—
H-380	63.0	2801	—	70.0	3144	—
H-4831	72.0	2881	—	80.5+	3184	—

165 gr. Jacketed Spire Pt.

Powder	Sugg. Starting Grains	Velocity fps	Pressure C.U.P.	Max. Load Grains	Velocity fps	Pressure C.U.P.
IMR-4064	56.0	2652	—	62.0	2985	—
IMR-4320	57.0	2680	—	63.0	2976	—
IMR-4350	66.0	2857	—	73.0	3194	—
H-380	61.0	2688	—	68.0	3030	—
H-4831	71.0	2762	—	79.0+	3105	—

180 gr. Jacketed Ptd. SP

Powder	Sugg. Starting Grains	Velocity fps	Pressure C.U.P.	Max. Load Grains	Velocity fps	Pressure C.U.P.
IMR-4064	53.0	2487	—	59.0	2793	—
IMR-4320	54.0	2487	—	60.0	2777	—
IMR-4350	64.0	2702	—	71.0	3067	—
H-380	59.0	2551	—	66.0	2873	—
H-4831	69.0	2652	—	77.0+	2994	—

200 gr. Jacketed RN

Powder	Sugg. Starting Grains	Velocity fps	Pressure C.U.P.	Max. Load Grains	Velocity fps	Pressure C.U.P.
IMR-4064	52.0	2392	—	58.0	2710	—
IMR-4320	53.0	2386	—	59.0	2688	—
IMR-4350	63.0	2659	—	70.0	2941	—
H-380	58.0	2506	—	64.0	2732	—
H-4831	68.0	2597	—	75.0+	2865	—

220 gr. Jacketed SP

Powder	Sugg. Starting Grains	Velocity fps	Pressure C.U.P.	Max. Load Grains	Velocity fps	Pressure C.U.P.
IMR-4064	50.0	2272	—	56.0	2570	—
IMR-4320	51.0	2252	—	57.0	2570	—
IMR-4350	61.0	2525	—	68.0+	2808	—
H-380	57.0	2427	—	63.0	2659	—
H-4831	66.0	2500	—	73.0+	2732	—

#311466**
151 gr. (#2 Alloy)

Powder	Sugg. Starting Grains	Velocity fps	Pressure C.U.P.	Max. Load Grains	Velocity fps	Pressure C.U.P.
Unique	15.0	1700	—	18.0	1893	—
2400	24.0	1845	—	27.0	2012	—
IMR-4227	24.0	1715	—	28.0	1949	—

#311291**
169 gr. (#2 Alloy)

Powder	Sugg. Starting Grains	Velocity fps	Pressure C.U.P.	Max. Load Grains	Velocity fps	Pressure C.U.P.
Unique	14.0	1569	—	17.0	1763	—
2400	22.0	1715	—	26.0	1919	—
IMR-4227	22.0	1600	—	27.0	1890	—

Note: Loads shown in shaded panels are maximum.
+ Designates a compressed powder charge.
** Indicates use of Rem. 9½ primers.

300 Weatherby Magnum

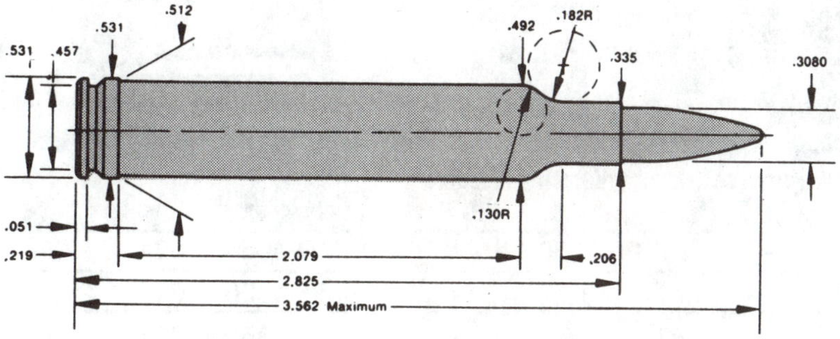

COMMENTS:

This cartridge uses a very large capacity case. Due to the relatively huge quantities of propellant being burned in a relatively small bore, barrels suffer from premature erosion, therefore short barrel life.

The use of a magnum primer is suggested to ensure proper ignition of the heavy powder charges used with jacketed bullets. However, use only standard primers with the listed cast bullet loads.

The listed data is for use in commercially made and chambered rifles. Custom chamber dimensions may be incompatible with the data to the point of unnecessary risk.

Of the listed propellants, Hodgdon H4831 is the best choice.

TEST COMPONENTS:

Cases .... Weatherby
Trim-to Length .... 2.815"
Primers .... Federal 215 and Remington 9½
Primer Size .... Large Rifle
Lyman Shell Holder .... No. 13
Jacketed Bullets Used .... Remington Ptd. SP, 110 gr.
Hornady Spire Pt., 130 gr.
Remington Ptd. SPCL, 150 gr.
Speer Spire Pt., 165 gr.
Remington Ptd. SPCL, 180 gr.
Speer RN, 200 gr.
Remington SPCL, 220 gr.
Cast Bullets Used .... (Sized to .308" dia.)
*Gas Check Bullets .... *#311291, 169 gr.

TEST SPECIFICATIONS: (Velocity Only)

Firearm Used .... Weatherby Mark V
Barrel Length .... 26"
Twist .... 1-12"
Groove Dia. .... .3075"

110 gr. Jacketed Ptd. SP

Powder	Sugg. Starting Grains	Velocity fps	Pressure C.U.P.	Max. Load Grains	Velocity fps	Pressure C.U.P.
IMR-4064	71.0	3360	—	79.0	3816	—
IMR-4320	74.0	3559	—	82.0	3937	—
IMR-4350	81.0	3440	—	90.0+	3906	—
H-380	78.0	3472	—	86.0	3831	—
H-4831	85.0	3205	—	94.0+	3559	—

130 gr. Jacketed Spire Pt.

Powder	Sugg. Starting Grains	Velocity fps	Pressure C.U.P.	Max. Load Grains	Velocity fps	Pressure C.U.P.
IMR-4064	66.0	3003	—	73.0	3390	—
IMR-4320	68.0	3144	—	76.0	3521	—
IMR-4350	77.0	3144	—	86.0	3636	—
H-380	73.0	3154	—	81.0	3521	—
H-4831	83.0	3076	—	92.0+	3509	—

Note: Loads shown in shaded panels are maximum.
\+ Designates a compressed powder charge.

300 Weatherby Magnum

150 gr. Jacketed Ptd. SP

Powder	Sugg. Starting Grains	Velocity fps	Pressure C.U.P.	Max. Load Grains	Velocity fps	Pressure C.U.P.
IMR-4064	64.0	2865	—	71.0	3230	—
IMR-4320	67.0	3076	—	74.0	3367	—
IMR-4350	75.0	3058	—	82.0	3401	—
H-380	70.0	3039	—	78.0	3322	—
H-4831	79.0	2849	—	88.0+	3333	—

165 gr. Jacketed Spire Pt.

Powder	Sugg. Starting Grains	Velocity fps	Pressure C.U.P.	Max. Load Grains	Velocity fps	Pressure C.U.P.
IMR-4064	62.0	2724	—	69.0	3003	—
IMR-4320	64.0	2857	—	71.0	3174	—
IMR-4350	72.0	2900	—	80.0	3285	—
H-380	69.0	2967	—	76.0	3215	—
H-4831	76.0	2754	—	85.0	3184	—

180 gr. Jacketed Ptd. SP

Powder	Sugg. Starting Grains	Velocity fps	Pressure C.U.P.	Max. Load Grains	Velocity fps	Pressure C.U.P.
IMR-4064	61.0	2604	—	68.0	2949	—
IMR-4320	63.0	2770	—	70.0	3048	—
IMR-4350	71.0	2793	—	79.0	3194	—
H-380	68.0	2577	—	75.0	3076	—
H-4831	74.0	2645	—	82.0	3067	—

200 gr. Jacketed RN

Powder	Sugg. Starting Grains	Velocity fps	Pressure C.U.P.	Max. Load Grains	Velocity fps	Pressure C.U.P.
IMR-4064	60.0	2750	—	66.0	2793	—
IMR-4320	60.0	2638	—	67.0	2865	—
IMR-4350	69.0	2676	—	76.0	3008	—
H-380	67.0	2724	—	73.0	2936	—
H-4831	71.0	2500	—	79.0	2865	—

220 gr. Jacketed SP

Powder	Sugg. Starting Grains	Velocity fps	Pressure C.U.P.	Max. Load Grains	Velocity fps	Pressure C.U.P.
IMR-4064	56.0	2369	—	62.0	2590	—
IMR-4320	60.0	2531	—	65.0	2717	—
IMR-4350	66.0	2506	—	73.0	2808	—
H-380	62.0	2518	—	69.0	2739	—
H-4831	69.0	2375	—	77.0	2754	—

#311291**
169 gr. (#2 Alloy)

Powder	Sugg. Starting Grains	Velocity fps	Pressure C.U.P.	Max. Load Grains	Velocity fps	Pressure C.U.P.
Unique	15.0	1572	—	19.0	1818	—
2400	23.0	1736	—	28.0	1976	—

Note: Loads shown in shaded panels are maximum.
\+ Designates a compressed powder charge.
** Indicates use of Rem. 9½ primers.

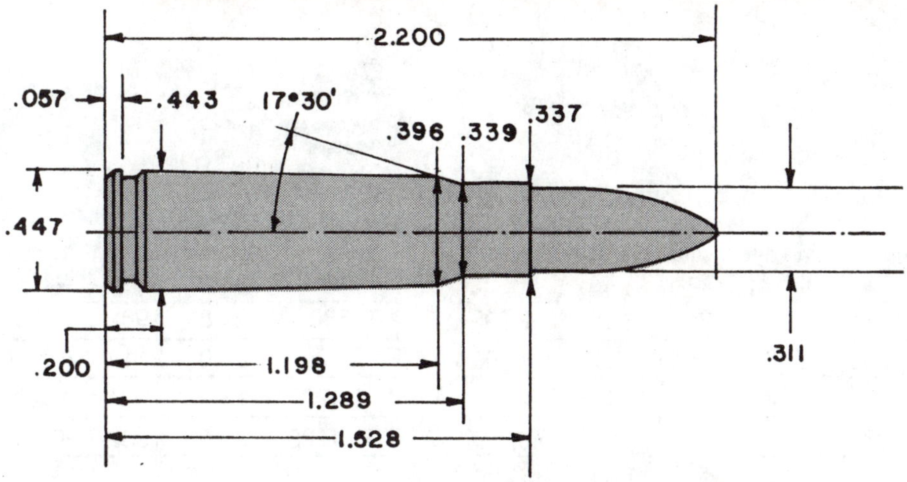

COMMENTS:

The Ruger Mini 30 was the final stamp of popularity for this cartridge which was originally designed for military style rifles. The Ruger rifles have used barrels with 0.308" groove diameters and are best loaded with the same diameter bullet. However, the standard bullet diameter for this cartridge, as loaded by most ammunition companies throughout the world, is 0.311". This is the proper diameter for the great quantity of SKS and AK47 rifles that are available in the surplus rifle market. We suggest slugging the bore on these rifles to verify their true bore size.

The 125 grain bullet weight seems ideal and a good first propellant choice would be Hercules Reloder 7. However many semi-auto firearms chambered for this cartridge are at their best when 5-shot groups at 100 yards get near 3 inches.

TEST COMPONENTS:

Cases	Winchester
Trim-to Length	1.515"
Primers	Winchester WLR
Primer Size	Large Rifle
Lyman Shell Holder	No. 3
Jacketed Bullets Used	Sierra HP #2110, 110 gr.
	Sierra SP #2120, 125 gr.
	Hornady SP #3020, 130 gr.
Cast Bullets Used	(Sized to .310" dia.)
*Gas Check Bullets	*#311359, 113 gr.
	#311410, 130 gr.

TEST SPECIFICATIONS:
(Velocity & Pressure)

Firearm Used	Universal Receiver
Barrel Length	20"
Twist	1-9½"
Groove Dia.	.310"

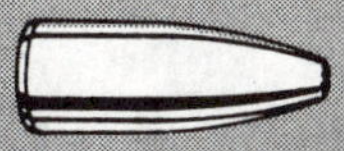

110 gr. Jacketed HP
2.200" OAL

Powder	Sugg. Starting Grains	Velocity fps	Pressure C.U.P.	Max. Load Grains	Velocity fps	Pressure C.U.P.
AA1680	24.0	2245	29,400	26.7	2533	42,800
RX7	27.0	2267	29,900	30.0	2501	35,900
IMR-4227	20.0	2100	28,000	22.8	2421	42,800
IMR-4198	22.6	2115	28,700	25.2	2416	40,200

Note: Loads shown in shaded panels are maximum.

7.62 x 39mm Russian

125 gr. Jacketed SP
2.200" OAL

Powder	Sugg. Starting Grains	Velocity fps	Pressure C.U.P.	Max. Load Grains	Velocity fps	Pressure C.U.P.
AA1680	22.8	2126	29,300	25.5	2383	42,200
RX7	26.5	2265	33,200	30.0+	2528	41,300
IMR-4227	19.0	1974	30,800	21.6	2264	42,700
IMR-4198	21.4	1980	27,300	24.2	2305	40,800

130 gr. Jacketed SP
2.200" OAL

Powder	Sugg. Starting Grains	Velocity fps	Pressure C.U.P.	Max. Load Grains	Velocity fps	Pressure C.U.P.
AA1680	21.8	1962	25,500	23.8	2259	40,100
RX7	25.0	2114	30,600	29.0+	2450	42,000
IMR-4227	18.3	1924	29,300	20.8	2195	42,500
IMR-4198	20.5	1932	28,200	23.4	2275	43,400

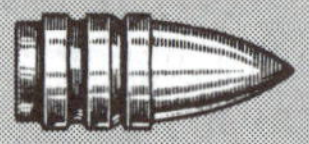

#311359
113 gr. (#2 Alloy) 2.060" OAL

Powder	Sugg. Starting Grains	Velocity fps	Pressure C.U.P.	Max. Load Grains	Velocity fps	Pressure C.U.P.
SR-4759	16.7	1978	29,000	18.7	2186	40,200
RX7	25.0	2189	28,100	30.0+	2561	41,100
AA2230	26.0	2092	32,000	32.0+	2445	44,200
Unique	8.0	1594	25,500	10.0	1816	36,500
IMR-4227	17.0	1906	25,800	19.2	2142	35,700
IMR-4198	19.8	1968	25,800	22.6	2273	39,000

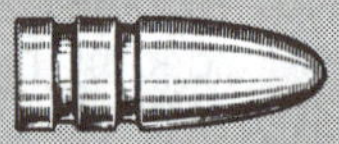

#311410
130 gr. (#2 Alloy) 2.200" OAL

Powder	Sugg. Starting Grains	Velocity fps	Pressure C.U.P.	Max. Load Grains	Velocity fps	Pressure C.U.P.
SR-4759	16.5	1935	29,900	19.0	2139	37,600
RX7	25.0	2164	30,600	29.0+	2442	38,800
AA2230	25.0	2015	30,700	30.0+	2280	39,600
Unique	8.0	1552	23,500	10.3	1786	36,900
IMR-4227	17.5	1883	25,800	19.5	2097	35,900
IMR-4198	20.0	1897	25,500	22.0	2134	34,200

Note: Loads shown in shaded panels are maximum.
\+ Designates a compressed powder charge.

7.62 x 54R Russian

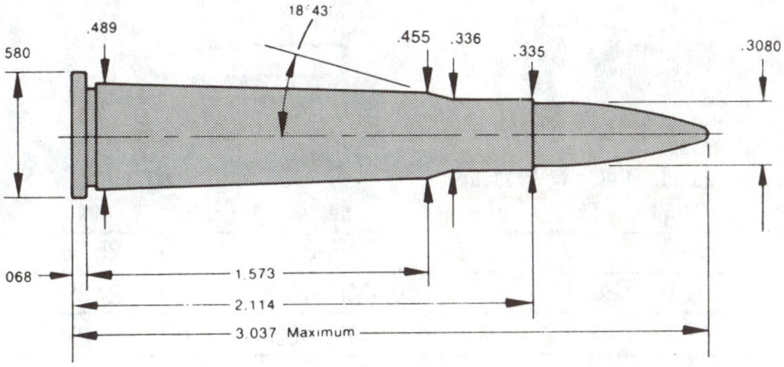

COMMENTS:

This is a large rimmed cartridge that should not be confused with the smaller and far more popular rimless 7.62 x 39 Russian. Bore and groove diameters of examined rifles have varied widely — from 0.308" to 0.312". For barrels with groove diameters no larger than 0.310" we suggest the use of 0.308" diameter bullets. Due to limitations of chamber size, larger diameter bullets are not suggested even though groove diameters are larger than 0.310 inch.

IMR 4350 was the best propellant in our tests with jacketed bullets. However, because of the groove diameter of the test rifle being 0.002" larger than the bullets used, accuracy was fair, at best.

TEST COMPONENTS:

Cases .... Norma
Trim-to Length .... 2.105"
Primers .... Remington 9½
Primer Size .... Large Rifle
Lyman Shell Holder .... No. 17
Jacketed Bullets Used .... Remington Ptd. SP, 110 gr.
Hornady Spire Pt., 130 gr.
Remington Ptd. SPCL, 150 gr.
Remington Ptd. SPCL, 180 gr.
Speer RN, 200 gr.
Remington SPCL, 220 gr.
Cast Bullets Used .... (Sized to .310" dia.)
*Gas Check Bullets .... *#311466, 151 gr.
*#311291, 169 gr.

TEST SPECIFICATIONS: (Velocity Only)

Firearm Used .... Westinghouse/Russian
Barrel Length .... 24"
Twist .... 1-10"
Groove Dia. .... .310"

110 gr. Jacketed Ptd. SP

Powder	Sugg. Starting Grains	Velocity fps	Pressure C.U.P.	Max. Load Grains	Velocity fps	Pressure C.U.P.
IMR-3031	45.0	2801	—	50.0	3134	—
IMR-4895	47.0	2724	—	52.0	3067	—
IMR-4064	49.0	2793	—	54.0+	3105	—
IMR-4320	50.0	2816	—	55.0	3105	—
IMR-4350	50.0	2352	—	56.0+	2710	—
H-4831	51.0	2105	—	57.0+	2444	—

Note: Loads shown in shaded panels are maximum.
+ Designates a compressed powder charge.

7.62 x 54R Russian

130 gr. Jacketed Spire Pt.

Powder	Sugg. Starting Grains	Velocity fps	Pressure C.U.P.	Max. Load Grains	Velocity fps	Pressure C.U.P.
IMR-3031	43.0	2638	—	48.0	2949	—
IMR-4895	45.0	2557	—	50.0	2881	—
IMR-4064	47.0	2631	—	52.0+	2967	—
IMR-4320	49.0	2702	—	54.0	3012	—
IMR-4350	50.0	2398	—	55.0+	2652	—
H-4831	51.0	2096	—	57.0+	2427	—

150 gr. Jacketed Ptd. SP

Powder	Sugg. Starting Grains	Velocity fps	Pressure C.U.P.	Max. Load Grains	Velocity fps	Pressure C.U.P.
IMR-3031	41.0	2518	—	46.0	2785	—
IMR-4895	43.0	2469	—	48.0	2762	—
IMR-4064	44.0	2427	—	49.0	2793	—
IMR-4320	46.0	2518	—	51.0	2849	—
IMR-4350	50.0	2439	—	55.0+	2695	—
H-4831	51.0	2155	—	57.0+	2457	—

180 gr. Jacketed Ptd. SP

Powder	Sugg. Starting Grains	Velocity fps	Pressure C.U.P.	Max. Load Grains	Velocity fps	Pressure C.U.P.
IMR-3031	39.0	2257	—	43.0	2506	—
IMR-4895	41.0	2277	—	45.0	2493	—
IMR-4064	43.0	2341	—	48.0	2659	—
IMR-4320	45.0	2409	—	50.0	2724	—
IMR-4350	50.0	2433	—	55.0+	2666	—
H-4831	51.0	2227	—	57.0+	2475	—

200 gr. Jacketed RN

Powder	Sugg. Starting Grains	Velocity fps	Pressure C.U.P.	Max. Load Grains	Velocity fps	Pressure C.U.P.
IMR-3031	37.0	2123	—	41.0	2358	—
IMR-4895	39.0	2136	—	43.0	2358	—
IMR-4064	41.0	2188	—	46.0	2493	—
IMR-4320	43.0	2298	—	48.0	2557	—
IMR-4350	48.0	2288	—	53.0+	2518	—
H-4831	50.0	2114	—	56.0+	2392	—

Note: Loads shown in shaded panels are maximum.
\+ Designates a compressed powder charge.

7.62 x 54R Russian

220 gr. Jacketed SP

Powder	Sugg. Starting Grains	Velocity fps	Pressure C.U.P.	Max. Load Grains	Velocity fps	Pressure C.U.P.
IMR-3031	35.0	1960	—	39.0	2188	—
IMR-4895	37.0	1949	—	41.0	2217	—
IMR-4064	40.0	2096	—	44.0	2309	—
IMR-4320	41.0	2127	—	46.0	2403	—
IMR-4350	46.0	2141	—	51.0+	2369	—
H-4831	50.0	2053	—	55.0+	2309	—

#311466
151 gr. (#2 Alloy)

Powder	Sugg. Starting Grains	Velocity fps	Pressure C.U.P.	Max. Load Grains	Velocity fps	Pressure C.U.P.
Unique	12.0	1592	—	15.0	1827	—
2400	23.5	1980	—	27.0	2197	—

#311291
169 gr. (#2 Alloy)

Powder	Sugg. Starting Grains	Velocity fps	Pressure C.U.P.	Max. Load Grains	Velocity fps	Pressure C.U.P.
Unique	11.0	1461	—	14.0	1680	—
2400	22.0	1865	—	24.0	1941	—

Note: Loads shown in shaded panels are maximum.
+ Designates a compressed powder charge.

7.65 Argentine Mauser
(7.65 x 53mm)

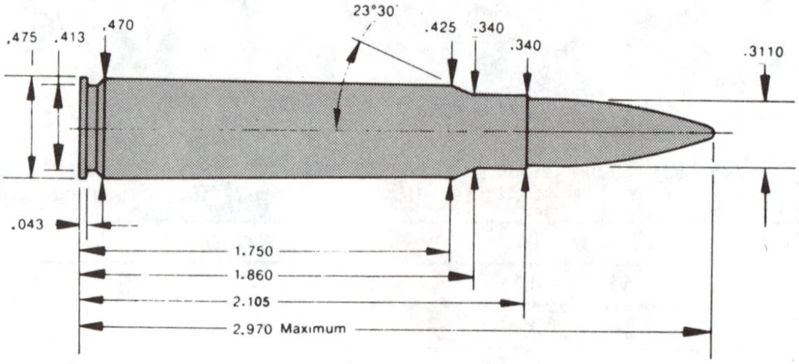

COMMENTS:

As with many types of foreign rifles, both military and commercial, internal barrel diameters vary widely for rifles chambered for this round. Slug your bore to determine if it is compatible with the nominal bullet diameter of 0.311 inch.

In barrels having a groove diameter of 0.311" cast bullet #311284 gave good results when shot as cast (0.311" diameter).

TEST COMPONENTS:

Cases .... Norma
Trim-to Length .... 2.100"
Primers .... Remington 9½ and Winchester 8½-120
Primer Size .... Large Rifle
Lyman Shell Holder .... No. 2
Jacketed Bullets Used .... Speer Spitzer, 150 gr.
Remington SPCL, 180 gr.
Remington SP, 215 gr.
Cast Bullets Used .... (Sized to .312" dia.)
*Gas Check Bullets .... *#311466, 151 gr.
*#311284, 210 gr.

TEST SPECIFICATIONS: (Velocity Only)

Firearm Used .... Argentine Mauser
Barrel Length .... 29¼"
Twist .... 1-9.8"
Groove Dia. .... .312"

150 gr. Jacketed Spitzer

Powder	Sugg. Starting Grains	Velocity fps	Pressure C.U.P.	Max. Load Grains	Velocity fps	Pressure C.U.P.
IMR-3031	40.0	2512	—	44.0	2739	—
IMR-4064	41.0	2331	—	46.0	2710	—
IMR-4895	41.0	2392	—	46.0	2732	—
IMR-4320	43.0	2457	—	48.0	2747	—
IMR-4350	45.0	2197	—	50.0+	2512	—
H-380	46.0	2531	—	51.0	2747	—
H-4831	48.0	2109	—	53.0+	2380	—

Note: Loads shown in shaded panels are maximum.
+ Designates a compressed powder charge.

7.65 Argentine Mauser

(7.65 x 53mm)

180 gr. Jacketed SP

Powder	Sugg. Starting Grains	Velocity fps	Pressure C.U.P.	Max. Load Grains	Velocity fps	Pressure C.U.P.
IMR-3031	38.0	2288	—	42.0	2531	—
IMR-4064	40.0	2298	—	44.0	2551	—
IMR-4895	40.0	2304	—	44.0	2551	—
IMR-4320	42.0	2040	—	47.0	2666	—
IMR-4350	44.0	2150	—	49.0+	2463	—
H-380	45.0	2364	—	50.0+	2544	—
H-4831	47.0	2049	—	52.0+	2320	—

215 gr. Jacketed SP

Powder	Sugg. Starting Grains	Velocity fps	Pressure C.U.P.	Max. Load Grains	Velocity fps	Pressure C.U.P.
IMR-3031	34.0	2000	—	38.0	2178	—
IMR-4064	36.0	1988	—	40.0+	2197	—
IMR-4895	36.0	2036	—	40.0	2227	—
IMR-4320	40.0	2159	—	44.0	2336	—
IMR-4350	41.0	1960	—	46.0+	2212	—
H-380	42.0	2092	—	47.0	2304	—
H-4831	46.0	1988	—	51.0+	2188	—

#311466**
151 gr. (#2 Alloy) 2.680" OAL

Powder	Sugg. Starting Grains	Velocity fps	Pressure C.U.P.	Max. Load Grains	Velocity fps	Pressure C.U.P.
Red Dot	10.0	1475	—	14.0	1780	—
700X	10.0	1475	—	14.0	1785	—
Green Dot	11.0	1530	—	14.5	1795	—
PB	11.0	1470	—	15.0	1770	—
Unique	12.0	1590	—	15.0	1820	—
SR-7625	11.5	1485	—	15.5	1775	—

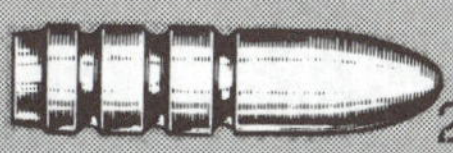

#311284**
210 gr. (#2 Alloy) 2.855" OAL

Powder	Sugg. Starting Grains	Velocity fps	Pressure C.U.P.	Max. Load Grains	Velocity fps	Pressure C.U.P.
Red Dot	9.0	1190	—	12.0	1415	—
700X	9.0	1190	—	11.0	1345	—
Green Dot	9.5	1215	—	12.0	1400	—
PB	9.5	1160	—	12.0	1340	—
Unique	9.5	1200	—	13.0	1445	—
SR-7625	10.0	1195	—	12.5	1365	—

Note: Loads shown in shaded panels are maximum.
\+ Designates a compressed powder charge.
** Indicates use of Win. 8½-120 primers.

303 British

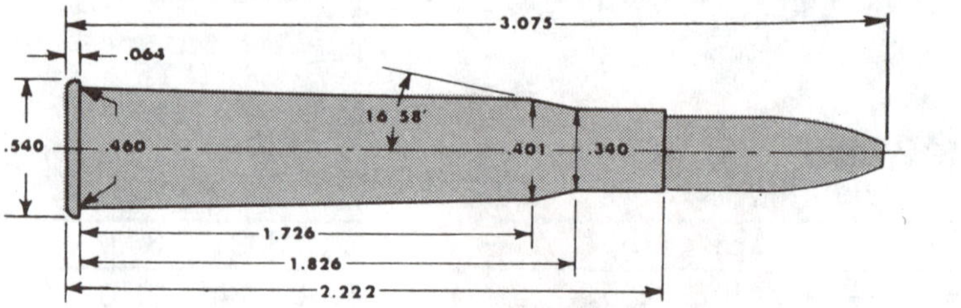

COMMENTS:

An extreme variation in groove diameters occurs in rifles chambered for this cartridge. Samples have been examined with groove measurements ranging from 0.308 to 0.317 inch. Naturally results will be poor if the barrel dimensions are incompatible with the jacketed or cast bullet diameter used.

Nominal bullet diameter is 0.312" with an honest plus or minus 0.001" being acceptable. Because of chamber limitations no attempt to use bullets larger than 0.313" should be made. A good cast bullet choice is #314299.

TEST COMPONENTS:

Cases .... Remington and Federal
Trim-to Length .... 2.212"
Primers .... Remington 9½ and Federal 210
Primer Size .... Large Rifle
Lyman Shell Holder .... No. 7
Jacketed Bullets Used .... Speer Spitzer, 150 gr.
Remington SPCL, 180 gr.
Remington SP, 215 gr.
Cast Bullets Used .... (Sized to .312" dia.)
*Gas Check Bullets *#311466, 151 gr.
*#314299, 200 gr.
*#311284, 210 gr.

TEST SPECIFICATIONS: (Velocity Only)

Firearm Used .... S.M.L.E. Mark III and Martini, Single Shot
Barrel Length .... S.M.L.E. 25"; Martini 22"
Twist .... 1-10"
Groove Dia. .... .312"

150 gr. Jacketed Spitzer

Powder	Sugg. Starting Grains	Velocity fps	Pressure C.U.P.	Max. Load Grains	Velocity fps	Pressure C.U.P.
IMR-3031	38.0	2288	—	42.0	2597	—
IMR-4064	40.0	2304	—	44.0	2577	—
IMR-4895	41.0	2409	—	45.0	2666	—
IMR-4320	41.0	2331	—	46.0	2672	—
H-380	42.0	2309	—	47.0	2570	—

180 gr. Jacketed SP

Powder	Sugg. Starting Grains	Velocity fps	Pressure C.U.P.	Max. Load Grains	Velocity fps	Pressure C.U.P.
IMR-3031	35.0	2044	—	39.0	2320	—
IMR-4064	38.0	2127	—	42.0	2386	—
IMR-4895	39.0	2207	—	43.0	2469	—
IMR-4320	40.0	2207	—	‡44.0	2469	—
H-380	41.0	2127	—	45.0	2369	—

Note: Loads shown in shaded panels are maximum.
‡ Designates a factory velocity duplication load.

303 British

215 gr. Jacketed SP

Powder	Sugg. Starting Grains	Velocity fps	Pressure C.U.P.	Max. Load Grains	Velocity fps	Pressure C.U.P.
IMR-3031	32.0	1840	—	36.0	2057	—
IMR-4064	35.0	1698	—	39.0	1996	—
IMR-4895	36.0	2000	—	40.0	2217	—
IMR-4320	37.0	2008	—	41.0	2217	—
H-380	40.0	2028	—	43.0	2150	—

#311466**
151 gr. (#2 Alloy) 2.788" OAL

Powder	Sugg. Starting Grains	Velocity fps	Pressure C.U.P.	Max. Load Grains	Velocity fps	Pressure C.U.P.
Red Dot	10.0	1430	—	13.5	1695	—
700X	10.0	1445	—	13.0	1660	—
Green Dot	10.5	1460	—	14.0	1715	—
PB	10.5	1410	—	13.0	1585	—
Unique	11.0	1475	—	15.0	1760	—
SR-7625	11.0	1425	—	13.5	1610	—

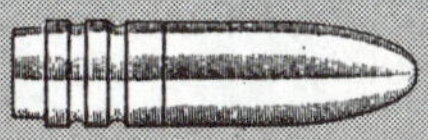

#314299**
200 gr. (#2 Alloy) 2.930" OAL

Powder	Sugg. Starting Grains	Velocity fps	Pressure C.U.P.	Max. Load Grains	Velocity fps	Pressure C.U.P.
Red Dot	10.0	1290	—	13.0	1490	—
700X	9.5	1240	—	12.5	1460	—
Green Dot	10.0	1280	—	13.5	1515	—
PB	10.0	1220	—	12.5	1390	—
Unique	10.5	1295	—	14.0	1530	—
SR-7625	10.5	1250	—	12.5	1380	—

#311284**
210 gr. (#2 Alloy) 3.005" OAL

Powder	Sugg. Starting Grains	Velocity fps	Pressure C.U.P.	Max. Load Grains	Velocity fps	Pressure C.U.P.
Red Dot	10.0	1245	—	12.5	1415	—
700X	9.0	1165	—	12.0	1375	—
Green Dot	9.5	1190	—	13.0	1425	—
PB	9.5	1130	—	12.0	1310	—
Unique	10.0	1210	—	13.5	1450	—
SR-7625	10.0	1170	—	12.0	1300	—

Note: Loads shown in shaded panels are maximum.
** Indicates use of Fed. 210 primers and Fed. cases.

32-20 Winchester
(32 WCF)

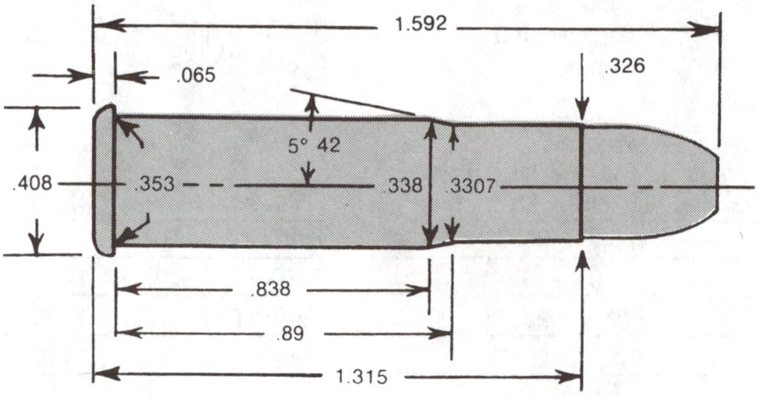

COMMENTS:

This was a dying cartridge until recent new lever action models, such as the Marlin 1894, gave it new life. This data is meant for this new crop of firearms and not older ones which may have deteriorated over the years. NEVER USE THIS DATA IN ANY FIREARMS DESIGNED FOR BLACK POWDER OR IN HANDGUNS.

Often factory ammo is encountered with cases shorter than the trim to length. These will cause no undue problems if they are segregated and crimping dies are adjusted to suit their unique length.

Ammo loaded for use in tubular magazines should only be loaded with blunt or flat nosed bullets for safety.

TEST COMPONENTS:

Cases .... Winchester
Trim-to Length .... 1.305"
Primers .... Remington 6$^1/_2$ and Winchester 6$^1/_2$-116
Primer Size .... Small Rifle
Lyman Shell Holder .... No. 10
Jacketed Bullets Used .... Remington SP, 100 gr.
Cast Bullets Used .... (Sized to .313" dia.)
*Gas Check Bullets .... *#311316, 115 gr.

TEST SPECIFICATIONS:
(Velocity Only)

Firearm Used .... Savage Sporter and Marlin 27S
Barrel Length .... 24"
Twist .... 1-22"
Groove Dia. .... .313"

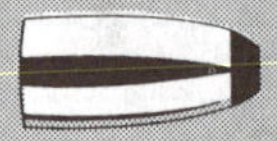

100 gr. Jacketed SP

Powder	Sugg. Starting Grains	Velocity fps	Pressure C.U.P.	Max. Load Grains	Velocity fps	Pressure C.U.P.
Unique	4.5	1169	—	6.0	1476	—
2400	9.0	1246	—	11.5	1700	—
IMR-4227	13.0	1720	—	15.0	1984	—

#311316**
115 gr. (#2 Alloy) 1.555" OAL

Powder	Sugg. Starting Grains	Velocity fps	Pressure C.U.P.	Max. Load Grains	Velocity fps	Pressure C.U.P.
Red Dot	3.2	1030	—	4.1	1215	—
700X	2.8	940	—	3.9	1195	—
Green Dot	3.5	1045	—	4.5	1255	—
PB	3.2	955	—	4.9	1295	—
Unique	4.2	1110	—	5.5	1390	—
SR-7625	3.5	985	—	5.5	1380	—

Note: Loads shown in shaded panels are maximum.
** Indicates use of Win. 6$^1/_2$-116 primers and Marlin 27S rifle.

32 Winchester Special

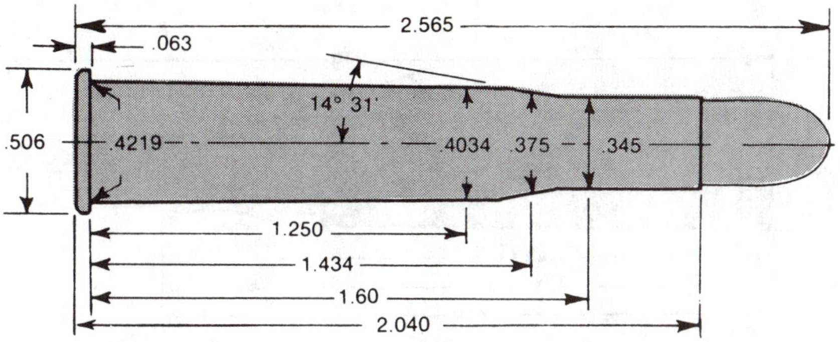

COMMENTS:

This is the ballistic twin to the 30-30 Winchester. It is the 30-30 case necked up to accept 0.321" bullets.

New cases which exceed the maximum case length have been encountered. Check all case lengths carefully before reloading.

Data is limited due to relatively low usage. IMR 4064 is a good first powder choice.

TEST COMPONENTS:

Cases .... Remington
Trim-to Length .... 2.035"
Primers .... Remington 9½
Primer Size .... Large Rifle
Lyman Shell Holder .... No. 6
Jacketed Bullets Used .... Remington HPCL, 170 gr.

TEST SPECIFICATIONS: (Velocity Only)

Firearm Used .... Winchester Model 94
Barrel Length .... 20"
Twist .... 1-16"
Groove Dia. .... .321"

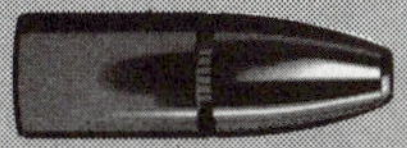

170 gr. Jacketed HP

Powder	Sugg. Starting Grains	Velocity fps	Pressure C.U.P.	Max. Load Grains	Velocity fps	Pressure C.U.P.
IMR-4198	24.0	1956	—	27.0	2150	—
IMR-3031	28.0	1912	—	31.0	2159	—
IMR-4064	28.0	1685	—	33.0	2070	—
IMR-4895	28.0	1792	—	33.0	2136	—
IMR-4320	31.0	1908	—	35.0	2169	—
RX7	28.0	2123	—	31.0	2283	—
BL-C(2)	31.0	1872	—	34.0	2040	—
H-380	36.0	1945	—	39.0	2109	—

Note: Loads shown in shaded panels are maximum.

8mm Mauser
(8 x 57mm JS)
(7.9 x 57mm)

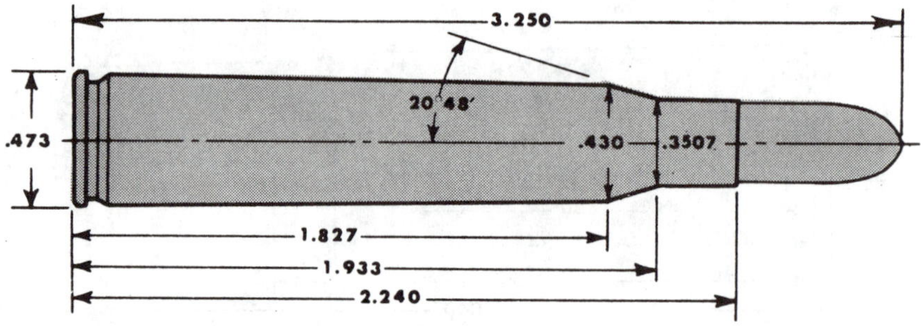

COMMENTS:

This data is not for use in rifles chambered for the 8 x 57mm J (or the rimmed version JR) cartridge which have a much smaller bore diameter of 0.318 inch. Use this data with the corresponding 0.323" bullets only in appropriate firearms.

IMR 4064 and 170 grain bullets will often bring the reloader to an accurate load with a minimum of testing.

TEST COMPONENTS:

Cases .... Winchester and Federal
Trim-to Length .... 2.235"
Primers .... Remington 9½ and Federal 210
Primer Size .... Large Rifle
Lyman Shell Holder .... No. 2
Jacketed Bullets Used .... Hornady SP #3230, 125 gr.
Speer SP #2277, 150 gr.
Hornady RN #3235, 170 gr.
Hornady SP #3238, 220 gr.
Cast Bullets Used .... (Sized to .323" dia.)
*Gas Check Bullets *#323470, 160 gr.

TEST SPECIFICATIONS: (Velocity & Pressure)

Firearm Used .... Mauser 98 and Universal Receiver
Barrel Length .... Mauser 23"; Universal Receiver 26"
Twist .... Mauser 1-9¼"; Universal Receiver 1-9½"
Groove Dia. .... .323"

125 gr. Jacketed SP**
2.875" OAL

Powder	Sugg. Starting Grains	Velocity fps	Pressure C.U.P.	Max. Load Grains	Velocity fps	Pressure C.U.P.
IMR-3031	46.0	2865	—	51.0+	3184	—
748	51.0	2813	32,800	58.0+	3063	35,100
IMR-4064	49.0	2840	—	54.0+	3125	—
IMR-4895	50.0	2898	—	53.0+	3194	—
IMR-4320	50.0	2824	—	55.0+	3105	—
H-380	51.0	2732	—	57.0+	2949	—
760	51.0	2493	27,500	58.0+	2739	32,400
H-414	50.1	2519	26,700	57.0+	2791	32,900
IMR-4350	50.0	2421	—	56.0+	2717	—

Note: Loads shown in shaded panels are maximum.
+ Designates a compressed powder charge.
** Used Win. cases and Rem. 9½ primers.

8mm Mauser
(8 x 57mm JS)
(7.9 x 57mm)

150 gr. Jacketed SP**
2.945" OAL

Powder	Sugg. Starting Grains	Velocity fps	Pressure C.U.P.	Max. Load Grains	Velocity fps	Pressure C.U.P.
IMR-3031	44.0	2604	—	49.0	2881	—
748	40.5	2392	33,700	46.0	2598	35,400
IMR-4064	48.0	2631	—	53.0+	2915	—
IMR-4895	48.0	2688	—	51.0+	2824	—
IMR-4320	48.0	2577	—	53.0+	2865	—
H-380	50.0	2570	—	56.0+	2840	—
760	47.5	2316	28,200	54.0	2605	36,800
H-414	47.5	2394	30,300	54.0+	2692	36,900
IMR-4350	50.0	2347	—	56.0+	2652	—

170 gr. Jacketed RN**
2.850" OAL

Powder	Sugg. Starting Grains	Velocity fps	Pressure C.U.P.	Max. Load Grains	Velocity fps	Pressure C.U.P.
RX7	27.2	1955	29,500	31.0	2140	34,400
IMR-3031	41.6	2331	—	47.0	2688	—
748	40.4	2187	28,100	46.0	2437	34,400
IMR-4064	46.0	2427	—	51.0+	2724	—
IMR-4895	46.0	2481	—	49.0+	2672	—
IMR-4320	46.0	2421	—	51.0+	2710	—
H-380	49.0	2433	—	54.0+	2631	—
760	43.1	2100	28,100	49.0	2302	35,000
IMR-4350	49.0	2237	—	54.0+	2518	—

220 gr. Jacketed SP**
3.000" OAL

Powder	Sugg. Starting Grains	Velocity fps	Pressure C.U.P.	Max. Load Grains	Velocity fps	Pressure C.U.P.
IMR-3031	33.4	1777	27,300	38.0	2047	36,900
748	36.0	1922	33,200	41.0	2157	36,800
IMR-4064	35.2	1817	29,300	40.0	2087	35,800
IMR-4895	34.3	1841	31,800	39.0	2081	36,000
IMR-4320	36.1	1867	35,200	41.0	2086	37,000
H-380	36.9	1810	27,000	42.0	2107	35,900
760	39.6	1798	27,200	45.0	2100	37,000
IMR-4350	39.6	1787	27,700	45.0+	2056	35,800

#323470***
160 gr. (#2 Alloy) 2.730" OAL

Powder	Sugg. Starting Grains	Velocity fps	Pressure C.U.P.	Max. Load Grains	Velocity fps	Pressure C.U.P.
Red Dot	10.0	1355	—	14.0	1645	—
700X	10.0	1360	—	14.0	1655	—
Green Dot	11.0	1405	—	15.0	1710	—
PB	11.5	1385	—	15.5	1670	—
Unique	13.0	1545	—	17.0	1820	—
SR-7625	12.0	1410	—	16.5	1725	—
SR-4759	20.5	1831	18,100	25.2	2139	30,200
IMR-4198	27.0	1925	16,700	35.0	2482	34,900
RX7	28.0	1917	16,900	40.0	2527	33,700
748	41.5	2326	25,400	45.0	2524	30,600
H-4895	31.0	1957	18,600	37.5	2330	28,200

Note: Loads shown in shaded panels are maximum.
\+ Designates a compressed powder charge.
** Used Win. cases and Rem. 9½ primers.
*** Used Federal cases and 210 primers.

8mm Remington Magnum

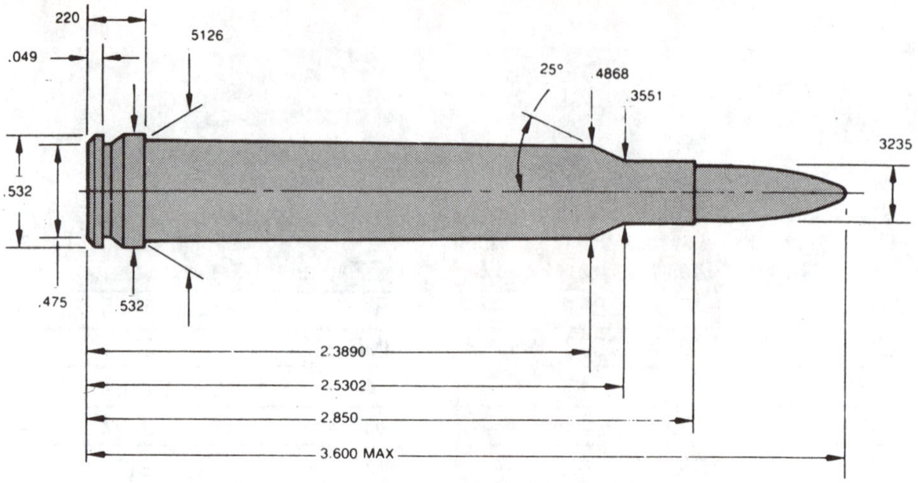

COMMENTS:

This was Remington's answer to the 338 Winchester Magnum. With its larger case (a necked down 375 H&H) it offered greater potential; but nevertheless it never became popular. This may be due to the fact that its hunting use is limited to the heaviest North American game. Or perhaps as some found the original factory bullets were really too soft for use on heavy rugged game.

In any event when properly loaded with a good premium grade bullet the 8mm Remington is a cartridge that can handle almost any situation to be encountered in North America. And it will do nicely for much of the African plains game.

With standard bullets it will perform well in non-demanding situations.

This round is at its best with a good 220 grain bullet and loaded with Hercules Reloder 22.

TEST COMPONENTS:

Cases .... Remington
Trim-to Length .... 2.840"
Primers .... CCI 250
Primer Size .... Large Rifle, Magnum
Lyman Shell Holder .... No. 13
Jacketed Bullets Used .... Speer SP #2277, 150 gr.
Hornady RN #3235, 170 gr.
Sierra SP #2410, 175 gr.
Hornady SP #3238, 220 gr.
Cast Bullets Used .... (Sized to .323" dia.)
*Gas Check Bullets .... *#323470, 160 gr.

TEST SPECIFICATIONS: (Velocity & Pressure)

Firearm Used .... Universal Receiver and Remington Model 700
Barrel Length .... 24"
Twist .... 1-10"
Groove Dia. .... .323"

150 gr. Jacketed SP
3.580" OAL

Powder	Sugg. Starting Grains	Velocity fps	Pressure C.U.P.	Max. Load Grains	Velocity fps	Pressure C.U.P.
IMR-4895	57.0	2821	35,200	64.0	3143	51,300
IMR-4320	57.5	2767	34,700	66.0	3126	52,300
H-380	65.0	2805	34,300	75.0	3198	53,600
760	71.0	3045	43,800	80.0	3294	52,900
H-414	68.0	3042	42,700	77.0	3272	51,700
IMR-4350	74.0	3102	41,800	80.0	3348	50,700
H-4831	80.0	3144	41,200	86.5	3382	50,000

Note: Loads shown in shaded panels are maximum.

8mm Remington Magnum

170 gr. Jacketed RN
3.440" OAL

Powder	Sugg. Starting Grains	Velocity fps	Pressure C.U.P.	Max. Load Grains	Velocity fps	Pressure C.U.P.
IMR-4320	59.7	2803	42,300	66.0	2976	50,600
H-380	67.0	2796	42,900	74.0	3049	52,400
760	73.0	2908	42,900	80.5	3144	52,000
H-414	71.5	2955	44,300	78.0	3181	53,000
IMR-4350	70.0	2919	42,900	76.5	3177	52,000
H-4831	78.0	2979	43,000	84.5	3257	52,300
RX19	81.0	3070	43,600	87.0+	3354	53,400
AA3100	76.5	2806	37,700	85.7+	3206	53,200
RX22	81.5	3004	40,600	87.3+	3296	51,300
AA8700	92.5	2908	41,400	102.4+	3155	50,200
IMR-7828	83.0	2977	43,100	90.0+	3261	52,100

175 gr. Jacketed SP
3.585" OAL

Powder	Sugg. Starting Grains	Velocity fps	Pressure C.U.P.	Max. Load Grains	Velocity fps	Pressure C.U.P.
H-380	67.0	2793	44,400	72.5	2997	52,800
H-414	68.8	2889	45,000	75.6	3099	52,400
IMR-4350	71.6	2871	41,100	77.5	3178	52,700
H-4831	75.3	2843	42,100	82.0	3131	53,000
RX19	80.4	3064	43,100	87.0+	3341	53,600
AA3100	77.0	2851	41,000	83.5+	3159	52,800
RX22	81.5	3034	42,500	87.2+	3305	52,200
AA8700	92.5	2920	42,500	102.4+	3196	51,800
IMR-7828	78.5	2812	39,100	86.3+	3216	52,500

220 gr. Jacketed SP
3.597" OAL

Powder	Sugg. Starting Grains	Velocity fps	Pressure C.U.P.	Max. Load Grains	Velocity fps	Pressure C.U.P.
H-380	60.0	2415	43,100	66.5	2622	52,000
H-414	60.0	2514	46,400	66.0	2678	52,600
IMR-4350	65.4	2579	43,800	‡73.0	2819	52,800
H-4831	65.7	2473	40,700	74.0	2734	51,600
H-870	81.0	2518	36,200	88.0+	2742	44,200
RX19	72.5	2712	42,600	79.5	2985	53,600
AA3100	72.5	2600	42,800	79.0+	2860	53,600
RX22	73.5	2673	40,600	79.5+	2939	51,800
AA8700	87.0	2662	43,100	94.5+	2885	50,400
IMR-7828	73.0	2596	40,300	80.0+	2888	50,800

#323470**
160 gr. (#2 Alloy) 3.398" OAL

Powder	Sugg. Starting Grains	Velocity fps	Pressure C.U.P.	Max. Load Grains	Velocity fps	Pressure C.U.P.
2400	24.0	1714	—	35.0	2313	—
SR-4759	23.0	1712	—	39.0	2469	—
IMR-4227	24.0	1729	—	40.0	2432	—
IMR-4198	26.0	1707	—	44.0	2471	—
RX7	29.0	1655	—	49.0	2465	—
H-4895	32.0	1711	—	50.0	2488	—

Note: Loads shown in shaded panels are maximum.
\+ Designates a compressed powder charge.
‡ Designates a factory velocity duplication load.
** Tested in Remington Model 700.

338 Winchester Magnum

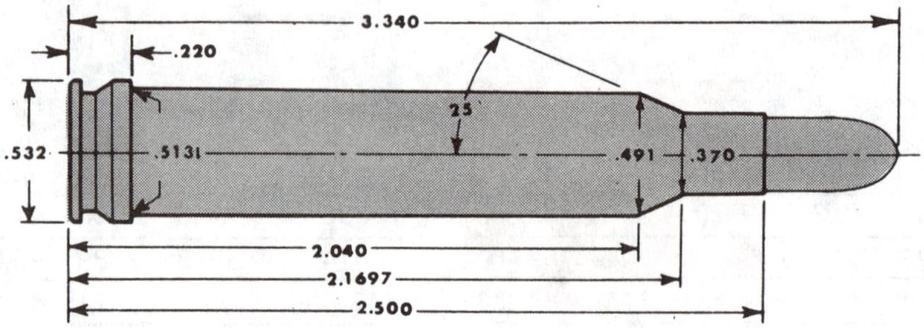

COMMENTS:

This cartridge has attained classic status in a relatively brief 34 years. It started slow but gradually gained a momentum equal to its ability to handle most of the world's game.

This round is best served with a premium style bullet but for non-demanding situations many use popularly priced bullets.

Propellants that have proven very accurate include IMR 4831, Hercules Reloder 22, and IMR 7828 — with all bullet weights. The shooter who really wants one cartridge for a wide variety of game, and who can see past the 30-06 mystique, would do well to learn to shoot a 338 Winchester Magnum. It has been used successfully on 30 pound duiker, 140 pound deer, giant Alaskan bear, and a lot of in between sized game around the world. It rivals the 375 H&H Magnum as being the most versatile of all big game cartridges. The 338 has the edge on lighter game whereas the 375 H&H can be pushed to the limit and used (carefully) for the likes of African cape buffalo.

TEST COMPONENTS:

Cases .... Winchester
Trim-to Length .... 2.490"
Primers .... Remington 9½ M
Primer Size .... Large Rifle, Magnum
Lyman Shell Holder .... No. 13
Jacketed Bullets Used .... Speer Spitzer #2405, 200 gr.
Hornady Spire Pt. #3320, 225 gr.
Hornady RN #3330, 250 gr.
Speer Semi-Spitzer #2411, 275 gr.

TEST SPECIFICATIONS: (Velocity Only)

Firearm Used .... Winchester Model 70
Barrel Length .... 24"
Twist .... 1-10"
Groove Dia. .... .336"

200 gr. Jacketed Spitzer
3.340" OAL

Powder	Sugg. Starting Grains	Velocity fps	Pressure C.U.P.	Max. Load Grains	Velocity fps	Pressure C.U.P.
IMR-3031	54.0	2620	—	60.5	2865	—
IMR-4895	58.0	2659	—	64.0	2898	—
IMR-4064	59.0	2665	—	65.0	2915	—
IMR-4320	59.0	2638	—	66.0	2915	—
IMR-4350	69.0	2777	—	75.0+	3058	—
H-380	63.0	2645	—	70.0	2881	—
H-4831	72.0	2652	—	80.0+	2923	—
RX15	58.0	2634	—	65.0	2927	—
RX19	69.0	2634	—	77.0+	2968	—
RX22	71.0	2644	—	79.0+	2932	—
IMR-4831	69.0	2664	—	77.5+	3023	—
AA3100	68.0	2513	—	76.0+	2805	—

Note: Loads shown in shaded panels are maximum.
+ Designates a compressed powder charge.

338 Winchester Magnum

225 gr. Jacketed Spire Pt.
3.310" OAL

Powder	Sugg. Starting Grains	Velocity fps	Pressure C.U.P.	Max. Load Grains	Velocity fps	Pressure C.U.P.
IMR-3031	52.0	2450	—	58.0	2638	—
IMR-4895	56.0	2469	—	62.5	2724	—
IMR-4064	57.0	2475	—	63.0	2717	—
IMR-4320	57.0	2469	—	63.0	2695	—
IMR-4350	67.0	2597	—	74.0+	2915	—
H-380	60.0	2457	—	67.0	2645	—
H-4831	70.0	2493	—	‡78.0+	2797	—
IMR-7828	68.0	2306	—	76.0+	2602	—
AA3100	67.0	2386	—	74.0+	2616	—
RX22	69.0	2462	—	77.0+	2747	—
IMR-4831	68.0	2524	—	75.5+	2801	—
RX19	68.0	2533	—	75.0+	2795	—

250 gr. Jacketed RN
3.300" OAL

Powder	Sugg. Starting Grains	Velocity fps	Pressure C.U.P.	Max. Load Grains	Velocity fps	Pressure C.U.P.
IMR-3031	50.0	2272	—	56.5	2469	—
IMR-4895	55.0	2352	—	61.5	2597	—
IMR-4064	56.0	2347	—	62.0	2577	—
IMR-4320	56.0	2352	—	62.0	2557	—
IMR-4350	65.0	2450	—	72.0+	2762	—
H-380	59.0	2352	—	66.0	2544	—
H-4831	69.0	2403	—	‡76.0+	2666	—
RX19	65.0	2297	—	72.0+	2591	—
AA3100	66.0	2267	—	72.0+	2509	—
IMR-7828	67.0	2225	—	74.0+	2523	—
IMR-4831	66.0	2367	—	73.5+	2686	—
RX22	68.0	2331	—	76.0+	2652	—

275 gr. Jacketed Semi-Spitzer
3.340" OAL

Powder	Sugg. Starting Grains	Velocity fps	Pressure C.U.P.	Max. Load Grains	Velocity fps	Pressure C.U.P.
IMR-3031	48.0	2145	—	53.0	2293	—
IMR-4895	54.0	2252	—	60.0	2469	—
IMR-4064	54.0	2227	—	60.0	2450	—
IMR-4320	54.0	2212	—	60.0	2415	—
IMR-4350	62.0	2331	—	69.0	2597	—
H-380	58.0	2237	—	64.0	2398	—
H-4831	66.0	2314	—	73.0+	2570	—
IMR-4831	64.0	2320	—	71.5+	2585	—
RX22	66.0	2321	—	73.0+	2553	—
IMR-7828	65.0	2151	—	71.0+	2400	—
AA3100	62.0	2135	—	68.0+	2355	—

Note: Loads shown in shaded panels are maximum.
+ Designates a compressed powder charge.
‡ Designates a factory velocity duplication load.

340 Weatherby Magnum

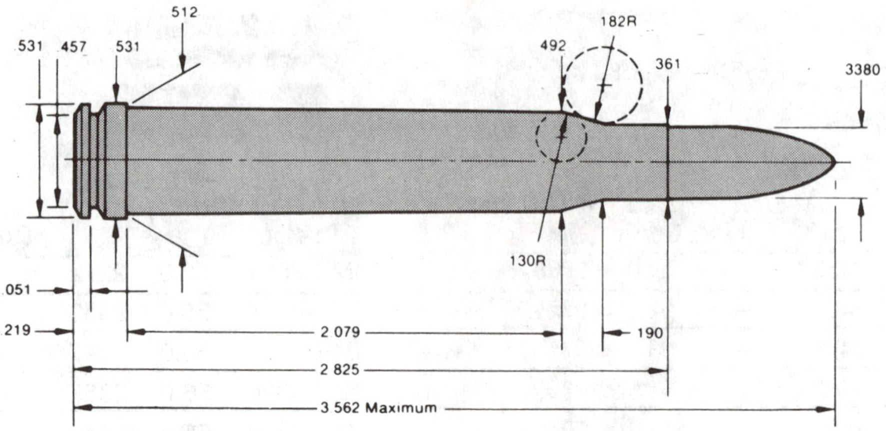

COMMENTS:

This and the 338 Winchester are the only two successful commercial cartridges to use the 0.338" bullet diameter. There is very little difference in the field performance of the two rounds. The 340 Weatherby, perhaps due to its proprietary status, never has become very popular, but is nonetheless every bit as satisfactory as the similar Winchester round.

The use of magnum style primers is urged for the loads shown. The data listed is intended only for rifles of Weatherby manufacture.

TEST COMPONENTS:

Cases .... Weatherby
Trim-to Length .... 2.815"
Primers .... Federal 215
Primer Size .... Large Rifle, Magnum
Lyman Shell Holder .... No. 13
Jacketed Bullets Used .... Speer Spitzer, 200 gr.
Hornady Spire Pt., 225 gr.
Hornady RN, 250 gr.
Speer Semi-Spitzer, 275 gr.

TEST SPECIFICATIONS: (Velocity Only)

Firearm Used .... Weatherby Mark V
Barrel Length .... 26"
Twist .... 1-10"
Groove Dia. .... .338"

200 gr. Jacketed Spitzer

Powder	Sugg. Starting Grains	Velocity fps	Pressure C.U.P.	Max. Load Grains	Velocity fps	Pressure C.U.P.
IMR-4064	65.0	2564	—	72.0	2890	—
IMR-4320	68.0	2710	—	76.0	3048	—
IMR-4350	76.0	2724	—	84.0	3144	—
H-380	72.0	2808	—	80.0	3039	—
H-4831	81.0	2604	—	90.0+	3021	—

Note: Loads shown in shaded panels are maximum.
+ Designates a compressed powder charge.

340 Weatherby Magnum

225 gr. Jacketed Spire Pt.

Powder	Sugg. Starting Grains	Velocity fps	Pressure C.U.P.	Max. Load Grains	Velocity fps	Pressure C.U.P.
IMR-4064	62.0	2409	—	69.0	2724	—
IMR-4320	65.0	2564	—	72.0	2832	—
IMR-4350	72.0	2538	—	80.0	2923	—
H-380	68.0	2575	—	76.0	2812	—
H-4831	78.0	2512	—	87.0+	2915	—

250 gr. Jacketed RN

Powder	Sugg. Starting Grains	Velocity fps	Pressure C.U.P.	Max. Load Grains	Velocity fps	Pressure C.U.P.
IMR-4064	59.0	2252	—	66.0	2551	—
IMR-4320	63.0	2427	—	70.0	2680	—
IMR-4350	70.0	2418	—	77.0	2739	—
H-380	66.0	2484	—	73.0	2672	—
H-4831	77.0	2439	—	85.0+	2793	—

275 gr. Jacketed Semi-Spitzer

Powder	Sugg. Starting Grains	Velocity fps	Pressure C.U.P.	Max. Load Grains	Velocity fps	Pressure C.U.P.
IMR-4064	56.0	2092	—	61.0	2314	—
IMR-4320	60.0	2283	—	67.0	2500	—
IMR-4350	66.0	2262	—	73.0	2544	—
H-380	63.0	2314	—	70.0	2481	—
H-4831	73.0	2293	—	81.0+	2617	—

Note: Loads shown in shaded panels are maximum.
+ Designates a compressed powder charge.

348 Winchester

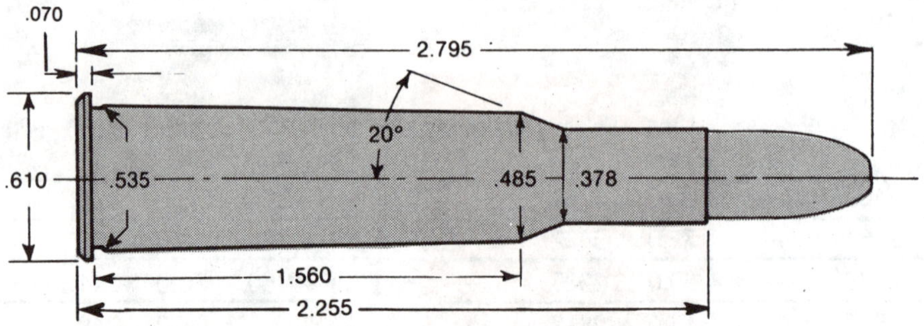

COMMENTS:

This cartridge is the only commercial success to use the 0.348" bullet diameter. The only rifle chambered for it was the Model 71 Winchester and when it died so did the 348. Browning gave the round a brief resurrection when it offered its Model 71 copy a few years ago.

Only blunt nosed or flat nosed bullets should be loaded for the tubular magazine of the Model 71. However, we think you would be hard pressed to find a 0.348" diameter bullet with any other shape, as bullet designs are limited.

TEST COMPONENTS:

Cases .... Remington
Trim-to Length .... 2.245"
Primers .... Remington 9½
Primer Size .... Large Rifle
Lyman Shell Holder .... No. 18
Jacketed Bullets Used .... Remington SPCL, 200 gr.

TEST SPECIFICATIONS: (Velocity Only)

Firearm Used .... Winchester Model 71
Barrel Length .... 24"
Twist .... 1-12"
Groove Dia. .... .3485"

200 gr. Jacketed SP

Powder	Sugg. Starting Grains	Velocity fps	Pressure C.U.P.	Max. Load Grains	Velocity fps	Pressure C.U.P.
IMR-3031	44.0	2262	—	49.0	2430	—
IMR-4895	49.0	2341	—	‡54.0	2506	—
IMR-4064	48.0	2222	—	53.6	2450	—
IMR-4320	50.0	2252	—	55.0	2481	—
IMR-4350	57.0	2262	—	63.0	2564	—
H-380	50.0	2188	—	57.0	2347	—
H-4831	61.0	2169	—	67.0+	2409	—

Note: Loads shown in shaded panels are maximum.
+ Designates a compressed powder charge.
‡ Designates a factory velocity duplication load.

9mm Luger (rifle data)

(9mm Parabellum)

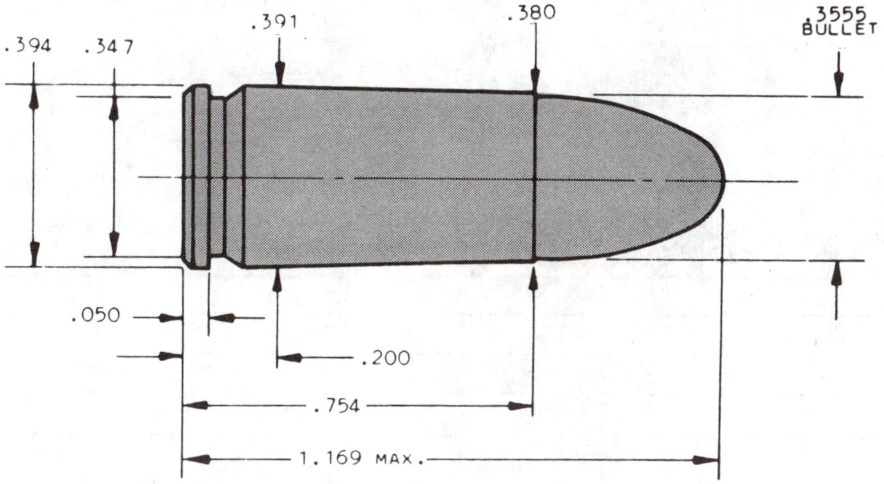

COMMENTS:

The 9mm has become the most popular autoloading pistol cartridge. This popularity has spread into small rifles chambered for the cartridge. The ballistics shown were obtained in a Marlin 16½" barrel.

A caution: do not go below the listed starting charges. Doing so may cause a bullet to be stuck in the rifle length barrel. If this occurs and another round is then fired the results could be disastrous. Stay alert and be sure every cartridge is properly loaded.

Hercules Unique with all jacketed and cast bullets would be the recommended first choice propellant.

TEST COMPONENTS:

Cases .... Federal
Trim-to Length .... .751"
Primers .... CCI 500
Primer Size .... Small Pistol
Lyman Shell Holder .... No. 12
Jacketed Bullets Used .... Sierra JHP #8100, 90 gr.
Sierra JHP #8110, 115 gr.
Sierra FMJ #8120, 125 gr.
Sierra FMJ #8345, 130 gr.
Cast Bullets Used .... (Sized to .355" dia.)
#356242, 120 gr.
#356634, 130 gr.
#356637, 147 gr.

TEST SPECIFICATIONS:
(Velocity Only)

Firearm Used .... Marlin Model 9
Barrel Length .... 16½"
Twist .... 1-10"
Groove Dia. .... .355"

90 gr. Jacketed HP
1.010" OAL

Powder	Sugg. Starting Grains	Velocity fps	Pressure C.U.P.	Max. Load Grains	Velocity fps	Pressure C.U.P.
AA5	5.2	1175	—	6.5	1572	—
Unique	5.0	1192	—	6.3	1537	—
SR-4756	5.0	1220	—	6.2	1510	—
Blue Dot	7.0	1264	—	8.9	1565	—

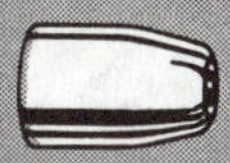

115 gr. Jacketed H.P.
1.014" OAL

Powder	Sugg. Starting Grains	Velocity fps	Pressure C.U.P.	Max. Load Grains	Velocity fps	Pressure C.U.P.
Unique	4.4	1054	—	5.8	1336	—
SR-4756	4.5	1037	—	6.2	1441	—
AA7	6.8	1074	—	8.5	1380	—
Blue Dot	6.3	1023	—	8.7	1421	—

Note: Loads shown in shaded panels are maximum.

9mm Luger (rifle data)
(9mm Parabellum)

125 gr. Full Metal Jacket
1.120" OAL

Powder	Sugg. Starting Grains	Velocity fps	Pressure C.U.P.	Max. Load Grains	Velocity fps	Pressure C.U.P.
Unique	4.1	816	—	5.3	1164	—
HS-6	5.4	898	—	6.8	1236	—
AA7	6.4	844	—	8.0	1188	—
Blue Dot	6.1	898	—	8.0	1336	—

130 gr. Full Metal Jacket
1.160" OAL

Powder	Sugg. Starting Grains	Velocity fps	Pressure C.U.P.	Max. Load Grains	Velocity fps	Pressure C.U.P.
Unique	3.5	675	—	4.8	1114	—
HS-6	5.0	716	—	6.5	1142	—
AA7	6.5	933	—	8.2	1224	—
Blue Dot	5.4	796	—	6.8	1136	—

#356242
120 gr. (#2 Alloy) 1.065" OAL

Powder	Sugg. Starting Grains	Velocity fps	Pressure C.U.P.	Max. Load Grains	Velocity fps	Pressure C.U.P.
Bullseye	3.4	1131	—	4.2	1305	—
Red Dot	3.0	1070	—	3.9	1263	—
700X	3.0	1062	—	3.7	1236	—
Unique	4.0	1214	—	5.0	1402	—
HS-6	5.0	1155	—	6.3	1384	—

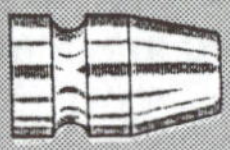

#356634
130 gr. (#2 Alloy) 1.016" OAL

Powder	Sugg. Starting Grains	Velocity fps	Pressure C.U.P.	Max. Load Grains	Velocity fps	Pressure C.U.P.
Unique	3.5	1069	—	4.0	1166	—
HS-6	5.1	1113	—	5.6+	1252	—
SR-4756	5.0	1331	—	5.6+	1385	—
AA7	6.8	1214	—	7.7	1341	—
Blue Dot	5.8	1174	—	6.5	1265	—

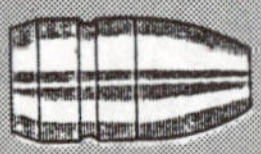

#356637
147 gr. (#2 Alloy) 1.058" OAL

Powder	Sugg. Starting Grains	Velocity fps	Pressure C.U.P.	Max. Load Grains	Velocity fps	Pressure C.U.P.
Unique	3.2	976	—	3.8	1108	—
HS-6	4.4	986	—	5.0	1108	—
SR-4756	3.1	935	—	3.6	1041	—
Blue Dot	5.1	1073	—	5.9	1144	—
WSF	3.4	1020	—	3.9	1104	—

Note: Loads shown in shaded panels are maximum.
+ Designates a compressed powder charge.

357 Magnum (rifle data)

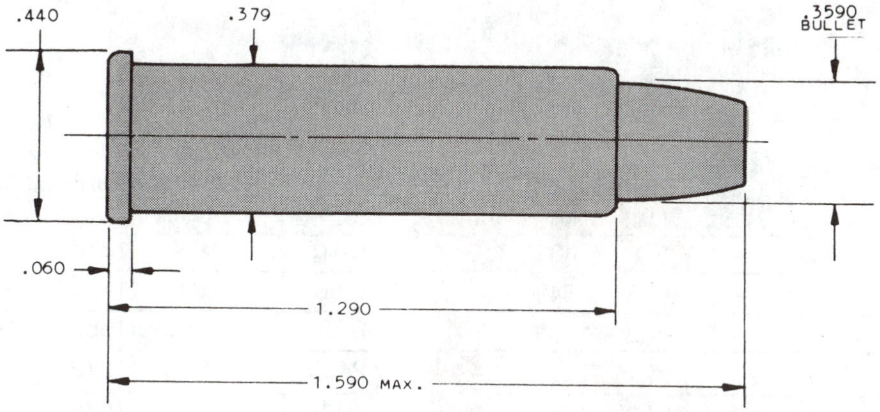

COMMENTS:

The Marlin rifles chambered for the 357 Magnum cartridge have become popular and hence this data entry. Hercules 2400 would be our suggestion for combining accuracy and punch.

Be sure to crimp bullets in place and use only blunt or flat point bullets in the tubular magazine of the Marlin. This data may, of course, be used in the limited production Ruger No. 1 single shot with any shape bullet.

TEST COMPONENTS:

Cases .... Federal
Trim-to Length .... 1.285"
Primers .... CCI 500
Primer Size .... Small Pistol
Lyman Shell Holder .... No. 1
Jacketed Bullets Used .... Hornady JHP #3570, 110 gr.
Hornady JHP #3571, 125 gr.
Speer JHP-SWC #4205, 146 gr.
Hornady JHP #3575, 158 gr.
Sierra JHC #8365, 170 gr.
Cast Bullets Used .... (Sized to .357" dia.)
*Gas Check Bullets .... *#358156, 155 gr.
#358429, 168 gr.

TEST SPECIFICATIONS:
(Velocity Only)

Firearm Used .... Marlin Model 1894
Barrel Length .... $18^1/_2$"
Twist .... 1-16"
Groove Dia. .... .357"

110 gr. Jacketed HP
1.590" OAL

Powder	Sugg. Starting Grains	Velocity fps	Pressure C.U.P.	Max. Load Grains	Velocity fps	Pressure C.U.P.
AA5	11.0	1948	—	13.0	2111	—
Unique	7.4	1543	—	10.0	1930	—
Blue Dot	10.0	1619	—	14.6	2098	—
2400	13.5	1709	—	20.3	2312	—
IMR-4227	14.8	1595	—	19.5	1972	—

125 gr. Jacketed HP
1.590" OAL

Powder	Sugg. Starting Grains	Velocity fps	Pressure C.U.P.	Max. Load Grains	Velocity fps	Pressure C.U.P.
Unique	7.0	1380	—	9.7	1818	—
Blue Dot	10.4	1511	—	13.3	1886	—
2400	13.0	1559	—	17.7	2009	—
H110	18.9	1964	—	21.5	2188	—
IMR-4227	15.6	1588	—	19.5	1955	—

Note: Loads shown in shaded panels are maximum.

357 Magnum (rifle data)

146 gr. Jacketed HP-SWC
1.590" OAL

Powder	Sugg. Starting Grains	Velocity fps	Pressure C.U.P.	Max. Load Grains	Velocity fps	Pressure C.U.P.
Unique	7.5	1424	—	8.1	1484	—
Blue Dot	9.0	1304	—	11.5	1643	—
AA9	13.0	1578	—	16.2	1883	—
2400	11.0	1270	—	15.1	1714	—
H110	13.0	1497	—	17.0	1746	—
IMR-4227	11.5	1144	—	16.2	1619	—

158 gr. Jacketed HP
1.590" OAL

Powder	Sugg. Starting Grains	Velocity fps	Pressure C.U.P.	Max. Load Grains	Velocity fps	Pressure C.U.P.
Unique	6.4	1218	—	8.3	1517	—
Blue Dot	9.6	1402	—	11.9	1669	—
AA9	13.0	1582	—	16.0	1838	—
2400	11.3	1372	—	14.9	1742	—
H110	13.7	1534	—	17.7	1922	—
IMR-4227	12.2	1339	—	16.1	1690	—

170 gr. Jacketed HC
1.590" OAL

Powder	Sugg. Starting Grains	Velocity fps	Pressure C.U.P.	Max. Load Grains	Velocity fps	Pressure C.U.P.
Herco	6.8	1190	—	7.5	1307	—
AA7	9.8	1348	—	11.0	1478	—
AA9	11.0	1355	—	12.6	1536	—
2400	10.5	1274	—	12.0	1456	—
H110	14.0	1502	—	15.5	1635	—
IMR-4227	12.0	1245	—	13.7	1377	—

#358156
155 gr. (#2 Alloy) 1.590" OAL

Powder	Sugg. Starting Grains	Velocity fps	Pressure C.U.P.	Max. Load Grains	Velocity fps	Pressure C.U.P.
Herco	5.0	1014	—	7.5	1366	—
AA7	10.0	1462	—	12.7	1749	—
Blue Dot	8.5	1337	—	10.5	1614	—
AA9	12.2	1585	—	15.2	1838	—
2400	10.6	1422	—	14.0	1740	—
IMR-4227	11.4	1270	—	15.2	1616	—

#358429
168 gr. (#2 Alloy) 1.553" OAL

Powder	Sugg. Starting Grains	Velocity fps	Pressure C.U.P.	Max. Load Grains	Velocity fps	Pressure C.U.P.
Herco	5.6	1099	—	7.1	1341	—
Blue Dot	8.3	1357	—	10.0	1579	—
AA9	10.8	1444	—	13.5	1713	—
2400	9.7	1317	—	13.5	1715	—
IMR-4227	9.8	1125	—	14.5	1596	—

Note: Loads shown in shaded panels are maximum.

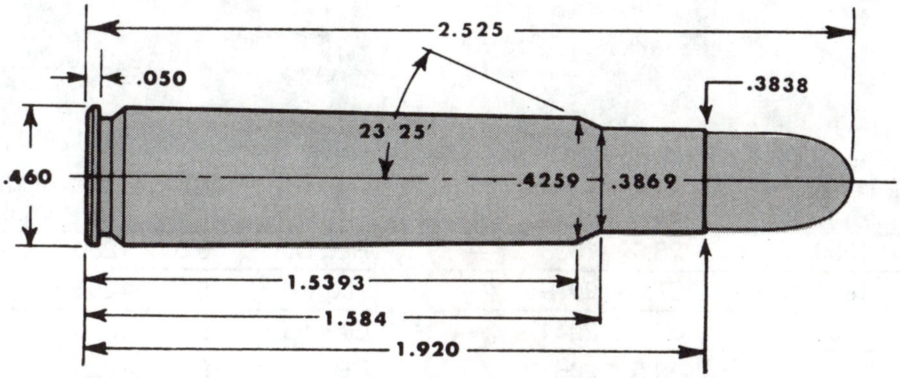

COMMENTS:

This cartridge has a strong but limited following as a short range deer round. It is generally chambered in lever action rifles. When loading for rifles having tubular magazines use only blunt or round nosed bullets. In these instances it is advisable to crimp bullets in place. RX7 is a good starting powder choice.

In multi-grooved barrels cast bullet velocity must be kept at or below 1,600 fps. for best results.

TEST COMPONENTS:

Cases .... Remington
Trim-to Length .... 1.910"
Primers .... Remington 9½ and Winchester 8½-120
Primer Size .... Large Rifle
Lyman Shell Holder .... No. 8 and 2
Jacketed Bullets Used .... Remington Ptd. SPCL, 150 gr.
Remington SPCL, 200 gr.
Cast Bullets Used .... (Sized to .357" dia.)
*Gas Check Bullets #358430, 195 gr.
*#358315, 206 gr.

TEST SPECIFICATIONS: (Velocity Only)

Firearm Used .... Marlin Model 336
Barrel Length .... 20"
Twist .... 1-16"
Groove Dia. .... .357"

150 gr. Jacketed Ptd. SP**

Powder	Sugg. Starting Grains	Velocity fps	Pressure C.U.P.	Max. Load Grains	Velocity fps	Pressure C.U.P.
IMR-4198	30.0	2000	—	33.5	2283	—
IMR-3031	36.0	2053	—	40.0+	2272	—
IMR-4064	39.0	2053	—	43.0+	2304	—
IMR-4320	40.0	2100	—	44.0+	2314	—
RX7	32.0	2262	—	35.0	2457	—
BL-C(2)	40.0	2036	—	45.0	2247	—
H-380	42.0	1956	—	47.0+	2127	—

Note: Loads shown in shaded panels are maximum.
\+ Designates a compressed powder charge.
** Used Winchester 8½-120 primers.

35 Remington

200 gr. Jacketed SP**

Powder	Sugg. Starting Grains	Velocity fps	Pressure C.U.P.	Max. Load Grains	Velocity fps	Pressure C.U.P.
IMR-4198	25.0	1669	—	28.0	1838	—
IMR-3031	34.0	1897	—	37.0+	2079	—
IMR-4064	35.0	1805	—	39.0+	2044	—
IMR-4320	35.0	1809	—	39.5	2061	—
RX7	30.0	2012	—	33.0	2188	—
BL-C(2)	37.0	1901	—	41.0	2114	—
H-380	40.0	1798	—	45.0+	1992	—

#358430
195 gr. (#2 Alloy) 2.215" OAL

Powder	Sugg. Starting Grains	Velocity fps	Pressure C.U.P.	Max. Load Grains	Velocity fps	Pressure C.U.P.
Red Dot	8.5	1280	—	11.0	1495	—
700X	8.0	1255	—	10.0	1415	—
Green Dot	9.0	1300	—	11.5	1515	—
PB	8.5	1225	—	11.5	1470	—
Unique	9.5	1320	—	12.0	1525	—
SR-7625	9.0	1230	—	11.5	1450	—
Herco	10.5	1360	—	13.0	1550	—
SR-4756	10.0	1275	—	13.0	1515	—

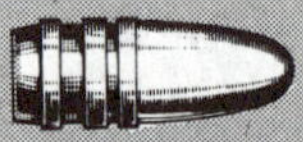

#358315
206 gr. (#2 Alloy) 2.500" OAL

Powder	Sugg. Starting Grains	Velocity fps	Pressure C.U.P.	Max. Load Grains	Velocity fps	Pressure C.U.P.
Red Dot	8.5	1240	—	11.0	1440	—
700X	8.0	1200	—	10.0	1370	—
Green Dot	9.0	1260	—	11.5	1455	—
PB	8.5	1170	—	11.5	1405	—
Unique	9.5	1265	—	12.0	1465	—
SR-7625	9.0	1185	—	11.5	1390	—
Herco	10.5	1315	—	13.0	1490	—
SR-4756	10.0	1235	—	13.0	1445	—

Note: Loads shown in shaded panels are maximum.
+ Designates a compressed powder charge.
** Used Winchester 8½-120 primers.

358 Winchester
(8.8 x 51 mm)

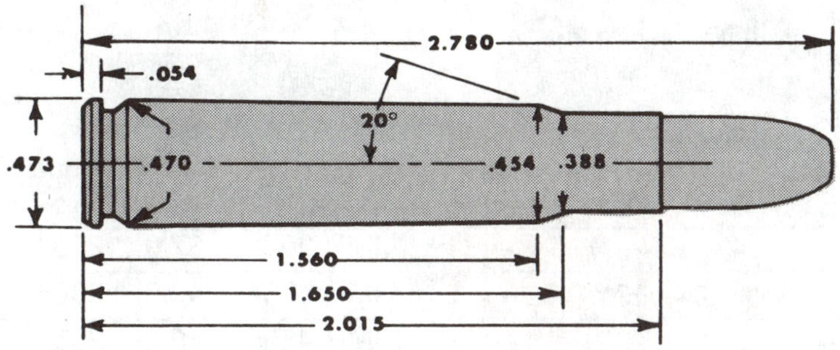

COMMENTS:

The 358 Winchester is the 308 case necked up to 35 caliber. It was designed as a medium range big game cartridge that could be chambered in short action rifles. While this idea certainly made sense, the cartridge never achieved the popularity it deserved.

The 358 Winchester would perform at its best with 225 or 250 grain bullets. IMR-3031 and IMR-4064 gave the best results ballistically.

TEST COMPONENTS:

Cases .... Winchester
Trim-to Length .... 2.005"
Primers .... Remington 9½ and Winchester 8½-120
Primer Size .... Large Rifle
Lyman Shell Holder .... No. 2
Jacketed Bullets Used .... Remington Ptd. SPCL, 150 gr.
Remington SPCL, 200 gr.
Remington Ptd. SPCL, 250 gr.
Hornady RN, 275 gr.
Cast Bullets Used .... (Sized to .359" dia.)
*Gas Check Bullets .... *#358315, 206 gr.

TEST SPECIFICATIONS:
(Velocity Only)

Firearm Used .... Savage Model 99F and Mannlicher-Schoenauer
Barrel Length .... Savage 22"; Mannlicher, 20"
Twist .... 1-12"
Groove Dia. .... Savage .359"; Mannlicher .362"

150 gr. Jacketed Ptd. SP

Powder	Sugg. Starting Grains	Velocity fps	Pressure C.U.P.	Max. Load Grains	Velocity fps	Pressure C.U.P.
IMR-3031	45.0	2506	—	50.5+	2808	—
IMR-4064	46.0	2272	—	51.0+	2570	—
IMR-4895	48.0	2506	—	53.5+	2747	—
IMR-4320	50.0	2544	—	55.0+	2717	—
RX7	40.0	2624	—	44.0	2801	—
BL-C(2)	52.0	2617	—	57.0+	2785	—
H-335	52.0	2717	—	57.0+	2801	—
H-380	50.0	2352	—	54.5+	2500	—

Note: Loads shown in shaded panels are maximum.
\+ Designates a compressed powder charge.

358 Winchester
(8.8 x 51 mm)

200 gr. Jacketed SP

Powder	Sugg. Starting Grains	Velocity fps	Pressure C.U.P.	Max. Load Grains	Velocity fps	Pressure C.U.P.
IMR-3031	43.0	2325	—	48.0+	2577	—
IMR-4064	46.0	2277	—	51.0+	2512	—
IMR-4895	46.0	2347	—	51.5+	2583	—
IMR-4320	47.0	2325	—	52.5+	2564	—
RX7	37.0	2272	—	‡42.0	2500	—
BL-C(2)	47.0	2304	—	52.0	2469	—
H-335	47.0	2352	—	52.0	2531	—
H-380	49.0	2212	—	54.0+	2375	—

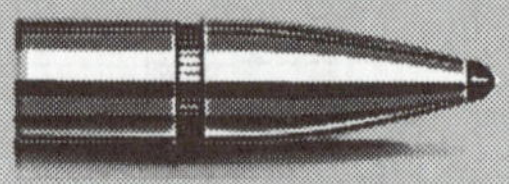

250 gr. Jacketed Ptd. SP

Powder	Sugg. Starting Grains	Velocity fps	Pressure C.U.P.	Max. Load Grains	Velocity fps	Pressure C.U.P.
IMR-3031	37.0	2008	—	41.5+	2252	—
IMR-4064	40.0	2008	—	44.5+	2217	—
IMR-4895	40.0	2053	—	44.5+	2252	—
IMR-4320	41.0	2040	—	45.5+	2227	—
RX7	31.0	1890	—	35.5	2118	—
BL-C(2)	41.0	2000	—	45.0	2197	—
H-335	41.0	2032	—	45.0	2252	—
H-380	42.0	1886	—	47.0+	2024	—

275 gr. Jacketed RN

Powder	Sugg. Starting Grains	Velocity fps	Pressure C.U.P.	Max. Load Grains	Velocity fps	Pressure C.U.P.
IMR-3031	35.0	1792	—	39.0	2040	—
IMR-4064	38.0	1838	—	42.0	2024	—
IMR-4895	38.0	1897	—	42.0	2079	—
IMR-4320	39.0	1858	—	43.0	2074	—
RX7	31.0	1831	—	34.0	1904	—
BL-C(2)	39.0	1792	—	43.0	1988	—
H-335	39.0	1818	—	43.0	2016	—
H-380	40.0	1738	—	44.0	1878	—

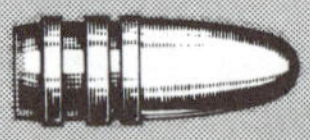

#358315**
206 gr. (#2 Alloy) 2.585" OAL

Powder	Sugg. Starting Grains	Velocity fps	Pressure C.U.P.	Max. Load Grains	Velocity fps	Pressure C.U.P.
Red Dot	9.0	1190	—	13.0	1505	—
700X	9.0	1205	—	12.5	1490	—
Green Dot	10.0	1310	—	13.0	1495	—
PB	10.0	1265	—	13.5	1465	—
Unique	11.0	1340	—	14.5	1550	—
SR-7625	11.0	1295	—	14.0	1480	—

Note: Loads shown in shaded panels are maximum.
\+ Designates a compressed powder charge.
‡ Designates a factory velocity duplication load.
** Indicates use of Win. 8½-120 primers and Mannlicher-Schoenauer rifle.

350 Remington Magnum

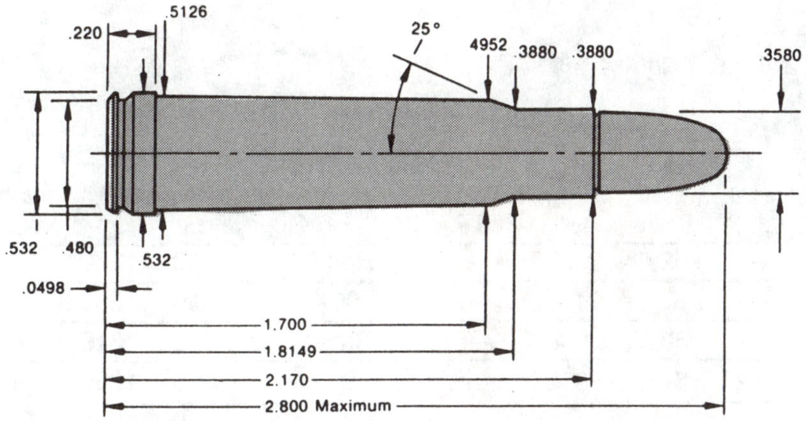

COMMENTS:

Either large rifle magnum or standard large rifle primers may be used with the listed dated for jacketed bullets. For cast bullets use only standard large rifle primers.

Due to length limitations of the rifle's magazine, bullets over 250 grains proved impractical in our testing. Even 250 grain bullets must be seated rather deeply into the case if they are to function through the magazine.

Suitable reloading powders were few due to the relatively small case capacity.

TEST COMPONENTS:

Cases .... Remington
Trim-to Length .... 2.160"
Primers .... Remington 9½ and 9½ M
Primer Size .... Large Rifle, and Large Rifle Magnum
Lyman Shell Holder .... No. 13
Jacketed Bullets Used .... Remington Ptd. SPCL, 150 gr.
Remington Ptd. SPCL, 200 gr.
Remington Ptd. SPCL, 250 gr.
Cast Bullets Used .... (Sized to .358" dia.)
*Gas Check Bullets *#358315, 206 gr.

TEST SPECIFICATIONS: (Velocity Only)

Firearm Used .... Remington Model 600 and 700
Barrel Length .... Model 600, 18½"; Model 700, 24"
Twist .... 1-16"
Groove Dia. .... .358"

150 gr. Jacketed Ptd. SP

Powder	Sugg. Starting Grains	Velocity fps	Pressure C.U.P.	Max. Load Grains	Velocity fps	Pressure C.U.P.
IMR-3031	52.0	2347	—	57.0	2732	—
IMR-4064	55.0	2331	—	61.0+	2710	—
IMR-4895	56.0	2493	—	62.0+	2849	—
IMR-4320	55.0	2314	—	61.0	2688	—

200 gr. Jacketed Ptd. SP

Powder	Sugg. Starting Grains	Velocity fps	Pressure C.U.P.	Max. Load Grains	Velocity fps	Pressure C.U.P.
IMR-3031	49.0	2262	—	54.0	2597	—
IMR-4064	53.0	2207	—	59.0+	2590	—
IMR-4895	54.0	2429	—	60.0+	2732	—
IMR-4320	53.0	2159	—	59.0	2544	—

Note: Loads shown in shaded panels are maximum.
\+ Designates a compressed powder charge.

350 Remington Magnum

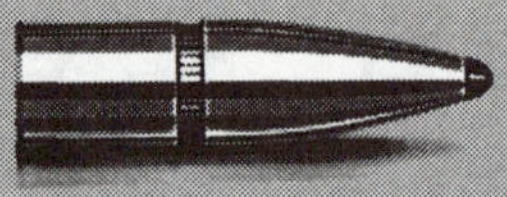

250 gr. Jacketed Ptd. SP

Powder	Sugg. Starting Grains	Velocity fps	Pressure C.U.P.	Max. Load Grains	Velocity fps	Pressure C.U.P.
IMR-3031	47.0	2155	—	52.0+	2403	—
IMR-4064	52.0	2169	—	57.0+	2415	—
IMR-4895	52.0	2293	—	58.0+	2538	—
IMR-4320	52.0	2192	—	57.0+	2392	—

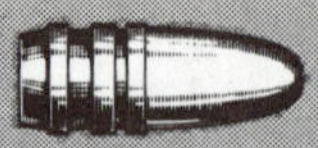

#358315
206 gr. (#2 Alloy) 2.760" OAL

Powder	Sugg. Starting Grains	Velocity fps	Pressure C.U.P.	Max. Load Grains	Velocity fps	Pressure C.U.P.
Red Dot	10.0	1270	—	14.5	1535	—
700X	10.0	1280	—	14.5	1540	—
Green Dot	11.0	1315	—	15.5	1620	—
PB	11.0	1250	—	16.0	1580	—
Unique	12.0	1345	—	17.0	1675	—
SR-7625	12.0	1305	—	17.0	1615	—

Note: Loads shown in shaded panels are maximum.
\+ Designates a compressed powder charge.

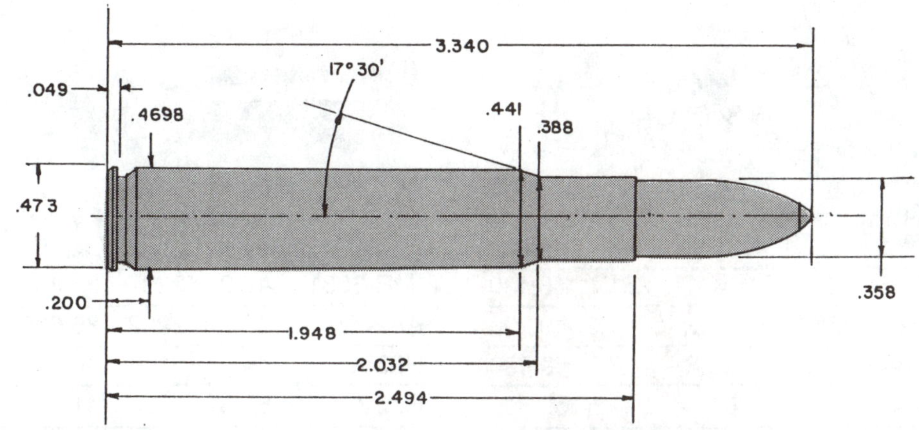

COMMENTS:

The 35 Whelen was around for many years as a wildcat cartridge. This data is not for use in such custom chambered rifles. Remington legitimized this round, basically a necked up 30-06, a few years ago when they added it to their standard cartridge offerings.

The 35 Whelen adds a bit of punch compared to its parent round—the 30-06. Still if one wants as much punch as possible without the short case life often associated with belted magnums, there is not a lot to choose from between the 30-06 and the 416 Rigby, except for the 35 Whelen. That's plenty of justification for the cartridge. IMR 4064 would be the best choice for loads with jacketed bullets.

Cast bullet #358009 with either IMR 4227 or IMR 4198 should give accurate results.

TEST COMPONENTS:

Cases .... Remington
Trim-to Length .... 2.484"
Primers .... Remington 9½
Primer Size .... Large Rifle
Lyman Shell Holder .... No. 2
Jacketed Bullets Used ...... Speer Flat Soft Point #2435, 180 gr.
Sierra RN #2800, 200 gr.
Nosler Spitzer #44800, 225 gr.
Hornady SP #3520, 250 gr.
Hornady HP (Pistol) #3575, 158 gr.
Cast Bullets Used .... (Sized to .358" dia.)
*Gas Check Bullets *#358315, 204 gr.
*#358009, 280 gr.
#358430, (Pistol) 195 gr.

TEST SPECIFICATIONS: (Velocity & Pressure)

Firearm Used .... Universal Receiver
Barrel Length .... 24"
Twist .... 1-16"
Groove Dia. .... .3575"

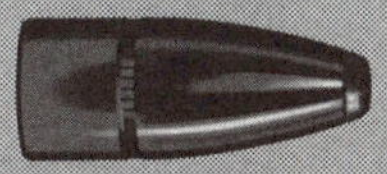

180 gr. Jacketed Flat Soft Point
3.045" OAL

Powder	Sugg. Starting Grains	Velocity fps	Pressure C.U.P.	Max. Load Grains	Velocity fps	Pressure C.U.P.
RX7	47.5	2645	40,300	53.0	2845	49,300
IMR-3031	54.0	2740	37,000	60.0+	2983	48,800
AA2230	55.0	2641	33,300	63.0+	2951	44,600
IMR-4895	56.0	2635	35,700	63.0+	2918	48,100
IMR-4064	57.5	2604	33,300	63.5+	2867	42,600
IMR-4320	57.5	2672	37,500	63.0+	2876	44,700

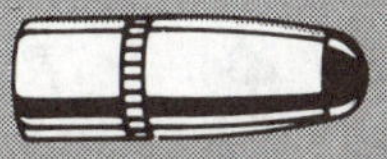

200 gr. Jacketed RN
3.045" OAL

Powder	Sugg. Starting Grains	Velocity fps	Pressure C.U.P.	Max. Load Grains	Velocity fps	Pressure C.U.P.
IMR-3031	52.0	2558	34,800	57.9+	2831	48,500
AA2230	55.0	2537	35,100	61.0+	2768	44,900
H335	57.0	2674	39,500	65.3+	2937	51,700
RX12	58.3	2574	33,300	63.7+	2796	42,700
IMR-4895	53.0	2469	32,700	61.0+	2791	48,400
IMR-4064	56.0	2482	32,400	63.0+	2839	48,700
IMR-4320	56.0	2532	36,500	61.5+	2744	44,400

Note: Loads shown in shaded panels are maximum.
+ Designates a compressed powder charge.

35 Whelen

225 gr. Jacketed Spitzer
3.340" OAL

Powder	Sugg. Starting Grains	Velocity fps	Pressure C.U.P.	Max. Load Grains	Velocity fps	Pressure C.U.P.
IMR-3031	49.0	2354	37,500	54.4+	2617	48,400
AA2460	50.0	2286	38,500	55.2	2563	49,200
H335	50.0	2349	38,700	55.8	2637	49,500
RX12	55.0	2457	39,800	61.5+	2722	50,800
IMR-4895	50.0	2288	32,400	58.0+	2648	48,600
IMR-4064	52.0	2364	35,300	60.0+	2689	49,600
IMR-4320	52.5	2401	37,500	59.7+	2693	51,900

250 gr. Jacketed SP
3.255" OAL

Powder	Sugg. Starting Grains	Velocity fps	Pressure C.U.P.	Max. Load Grains	Velocity fps	Pressure C.U.P.
IMR-3031	46.0	2237	35,400	51.3+	2485	48,200
AA2460	49.0	2252	36,500	56.0	2520	49,100
H335	48.0	2186	35,100	55.7	2514	48,200
RX12	52.5	2323	41,500	58.2+	2578	51,200
IMR-4895	47.4	2169	33,300	55.3+	2511	50,100
IMR-4064	49.3	2260	37,800	56.7+	2529	49,300
IMR-4320	48.3	2198	34,800	56.0+	2501	49,400

158 gr. Jacketed HP
(Pistol Bullet) 2.787" OAL

Powder	Sugg. Starting Grains	Velocity fps	Pressure C.U.P.	Max. Load Grains	Velocity fps	Pressure C.U.P.
2400	26.7	1405	23,300	31.0	1546	40,800

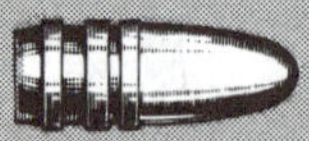

#358315
204 gr. (#2 Alloy) 3.045" OAL

Powder	Sugg. Starting Grains	Velocity fps	Pressure C.U.P.	Max. Load Grains	Velocity fps	Pressure C.U.P.
SR-4759	31.0	2164	32,600	37.3	2416	45,900
IMR-3031	50.0	2558	34,600	58.5	2860	49,000
IMR-4227	28.4	2064	28,200	31.4	2230	40,700
IMR-4198	38.0	2230	25,800	48.0+	2678	46,100

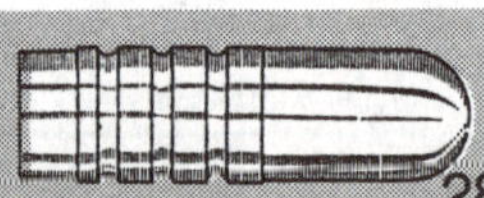

#358009
280 gr. (#2 Alloy) 3.040" OAL

Powder	Sugg. Starting Grains	Velocity fps	Pressure C.U.P.	Max. Load Grains	Velocity fps	Pressure C.U.P.
SR-4759	25.5	1654	25,300	31.6	1968	47,700
IMR-3031	39.0	1901	24,000	48.5+	2349	47,700
IMR-4227	24.0	1651	25,800	28.0	1825	36,100
IMR-4198	30.0	1753	23,500	40.0	2154	41,700

#358430 (Pistol Bullet)
195 gr. (#2 Alloy) 2.805" OAL

Powder	Sugg. Starting Grains	Velocity fps	Pressure C.U.P.	Max. Load Grains	Velocity fps	Pressure C.U.P.
SR-4759	27.0	1935	20,200	34.7	2337	42,400
IMR-3031	39.0	2094	20,500	50.0+	2594	36,300
2400	22.5	1815	17,800	29.0	2255	43,600
Unique	13.0	1546	20,600	19.2	1894	38,100
IMR-4198	33.0	2060	21,800	44.0	2553	40,200

Note: Loads shown in shaded panels are maximum.
\+ Designates a compressed powder charge.

358 Norma Magnum

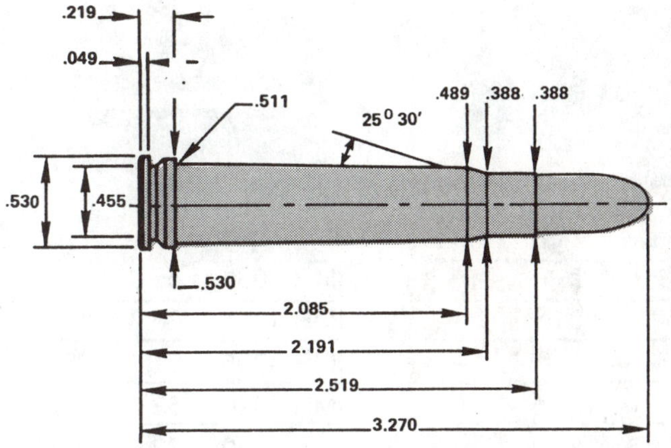

COMMENTS:

The data shown is intended only for use in commercial rifles chambered by the manufacturer for this round. Such rifles have the necessary free bore to be compatible with our loads.

Magnum primers for jacketed bullets and standard primers for cast bullets are suggested.

TEST COMPONENTS:

Cases .... Norma
Trim-to Length .... 2.505"
Primers .... Remington 9½ and 9½ M
Primer Size .... Large Rifle and Large Rifle Magnum
Lyman Shell Holder .... No. 2
Jacketed Bullets Used .... Remington Ptd. SPCL, 150 gr.
Remington SPCL, 200 gr.
Remington Ptd. SPCL, 250 gr.
Hornady RN, 275 gr.
Cast Bullets Used .... (Sized to .358" dia.)
*Gas Check Bullets .... *#358315, 206 gr.

TEST SPECIFICATIONS:
(Velocity Only)

Firearm Used .... Schultz & Larsen
Barrel Length .... 24"
Twist .... 1-12"
Groove Dia. .... .358"

150 gr. Jacketed Ptd. SP

Powder	Sugg. Starting Grains	Velocity fps	Pressure C.U.P.	Max. Load Grains	Velocity fps	Pressure C.U.P.
IMR-3031	65.0	3021	—	72.0+	3378	—
IMR-4064	70.0	2994	—	77.0+	3322	—
IMR-4320	73.0	3174	—	81.5+	3546	—
IMR-4350	72.0	2577	—	79.5+	2941	—
H-380	74.0	2949	—	82.0+	3215	—

Note: Loads shown in shaded panels are maximum.
\+ Designates a compressed powder charge.

358 Norma Magnum

200 gr. Jacketed SP

Powder	Sugg. Starting Grains	Velocity fps	Pressure C.U.P.	Max. Load Grains	Velocity fps	Pressure C.U.P.
IMR-3031	59.0	2617	—	65.0	2941	—
IMR-4064	64.0	2688	—	71.0+	3030	—
IMR-4895	65.0	2793	—	72.0	3115	—
IMR-4320	66.0	2777	—	73.0	3095	—
IMR-4350	72.0	2710	—	79.5+	2941	—
H-380	70.0	2710	—	77.0+	2932	—

250 gr. Jacketed Ptd. SP

Powder	Sugg. Starting Grains	Velocity fps	Pressure C.U.P.	Max. Load Grains	Velocity fps	Pressure C.U.P.
IMR-3031	55.0	2358	—	61.0	2652	—
IMR-4064	60.0	2433	—	67.0+	2739	—
IMR-4895	61.0	2518	—	67.6+	2824	—
IMR-4320	62.0	2525	—	68.0+	2793	—
IMR-4350	72.0	2617	—	79.5+	2857	—
H-380	67.0	2525	—	74.0+	2710	—

275 gr. Jacketed RN

Powder	Sugg. Starting Grains	Velocity fps	Pressure C.U.P.	Max. Load Grains	Velocity fps	Pressure C.U.P.
IMR-3031	52.0	2083	—	58.0	2392	—
IMR-4064	58.0	2267	—	64.0	2528	—
IMR-4895	59.0	2336	—	65.0	2610	—
IMR-4320	59.0	2288	—	65.0	2557	—
IMR-4350	68.0	2409	—	76.0	2688	—
H-380	63.0	2267	—	70.0	2531	—

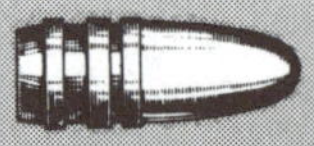

#358315**
206 gr. (#2 Alloy)

Powder	Sugg. Starting Grains	Velocity fps	Pressure C.U.P.	Max. Load Grains	Velocity fps	Pressure C.U.P.
Unique	16.0	1604	—	21.0	1865	—

Note: Loads shown in shaded panels are maximum.
\+ Designates a compressed powder charge.
** Indicates use of Rem. 9½ primers

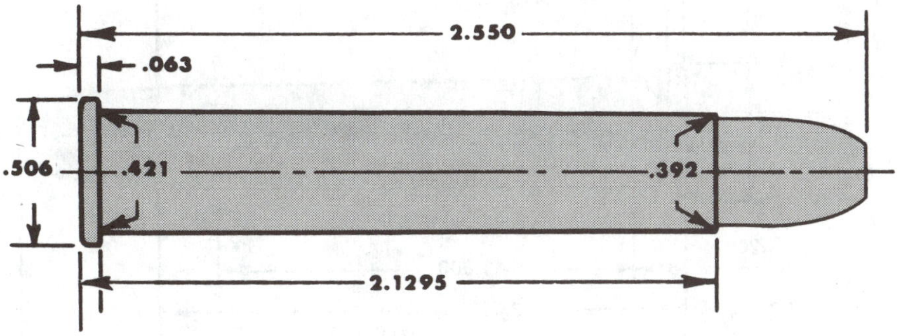

COMMENTS:

Originally developed for early Winchester lever action rifles, this cartridge made the transition from black to smokeless powder loadings. Many rifles chambered for this cartridge have developed excessive headspace and should no longer be used. Therefore this data is intended only for more modern firearms such as the Winchester 94 commemorative so chambered.

TEST COMPONENTS:

Cases .... Remington
Trim-to Length .... 2.118"
Primers .... Remington 9½
Primer Size .... Large Rifle
Lyman Shell Holder .... No. 6
Jacketed Bullets Used .... Remington FP, 255 gr.
Cast Bullets Used .... (Sized to .379" dia.) #375248, 249 gr.

TEST SPECIFICATIONS: (Velocity Only)

Firearm Used .... Winchester Model 94 and Stevens Model 44½
Barrel Length .... Winchester 26"; Stevens 29"
Twist .... 1-18"
Groove Dia. .... Winchester .377"; Stevens .379"

255 gr. Jacketed FP

Powder	Sugg. Starting Grains	Velocity fps	Pressure C.U.P.	Max. Load Grains	Velocity fps	Pressure C.U.P.
IMR-4227	20.0	1275	—	22.0	1374	—
IMR-4198	21.0	1228	—	‡23.0	1322	—
IMR-3031	31.0	1271	—	35.0	1805	—
2400	17.0	1221	—	19.0	1360	—
RX7	27.0	1219	—	30.0	1543	—

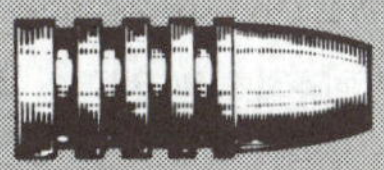

#375248**
249 gr. (#2 Alloy) 2.550" OAL

Powder	Sugg. Starting Grains	Velocity fps	Pressure C.U.P.	Max. Load Grains	Velocity fps	Pressure C.U.P.
Red Dot	6.0	945	—	8.5	1145	—
700X	6.0	950	—	8.5	1150	—
Green Dot	6.5	970	—	9.0	1160	—
PB	6.5	910	—	9.5	1140	—
Unique	7.0	975	—	9.5	1170	—

Note: Loads shown in shaded panels are maximum.
‡ Designates a factory velocity duplication load.
** Tested in a Stevens Model 44½.

375 Winchester

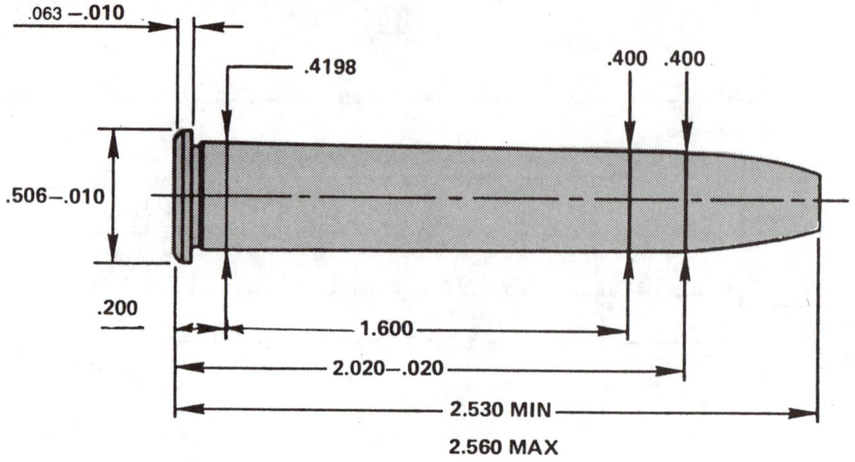

COMMENTS:

This cartridge was the first attempt to beef up the ballistics that could be obtained with a 94 Winchester carbine. Basically it is the rebirth of the 38-55 loaded to higher pressure levels. The rifle was called the Big Bore.

Never put a 375 Winchester round into a 38-55 Winchester. Nor should you attempt to use 38-55 ammo in the 375 Winchester. Incompatible neck diameters can cause serious hazard.

The round ball load listed must be used in a single shot mode. Seat the ball slightly past its full diameter.

TEST COMPONENTS:

Cases .... Winchester
Trim-to Length .... 2.010"
Primers .... Winchester 8½-120
Primer Size .... Large Rifle
Lyman Shell Holder .... No. 6
Jacketed Bullets Used .... Hornady FP #3705, 220 gr.
Barnes FP (.375" dia.) 255 gr.
Cast Bullets Used .... (Sized to .376" dia.)
*Gas Check Bullets #375248, 248 gr.
*#375449, 264 gr.
#375 Round Ball, 79 gr.

TEST SPECIFICATIONS: (Velocity & Pressure)

Firearm Used .... Winchester Big Bore 94 and Universal Receiver
Barrel Length .... Winchester, 20"; Universal Receiver, 24"
Twist .... 1-12"
Groove Dia. .... .376"

220 gr. Jacketed FP
2.560" OAL

Powder	Sugg. Starting Grains	Velocity fps	Pressure C.U.P.	Max. Load Grains	Velocity fps	Pressure C.U.P.
IMR-4198	29.0	1949	44,100	33.0	2145	50,900
RX7	34.0	2031	31,000	38.0+	2259	42,100
IMR-3031	30.3	1732	31,700	35.0+	1983	40,700
748	38.0	1705	25,200	41.5+	1879	30,800
H-4895	34.3	1868	30,000	38.0+	2069	36,600
AA2230	37.5	1951	34,900	42.0+	2178	44,500
AA2460	38.0	1991	33,800	42.0+	2162	41,700
RX12	37.5	1891	29,500	41.0+	2072	33,500

Note: Loads shown in shaded panels are maximum.
\+ Designates a compressed powder charge.

255 gr. Jacketed FP
2.560" OAL

Powder	Sugg. Starting Grains	Velocity fps	Pressure C.U.P.	Max. Load Grains	Velocity fps	Pressure C.U.P.
IMR-4198	23.0	1542	34,500	25.7+	1743	48,700
RX7	25.2	1599	30,000	29.0+	1781	33,700
IMR-3031	26.8	1547	35,800	32.0+	1771	40,900
H-322	29.4	1736	36,500	34.0+	1940	45,800
H-4895	30.0	1632	27,300	34.0+	1838	35,500
AA2230	35.5	1844	37,700	40.5+	2045	48,400
AA2460	34.5	1796	37,800	39.0+	1977	47,100
RX12	33.5	1749	30,700	38.0+	1933	37,900

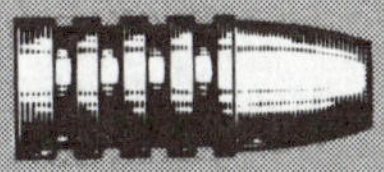

#375248‡
248 gr. (#2 Alloy) 2.450" OAL

Powder	Sugg. Starting Grains	Velocity fps	Pressure C.U.P.	Max. Load Grains	Velocity fps	Pressure C.U.P.
2400	20.0	1595	—	24.0	1897	—
SR-4759	21.0	1606	—	26.0+	1866	—
IMR-4227	22.0	1617	—	26.0	1925	—
IMR-4198	24.0	1581	—	28.5+	1907	—
RX7	27.5	1614	—	32.0+	1811	—
IMR-3031	27.0	1563	—	31.4+	1771	—

#375449‡
264 gr. (#2 Alloy) 2.560" OAL

Powder	Sugg. Starting Grains	Velocity fps	Pressure C.U.P.	Max. Load Grains	Velocity fps	Pressure C.U.P.
2400	19.5	1503	—	24.3	1806	—
SR-4759	20.0	1502	—	24.0	1738	—
IMR-4227	21.0	1520	—	24.8	1807	—
IMR-4198	23.0	1496	—	25.5	1769	—
RX7	25.0	1507	—	32.0+	1762	—
IMR-3031	26.2	1484	—	32.0+	1770	—

#375 Round Ball‡
79 gr. (#2 Alloy .377" dia., as-cast) 2.178" OAL

Powder	Sugg. Starting Grains	Velocity fps	Pressure C.U.P.	Max. Load Grains	Velocity fps	Pressure C.U.P.
Unique	8.0	1542	—	10.5	1851	—
SR-4759	16.0	1551	—	19.0	1800	—

Note: Loads shown in shaded panels are maximum.
\+ Designates a compressed powder charge.
‡ Tested in a Winchester, Big Bore 94.

375 H&H Magnum
(375 Holland and Holland)

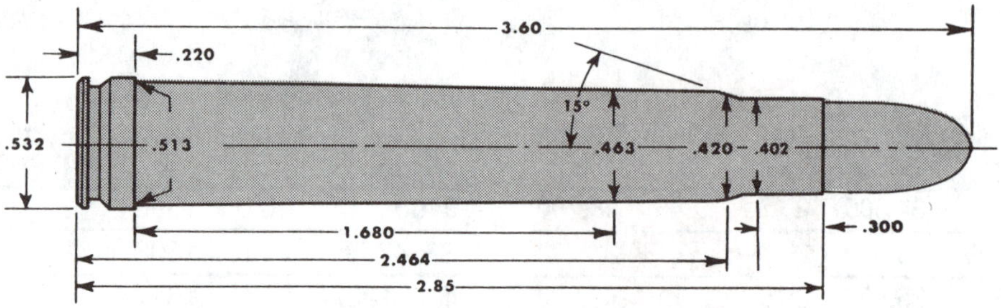

COMMENTS:

This cartridge requires a long magazine and action. This requirement has made it vulnerable to the shorter 338 Winchester Magnum's challenge as the most versatile of world wide hunting cartridges. The 375 H&H also kicks harder than the 338. Nevertheless it remains one of the two popular cartridges that truly are universal in application.

The 375 H&H with the right bullets and good reloads is also an exceptionally accurate cartridge. With a capable rifle and shooter, one inch groups are not at all uncommon.

For use on dangerous game crimped bullets will give the needed reliability of no magazine or feeding jams. Do not attempt to crimp bullets without a cannelure (such as the 235-grain Speer).

IMR 4064 is a very popular choice with all bullet weights. At least once it was a factory selected loading powder. However, of the listed powders IMR 4350 is perhaps the best general selection. It is as accurate as IMR 4064 and offers a bit more velocity.

TEST COMPONENTS:

Cases .... Winchester
Trim-to Length .... 2.840"
Primers .... Remington 9½ M and Winchester 8½-120
Primer Size .... Large Rifle and Large Rifle, Magnum
Lyman Shell Holder .... No. 13
Jacketed Bullets Used .... Speer Semi-Spitzer, 235 gr.
Remington SP, 270 gr.
Speer Semi-Ptd., 285 gr.
Hornady RN, 300 gr.
Cast Bullets Used .... (Sized to .375" dia.)
#375248, 249 gr.

TEST SPECIFICATIONS:
(Velocity Only)

Firearm Used .... Winchester Model 70
Barrel Length .... 24"
Twist .... 1-12"
Groove Dia. .... .375"

235 gr. Jacketed Semi-Spitzer

Powder	Sugg. Starting Grains	Velocity fps	Pressure C.U.P.	Max. Load Grains	Velocity fps	Pressure C.U.P.
IMR-3031	64.0	2597	—	71.0	2849	—
IMR-4895	66.0	2577	—	73.0	2840	—
IMR-4064	67.0	2551	—	74.0	2808	—
IMR-4320	68.0	2583	—	76.0	2865	—
IMR-4350	79.0	2672	—	88.0+	2898	—
H-380	72.0	2512	—	80.0	2754	—
H-4831	80.0	2403	—	88.0+	2604	—

Note: Loads shown in shaded panels are maximum.
+ Designates a compressed powder charge.

375 H&H Magnum

(375 Holland and Holland)

270 gr. Jacketed SP

Powder	Sugg. Starting Grains	Velocity fps	Pressure C.U.P.	Max. Load Grains	Velocity fps	Pressure C.U.P.
IMR-3031	59.0	2369	—	66.5	2631	—
IMR-4895	63.0	2398	—	‡70.0	2680	—
IMR-4064	64.0	2389	—	71.0	2666	—
IMR-4320	65.0	2427	—	72.0	2680	—
IMR-4350	76.0	2570	—	78.5+	2700	—
H-380	70.0	2415	—	78.0	2672	—
H-4831	80.0	2409	—	88.0+	2631	—

285 gr. Jacketed Semi-Ptd.

Powder	Sugg. Starting Grains	Velocity fps	Pressure C.U.P.	Max. Load Grains	Velocity fps	Pressure C.U.P.
IMR-3031	58.0	2217	—	64.5	2465	—
IMR-4895	61.0	2227	—	68.0	2512	—
IMR-4064	62.0	2202	—	69.5	2500	—
IMR-4320	63.0	2237	—	70.5	2525	—
IMR-4350	75.0	2439	—	78.0+	2650	—
H-380	69.0	2262	—	76.0	2512	—
H-4831	79.0	2288	—	87.0+	2525	—

300 gr. Jacketed RN

Powder	Sugg. Starting Grains	Velocity fps	Pressure C.U.P.	Max. Load Grains	Velocity fps	Pressure C.U.P.
IMR-3031	57.0	2150	—	63.0	2386	—
IMR-4895	60.0	2188	—	67.0	2457	—
IMR-4064	61.0	2141	—	68.5	2439	—
IMR-4320	62.0	2212	—	69.5	2493	—
IMR-4350	74.0	2369	—	78.0+	2617	—
H-380	67.0	2222	—	74.0	2444	—
H-4831	77.0	2183	—	85.0+	2444	—

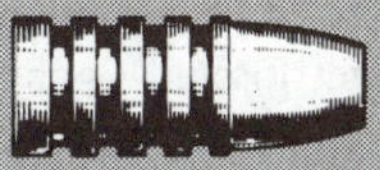

#375248**
249 gr. (#2 Alloy) 3.395" OAL

Powder	Sugg. Starting Grains	Velocity fps	Pressure C.U.P.	Max. Load Grains	Velocity fps	Pressure C.U.P.
Red Dot	10.5	1150	—	15.0	1420	—
700X	10.0	1115	—	15.0	1410	—
Green Dot	11.5	1195	—	16.0	1450	—
PB	11.0	1105	—	16.0	1380	—
Unique	13.0	1260	—	18.0	1525	—
Herco	15.0	1335	—	20.0	1590	—
SR-4756	18.0	1430	—	24.5	1720	—
Blue Dot	17.0	1395	—	24.5	1750	—

Note: Loads shown in shaded panels are maximum.
\+ Designates a compressed powder charge.
‡ Designates a factory velocity duplication load.
** Indicates use of Win. 8½-120 primers

378 Weatherby Magnum

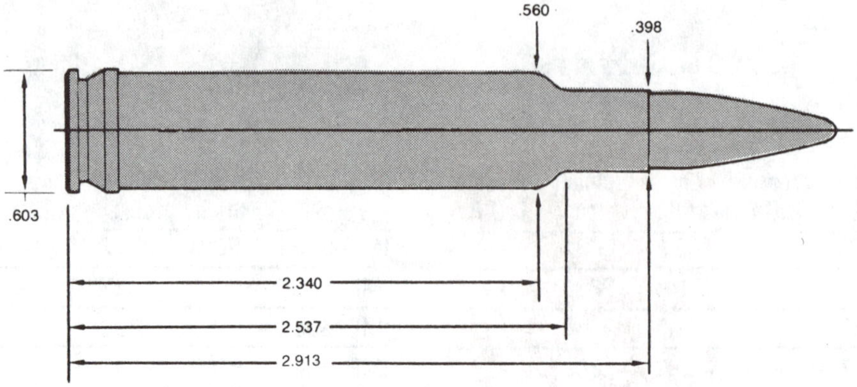

COMMENTS:

This is one of the world's most powerful cartridges. It uses the 416 Rigby case necked down to accept a 0.375" bullet. Its recoil is in proportion to its enormous power. Few shooters can handle this cartridge unless the rifle is equipped with a muzzle break. Due to extremely heavy recoil, the crimping of all bullets is urged.

With a muzzle break and good reloads it is not difficult to get groups approaching one inch at 100 yards. Very strong premium grade bullets are required to withstand impact forces. Lesser bullets will simply disintegrate before penetration. Popular priced bullets are satisfactory for range practice. Shooters will need practice to avoid flinching when using this round. It is the smallest diameter round that many serious African professional hunters will recommend for cape buffalo or elephant.

Magnum primers are essential to igniting the heavy powder charges. The Federal 215 primer is exclusively recommended by Weatherby.

The accompanying data is for use only in Weatherby Mark V rifles.

TEST COMPONENTS:

Cases .... Weatherby
Trim-to Length .... 2.903"
Primers .... Federal 215 and Remington 9½
Primer Size .... Large Rifle and Large Rifle Magnum
Lyman Shell Holder .... No. 17
Jacketed Bullets Used .... Remington SP, 270 gr.
Hornady RN, 300 gr.
Cast Bullets Used .... (Sized to .375" dia.)
*Gas Check Bullets #375248, 249 gr.
*#375449, 267 gr.

TEST SPECIFICATIONS:
(Velocity Only)

Firearm Used .... Weatherby Mark V
Barrel Length .... 26"
Twist .... 1-12"
Groove Dia. .... .375"

270 gr. Jacketed SP

Powder	Sugg. Starting Grains	Velocity fps	Pressure C.U.P.	Max. Load Grains	Velocity fps	Pressure C.U.P.
IMR-4320	86.0	2739	—	96.0	2941	—
IMR-4350	98.0	2695	—	108.0	3076	—
H-4831	105.0	2717	—	118.0+	3086	—

300 gr. Jacketed RN

Powder	Sugg. Starting Grains	Velocity fps	Pressure C.U.P.	Max. Load Grains	Velocity fps	Pressure C.U.P.
IMR-4320	85.0	2481	—	91.0	2695	—
IMR-4350	93.0	2518	—	103.0	2832	—
H-4831	105.0	2672	—	112.0+	2923	—

Note: Loads shown in shaded panels are maximum.
+ Designates a compressed powder charge.

378 Weatherby Magnum

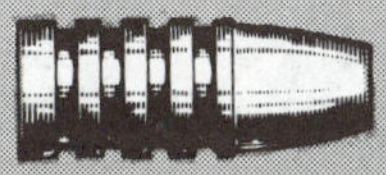

#375248**
249 gr. (#2 Alloy)

Powder	Sugg. Starting Grains	Velocity fps	Pressure C.U.P.	Max. Load Grains	Velocity fps	Pressure C.U.P.
Unique	25.0	1760	—	29.0	1904	—

#375449**
267 gr. (#2 Alloy)

Powder	Sugg. Starting Grains	Velocity fps	Pressure C.U.P.	Max. Load Grains	Velocity fps	Pressure C.U.P.
Unique	24.0	1685	—	28.0	1831	—

Note: Loads shown in shaded panels are maximum.
** Indicates use of Rem. $9^{1}/_{2}$ primers.

41 Remington Magnum
(rifle data)

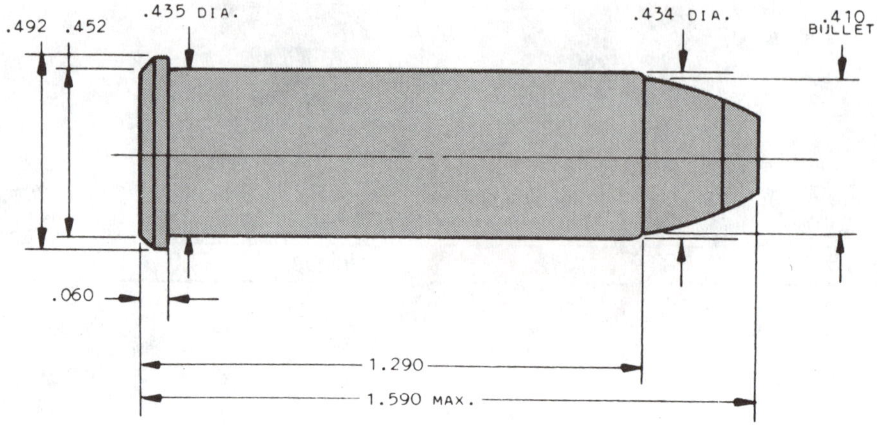

COMMENTS:

We suggest Hercules 2400 with all cast and jacketed bullets as being a good place to begin loading. Used in the Marlin 1894 lever action, this cartridge is popular with shooters who own handguns of the same caliber.

TEST COMPONENTS:

Cases .... Winchester
Trim-to Length .... 1.285"
Primers .... Remington 2½
Primer Size .... Large Pistol
Lyman Shell Holder .... No. 30
Jacketed Bullets Used .... Speer JHP-SWC #4405, 200 gr.
Hornady HP #4100, 210 gr.
Sierra FPJ #8530, 220 gr.
Cast Bullets Used .... (Sized to .410" dia.)
*Gas Check Bullets *#410610, 215 gr.
#410459, 220 gr.

TEST SPECIFICATIONS:
(Velocity Only)

Firearm Used .... Marlin 1894S Carbine
Barrel Length .... 20"
Twist .... 1-18½"
Groove Dia. .... .409"

200 gr. Jacketed HP
1.590" OAL

Powder	Sugg. Starting Grains	Velocity fps	Pressure C.U.P.	Max. Load Grains	Velocity fps	Pressure C.U.P.
Blue Dot	12.5	1406	—	15.4	1679	—
2400	15.3	1467	—	20.0	1827	—
H-110	19.6	1613	—	23.4	1896	—
IMR-4227	18.0	1370	—	21.0	1663	—

210 gr. Jacketed HP
1.585" OAL

Powder	Sugg. Starting Grains	Velocity fps	Pressure C.U.P.	Max. Load Grains	Velocity fps	Pressure C.U.P.
Blue Dot	12.6	1416	—	15.8	1675	—
2400	15.3	1449	—	20.0	1816	—
H-110	19.0	1579	—	22.9	1846	—
IMR-4227	17.8	1366	—	21.0	1641	—

Note: Loads shown in shaded panels are maximum.

41 Remington Magnum
(rifle data)

220 gr. Jacketed Full Profile
1.590" OAL

Powder	Sugg. Starting Grains	Velocity fps	Pressure C.U.P.	Max. Load Grains	Velocity fps	Pressure C.U.P.
Blue Dot	11.0	1198	—	14.2	1549	—
2400	12.9	1167	—	17.4	1599	—
H-110	15.7	1338	—	19.4	1596	—
IMR-4227	15.3	1139	—	19.7	1534	—

#410610
215 gr. (#2 Alloy) 1.575" OAL

Powder	Sugg. Starting Grains	Velocity fps	Pressure C.U.P.	Max. Load Grains	Velocity fps	Pressure C.U.P.
Unique	7.8	1253	—	10.4	1552	—
Blue Dot	11.4	1335	—	14.6	1620	—
2400	14.2	1405	—	18.8	1754	—

#410459
220 gr. (#2 Alloy) 1.700" OAL

Powder	Sugg. Starting Grains	Velocity fps	Pressure C.U.P.	Max. Load Grains	Velocity fps	Pressure C.U.P.
Unique	8.1	1256	—	11.2	1574	—
Blue Dot	13.0	1473	—	16.0	1715	—
2400	15.7	1531	—	20.1	1834	—

Note: Loads shown in shaded panels are maximum.

416 Remington Magnum

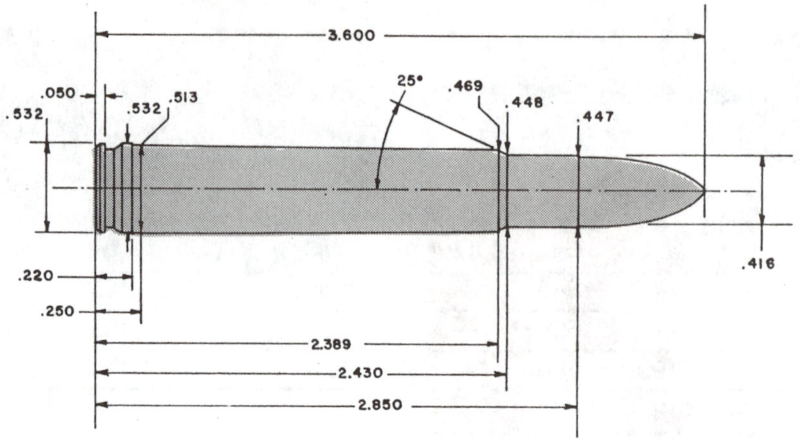

COMMENTS:

This cartridge is a necked up 8mm Remington Magnum. It is ideal for big game applications from brown bear to elephant.

The 416 Remington Magnum is not much different from the wildcat 416 Hoffman. It offers all the power of the 416 Rigby (albeit at higher chamber pressure) and the advantage of more shells in the magazine because of its narrower case diameter.

The 400 grain bullet offers the most punch, but lighter bullets better fit the needs of most North American hunting. Hercules Reloder 15 is the single best propellant and 1"groups are possible with it and the 400 grain bullets.

TEST COMPONENTS:

Cases .... Remington
Trim-to Length .... 2.840"
Primers .... Remington 9½ M
Primer Size .... Large Rifle, Magnum
Lyman Shell Holder .... No. 13
Jacketed Bullets Used .... Barnes "X" #41682, 325 gr.
Barnes Solid R.N. #41628, 350 gr.
Barnes Solid R.N. #41660, 400 gr.

TEST SPECIFICATIONS: (Velocity & Pressure)

Firearm Used .... Universal Receiver
Barrel Length .... 24"
Twist .... 1-14"
Groove Dia. .... .4165"

325 gr. Jacketed "X" Spire Pt.
3.600" OAL

Powder	Sugg. Starting Grains	Velocity fps	Pressure C.U.P.	Max. Load Grains	Velocity fps	Pressure C.U.P.
IMR-3031	74.5	2466	40,300	83.0+	2708	53,100
IMR-4895	75.5	2450	39,700	84.4+	2699	53,100
IMR-4064	77.0	2426	38,200	85.7+	2702	51,500
IMR-4320	76.0	2428	39,800	84.2+	2651	51,400
RX15	80.5	2552	42,200	88.5+	2753	52,300
H-380	81.0	2389	38,300	91.0+	2644	48,500

Note: Loads shown in shaded panels are maximum.
+ Designates a compressed powder charge.

416 Remington Magnum

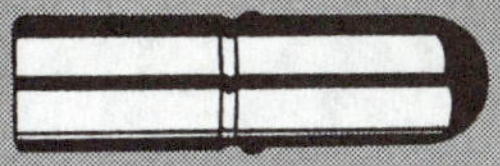

350 gr. Jacketed R.N. Solid
3.590" OAL

Powder	Sugg. Starting Grains	Velocity fps	Pressure C.U.P.	Max. Load Grains	Velocity fps	Pressure C.U.P.
IMR-3031	74.0	2389	41,200	82.5+	2608	52,300
IMR-4895	75.0	2370	41,300	84.2+	2610	52,800
IMR-4064	76.5	2364	39,700	85.5+	2613	52,400
IMR-4320	75.0	2341	40,300	84.0+	2565	49,900
RX15	79.0	2442	41,800	88.0+	2668	52,500
H-380	81.0	2329	39,800	91.0+	2567	48,700

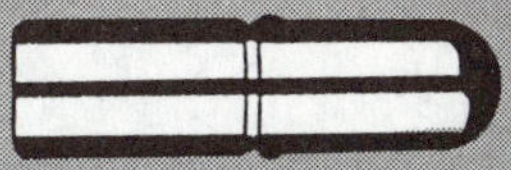

400 gr. Jacketed R.N. Solid
3.590" OAL

Powder	Sugg. Starting Grains	Velocity fps	Pressure C.U.P.	Max. Load Grains	Velocity fps	Pressure C.U.P.
IMR-3031	67.5	2149	40,700	75.2+	2353	52,500
IMR-4895	69.0	2151	40,000	77.3+	2376	52,900
IMR-4064	71.0	2156	39,800	78.8+	2385	52,000
IMR-4320	70.0	2106	39,000	77.2+	2320	50,800
RX15	71.5	2155	38,000	80.0+	2415	51,600
H-380	78.0	2200	42,200	87.0+	2403	50,400

Note: Loads shown in shaded panels are maximum.
\+ Designates a compressed powder charge.

416 Rigby

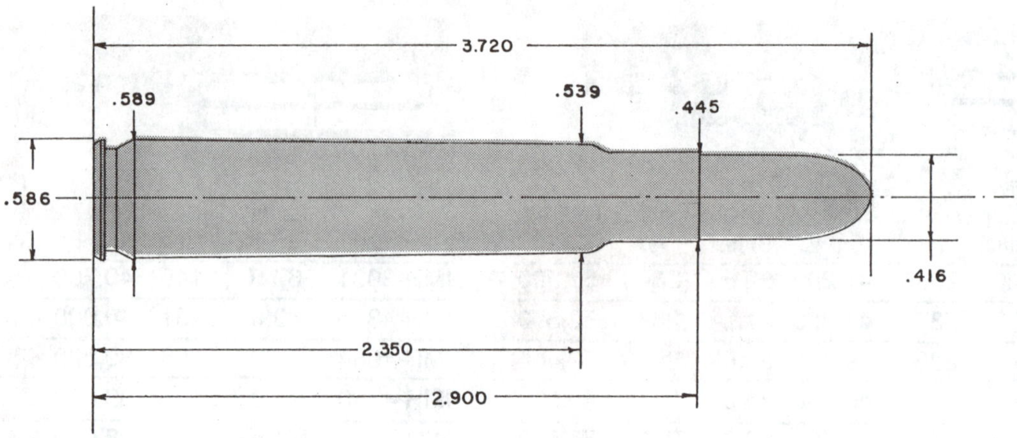

COMMENTS:

The 416 Rigby is a standard all around African cartridge, and has been so for most of the century. It is a large cartridge and as such it reaches its enormous power with relatively low chamber pressure. This in turn means easy case extraction even under adverse conditions.

As in the 416 Remington, the 400 grain bullet seems most accurate and will get the most power out of the round. Groups to 1$^1/_2$" are not uncommon. The Rigby case, due to its huge volume needs slow burning powders to fill it so as to get uniform ballistics. IMR 4831, and IMR 7828 are top selections.

With lighter bullets such as Barnes "X" series reloaders can obtain about a three hundred feet per second more velocity than with standard 400 gr. factory loads. This is possible at a slight increase in chamber pressure, though pressure still stays well within the limits of safety.

TEST COMPONENTS:

Cases .... Federal
Trim-to Length .... 2.890"
Primers .... Federal 215
Primer Size .... Large Rifle, Magnum
Lyman Shell Holder .... No. 17
Jacketed Bullets Used .... Barnes "X" #41682, 325 gr.
Barnes Solid R.N. #41628, 350 gr.
Barnes Solid R.N. #41660, 400 gr.

TEST SPECIFICATIONS:
(Velocity Only)

Firearm Used .... Ruger M/77
Barrel Length .... 24"
Twist .... 1-16$^1/_2$"
Groove Dia. .... .4165"

325 gr. Jacketed "X" Spire Pt.
3.720" OAL

Powder	Sugg. Starting Grains	Velocity fps	Pressure C.U.P.	Max. Load Grains	Velocity fps	Pressure C.U.P.
H-380	88.0	2467	—	98.0	2723	—
760	90.0	2498	—	100.0	2762	—
IMR-4350	97.5	2619	—	108.0+	2861	—
IMR-4831	100.0	2592	—	111.0+	2898	—
AA3100	103.0	2550	—	114.0+	2757	—
RX22	104.0	2629	—	115.0+	2871	—
IMR-7828	105.0	2497	—	116.0+	2768	—

Note: Loads shown in shaded panels are maximum.
\+ Designates a compressed powder charge.

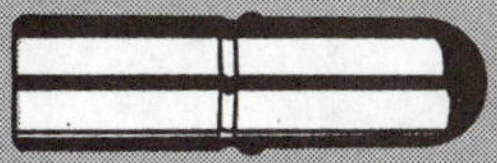

350 gr. Jacketed R.N. Solid
3.625" OAL

Powder	Sugg. Starting Grains	Velocity fps	Pressure C.U.P.	Max. Load Grains	Velocity fps	Pressure C.U.P.
760	89.0	2454	—	99.0	2601	—
IMR-4350	93.0	2408	—	104.0+	2753	—
RX19	98.0	2417	—	110.0+	2749	—
IMR-4831	98.0	2467	—	109.0+	2738	—
AA3100	101.0	2450	—	112.0+	2657	—
RX22	102.0	2404	—	113.0+	2713	—
IMR-7828	103.0	2399	—	114.0+	2644	—

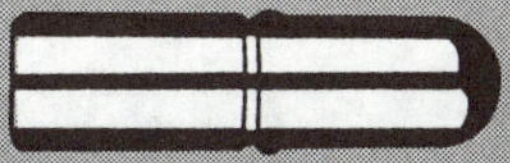

400 gr. Jacketed R.N. Solid
3.625" OAL

Powder	Sugg. Starting Grains	Velocity fps	Pressure C.U.P.	Max. Load Grains	Velocity fps	Pressure C.U.P.
760	85.0	2261	—	96.0	2508	—
IMR-4350	90.0	2373	—	101.0+	2585	—
RX19	97.0	2394	—	108.0+	2594	—
IMR-4831	95.0	2397	—	106.0+	2603	—
AA3100	100.0	2358	—	109.0+	2523	—
RX22	100.0	2384	—	110.0+	2575	—
IMR-7828	103.0	2376	—	111.0+	2541	—

Note: Loads shown in shaded panels are maximum.
+ Designates a compressed powder charge.

416 Weatherby Magnum

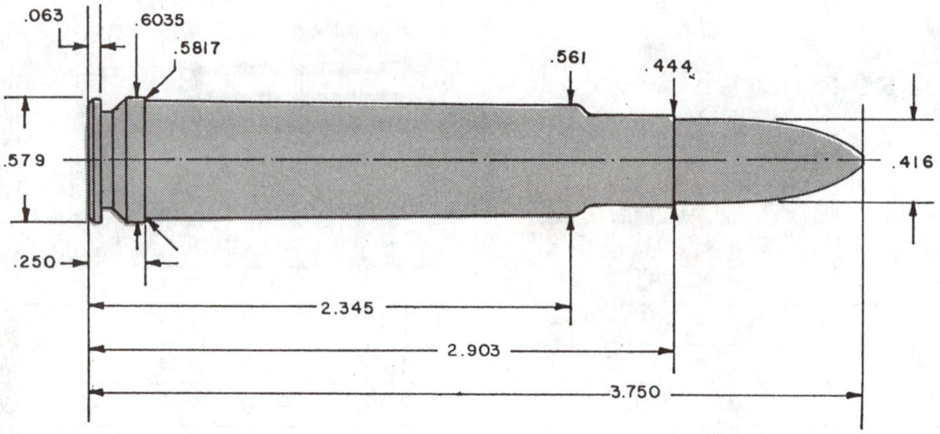

COMMENTS:

This case is the 416 Rigby with a belt added and loaded to substantially higher pressure. The much higher chamber pressures result in almost 300 feet per second additional velocity with the 400 grain bullet compared to the 2400 feet per second velocity of factory 416 Rigby ammo. Recoil may exceed most shooters' ability. Of course the 416 Weatherby can easily be loaded down to Rigby ballistics which makes it a manageable round.

The slowest burning powders are ideal. For Rigby level ballistics IMR 4831 seems best. For full power loads, IMR 7828 is the wisest selection, though all the listed propellants are quite satisfactory. This is enough cartridge to make almost anyone believe in muzzle brakes. Even then full power loads are nothing for the timid shooter. Such loads would undoubtedly give an advantage when hunting elephant as compared to the Remington or Rigby 416's, but for almost any other purpose they might be overly potent.

TEST COMPONENTS:

Cases .... Weatherby
Trim-to Length .... 2.898"
Primers .... Federal 215
Primer Size .... Large Rifle, Magnum
Lyman Shell Holder .... No. 17
Jacketed Bullets Used .... Barnes "X" #41682, 325 gr.
Barnes Solid R.N. #41628, 350 gr.
A-Square Dead-Tough SP, 400 gr.

TEST SPECIFICATIONS:
(Velocity Only)

Firearm Used .... Weatherby Mark V
Barrel Length .... 24"
Twist .... 1-14"
Groove Dia. .... .416"

325 gr. Jacketed "X" Spire Pt.
3.700" OAL

Powder	Sugg. Starting Grains	Velocity fps	Pressure C.U.P.	Max. Load Grains	Velocity fps	Pressure C.U.P.
760	98.0	2677	—	109.0	2800	—
IMR-4350	102.0	2627	—	113.0+	2914	—
RX19	108.0	2613	—	120.0+	2902	—
IMR-4831	105.0	2603	—	117.0+	2877	—
AA 3100	107.0	2551	—	119.0+	2767	—
RX22	111.0	2586	—	124.0+	2887	—
IMR-7828	110.0	2536	—	122.0+	2819	—

Note: Loads shown in shaded panels are maximum.
+ Designates a compressed powder charge.

416 Weatherby Magnum

350 gr. Jacketed R.N. Solid
3.640" OAL

Powder	Sugg. Starting Grains	Velocity fps	Pressure C.U.P.	Max. Load Grains	Velocity fps	Pressure C.U.P.
760	96.0	2495	—	107.0	2708	—
IMR-4350	98.0	2464	—	109.0+	2783	—
RX19	106.0	2533	—	118.0+	2822	—
IMR-4831	102.0	2460	—	114.0+	2766	—
AA 3100	106.0	2469	—	118.0+	2687	—
RX22	109.0	2477	—	121.0+	2756	—
IMR-7828	107.0	2445	—	119.0+	2727	—

400 gr. Jacketed Soft Pt.
3.600" OAL

Powder	Sugg. Starting Grains	Velocity fps	Pressure C.U.P.	Max. Load Grains	Velocity fps	Pressure C.U.P.
IMR-4350	94.0	2261	—	104.0+	2609	—
RX19	101.0	2307	—	112.0+	2686	—
IMR-4831	100.0	2309	—	111.0+	2659	—
RX22	104.0	2286	—	116.0+	2685	—
IMR-7828	104.0	2290	—	116.0+	2635	—

Note: Loads shown in shaded panels are maximum.
+ Designates a compressed powder charge.

44-40 Winchester
(44 WCF)

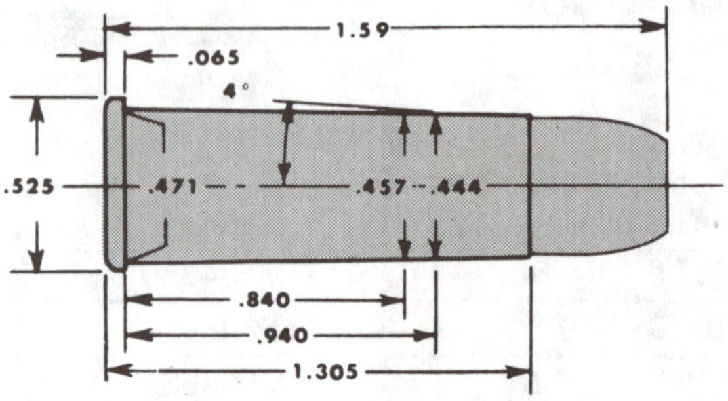

COMMENTS:

These loads should not be used in handguns or in rifles that were designed for black powder.

Individual tolerances vary greatly in rifles chambered for this cartridge. Therefore , extreme care should be used in working up maximum loads. CAUTION: THE RELATIVE STRENGTH OF 44/40 FIREARMS VARIES GREATLY WITH ACTION DESIGN AND DATE OF MANUFACTURE. As a reference we have categorized actions into two groups:

GROUP I (Weaker Actions)	GROUP 2 (Strong Actions)
Winchester Model 1873 Lever Action	Winchester Model 1892 Lever Action
Whitney-Kennedy Lever Action	Marlin Model 1889 Lever Action
Colt-Burgess Lever Action	Marlin Model 1894 Lever Action
Marlin Model 1888 Lever Action	Remington-Keene Bolt Action
Colt "Lightning" Slide Action	Remington Model 14½ Slide Action
Euroarms Replica 1873 Lever Action	Winchester Single Shot Rifles
Navy Arms Replica 1873 Lever Action	Remington No. 1 Rolling Block S.S.
Remington No. 2 Rolling-Block S.S.	Remington "Baby Carbine" S.S.
Ballard No. 2 Single Shot	Stevens Model 44½ Single Shot
Stevens Model 44 Single Shot	

For Group 1 we recommend not to exceed 13,700 CUP maximum pressure and for Group 2 not to exceed 22,000 CUP maximum pressure. Our data is divided up into loads for each category. In all cases use caution in working up to maximum loads.

Salvaged cases should be examined to determine if they are of the old balloon head construction. These cases should be considered unsafe for smokeless powder loads. Due to variations in groove diameters, it is recommended that you slug your barrel before reloading.

TEST COMPONENTS:

Cases .... Winchester
Trim-to Length .... 1.300"
Primers .... Winchester 7-111
Primer Size .... Large Pistol
Lyman Shell Holder .... No. 14B
Jacketed Bullets Used .... Speer HP, 200 gr.
Cast Bullets Used .... (Sized to .429" dia.)
*Gas Check Bullets .... #427098, 205 gr.
*#429434, 220 gr.

TEST SPECIFICATIONS:

Firearm Used .... Special Pressure Test Barrel
Barrel Length .... 24"
Twist .... 1-36"
Groove Dia. .... .429"

Note: Loads shown in shaded panels are maximum.
‡ Designates a factory velocity duplication load.

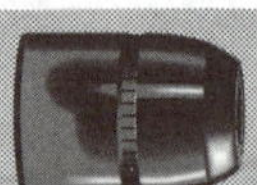

200 gr. Jacketed HP
1.600" OAL

Group 1

Powder	Sugg. Starting Grains	Velocity fps	Pressure C.U.P.	Max. Load Grains	Velocity fps	Pressure C.U.P.
Unique	7.6	865	10,600	8.5	981	13,300
2400	14.4	980	6,600	16.0	1183	11,900
IMR-4227	16.5	895	4,800	‡18.5	1212	11,600

Group 2

Powder	Sugg. Starting Grains	Velocity fps	Pressure C.U.P.	Max. Load Grains	Velocity fps	Pressure C.U.P.
Unique	8.5	981	13,300	10.2	1282	19,600
2400	18.0	1380	14,600	20.0	1638	19,000
IMR-4227	‡18.5	1212	11,600	20.5	1455	19,600

44-40 Winchester
(44 WCF)

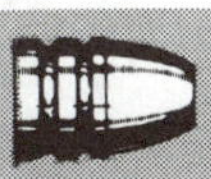

#427098
205 gr. (#2 Alloy) 1.600" OAL

Group 1

Powder	Sugg. Starting Grains	Velocity fps	Pressure C.U.P.	Max. Load Grains	Velocity fps	Pressure C.U.P.
Red Dot	5.5	980	7,500	6.5	1125	13,600
Green Dot	7.0	1059	8,700	‡8.0	1194	12,900
Unique	8.0	1134	8,700	9.0	1278	13,700

Group 2

Powder	Sugg. Starting Grains	Velocity fps	Pressure C.U.P.	Max. Load Grains	Velocity fps	Pressure C.U.P.
Red Dot	7.0	1180	17,200	8.0	1290	21,900
Green Dot	‡8.0	1194	12,900	9.0	1340	19,300
Unique	9.5	1319	15,200	10.5	1410	19,700

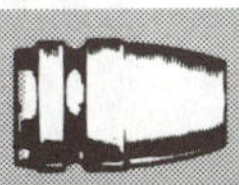

#429434
220 gr. (#2 Alloy) 1.600" OAL

Group 1

Powder	Sugg. Starting Grains	Velocity fps	Pressure C.U.P.	Max. Load Grains	Velocity fps	Pressure C.U.P.
700X	5.2	878	9,200	6.2	1017	13,700
Unique	6.5	919	7,900	7.5	1062	12,800
SR-7625	6.8	902	6,200	7.8	1059	10,800

Group 2

Powder	Sugg. Starting Grains	Velocity fps	Pressure C.U.P.	Max. Load Grains	Velocity fps	Pressure C.U.P.
700X	6.5	1049	15,300	7.5	1169	20,600
Unique	7.5	1062	12,800	8.5	1187	18,900
SR-7625	7.8	1059	10,800	9.0	1229	21,600

Note: Loads shown in shaded panels are maximum.
‡ Designates a factory velocity duplication load.

44 Remington Magnum
(rifle data)

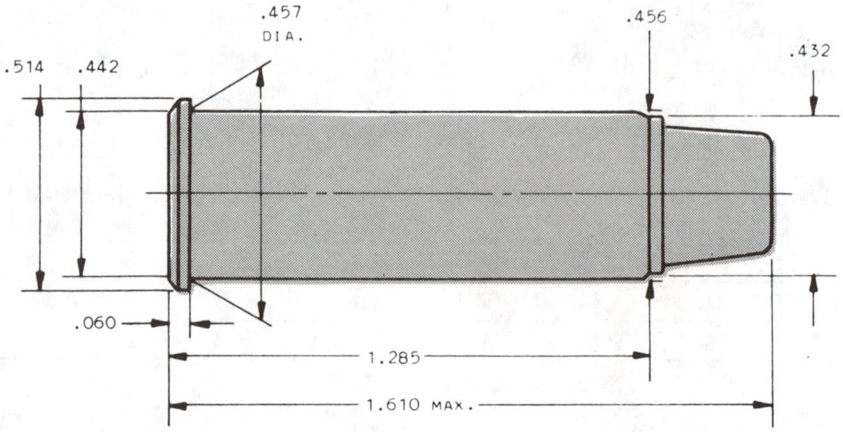

COMMENTS:

Pressure limits for this cartridge are the same in rifles or handguns. High primers are an extraordinary danger in semi-automatic rifles. Be extra careful. Cast bullets tend to foul the gas piston in these same firearms and hence are not recommended for use in the semi-autos.

In rifles with shallow multi-groove rifling cast bullet velocity must be held to 1,600 fps. or less. The use of hard alloys and maximum sizing diameters will also help accuracy in these rifles.

Use the exact listed overall length to get correct seating depth with cast bullets. Bullet #429640 may not feed from the magazine in some Marlin rifles.

Use only flat nosed bullets in tubular magazines.

TEST COMPONENTS:

Cases .... Remington
Trim-to Length .... 1.280"
Primers .... Remington 2½
Primer Size .... Large Pistol
Lyman Shell Holder .... No. 7
Jacketed Bullets Used .... Speer HP, 225 gr.
Remington SP, 240 gr.
Cast Bullets Used .... (Sized to .429" dia.)
*Gas Check Bullets *#429215, 215 gr.
#429421, 245 gr.
*#429244, 250 gr.
*#429640, 275 gr.

TEST SPECIFICATIONS:
(Velocity Only)

Firearm Used .... Marlin 336 and Ruger .44 Carbine
Barrel Length .... Marlin 20"; Ruger 18½"
Twist .... 1-38"
Groove Dia. .... Marlin .431"; Ruger .429"

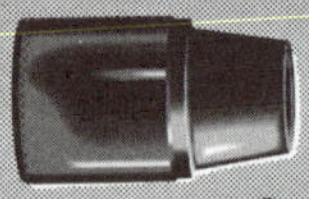

225 gr. Jacketed HP††

Powder	Sugg. Starting Grains	Velocity fps	Pressure C.U.P.	Max. Load Grains	Velocity fps	Pressure C.U.P.
SR-4756	14.0	1519	—	16.5+	1680	—
IMR-4227	23.0	1225	—	26.0+	1760	—
H-110	22.0	1666	—	25.0+	1872	—
Herco	12.0	1396	—	14.5+	1604	—
2400	21.0	1689	—	23.0+	1833	—

240 gr. Jacketed SP††

Powder	Sugg. Starting Grains	Velocity fps	Pressure C.U.P.	Max. Load Grains	Velocity fps	Pressure C.U.P.
SR-4756	13.0	1512	—	15.6	1602	—
IMR-4227	22.0	1536	—	‡25.0+	1742	—
H-110	22.0	1630	—	24.0+	1779	—
Herco	11.0	1344	—	13.5	1530	—
2400	20.0	1547	—	22.0+	1726	—

Note: Loads shown in shaded panels are maximum.
\+ Designates a compressed powder charge.
‡ Designates a factory velocity duplication load.
†† Tested in a Marlin 336

44 Remington Magnum
(rifle data)

#429215†††
215 gr. (#2 Alloy) 1.645" OAL

Powder	Sugg. Starting Grains	Velocity fps	Pressure C.U.P.	Max. Load Grains	Velocity fps	Pressure C.U.P.
Bullseye	5.0	935	—	7.5	1210	—
Unique	8.5	1205	—	12.5	1595	—
Herco	11.0	1365	—	15.5	1685	—
SR-4756	12.0	1425	—	15.0	1660	—
Blue Dot	14.0	1370	—	19.5	1825	—
2400	18.0	1365	—	23.5	1735	—
IMR-4227	20.0	1265	—	25.5	1695	—

#429421†††
245 gr. (#2 Alloy) 1.710" OAL

Powder	Sugg. Starting Grains	Velocity fps	Pressure C.U.P.	Max. Load Grains	Velocity fps	Pressure C.U.P.
Bullseye	5.0	915	—	7.5	1180	—
Unique	8.5	1195	—	12.0	1495	—
Herco	11.0	1345	—	14.5	1565	—
SR-4756	11.0	1305	—	13.0	1470	—
Blue Dot	14.0	1410	—	17.5	1665	—
2400	18.0	1375	—	22.5	1680	—
IMR-4227	20.0	1310	—	24.0	1610	—

#429244†††
250 gr. (#2 Alloy) 1.680" OAL

Powder	Sugg. Starting Grains	Velocity fps	Pressure C.U.P.	Max. Load Grains	Velocity fps	Pressure C.U.P.
Bullseye	5.0	910	—	7.5	1160	—
Unique	8.5	1180	—	12.0	1485	—
Herco	11.0	1320	—	14.5	1625	—
SR-4756	11.0	1270	—	12.5	1405	—
Blue Dot	14.0	1410	—	17.5	1655	—
2400	18.0	1370	—	22.0	1635	—
IMR-4227	20.0	1300	—	24.0	1595	—

#429640††
275 gr. (#2 Alloy) 1.700" OAL

Powder	Sugg. Starting Grains	Velocity fps	Pressure C.U.P.	Max. Load Grains	Velocity fps	Pressure C.U.P.
Unique	9.5	1245	—	11.5	1380	—
Herco	10.2	1188	—	12.0	1337	—
SR-4756	10.4	1251	—	11.8	1348	—
800X	10.5	1308	—	12.8	1464	—
Blue Dot	13.2	1317	—	16.7	1568	—
AA9	18.0	1495	—	20.2	1619	—
2400	17.6	1468	—	20.4	1656	—
IMR-4227	19.5	1400	—	23.0	1615	—

Note: Loads shown in shaded panels are maximum.
\+ Designates a compressed powder charge.
†† Tested in a Marlin 336.
††† Tested in a Ruger .44 Carbine.

444 Marlin

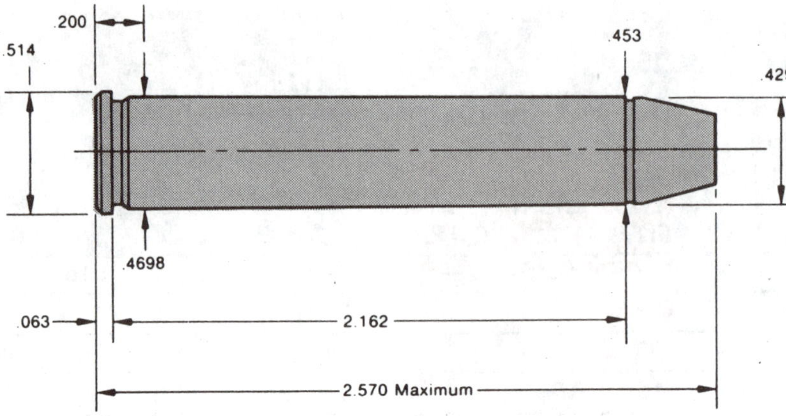

COMMENTS:

Many of the listed jacketed bullet loads are heavily compressed. As a result, crimping die as well as the bullet seating screw adjustments will need to be precise or crushed cases will occur. Case neck and bullet cannelure must align precisely to prevent this problem. If the load must be excessively compressed with your lots of components this precision will not be possible.

The shallow groove rifling commonly used in these rifles requires hard bullet alloys and velocities at or below 1,600 fps. when using cast bullets if accuracy is to be acceptable. A maximum diameter sizing die will also be beneficial.

In order to seat cast bullets properly the listed maximum overall cartridge length must be exceeded. Our data takes this into consideration. Use the length listed with each specific bullet. Bullet #429421 is crimped on the forward edge of the driving band.

TEST COMPONENTS:

Cases .... Remington
Trim-to Length .... 2.220"
Primers .... Remington 9½
Primer Size .... Large Rifle
Lyman Shell Holder .... No. 14B
Jacketed Bullets Used .... Speer HP, 225 gr.
Remington SP, 240 gr.
Cast Bullets Used .... (Sized to .431" dia.)
*Gas Check Bullets: *#429215, 215 gr.
#429360, 232 gr.
#429421, 245 gr.
*#429244, 250 gr.

TEST SPECIFICATIONS:
(Velocity Only)

Firearm Used .... Marlin Model 444
Barrel Length .... 22"
Twist .... 1-38"
Groove Dia. .... .431"

Note: Loads shown in shaded panels are maximum.
+ Designates a compressed powder charge.
‡ Designates a factory velocity duplication load.

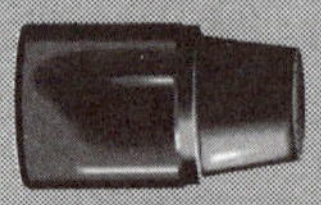

225 gr. Jacketed HP

Powder	Sugg. Starting Grains	Velocity fps	Pressure C.U.P.	Max. Load Grains	Velocity fps	Pressure C.U.P.
IMR-4198	46.0	2288	—	49.0+	2409	—
IMR-3031	50.0	2032	—	55.0+	2237	—
IMR-4895	52.0	2036	—	57.0+	2217	—

240 gr. Jacketed SP

Powder	Sugg. Starting Grains	Velocity fps	Pressure C.U.P.	Max. Load Grains	Velocity fps	Pressure C.U.P.
IMR-4198	42.0	2096	—	‡47.0+	2341	—
IMR-3031	50.0	2053	—	55.0+	2242	—
IMR-4895	52.0	2032	—	57.0+	2232	—

#429215
215 gr. (#2 Alloy) 2.595" OAL

Powder	Sugg. Starting Grains	Velocity fps	Pressure C.U.P.	Max. Load Grains	Velocity fps	Pressure C.U.P.
Red Dot	10.5	1335	—	16.0	1685	—
700X	10.0	1295	—	15.5	1655	—
Green Dot	11.0	1325	—	16.5	1685	—
Unique	12.0	1335	—	18.0	1735	—
SR-7625	13.0	1405	—	18.0	1705	—
SR-4756	14.0	1350	—	19.0	1725	—
Blue Dot	15.0	1295	—	20.0	1690	—

#429360
232 gr. (#2 Alloy) 2.575" OAL

Powder	Sugg. Starting Grains	Velocity fps	Pressure C.U.P.	Max. Load Grains	Velocity fps	Pressure C.U.P.
Red Dot	10.5	1295	—	15.5	1630	—
700X	10.0	1260	—	15.0	1600	—
Green Dot	11.0	1290	—	16.0	1615	—
Unique	12.0	1300	—	17.5	1660	—
SR-7625	13.0	1380	—	17.5	1650	—
SR-4756	14.0	1365	—	18.5	1680	—
Blue Dot	15.0	1310	—	19.5	1605	—

#429421
245 gr. (#2 Alloy) 2.550" OAL

Powder	Sugg. Starting Grains	Velocity fps	Pressure C.U.P.	Max. Load Grains	Velocity fps	Pressure C.U.P.
Red Dot	10.0	1225	—	15.0	1555	—
700X	9.5	1200	—	14.5	1535	—
Green Dot	10.5	1235	—	15.5	1570	—
Unique	11.5	1270	—	17.0	1620	—
SR-7625	12.5	1310	—	17.0	1590	—
SR-4756	13.5	1305	—	18.0	1590	—
Blue Dot	14.5	1240	—	19.0	1570	—

#429244
250 gr. (#2 Alloy) 2.620" OAL

Powder	Sugg. Starting Grains	Velocity fps	Pressure C.U.P.	Max. Load Grains	Velocity fps	Pressure C.U.P.
Red Dot	10.0	1205	—	15.0	1520	—
700X	9.5	1175	—	14.5	1500	—
Green Dot	10.5	1215	—	15.5	1525	—
Unique	11.5	1225	—	17.0	1590	—
SR-7625	12.5	1285	—	17.0	1560	—
SR-4756	13.5	1230	—	18.0	1560	—
Blue Dot	14.5	1240	—	19.0	1540	—

Note: Loads shown in shaded panels are maximum.

45 Automatic (rifle data)

(45 ACP)

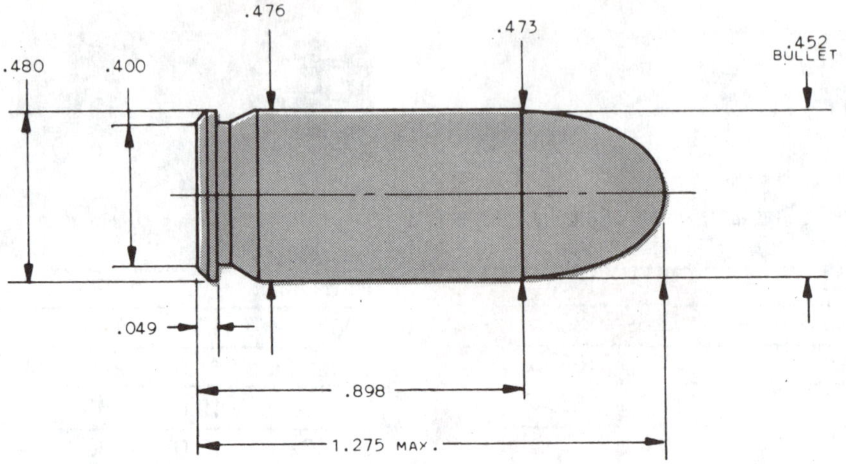

COMMENTS:

This cartridge headspaces on the case mouth. Adhere to the listed maximum and trim-to case lengths exactly.

Do not go below the starting load as in so doing a bullet may become lodged in the barrel. Should the gun be fired in this condition a serious accident could result. For the same reason be doubly certain that every case has the correct powder charge.

TEST COMPONENTS:

Cases .... Federal
Trim-to Length .... .895"
Primers .... CCI 300
Primer Size .... Large Pistol
Lyman Shell Holder .... No. 2
Jacketed Bullets Used .... Hornady JHP #4510, 185 gr.
Speer JHP #4477, 200 gr.
Speer JHP #4479, 225 gr.
Sierra FMJ #8815, 230 gr.
Sierra JHC #8820, 240 gr.
Cast Bullets Used .... (Sized to .451" dia.)
*Gas Check Bullets #452460, 200 gr.
#452374, 225 gr.
*#452484, 250 gr.

TEST SPECIFICATIONS:
(Velocity Only)

Firearm Used .... Marlin Model 45 Carbine
Barrel Length .... 16 1/2"
Twist .... 1-16"
Groove Dia. .... .451"

185 gr. Jacketed HP
1.175" OAL

Powder	Sugg. Starting Grains	Velocity fps	Pressure C.U.P.	Max. Load Grains	Velocity fps	Pressure C.U.P.
Bullseye	4.5	756	—	6.0	1041	—
Red Dot	4.5	818	—	6.1	1095	—
231	5.0	771	—	6.1	986	—
700X	4.5	768	—	5.5	987	—
AA5	7.4	911	—	9.2	1245	—
Unique	5.8	829	—	7.8	1206	—

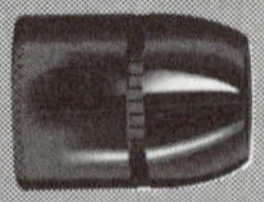

200 gr. Jacketed HP
1.178" OAL

Powder	Sugg. Starting Grains	Velocity fps	Pressure C.U.P.	Max. Load Grains	Velocity fps	Pressure C.U.P.
231	5.0	667	—	6.5	1009	—
AA5	6.8	715	—	8.5	1099	—
Unique	6.0	827	—	6.5	911	—
SR-4756	6.5	640	—	7.4	955	—

Note: Loads shown in shaded panels are maximum.

45 Automatic (rifle data)

(45 ACP)

225 gr. Jacketed HP
1.243" OAL

Powder	Sugg. Starting Grains	Velocity fps	Pressure C.U.P.	Max. Load Grains	Velocity fps	Pressure C.U.P.
Unique	6.0	765	—	6.7	906	—
HS-7	9.0	783	—	10.3	984	—
Blue Dot	8.5	738	—	10.7	1099	—

230 gr. Jacketed FMJ
1.275" OAL

Powder	Sugg. Starting Grains	Velocity fps	Pressure C.U.P.	Max. Load Grains	Velocity fps	Pressure C.U.P.
Unique	5.5	733	—	6.8	974	—
SR-4756	6.0	597	—	7.0	894	—
Blue Dot	8.5	787	—	10.2	1034	—

240 gr. Jacketed Hollow Cavity
1.185" OAL

Powder	Sugg. Starting Grains	Velocity fps	Pressure C.U.P.	Max. Load Grains	Velocity fps	Pressure C.U.P.
Unique	5.5	712	—	6.6	957	—
SR-4756	6.0	764	—	6.9	911	—
Blue Dot	7.5	701	—	9.3	948	—

#452460
200 gr. (#2 Alloy) 1.161" OAL

Powder	Sugg. Starting Grains	Velocity fps	Pressure C.U.P.	Max. Load Grains	Velocity fps	Pressure C.U.P.
Red Dot	4.0	913	—	5.3	1133	—
231	4.0	819	—	6.0	1138	—
700X	4.0	896	—	5.2	1111	—
AA5	6.4	939	—	8.0	1223	—
Unique	5.0	955	—	7.5	1314	—
HS-6	6.0	761	—	8.7	1124	—
SR-4756	5.3	880	—	7.3	1189	—
Blue Dot	7.1	767	—	10.6	1275	—

#452374
225 gr. (#2 Alloy) 1.272" OAL

Powder	Sugg. Starting Grains	Velocity fps	Pressure C.U.P.	Max. Load Grains	Velocity fps	Pressure C.U.P.
Red Dot	4.3	923	—	5.3	1061	—
231	5.0	942	—	5.8	1055	—
700X	4.0	847	—	5.0	1019	—
AA5	6.6	931	—	8.2	1207	—
Unique	5.5	916	—	7.3	1196	—
HS-6	6.2	770	—	8.6	1087	—
SR-4756	5.5	803	—	7.5	1188	—
Blue Dot	7.5	869	—	10.7	1262	—

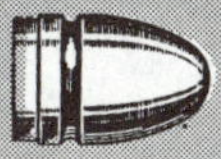

#452484
250 gr. (#2 Alloy) 1.272" OAL

Powder	Sugg. Starting Grains	Velocity fps	Pressure C.U.P.	Max. Load Grains	Velocity fps	Pressure C.U.P.
700X	4.0	830	—	4.5	903	—
Unique	5.0	811	—	6.0	1021	—
HS-7	8.0	869	—	9.3	1027	—
Blue Dot	7.0	764	—	8.5	963	—

Note: Loads shown in shaded panels are maximum.

45 Colt (rifle data)

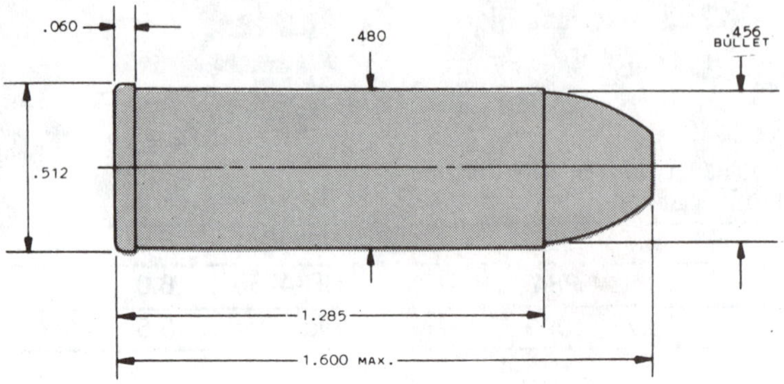

COMMENTS:

This cartridge is often erroneously called the 45 Long Colt. Care must be taken when sizing as the tiny rim of this cartridge does not afford a good purchase for the shell holder. If undue force is needed to withdraw the case, the rim may tear away leaving the case jammed in the die.

Unique is highly favored with all bullets for accuracy with this cartridge.

TEST COMPONENTS:

Cases .... Winchester
Trim-to Length .... 1.280"
Primers .... Winchester #7-111, WLP
Primer Size .... Large Pistol
Lyman Shell Holder No.11
Jacketed Bullets Used .... Sierra JHP #8800, 185 gr.
Speer JHP #4477, 200 gr.
Speer JHP #4479, 225 gr.
Sierra JHC #8820, 240 gr.
Hornady JHP #4520, 250 gr.
Cast Bullets Used .... (Sized to .452" dia.)
#454190, 250 gr.
#454424, 250 gr.

TEST SPECIFICATIONS:
(Velocity Only)

Firearm Used .... Winchester Model 94AE
Barrel Length .... 16"
Twist .... 1-38"
Groove Dia. .... .451"

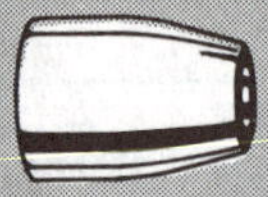

185 gr. Jacketed HP
1.520" OAL

Powder	Sugg. Starting Grains	Velocity fps	Pressure C.U.P.	Max. Load Grains	Velocity fps	Pressure C.U.P.
700X	6.0	946	—	7.0	1108	—
AA5	13.0	1416	—	13.7	1493	—
Unique	7.5	944	—	10.0	1263	—
HS-7	13.0	1121	—	14.5	1238	—
Blue Dot	12.0	979	—	13.5	1116	—
IMR-4227	16.5	843	—	20.0	1145	—
AA N100	6.0	874	—	7.8	1196	—

Note: Loads shown in shaded panels are maximum

45 Colt (rifle data)

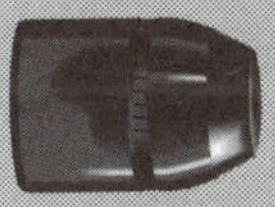

200 gr. Jacketed HP
1.557" OAL

Powder	Sugg. Starting Grains	Velocity fps	Pressure C.U.P.	Max. Load Grains	Velocity fps	Pressure C.U.P.
AA5	12.6	1370	—	13.3	1461	—
Unique	8.0	1035	—	11.2	1391	—
HS-6	10.3	896	—	13.3	1294	—
SR-4756	9.5	892	—	11.8	1354	—
Blue Dot	11.6	941	—	12.8	1047	—
SR-7625	7.0	711	—	9.5	1173	—
AA N100	7.3	1041	—	7.7	1141	—

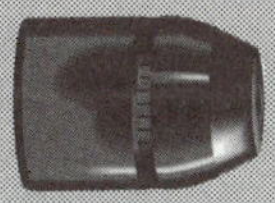

225 gr. Jacketed HP
1.557" OAL

Powder	Sugg. Starting Grains	Velocity fps	Pressure C.U.P.	Max. Load Grains	Velocity fps	Pressure C.U.P.
AA5	11.2	1170	—	11.8	1258	—
Unique	8.3	1049	—	10.8	1291	—
HS-6	11.6	1096	—	13.6	1265	—
IMR-4227	16.5	953	—	21.0	1265	—
AA N100	6.3	770	—	6.9	1012	—
Blue Dot	12.4	1076	—	15.4	1310	—
SR-7625	7.0	741	—	9.3	1151	—

240 gr. Jacketed HC
1.575" OAL

Powder	Sugg. Starting Grains	Velocity fps	Pressure C.U.P.	Max. Load Grains	Velocity fps	Pressure C.U.P.
AA 5	10.9	1148	—	11.5	1210	—
Unique	7.5	952	—	9.5	1176	—
HS-7	12.0	1041	—	14.0	1185	—
2400	15.0	973	—	18.5	1321	—
IMR-4227	16.0	842	—	20.5	1215	—

250 gr. Jacketed HP
1.590" OAL

Powder	Sugg. Starting Grains	Velocity fps	Pressure C.U.P.	Max. Load Grains	Velocity fps	Pressure C.U.P.
AA 5	10.9	1124	—	11.5	1217	—
Unique	7.5	878	—	9.3	1102	—
HS-7	11.7	991	—	14.3	1187	—
2400	15.2	1054	—	17.5	1236	—
IMR-4227	17.0	949	—	20.3	1198	—

#454190
250 gr. (#2 Alloy) 1.600" OAL

Powder	Sugg. Starting Grains	Velocity fps	Pressure C.U.P.	Max. Load Grains	Velocity fps	Pressure C.U.P.
Red Dot	4.5	763	—	6.5	1002	—
231	6.5	967	—	7.4	1069	—
700X	4.5	816	—	6.0	992	—
AA 5	10.5	1160	—	11.0	1237	—
Unique	6.0	809	—	9.0	1157	—
SR-7625	6.0	767	—	8.5	1106	—
AA N100	6.0	961	—	6.5	1020	—

#452424
250 gr. (#2 Alloy) 1.575" OAL

Powder	Sugg. Starting Grains	Velocity fps	Pressure C.U.P.	Max. Load Grains	Velocity fps	Pressure C.U.P.
Red Dot	4.5	746	—	6.0	947	—
700X	4.5	809	—	6.0	993	—
AA 5	10.0	1119	—	10.7	1204	—
Unique	6.0	836	—	8.5	1110	—
HS-6	7.8	704	—	9.5	911	—
SR-7625	6.0	775	—	8.0	1054	—
AA N100	5.5	912	—	6.2	997	—

Note: Loads shown in shaded panels are maximum.

45-70 Government

For 1873 Springfield

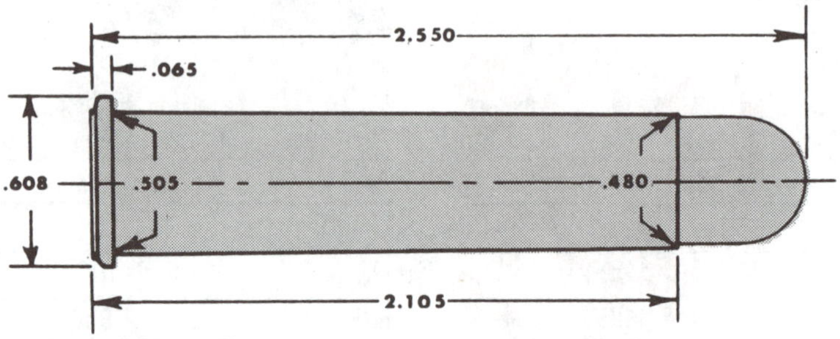

COMMENTS:

Never reload old brass as they may be of the weak balloon head construction or may have been fired at one time with mercuric primers. Never use an older 45-70 that has not at first been carefully checked by a gunsmith and found to be suitable for shooting.

These loads are intended for "Trap Door Springfields", reproductions of these 1873 to 1884 Springfields, Remington Rolling blocks, Sharps rifles and replicas of same, as well as the H&R "Shikari".

WADS: 100% Polyester fiber wads in the form of 1/4" thick batting were used in conjunction with some cast bullet loads as indicated. The Polyester fiber should be cut into 5/8" squares weighing about a 1/2 grain each. Do not use wads in any load unless specifically listed in the data.

TEST COMPONENTS:

Cases .... Remington
Trim-to Length .... 2.100"
Primers .... Remington 9½
Primer Size .... Large Rifle
Lyman Shell Holder .... No. 17
Jacketed Bullets Used .... Hornady Short Jacket, 300 gr.
Remington SP, 405 gr.
Cast Bullets Used .... (Sized to .457" dia.)
#457191, 293 gr.
#457122 HP, 322 gr.
#457124, 385 gr.
#457193, 420 gr.

TEST SPECIFICATIONS: (Velocity & Pressure)

Firearm Used .... Springfield and Universal Receiver
Barrel Length .... Springfield, 32½"; Universal Receiver 24"
Twist .... Springfield 1-22"; Universal Receiver 1-18"
Groove Dia. .... .457"

300 gr. Jacketed (Short Jacket)

Powder	Sugg. Starting Grains	Velocity fps	Pressure C.U.P.	Max. Load Grains	Velocity fps	Pressure C.U.P.
2400	24.0	1426	—	30.0	1715	—
RX7	33.0	1408	—	38.0	1639	—
IMR-4227	27.0	1459	—	32.0	1647	—
IMR-4198	31.0	1447	—	36.0	1647	—
IMR-3031	38.0	1436	—	43.0	1582	—

Note: Loads shown in shaded panels are maximum.

45-70 Government
For 1873 Springfield

405 gr. Jacketed SP

Powder	Sugg. Starting Grains	Velocity fps	Pressure C.U.P.	Max. Load Grains	Velocity fps	Pressure C.U.P.
Unique	11.0	890	—	15.0	1223	—
2400	22.0	1124	—	27.0	1394	—
RX7	29.0	1211	—	34.0	1390	—
IMR-4227	25.0	1159	—	30.0	1377	—
IMR-4198	28.0	1130	—	33.0	1353	—
IMR-3031	34.0	1137	—	39.0	1356	—

#457191**
293 gr. (#2 Alloy) 2.550" OAL

Powder	Sugg. Starting Grains	Velocity fps	Pressure C.U.P.	Max. Load Grains	Velocity fps	Pressure C.U.P.
SR-4759	26.0	1311	9,400	30.0	1602	14,600
IMR-4198	28.5§	1327	9,900	34.5§	1613	16,600
IMR-3031	34.0§	1269	10,500	44.0§	1577	15,300

#457122 HP**
322 gr. (#2 Alloy) 2.550" OAL

Powder	Sugg. Starting Grains	Velocity fps	Pressure C.U.P.	Max. Load Grains	Velocity fps	Pressure C.U.P.
SR-4759	19.0§	1043	8,100	25.0§	1412	16,400
IMR-4198	23.0§	1088	7,700	32.0§	1480	17,000
RX7	47.0§	1687	13,100	51.0	1813	15,800
IMR-3031	29.0§	1056	7,600	37.0§	1435	16,800

#457124**
385 gr. (#2 Alloy) 2.540" OAL

Powder	Sugg. Starting Grains	Velocity fps	Pressure C.U.P.	Max. Load Grains	Velocity fps	Pressure C.U.P.
SR-4759	19.0§	1035	9,500	23.3§	1298	15,400
IMR-4198	23.0§	1039	7,400	30.0§	1363	16,100
RX7	37.0§	1300	7,200	48.0	1764	16,200
IMR-3031	30.0§	1090	8,900	39.0§	1403	17,000

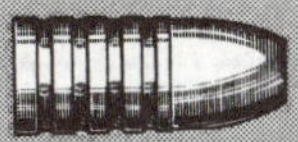

#457193**
420 gr. (#2 Alloy) 2.550" OAL

Powder	Sugg. Starting Grains	Velocity fps	Pressure C.U.P.	Max. Load Grains	Velocity fps	Pressure C.U.P.
SR-4759	18.0§	942	9,000	23.0§	1207	15,700
IMR-4198	21.5§	952	6,800	28.5§	1267	13,900
RX7	33.0§	1198	7,800	41.0	1467	13,300
IMR-3031	34.0	1161	11,100	38.5	1352	16,000

Note: Loads shown in shaded panels are maximum.
§ Designates the use of a 1/2 gr. Polyester fiber wad, 5/8" square x 1/4" thick over the powder.
** Tested in a Universal Receiver.

45-70 Government

For 1886 Winchester and 1895 Marlin only

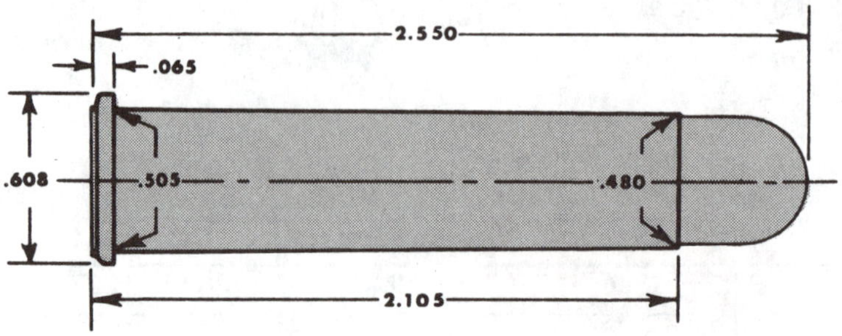

COMMENTS:

Never reload old brass as they may be of the weak balloon head construction or may have been fired at one time with mercuric primers. Never use an older 45-70 that has not been carefully checked by a gunsmith and found to be suitable for shooting.

This data is not safe in "Trap Door" Springfields and other similar strength rifles, some of which are listed in the previous data listing. These loads are intended for Winchester and Marlin rifles of the specified model, both old and new.

WADS: 100% Polyester fiber wads in the form of 1/4" thick batting were used in conjunction with some cast bullet loads as indicated. The Polyester fiber should be cut into 5/8" squares weighing about a 1/2 grain each. Do not use wads in any load unless specifically listed in the data.

The shallow groove rifling commonly used in the new Marlin rifles requires hard bullet alloys and velocities at or below 1,600 fps. when using cast bullets if accuracy is to be acceptable. A maximum diameter sizing die will also be beneficial. Cast bullet #457643 is a 400 grain bullet designed specifically for lever action rifles.

TEST COMPONENTS:

Cases .... Remington
Trim-to Length .... 2.100"
Primers .... Remington 9 1/2
Primer Size .... Large Rifle
Lyman Shell Holder .... No. 17
Jacketed Bullets Used .... Hornady Short Jacket, 300 gr.
Remington SP, 405 gr.
Cast Bullets Used .... (Sized to .457" dia.)
#457191, 293 gr.
#457122 HP, 322 gr.
#457643, 400 gr.
#457193, 420 gr.

TEST SPECIFICATIONS: (Velocity & Pressure)

Firearm Used .... Winchester 1886 and Universal Receiver
Barrel Length .... Winchester 26"; Universal Receiver 24"
Twist .... Winchester 1-22"; Universal Receiver 1-18"
Groove Dia. .... .457"

300 gr. Jacketed (Short Jacket)

Powder	Sugg. Starting Grains	Velocity fps	Pressure C.U.P.	Max. Load Grains	Velocity fps	Pressure C.U.P.
2400	30.0	1658	—	34.0	1870	—
RX7	38.0	1428	—	45.0	1811	—
IMR-4227	32.0	1612	—	37.0	1845	—
IMR-4198	36.0	1533	—	41.0	1779	—
IMR-3031	43.0	1557	—	50.0	1831	—
IMR-4320	47.0	1524	—	54.0	1782	—

Note: Loads shown in shaded panels are maximum.

45-70 Government

For 1886 Winchester and 1895 Marlin only

405 gr. Jacketed SP

Powder	Sugg. Starting Grains	Velocity fps	Pressure C.U.P.	Max. Load Grains	Velocity fps	Pressure C.U.P.
2400	27.0	1355	—	32.0	1582	—
RX7	34.0	1200	—	44.0	1590	—
IMR-4227	30.0	1380	—	35.0	1618	—
IMR-4198	‡33.0	1336	—	38.0	1557	—
IMR-3031	38.5	1312	—	48.0	1736	—
IMR-4320	45.0	1342	—	51.0	1647	—

#457191**
293 gr. (#2 Alloy) 2.550" OAL

Powder	Sugg. Starting Grains	Velocity fps	Pressure C.U.P.	Max. Load Grains	Velocity fps	Pressure C.U.P.
Unique	12.0§	1200	13,200	15.0§	1416	24,500
2400	24.4§	1524	15,300	30.5§	1842	26,900
SR-4759	30.0	1602	14,600	34.0	1802	21,000
IMR-4227	25.4§	1495	11,400	31.0§	1778	24,800
IMR-4198	34.5§	1613	16,600	49.7	2065	24,600
IMR-3031	51.0	1695	16,500	59.0+	2065	27,400

#457122 HP**
322 gr. (#2 Alloy) 2.550" OAL

Powder	Sugg. Starting Grains	Velocity fps	Pressure C.U.P.	Max. Load Grains	Velocity fps	Pressure C.U.P.
Unique	9.5§	963	9,600	14.5§	1295	24,400
2400	21.0§	1293	14,200	28.0§	1640	25,000
SR-4759	25.0§	1412	16,400	31.5§	1719	28,000
IMR-4227	22.1§	1254	13,800	30.0§	1639	27,000
IMR-4198	33.5§	1539	18,800	43.0	1867	28,000
RX7	50.0§	1842	16,200	58.5+	2132	27,000
IMR-3031	47.0§	1723	22,000	54.0	1934	28,000

#457643**
400 gr. (#2 Alloy) 2.530" OAL

Powder	Sugg. Starting Grains	Velocity fps	Pressure C.U.P.	Max. Load Grains	Velocity fps	Pressure C.U.P.
Unique	13.7	1154	17,100	16.5	1286	27,300
2400	21.0§	1322	17,800	24.6§	1497	26,300
SR-4759	22.5§	1353	17,100	25.3§	1497	27,700
IMR-4227	24.5§	1373	20,500	27.3§	1490	26,100
IMR-4198	28.5§	1375	16,600	32.3§	1563	25,400
RX7	40.0	1542	14,000	48.0+	1875	26,900
IMR-3031	41.5	1414	15,700	49.0+	1715	26,300
AA 2230	43.5	1515	16,300	50.4	1762	27,400

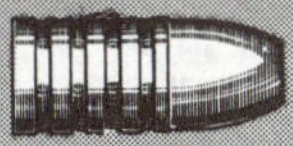

#457193**
420 gr. (#2 Alloy) 2.550" OAL

Powder	Sugg. Starting Grains	Velocity fps	Pressure C.U.P.	Max. Load Grains	Velocity fps	Pressure C.U.P.
Unique	11.0§	1006	16,800	13.5§	1163	27,200
2400	21.0§	1252	17,700	25.7§	1490	27,500
SR-4759	23.0§	1207	15,700	28.5§	1418	24,500
IMR-4227	21.0§	1163	14,700	26.2§	1418	24,200
IMR-4198	28.5§	1267	13,900	41.5	1737	28,300
RX7	40.0	1493	13,500	48.5	1828	25,000
IMR-3031	38.5	1352	16,000	47.0	1684	25,000

Note: Loads shown in shaded panels are maximum.
\+ Designates a compressed powder charge.
‡ Designates a factory velocity duplication load.
§ Designates the use of a 1/2 gr. Polyester fiber wad, 5/8" square x 1/4" thick over the powder.
** Tested in a Universal Receiver.

45-70 Government

FOR RUGER No. 1 and No. 3 ONLY

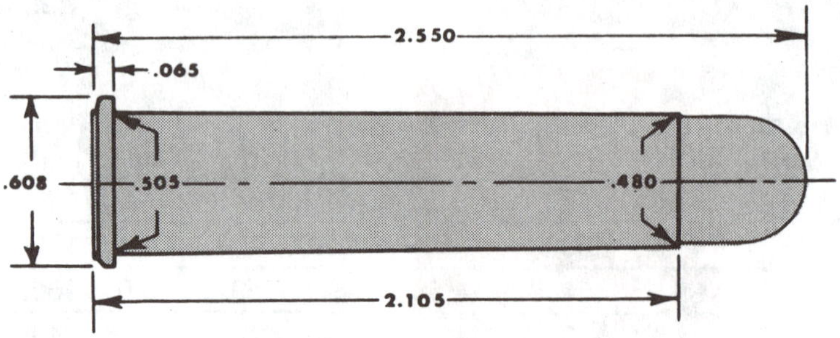

COMMENTS:

This data is not safe in "Trap Door" Springfields and other similar strength rifles, some of which are listed in the previous data listing. Nor are they intended for the lever action Winchester and Marlin rifles. These loads are intended only for the Ruger single shot rifles and must be loaded only in newly manufactured cases that have not been repeatedly fired.

The listed loads often exceed the SAAMI maximum cartridge overall length of 2.550 inches. The Ruger chamber allows for the increased lengths shown in the data.

WADS: 100% Polyester fiber wads in the form of 1/4" thick batting were used in conjunction with some cast bullet loads as indicated. The Polyester fiber should be cut into 5/8" squares weighing about a 1/2 grain each. Do not use wads in any load unless specifically listed in the data.

With heavily compressed loads it is advisable to crimp bullets so that they will not "grow" in length after assembly.

TEST COMPONENTS:

Cases .... Remington
Trim-to Length .... 2.105"
Primers .... Remington 9½
Primer Size .... Large Rifle
Lyman Shell Holder .... No. 17
Jacketed Bullets Used .... Sierra FN #8900, 300 gr.
Speer FN #2479, 400 gr.
Winchester SP #4540, 405 gr.
Hornady RN, 500 gr.
Cast Bullets Used .... (Sized to .457" dia.)
*Gas Check Bullets #457122 HP, 322 gr.
#457124, 385 gr.
#457193, 420 gr.
*#457406, 451 gr.
#457125, 500 gr.

TEST SPECIFICATIONS:
(Velocity & Pressure)

Firearm Used .... Universal Receiver (Ruger Chamber)
Barrel Length .... 28"
Twist .... 1-20"
Groove Dia. .... .456"

300 gr. Jacketed FN
2.550" OAL

Powder	Sugg. Starting Grains	Velocity fps	Pressure C.U.P.	Max. Load Grains	Velocity fps	Pressure C.U.P.
IMR-4198	49.0	2047	24,000	55.0+	2347	35,200
IMR-3031	53.0	1753	16,400	60.0+	2076	25,000
IMR-4227	32.0	1702	22,800	38.0	1974	36,300
SR-4759	35.0	1786	23,000	42.0	2045	37,000
2400	31.0	1703	22,600	35.0	1946	35,100
RX7	50.0	2113	25,100	63.0+	2405	30,400
H-322	55.0	1784	13,900	62.0+	2076	22,400

Note: Loads shown in shaded panels are maximum.
\+ Designates a compressed powder charge.

45-70 Government

FOR RUGER No. 1 and No. 3 ONLY

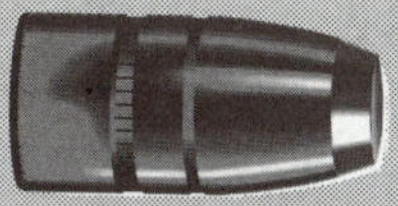

400 gr. Jacketed FN
2.640" OAL

Powder	Sugg. Starting Grains	Velocity fps	Pressure C.U.P.	Max. Load Grains	Velocity fps	Pressure C.U.P.
IMR-4198	39.0	1534	17,700	48.2	1957	34,000
IMR-3031	50.0	1696	21,500	57.5+	2039	37,400
IMR-4320	51.5	1610	21,000	60.5	1963	38,100
SR-4759	32.0	1501	24,400	39.0	1771	39,700
H-4895	50.0	1570	14,900	59.0+	1934	27,200
RX7	49.5	1872	26,900	57.0	2113	37,000
748	49.0	1479	12,600	61.0	1886	25,000

500 gr. Jacketed RN
2.930" OAL

Powder	Sugg. Starting Grains	Velocity fps	Pressure C.U.P.	Max. Load Grains	Velocity fps	Pressure C.U.P.
IMR-4198	35.0	1440	27,500	43.0	1657	38,400
IMR-3031	44.0	1444	22,300	53.1+	1742	32,700
H-4895	50.0	1646	28,100	57.0	1879	39,000
RX7	37.0	1487	31,200	45.5	1738	39,100

#457124
385 gr. (#2 Alloy) 2.723" OAL

Powder	Sugg. Starting Grains	Velocity fps	Pressure C.U.P.	Max. Load Grains	Velocity fps	Pressure C.U.P.
Unique	10.0§	939	10,300	17.5§	1411	38,500
2400	26.5§	1517	20,400	33.5§	1853	37,900
SR-4759	30.0	1478	16,300	39.0	1880	36,700
IMR-4227	27.0§	1458	18,300	34.0§	1769	34,300
IMR-4198	41.5§	1832	25,400	50.5	2135	37,500
IMR-3031	50.0	1714	17,800	60.0+	2117	33,000
FFg	52.5§	1087	7,700	70.0+	1328	14,400
Pyrodex CTG	39.4	1054	7,400	52.5+	1198	9,000

405 gr. Jacketed SP
2.735" OAL

Powder	Sugg. Starting Grains	Velocity fps	Pressure C.U.P.	Max. Load Grains	Velocity fps	Pressure C.U.P.
IMR-4198	41.0	1761	30,800	50.0	2009	39,500
IMR-3031	50.0	1698	20,800	58.0+	2048	36,500
IMR-4320	53.0	1668	23,600	62.0	1976	37,400
H-4895	50.0	1537	14,900	59.0+	1877	24,400
RX7	51.0	1895	26,700	60.0	2157	37,500

#457122 HP
322 gr. (#2 Alloy) 2.610" OAL

Powder	Sugg. Starting Grains	Velocity fps	Pressure C.U.P.	Max. Load Grains	Velocity fps	Pressure C.U.P.
Unique	12.5§	1171	16,900	17.5§	1472	38,700
2400	27.0	1608	21,200	34.5	1958	40,000
SR-4759	30.5	1635	20,600	39.5	1973	38,300
IMR-4227	29.0§	1635	23,500	35.0§	1893	37,700
IMR-4198	43.0§	1930	26,900	51.0	2220	37,600
IMR-3031	52.0	1865	21,500	60.0+	2179	33,700
FFg	40.0§	1041	8,400	70.0+	1399	14,700
Pyrodex CTG	46.5	1153	7,100	62.0+	1445	12,800

Note: Loads shown in shaded panels are maximum.
\+ Designates a compressed powder charge.
§ Designates the use of a 1/2 gr. Polyester fiber wad, 5/8" square x 1/4" thick over the powder.

45-70 Government

FOR RUGER No. 1 and No. 3 ONLY

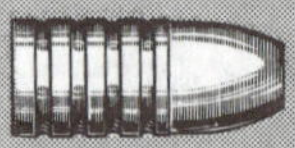

#457193
420 gr. (#2 Alloy) 2.680" OAL

Powder	Sugg. Starting Grains	Velocity fps	Pressure C.U.P.	Max. Load Grains	Velocity fps	Pressure C.U.P.
Unique	13.5§	1113	22,800	16.5§	1275	38,200
2400	26.3§	1451	23,800	32.0§	1698	37,100
SR-4759	28.0§	1370	19,300	35.0	1682	36,400
IMR-4227	27.8§	1457	24,200	34.0§	1684	37,700
IMR-4198	37.6§	1662	25,600	47.0	1932	36,600
RX7	46.0§	1687	17,800	56.0+	2058	34,700
IMR-3031	44.8	1599	20,200	56.0+	1996	35,600
FFg	63.0	1185	12,500	70.0+	1268	16,400
Pyrodex CTG	58.0+	1084	9,100	65.0+	1161	10,100

#457406
451 gr. (#2 Alloy) 2.650" OAL

Powder	Sugg. Starting Grains	Velocity fps	Pressure C.U.P.	Max. Load Grains	Velocity fps	Pressure C.U.P.
Unique	11.0§	950	17,100	15.2§	1209	40,000
2400	18.5§	1355	25,100	27.5§	1517	34,400
SR-4759	24.0§	1305	19,900	32.0	1635	39,400
IMR-4227	24.0§	1267	20,400	31.5§	1568	36,600
IMR-4198	33.0§	1445	24,100	43.0	1833	38,400
IMR-3031	36.0§	1304	15,300	44.0+	1590	29,100
FFg	47.3§	952	7,900	63.0+	1168	15,000
Pyrodex CTG	38.2§	927	7,600	51.0+	1096	10,200

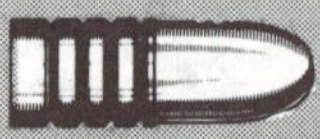

#457125
500 gr. (#2 Alloy) 2.878" OAL

Powder	Sugg. Starting Grains	Velocity fps	Pressure C.U.P.	Max. Load Grains	Velocity fps	Pressure C.U.P.
Unique	10.0§	888	14,100	14.8§	1153	33,700
2400	24.0§	1350	22,100	28.0§	1604	38,500
SR-4759	23.0§	1237	15,400	33.0	1635	36,600
IMR-4227	22.0§	1169	16,400	30.0§	1532	35,000
IMR-4198	33.0§	1477	22,100	43.5	1797	34,900
IMR-3031	42.0§	1548	22,300	52.0+	1860	34,700
FFg	45.7§	926	7,800	61.0+	1125	12,300
Pyrodex CTG	48.8§	1098	10,400	65.0+	1282	19,800

Note: Loads shown in shaded panels are maximum.
+ Designates a compressed powder charge.
§ Designates the use of a 1/2 gr. Polyester fiber wad, 5/8" square x 1/4" thick over the powder.

458 Winchester Magnum

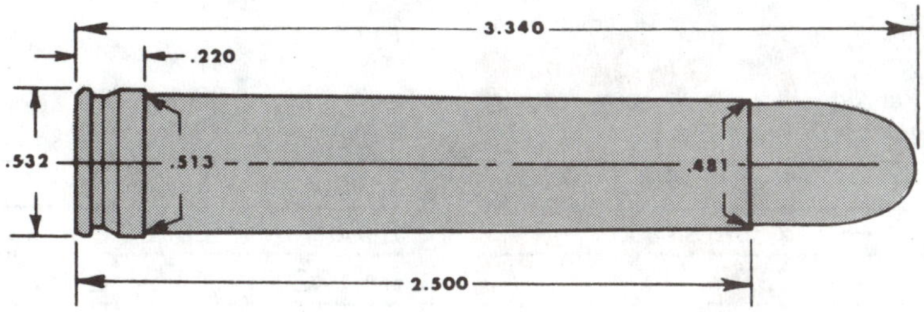

COMMENTS:

The heavy compression of most powders in this case's limited space can lead to numerous loading difficulties. The use of Hercules Reloder 7 will minimize this difficulty when loading the long 500 grain bullets. With a capable rifle and shooter, one inch groups are not uncommon.

Due to heavy recoil crimp all cartridges (except when the bullets do not have a cannelure).

Reloading die adjustments are critical. Case mouth and bullet cannelure must align precisely or the case will be crushed.

TEST COMPONENTS:

Cases .... Winchester
Trim-to Length .... 2.495"
Primers .... Winchester 8½-120
Primer Size .... Large Rifle
Lyman Shell Holder .... No. 13
Jacketed Bullets Used .... Remington SP, 405 gr.
Remington MC, 500 gr.
Cast Bullets Used .... (Sized to .459" dia.)
*Gas Check Bullets #457191, 292 gr.
#457124, 385 gr.
*#457406, 482 gr.

TEST SPECIFICATIONS:
(Velocity Only)

Firearm Used .... Winchester Model 70 and Ruger #1
Barrel Length .... Winchester 22"; Ruger 24"
Twist .... 1-14"
Groove Dia. .... Winchester .457"; Ruger .459"

405 gr. Jacketed SP††

Powder	Sugg. Starting Grains	Velocity fps	Pressure C.U.P.	Max. Load Grains	Velocity fps	Pressure C.U.P.
IMR-3031	64.0	2036	—	71.0+	2237	—
IMR-4064	67.0	1949	—	74.0+	2127	—
IMR-4895	67.0	1988	—	75.0+	2217	—
IMR-4320	69.0	1988	—	77.0+	2247	—
IMR-4350	72.0	1733	—	80.0+	1923	—
BL-C(2)	71.0	1733	—	77.0+	1919	—
H-335	74.0	1815	—	80.0+	1984	—
H-380	72.0	1897	—	80.0+	2016	—

Note: Loads shown in shaded panels are maximum.
\+ Designates a compressed powder charge.
†† Tested in a Winchester Mod. 70.

458 Winchester Magnum

500 gr. Jacketed MC††

Powder	Sugg. Starting Grains	Velocity fps	Pressure C.U.P.	Max. Load Grains	Velocity fps	Pressure C.U.P.
IMR-3031	62.0	1890	—	‡69.0+	2053	—
IMR-4064	64.0	1785	—	71.0+	1968	—
IMR-4895	66.0	1886	—	73.0+	2066	—
IMR-4320	67.0	1858	—	75.0+	2066	—
IMR-4350	69.0	1630	—	77.0+	1855	—
BL-C(2)	70.0	1736	—	75.0+	1912	—
H-335	73.0	1821	—	78.0+	2016	—
H-380	71.0	1779	—	76.0+	1855	—
RX7	60.0	1709	—	65.0+	1937	—

#457191†††
292 gr. (#2 Alloy) 2.995" OAL

Powder	Sugg. Starting Grains	Velocity fps	Pressure C.U.P.	Max. Load Grains	Velocity fps	Pressure C.U.P.
Unique	19.0	1460	—	23.0	1625	—
2400	28.0	1585	—	33.0	1765	—
IMR-4227	44.5	2100	—	50.5	2305	—
IMR-4198	54.0	2135	—	58.0	2205	—
IMR-3031	60.0	1885	—	69.0	2220	—
IMR-4064	64.0	1865	—	72.0	2110	—

#457124†††
385 gr. (#2 Alloy) 3.012" OAL

Powder	Sugg. Starting Grains	Velocity fps	Pressure C.U.P.	Max. Load Grains	Velocity fps	Pressure C.U.P.
Unique	18.0	1315	—	22.0	1510	—
2400	25.0	1310	—	31.0	1565	—
IMR-4227	44.0	1935	—	50.0	2090	—
IMR-4198	53.0	2010	—	57.0	2125	—
IMR-3031	59.0	1900	—	68.0+	2210	
IMR-4064	62.0	1860	—	70.0+	2115	—

#457406†††
482 gr. (#2 Alloy) 3.090" OAL

Powder	Sugg. Starting Grains	Velocity fps	Pressure C.U.P.	Max. Load Grains	Velocity fps	Pressure C.U.P.
Unique	17.0	1175	—	21.0	1320	—
2400	24.0	1240	—	30.0	1480	—
IMR-4227	41.0	1700	—	47.0	1855	—
IMR-4198	49.0	1830	—	53.0	1940	—
IMR-3031	57.0	1850	—	66.0+	2120	—
IMR-4064	60.0	1805	—	68.0+	2060	—

Note: Loads shown in shaded panels are maximum.
+ Designates a compressed powder charge.
‡ Designates a factory velocity duplication load.
†† Tested in a Winchester Mod. 70.
††† Tested in a Ruger #1.

460 Weatherby Magnum

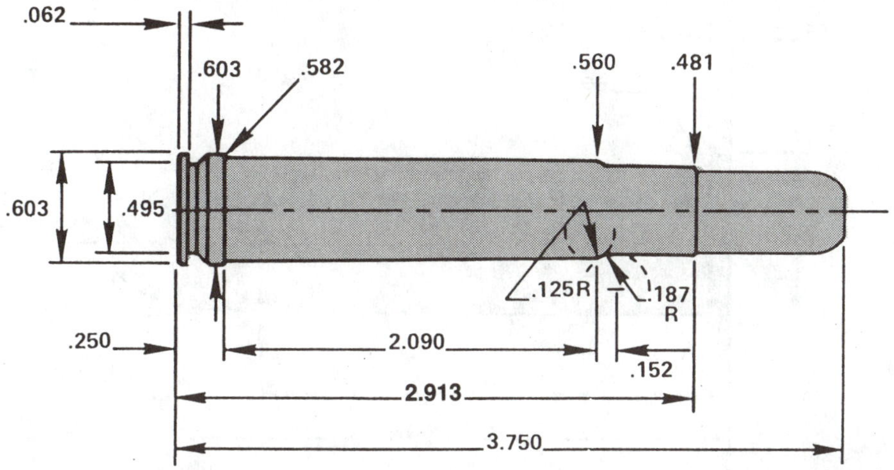

COMMENTS:

The data listed for this cartridge was obtained in a Weatherby rifle and is intended for Weatherby rifles only. The free-boring built into these arms allows higher velocities at safe working pressures.

The exclusive use of Federal 215 large rifle primers is recommended by Weatherby Inc. for all jacketed bullet loads.

Due to the heavy recoil of these rifles, all bullets should be crimped in place.

TEST COMPONENTS:

Cases .... Weatherby
Trim-to Length .... 2.903"
Primers .... Federal 215
Primer Size .... Large Rifle
Lyman Shell Holder .... No. 17
Jacketed Bullets Used .... Remington MC, 500 gr.

TEST SPECIFICATIONS: (Velocity Only)

Firearm Used .... Weatherby Mark V
Barrel Length .... 26"
Twist .... 1-16"
Groove Dia. .... .457"

500 gr. Jacketed MC

Powder	Sugg. Starting Grains	Velocity fps	Pressure C.U.P.	Max. Load Grains	Velocity fps	Pressure C.U.P.
IMR-3031	86.0	2183	—	96.0	2444	—
IMR-4064	96.0	2320	—	106.0	2538	—
IMR-4350	110.0	2392	—	121.0+	2624	—
H-380	100.0	2227	—	112.0	2481	—
H-4831	115.0	2262	—	128.0+	2481	—

Note: Loads shown in shaded panels are maximum.
\+ Designates a compressed powder charge.

50-70 Government

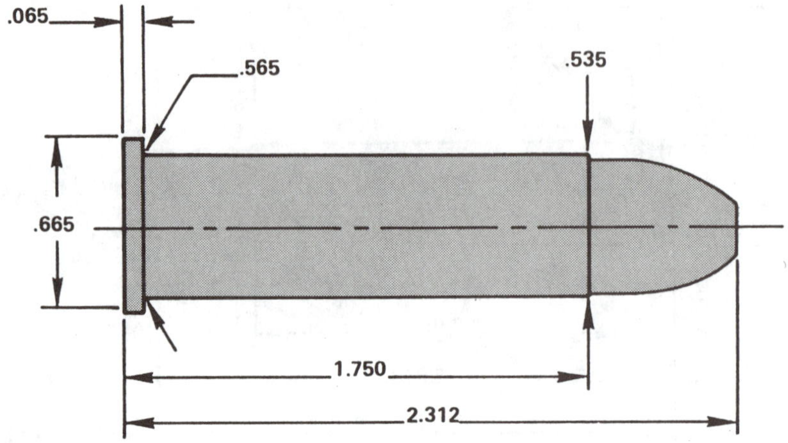

COMMENTS:

Caution: The load data listed is for use only in modern solid head brass cases and in conjunction with the use of new modern replica rifles.

Wads: 100% Polyester fiber wads in the form of 1/4" batting were used in conjunction with cast bullet loads, where indicated. This material can be purchased in most yard-goods stores. It should be cut into 5/8 inch squares, each weighing about 1/2-grain.

When developing a load, if a wad is desired, it should be used from the beginning as the charge weight is increased. It should never be added as an afterthought once a maximum load has been established since its presence could result in a pressure increase of 2000 CUP or more.

TEST COMPONENTS:

Cases	BELL
Trim-to Length	1.740"
Primers	Federal 215
Primer Size	Large Rifle, Magnum
Lyman Shell Holder	No. 22
Cast Bullets Used	(Sized to .512" dia.) #515141, 422 gr. #515142, 498 gr.

TEST SPECIFICATIONS:
(Velocity Only)

Firearm Used	Shiloh Sharps
Barrel Length	22"
Twist	1-48"
Groove Dia.	.511"

#515141
422 gr. (#2 Alloy) 2.312" OAL

Powder	Sugg. Starting Grains	Velocity fps	Pressure C.U.P.	Max. Load Grains	Velocity fps	Pressure C.U.P.
Unique	11.0§	1003	—	14.8§	1211	—
SR-4759	22.0§	1114	—	26.5§	1313	—
IMR-4227	22.0§	1102	—	27.3§	1333	—
IMR-4198	25.5§	1129	—	30.0§	1305	—
FFg	50.0	1050	—	70.0+	1238	—
Pyrodex CTG	37.0	1037	—	52.5	1252	—

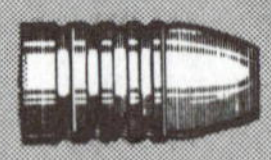

#515142
498 gr. (#2 Alloy) 2.223" OAL

Powder	Sugg. Starting Grains	Velocity fps	Pressure C.U.P.	Max. Load Grains	Velocity fps	Pressure C.U.P.
Unique	10.5§	940	—	13.8§	1114	—
SR-4759	18.5§	942	—	24.0§	1215	—
IMR-4227	19.0§	950	—	24.5§	1202	—
IMR-4198	22.3§	950	—	28.5§	1188	—
IMR-3031	27.5§	865	—	31.5§	1005	—
FFg	43.0	913	—	55.0+	1054	—
Pyrodex CTG	32.0	892	—	42.0	1074	—

Note: Loads shown in shaded panels are maximum.
\+ Designates a compressed powder charge.
§ Designates the use of a 1/2 gr. Polyester fiber wad, 5/8" square x 1/4" thick over the powder.

25 Automatic

(25 ACP)
(6.35 mm)

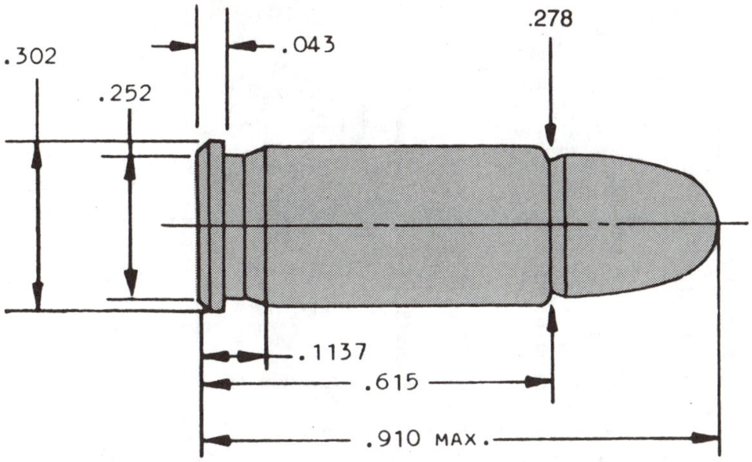

COMMENTS:

There are relatively few bullets to select from for this tiny cartridge. When using cast bullet #252435 use a hard bullet alloy and size to 0.251" diameter.

The 25 Automatic is a semi-rimmed case and headspaces on its rim. Cases can therefore be roll crimped.

Hercules Bullseye, Hodgdon HP-38, or Winchester 231 are the preferred powders.

TEST COMPONENTS:

Cases .... Remington
Trim-to Length .... .610"
Primers .... Winchester 1½-108
Primer Size .... Small Pistol
Lyman Shell Holder .... No. 32
Jacketed Bullets Used .... Hornady FMC #3545, 50 gr.
Cast Bullets Used .... (Sized to .251" dia.) #252435, 51 gr.

TEST SPECIFICATIONS: (Velocity & Pressure)

Firearm Used .... Universal Receiver
Barrel Length .... 2"
Twist .... 1-16"
Groove Dia. .... .251"

50 gr. Jacketed FMC
.900" OAL

Powder	Sugg. Starting Grains	Velocity fps	Pressure C.U.P.	Max. Load Grains	Velocity fps	Pressure C.U.P.
Bullseye	1.0	558	11,000	1.4	754	17,300
231	1.0	563	11,800	1.4	752	17,800
HP-38	0.8	565	12,200	1.1	710	17,500
Red Dot	1.0	562	11,800	1.3	749	18,100
700X	0.8	556	11,600	1.1	743	18,000
Unique	1.2	544	11,000	1.6	740	16,700

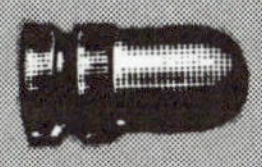

#252435
51 gr. (#2 Alloy) .910" OAL

Powder	Sugg. Starting Grains	Velocity fps	Pressure C.U.P.	Max. Load Grains	Velocity fps	Pressure C.U.P.
Bullseye	0.9	591	9,700	1.2	701	14,200
231	0.9	604	10,100	1.2	708	14,700
Red Dot	0.8	566	9,500	1.1	710	13,700
700X	0.7	580	9,200	0.9	677	12,800
Green Dot	1.0	585	9,800	1.3	712	14,100
PB	0.9	591	9,200	1.2	739	13,600
Unique	1.2	561	8,700	1.6	727	14,000

Note: Loads shown in shaded panels are maximum.

30 Luger
(7.65mm Luger)
(7.65mm Parabellum)

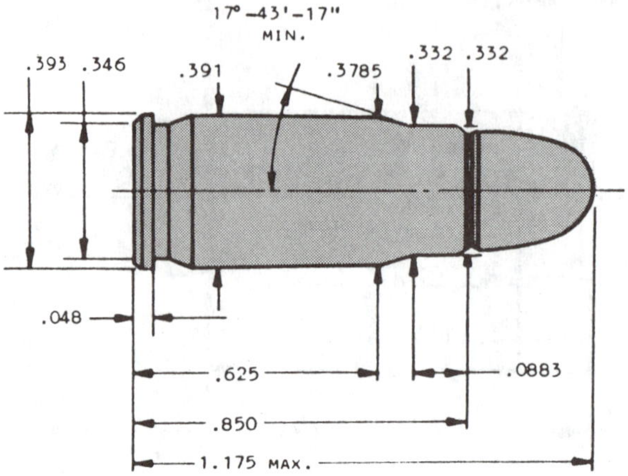

COMMENTS:

Groove diameter variations are the norm in firearms chambered for this cartridge. In some handguns the starting load may not function the action. In such an instance work up the load cautiously.

For best results with cast bullets, size bullets to, or 0.001" above groove diameter. Cast bullet #313249 should be crimped on the leading edge of the first driving band.

TEST COMPONENTS:

Cases .... Remington
Trim-to Length .... .845"
Primers .... Remington 1 1/2
Primer Size .... Small Pistol
Lyman Shell Holder .... No. 12
Jacketed Bullets Used .... Remington MC #B22764, 93 gr.
Cast Bullets Used .... (Sized to .309" dia.) #313249, 84 gr.

TEST SPECIFICATIONS: (Velocity Only)

Firearm Used .... Walther P-38
Barrel Length .... 4 15/16"
Twist .... 1-9.85"
Groove Dia. .... .3085"

93 gr. Jacketed MC
1.175" OAL

Powder	Sugg. Starting Grains	Velocity fps	Pressure C.U.P.	Max. Load Grains	Velocity fps	Pressure C.U.P.
700X	3.6	1065	—	4.1	1171	—
SR-4756	5.0	1038	—	5.5	1161	—
Bullseye	3.5	968	—	4.0	1067	—
Unique	4.5	1003	—	5.0	1115	—
231	3.7	935	—	4.2	1046	—
HP-38	3.5	1028	—	4.0	1115	—

#313249
84 gr. (#2 Alloy) 1.125" OAL

Powder	Sugg. Starting Grains	Velocity fps	Pressure C.U.P.	Max. Load Grains	Velocity fps	Pressure C.U.P.
700X	3.7	1143	—	4.2	1204	—
PB	4.0	1105	—	4.5	1177	—
SR-4756	4.7	966	—	5.6	1182	—
Bullseye	3.8	1106	—	4.5	1244	—
Red Dot	3.8	1116	—	4.3	1201	—
Green Dot	4.2	1111	—	4.7	1210	—
Unique	4.6	1110	—	5.2	1226	—
231	3.9	1054	—	4.4	1144	—

Note: Loads shown in shaded panels are maximum.

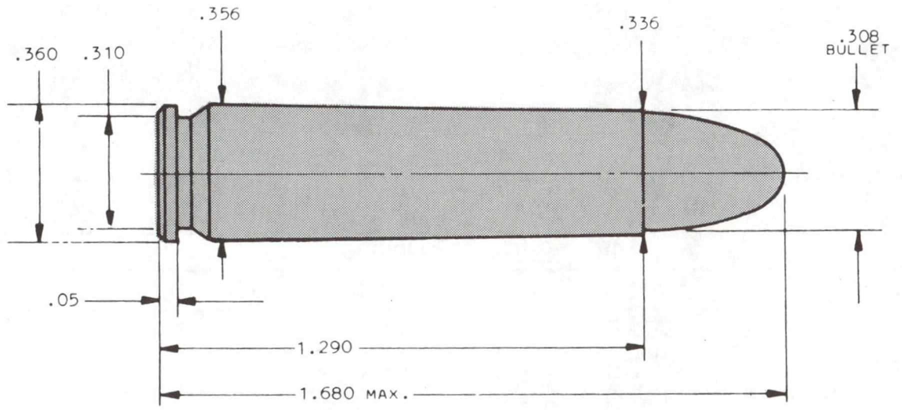

COMMENTS:

This cartridge headspaces from the mouth and therefore case trimming must be uniform and accurate. Do not reduce cases below the trim-to length. Additionally do not crimp bullets as this will prevent the case from properly headspacing on its mouth.

Best performance is with IMR 4227 with both jacketed and cast bullets.

Ballistics with this cartridge are too light for any big game applications.

TEST COMPONENTS:

Cases .... Federal
Trim-to Length .... 1.286"
Primers .... CCI 450 (jacketed bullets only)
Remington 6½ (cast bullets only)
Primer Size .... Small Rifle Magnum and Small Rifle
Lyman Shell Holder .... No. 19
Jacketed Bullets Used .... Remington MC #B22764, 93 gr.
Speer Plinker #1805, 100 gr.
Hornady HP, 110 gr.
Cast Bullets Used .... (Sized to .308" dia.)
*Gas Check Bullets .... *#311359, 120 gr.

TEST SPECIFICATIONS:
(Velocity Only)

Firearm Used .... Ruger Blackhawk
Barrel Length .... 7½"
Twist .... 1-20"
Groove Dia. .... .308"

93 gr. Jacketed MC
1.595" OAL

Powder	Sugg. Starting Grains	Velocity fps	Pressure C.U.P.	Max. Load Grains	Velocity fps	Pressure C.U.P.
SR-4759	9.5	948	—	12.5	1216	—
IMR-4227	10.0	1076	—	15.0	1491	—
Blue Dot	6.5	922	—	8.5	1284	—
2400	9.5	1078	—	12.5	1390	—
H-110	10.0	1176	—	15.5	1456	—

100 gr. Jacketed RN
1.600" OAL

Powder	Sugg. Starting Grains	Velocity fps	Pressure C.U.P.	Max. Load Grains	Velocity fps	Pressure C.U.P.
2400	9.5	1147	—	12.5	1412	—
SR-4759	9.0	918	—	12.0+	1248	—
IMR-4227	10.0	1112	—	15.0+	1438	—
H-110	10.0	1124	—	14.5	1416	—

Note: Loads shown in shaded panels are maximum.
\+ Designates a compressed powder charge.

30 M1 Carbine
(revolver data)

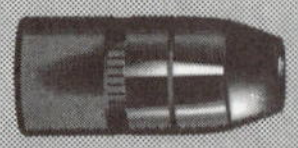

110 gr. Jacketed HP
1.660" OAL

Powder	Sugg. Starting Grains	Velocity fps	Pressure C.U.P.	Max. Load Grains	Velocity fps	Pressure C.U.P.
2400	10.0	1161	—	12.0	1348	—
SR-4759	9.0	851	—	12.0+	1172	—
IMR-4227	11.0	1188	—	14.0	1377	—
H-110	9.5	1054	—	14.0	1369	—

#311359
120 gr. (#2 Alloy) 1.680" OAL

Powder	Sugg. Starting Grains	Velocity fps	Pressure C.U.P.	Max. Load Grains	Velocity fps	Pressure C.U.P.
Unique	5.5	1138	—	6.5	1335	—
Blue Dot	6.8	1066	—	8.8	1339	—
2400	11.0	1236	—	12.5+	1436	—
SR-7625	4.6	1027	—	5.6	1162	—
SR-4756	5.5	1050	—	6.5	1220	—
IMR-4227	11.5	1169	—	13.3+	1388	—
H-110	10.0	1055	—	13.5	1358	—

Note: Loads shown in shaded panels are maximum.
+ Designates a compressed powder charge.

32 Automatic
(32 ACP)
(7.65mm Browning)

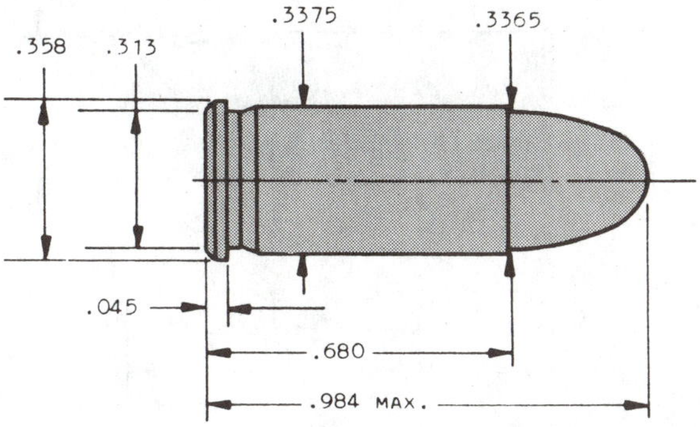

COMMENTS:

The loads listed may be used with 0.311" jacketed bullets in pistols which have a groove diameter as small as 0.309 inches. Groove diameter variations are extensive in handguns chambered for this cartridge. Chamber limits make it unsafe to attempt to use larger diameter bullets, regardless of groove diameter measurement.

The very fast powders, such as Hercules Bullseye, Hodgdon HP-38, and Winchester 231 are best for this tiny cartridge.

TEST COMPONENTS:

Cases .... Remington; Winchester
Trim-to Length .... .672"
Primers .... Remington 1½; Winchester 1½-108
Primer Size .... Small Pistol
Lyman Shell Holder .... No. 23
Jacketed Bullets Used .... Remington MC #B22816, 71 gr.
Cast Bullets Used .... (Sized to .309" dia.) #311252, 77 gr.

TEST SPECIFICATIONS: (Velocity & Pressure)

Firearm Used .... Mauser HSc; also Special Pressure Test Barrel
Barrel Length .... 3" (Mauser); 3¼" (Special Pressure Barrel)
Twist .... 1-16"
Groove Dia. .... .309"

71 gr. Jacketed MC
.984" OAL

Powder	Sugg. Starting Grains	Velocity fps	Pressure C.U.P.	Max. Load Grains	Velocity fps	Pressure C.U.P.
**700X	1.7	706	9,500	‡2.2	902	14,700
Bullseye	1.5	603	—	2.2	825	—
Red Dot	1.7	705	—	2.5	937	—
Unique	2.0	571	—	3.1	945	—
**HP-38	1.7	733	11,100	2.2	883	14,900

#311252
77 gr. (#2 Alloy) .975" OAL

Powder	Sugg. Starting Grains	Velocity fps	Pressure C.U.P.	Max. Load Grains	Velocity fps	Pressure C.U.P.
700X	1.5	750	—	1.9	895	—
PB	1.6	665	—	2.2	870	—
SR-7625	1.7	675	—	2.3	860	—
Bullseye	1.5	685	—	2.0	830	—
Red Dot	1.5	695	—	2.3	940	—
Green Dot	1.6	705	—	2.4	895	—
Unique	1.8	625	—	2.6	860	—
**231	1.5	620	7,200	2.5	888	15,000
**HP-38	1.5	688	9,200	2.0	824	13,800

Note: Loads shown in shaded panels are maximum.
‡ Designates a factory velocity duplication load.
** Designates Winchester cases, primers, and special pressure test barrel.

32 Smith and Wesson

(32 S&W)

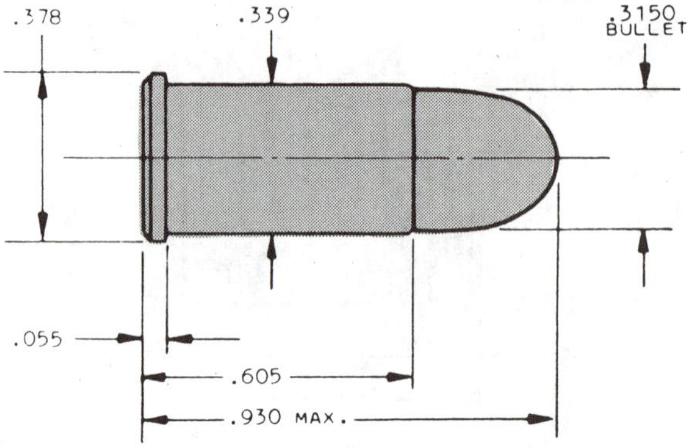

COMMENTS:

The data listed is intended only for use in solid frame revolvers. We no longer suggest the use of reloads in revolvers of the top break design due to their age and relative weakness.

Due to wide internal barrel dimensions it is wise to slug your bore before beginning to reload. Size cast bullets to groove diameter or 0.001" larger. Then make up a dummy round to ensure that your revolver chambers will easily accept ammo with bullets of the chosen diameter.

The fastest burning propellants seem to work best. As with other small capacity cases, begin with either Hercules Bullseye, Hodgdon HP-38 or Winchester 231.

TEST COMPONENTS:

Cases .... Remington; Norma (cast bullets only)
Trim-to Length .... .600"
Primers .... Remington $1^1/_2$; Winchester $1^1/_2$-108 (cast bullets only)
Primer Size .... Small Pistol
Lyman Shell Holder .... No. 9
Jacketed Bullets Used .... Remington MC #B22816, 71 gr.; Remington SP #B22810, 100 gr.
Cast Bullets Used .... (Sized to .310" dia.) #311252, 77 gr.

TEST SPECIFICATIONS:
(Velocity Only)

Firearm Used .... Smith & Wesson Model 31
Barrel Length .... 4"
Twist .... 1-$18^3/_4$"
Groove Dia. .... .312"

71 gr. Jacketed MC
.897" OAL

Powder	Sugg. Starting Grains	Velocity fps	Pressure C.U.P.	Max. Load Grains	Velocity fps	Pressure C.U.P.
700X	2.0	562	—	2.5	722	—
SR-7625	2.4	522	—	2.9	674	—
Bullseye	2.0	515	—	2.5	671	—
Unique	2.7	482	—	3.2+	650	—
231	2.4	568	—	2.9	711	—
HP-38	2.0	562	—	2.5	710	—

Note: Loads shown in shaded panels are maximum.
\+ Designates a compressed powder charge.

32 Smith and Wesson

(32 S&W)

100 gr. Jacketed SP
.897" OAL

Powder	Sugg. Starting Grains	Velocity fps	Pressure C.U.P.	Max. Load Grains	Velocity fps	Pressure C.U.P.
700X	2.0	625	—	2.5	787	—
SR-7625	2.0	536	—	2.5	682	—
Bullseye	1.9	578	—	2.5	736	—
Unique	2.5	556	—	3.0	712	—
Trap 100	2.9	595	—	3.5+	695	—

#311252
77 gr. (#2 Alloy) .860" OAL

Powder	Sugg. Starting Grains	Velocity fps	Pressure C.U.P.	Max. Load Grains	Velocity fps	Pressure C.U.P.
700X	1.3	598	—	1.8	758	—
PB	1.4	523	—	2.1	803	—
Bullseye	1.1	478	—	1.5	578	—
Red Dot	1.4	578	—	1.9	763	—
Green Dot	1.5	553	—	2.0	723	—
Unique	1.6	493	—	2.5	743	—
**231	1.9	641	—	2.5	800	—
**HP-38	1.4	642	—	2.2	821	—

Note: Loads shown in shaded panels are maximum.
+ Designates a compressed powder charge.
** Designates the use of Remington cases and primers.

32 Smith & Wesson Long

(32 Colt New Police) (32 S&W Long)

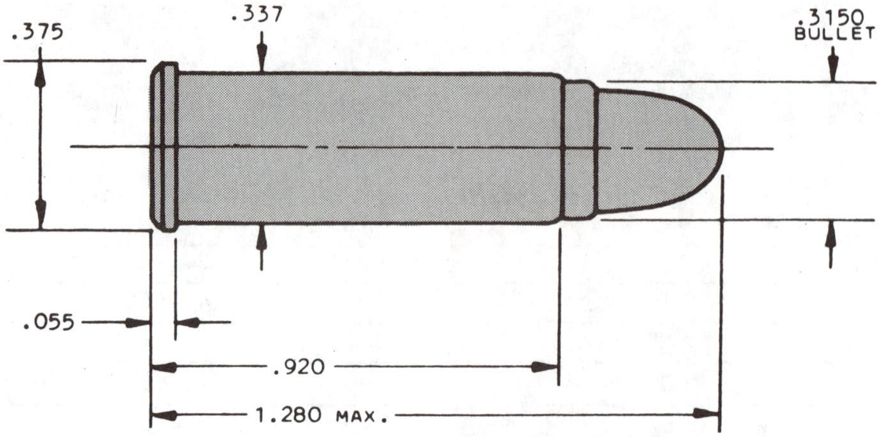

COMMENTS:

The data listed is intended only for use in solid frame revolvers. We do not recommend the use of reloads in revolvers of the top break design due to their age and relative weakness.

Due to wide internal barrel dimensions it is wise to slug your bore before beginning to reload. Size cast bullets to groove diameter or 0.001" larger. Then make up a dummy round to ensure that your revolver chambers will easily accept ammo with bullets of the chosen diameter.

The fastest burning propellants seem to work best. As with other small capacity cases, begin with either Hercules Bullseye or Winchester 231.

TEST COMPONENTS:

Cases .... Remington
Trim-to Length .... .915"
Primers .... Remington 1 1/2
Primer Size .... Small Pistol
Lyman Shell Holder .... No. 9
Jacketed Bullets Used .... Remington SP #B22810, 100 gr.
Cast Bullets Used .... (Sized to .312" dia.) #311252, 77 gr.

TEST SPECIFICATIONS: (Velocity Only)

Firearm Used .... Smith & Wesson Model 31
Barrel Length .... 4"
Twist .... 1-18 3/4"
Groove Dia. .... .312"

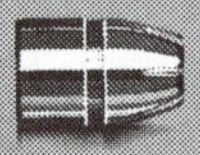

100 gr. Jacketed SP
1.280" OAL

Powder	Sugg. Starting Grains	Velocity fps	Pressure C.U.P.	Max. Load Grains	Velocity fps	Pressure C.U.P.
SR-4756	3.7	677	—	4.1	761	—
Bullseye	2.4	624	—	3.0	781	—
Unique	3.5	674	—	4.0	779	—

#311252
77 gr. (#2 Alloy) 1.115" OAL

Powder	Sugg. Starting Grains	Velocity fps	Pressure C.U.P.	Max. Load Grains	Velocity fps	Pressure C.U.P.
Bullseye	1.5	513	—	3.0	958	—
Red Dot	1.8	588	—	3.3	1028	—
Green Dot	1.9	563	—	3.4	993	—
Unique	2.5	608	—	4.0	943	—
700X	1.8	598	—	3.3	1098	—
SR-7625	2.4	583	—	3.6	988	—
231	2.0	677	—	3.5	1043	—
Trap 100	1.7	645	—	3.5	1003	—

Note: Loads shown in shaded panels are maximum.

32 H & R Magnum

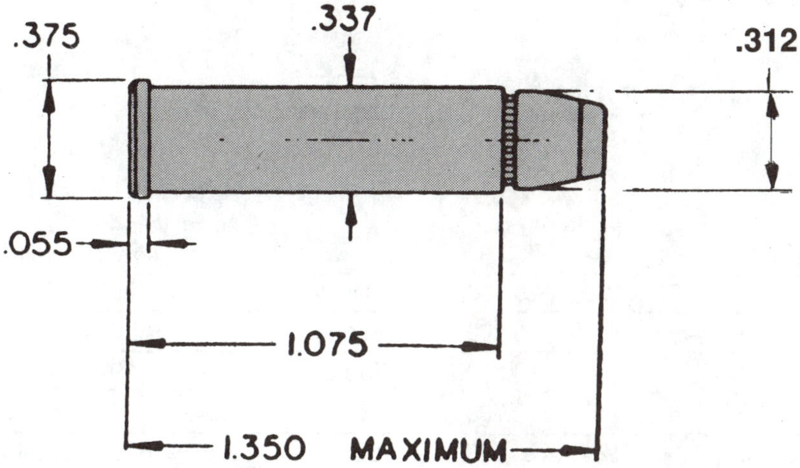

COMMENTS:

Revolvers chambered for this cartridge will also fire 32 S&W, or 32 S&W Long ammo. However the reverse is not true.

Do not load 32 H&R data into any case except those so headstamped.

Hercules Bullseye, Hodgdon HP-38, and Winchester 231 will provide the best accuracy.

Cast bullet #313631 is an excellent choice for small game.

TEST COMPONENTS:

Cases .... Federal
Trim-to Length .... 1.070"
Primers .... Federal 100
Primer Size .... Small Pistol
Lyman Shell Holder No.9
Jacketed Bullets Used .... Sierra FMJ #8100, 71 gr.
Hornady JHP #3205, 85 gr.
Sierra JHC #8030, 90 gr.
Cast Bullets Used .... (Sized to .313" dia.)
*Gas Check Bullets #313249, 85 gr.
#313492, 95 gr.
*#313631, 100 gr.

TEST SPECIFICATIONS: (Velocity & Pressure)

Firearm Used .... Universal Receiver
Barrel Length .... 5"
Twist .... 1-16"
Groove Dia. .... .312"

71 gr. FMJ
1.350" OAL

Powder	Sugg. Starting Grains	Velocity fps	Pressure C.U.P.	Max. Load Grains	Velocity fps	Pressure C.U.P.
Bullseye	3.5	901	12,600	4.3	1112	19,000
Unique	4.7	986	15,900	5.5	1140	20,300
Blue Dot	6.5	936	14,300	7.8	1188	20,600
SR-4756	4.7	900	12,400	5.5	1141	19,200
800X	4.8	936	12,200	6.0	1174	19,400
231	3.5	817	13,400	4.5	1132	20,000
HP38	3.0	902	14,900	3.9	1104	20,100
AA 5	5.3	951	14,300	6.2	1160	19,400

Note: Loads shown in shaded panels are maximum.

32 H & R Magnum

85 gr. Jacketed HP
1.310" OAL

Powder	Sugg. Starting Grains	Velocity fps	Pressure C.U.P.	Max. Load Grains	Velocity fps	Pressure C.U.P.
Bullseye	2.8	805	13,700	3.8	1039	20,000
Unique	3.5	791	13,300	4.8	1063	20,800
Blue Dot	5.5	798	12,700	6.7	1044	19,200
SR-4756	3.5	641	11,500	4.8	1073	20,700
800X	3.7	758	13,600	5.0	1005	20,500
231	2.8	773	13,300	4.0	1054	20,800
HP 38	2.3	755	13,700	3.5	1022	21,000
AA 5	4.0	788	13,000	5.4	1089	20,600

90 gr. Jacketed HC
1.315" OAL

Powder	Sugg. Starting Grains	Velocity fps	Pressure C.U.P.	Max. Load Grains	Velocity fps	Pressure C.U.P.
Bullseye	2.5	729	12,900	3.7	1009	20,600
Unique	3.2	656	11,600	4.5	973	20,000
Blue Dot	5.0	719	11,800	6.5	1014	19,700
SR-4756	3.2	549	9,800	4.6	1015	20,500
800X	3.8	805	13,400	5.0	1057	20,600
231	2.5	679	12,500	3.8	980	20,600
HP38	2.0	612	12,100	3.3	941	20,600
AA 5	4.0	723	11,900	5.4	1064	20,700

#313249
85 gr. (#2 Alloy) 1.350" OAL

Powder	Sugg. Starting Grains	Velocity fps	Pressure C.U.P.	Max. Load Grains	Velocity fps	Pressure C.U.P.
Bullseye	2.5	803	12,400	4.0	1115	20,900
Red Dot	2.5	801	12,800	3.8	1110	20,400
Unique	3.0	747	10,800	4.5	1092	20,100
Herco	3.5	845	12,500	4.9	1080	19,300
Blue Dot	5.0	824	12,700	6.5	1105	19,000
SR-7625	2.7	709	9,700	4.3	1083	19,800
SR-4756	3.0	659	8,000	4.6	1091	18,900
800X	3.8	889	12,400	5.2	1136	19,300
231	2.5	777	12,500	4.1	1101	20,400
HP38	2.0	738	10,900	3.5	1082	20,700
AA 5	4.0	820	11,500	5.4	1124	19,200

#313492
95 gr. (#2 Alloy) 1.260" OAL

Powder	Sugg. Starting Grains	Velocity fps	Pressure C.U.P.	Max. Load Grains	Velocity fps	Pressure C.U.P.
Bullseye	2.5	825	12,100	3.7	1072	19,300
Red Dot	2.3	723	9,400	3.7	1081	19,900
Unique	3.0	751	10,200	4.4	1075	19,500
Herco	3.5	812	11,600	4.8	1122	19,800
Blue Dot	5.0	810	11,100	6.5	1094	19,000
SR-7625	3.0	793	10,300	4.4	1103	20,900
SR-4756	3.2	728	9,300	4.6	1113	20,100
800X	3.7	855	12,200	5.1	1113	19,900
231	2.5	772	10,900	3.9	1071	19,400
HP38	2.0	746	10,700	3.4	1057	19,700
AA 5	3.8	797	11,000	5.2	1123	20,500

Note: Loads shown in shaded panels are maximum.

32 H & R Magnum

#313631
100 gr. (#2 Alloy) 1.325" OAL

Powder	Sugg. Starting Grains	Velocity fps	Pressure C.U.P.	Max. Load Grains	Velocity fps	Pressure C.U.P.
Bullseye	2.0	687	11,000	3.5	1023	19,800
Red Dot	2.0	660	10,300	3.5	1027	20,400
Unique	2.7	729	12,100	4.2	1035	20,900
Herco	2.8	700	11,400	4.2	1013	19,700
Blue Dot	4.0	690	10,800	5.4	1015	19,700
SR-7625	2.3	671	10,300	3.7	1026	20,800
SR-4756	2.5	594	8,200	4.0	1034	20,800
800X	3.0	754	12,300	4.3	999	19,700
231	2.2	697	11,300	3.7	1031	20,000
HP38	1.8	673	10,700	3.3	1018	19,900
AA 5	3.2	699	11,100	4.6	974	20,900

Note: Loads shown in shaded panels are maximum.

38 Smith and Wesson

(38 S&W) (38 Colt New Police) (380 MK1)

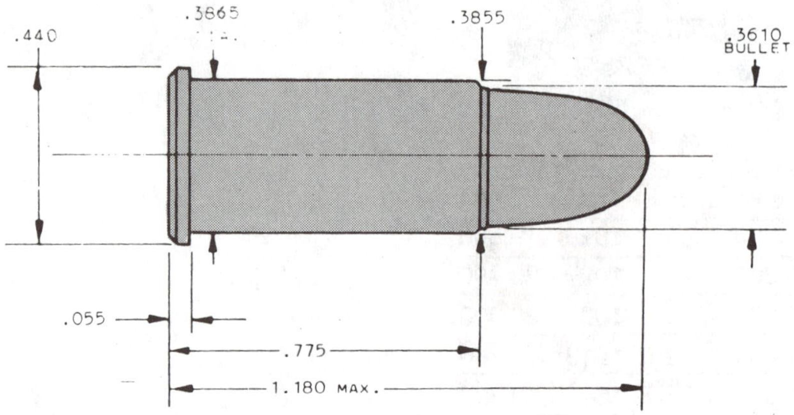

COMMENTS:

The data listed is intended only for use in solid frame revolvers. We do not recommend the use of reloads in revolvers of the top break design due to their age and relative weakness.

Because of wide internal barrel dimensions it is wise to slug your bore before beginning to reload. Size cast bullets to groove diameter or 0.001" larger. Then make up a dummy round to ensure that your revolver chambers will easily accept ammo with bullets of the chosen diameter.

The fastest burning propellants seem to work best. So, as with other small capacity cases, begin with either Hercules Bullseye or Winchester 231.

TEST COMPONENTS:

Cases .... Remington; Winchester
Trim-to Length .... .765"
Primers .... Winchester 1½-108
Primer Size .... Small Pistol
Lyman Shell Holder .... No. 21
Cast Bullets Used .... (Sized to .360" dia.)
#356242, 121 gr.
#358480, 133 gr.
#358311, 158 gr.

TEST SPECIFICATIONS: (Velocity & Pressure)

Firearm Used .... Special Pressure Test Barrel; Smith & Wesson Model 33
Barrel Length .... 4"
Twist .... 1-18¾"
Groove Dia. .... .361"

#356242
121 gr. (#2 Alloy) 1.125" OAL

Powder	Sugg. Starting Grains	Velocity fps	Pressure C.U.P.	Max. Load Grains	Velocity fps	Pressure C.U.P.
Bullseye	2.0	560	—	3.0	805	—
Unique	3.5	700	—	4.5	945	—
700X	2.3	680	—	3.1	895	—
PB	2.6	575	—	3.4	825	—
SR-7625	2.7	580	—	3.5	830	—
231	2.0	610	6,900	2.8	798	12,400
Red Dot	2.4	650	—	3.2	855	—
Green Dot	2.5	615	—	3.3	815	—

Note: Loads shown in shaded panels are maximum.

38 Smith and Wesson

(38 S&W) (38 Colt New Police) (380 MK1)

#358480
133 gr. (#2 Alloy) 1.062" OAL

Powder	Sugg. Starting Grains	Velocity fps	Pressure C.U.P.	Max. Load Grains	Velocity fps	Pressure C.U.P.
Bullseye	2.0	595	—	2.9	820	—
Red Dot	2.4	705	—	3.1	865	—
Green Dot	2.5	680	—	3.2	830	—
Unique	3.5	790	—	4.2	920	—
700X	2.3	735	—	2.9	875	—
PB	2.6	670	—	3.2	830	—
SR-7625	2.7	675	—	3.3	850	—
231	1.9	568	6,400	2.8	786	12,300

#358311
158 gr. (#2 Alloy) 1.150" OAL

Powder	Sugg. Starting Grains	Velocity fps	Pressure C.U.P.	Max. Load Grains	Velocity fps	Pressure C.U.P.
Bullseye	1.8	545	—	2.6	730	—
Red Dot	2.1	605	—	2.8	770	—
Green Dot	2.2	595	—	2.9	745	—
Unique	3.2	715	—	3.7	805	—
700X	2.1	665	—	2.6	785	—
PB	2.3	620	—	2.8	730	—
SR-7625	2.4	660	—	2.9	770	—
HS-7	3.6	619	8,100	4.5	749	11,900
231	1.5	486	4,600	2.5	727	12,400

Note: Loads shown in shaded panels are maximum.

380 Automatic(380 ACP)

(9mm Corto) (9mm Kurtz) (9mm Short)

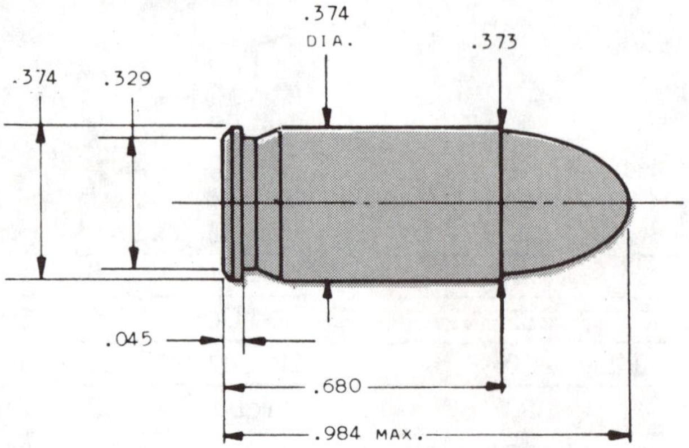

COMMENTS:

Groove diameters have been measured as small as 0.355" to as large as 0.362 inch. Such variations can play havoc with accuracy. Because of chamber restrictions it is generally not safe to load bullets of a diameter larger than 0.355 inch.

Heavy (long) bullets of 0.355" diameter may cause case bulging due to the internal case taper. This may be ignored so long as the outside diameter of the case at the bulge does not exceed 0.374 inch.

Winchester 231 is an outstanding performer with all weight bullets in this cartridge. Cast bullet #356632 is a good all around choice for this caliber.

TEST COMPONENTS:

Cases .... Federal; Super Vel (cast bullets only)
Trim-to Length .... .677"
Primers .... Federal 100;
Winchester 1 1/2-108 (cast bullets only)
Primer Size .... Small Pistol
Lyman Shell Holder .... No. 26
Jacketed Bullets Used .... Sierra HC #8100, 90 gr.
Sierra FMJ #8105, 95 gr.
Speer HP #3983, 100 gr.
Hornady HP, 115 gr.
Cast Bullets Used .... (Sized to .355" dia.)
#356242, 92 gr.
#356632, 100 gr.
#356242, 121 gr.

TEST SPECIFICATIONS: (Velocity & Pressure)

Firearm Used .... Universal Receiver
Barrel Length .... 3 3/4"
Twist .... 1-16"
Groove Dia. .... .355"

90 gr. Jacketed HC
.925" OAL

Powder	Sugg. Starting Grains	Velocity fps	Pressure C.U.P.	Max. Load Grains	Velocity fps	Pressure C.U.P.
700X	1.8	704	9,100	2.5	959	15,900
SR-7625	2.2	583	8,100	3.1	941	15,200
Bullseye	2.0	713	9,100	2.9	970	15,500
Red Dot	1.9	691	8,400	2.8	983	15,700
Unique	2.5	709	8,700	3.4	967	15,500
231	2.1	681	8,100	3.2	992	16,000

Note: Loads shown in shaded panels are maximum.

380 Automatic(380 ACP)

(9mm Corto) (9mm Kurtz) (9mm Short)

95 gr. FMJ
.900" OAL

Powder	Sugg. Starting Grains	Velocity fps	Pressure C.U.P.	Max. Load Grains	Velocity fps	Pressure C.U.P.
Bullseye	2.0	724	9,400	2.8	953	15,700
AA 2	1.9	694	9,900	2.9	922	15,600
Red Dot	2.0	712	9,600	2.7	938	15,600
231	2.1	704	9,700	2.9	939	15,800
700X	1.7	693	9,500	2.5	935	15,700
AA 5	3.0	671	9,100	3.9	940	16,000
Unique	2.3	706	9,600	3.1	952	16,000

115 gr. Jacketed HP
.956" OAL

Powder	Sugg. Starting Grains	Velocity fps	Pressure C.U.P.	Max. Load Grains	Velocity fps	Pressure C.U.P.
700X	1.6	658	9,100	2.5	921	15,900
SR-7625	2.2	667	8,700	3.1	928	15,800
Bullseye	2.0	711	9,900	2.9	937	15,100
Red Dot	1.8	670	8,700	2.8	945	15,500
Unique	2.4	718	9,500	3.4	973	15,900
231	2.0	664	8,300	3.1	934	15,300
HS-6	3.1	677	8,700	4.5	945	15,200

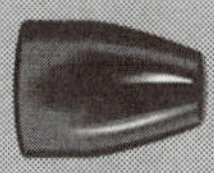

100 gr. Jacketed HP
.955" OAL

Powder	Sugg. Starting Grains	Velocity fps	Pressure C.U.P.	Max. Load Grains	Velocity fps	Pressure C.U.P.
700X	1.9	704	9,100	2.5	889	14,700
SR-7625	2.4	694	8,000	3.3	960	14,900
Bullseye	2.0	695	7,700	3.1	988	15,400
Red Dot	1.8	651	6,500	3.1	1026	15,900
Unique	2.6	725	8,600	3.6	1013	15,200
231	2.2	693	6,600	3.5	1024	15,600

#356242
92 gr. (#2 Alloy) .980" OAL

Powder	Sugg. Starting Grains	Velocity fps	Pressure C.U.P.	Max. Load Grains	Velocity fps	Pressure C.U.P.
700X	1.9	695	11,100	2.5	900	15,800
PB	2.2	645	10,300	2.8	860	15,200
**SR-4756	2.8	633	6,900	4.1	970	15,600
Bullseye	2.4	750	11,800	3.0	915	16,000
Red Dot	2.0	690	11,000	2.6	880	15,200
Green Dot	2.1	655	10,000	3.0	910	16,000
Unique	3.1	755	11,800	3.7	920	16,000
**Blue Dot	4.0	667	7,000	5.5	961	14,600
**231	2.3	701	7,900	3.5	1000	15,600

Note: Loads shown in shaded panels are maximum.
** Designates use of Federal cases.

380 Automatic (380ACP)

(9mm Corto) (9mm Kurtz) (9mm Short)

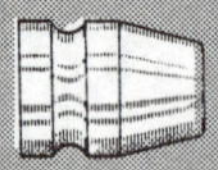

#356632
100 gr. (#2 Alloy) .885" OAL

Powder	Sugg. Starting Grains	Velocity fps	Pressure C.U.P.	Max. Load Grains	Velocity fps	Pressure C.U.P.
Bullseye	1.9	745	9,300	2.7	957	15,100
AA 2	2.0	736	9,600	3.4	977	15,400
Red Dot	1.8	745	9,200	2.7	991	15,700
231	2.0	739	9,200	3.1	1009	15,900
700X	1.7	713	9,200	2.5	959	14,700
AA 5	2.8	771	9,900	3.6	973	15,300
Unique	2.4	777	9,900	3.1	1001	15,700
SR-4756	2.4	724	9,400	3.3	981	15,600
Blue Dot	3.3	715	8,900	5.0	988	15,800

#356242
121 gr. (#2 Alloy) .980" OAL

Powder	Sugg. Starting Grains	Velocity fps	Pressure C.U.P.	Max. Load Grains	Velocity fps	Pressure C.U.P.
700X	1.4	610	11,300	1.9	750	15,800
PB	1.7	605	10,800	2.2	750	15,800
**SR-4756	2.3	662	7,800	3.5	927	15,800
Bullseye	1.6	600	10,500	2.1	750	15,200
Red Dot	1.6	615	11,000	2.1	770	15,800
Green Dot	1.7	620	10,800	2.3	760	15,400
Unique	2.2	630	11,300	2.8	780	15,600
**Blue Dot	3.7	732	9,000	5.3	948	15,900
**231	2.1	718	9,200	3.2	946	15,800

Note: Loads shown in shaded panels are maximum.
** Designates use of Federal cases.

9mm Luger
(9mm Parabellum)

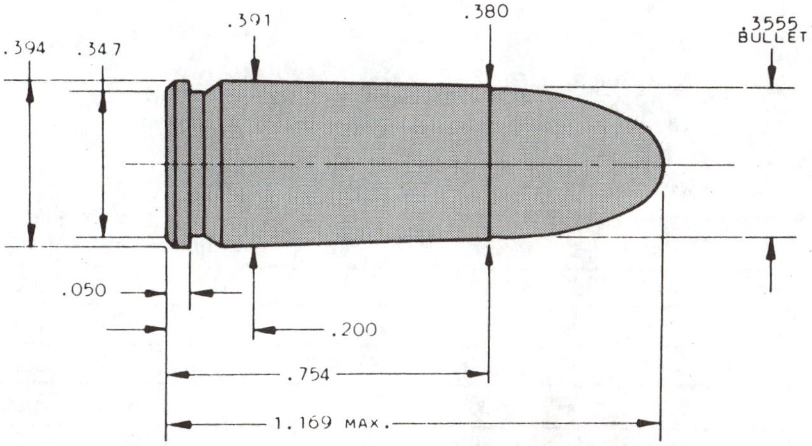

COMMENTS:

The most common diameter for this cartridge is 0.354 inch. However, the industry standard has recently been established at 0.355 inch. All barrels of 0.354" to 0.356" are correctly matched with a bullet diameter of 0.355 inch. Barrels with larger groove diameters will often provide poor accuracy with a 0.355" bullet but chamber restrictions prevent the safe use of larger diameter bullets.

This cartridge headspaces from the mouth and therefore cases trimming must be uniform and accurate. Do not reduce cases below the trim-to length. Additionally do not roll crimp bullets as this will prevent the case from properly headspacing on its mouth. A modest taper crimp may be employed if found necessary.

The most popular cast bullet for this round has been #356402. However, there are many good choices based upon intended use. Bullet #356637 (147 gr.) closely duplicates the U.S. Military subsonic bullet design and has proven to be very accurate.

Hercules Bullseye, Winchester 231, and Hercules Unique are very good popular propellants for this cartridge.

With any handgun cartridge, it is important not to seat bullets to a shorter length than specified in the data. This is especially important with 9mm loads. Pressures can be raised dramatically with deep seating.

TEST COMPONENTS:

Cases .... Federal; Winchester
Trim-to Length .... .751"
Primers .... CCI 500; Winchester 1½-108
Primer Size .... Small Pistol
Lyman Shell Holder .... No. 12
Jacketed Bullets Used .... Sierra JHP #8100, 90 gr.
Sierra MC #8105, 95 gr.
Hornady FMJ #3552, 100 gr.
Hornady HP #3554, 115 gr.
Speer SP #4005, 125 gr.
Sierra FMJ #8345, 130 gr.
Cast Bullets Used .... (Sized to .356" dia.)
#356242, 92 gr.
#356632, 100 gr.
#358345, 115 gr.
#356402, 121 gr.
#356242, 121 gr.
#356634, 130 gr.
#356637, 147 gr.

TEST SPECIFICATIONS: (Velocity & Pressure)

Firearm Used .... Universal Receiver
Barrel Length .... 4"
Twist .... 1-10"
Groove Dia. .... .355"

90 gr. Jacketed HP
1.010" OAL

Powder	Sugg. Starting Grains	Velocity fps	Pressure C.U.P.	Max. Load Grains	Velocity fps	Pressure C.U.P.
700X	3.6	1027	25,000	4.5	1292	30,500
SR-4756	5.0	1028	24,000	6.2	1352	31,900
Bullseye	4.1	971	21,600	5.2	1376	30,900
Unique	5.0	1089	26,900	6.3	1343	32,600
Blue Dot	7.0	1072	24,800	8.9	1341	31,500
HS-6	5.9	858	20,200	7.4	1333	31,800
AA 5	5.2	835	16,600	6.5	1347	30,800
Clays	3.0	982	22,000	3.7+	1213	30,800
231	4.0	1156	20,600	5.2	1411	30,400

Note: Loads shown in shaded panels are maximum.
+ Designates a compressed powder charge.

9mm Luger
(9mm Parabellum)

95 gr. Jacketed MC
1.005" OAL

Powder	Sugg. Starting Grains	Velocity fps	Pressure C.U.P.	Max. Load Grains	Velocity fps	Pressure C.U.P.
700X	3.2	893	19,600	4.5	1301	31,400
SR-4756	4.5	983	20,700	5.8	1306	32,700
Bullseye	3.5	893	19,400	5.0	1307	30,600
Unique	4.5	1007	22,000	5.8	1399	32,000
Blue Dot	6.8	1101	22,900	8.9+	1457	32,300
HS-6	5.8	952	20,100	7.0	1315	29,600
AA 5	5.2	915	18,100	6.5	1394	31,000
231	3.7	1068	19,200	5.1	1374	32,400

100 gr. FMJ
1.105" OAL

Powder	Sugg. Starting Grains	Velocity fps	Pressure C.U.P.	Max. Load Grains	Velocity fps	Pressure C.U.P.
700 X	3.5	870	20,200	4.4	1171	31,500
231	4.2	921	21,400	5.3	1216	32,300
SR-4756	4.5	681	13,000	5.7	1225	30,400
Bullseye	4.0	976	22,000	5.0	1263	31,700
Unique	4.5	931	20,000	5.6	1256	31,400
Blue Dot	6.5	973	19,800	7.8	1252	30,200
HS-6	6.0	801	15,900	7.4	1206	30,100
AA 7	8.0	884	18,600	9.6	1271	31,000

115 gr. Jacketed HP
1.090" OAL

Powder	Sugg. Starting Grains	Velocity fps	Pressure C.U.P.	Max. Load Grains	Velocity fps	Pressure C.U.P.
700X	3.1	918	20,200	4.5	1171	31,700
SR-4756	4.5	946	18,600	‡6.2+	1147	28,100
Bullseye	3.5	945	18,700	4.8	1184	31,700
Unique	4.4	996	20,600	5.8	1233	30,700
Blue Dot	6.3	910	17,100	8.7+	1211	29,200
HS-6	5.7	956	17,400	7.2	1193	30,400
AA 7	6.8	871	20,700	8.5	1158	30,400
231	3.5	959	20,700	4.9	1253	32,100

125 gr. Jacketed SP
1.120" OAL

Powder	Sugg. Starting Grains	Velocity fps	Pressure C.U.P.	Max. Load Grains	Velocity fps	Pressure C.U.P.
700X	3.0	835	19,300	4.4	1095	32,000
Bullseye	3.3	876	18,700	4.8	1142	32,500
Unique	4.1	859	18,200	5.3+	1118	31,100
Blue Dot	6.1	895	18,200	8.0+	1190	32,600
HS-6	5.4	901	19,300	6.8	1144	31,100
AA 5	4.4	785	18,900	5.5	1050	29,700
AA 7	6.4	825	20,200	8.0	1102	29,800
Clays	2.6	739	19,500	3.7+	974	32,300
231	3.2	910	17,800	4.6	1099	31,500

Note: Loads shown in shaded panels are maximum.
\+ Designates a compressed powder charge.
‡ Designates a factory velocity duplication load.

9mm Luger
(9mm Parabellum)

130 gr. FMJ
1.160" OAL

Powder	Sugg. Starting Grains	Velocity fps	Pressure C.U.P.	Max. Load Grains	Velocity fps	Pressure C.U.P.
700X	3.0	803	21,700	4.0	991	32,500
231	4.0	834	24,000	5.0	1058	32,100
SR-4756	4.0	801	20,200	5.1	1063	32,000
Bullseye	3.6	780	21,500	4.5	1002	31,000
Unique	3.5	710	18,200	4.8	1014	31,200
Blue Dot	5.4	852	21,700	6.8	1114	31,700
HS-6	5.0	752	18,000	6.5	1112	32,500
AA 7	6.5	833	20,500	8.2	1119	32,600
Clays	2.5	690	19,200	3.6+	933	31,500

#356242
92 gr. (#2 Alloy) 1.045" OAL

Powder	Sugg. Starting Grains	Velocity fps	Pressure C.U.P.	Max. Load Grains	Velocity fps	Pressure C.U.P.
700X	4.1	1125	22,800	5.1	1285	31,000
PB	4.8	1115	19,800	6.0	1360	33,000
SR-7625	5.0	1180	23,400	5.9	1310	30,500
SR-4756	5.3	1168	22,400	7.2	1380	30,500
Bullseye	4.0	1060	19,800	5.2	1315	32,000
Red Dot	4.1	1120	22,800	5.5	1340	32,500
Green Dot	4.3	1095	20,400	6.0	1325	31,000
Unique	5.0	1065	19,800	6.8	1325	31,500
HS-6	5.8	1011	19,500	7.2	1356	29,800
231	3.8	1093	20,500	5.1	1455	32,100

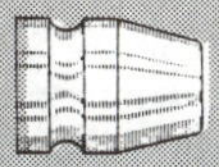

#356632
100 gr. (#2 Alloy) .976" OAL

Powder	Sugg. Starting Grains	Velocity fps	Pressure C.U.P.	Max. Load Grains	Velocity fps	Pressure C.U.P.
Bullseye	3.3	1157	21,100	4.4	1395	32,400
Unique	4.2	1193	28,300	4.7	1329	31,200
700X	3.3	1072	24,100	4.0	1300	30,800
SR-4756	3.9	1215	24,100	5.3	1415	32,700
HS-6	6.0	1205	27,600	6.6	1398	31,700
WSL	3.7	1089	26,900	4.3	1307	32,200
AA 2	4.0	1109	26,300	4.5	1259	32,200
AA 5	5.1	1129	24,800	5.7	1349	30,800
Clays	2.4	873	20,000	3.3	1168	31,500
231	3.2	1039	19,000	4.5	1333	31,000

#358345
115 gr. (#2 Alloy) .997" OAL

Powder	Sugg. Starting Grains	Velocity fps	Pressure C.U.P.	Max. Load Grains	Velocity fps	Pressure C.U.P.
**700X	3.7	1045	25,200	4.4	1160	32,500
**PB	3.7	995	23,400	4.5	1130	32,000
**SR-7625	3.8	1005	24,000	4.5	1130	31,000
SR-4756	4.2	1005	18,800	6.1	1284	32,000
**Bullseye	3.8	1035	22,800	4.9	1200	31,000
**Red Dot	3.9	1055	24,600	4.9	1200	32,000
**Green Dot	4.0	1030	24,000	5.0	1200	32,500
**Unique	4.5	1025	22,200	5.7	1210	31,500
Blue Dot	5.8	983	16,800	8.4	1339	32,300
231	2.9	1040	18,000	4.7	1443	31,600
HS-6	5.6	1000	23,800	7.0	1326	31,200

Note: Loads shown in shaded panels are maximum.
+ Designates a compressed powder charge.
** Designates use of Winchester cases and primers.

9mm Luger
(9mm Parabellum)

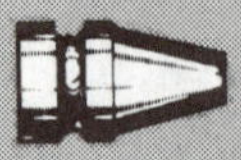

#356402
121 gr. (#2 Alloy) 1.110" OAL

Powder	Sugg. Starting Grains	Velocity fps	Pressure C.U.P.	Max. Load Grains	Velocity fps	Pressure C.U.P.
**700X	3.3	975	25,200	4.0	1080	33,000
**PB	3.4	915	20,400	4.3	1095	31,500
**SR-7625	3.5	935	21,000	4.3	1080	31,500
SR-4756	4.0	925	17,500	5.5	1198	32,900
**Bullseye	3.7	950	21,600	4.4	1125	32,000
**Red Dot	3.8	1020	26,400	4.5	1130	32,500
**Green Dot	3.9	970	22,200	4.8	1165	32,500
**Unique	4.3	940	21,000	5.3	1135	31,000
Blue Dot	5.9	973	18,200	8.0	1253	31,600
Clays	2.2	784	18,700	3.1	1006	31,900
231	2.9	963	17,000	4.4	1264	29,600
HS-6	5.0	910	20,000	6.2	1163	29,300

#356242
121 gr. (#2 Alloy) 1.065" OAL

Powder	Sugg. Starting Grains	Velocity fps	Pressure C.U.P.	Max. Load Grains	Velocity fps	Pressure C.U.P.
700X	3.0	940	26,000	3.7	1080	31,600
PB	3.2	872	24,500	4.1	1106	31,400
SR-7625	3.0	852	21,400	3.9	1095	31,300
Bullseye	3.4	939	23,600	4.2	1175	32,500
Red Dot	3.0	919	25,100	3.9	1140	32,000
Green Dot	3.5	966	26,200	4.4	1152	32,800
Unique	4.0	1013	26,500	5.0	1194	32,800
HS-6	5.0	904	23,000	6.3	1182	31,700
231	3.0	1010	20,000	4.1	1148	30,200

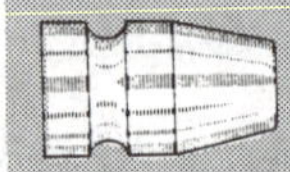

#356634
130 gr. (#2 Alloy) 1.016" OAL

Powder	Sugg. Starting Grains	Velocity fps	Pressure C.U.P.	Max. Load Grains	Velocity fps	Pressure C.U.P.
Bullseye	3.4	977	27,200	3.7	1049	31,900
Unique	3.5	972	28,000	4.0	1084	32,200
Blue Dot	5.8	969	26,900	6.5	1089	30,100
700X	3.2	927	27,100	3.7	1080	32,400
SR-4756	5.0	982	28,400	5.6	1097	31,600
HS-6	5.1	997	28,300	5.6	1060	31,000
WSF	3.8	888	24,400	4.4	1047	31,700
AA 7	6.8	1016	28,700	7.7	1191	32,700
Clays	2.2	787	22,200	3.2+	1021	31,500
231	2.6	990	19,700	4.1	1133	31,600

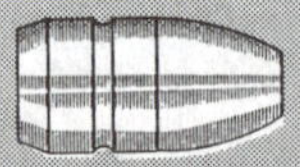

#356637
147 gr. (#2 Alloy) 1.058" OAL

Powder	Sugg. Starting Grains	Velocity fps	Pressure C.U.P.	Max. Load Grains	Velocity fps	Pressure C.U.P.
Bullseye	2.8	780	24,900	3.5	942	31,600
Unique	3.2	810	25,600	3.8	956	31,800
Blue Dot	5.1	864	26,700	5.9	980	31,200
700X	2.7	803	26,500	3.3	921	32,200
SR-4756	3.1	769	25,000	3.6	896	30,700
HS-6	4.4	846	26,700	5.0	968	31,900
WSF	3.4	867	26,900	3.9	961	32,100
Clays	1.9	669	21,000	2.8+	873	31,700

Note: Loads shown in shaded panels are maximum.
** Designates use of Winchester cases and primers.
+ Designates a compressed powder charge.

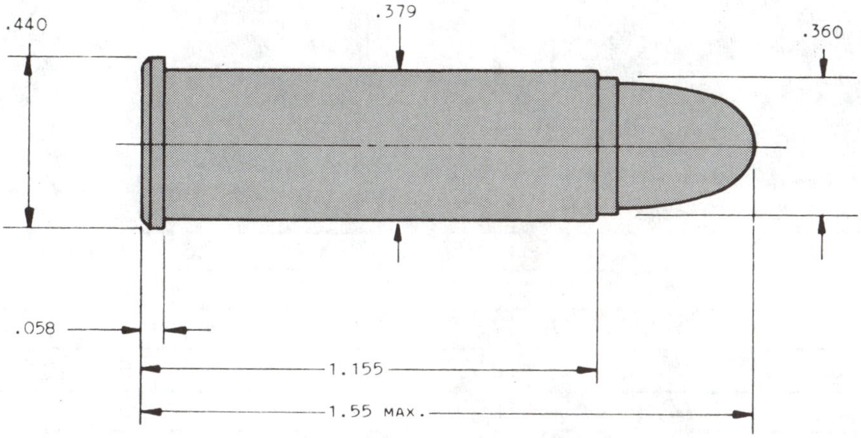

COMMENTS:

Loads designated with a +P symbol are higher pressure loads and must not be used in standard 38 Special revolvers unless the firearm manufacturer has suggested the use of such ammunition in your specific revolver. This caution should not be ignored. Under no circumstance should +P (plus P) ammo be used in revolvers with aluminum frames and/or aluminum cylinders.

Do not use charge weights below the starting load to prevent bullet jackets from half jacket bullets from becoming lodged in the barrel (the lead cores may exit the muzzle and strike the target). To crimp half jacket bullets, form the crimp at the junction of the jacket and exposed lead nose.

For target (mid-range) loads we suggest bullet #358495 with the suggested starting grains load. Bullet #358311 duplicates the factory lead round nose 158 grain bullet very closely. Bullet #358429 closely duplicates the factory 158 grain semi-wadcutter. This is the Elmer Keith design bullet and makes an excellent choice for hunting.

Handguns can vary in groove diameter and it is wise to slug your barrel before sizing cast bullets.

Hercules Bullseye and Winchester 231 consistently provide the best accuracy with all bullets.

TEST COMPONENTS:

Cases .... Federal
Trim-to Length .... 1.149"
Primers .... CCI 500
Primer Size .... Small Pistol
Lyman Shell Holder .... No. 1
Jacketed Bullets Used .... Remington SJHP #B22944, 95 gr.
Winchester JHP #B3573HP, 110 gr.
Hornady HP, 125 gr.
Speer JHP #4203, 140 gr.
Speer HJHP #4205, 146 gr.
Hornady HP, 158 gr.
Speer HJSP #4223, 160 gr.
Cast Bullets Used .... (Sized to .357" dia.)
* Gas Check Bullets .... #356242, 92 gr.
#356242, 121 gr.
#358495, 141 gr.
#358212, 146 gr.
*#358156, 155 gr.
#358311, 158 gr.
#357446, 162 gr.
#358429, 168 gr.

TEST SPECIFICATIONS:
(Velocity & Pressure)

Firearm Used .... Universal Receiver and Special Vented Barrel
Barrel Length .... 4"
Twist .... 1-18¾"
Groove Dia. .... .357"

95 gr. Jacketed HP
1.470" OAL

Powder	Sugg. Starting Grains	Velocity fps	Pressure C.U.P.	Max. Load Grains	Velocity fps	Pressure C.U.P.
700X	3.8	737	10,100	4.8	995	16,800
SR-7625	4.5	614	6,800	6.6	1048	16,500
SR-4756	5.5	693	9,400	7.3	1036	16,800
Bullseye	3.2	613	7,900	5.4	1042	16,000
				‡(+P)5.8	1095	17,300
Unique	4.5	671	9,100	6.6	1046	17,000
				(+P)7.1	1119	17,400
231	4.5	711	8,100	6.0	1016	15,300
				(+P)6.2	1060	18,400
HS-6	6.1	660	7,900	9.0	1072	16,500

Note: Loads shown in shaded panels are maximum.
‡ Designates a factory velocity duplication load.
(+P) Designates higher pressure loads. See COMMENTS.

38 Special
(38 Smith & Wesson Special)

110 gr. Jacketed HP
1.478" OAL

Powder	Sugg. Starting Grains	Velocity fps	Pressure C.U.P.	Max. Load Grains	Velocity fps	Pressure C.U.P.
700X	3.7	596	7,300	5.1	955	16,800
SR-4756	5.8	675	9,100	7.3	980	16,400
				(+P)7.5	1027	17,500
Bullseye	3.4	528	6,300	5.4	946	16,600
				(+P)5.6	981	18,000
Unique	4.7	580	7,600	6.7	967	16,600
				(+P)7.0	1026	18,200
Blue Dot	8.0	709	8,800	9.7	986	15,700
231	4.5	653	10,400	6.0	947	16,800
				(+P)6.3	985	18,400
HS-6	6.5	593	7,700	9.0	978	16,200
				(+P)9.5	1043	18,000

125 gr. Jacketed HP
1.470" OAL

Powder	Sugg. Starting Grains	Velocity fps	Pressure C.U.P.	Max. Load Grains	Velocity fps	Pressure C.U.P.
700X	2.8	473	8,100	4.1	814	16,500
				(+P)4.5	877	17,300
SR-4756	4.8	535	7,400	6.3	874	16,200
Bullseye	3.2	568	10,200	4.4	860	16,500
				(+P)5.0	920	18,300
Unique	4.0	498	8,000	6.0	895	16,700
Blue Dot	7.4	703	11,400	8.7	893	16,500
231	3.9	589	10,600	5.1	820	15,600
				(+P)5.4	882	18,300
HS-6	5.6	600	9,400	7.4	862	16,100
AA 2	4.3	751	11,800	5.2	899	16,200
Clays	3.2	719	11,800	3.8	870	15,900

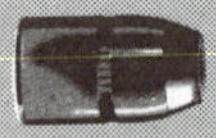

140 gr. Jacketed HP
1.470" OAL

Powder	Sugg. Starting Grains	Velocity fps	Pressure C.U.P.	Max. Load Grains	Velocity fps	Pressure C.U.P.
SR-7625	4.2	505	10,800	5.1	715	16,100
Bullseye	3.5	533	11,900	4.3	714	16,500
Unique	4.3	537	11,200	5.3	709	15,900
Blue Dot	7.6	662	13,000	8.4	824	16,800
231	3.9	496	12,100	4.7	684	15,800
HS-6	5.6	526	11,200	7.2	777	16,700

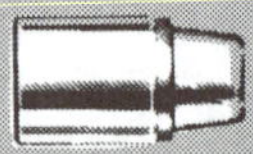

146 gr. Jacketed HP
1.408" OAL

Powder	Sugg. Starting Grains	Velocity fps	Pressure C.U.P.	Max. Load Grains	Velocity fps	Pressure C.U.P.
SR-7625	4.7	694	15,100	4.9	740	16,800
SR-4756	5.0	633	13,400	5.6	778	16,600
Unique	4.3	741	11,200	5.1	745	16,800
				(+P)5.4	795	17,600
Herco	5.0	711	15,100	5.2	744	15,900
Blue Dot	6.7	595	12,500	7.4	772	16,600
2400	9.0	681	13,400	10.0	911	16,300
HS-6	6.2	691	14,800	6.7	782	16,000

Note: Loads shown in shaded panels are maximum.
(+P) Designates higher pressure loads.
See COMMENTS.

158 gr. Jacketed HP
1.480" OAL

Powder	Sugg. Starting Grains	Velocity fps	Pressure C.U.P.	Max. Load Grains	Velocity fps	Pressure C.U.P.
SR-4756	4.9	611	13,800	5.4	705	16,800
Unique	4.7	645	14,400	5.0	710	16,600
				(+P)5.2	761	18,200
Herco	4.7	619	14,000	5.2	708	16,800
Blue Dot	6.0	546	12,200	7.1	721	16,300
2400	8.4	645	13,500	9.4	745	16,400
HS-6	5.5	572	12,400	6.5	726	16,600
AA 5	5.0	674	12,100	5.8	821	16,300
Clays	2.8	648	14,500	3.2	712	16,600

160 gr. Jacketed SP
1.430" OAL

Powder	Sugg. Starting Grains	Velocity fps	Pressure C.U.P.	Max. Load Grains	Velocity fps	Pressure C.U.P.
IMR-4227	9.6	747	15,300	10.0	793	16,700
Unique	4.3	625	11,600	5.2	769	16,600
				(+P)5.3	808	17,800
Blue Dot	6.4	621	11,900	7.2	786	16,500
2400	9.0	710	14,700	9.6	798	16,900
HS-6	5.8	653	13,900	6.4	742	16,500

#356242
92 gr. (Linotype) 1.456" OAL

Powder	Sugg. Starting Grains	Velocity fps	Pressure C.U.P.	Max. Load Grains	Velocity fps	Pressure C.U.P.
700X	3.0	675	5,700	5.0	1107	15,700
				(+P)5.3	1168	18,300
Bullseye	3.3	765	7,600	5.3	1110	15,700
				(+P)5.6	1168	17,600
Red Dot	3.8	830	8,800	5.2	1108	16,300
				(+P)5.5	1154	17,700
231	3.5	689	5,900	5.7	1104	16,300
				(+P)6.0	1163	17,800
HP38	3.6	836	9,100	5.0	1098	16,500

#356242
121 gr. (Linotype) 1.450" OAL

Powder	Sugg. Starting Grains	Velocity fps	Pressure C.U.P.	Max. Load Grains	Velocity fps	Pressure C.U.P.
700X	2.5	647	7,600	3.9	940	16,200
				(+P)4.2	987	17,900
SR-7625	4.0	788	10,600	5.0	988	16,200
Bullseye	2.8	690	8,600	4.6	1001	16,400
				(+P)4.9	1045	18,400
Red Dot	3.3	749	10,100	4.6	974	16,200
				(+P)5.0	1047	18,400
Unique	4.0	725	9,000	5.7	1012	16,200
231	3.4	693	8,800	5.1	986	16,600
				(+P)5.4	1036	17,900
HP38	3.0	725	9,600	4.3	961	16,700
				(+P)4.8	1045	18,300

Note: Loads shown in shaded panels are maximum.
(+P) Designates higher pressure loads.
See COMMENTS.

38 Special

(38 Smith & Wesson Special)

#358495
141 gr. (Linotype) 1.310" OAL

Powder	Sugg. Starting Grains	Velocity fps	Pressure C.U.P.	Max. Load Grains	Velocity fps	Pressure C.U.P.
700X	2.2	636	8,300	4.0	994	16,600
SR-7625	3.1	683	9,100	4.3	905	15,600
Bullseye	3.0	778	11,000	4.1	973	16,800
Red Dot	2.8	706	9,500	4.2	961	16,300
Unique	3.8	757	10,300	5.3	1013	16,700
231	3.0	667	8,500	4.9	1001	16,800
HP38	2.7	683	9,000	4.5	1001	16,800
AA 2	2.7	788	12,200	3.5	934	16,300
Clays	2.2	769	11,600	2.8	905	16,200

#358212
146 gr. (Linotype) 1.515" OAL

Powder	Sugg. Starting Grains	Velocity fps	Pressure C.U.P.	Max. Load Grains	Velocity fps	Pressure C.U.P.
700X	2.8	674	9,400	4.0	881	15,900
SR-7625	3.5	594	6,900	5.0	899	16,100
Bullseye	2.7	612	8,000	4.3	906	15,800
Red Dot	2.8	650	8,800	4.3	917	16,400
Unique	3.7	652	9,000	5.4	928	15,900
231	3.2	617	7,900	5.0	930	16,700
HP38	3.0	654	9,300	4.4	900	16,500

#358156
155 gr. (Linotype) 1.460" OAL

Powder	Sugg. Starting Grains	Velocity fps	Pressure C.U.P.	Max. Load Grains	Velocity fps	Pressure C.U.P.
PB	3.0	584	8,700	4.2	825	16,200
SR-4756	4.2	630	9,500	5.5	898	16,700
Bullseye	2.8	638	10,000	4.0	858	16,000
				(+P)4.4	915	18,100
Unique	3.4	601	8,900	5.1	895	16,100
				(+P)5.4	954	18,000
Blue Dot	6.0	686	10,400	7.2	867	15,700
231	3.5	680	11,100	4.7	885	16,800
				(+P)5.0	934	18,000
HS-6	4.4	564	8,000	6.7	894	16,600
HP38	2.8	642	10,200	3.9	854	16,300
				(+P)4.1	882	17,800

#358311
158 gr. (Linotype) 1.550" OAL

Powder	Sugg. Starting Grains	Velocity fps	Pressure C.U.P.	Max. Load Grains	Velocity fps	Pressure C.U.P.
PB	3.0	604	7,900	4.5	871	16,100
SR-4756	4.2	620	8,700	5.5	874	16,000
Bullseye	2.8	656	10,000	4.2	880	16,200
				(+P)4.5	931	18,200
Unique	3.4	611	8,300	5.1	896	16,400
				(+P)5.4	946	18,000
Blue Dot	5.9	673	9,100	7.8	919	16,400
231	3.5	675	9,600	4.9	906	16,800
				(+P)5.2	956	18,500
HS-6	4.5	610	8,200	6.7	901	16,200
HP38	2.6	621	8,800	4.3	902	16,900
				(+P)4.5	944	18,400

Note: Loads shown in shaded panels are maximum.
(+P) Designates higher pressure loads.
See COMMENTS.

#357446
162 gr. (Linotype) 1.457" OAL

Powder	Sugg. Starting Grains	Velocity fps	Pressure C.U.P.	Max. Load Grains	Velocity fps	Pressure C.U.P.
PB	3.4	646	9,500	4.6	877	16,300
SR-4756	4.6	694	10,400	5.8	921	16,800
Bullseye	3.0	680	10,100	4.2	889	16,900
Unique	3.5	639	8,900	5.2	914	16,200
Blue Dot	5.8	675	9,500	7.7	948	16,500
231	3.4	702	10,400	4.6	896	16,000
HS-6	5.1	690	9,600	6.7	922	16,100

#358429
168 gr. (Linotype) 1.537" OAL

Powder	Sugg. Starting Grains	Velocity fps	Pressure C.U.P.	Max. Load Grains	Velocity fps	Pressure C.U.P.
SR-7625	3.4	621	9,700	4.6	821	16,300
SR-4756	4.2	645	10,000	5.4	836	16,400
Bullseye	2.6	597	9,300	4.2	850	16,800
Unique	3.8	673	11,000	5.0	860	16,700
Blue Dot	5.5	647	10,000	7.3	869	16,500
231	3.0	599	9,300	4.6	855	16,900
HS-6	5.0	677	11,000	6.4	856	16,800
AA 5	4.3	755	11,500	5.1	894	16,100
Clays	2.3	722	11,500	2.9	824	15,900

Note: Loads shown in shaded panels are maximum.
(+P) Designates higher pressure loads.
See COMMENTS.

357 Magnum
(357 S&W Magnum)

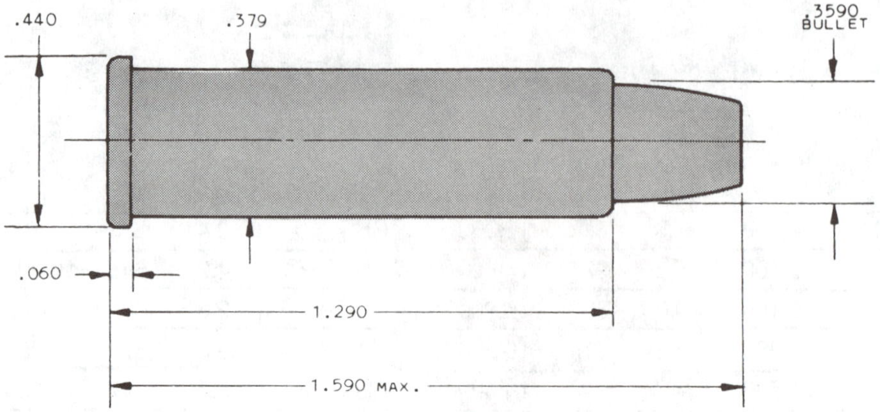

COMMENTS:

Never use 357 Magnum Loads in 38 Special cases as very dangerous pressure will result.

Handguns can vary in groove diameter and it is wise to slug your barrel before sizing cast bullets.

In order to maintain a maximum overall cartridge length of 1.590", it is sometimes necessary to crimp cast bullets on the forward edge of the first driving band.

When using half jacketed bullets velocities must be kept above 750 fps. to prevent bullet jackets from becoming lodged in the barrel (the lead cores may exit the muzzle and strike the target).

To crimp half jacket bullets, form the crimp at the junction of the jacket and exposed lead nose.

For target (mid-range) loads we recommend bullet #358495 with the suggested starting grains load. Bullet #358156 is extremely popular for heavy loads. Bullet #358429 closely duplicates the factory 158 grain semi-wadcutter. This is the Elmer Keith design bullet and makes an excellent choice for hunting.

For light loads Hercules Bullseye and Winchester 231 are best. Heavy loads work well with a wide range of powders but you might want to try 2400 for your first accuracy tests.

TEST COMPONENTS:

Cases .... Federal
Trim-to Length .... 1.285"
Primers .... CCI 550
Primer Size .... Small Pistol Magnum
Lyman Shell Holder .... No. 1
Jacketed Bullets Used .... Winchester JHP #B3573, 110 gr.
Hornady HP, 125 gr.
Speer JHP #4203, 140 gr.
Speer JHP, #4205, 146 gr.
Sierra HC #8330, 150 gr.
Hornady HP, 158 gr.
Speer JSP #4223, 160 gr.
Hornady JTC SIL #3577, 180 gr.

Cast Bullets Used .... (Size to .357" dia.)
*Gas Check Bullets
#356242, 92 gr.
#358345, 115 gr.
#356242, 121 gr.
#356402, 121 gr.
#358480, 133 gr.
#358495, 141 gr.
#358477, 150 gr.
*#358156, 155 gr.
#358311, 158 gr.
#358429, 168 gr.

Note: Loads shown in shaded panels are maximum.
\+ Designates a compressed powder charge.

TEST SPECIFICATIONS:
(Velocity & Pressure)

Firearm Used .... Universal Receiver and Special Vented Barrel
Barrel Length .... 4"
Twist .... 1-18 3/4"
Groove Dia. .... .356"

110 gr. Jacketed HP
1.590" OAL

Powder	Sugg. Starting Grains	Velocity fps	Pressure C.U.P.	Max. Load Grains	Velocity fps	Pressure C.U.P.
IMR-4227	14.8	1100	20,700	19.5+	1383	30,100
Unique	7.4	1052	17,300	10.0	1392	36,400
Blue Dot	10.0	930	12,400	14.6	1545	39,400
2400	13.5	1045	17,400	20.3+	1561	39,100
AA 5	11.0	1211	22,500	13.0	1547	42,000
Bullseye	7.0	1030	20,500	9.7	1471	40,300

357 Magnum

(357 S&W Magnum)

125 gr. Jacketed HP
1.590" OAL

Powder	Sugg. Starting Grains	Velocity fps	Pressure C.U.P.	Max. Load Grains	Velocity fps	Pressure C.U.P.
IMR-4227	15.6	1232	31,200	19.5+	1427	37,400
Unique	7.0	990	18,900	9.7	1359	41,600
Blue Dot	10.4	1113	20,000	13.3	1402	36,700
2400	13.0	1159	24,900	‡17.7	1478	40,600
HS-7	10.9	1098	21,400	14.4	1403	39,200
Bullseye	6.5	864	21,000	8.6	1315	38,400

140 gr. Jacketed HP
1.590" OAL

Powder	Sugg. Starting Grains	Velocity fps	Pressure C.U.P.	Max. Load Grains	Velocity fps	Pressure C.U.P.
IMR-4227	15.0	1106	28,800	17.8	1372	41,400
Unique	7.0	949	20,800	8.9	1233	40,000
Blue Dot	9.2	917	17,900	11.9	1272	35,200
2400	12.5	1047	26,500	16.5	1343	40,300
H110	15.2	1189	25,800	19.5	1427	37,000
AA 5	9.0	852	21,800	11.0	1219	38,800
Bullseye	5.7	705	20,300	8.0	1116	39,200

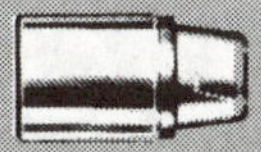

146 gr. Jacketed HP
1.590" OAL

Powder	Sugg. Starting Grains	Velocity fps	Pressure C.U.P.	Max. Load Grains	Velocity fps	Pressure C.U.P.
IMR-4227	11.5	914	21,800	16.2	1253	40,600
Unique	6.3	928	25,900	8.1	1200	41,100
Blue Dot	9.0	943	21,300	11.5	1273	39,700
2400	11.0	895	20,800	15.1	1242	39,100
H110	13.0	1064	22,300	17.0	1380	41,900
Bullseye	5.7	759	23,300	7.5	1077	39,900

150 gr. Jacketed HC
1.590" OAL

Powder	Sugg. Starting Grains	Velocity fps	Pressure C.U.P.	Max. Load Grains	Velocity fps	Pressure C.U.P.
IMR-4227	12.5	896	20,700	16.8	1251	39,200
Unique	6.5	878	20,400	8.7	1204	41,000
2400	12.0	958	22,800	15.6	1311	40,700
H110	13.9	1103	22,800	17.9	1359	38,400
AA 9	13.0	794	18,100	16.0	1277	36,700
Bullseye	5.8	692	23,300	7.5	1029	38,800

Note: Loads shown in shaded panels are maximum.
\+ Designates a compressed powder charge.
‡ Designates factory velocity duplication load.

357 Magnum

(357 S&W Magnum)

158 gr. Jacketed HP
1.590" OAL

Powder	Sugg. Starting Grains	Velocity fps	Pressure C.U.P.	Max. Load Grains	Velocity fps	Pressure C.U.P.
IMR-4227	12.2	944	21,500	‡16.1	1249	41,400
Unique	6.4	887	18,900	8.3	1185	38,300
Blue Dot	9.6	995	20,100	11.9	1319	40,900
2400	11.3	938	20,700	14.9	1279	41,800
H110	13.7	1087	22,300	17.7	1387	42,000
AA 9	13.0	811	19,600	16.0	1243	39,400

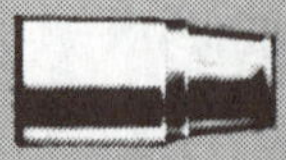

160 gr. Jacketed SP
1.590" OAL

Powder	Sugg. Starting Grains	Velocity fps	Pressure C.U.P.	Max. Load Grains	Velocity fps	Pressure C.U.P.
IMR-4227	11.5	897	20,600	15.4	1242	40,900
Unique	6.1	858	19,400	8.2	1178	40,900
Blue Dot	9.2	996	20,100	11.2	1260	40,900
2400	11.4	870	23,600	14.8	1273	40,300
H110	12.8	1029	20,700	16.8	1398	42,000
AA 9	12.5	801	21,300	15.5	1234	38,700

180 gr. JTC Sil
1.585" OAL

Powder	Sugg. Starting Grains	Velocity fps	Pressure C.U.P.	Max. Load Grains	Velocity fps	Pressure C.U.P.
IMR-4227	11.7	602	22,900	14.6	933	40,000
2400	10.0	597	23,100	12.6	949	40,500
HS-7	9.0	616	24,400	11.2	892	40,300
H110	12.0	655	20,700	15.0	1047	39,100
AA 9	11.2	654	19,700	14.0	1059	40,800

#356242
92 gr. (Linotype) 1.585" OAL

Powder	Sugg. Starting Grains	Velocity fps	Pressure C.U.P.	Max. Load Grains	Velocity fps	Pressure C.U.P.
700X	6.0	1206	19,700	8.3	1581	41,500
SR-7625	7.5	1262	20,700	9.8	1613	40,900
Bullseye	7.5	1365	25,900	9.5	1630	39,900
231	8.1	1329	25,600	10.1	1629	41,200
AA 5	10.0	1106	14,800	12.5	1642	30,900

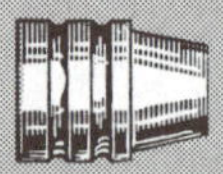

#358345
115 gr. (Linotype) 1.465" OAL

Powder	Sugg. Starting Grains	Velocity fps	Pressure C.U.P.	Max. Load Grains	Velocity fps	Pressure C.U.P.
SR-7625	5.4	1036	17,600	8.3	1446	41,900
SR-4756	6.8	1137	19,600	12.0	1626	41,900
Unique	7.5	1233	24,300	9.2	1489	41,400
Blue Dot	9.4	1134	20,400	12.8	1586	41,200
231	6.4	1166	23,900	8.4	1442	40,600
Bullseye	6.0	1050	18,800	8.2	1502	39,200

#356402
121 gr. (Linotype) 1.590" OAL

Powder	Sugg. Starting Grains	Velocity fps	Pressure C.U.P.	Max. Load Grains	Velocity fps	Pressure C.U.P.
SR-7625	7.0	1205	23,000	8.8	1429	41,600
Unique	7.2	1134	15,800	9.4	1482	41,300
Blue Dot	10.8	1156	17,000	13.0	1564	41,400
2400	13.5	1140	17,300	19.5+	1565	37,200
Bullseye	6.5	994	18,800	8.3	1390	37,600

Note: Loads shown in shaded panels are maximum.
\+ Designates a compressed powder charge.
‡ Designates factory velocity duplication load.

357 Magnum

(357 S&W Magnum)

#356242
121 gr. (Linotype) 1.585" OAL

Powder	Sugg. Starting Grains	Velocity fps	Pressure C.U.P.	Max. Load Grains	Velocity fps	Pressure C.U.P.
SR-7625	5.5	974	17,500	8.0	1335	41,100
Unique	6.9	1088	21,100	9.1	1409	42,000
2400	13.0	1178	25,700	16.4	1494	41,700
HS-7	10.0	1053	18,400	13.2	1436	40,300
AA 5	8.6	989	19,600	10.8	1450	37,300
Bullseye	5.7	1045	16,500	8.0	1320	35,800

#358480
133 gr. (Linotype) 1.508" OAL

Powder	Sugg. Starting Grains	Velocity fps	Pressure C.U.P.	Max. Load Grains	Velocity fps	Pressure C.U.P.
IMR-4227	12.0	1045	21,700	17.4	1421	40,200
Unique	6.2	1006	18,300	8.6	1356	40,300
Blue Dot	10.0	1162	18,800	11.6	1426	41,000
2400	11.8	1058	15,500	16.8	1501	40,200
AA 5	8.0	1013	24,800	10.0	1391	38,700
Bullseye	5.0	909	16,900	7.6	1316	40,600

#358495
141 gr. (Linotype) 1.435" OAL

Powder	Sugg. Starting Grains	Velocity fps	Pressure C.U.P.	Max. Load Grains	Velocity fps	Pressure C.U.P.
IMR-4227	12.3	1084	22,200	16.3+	1365	39,600
Unique	5.8	960	14,600	7.9	1289	40,300
Blue Dot	9.6	1134	21,800	11.2	1382	40,600
2400	10.5	1015	18,200	14.6	1376	39,700
HS-7	9.8	1130	23,200	11.8	1356	40,700
AA 5	7.4	863	22,200	9.2	1263	38,700
Bullseye	4.8	730	18,000	6.8	1157	40,600

#358477
150 gr. (Linotype) 1.510" OAL

Powder	Sugg. Starting Grains	Velocity fps	Pressure C.U.P.	Max. Load Grains	Velocity fps	Pressure C.U.P.
IMR-4227	11.8	1044	21,700	16.0+	1333	40,200
Herco	6.2	972	17,700	7.8	1225	41,900
Blue Dot	8.2	950	13,000	10.8	1356	41,200
2400	11.0	998	17,900	15.0	1362	41,400
H110	12.4	1075	17,300	17.7	1459	41,200
AA 7	10.0	860	20,900	12.3	1255	37,700
AA 9	12.0	903	21,500	15.0	1371	39,200
Bullseye	4.6	772	17,300	7.0	1114	36,900

Note: Loads shown in shaded panels are maximum.
\+ Designates a compressed powder charge.

357 Magnum

(357 S&W Magnum)

#358156
155 gr. (Linotype) 1.590" OAL

Powder	Sugg. Starting Grains	Velocity fps	Pressure C.U.P.	Max. Load Grains	Velocity fps	Pressure C.U.P.
IMR-4227	11.4	973	24,100	15.2	1254	41,300
Herco	5.0	785	16,300	7.5	1151	41,000
Blue Dot	8.5	975	21,500	10.5	1277	40,800
2400	10.6	999	24,900	14.0	1299	41,900
H110	11.6	1037	21,800	15.7	1363	40,300
AA 7	10.0	739	18,600	12.7	1171	40,100
AA 9	12.2	801	19,800	15.2	1242	39,000
Bullseye	4.8	828	18,000	7.0	1122	39,400

#358311
158 gr. (Linotype) 1.590" OAL

Powder	Sugg. Starting Grains	Velocity fps	Pressure C.U.P.	Max. Load Grains	Velocity fps	Pressure C.U.P.
IMR-4227	11.8	977	19,600	17.0+	1345	40,600
Herco	6.3	963	18,700	7.9	1203	41,000
Blue Dot	8.2	888	11,600	10.9	1316	39,200
2400	11.4	1024	20,200	15.5	1344	39,700
H110	13.0	1115	19,100	18.3	1460	40,100
AA 7	10.0	769	16,800	12.2	1125	36,000
AA 9	12.0	978	23,600	15.0	1339	42,000
Bullseye	4.9	821	17,500	6.9	1119	41,100

#358429
168 gr. (Linotype) 1.553" OAL

Powder	Sugg. Starting Grains	Velocity fps	Pressure C.U.P.	Max. Load Grains	Velocity fps	Pressure C.U.P.
IMR-4227	9.8	835	14,100	14.5+	1233	40,800
Herco	5.6	885	17,900	7.1	1104	39,000
Blue Dot	8.3	970	18,400	10.0	1233	39,200
2400	9.7	879	15,900	13.5	1242	41,100
H110	11.8	1037	18,900	15.7	1318	39,800
AA 9	10.8	767	21,100	13.5	1172	38,800
Bullseye	4.1	813	16,500	6.1	976	39,300

Note: Loads shown in shaded panels are maximum.
+ Designates a compressed powder charge.

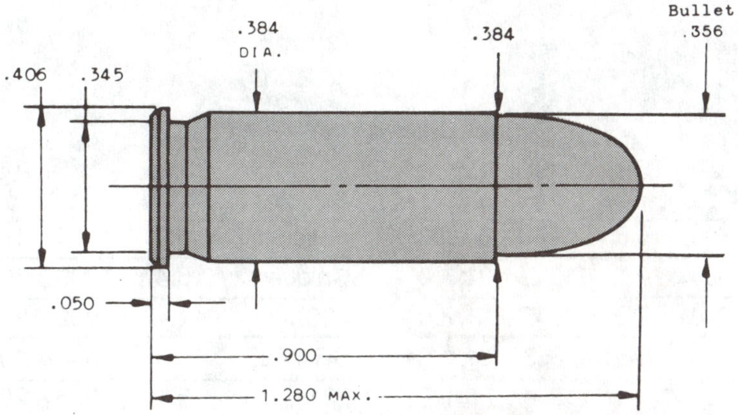

COMMENTS:

Use this data only in pistols specifically marked 38 Super by the manufacturer. This data would be extremely dangerous if used in a pistol marked 38 Automatic. The 38 Automatic uses the same size case but loaded to much lower chamber pressures.

This has traditionally been a semi-rimless case that headspaces on the small rim. Some modern pistols and replacement barrels headspace on the case mouth. If your handgun headspaces on the case mouth do not roll crimp.

38 Super has become extremely popular with IPSC shooters due to its lower recoil which allows for faster recovery time.

TEST COMPONENTS:

Cases .... Remington
Trim-to Length .... .895"
Primers .... CCI 500; Winchester 1½-108 (cast bullets only)
Primer Size .... Small Pistol
Lyman Shell Holder .... No. 12
Jacketed Bullets Used .... Sierra HC #8100, 90 gr.
Speer HP #3983, 100 gr.
Hornady HP, 115 gr.
Hornady FMJ/FP #3556, 124 gr.
Speer SP #4005, 125 gr.
Sierra FMJ #8345, 130 gr.
Cast Bullets Used .... (Sized to .355" dia.)
#356242, 95 gr.
#356242, 121 gr.
#356402, 121 gr.
#358480, 133 gr.
#356637, 147 gr.
#358311, 158 gr.

TEST SPECIFICATIONS:
(Velocity & Pressure)

Firearm Used .... Universal Receiver
Barrel Length .... 5"
Twist .... 1-16"
Groove Dia. .... .355"

NOTE: Loads shown in shade panels are maximum.

90 gr. Jacketed HC
1.125" OAL

Powder	Sugg. Starting Grains	Velocity fps	Pressure C.U.P.	Max. Load Grains	Velocity fps	Pressure C.U.P.
Bullseye	5.0	1222	17,600	6.8	1472	30,600
Unique	6.0	1221	19,400	7.1	1409	30,000
Herco	6.0	1193	17,500	8.2	1544	30,500
PB	5.5	1193	20,400	7.3	1482	29,900
SR-7625	6.0	1203	21,200	7.8	1515	30,600
231	5.5	1198	20,500	6.9	1383	30,400
HS-6	7.5	1210	20,000	9.4	1550	30,500

38 Super Automatic

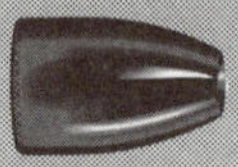

100 gr. Jacketed HP
1.185" OAL

Powder	Sugg. Starting Grains	Velocity fps	Pressure C.U.P.	Max. Load Grains	Velocity fps	Pressure C.U.P.
Bullseye	5.3	1158	21,100	6.5	1374	30,800
Unique	5.8	1126	18,800	7.1	1343	31,000
Herco	6.1	1141	19,600	7.9	1397	31,100
700X	4.6	1136	21,800	6.5	1358	31,400
PB	5.3	1153	21,100	7.0	1333	30,400
231	5.5	1101	18,700	6.5	1303	28,900
HS-6	7.5	1172	17,800	9.3	1433	30,800

115 gr. Jacketed HP
1.265" OAL

Powder	Sugg. Starting Grains	Velocity fps	Pressure C.U.P.	Max. Load Grains	Velocity fps	Pressure C.U.P.
Bullseye	5.0	1084	19,100	6.2	1182	31,300
Unique	5.4	1039	19,900	6.9	1186	30,700
700X	4.4	1032	20,200	6.0	1202	32,600
SR-7625	5.3	1042	19,200	6.6	1195	31,900
231	5.1	1026	20,100	6.6	1173	31,200
HS-6	7.5	1114	20,500	9.0	1316	31,400
Herco	5.6	1039	19,600	6.9	1200	30,600

124 gr. FMJ/FP
1.165" OAL

Powder	Sugg. Starting Grains	Velocity fps	Pressure C.U.P.	Max. Load Grains	Velocity fps	Pressure C.U.P.
Bullseye	4.0	842	19,500	5.1	1125	29,600
231	4.4	879	22,400	5.5	1132	31,300
SR-4756	4.7	737	12,900	5.9	1156	31,000
Unique	4.5	776	16,600	5.8	1123	31,200
Herco	4.8	801	18,200	6.0	1123	30,900
Blue Dot	6.5	904	20,400	8.3	1248	31,800
HS-7	7.0	816	16,000	8.8	1212	29,900
AA 5	5.7	908	19,300	6.7	1175	29,400

125 gr. Jacketed SP
1.280" OAL

Powder	Sugg. Starting Grains	Velocity fps	Pressure C.U.P.	Max. Load Grains	Velocity fps	Pressure C.U.P.
Unique	5.5	1028	20,800	7.0	1134	32,100
Herco	5.8	995	19,000	7.0	1137	30,900
Blue Dot	8.0	1090	20,300	10.1	1289	31,500
SR-7625	4.6	970	20,100	6.5	1095	31,800
HS-7	8.6	1070	20,600	11.0	1240	31,800
SR-4756	5.4	1039	20,600	‡7.7	1248	30,600

Note: Loads shown in shaded panels are maximum.
‡ Designates factory velocity duplication load.

38 Super Automatic

130 gr. FMJ
1.280" OAL

Powder	Sugg. Starting Grains	Velocity fps	Pressure C.U.P.	Max. Load Grains	Velocity fps	Pressure C.U.P.
Bullseye	4.3	855	19,800	5.4	1095	30,300
231	4.5	847	21,300	5.7	1100	31,500
SR-4756	5.0	747	13,500	6.1	1083	28,000
Unique	4.8	771	16,800	6.1	1089	31,300
Herco	5.0	766	17,900	6.2	1050	31,200
Blue Dot	6.6	790	15,700	8.3	1183	31,600
HS-7	8.0	935	21,300	9.0	1145	29,900
AA7	7.8	814	14,600	9.4	1150	30,000
WST	4.5	841	22,000	5.2	977	30,800
WSL	4.7	920	24,100	5.4	1050	30,500

#356242
92 gr. (#2 Alloy) 1.175" OAL

Powder	Sugg. Starting Grains	Velocity fps	Pressure C.U.P.	Max. Load Grains	Velocity fps	Pressure C.U.P.
Bullseye	4.2	1022	19,500	5.6	1261	30,900
Unique	5.1	1061	20,400	6.3	1267	30,500
PB	4.3	1009	18,900	5.7	1222	30,500
SR-4756	5.9	1089	19,700	7.1	1263	30,300
231	4.7	1039	20,300	6.3	1282	31,600
HP38	4.1	1026	20,600	5.8	1246	30,500

#356242
121 gr. (#2 Alloy) 1.160" OAL

Powder	Sugg. Starting Grains	Velocity fps	Pressure C.U.P.	Max. Load Grains	Velocity fps	Pressure C.U.P.
Bullseye	3.9	970	17,600	5.4	1214	29,200
Unique	4.9	1031	20,000	6.4	1281	28,800
700X	3.7	979	19,800	5.3	1206	29,200
SR-7625	4.2	978	18,100	5.8	1230	29,300
231	4.3	988	17,700	5.8	1233	29,600
HP38	3.9	975	17,800	5.6	1230	29,400

Note: Loads shown in shaded panels are maximum.

38 Super Automatic

#356402
121 gr. (#2 Alloy) 1.230" OAL

Powder	Sugg. Starting Grains	Velocity fps	Pressure C.U.P.	Max. Load Grains	Velocity fps	Pressure C.U.P.
Bullseye	4.0	1005	18,600	5.3	1235	30,000
Unique	5.1	1075	20,400	6.0	1255	31,000
700X	4.0	1025	20,600	5.5	1252	29,900
SR-7625	5.1	1099	20,600	6.6	1274	29,900
231	5.0	1092	20,700	6.5	1307	30,800
HP38	4.4	1063	20,900	5.9	1267	29,700

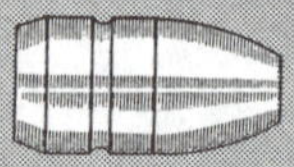

#356637
147 gr. (#2 Alloy) 1.224" OAL

Powder	Sugg. Starting Grains	Velocity fps	Pressure C.U.P.	Max. Load Grains	Velocity fps	Pressure C.U.P.
Bullseye	4.3	793	24,300	4.8	940	31,300
Unique	4.5	791	22,900	5.2	954	31,300
Herco	4.9	830	25,200	5.7	928	31,200
Blue Dot	6.6	869	24,000	7.6	1061	31,200
HS-7	6.7	877	23,900	7.7	1048	31,400
AA 7	7.5	862	22,200	8.9	1100	31,200
WST	4.2	894	26,400	5.5	969	31,100
WSL	3.7	800	24,100	4.3	914	30,400

#358480
133 gr. (#2 Alloy) 1.115" OAL

Powder	Sugg. Starting Grains	Velocity fps	Pressure C.U.P.	Max. Load Grains	Velocity fps	Pressure C.U.P.
SR-4756	5.0	975	16,900	7.0	1237	29,400
Bullseye	4.0	1000	19,800	5.3	1180	31,500
Unique	5.1	1045	21,000	6.0	1200	30,000
Blue Dot	6.8	1042	18,500	8.9	1346	30,100
HS-6	5.9	1026	18,000	7.8	1303	30,200

#358311
158 gr. (#2 Alloy) 1.245" OAL

Powder	Sugg. Starting Grains	Velocity fps	Pressure C.U.P.	Max. Load Grains	Velocity fps	Pressure C.U.P.
SR-4756	4.9	927	18,900	5.8	1085	29,800
Unique	4.4	958	20,100	5.1	1058	28,100
Herco	4.5	920	21,000	5.4	1058	29,000
Blue Dot	6.5	987	20,900	7.6	1133	28,400
HS-7	6.8	956	20,200	8.3	1100	28,200

Note: Loads shown in shaded panels are maximum.

38-40 Winchester
(38 WCF)

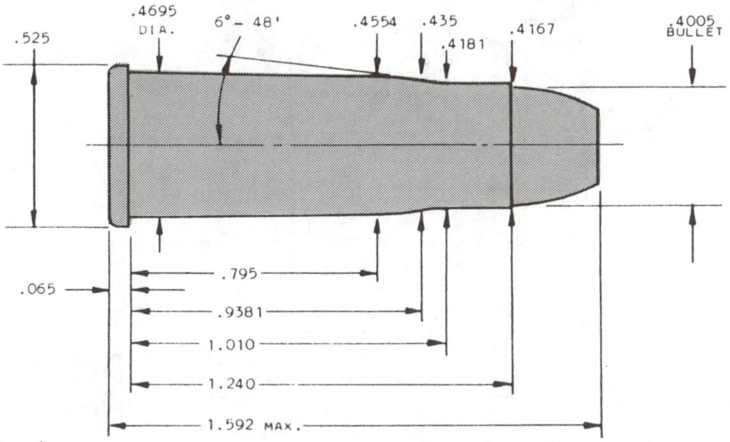

COMMENTS:

This is a very old cartridge and care should be taken not to use the data listed in any handgun which has deteriorated. Additionally this data should not be used in conjunction with any firearm designed for black powder cartridges.

Individual tolerances vary greatly in firearms chambered for this cartridge, even among the same make and model. Therefore exercise extreme caution when developing loads as you work up from the sugg. starting grains listing.

Never load old brass as it may be of the weaker balloon head types or have been fired at one time with corrosive primers.

TEST COMPONENTS:

Cases .... Winchester
Trim-to Length .... 1.300"
Primers .... Winchester 7-111
Primer Size .... Large Pistol
Lyman Shell Holder .... No. 14B
Cast Bullets Used .... (Sized to .401" dia.)
#401043, 172 gr.

TEST SPECIFICATIONS: (Velocity Only)

Firearm Used .... Colt S.A.A.
Barrel Length .... 7$^1/_2$"
Twist .... 1-36"
Groove Dia. .... .400"

#401043
172 gr. (#2 Alloy) 1.592" OAL

Powder	Sugg. Starting Grains	Velocity fps	Pressure C.U.P.	Max. Load Grains	Velocity fps	Pressure C.U.P.
Bullseye	4.0	740	—	5.9	965	—
Unique	7.0	830	—	10.0	1105	—

Note: Loads shown in shaded panels are maximum.

40 Smith & Wesson
(40 S&W)

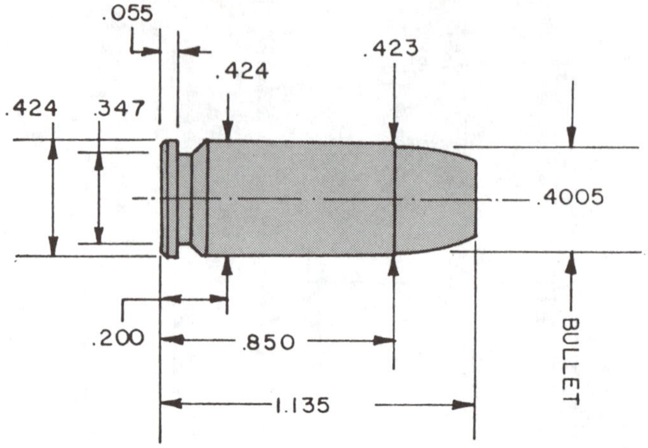

COMMENTS:

The 40 S&W has quickly become a very popular cartridge as it is a good choice between the 9mm and 45 ACP. It has lighter recoil and greater magazine capacity than the 45 ACP and is more powerful than the 9mm. Accuracy in most handguns is superb, especially with Hercules Bullseye, Winchester 231 and Accurate Arms No. 5.

This cartridge headspaces from the mouth and therefore case trimming must be uniform and accurate. Do not reduce cases below the trim-to length. Additionally do not roll crimp bullets as this will prevent the case from properly headspacing on its mouth. A modest taper crimp may be employed if found necessary.

It is important not to seat bullets to a shorter length than listed in any handgun cartridge; but especially so in this one. Pressures can be raised dramatically with deep seating.

TEST COMPONENTS:

Cases .... Winchester
Trim-to Length .... .847"
Primers .... Winchester 1 1/2-108
Primer Size .... Small Pistol
Lyman Shell Holder .... No. 15
Jacketed Bullets Used .... Sierra JHP #8430, 150 gr.
Nosler JHP #44844, 170 gr.
Sierra JHP #8460, 180 gr.
Sierra FPJ #8480, 190 gr.
Cast Bullets Used .... (Sized to .401" dia.)
#401043, 175 gr.
#401638, 175 gr.
#401633, 200 gr.

TEST SPECIFICATIONS:
(Velocity & Pressure)

Firearm Used .... Universal Receiver
Barrel Length .... 4"
Twist .... 1-16"
Groove Dia. .... .401"

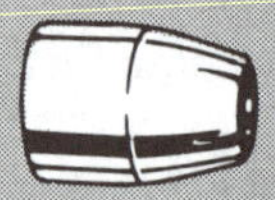

150 gr. Jacketed HP
1.100" OAL

Powder	Sugg. Starting Grains	Velocity fps	Pressure C.U.P.	Max. Load Grains	Velocity fps	Pressure C.U.P.
Unique	5.5	982	17,800	6.6	1132	23,300
Blue Dot	8.6	999	18,800	10.0	1140	23,100
231	5.0	799	17,900	6.2	1054	22,800
WST	5.5	968	19,100	6.4	1096	23,400
WSL	4.9	972	18,900	5.5	1076	22,600
800X	6.9	999	18,500	8.4	1109	23,200
HS-6	7.6	990	17,100	8.6	1093	22,300
AA 5	6.9	998	18,400	7.7	1101	23,300
Bullseye	4.9	740	18,200	5.7	890	22,500
Clays	3.5	856	17,100	4.4	1025	21,600

Note: Loads shown in shaded panels are maximum.

40 Smith & Wesson

(40 S&W)

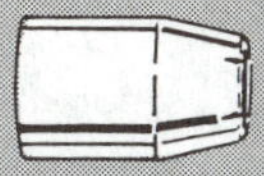

170 gr. Jacketed HP
1.095" OAL

Powder	Sugg. Starting Grains	Velocity fps	Pressure C.U.P.	Max. Load Grains	Velocity fps	Pressure C.U.P.
Unique	5.2	817	17,100	6.1	1016	22,700
Blue Dot	7.5	944	17,200	9.0	1018	23,400
231	4.6	874	17,200	5.8	1042	22,500
WST	4.5	697	17,600	5.7	998	22,900
WSL	4.5	814	17,900	5.3	908	23,500
800X	6.4	817	18,000	7.9	1005	23,200
HS-6	7.0	943	17,500	8.1	1004	22,400
AA 5	6.0	839	17,800	6.9	1000	23,500
Bullseye	4.4	641	17,900	5.4	819	22,300

180 gr. Jacketed HP
1.115" OAL

Powder	Sugg. Starting Grains	Velocity fps	Pressure C.U.P.	Max. Load Grains	Velocity fps	Pressure C.U.P.
Unique	4.9	839	18,800	5.8	916	23,100
Blue Dot	7.3	754	18,000	8.8	990	23,300
WSL	4.2	707	18,200	‡5.2	979	23,300
WSF	5.0	853	17,700	6.2	993	23,400
800X	5.9	906	18,700	7.4	1013	23,400
HS-6	6.9	860	19,600	8.0	1048	23,500
Bullseye	4.4	628	18,200	5.6	844	23,500
Clays	3.4	551	17,900	4.0	912	21,900

190 gr. Jacketed FP
1.135" OAL

Powder	Sugg. Starting Grains	Velocity fps	Pressure C.U.P.	Max. Load Grains	Velocity fps	Pressure C.U.P.
Bullseye	3.9	520	16,200	5.0	722	23,100
231	4.2	540	16,800	5.3	723	23,200
AA 5	5.5	545	16,000	6.5	745	23,200
Unique	4.5	535	15,900	5.3	720	23,100
HS-6	6.4	594	16,200	7.2	734	22,300
800X	5.2	573	17,200	6.3	737	22,200
Blue Dot	6.7	567	16,800	8.1	724	22,600
WSL	3.8	510	15,800	5.0	736	22,800
WSF	4.9	530	16,100	5.8	702	22,700

#401043
175 gr. (#2 Alloy) 1.125" OAL

Powder	Sugg. Starting Grains	Velocity fps	Pressure C.U.P.	Max. Load Grains	Velocity fps	Pressure C.U.P.
Unique	4.8	911	15,900	5.8	1023	23,300
Blue Dot	7.3	897	17,900	9.0	1123	23,500
231	4.3	863	15,400	5.8	977	22,500
WST	4.8	944	18,300	6.3	1094	22,900
WSL	4.1	891	17,400	4.9	1049	22,700
800X	5.6	853	16,500	7.4	974	22,400
HS-6	7.0	940	17,000	8.2	1047	23,300
AA 5	6.1	903	19,300	6.9	1066	23,000
Bullseye	4.2	665	17,100	5.1	812	22,000

Note: Loads shown in shaded panels are maximum.
‡ Designates a factory velocity duplication load.

40 Smith & Wesson
(40 S&W)

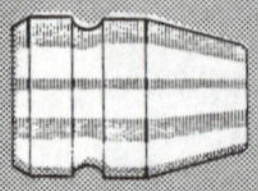

#401638
175 gr. (#2 Alloy) 1.100" OAL

Powder	Sugg. Starting Grains	Velocity fps	Pressure C.U.P.	Max. Load Grains	Velocity fps	Pressure C.U.P.
Unique	4.7	898	18,600	5.8	971	23,200
Blue Dot	6.9	979	18,100	8.6	1059	23,200
231	4.3	796	16,000	5.8	970	22,700
WSL	4.3	957	18,300	5.1	992	23,000
800X	5.8	878	17,000	7.6	996	22,900
HS-6	7.0	976	18,600	8.4	1129	22,800
AA 5	6.0	844	18,600	6.9	1047	22,600
Bullseye	4.0	654	16,900	5.0	842	23,100
Clays	3.3	641	17,800	4.0	992	20,900

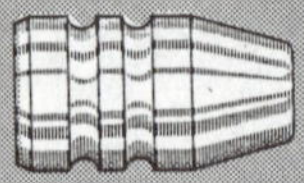

#401633
200 gr. (#2 Alloy) 1.115" OAL

Powder	Sugg. Starting Grains	Velocity fps	Pressure C.U.P.	Max. Load Grains	Velocity fps	Pressure C.U.P.
Unique	4.3	902	18,900	5.3	953	23,100
231	4.0	867	17,300	5.5	945	22,600
WSF	4.7	909	18,700	5.6	1043	23,400
800X	4.9	814	17,500	6.4	992	22,500
HS-6	5.6	902	18,100	6.9	996	23,100
AA 5	4.6	878	17,000	5.3	990	22,600

Note: Loads shown in shaded panels are maximum.

10mm Automatic

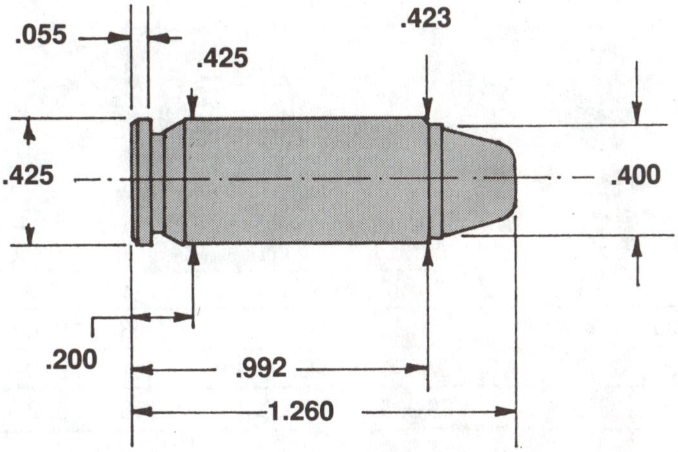

COMMENTS:

This is a high performance semi-auto pistol cartridge with applications in law enforcement and hunting.

The 10mm Auto cartridge headspaces on the mouth and therefore case trimming must be uniform and accurate. Do not reduce cases below the recommended trim-to length. Additionally do not roll crimp bullets as this will prevent the case from properly headspacing on its mouth. A modest taper crimp is also suggested.

It is important not to seat bullets to a shorter length than listed in any handgun cartridge; but especially so in this one. Pressures can be raised dramatically with deep seating.

Winchester 231 frequently provided the best accuracy though not the highest velocities when used in this cartridge. Cast bullet #401633 is an excellent choice for hunting.

TEST COMPONENTS:

Cases .... Norma
Trim-to Length .... .989"
Primers .... CCI 300
Primer Size .... Large Pistol
Lyman Shell Holder .... No. 15
Jacketed Bullets Used .... Sierra JHP #8430, 150 gr.
Nosler JHP #44844, 170 gr.
Sierra JHP #8460, 180 gr.
Sierra FPJ #8480, 190 gr.
Hornady JHP #40060, 200 gr.
Cast Bullets Used .... (Sized to .401" dia.)
#401043, 175 gr.
#401638, 175 gr.
#401633, 200 gr.

TEST SPECIFICATIONS: (Velocity & Pressure)

Firearm Used .... Universal Receiver
Barrel Length .... 5"
Twist .... 1-16"
Groove Dia. .... .400"

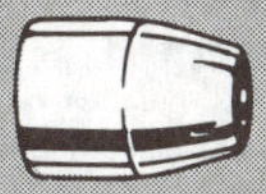

150 gr. Jacketed HP
1.250" OAL

Powder	Sugg. Starting Grains	Velocity fps	Pressure C.U.P.	Max. Load Grains	Velocity fps	Pressure C.U.P.
Bullseye	5.5	1049	20,900	7.0	1263	28,500
Red Dot	4.7	864	19,300	6.5	1243	28,800
Unique	6.5	994	21,800	8.1	1265	29,500
700X	5.0	937	20,700	6.8	1282	30,000
800X	7.0	1029	20,200	9.0	1301	28,900
231	5.7	1053	20,700	7.5	1292	29,800
HS-6	8.3	959	17,900	9.9	1278	27,200
AA 5	7.0	986	18,100	9.0	1278	29,800

Note: Loads shown in shaded panels are maximum.

10mm Automatic

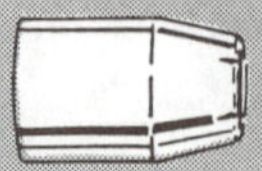

170 gr. Jacketed HP
1.250" OAL

Powder	Sugg. Starting Grains	Velocity fps	Pressure C.U.P.	Max. Load Grains	Velocity fps	Pressure C.U.P.
Bullseye	5.0	930	18,800	6.4	1068	28,100
Unique	6.3	1025	21,800	7.2	1158	27,100
Blue Dot	8.5	989	18,800	10.5	1099	27,800
700X	5.0	1001	21,000	6.2	1168	29,300
800X	6.6	1003	18,800	8.7	1146	28,500
231	5.2	963	21,000	6.8	1129	27,600
HS-6	7.9	1020	19,500	9.6	1121	27,900
AA 5	7.0	766	18,000	8.3	1197	28,300

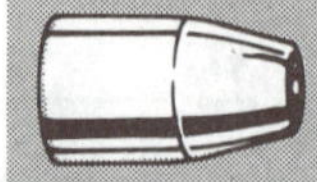

180 gr. Jacketed HP
1.250" OAL

Powder	Sugg. Starting Grains	Velocity fps	Pressure C.U.P.	Max. Load Grains	Velocity fps	Pressure C.U.P.
Bullseye	5.5	998	21,000	6.5	1088	29,300
Unique	5.8	1018	21,800	6.7	1064	28,100
Blue Dot	8.1	939	18,000	10.3	1133	30,000
700X	4.5	892	17,300	5.8	1053	28,800
800X	6.8	831	20,300	8.4	1073	28,000
231	5.1	917	20,300	6.7	1074	26,900
HS-6	8.1	1057	21,000	9.6	1225	28,700
AA 7	9.5	984	18,000	11.5	1087	28,300

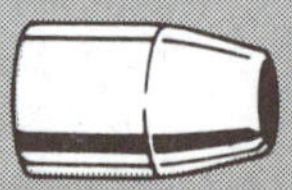

190 gr. Jacketed FP
1.250" OAL

Powder	Sugg. Starting Grains	Velocity fps	Pressure C.U.P.	Max. Load Grains	Velocity fps	Pressure C.U.P.
Bullseye	4.7	879	18,400	6.1	1088	28,200
231	5.5	962	20,700	6.7	1086	29,700
700X	4.8	951	20,300	5.8	1086	29,800
Unique	5.5	896	19,200	6.5	1001	27,600
HS-6	7.7	744	20,700	9.2	1037	30,000
800X	6.4	987	22,200	8.5	1189	29,300
AA 7	9.0	932	20,600	10.5	1106	28,000
Blue Dot	8.3	968	19,200	9.9	1175	28,600
Clays	4.3	878	22,900	5.6	1071	29,800

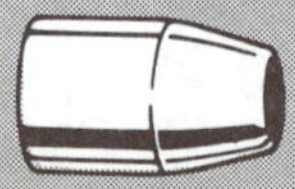

200 gr. Jacketed HP
1.250" OAL

Powder	Sugg. Starting Grains	Velocity fps	Pressure C.U.P.	Max. Load Grains	Velocity fps	Pressure C.U.P.
Bullseye	4.8	910	20,3600	5.7	1059	26,500
Unique	4.8	836	17,700	6.1	1071	29,000
Blue Dot	8.0	1015	22,900	9.6	1148	29,900
700X	4.4	936	19,900	5.8	1080	29,500
800X	5.9	963	22,600	7.6	1119	29,200
231	5.3	947	21,400	6.5	1104	27,800
HS-6	7.5	987	22,600	8.6	1126	29.500
AA 7	9.0	987	19,900	10.6	1145	28,200

Note: Loads shown in shaded panels are maximum.

10mm Automatic

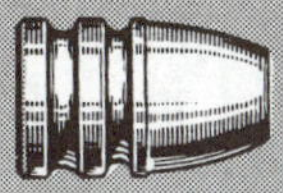

#401043
175 gr. (#2 Alloy) 1.260" OAL

Powder	Sugg. Starting Grains	Velocity fps	Pressure C.U.P.	Max. Load Grains	Velocity fps	Pressure C.U.P.
Bullseye	5.5	1069	22,200	6.5	1199	28,300
Red Dot	5.5	1087	24,500	6.2	1194	29,300
Unique	6.1	1097	23,300	7.0	1224	28,800
Blue Dot	9.0	1100	21,100	10.4	1245	28,100
700X	5.2	1102	24,700	5.8	1170	28,200
800X	7.0	1090	20,300	8.2	1242	27,600
231	5.5	1023	20,300	6.7	1189	28,800
HS-6	8.0	1071	20,300	9.4	1236	29,000
AA 5	7.0	1038	18,100	8.2	1236	28,400

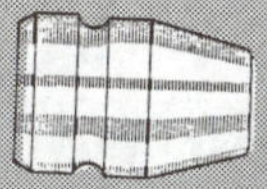

#401638
175 gr. (#2 Alloy) 1.253" OAL

Powder	Sugg. Starting Grains	Velocity fps	Pressure C.U.P.	Max. Load Grains	Velocity fps	Pressure C.U.P.
Bullseye	5.9	1115	23,500	6.4	1197	28,400
Red Dot	5.4	1113	23,900	6.0	1152	28,700
Unique	6.1	1104	22,900	6.8	1180	28,900
Blue Dot	9.3	1175	23,800	10.4	1275	29,900
800X	7.7	1092	23,300	8.6	1269	28,300
231	5.9	1115	24,000	6.5	1181	29,400
WSL	5.4	1096	24,300	6.0	1154	28,500
WSF	6.2	1094	22,800	7.0	1176	28,600
AA 5	7.2	1138	24,000	7.6	1196	28,700

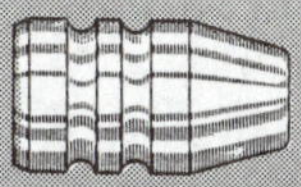

#401633
200 gr. (#2 Alloy) 1.253" OAL

Powder	Sugg. Starting Grains	Velocity fps	Pressure C.U.P.	Max. Load Grains	Velocity fps	Pressure C.U.P.
Bullseye	5.0	1034	24,300	5.6	1102	29,300
Red Dot	5.1	1034	24,800	5.9	1150	29,800
Unique	5.7	1031	24,200	6.4	1154	29,600
Blue Dot	8.1	1093	24,500	9.0	1173	29,400
800X	6.8	1101	23,300	7.7	1181	28,400
231	5.3	1045	23,300	6.0	1108	28,200
WSL	4.7	1025	23,000	5.5	1104	29,500
WSF	5.8	1036	22,600	6.4	1117	28,600
AA 7	9.1	1084	22,800	10.2	1184	28,800

Note: Loads shown in shaded panels are maximum.

41 Remington Magnum

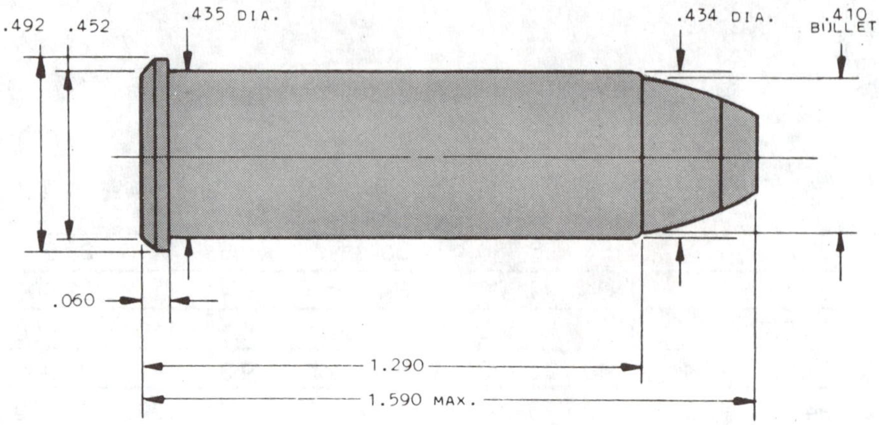

COMMENTS:

The 41 Magnum has a small but loyal following. For hunting use, the 41 Magnum has a flatter trajectory than the 44 Magnum.

We suggest Hercules 2400 with all cast and jacketed bullets as being a good place to start your loading.

Loads shown with cast bullet #410459 exceeds the standard maximum cartridge length of 1.590 inches. However, most modern revolvers have cylinders which will accept cartridges of 1.700 inches, as shown for this bullet. These loads can also be used in the T/C Contender.

TEST COMPONENTS:

Cases .... Winchester
Trim-to Length .... 1.285"
Primers .... Remington 2½
Primer Size .... Large Pistol
Lyman Shell Holder .... No. 30
Jacketed Bullets Used .... Sierra HC #8500, 170 gr.
Speer JHP #4405, 200 gr.
Hornady HP, 210 gr.
Speer JSP #4417, 220 gr.
Cast Bullets Used .... (Sized to .410" dia.)
*Gas Check Bullets .... *#410610, 215 gr.
#410459, 220 gr.

TEST SPECIFICATIONS:
(Velocity & Pressure)

Firearm Used .... Universal Receiver and Special Vented Barrel
Barrel Length .... 4"
Twist .... 1-18¾"
Groove Dia. .... .409"

170 gr. Jacketed HC
1.580" OAL

Powder	Sugg. Starting Grains	Velocity fps	Pressure C.U.P.	Max. Load Grains	Velocity fps	Pressure C.U.P.
SR-4756	10.6	1052	13,300	13.0	1360	36,700
IMR-4227	20.5	1131	20,000	23.5+	1340	33,000
2400	17.5	1040	15,000	22.0+	1311	28,600
Unique	9.8	1056	13,700	12.4	1378	36,100
H110	22.5	1217	15,200	26.3+	1507	33,700

Note: Loads shown in shaded panels are maximum.
+ Designates a compressed powder charge.

41 Remington Magnum

200 gr. Jacketed HP
1.590" OAL

Powder	Sugg. Starting Grains	Velocity fps	Pressure C.U.P.	Max. Load Grains	Velocity fps	Pressure C.U.P.
IMR-4227	18.0	1043	21,200	21.0+	1254	35,800
Blue Dot	12.5	1018	15,900	15.4	1302	35,200
2400	15.3	966	15,300	20.0	1277	36,100
Unique	8.2	984	14,100	10.7	1218	35,400
H110	19.6	1164	20,800	23.4	1421	37,000

210 gr. Jacketed HP
1.585" OAL

Powder	Sugg. Starting Grains	Velocity fps	Pressure C.U.P.	Max. Load Grains	Velocity fps	Pressure C.U.P.
IMR-4227	17.8	1016	19,100	21.0+	1232	36,500
Blue Dot	12.6	1029	16,800	15.8	1304	37,000
2400	15.3	966	14,600	20.0	1233	36,100
Unique	8.6	1011	15,600	11.0	1196	34,700
H110	19.0	1105	19,800	‡22.9	1349	35,300

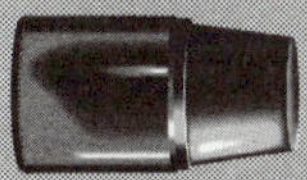

220 gr. Jacketed SP
1.590" OAL

Powder	Sugg. Starting Grains	Velocity fps	Pressure C.U.P.	Max. Load Grains	Velocity fps	Pressure C.U.P.
IMR-4227	15.3	900	15,500	19.7	1190	36,900
Blue Dot	11.0	996	14,900	14.2	1219	34,900
2400	12.9	876	17,700	17.4	1172	36,700
Unique	7.5	810	13,100	10.0	1139	34,800
H110	15.7	1031	20,000	19.4	1261	36,000

#410610
215 gr. (Linotype) 1.575" OAL

Powder	Sugg. Starting Grains	Velocity fps	Pressure C.U.P.	Max. Load Grains	Velocity fps	Pressure C.U.P.
SR-4756	8.0	909	17,100	10.8	1213	36,000
IMR-4227	14.9	956	20,100	19.2	1221	35,200
Unique	7.8	979	19,800	10.4	1214	36,100
Blue Dot	11.4	1003	20,200	14.6	1170	35,800
2400	14.2	835	19,600	18.8	1272	37,000
HS-7	11.5	1007	20,900	14.5	1249	36,800

#410459
220 gr. (Linotype) 1.700" OAL

Powder	Sugg. Starting Grains	Velocity fps	Pressure C.U.P.	Max. Load Grains	Velocity fps	Pressure C.U.P.
SR-4756	9.8	1061	20,900	11.6	1246	33,800
IMR-4227	17.3	1048	22,100	21.5+	1280	35,500
Unique	8.1	970	18,000	11.2	1261	36,200
Blue Dot	13.0	1100	21,700	16.0	1359	36,000
2400	15.7	1022	21,100	20.1	1330	35,400
HS-7	13.0	1061	21,400	15.9	1313	36,900

Note: Loads shown in shaded panels are maximum.
\+ Designates a compressed powder charge.
‡ Designates a factory velocity duplication load.

44-40 Winchester
(44 WCF)

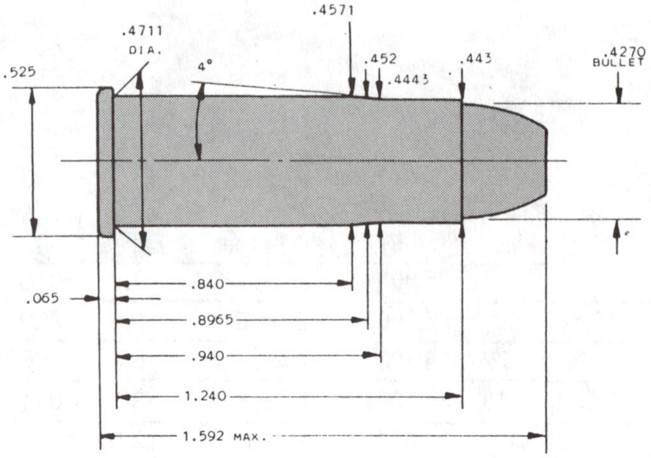

COMMENTS:

This is a very old cartridge and care should be taken not to use the data listed in any handgun which has deteriorated. Additionally this data should not be used in conjunction with any firearm designed for black powder cartridges.

Individual tolerances vary greatly in firearms chambered for this cartridge, even among the same make and model. Therefore exercise extreme caution when developing loads as you work up from the suggested starting grains listing.

Never load old brass as it may be of the weaker balloon head type or have been fired at one time with corrosive primers.

TEST COMPONENTS:

Cases .... Winchester
Trim-to Length .... 1.300"
Primers .... Winchester 7-111
Primer Size .... Large Pistol
Lyman Shell Holder .... No. 14B
Cast Bullets Used .... (Sized to .427" dia.) #427098, 205 gr.

TEST SPECIFICATIONS:
(Velocity Only)

Firearm Used .... Colt S.A.A.
Barrel Length .... 7 1/2"
Twist .... 1-20"
Groove Dia. .... .426"

#427098
205 gr. (#2 Alloy) 1.592" OAL

Powder	Sugg. Starting Grains	Velocity fps	Pressure C.U.P.	Max. Load Grains	Velocity fps	Pressure C.U.P.
Bullseye	4.0	695	—	6.6	945	—
Unique	6.0	750	—	9.0	1095	—

Note: Loads shown in shaded panels are maximum.

44 Smith & Wesson Special
(44 Special)

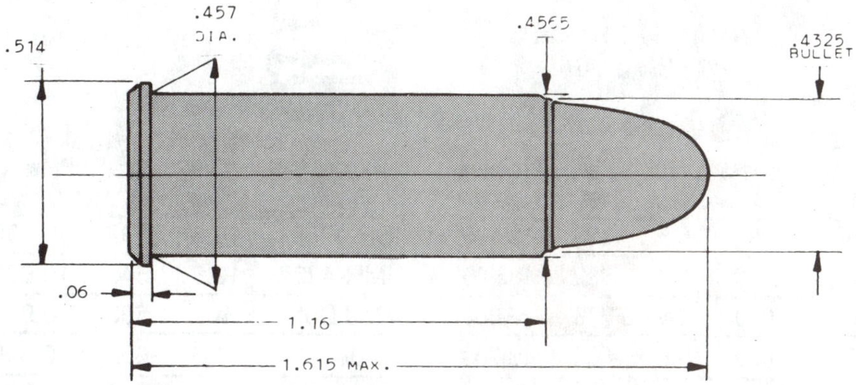

COMMENTS:

When using half jacketed bullets do not use charge weights below the starting load to prevent bullet jackets from becoming lodged in the barrel (the lead cores may exit the muzzle and strike the target). To crimp half jacket bullets form the crimp at the junction of the jacket and exposed lead nose.

Due to wide internal barrel dimensions it is wise to slug your bore before beginning to reload. Size cast bullets to match or 0.001" larger than groove diameter. Then make up a dummy round to ensure that your revolver chambers will easily accept ammo with bullets of the chosen diameter.

Bullet #429421, the Elmer Keith design, is extremely popular in this cartridge.

This cartridge produces accuracy with a great many powders. We would suggest starting with Winchester 231 for light loads or either IMR 4227 or Hercules 2400 for heavy loads.

TEST COMPONENTS:

Cases .... Winchester
Trim-to Length .... 1.152"
Primers .... CCI 300
Primer Size .... Large Pistol
Lyman Shell Holder .... No. 7
Jacketed Bullets Used .... Sierra HC #8600, 180 gr.
Hornady HP, 200 gr.
Speer JHP #4435, 225 gr.
Remington SP #B22906, 240 gr.
Cast Bullets Used .... (Sized to .429" dia.)
*Gas Check Bullets
#427098, 205 gr.
*#429215, 215 gr.
#429360, 232 gr.
#429421, 245 gr.
*#429244, 250 gr.

TEST SPECIFICATIONS: (Velocity & Pressure)

Firearm Used .... Universal Receiver and Special Vented Barrel
Barrel Length .... 4"
Twist .... 1-20"
Groove Dia. .... .429"

180 gr. Jacketed HC
1.470" OAL

Powder	Sugg. Starting Grains	Velocity fps	Pressure C.U.P.	Max. Load Grains	Velocity fps	Pressure C.U.P.
SR-4759	14.6	705	10,100	16.3+	869	12,800
IMR-4227	15.1	685	9,400	16.8	862	13,100
Red Dot	6.4	793	11,600	7.1	850	13,700
2400	14.0	674	8,100	15.8	851	13,300
231	6.5	689	9,300	7.5	832	13,600

Note: Loads shown in shaded panels are maximum.
\+ Designates a compressed powder charge.

44 Smith & Wesson Special
(44 Special)

200 gr. Jacketed HP
1.495" OAL

Powder	Sugg. Starting Grains	Velocity fps	Pressure C.U.P.	Max. Load Grains	Velocity fps	Pressure C.U.P.
SR-4759	13.9	665	11,400	15.5	828	13,000
IMR-4227	14.3	657	10,200	15.8	882	13,900
Red Dot	5.5	660	9,000	6.3	776	13,800
2400	12.0	599	8,000	13.5	838	13,700
231	6.4	707	11,100	7.1	797	13,800

225 gr. Jacketed HP
1.615" OAL

Powder	Sugg. Starting Grains	Velocity fps	Pressure C.U.P.	Max. Load Grains	Velocity fps	Pressure C.U.P.
SR-4759	12.3	597	9,200	14.0	763	14,000
IMR-4227	12.3	640	10,600	14.0	749	13,400
Red Dot	4.9	590	10,900	5.7	767	13,900
2400	11.6	653	10,600	13.3	790	13,800
231	5.0	585	11,300	5.8	729	13,900

240 gr. Jacketed SP
1.475" OAL

Powder	Sugg. Starting Grains	Velocity fps	Pressure C.U.P.	Max. Load Grains	Velocity fps	Pressure C.U.P.
SR-4759	12.8	607	10,400	14.2+	769	13,300
IMR-4227	13.3	616	11,100	14.8	711	13,300
Red Dot	4.8	571	10,500	5.6	713	13,600
2400	12.1	604	10,300	13.5	723	13,000
231	5.3	581	11,000	6.1	616	13,400

#427098
205 gr. (Linotype) 1.537" OAL

Powder	Sugg. Starting Grains	Velocity fps	Pressure C.U.P.	Max. Load Grains	Velocity fps	Pressure C.U.P.
SR-7625	6.8	761	9,700	7.7	838	12,600
SR-4759	13.2	651	9,200	15.2+	798	12,100
Unique	7.0	736	10,500	8.2	869	14,000
Blue Dot	10.5	747	10,000	12.0	855	13,300
2400	13.3	734	10,800	15.2	870	13,500
HS-7	10.0	732	10,100	11.6	874	13,600

#429215
215 gr. (Linotype) 1.500" OAL

Powder	Sugg. Starting Grains	Velocity fps	Pressure C.U.P.	Max. Load Grains	Velocity fps	Pressure C.U.P.
SR-7625	6.2	732	10,900	6.8	801	13,500
Unique	6.3	729	9,500	7.7	884	13,600
Blue Dot	9.5	689	8,800	11.0	885	13,400
2400	13.0	792	11,300	14.5	905	14,000
HS-7	9.0	659	8,100	11.0	841	12,600

#429360
232 gr. (Linotype) 1.525" OAL

Powder	Sugg. Starting Grains	Velocity fps	Pressure C.U.P.	Max. Load Grains	Velocity fps	Pressure C.U.P.
SR-4756	6.7	666	9,800	8.0	774	13,400
SR-4759	12.5	647	10,300	14.3	803	13,500
Unique	6.0	604	10,400	7.1	781	13,500
Blue Dot	9.0	646	10,300	10.0	751	13,100
2400	12.0	732	10,500	13.8	848	13,800
HS-7	9.5	729	10,500	11.0	853	13,900

Note: Loads shown in shaded panels are maximum.
+ Designates a compressed powder charge.

44 Smith & Wesson Special

(44 Special)

#429421

245 gr. (Linotype) 1.571" OAL

Powder	Sugg. Starting Grains	Velocity fps	Pressure C.U.P.	Max. Load Grains	Velocity fps	Pressure C.U.P.
SR-4756	6.8	661	9,000	8.1	780	13,800
SR-4759	11.9	652	10,500	13.4	789	14,000
Unique	6.0	649	10,300	‡6.9	767	13,300
Blue Dot	9.0	710	10,800	10.0	807	13,200
2400	11.4	704	10,400	13.2	797	13,800
HS-7	9.1	728	11,300	10.4	828	14,000

#429244

250 gr. (Linotype) 1.550" OAL

Powder	Sugg. Starting Grains	Velocity fps	Pressure C.U.P.	Max. Load Grains	Velocity fps	Pressure C.U.P.
SR-4756	6.2	584	8,400	7.5	765	14,000
Unique	5.7	635	10,200	6.6	747	13,600
2400	10.5	602	8,500	12.3	742	12,700
HS-7	8.5	606	8,700	10.0	743	13,400
SR-4759	11.6	658	11,600	13.1	752	13,200

Note: Loads shown in shaded panels are maximum.
‡ Designates a factory velocity duplication load.

44 Remington Magnum
(44 Magnum)

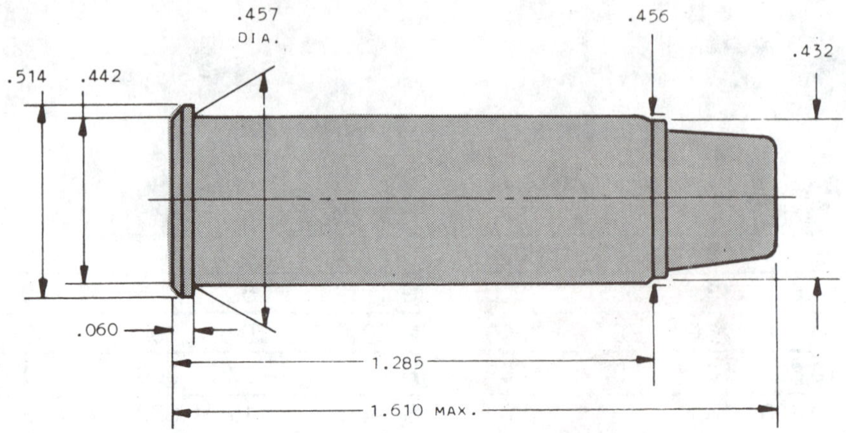

COMMENTS:

Pressure limits for this cartridge are the same in rifles or handguns. Heavy recoil in handguns demands a substantial bullet crimp.

Loads shown with cast bullet #429215, #429303, #429244, #429360, #429421, and #429640 all exceed the standard maximum cartridge length. However, most modern 44 Magnum revolvers have cylinders which will accept the cartridge lengths listed in the data. These can also be used in the Thompson Center Contender.

Do not use charge weights below the starting load when using half-jacketed bullets as the jacket may become lodged in the barrel (the lead cores may exit the muzzle and strike the target). To crimp half jacket bullets, form the crimp at the junction of the jacket and exposed lead nose.

Bullet #429348 at starting charges is suggested for target use. Bullet #429215 is extremely popular for all around use. An excellent choice hunting or silhouettes is #429640.

IMR 4227 or Hercules 2400 are suggested for maximum accuracy with full power loads.

TEST COMPONENTS:

Cases .... Remington
Trim-to Length .... 1.280"
Primers .... CCI 300
Primer Size .... Large Pistol
Lyman Shell Holder .... No. 7
Jacketed Bullets Used .... Sierra HC #8600, 180 gr.
Hornady HP, 200 gr.
Speer JHP #4435, 225 gr.
Sierra HC #8610, 240 gr.
Speer JHP #4453, 240 gr.
Hornady FP, 265 gr.
Cast Bullets Used .... (Sized to .429" dia.)
*Gas Check Bullets
#429348, 180 gr.
*#429303, 200 gr.
*#429215, 215 gr.
#429360, 232 gr.
#429421, 245 gr.
*#429244, 245 gr.
*#429640, 275 gr.

TEST SPECIFICATIONS:
(Velocity & Pressure)

Firearm Used .... Universal Receiver and Special Vented Barrel
Barrel Length .... 4"
Twist .... 1-20"
Groove Dia. .... .429"

180 gr. Jacketed HC
1.610" OAL

Powder	Sugg. Starting Grains	Velocity fps	Pressure C.U.P.	Max. Load Grains	Velocity fps	Pressure C.U.P.
SR-4756	11.4	970	17,400	13.9	1262	34,400
IMR-4227	24.2	1136	27,600	27.0	1317	34,800
Unique	12.0	1061	25,100	13.6	1301	35,700
Blue Dot	17.8	1126	25,800	20.2	1404	36,400
2400	21.3	1103	24,800	25.5	1370	36,400
231	10.2	1077	22,100	13.0	1267	36,300
HS-7	17.8	1111	24,200	20.6	1341	36,300
H110	28.7	1277	23,300	32.3+	1510	36,500
AA 5	12.8	976	16,400	16.0	1344	36,000

Note: Loads shown in shaded panels are maximum.
+ Designates a compressed powder charge.

44 Remington Magnum
(44 Magnum)

200 gr. Jacketed HP
1.610" OAL

Powder	Sugg. Starting Grains	Velocity fps	Pressure C.U.P.	Max. Load Grains	Velocity fps	Pressure C.U.P.
SR-4756	11.0	894	18,500	13.6	1164	36,300
IMR-4227	22.8	1000	22,800	26.4	1261	36,600
Unique	10.5	905	21,100	13.2	1148	35,100
Blue Dot	15.7	942	19,600	18.7	1245	35,100
2400	19.5	972	23,100	23.6	1227	35,000
HS-7	16.0	939	19,600	19.4	1209	34,700
H110	24.5	1059	21,300	29.9	1378	35,800
AA 7	18.0	1108	18,200	21.0	1260	35,200
800X	12.0	861	20,200	14.9	1239	36,100

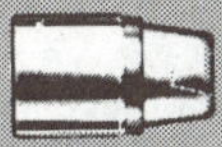

225 gr. Jacketed HP
1.610" OAL

Powder	Sugg. Starting Grains	Velocity fps	Pressure C.U.P.	Max. Load Grains	Velocity fps	Pressure C.U.P.
SR-4756	10.2	865	20,800	12.6	1123	36,500
IMR-4227	20.4	1010	24,800	24.0	1205	35,800
Unique	10.0	922	23,300	12.1	1127	35,500
Blue Dot	14.5	993	24,100	17.4	1210	35,700
2400	17.7	935	22,300	22.0	1199	36,000
HS-7	14.8	916	21,400	18.2	1179	35,600
H110	21.3	1028	23,300	26.0	1276	36,500
AA 7	15.5	836	23,300	19.0	1227	37,300
800X	12.0	958	27,700	14.7	1178	36,300

240 gr. Jacketed HC
1.610" OAL

Powder	Sugg. Starting Grains	Velocity fps	Pressure C.U.P.	Max. Load Grains	Velocity fps	Pressure C.U.P.
IMR-4227	19.3	921	22,600	23.1	1177	36,800
Unique	10.0	886	23,800	12.1	1101	35,600
Blue Dot	13.8	916	23,500	16.8	1141	35,800
2400	17.3	878	22,100	21.5	1160	36,800
HS-7	14.2	878	22,000	17.6	1132	36,700
H110	19.3	931	22,100	23.9	1191	35,100
SR-4756	9.9	854	23,700	12.3	1072	35,800
AA 9	19.0	829	22,300	22.2	1135	33,500
800X	11.0	776	23,700	13.7	1079	37,300

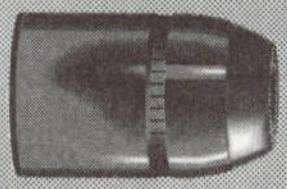

240 gr. Jacketed HP
1.610" OAL

Powder	Sugg. Starting Grains	Velocity fps	Pressure C.U.P.	Max. Load Grains	Velocity fps	Pressure C.U.P.
IMR-4227	19.3	884	23,100	23.1	1109	36,000
Unique	10.0	847	24,200	12.2	1042	37,000
Blue Dot	13.0	803	19,600	16.2	1068	35,300
2400	17.4	840	21,600	21.6	1119	35,500
HS-7	14.0	929	21,700	17.6	1107	37,000
H110	19.6	937	20,800	24.5	1177	36,000
AA 9	19.0	815	22,100	22.0	1195	37,100
800X	11.2	807	24,700	13.9	1068	36,500

Note: Loads shown in shaded panels are maximum.

44 Remington Magnum
(44 Magnum)

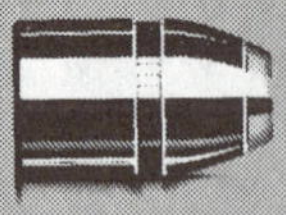

265 gr. Jacketed FP
1.610" OAL

Powder	Sugg. Starting Grains	Velocity fps	Pressure C.U.P.	Max. Load Grains	Velocity fps	Pressure C.U.P.
IMR-4227	18.5	870	23,800	22.0	1088	34,800
Blue Dot	13.0	830	21,600	16.0	1086	35,100
2400	16.2	794	21,100	20.2	1096	35,200
H110	17.9	871	19,700	22.8	1113	34,800
AA 9	18.0	823	25,300	21.3	1066	35,600
800X	10.7	745	25,400	13.4	999	36,700

#429348
180 gr.(Linotype)1.420" OAL

Powder	Sugg. Starting Grains	Velocity fps	Pressure C.U.P.	Max. Load Grains	Velocity fps	Pressure C.U.P.
SR-4756	11.9	1096	23,300	15.8	1359	36,900
IMR-4227	22.7	1065	21,700	27.0+	1394	35,900
Unique	11.0	1079	22,500	13.5	1343	36,500
Blue Dot	16.1	1110	22,600	20.4	1408	35,800
2400	20.3	1190	22,000	25.3+	1410	36,100
231	9.0	1002	22,000	12.6	1315	36,700
HS-6	14.2	1082	23,200	17.3	1352	36,400
Herco	11.7	1049	22,800	14.6	1344	36,400

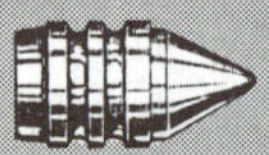

#429303
200 gr. (Linotype) 1.692" OAL

Powder	Sugg. Starting Grains	Velocity fps	Pressure C.U.P.	Max. Load Grains	Velocity fps	Pressure C.U.P.
SR-4756	12.3	1043	22,600	15.5	1289	36,700
IMR-4227	22.0	1038	22,100	24.5+	1201	29,600
Unique	11.0	1004	22,100	13.2	1265	36,700
Herco	11.4	1035	24,300	13.8	1245	36,700
Blue Dot	15.8	1062	22,800	19.8	1363	36,800
2400	19.0	991	20,400	24.0+	1314	34,600
231	9.1	913	19,600	12.6	1231	36,900
HS-6	14.2	1020	22,000	17.3	1295	36,900

#429215
215 gr. (Linotype) 1.645" OAL

Powder	Sugg. Starting Grains	Velocity fps	Pressure C.U.P.	Max. Load Grains	Velocity fps	Pressure C.U.P.
SR-4756	12.2	1036	24,900	15.2	1228	36,100
IMR-4227	21.8	1017	22,100	25.5+	1227	32,800
Unique	10.0	930	21,000	13.2	1221	36,000
Herco	11.4	984	22,800	13.5	1205	36,000
Blue Dot	15.7	1032	23,500	19.5	1314	37,000
2400	20.0	986	21,800	25.0	1310	37,000
231	9.3	914	21,000	12.8	1191	37,000
HS-6	13.5	985	21,700	16.5	1238	36,000
AA 7	16.0	776	16,400	20.2	1277	37,000
800X	12.0	932	23,900	15.1	1274	36,200

Note: Loads shown in shaded panels are maximum.
\+ Designates a compressed powder charge.

44 Remington Magnum
(44 Magnum)

#429360
232 gr. (Linotype) 1.655" OAL

Powder	Sugg. Starting Grains	Velocity fps	Pressure C.U.P.	Max. Load Grains	Velocity fps	Pressure C.U.P.
SR-4756	9.9	887	16,000	13.3	1145	36,100
IMR-4227	20.2	961	21,900	24.5	1236	36,100
Unique	9.5	896	16,100	12.7	1159	36,200
Herco	10.7	937	23,300	13.5	1163	36,500
Blue Dot	14.4	952	20,400	18.2	1240	36,300
2400	18.5	998	22,700	23.0	1256	36,700
HS-6	12.5	921	20,800	16.2	1177	36,600
AA 7	15.5	881	23,500	19.0	1233	35,300
AA 9	19.0	913	20,700	22.5	1261	33,400
800X	11.0	923	25,000	13.7	1222	36,000

#429421
245 gr. (Linotype) 1.710" OAL

Powder	Sugg. Starting Grains	Velocity fps	Pressure C.U.P.	Max. Load Grains	Velocity fps	Pressure C.U.P.
SR-4756	11.4	927	20,500	13.1	1125	36,500
IMR-4227	20.2	938	21,300	24.0+	1196	34,600
Unique	9.8	912	20,800	13.0	1147	36,800
Herco	10.9	918	22,900	13.1	1090	34,600
Blue Dot	15.0	941	20,000	18.4	1196	35,300
2400	19.4	974	23,500	23.4	1232	36,000
HS-7	15.2	931	20,900	18.2	1171	35,300
AA 9	17.5	711	15,100	21.7	1229	37,000
800X	10.6	812	20,700	13.3	1113	34,200

#429244
245 gr. (Linotype) 1.680" OAL

Powder	Sugg. Starting Grains	Velocity fps	Pressure C.U.P.	Max. Load Grains	Velocity fps	Pressure C.U.P.
SR-4756	10.5	892	21,500	12.2	1119	35,300
IMR-4227	20.2	903	20,600	24.0+	1135	33,700
Unique	8.9	805	17,500	12.1	1100	36,400
Herco	11.0	927	24,500	12.4	1085	35,400
Blue Dot	14.0	911	20,300	17.4	1160	34,800
2400	18.2	915	21,600	22.2	1165	35,300
HS-7	14.5	921	22,600	17.5	1160	37,000
AA 9	17.5	750	17,000	21.7	1223	34,600
800X	10.6	842	25,300	13.2	1075	36,600

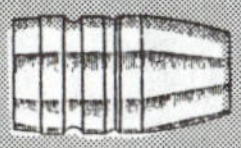

#429640
275 gr. (Linotype) 1.700" OAL

Powder	Sugg. Starting Grains	Velocity fps	Pressure C.U.P.	Max. Load Grains	Velocity fps	Pressure C.U.P.
SR-4756	10.4	722	27,800	11.8	871	36,900
800X	10.5	776	25,400	12.8	967	36,600
IMR 4227	19.5	726	23,300	23.0	1027	36,100
Unique	9.5	700	23,300	11.5	916	36,100
Herco	10.2	667	23,300	12.0	892	36,400
Blue Dot	13.2	685	22,000	16.7	991	36,600
2400	17.6	879	27,700	20.4	1129	37,000
AA 9	18.0	812	23,800	20.2	1048	35,800

Note: Loads shown in shaded panels are maximum.
+ Designates a compressed powder charge.

45 Automatic
(45 ACP)

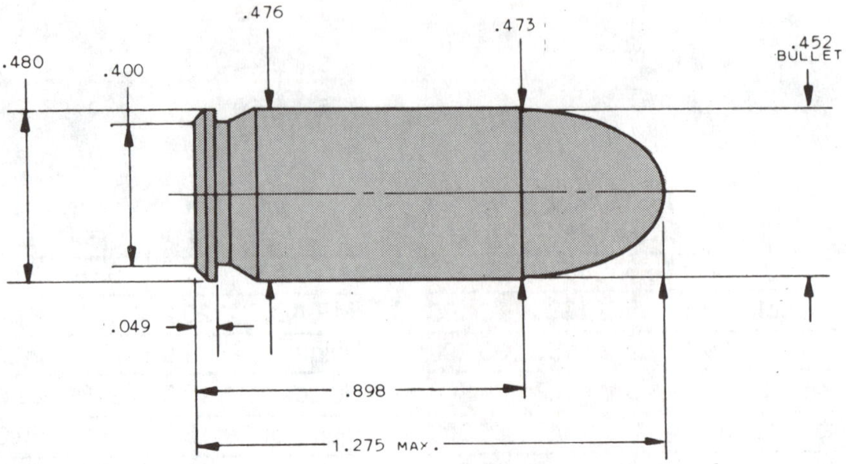

COMMENTS:

This cartridge headspaces on the case mouth. Adhere to the listed maximum and trim-to case lengths exactly. Do not roll crimp as doing so will alter headspacing. A taper crimp is suggested.

While groove diameters will vary from 0.450" to 0.453", chamber dimensions suggest that bullets should not exceed 0.451" for safety.

Some semi-auto pistols will not function with any but round nose bullets unless they are expertly altered by a qualified gunsmith. Bullet #452374 is a good choice for most unaltered pistols.

Hercules Bullseye, Winchester 231, and Hercules Unique are the unquestionable accuracy champs with most bullets in this cartridge.

TEST COMPONENTS:

Cases .... Federal; Remington (cast bullets only)
Trim-to Length .... .895"
Primers .... CCI 300; Remington 2½ (cast bullets only)
Primer Size .... Large Pistol
Lyman Shell Holder .... No. 2
Jacketed Bullets Used .... Remington
Remington MCWC #B22890, 185 gr.
Hornady HP, 185 gr.
Speer HP #4477, 200 gr.
Speer HP #4479, 225 gr.
Remington MC #B22892, 230 gr.
Cast Bullets Used .... (Sized to .450" dia.)
#452389, 185 gr.
#452488, 195 gr.
#452460, 200 gr.
#452630, 200 gr.
#452374, 225 gr.

TEST SPECIFICATIONS: (Velocity & Pressure)

Firearm Used .... Universal Receiver
Barrel Length .... 5"
Twist .... 1-16"
Groove Dia. .... .450"

185 gr. Jacketed MCWC
1.135" OAL

Powder	Sugg. Starting Grains	Velocity fps	Pressure C.U.P.	Max. Load Grains	Velocity fps	Pressure C.U.P.
700X	3.4	675	7,900	5.0	967	17,800
SR-7625	4.4	706	9,100	6.5	972	16,800
SR-4756	5.7	666	7,900	7.7	1002	17,000
Bullseye	3.5	678	8,000	5.6	975	16,600
Unique	4.7	661	7,700	7.5	1055	18,000
Blue Dot	7.3	666	8,000	10.2	1015	17,800
231	4.1	680	9,100	6.1	981	16,600
HS-6	6.6	689	8,000	9.1	1038	18,000

Note: Loads shown in shaded panels are maximum.

45 Automatic

(45 ACP)

185 gr. Jacketed HP
1.175" OAL

Powder	Sugg. Starting Grains	Velocity fps	Pressure C.U.P.	Max. Load Grains	Velocity fps	Pressure C.U.P.
700X	3.5	612	7,500	5.5	933	17,800
SR-7625	4.9	633	7,100	7.0	960	17,900
Bullseye	3.5	610	7,300	6.0	976	18,000
Unique	4.8	630	7,800	7.8	985	17,800
231	4.4	640	7,500	‡6.1	937	15,700
PB	4.6	660	9,100	6.9	968	17,700
Red Dot	3.5	607	7,200	6.1	942	17,600
HP38	3.5	599	7,500	5.8	960	17,900
AA5	7.4	720	10,700	9.2	1008	17,700
Clays	4.1	766	13,500	4.6	841	16,600

200 gr. Jacketed HP
1.178" OAL

Powder	Sugg. Starting Grains	Velocity fps	Pressure C.U.P.	Max. Load Grains	Velocity fps	Pressure C.U.P.
700X	3.3	597	7,400	4.9	871	17,300
SR-7625	4.5	601	7,300	6.7	935	17,400
SR-4756	5.4	588	7,100	7.4	961	18,000
Bullseye	3.4	604	7,100	5.7	942	17,800
Unique	4.0	604	7,300	6.5	927	17,700
Blue Dot	7.2	620	7,100	10.2	956	17,500
231	4.0	636	6,800	6.5	970	17,800
HS-6	5.7	566	6,500	9.0	951	17,600
AA5	6.8	623	9,400	8.5	900	16,800
Clays	3.9	713	13,100	4.4	790	16,500

225 gr. Jacketed HP
1.243" OAL

Powder	Sugg. Starting Grains	Velocity fps	Pressure C.U.P.	Max. Load Grains	Velocity fps	Pressure C.U.P.
700X	3.1	549	7,500	4.8	796	17,200
SR-7625	4.4	587	7,400	6.3	844	17,400
SR-4756	4.8	552	6,700	7.3	882	17,300
Bullseye	3.1	554	7,300	5.3	840	17,400
Unique	4.1	548	6,800	6.7	883	17,500
Blue Dot	7.2	602	6,600	10.7	925	17,700
HS-7	6.7	564	6,800	10.3	884	17,600

230 gr. Jacketed MC
1.275" OAL

Powder	Sugg. Starting Grains	Velocity fps	Pressure C.U.P.	Max. Load Grains	Velocity fps	Pressure C.U.P.
700X	3.6	667	11,200	4.8	867	17,900
SR-7625	4.9	678	10,800	‡5.9	838	15,800
SR-4756	6.0	695	11,200	7.0	866	17,400
Bullseye	3.8	662	10,400	5.3	878	17,100
Unique	5.0	677	10,000	6.8	916	16,900
Blue Dot	7.5	662	8,700	10.2	923	16,900
HS-7	8.1	671	10,000	10.7	914	17,800

Note: Loads shown in shaded panels are maximum.
‡ Designates a factory velocity duplication load.

45 Automatic

(45 ACP)

#452389
185 gr. (#2 Alloy) 1.080" OAL

Powder	Sugg. Starting Grains	Velocity fps	Pressure C.U.P.	Max. Load Grains	Velocity fps	Pressure C.U.P.
700X	4.0	760	8,800	5.5	1010	17,500
PB	4.6	740	8,300	5.5	920	13,500
SR-7625	5.5	825	9,800	6.5	1030	17,400
SR-4756	5.3	663	8,900	7.9	988	17,600
Bullseye	3.5	660	6,000	5.0	890	12,300
Red Dot	4.0	735	8,100	5.5	970	14,800
Green Dot	4.5	745	8,300	6.0	965	14,800
Unique	5.0	684	6,900	7.5	995	15,500
Blue Dot	7.0	636	7,600	9.2+	844	13,200
231	3.9	662	7,900	6.0	958	17,500
HS-6	5.6	613	7,500	9.1	1034	17,100

#452488
195 gr. (#2 Alloy) 1.120" OAL

Powder	Sugg. Starting Grains	Velocity fps	Pressure C.U.P.	Max. Load Grains	Velocity fps	Pressure C.U.P.
700X	4.0	735	9,600	5.2	950	16,700
PB	4.6	765	10,100	5.5	910	14,800
SR-7625	5.0	780	10,300	6.2	945	16,200
SR-4756	5.3	639	8,600	7.8	977	17,500
Bullseye	3.5	680	7,700	5.0	895	13,900
Red Dot	4.0	730	9,000	5.5	950	16,200
Green Dot	4.5	740	9,400	6.0	955	15,800
Unique	5.0	695	7,900	7.5	1010	16,800
Blue Dot	7.0	646	8,200	10.7	1033	17,900
231	3.9	642	8,500	5.7	962	16,900
HS-6	6.1	652	8,300	9.0	1027	17,900

#452460
200 gr. (#2 Alloy) 1.161" OAL

Powder	Sugg. Starting Grains	Velocity fps	Pressure C.U.P.	Max. Load Grains	Velocity fps	Pressure C.U.P.
700X	4.0	745	10,300	5.2	940	17,300
PB	4.6	725	9,000	5.5	880	14,200
SR-7625	5.0	735	9,000	6.2	945	17,000
**SR-4756	5.3	704	9,100	7.3	993	17,000
Bullseye	3.5	645	6,900	5.0	860	13,400
Red Dot	4.0	695	8,400	5.3	895	14,700
Green Dot	4.5	715	9,000	5.8	895	14,400
Unique	5.0	670	7,700	7.5	980	16,600
**Blue Dot	7.1	701	8,900	10.6	1012	17,200
**231	4.0	694	9,200	6.0	987	18,000
**HS-6	6.0	690	8,600	8.7	1016	17,700
**AA5	6.4	686	10,000	8.0	970	17,000

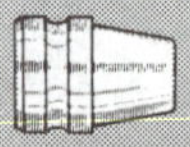

#452630 **
200 gr. (#2 Alloy) 1.235" OAL

Powder	Sugg. Starting Grains	Velocity fps	Pressure C.U.P.	Max. Load Grains	Velocity fps	Pressure C.U.P.
Bullseye	4.9	840	12,900	6.2	918	17,600
231	5.2	739	12,300	6.5	923	17,400
WST	4.8	825	14,700	5.6	933	17,900
Red Dot	4.4	713	12,400	5.5	894	17,600
700X	4.4	676	10,000	5.5	884	17,500
Green Dot	5.1	727	11,900	6.4	894	16,900
PB	5.5	728	11,800	6.7	900	16,900
Unique	6.0	717	11,400	7.5	913	17,400
SR-7625	5.4	653	8,600	6.7	900	16,800
HS-6	7.6	649	9,200	9.5	937	17,600
SR-4756	6.4	660	9,300	8.0	933	16,700
Blue Dot	8.5	679	9,800	10.6	972	17,300
AA5	7.0	697	10,500	8.5	927	16,800
Clays	3.9	701	10,200	4.4	839	15,500

Note: Loads shown in shaded panels are maximum.
\+ Designates a compressed powder charge.
** Designates use of Federal cases.

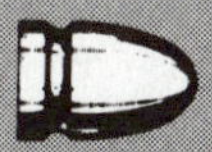

#452374
225 gr. (#2 Alloy) 1.272" OAL

Powder	Sugg. Starting Grains	Velocity fps	Pressure C.U.P.	Max. Load Grains	Velocity fps	Pressure C.U.P.
700X	4.0	695	10,500	5.0	855	17,300
PB	4.8	705	10,300	5.5	810	14,200
SR-7625	5.0	675	9,000	6.0	850	15,200
**SR-4756	5.5	662	8,800	7.5	955	18,000
Bullseye	4.0	680	10,100	5.0	815	14,400
Red Dot	4.3	705	11,000	5.3	835	15,300
Green Dot	4.8	725	11,400	5.8	845	15,100
Unique	5.5	695	10,100	7.3	905	16,500
**Blue Dot	7.5	660	8,600	10.7	964	17,300
**231	4.0	661	9,200	5.8	902	17,500
**HS-6	6.2	664	8,400	8.6	921	16,600
**AA5	6.6	625	9,200	8.2	874	17,300
**WSL	4.9	803	14,500	5.5	881	17,800
**Clays	3.6	683	10,600	4.3	818	17,100

Note: Loads shown in shaded panels are maximum.
** Designates use of Federal cases.

45 Auto Rim

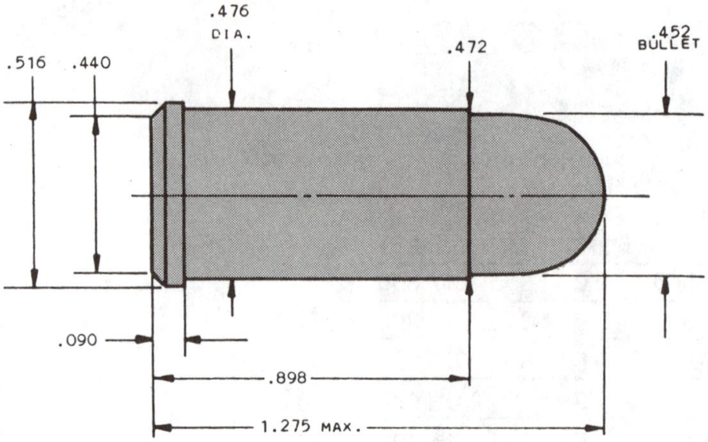

COMMENTS:

The 45 Auto Rim was designed for use in revolvers that were chambered for the 45 Automatic. The 45 Auto Rim cartridge did away with the need to use the cumbersome half moon clips in such revolvers. The rim of this case both headspaced the round and afforded a purchase for the revolvers extractor. Hercules Bullseye, Winchester 231 or Hercules Unique should give the best results.

TEST COMPONENTS:

Cases .... Remington
Trim-to Length .... .894"
Primers .... Remington 2½
Primer Size .... Large Pistol
Lyman Shell Holder .... No. 14A
Jacketed Bullets Used .... Hornady HP, 185 gr.
Speer HP #4477, 200 gr.
Speer HP #4479, 225 gr.
Cast Bullets Used .... (Sized to .451" dia.)
#452389, 185 gr.
#452460, 200 gr.
#452374, 225 gr.

TEST SPECIFICATIONS: (Velocity & Pressure)

Firearm Used .... Special Test Barrel; Smith & Wesson Model 25 (cast bullets only)
Barrel Length .... 4½" and 6½"
Twist .... 1-16"
Groove Dia. .... .451"

185 gr. Jacketed HP
1.236" OAL

Powder	Sugg. Starting Grains	Velocity fps	Pressure C.U.P.	Max. Load Grains	Velocity fps	Pressure C.U.P.
700X	4.5	713	8,200	5.2	838	11,900
SR-7625	6.0	709	7,500	7.2	849	13,700
Bullseye	4.6	722	8,600	5.8	915	14,800
Red Dot	4.8	711	8,600	6.4	933	14,700
Unique	5.0	585	5,100	7.6	905	13,200
231	4.4	618	6,400	6.1	889	13,700

Note: Loads shown in shaded panels are maximum.

200 gr. Jacketed HP
1.180" OAL

Powder	Sugg. Starting Grains	Velocity fps	Pressure C.U.P.	Max. Load Grains	Velocity fps	Pressure C.U.P.
700X	3.4	491	5,000	4.9	799	13,400
SR-7625	4.8	608	7,200	6.0	809	12,400
Bullseye	4.0	614	7,800	5.4	838	13,700
Red Dot	4.0	615	7,700	5.4	852	14,600
Unique	5.8	719	10,100	7.1	882	14,500
231	4.0	570	7,800	5.7	852	14,800
HS-6	6.1	558	6,400	8.7	846	13,600

225 gr. Jacketed HP
1.255" OAL

Powder	Sugg. Starting Grains	Velocity fps	Pressure C.U.P.	Max. Load Grains	Velocity fps	Pressure C.U.P.
SR-7625	4.4	518	6,100	6.2	812	14,900
Unique	5.0	609	8,200	7.0	834	14,500
Blue Dot	8.5	718	10,200	10.0	864	14,500
231	3.7	496	6,500	5.2	773	13,800
HS-6	6.7	634	8,800	8.3	812	13,600

#452389
185 gr. (#2 Alloy) 1.118" OAL

Powder	Sugg. Starting Grains	Velocity fps	Pressure C.U.P.	Max. Load Grains	Velocity fps	Pressure C.U.P.
700X	4.0	655	—	5.3	905	—
PB	4.5	620	—	5.8	855	—
SR-7625	5.0	615	—	6.3	810	—
Bullseye	3.0	505	—	4.8	780	—
Red Dot	4.0	650	—	5.3	885	—
Green Dot	4.5	650	—	5.8	855	—
Unique	5.0	600	—	7.0	870	—
231	4.0	735	7,200	5.7	972	13,400
HS-6	5.5	674	5,300	8.8	1030	14,800

#452460
200 gr. (#2 Alloy) 1.161" OAL

Powder	Sugg. Starting Grains	Velocity fps	Pressure C.U.P.	Max. Load Grains	Velocity fps	Pressure C.U.P.
700X	4.0	675	—	5.0	875	—
PB	4.5	650	—	5.5	840	—
SR-7625	5.0	660	—	6.1	850	—
Bullseye	3.0	510	—	4.7	775	—
Red Dot	4.0	645	—	5.2	845	—
Green Dot	4.5	660	—	5.6	840	—
Unique	5.0	610	—	6.8	845	—
231	4.0	694	7,200	5.5	936	14,100

#452374
225 gr. (#2 Alloy) 1.265" OAL

Powder	Sugg. Starting Grains	Velocity fps	Pressure C.U.P.	Max. Load Grains	Velocity fps	Pressure C.U.P.
700X	4.0	625	—	4.8	770	—
PB	4.5	565	—	5.4	735	—
SR-7625	5.0	560	—	5.9	750	—
Bullseye	3.0	475	—	4.5	690	—
Red Dot	4.0	625	—	5.0	760	—
Green Dot	4.5	615	—	5.4	750	—
Unique	5.0	550	—	6.6	785	—
231	4.0	660	6,900	5.5	873	13,800
HS-6	7.3	799	9,400	8.6	922	13,800

Note: Loads shown in shaded panels are maximum.

45 Colt

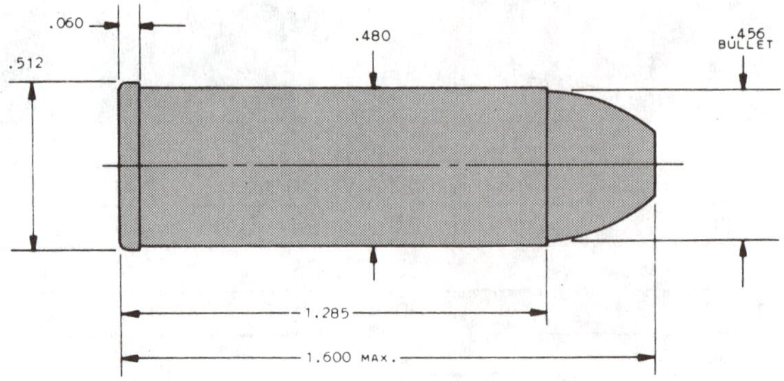

COMMENTS:

This cartridge is often mistakenly called the 45 Long Colt. Care must be taken when sizing as the tiny rim of this cartridge does not afford a good purchase for the shell holder. If undue force is needed to withdraw the case, the rim may tear away leaving the case jammed in the die.

Hercules Unique is highly favored with all bullets for accuracy with this cartridge.

Pre-World War II revolvers normally have a groove diameter of 0.454 inch. Later handguns are nominally equipped with barrels having a groove diameter of 0.451 inch. If in doubt about your handgun, slug the bore and size bullets accordingly. The first step of the expanding plug used must be 0.001" to 0.0015" under actual bullet diameter or bullets will not be tight in the case neck.

Bullet #452424, the Elmer Keith bullet, has replaced bullet #454424 in all our data. The bullets are identical except that #452424 is a few thousandths smaller in diameter. This is our most popular bullet in this caliber. Carefully control seating depths of cast bullets to the listed length. Crimp #452424 on the front edge of the forward driving band. Number 454190 should be used for larger 0.454" groove diameter revolvers and should be crimped on the ogive.

TEST COMPONENTS:

Cases .... Remington; Winchester
Trim-to Length .... 1.280"
Primers .... Winchester 7-111; Remington 2½
Primer Size .... Large Pistol
Lyman Shell Holder .... No. 11
Jacketed Bullets Used .... Sierra HC #8800, 185 gr.
Speer HP #4477, 200 gr.
Speer HP #4479, 225 gr.
Sierra HC #8820, 240 gr.
Hornady HP, 250 gr.
Cast Bullets Used .... (Sized to .452" dia.)
#454190, 250 gr.
#452424, 255 gr.
#454 Round Ball, 142 gr.

TEST SPECIFICATIONS:
(Velocity Only)

Firearm Used .... Ruger Blackhawk
Barrel Length .... 7½"
Twist .... 1-16"
Groove Dia. .... .451"

185 gr. Jacketed HC
1.520" OAL

Powder	Sugg. Starting Grains	Velocity fps	Pressure C.U.P.	Max. Load Grains	Velocity fps	Pressure C.U.P.
Unique	7.5	615	—	‡10.0	934	—
Blue Dot	12.0	692	—	13.5	826	—
700X	6.0	658	—	7.0	884	—
IMR-4227	16.5	711	—	20.0	875	—
HS-7	13.0	700	—	14.5	908	—

200 gr. Jacketed HP
1.557" OAL

Powder	Sugg. Starting Grains	Velocity fps	Pressure C.U.P.	Max. Load Grains	Velocity fps	Pressure C.U.P.
SR-4756	9.5	679	—	11.8	950	—
Unique	8.0	683	—	11.2	994	—
Blue Dot	11.6	560	—	12.8	835	—
HS-6	10.3	740	—	13.3	993	—
SR-7625	7.0	581	—	9.5	894	—

Note: Loads shown in shaded panels are maximum.
‡ Designates a factory velocity duplication load.

225 gr. Jacketed HP
1.557" OAL

Powder	Sugg. Starting Grains	Velocity fps	Pressure C.U.P.	Max. Load Grains	Velocity fps	Pressure C.U.P.
SR-7625	7.0	482	—	9.3	810	—
IMR-4227	16.5	581	—	21.0	914	—
Unique	8.3	696	—	10.8	926	—
Blue Dot	12.4	624	—	15.4	947	—
HS-6	11.6	674	—	13.6	914	—

250 gr. Jacketed HP
1.590" OAL

Powder	Sugg. Starting Grains	Velocity fps	Pressure C.U.P.	Max. Load Grains	Velocity fps	Pressure C.U.P.
Unique	7.5	625	—	9.3	801	—
2400	15.2	696	—	17.5	814	—
SR-4759	16.0	682	—	18.5	807	—
IMR-4227	17.0	646	—	20.3	862	—
HS-7	11.7	654	—	14.3	837	—

#452424
255 gr. (#2 Alloy) 1.575" OAL

Powder	Sugg. Starting Grains	Velocity fps	Pressure C.U.P.	Max. Load Grains	Velocity fps	Pressure C.U.P.
++Red Dot	4.5	550	—	6.0	780	—
++Green Dot	4.5	500	—	6.5	765	—
++Unique	6.0	590	—	8.5	845	—
++700X	4.5	535	—	6.0	785	—
++PB	5.0	530	—	7.0	750	—
++SR-7625	6.0	555	—	8.0	835	—
HS-6	7.8	630	—	9.5	761	—

240 gr. Jacketed HC
1.575" OAL

Powder	Sugg. Starting Grains	Velocity fps	Pressure C.U.P.	Max. Load Grains	Velocity fps	Pressure C.U.P.
Unique	7.5	633	—	9.5	851	—
2400	15.0	671	—	18.5	899	—
SR-4759	15.0	619	—	18.5	819	—
IMR-4227	16.0	610	—	20.5	904	—
HS-7	12.0	687	—	14.0	865	—

#454190
250 gr. (#2 Alloy) 1.600" OAL

Powder	Sugg. Starting Grains	Velocity fps	Pressure C.U.P.	Max. Load Grains	Velocity fps	Pressure C.U.P.
++Red Dot	4.5	535	—	6.5	835	—
++Green Dot	4.5	505	—	7.0	835	—
++Unique	6.0	595	—	‡9.0	875	—
++700X	4.5	535	—	6.0	800	—
++PB	5.0	545	—	7.5	820	—
++SR-7625	6.0	555	—	8.5	890	—
231	6.5	762	—	7.4	931	—

454 Round Ball
142 gr. (#2 Alloy) 1.440" OAL

Powder	Sugg. Starting Grains	Velocity fps	Pressure C.U.P.	Max. Load Grains	Velocity fps	Pressure C.U.P.
Red Dot	—	—	—	4.3	536	—
Unique	—	—	—	6.5	521	—
SR-7625	—	—	—	7.2	602	—

Note: Loads shown in shaded panels are maximum.
++ Designates use of Winchester cases and primers.
‡ Designates a factory velocity duplication load.

Thompson/Center Contender/ Remington XP-100 Data

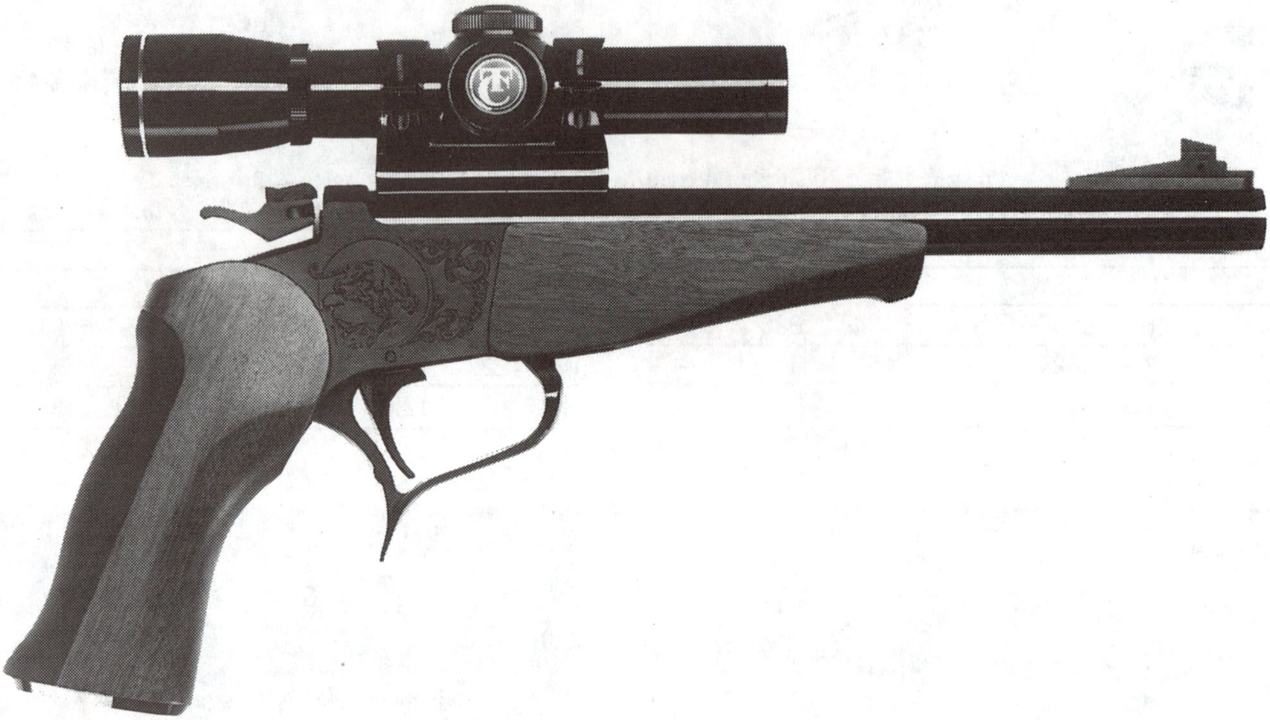

Introduction

The load data found herein has been reviewed by Thompson/Center. This data is expressly for use only in the T/C Contender or the Remington XP-100. *Do not use these loads in any other handgun.*

The T/C Contender has become extremely popular with handgun hunters and silhouette shooters. It has been chambered for a wide variety of commercial cartridges, including many that were generally up till then rifle rounds. These range from the 22 Hornet to the 45-70 Government.

Maximum overall cartridge lengths as suggested by SAAMI may be exceeded in the Contender. Use the exact cartridge overall length shown for each specific bullet in this data section.

When using bottleneck rimmed cases such as the 30-30 Winchester, accuracy may be improved by avoiding full length resizing of cases. Once cases have been fire formed it is best to neck size only so as to allow rimmed cartridges to headspace off the shoulder.

The data in this section cannot be used without a full understanding of what it represents or without complete reloading knowledge. Any attempts to use this data without the necessary cautions, procedures and background knowledge could be extremely dangerous causing personal and property injury, even death.

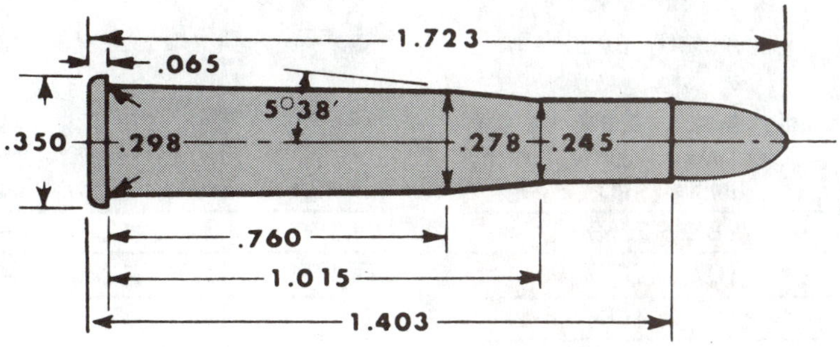

COMMENTS:

The Hornet is a fine short range varmint cartridge for the Contender and noise level is kept to a minimum. Trajectory is quite flat for 100 yard shooting.

IMR 4227 is the first choice for accuracy. Do not forget to use thin jacketed bullets expressly designed for the Hornet.

TEST COMPONENTS:

Cases .... Winchester
Trim-to Length .... 1.393"
Primers .... Remington 6½
Primer Size .... Small Rifle
Lyman Shell Holder .... No. 4
Jacketed Bullets Used .... Sierra SP #1200, 40 gr.
Hornady SP #2230, 45 gr.
Hornady HP #2249, 52 gr.
Cast Bullets Used .... (Sized to .224" dia.)
*Gas Check Bullets *#225415, 49 gr.
*#225462, 61 gr.

TEST SPECIFICATIONS: (Velocity Only)

Firearm Used .... Thompson/Center Contender
Barrel Length .... 10"
Twist .... 1-14"
Groove Dia. .... .224"

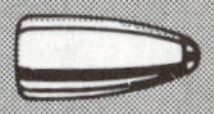

40 gr. Jacketed SP
1.722" OAL

Powder	Sugg. Starting Grains	10" Bbl. Velocity fps	14" Bbl. Velocity fps	Max. Load Grains	10" Bbl. Velocity fps	14" Bbl. Velocity fps
2400	7.2	1919	—	8.8	2402	—
IMR-4227	10.0	2105	—	11.7+	2348	—
680	11.2	2078	—	13.8+	2440	—

45 gr. Jacketed Spire Pt.
1.722" OAL

Powder	Sugg. Starting Grains	10" Bbl. Velocity fps	14" Bbl. Velocity fps	Max. Load Grains	10" Bbl. Velocity fps	14" Bbl. Velocity fps
2400	7.0	1920	—	8.3	2238	—
IMR-4227	10.0	2062	—	11.3	2280	—
680	11.0	2077	—	12.5	2240	—

Note: Loads shown in shaded panels are maximum.
+ Designates a compressed powder charge.

22 Hornet

52 gr. Jacketed HP
1.860" OAL

Powder	Sugg. Starting Grains	10" Bbl. Velocity fps	14" Bbl. Velocity fps	Max. Load Grains	10" Bbl. Velocity fps	14" Bbl. Velocity fps
2400	6.7	1709	—	8.0	2063	—
IMR-4227	9.5	1922	—	10.5	2077	—
680	8.5	1753	—	11.5	1986	—

#225415
49 gr. (#2 Alloy) 1.694" OAL

Powder	Sugg. Starting Grains	10" Bbl. Velocity fps	14" Bbl. Velocity fps	Max. Load Grains	10" Bbl. Velocity fps	14" Bbl. Velocity fps
Unique	3.0	1270	—	4.6	1732	—
SR-4759	5.6	1230	—	7.5	1639	—

#225462
61 gr. (#2 Alloy) 1.610" OAL

Powder	Sugg. Starting Grains	10" Bbl. Velocity fps	14" Bbl. Velocity fps	Max. Load Grains	10" Bbl. Velocity fps	14" Bbl. Velocity fps
Unique	2.5	1074	—	4.5	1624	—
SR-4759	5.6	1216	—	7.5	1619	—

Note: Loads shown in shaded panels are maximum.

221 Remington Fireball

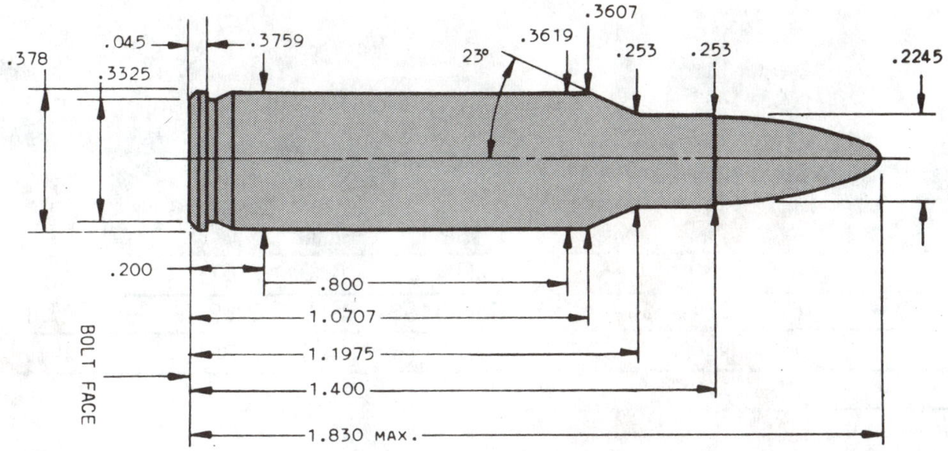

COMMENTS:

This is a fine cartridge with much longer useful case reloading life than the 22 Hornet. It will actually best the Hornet in most performance categories. Originally designed for the Remington bolt-action XP-100 pistol this cartridge is excellent for varmint hunting, particularly when loaded with 52 grain bullets. IMR 4227 and IMR 4198 are the best powder choices.

TEST COMPONENTS:

Cases .... Remington
Trim-to Length .... 1.390"
Primers .... Remington 7½
Primer Size .... Small Rifle, Magnum
Lyman Shell Holder .... No. 26
Jacketed Bullets Used .... Sierra SP #1200, 40 gr.
Hornady SP #2230, 45 gr.
Hornady HP #2249, 52 gr.
Cast Bullets Used .... (Sized to .224" dia.)
*Gas Check Bullets .... *#225415, 49 gr.
*#225462, 61 gr.

TEST SPECIFICATIONS: (Velocity Only)

Firearm Used .... Thompson/Center Contender
Barrel Length .... 10"
Twist .... 1-14"
Groove Dia. .... .224"

40 gr. Jacketed SP
1.725" OAL

Powder	Sugg. Starting Grains	10" Bbl. Velocity fps	14" Bbl. Velocity fps	Max. Load Grains	10" Bbl. Velocity fps	14" Bbl. Velocity fps
IMR-4227	13.0	2284	—	16.5	2722	—
IMR-4198	16.0	2254	—	18.0+	2541	—
RX7	15.0	1605	—	18.0	2287	—

Note: Loads shown in shaded panels are maximum.
\+ Designates a compressed powder charge.

221 Remington Fireball

45 gr. Jacketed Spire Pt.
1.820" OAL

Powder	Sugg. Starting Grains	10" Bbl. Velocity fps	14" Bbl. Velocity fps	Max. Load Grains	10" Bbl. Velocity fps	14" Bbl. Velocity fps
2400	12.5	2212	—	14.7	2611	—
IMR-4227	13.0	2226	—	16.0	2641	—
IMR-4198	16.0	2251	—	18.0+	2549	—
RX7	15.0	1748	—	17.0	2111	—

52 gr. Jacketed HP
1.820" OAL

Powder	Sugg. Starting Grains	10" Bbl. Velocity fps	14" Bbl. Velocity fps	Max. Load Grains	10" Bbl. Velocity fps	14" Bbl. Velocity fps
IMR-4227	12.0	1987	—	15.5	2516	—
IMR-4198	16.0	2269	—	17.8+	2497	—
RX7	14.5	1735	—	16.5	2096	—

#225415
49 gr. (#2 Alloy) 1.665" OAL

Powder	Sugg. Starting Grains	10" Bbl. Velocity fps	14" Bbl. Velocity fps	Max. Load Grains	10" Bbl. Velocity fps	14" Bbl. Velocity fps
Unique	4.3	1444	—	6.0	1751	—
2400	11.0	1924	—	14.0	2469	—
SR-4759	10.5	1779	—	13.5	2226	—

#225462
61 gr. (#2 Alloy) 1.605" OAL

Powder	Sugg. Starting Grains	10" Bbl. Velocity fps	14" Bbl. Velocity fps	Max. Load Grains	10" Bbl. Velocity fps	14" Bbl. Velocity fps
Unique	5.0	1453	—	6.0	1658	—
2400	10.0	1843	—	13.0	2274	—
SR-4759	9.5	1640	—	12.5	2038	—

Note: Loads shown in shaded panels are maximum.
+ Designates a compressed powder charge.

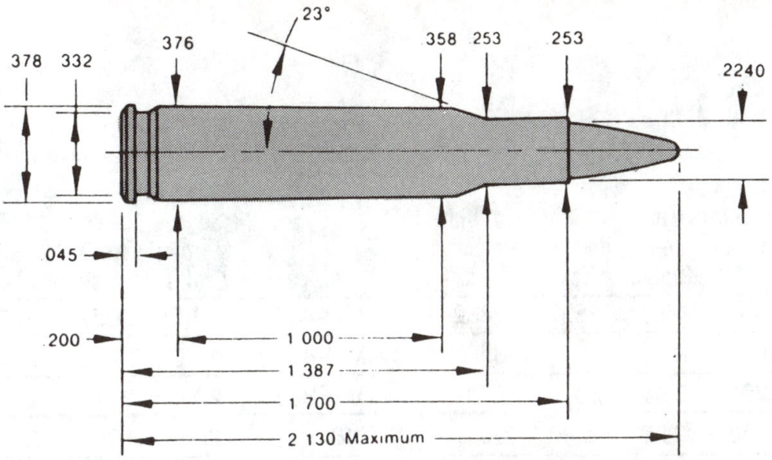

COMMENTS:

The 222 Remington is slightly more accurate than the 223 but it has been all but displaced by the latter.

The best jacketed bullet weights are 50 to 53 grains and the first powder choice might be IMR 4198 although the 222 Remington has shot spectacular groups with all of the listed powders.

The 222 Remington requires jacketed bullets designed for modest velocity expansion. These often carry names that suggest rapid expansion, i.e. Blitz.

Hornet style bullets may be used in the short handgun barrel lengths if velocities do not exceed 2,600 fps.

TEST COMPONENTS:

Cases .... Remington
Trim-to Length .... 1.690"
Primers .... Remington 7½
Primer Size .... Small Rifle, Magnum
Lyman Shell Holder .... No. 26
Jacketed Bullets Used .... Sierra SP #1200, 40 gr.
Hornady SP #2230, 45 gr.
Sierra SP #1340, 50 gr.
Hornady HP #2249, 52 gr.
Cast Bullets Used .... (Sized to .224" dia.)
*Gas Check Bullets .... *#225415, 49 gr.
*#225462, 61 gr.

TEST SPECIFICATIONS:
(Velocity Only)

Firearm Used .... Thompson/Center Contender
Barrel Length .... 10" and 14"
Twist .... 1-14"
Groove Dia. .... .224"

40 gr. Jacketed SP
1.995" OAL

Powder	Sugg. Starting Grains	10" Bbl. Velocity fps	14" Bbl. Velocity fps	Max. Load Grains	10" Bbl. Velocity fps	14" Bbl. Velocity fps
IMR-4227	15.0	2291	2671	17.0	2632	2915
IMR-4198	18.0	2319	2592	22.0+	2792	3163
IMR-3031	20.0	2057	2227	23.5+	2456	2794
748	22.0	1947	2262	25.0	2321	2661
BL-C(2)	23.5	2187	2534	27.5+	2585	3002

Note: Loads shown in shaded panels are maximum.
\+ Designates a compressed powder charge.

222 Remington

45 gr. Jacketed Spire Pt.
2.095" OAL

Powder	Sugg. Starting Grains	10" Bbl. Velocity fps	14" Bbl. Velocity fps	Max. Load Grains	10" Bbl. Velocity fps	14" Bbl. Velocity fps
IMR-4227	14.0	2126	2491	16.5	2567	2832
IMR-4198	18.0	2259	2625	21.5+	2750	3065
IMR-3031	20.0	1968	2200	23.0+	2410	2707
748	21.5	1895	2228	24.5	2290	2581
BL-C(2)	23.0	2163	2426	27.0+	2529	2917

50 gr. Jacketed SP
2.125" OAL

Powder	Sugg. Starting Grains	10" Bbl. Velocity fps	14" Bbl. Velocity fps	Max. Load Grains	10" Bbl. Velocity fps	14" Bbl. Velocity fps
IMR-4227	13.5	2057	2297	16.0	2465	2658
IMR-4198	18.0	2260	2557	20.5	2634	2836
IMR-3031	20.0	2019	2230	23.0+	2370	2674
748	22.0	1997	2294	25.0+	2236	2631
BL-C(2)	23.0	2225	2449	26.3+	2480	2826

52 gr. Jacketed HP
2.125" OAL

Powder	Sugg. Starting Grains	10" Bbl. Velocity fps	14" Bbl. Velocity fps	Max. Load Grains	10" Bbl. Velocity fps	14" Bbl. Velocity fps
IMR-4227	14.0	2140	2378	15.5	2443	2624
IMR-4198	17.1	2212	2428	19.5	2499	2829
IMR-3031	20.0	2065	2298	23.0+	2430	2710
748	21.5	1927	2258	24.5	2298	2593
BL-C(2)	22.0	2069	2398	25.0+	2387	2715

#225415
49 gr. (#2 Alloy) 2.120" OAL

Powder	Sugg. Starting Grains	10" Bbl. Velocity fps	14" Bbl. Velocity fps	Max. Load Grains	10" Bbl. Velocity fps	14" Bbl. Velocity fps
Unique	6.0	1605	1711	8.0	1943	2076
SR-4759	11.0	1674	1899	13.0	1959	2195
IMR-4227	11.0	1603	1915	13.0	1970	2263
748	14.5	1347	1556	18.0	1684	1793

#225462
61 gr. (#2 Alloy) 1.930" OAL

Powder	Sugg. Starting Grains	10" Bbl. Velocity fps	14" Bbl. Velocity fps	Max. Load Grains	10" Bbl. Velocity fps	14" Bbl. Velocity fps
Unique	5.7	1482	1573	8.0	1843	1938
SR-4759	9.5	1455	1617	12.2	1869	2051
IMR-4227	10.0	1521	1758	12.0	1861	2059
748	13.5	1256	1455	17.5	1584	1903

Note: Loads shown in shaded panels are maximum.
+ Designates a compressed powder charge.

223 Remington
(5.56mm)

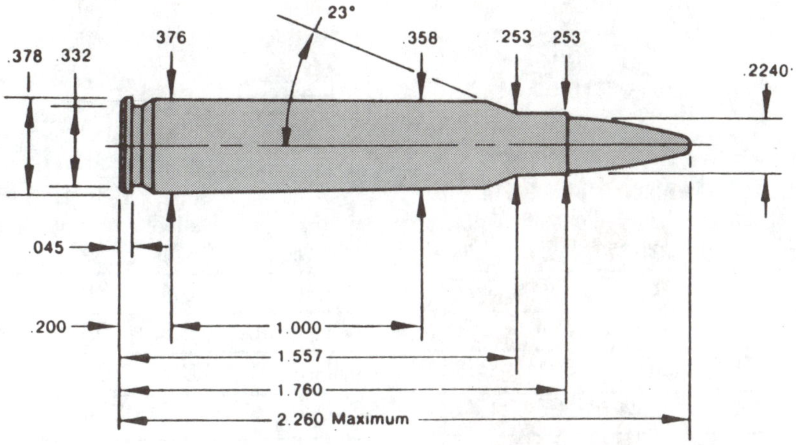

COMMENTS:

This is the current official cartridge of our armed forces. It is extremely accurate when fired in a suitable handgun and has a good range capability for varmint hunting.

Propellant choice can include numerous types which all deliver balanced accuracy and velocity. A good place to start is Winchester 748.

TEST COMPONENTS:

Cases .... Remington
Trim-to Length .... 1.750"
Primers .... Remington 7 1/2
Primer Size .... Small Rifle, Magnum
Lyman Shell Holder .... No. 26
Jacketed Bullets Used .... Hornady SP #2230, 45 gr.
Sierra SP #1340, 50 gr.
Hornady HP #2249, 52 gr.
Cast Bullets Used .... (Sized to .224" dia.)
*Gas Check Bullets — *#225415, 49 gr.
*#225462, 61 gr.

TEST SPECIFICATIONS:
(Velocity Only)

Firearm Used .... Thompson/Center Contender
Barrel Length .... 10" and 14"
Twist .... 1-14"
Groove Dia. .... .224"

45 gr. Jacketed Spire Pt.
2.170" OAL

Powder	Sugg. Starting Grains	10" Bbl. Velocity fps	14" Bbl. Velocity fps	Max. Load Grains	10" Bbl. Velocity fps	14" Bbl. Velocity fps
IMR-4227	14.0	2066	2340	17.0	2529	2690
IMR-4198	19.0	2315	2602	21.0	2504	2827
IMR-3031	22.0	2083	2323	24.5+	2346	2747
748	25.0	2029	2378	28.0+	2395	2714

Note: Loads shown in shaded panels are maximum.
\+ Designates a compressed powder charge.

223 Remington
(5.56mm)

50 gr. Jacketed SP
2.235" OAL

Powder	Sugg. Starting Grains	10" Bbl. Velocity fps	14" Bbl. Velocity fps	Max. Load Grains	10" Bbl. Velocity fps	14" Bbl. Velocity fps
IMR-4227	13.5	1997	2211	16.5	2430	2679
IMR-4198	17.0	2028	2290	20.0	2364	2668
IMR-3031	21.0	1991	2297	24.0	2321	2584
748	25.0	2067	2383	28.0+	2336	2712

52 gr. Jacketed HP
2.250" OAL

Powder	Sugg. Starting Grains	10" Bbl. Velocity fps	14" Bbl. Velocity fps	Max. Load Grains	10" Bbl. Velocity fps	14" Bbl. Velocity fps
IMR-4227	13.4	1886	2182	16.4	2384	2629
IMR-4198	16.8	1957	2238	19.8	2382	2652
IMR-3031	20.8	1980	2200	23.8	2358	2595
748	24.5	2042	2341	27.5	2313	2656

#225415
49 gr. (#2 Alloy) 2.060" OAL

Powder	Sugg. Starting Grains	10" Bbl. Velocity fps	14" Bbl. Velocity fps	Max. Load Grains	10" Bbl. Velocity fps	14" Bbl. Velocity fps
Unique	6.0	1519	1703	8.0	1850	2029
SR-4759	11.0	1565	1834	13.0	1876	2094
IMR-4227	10.5	1518	1745	12.5	1863	2086
748	21.0	1633	1839	23.0	1816	2080

#225462
61 gr. (#2 Alloy) 2.090" OAL

Powder	Sugg. Starting Grains	10" Bbl. Velocity fps	14" Bbl. Velocity fps	Max. Load Grains	10" Bbl. Velocity fps	14" Bbl. Velocity fps
Unique	6.5	1517	1652	8.5	1809	1947
SR-4759	11.0	1587	1792	13.0	1824	2028
IMR-4227	10.5	1488	1729	12.5	1811	2037
748	18.5	1544	1823	20.5	1826	2039

Note: Loads shown in shaded panels are maximum.
+ Designates a compressed powder charge.

7mm BR
(7mm Remington Benchrest)

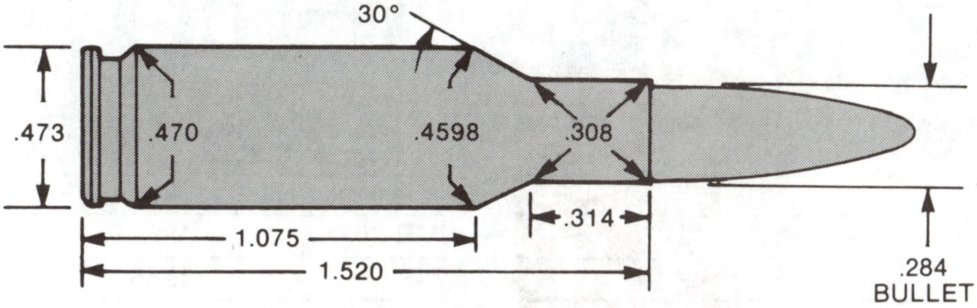

COMMENTS:

This is an extremely accurate cartridge primarily used in silhouette shooting. It has also been effectively used on deer sized game at moderate ranges. Since our last handbook this cartridge has become available in factory loaded rounds and as factory formed brass. Thus it is a fine choice for the reloading handgunner who does not wish to be burdened with case forming operations.

We strongly recommend the use of factory formed cases with this data. If reloader formed cases are used only the special Remington case with the small rifle primer will be safe with this data. Winchester 748 is a good powder choice.

Excellent accuracy can be attained with cast bullet #287641 which has a bore riding nose design.

TEST COMPONENTS:

Cases .... Remington Basic
Trim-to Length .... 1.510"
Primers .... Remington 7½
Primer Size .... Small Rifle, Magnum
Lyman Shell Holder .... No. 2
Jacketed Bullets Used .... Speer HP #1617, 115 gr.
Hornady SP #2810, 120 gr.
Speer SP #1623, 130 gr.
Nosler SP #16325, 140 gr.
Hornady BTSP #2845, 162 gr.
Cast Bullets Used .... (Sized to .284" dia.)
*Gas Check Bullets .... *#287405, 150 gr.
*#287641, 160 gr.

TEST SPECIFICATIONS:
(Velocity Only)

Firearm Used .... Remington XP-100
Barrel Length .... 14⅞"
Twist .... 1-10"
Groove Dia. .... .284"

115 gr. Jacketed HP
2.225" OAL

Powder	Sugg. Starting Grains	Velocity fps	Pressure C.U.P.	Max. Load Grains	Velocity fps	Pressure C.U.P.
IMR-4198	23.3	2122	—	26.5+	2396	—
RX7	23.7	2076	—	27.0	2360	—
IMR-3031	26.4	1949	—	30.0+	2345	—
748	29.0	2040	—	33.0+	2314	—
H-335	27.2	2075	—	31.0	2342	—
BL-C(2)	28.6	2060	—	32.5+	2304	—

Note: Loads shown in shaded panels are maximum.
+ Designates a compressed powder charge.

7mm BR

(7mm Remington Benchrest)

120 gr. Jacketed SP
2.240" OAL

Powder	Sugg. Starting Grains	Velocity fps	Pressure C.U.P.	Max. Load Grains	Velocity fps	Pressure C.U.P.
IMR-4198	22.0	2051	—	25.0+	2290	—
RX7	21.6	1927	—	24.5	2197	—
IMR-3031	25.0	1793	—	28.5+	2151	—
748	27.7	1952	—	31.5+	2228	—
H-335	26.8	2004	—	30.5	2277	—
BL-C(2)	27.3	1913	—	31.0	2187	—

130 gr. Jacketed SP
2.240" OAL

Powder	Sugg. Starting Grains	Velocity fps	Pressure C.U.P.	Max. Load Grains	Velocity fps	Pressure C.U.P.
IMR-4198	22.0	2049	—	25.0+	2279	—
IMR-3031	24.8	1857	—	28.2+	2182	—
748	27.2	1966	—	31.0	2194	—
H-335	24.3	1874	—	27.7	2156	—
BL-C(2)	26.8	1997	—	30.5+	2206	—

140 gr. Jacketed SP
2.275" OAL

Powder	Sugg. Starting Grains	Velocity fps	Pressure C.U.P.	Max. Load Grains	Velocity fps	Pressure C.U.P.
IMR-4198	20.2	1877	—	23.0+	2071	—
IMR-3031	23.5	1776	—	26.7+	2069	—
748	25.5	1889	—	29.0+	2089	—
H-335	23.5	1853	—	26.8	2069	—
BL-C(2)	26.4	1944	—	30.0+	2129	—
IMR-4895	24.2	1821	—	27.5+	2073	—

162 gr. Jacketed BTSP
2.325" OAL

Powder	Sugg. Starting Grains	Velocity fps	Pressure C.U.P.	Max. Load Grains	Velocity fps	Pressure C.U.P.
IMR-3031	22.0	1682	—	25.0+	1910	—
748	24.6	1802	—	28.0+	1931	—
H-335	22.7	1721	—	25.8+	1943	—
BL-C(2)	24.6	1724	—	28.0+	1979	—
IMR-4895	22.8	1715	—	26.0+	1970	—

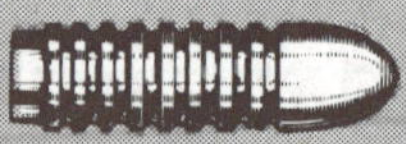

#287405
150 gr. (#2 Alloy) 2.055" OAL

Powder	Sugg. Starting Grains	Velocity fps	Pressure C.U.P.	Max. Load Grains	Velocity fps	Pressure C.U.P.
SR-4759	17.6	1848	—	20.0+	1988	—
IMR-4227	16.7	1765	—	19.0	1969	—
IMR-4198	18.4	1748	—	21.0+	1958	—
RX7	19.3	1749	—	22.0+	1964	—
748	24.6	1768	—	28.0	1955	—

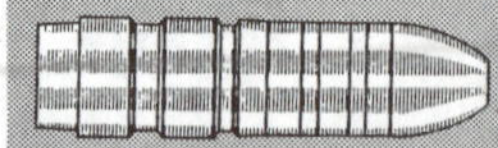

#287641
160 gr. (#2 Alloy) 2.100" OAL

Powder	Sugg. Starting Grains	Velocity fps	Pressure C.U.P.	Max. Load Grains	Velocity fps	Pressure C.U.P.
SR-4759	14.6	1499	—	17.0	1710	—
IMR-4227	15.5	1456	—	18.0	1652	—
IMR-4198	17.8	1508	—	20.5+	1732	—
RX7	18.5	1309	—	21.5+	1595	—
748	24.0	1393	—	27.5	1638	—

Note: Loads shown in shaded panels are maximum.
+ Designates a compressed powder charge.

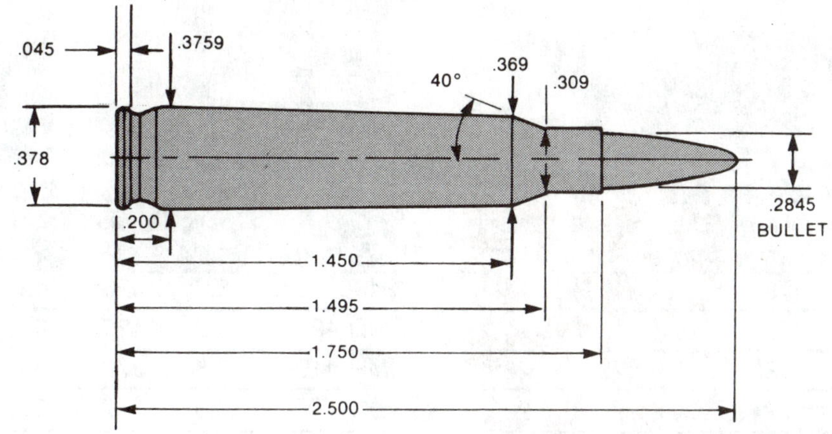

COMMENTS:

This cartridge is formed by running a 223 Remington case into a special Lyman 7mm TCU full length sizing, decapping, expanding die. The special die forms the proper case neck dimensions without additional tooling.

Case neck splits will occur at times. The use of new cases will minimize the splitting problem. IMR 4227 would be the recommended powder choice.

TEST COMPONENTS:

Cases .... Federal
Trim-to Length .... 1.740"
Primers .... Remington 6½
Primer Size .... Small Rifle
Lyman Shell Holder .... No. 26
Jacketed Bullets Used .... Speer HP #1617, 115 gr.
Hornady SP #2810, 120 gr.
Speer SP #1623, 130 gr.
Cast Bullets Used .... (Sized to .284" dia.)
*Gas Check Bullets *#287346, 135 gr.
*#287405, 150 gr.
*#287641, 160 gr.

TEST SPECIFICATIONS: (Velocity Only)

Firearm Used .... Thompson/Center Contender
Barrel Length .... 10" and 14"
Twist .... 1-10"
Groove Dia. .... .283"

115 gr. Jacketed HP
2.300" OAL

Powder	Sugg. Starting Grains	10" Bbl. Velocity fps	14" Bbl. Velocity fps	Max. Load Grains	10" Bbl. Velocity fps	14" Bbl. Velocity fps
2400	15.7	1685	1893	20.0	2106	2290
IMR-4227	16.5	1700	1863	21.0	2078	2233
IMR-4198	19.3	1599	1815	24.0+	2047	2242
RX7	21.3	1622	1844	26.5+	1987	2270

Note: Loads shown in shaded panels are maximum.
\+ Designates a compressed powder charge.

7mm TCU

120 gr. Jacketed Spire Pt.
2.467" OAL

Powder	Sugg. Starting Grains	10" Bbl. Velocity fps	14" Bbl. Velocity fps	Max. Load Grains	10" Bbl. Velocity fps	14" Bbl. Velocity fps
2400	15.2	1523	1759	19.5	1999	2221
IMR-4227	16.0	1444	1780	20.5	1957	2131
IMR-4198	18.8	1564	1788	23.5+	2012	2211
RX7	20.8	1719	1924	26.0+	2108	2223

130 gr. Jacketed SP
2.500" OAL

Powder	Sugg. Starting Grains	10" Bbl. Velocity fps	14" Bbl. Velocity fps	Max. Load Grains	10" Bbl. Velocity fps	14" Bbl. Velocity fps
2400	13.6	1426	1493	18.0	1802	1993
IMR-4227	15.2	1425	1633	20.0	1906	2089
IMR-4198	16.8	1372	1530	22.0	1864	2004
RX7	19.6	1627	1825	24.5	2036	2154

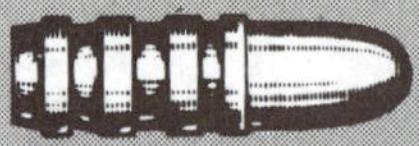

#287346
135 gr. (#2 Alloy) 2.300" OAL

Powder	Sugg. Starting Grains	10" Bbl. Velocity fps	14" Bbl. Velocity fps	Max. Load Grains	10" Bbl. Velocity fps	14" Bbl. Velocity fps
H-110	14.4	1470	1676	18.0	1767	1944
SR-4759	15.5	1533	1735	20.0+	1890	2071
IMR-4227	15.5	1461	1740	20.0	1880	2113
IMR-4198	17.5	1464	1752	23.0	1926	2098
RX7	18.0	1474	1516	23.0	1885	1927

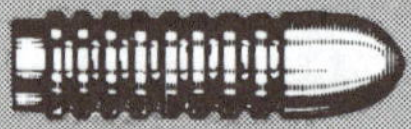

#287405
150 gr. (#2 Alloy) 2.300" OAL

Powder	Sugg. Starting Grains	10" Bbl. Velocity fps	14" Bbl. Velocity fps	Max. Load Grains	10" Bbl. Velocity fps	14" Bbl. Velocity fps
H-110	13.6	1392	1532	17.0	1688	1882
SR-4759	14.5	1428	1612	19.5+	1820	1996
IMR-4227	14.5	1376	1532	19.5	1797	1958
IMR-4198	16.5	1405	1561	22.0	1874	2004
RX7	17.2	1427	1480	23.0	1778	1889

#287641
160 gr. (#2 Alloy) 2.392" OAL

Powder	Sugg. Starting Grains	10" Bbl. Velocity fps	14" Bbl. Velocity fps	Max. Load Grains	10" Bbl. Velocity fps	14" Bbl. Velocity fps
SR-4759	14.0	1449	1565	19.0+	1826	1926
IMR-4227	14.0	1436	1485	19.0	1779	1902
IMR-4198	16.0	1361	1494	21.5+	1812	1953
RX7	16.7	1340	1462	22.5+	1758	1907

Note: Loads shown in shaded panels are maximum.
+ Designates a compressed powder charge.

7-30 Waters

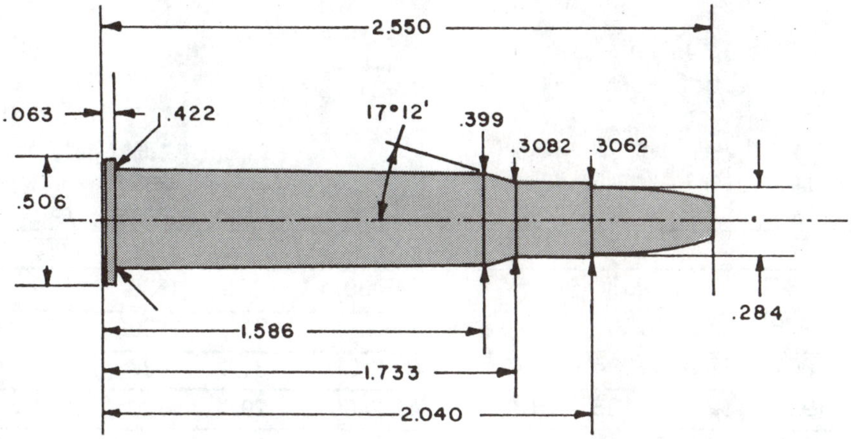

COMMENTS:

This is a modified version of the 30-30 Win. that has gained some popularity, perhaps because of the recent trend toward 7mm cartridges and bullets.

The cartridge is suitable for light big game at modest ranges.

The Thompson/Center Contender, being a single shot design, allows the use of Spitzer bullets. Good powder choices are Winchester 748 and IMR 3031.

TEST COMPONENTS:

Cases .... Federal
Trim-to Length .... 2.030"
Primers .... CCI 200
Primer Size .... Large Rifle
Lyman Shell Holder .... No. 6
Jacketed Bullets Used .... Sierra SP #1900,120 gr.
Sierra SP #7250, 130 gr.
Sierra SBT #1905, 140 gr.
Sierra SBT #1913, 150 gr.
Sierra HPBT #1925, 160 gr.
Cast Bullets Used .... (Sized to .285" dia.)
*Gas Check Bullets .... *287346, 135 gr.

TEST SPECIFICATIONS:
(Velocity Only)

Firearm Used .... Thompson/Center Contender
Barrel Length .... 10" and 14"
Twist .... 1-9"
Groove Dia. .... .284"

120 gr. Jacketed SP
2.550" OAL

Powder	Sugg. Starting Grains	10" Bbl. Velocity fps	14" Bbl. Velocity fps	Max. Load Grains	10" Bbl. Velocity fps	14" Bbl. Velocity fps
RX7	27.5	1873	2043	31.0	2127	2339
IMR-3031	30.5	1811	2063	34.5+	2096	2350
748	34.0	1788	2073	39.5+	2159	2430
BL-C(2)	32.3	1786	1955	36.8	2072	2316
AA 2460	29.5	1683	1974	33.8	2032	2267
H-335	31.0	1838	2096	34.8	2167	2398
IMR-4895	31.8	1834	2068	35.8+	2083	2312

Note: Loads shown in shaded panels are maximum.
\+ Designates a compressed powder charge.

7-30 Waters

130 gr. Jacketed SP
2.550" OAL

Powder	Sugg. Starting Grains	10" Bbl. Velocity fps	14" Bbl. Velocity fps	Max. Load Grains	10" Bbl. Velocity fps	14" Bbl. Velocity fps
RX7	25.7	1712	1872	29.7	1993	2189
IMR-3031	29.5	1756	2081	33.6+	2036	2254
748	31.0	1579	1778	35.0+	1901	2171
BL-C(2)	31.7	1695	1897	35.9	1984	2212
AA 2460	28.3	1849	1918	32.3	2108	2124
H-335	29.5	1734	1912	33.5	2012	2249
IMR-4895	30.0	1694	1950	34.6+	1962	2201

140 gr. Jacketed SBT
2.550" OAL

Powder	Sugg. Starting Grains	10" Bbl. Velocity fps	14" Bbl. Velocity fps	Max. Load Grains	10" Bbl. Velocity fps	14" Bbl. Velocity fps
RX7	25.0	1714	1795	28.8	1966	2125
IMR-3031	27.5	1636	1856	32.6+	1966	2175
748	31.5	1689	1920	34.5	1905	2134
BL-C(2)	31.4	1687	1920	35.0	1920	2141
AA 2460	29.0	1720	1914	32.6	1950	2152
H-335	28.7	1614	1818	33.0	1989	2190
IMR-4895	28.8	1658	1839	33.6+	1947	2173

150 gr. Jacketed SBT
2.550" OAL

Powder	Sugg. Starting Grains	10" Bbl. Velocity fps	14" Bbl. Velocity fps	Max. Load Grains	10" Bbl. Velocity fps	14" Bbl. Velocity fps
RX7	23.5	1681	1790	27.4	1875	2013
IMR-3031	26.2	1746	1850	30.7+	1899	2060
748	31.6	1731	1927	34.8+	1940	2180
BL-C(2)	30.3	1656	1858	33.7	1863	2070
AA 2460	26.7	1657	1806	31.6	1845	2024
H-335	27.5	1693	1865	32.0	1913	2129
IMR-4895	28.2	1760	1831	32.5+	1930	2112

160 gr. Jacketed HPBT
2.550" OAL

Powder	Sugg. Starting Grains	10" Bbl. Velocity fps	14" Bbl. Velocity fps	Max. Load Grains	10" Bbl. Velocity fps	14" Bbl. Velocity fps
RX7	22.4	1585	1653	26.5	1825	1938
IMR-3031	23.6	1545	1688	27.8+	1772	1963
748	31.5	1711	1875	34.8+	1898	2111
BL-C(2)	29.7	1649	1789	33.3	1829	2021
AA 2460	26.7	1646	1786	30.8	1813	1960
H-335	27.0	1582	1768	31.4	1867	2029
IMR-4895	25.8	1599	1787	29.7+	1856	1968

#287346
135 gr. (#2 Alloy) 2.475" OAL

Powder	Sugg. Starting Grains	10" Bbl. Velocity fps	14" Bbl. Velocity fps	Max. Load Grains	10" Bbl. Velocity fps	14" Bbl. Velocity fps
SR-4759	16.5	1443	1592	20.0	1800	1916
RX7	23.0	1687	1901	26.9	1915	2069
IMR-3031	25.5	1565	1793	31.0+	1902	2139
748	29.5	1556	1827	33.5+	1891	2123
AA 2460	26.0	1574	1847	32.0	1915	2122
H-335	26.2	1665	1811	32.0	1912	2161

Note: Loads shown in shaded panels are maximum.
\+ Designates a compressed powder charge.

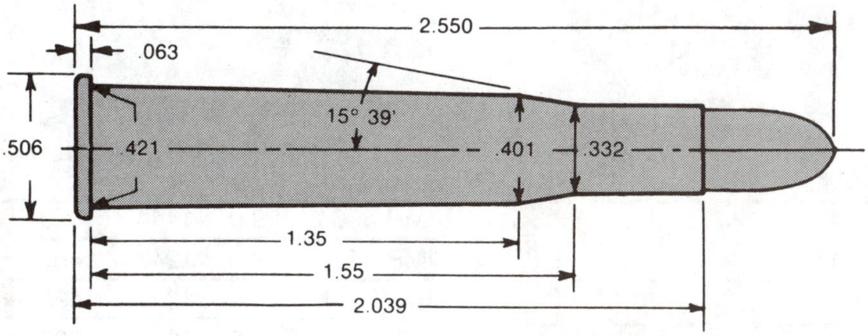

COMMENTS:

When used in the Contender this is an extremely accurate cartridge.

The 110 grain to 130 grain bullets are for varmints and plinking.

This cartridge works very well with a number of powders including IMR-3031, Winchester 748 and Reloder 7.

It is also very accurate with cast bullets, particularly #311291.

TEST COMPONENTS:

Cases .... Winchester
Trim-to Length .... 2.029"
Primers .... Winchester 8½-120
Primer Size .... Large Rifle
Lyman Shell Holder .... No. 6
Jacketed Bullets Used .... Hornady SP #3010, 110 gr.
Sierra SP #2120, 125 gr.
Hornady SP #3020, 130 gr.
Hornady SP #3031, 150 gr.
Cast Bullets Used .... (Sized to .309" dia.)
*Gas Check Bullets .... *#311359, 115 gr.
*#311466, 152 gr.
*#311291, 169 gr.

TEST SPECIFICATIONS:
(Velocity Only)

Firearm Used .... Thompson/Center Contender
Barrel Length .... 10" and 14"
Twist .... 1-14"
Groove Dia. .... .3087"

110 gr. Jacketed Spire Pt.
2.530" OAL

Powder	Sugg. Starting Grains	10" Bbl. Velocity fps	14" Bbl. Velocity fps	Max. Load Grains	10" Bbl. Velocity fps	14" Bbl. Velocity fps
IMR-4198	25.0	1773	2038	29.0	2128	2346
IMR-3031	28.5	1572	1854	33.5	1923	2166
748	35.5	1614	1892	41.0+	1905	2222

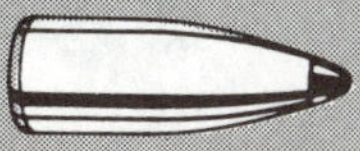

125 gr. Jacketed SP
2.600" OAL

Powder	Sugg. Starting Grains	10" Bbl. Velocity fps	14" Bbl. Velocity fps	Max. Load Grains	10" Bbl. Velocity fps	14" Bbl. Velocity fps
IMR-4198	24.0	1593	1909	26.0	1969	2159
IMR-3031	28.0	1591	1893	32.0	1946	2239
748	35.5	1653	1907	41.0+	1897	2210

Note: Loads shown in shaded panels are maximum.
\+ Designates a compressed powder charge.

30-30 Winchester

130 gr. Jacketed Spire Pt.
2.620" OAL

Powder	Sugg. Starting Grains	10" Bbl. Velocity fps	14" Bbl. Velocity fps	Max. Load Grains	10" Bbl. Velocity fps	14" Bbl. Velocity fps
IMR-4198	23.0	1552	1768	25.5	1894	2095
IMR-3031	27.5	1533	1779	31.5	1936	2183
748	35.0	1549	1851	40.5+	1909	2166

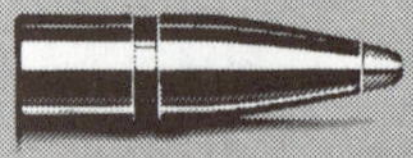

150 gr. Jacketed Spire Pt.
2.745" OAL

Powder	Sugg. Starting Grains	10" Bbl. Velocity fps	14" Bbl. Velocity fps	Max. Load Grains	10" Bbl. Velocity fps	14" Bbl. Velocity fps
IMR-4064	29.0	1444	1699	31.5	1719	1950
IMR-4198	19.5	1243	1457	23.5	1578	1749
RX7	21.0	1341	1595	28.0	1680	1934
IMR-3031	25.5	1361	1602	28.5	1627	1847
748	33.0	1534	1779	38.5+	1832	2077

#311359
115 gr. (#2 Alloy) 2.440" OAL

Powder	Sugg. Starting Grains	10" Bbl. Velocity fps	14" Bbl. Velocity fps	Max. Load Grains	10" Bbl. Velocity fps	14" Bbl. Velocity fps
Unique	7.5	1175	1303	11.4	1598	1740
SR-4759	18.0	1575	1754	20.0	1844	2027
RX7	20.0	1385	1656	30.5	1700	2039
IMR-3031	26.0	1499	1805	32.0	1850	2116
748	24.5	1209	1531	36.5	1697	1981

#311466
152 gr. (#2 Alloy) 2.540" OAL

Powder	Sugg. Starting Grains	10" Bbl. Velocity fps	14" Bbl. Velocity fps	Max. Load Grains	10" Bbl. Velocity fps	14" Bbl. Velocity fps
Unique	7.5	1194	1193	11.0	1397	1525
SR-4759	14.4	1144	1313	18.0	1478	1650
RX7	22.5	1468	1785	30.0	1806	2068
IMR-3031	25.0	1362	1657	30.0	1845	1963
748	27.0	1287	1563	36.5	1709	1986

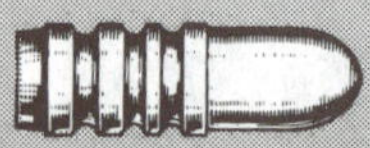

#311291
169 gr. (#2 Alloy) 2.540" OAL

Powder	Sugg. Starting Grains	10" Bbl. Velocity fps	14" Bbl. Velocity fps	Max. Load Grains	10" Bbl. Velocity fps	14" Bbl. Velocity fps
Unique	7.0	995	1065	10.6	1287	1416
SR-4759	14.8	1194	1395	18.5	1498	1649
RX7	19.0	1269	1761	28.0	1509	1936
IMR-3031	22.5	1259	1439	28.5	1668	1927
748	23.8	1185	1429	37.3+	1781	2025

Note: Loads shown in shaded panels are maximum.
+ Designates a compressed powder charge.

32-20 Winchester

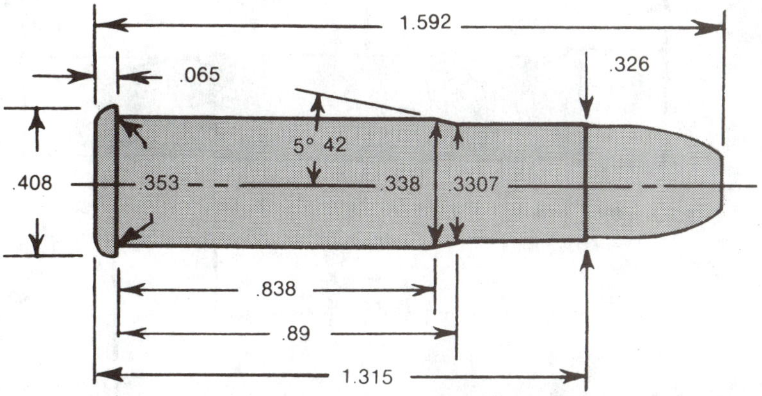

COMMENTS:

Hercules Unique will be a fine first powder selection.

The 32-20 in the T/C Contender is a good small game cartridge. It is also popular as a silhouette cartridge.

TEST COMPONENTS:

Cases .... Winchester
Trim-to Length .... 1.305"
Primers .... Remington 1½
Primer Size .... Small Pistol
Lyman Shell Holder .... No. 10
Jacketed Bullets Used .... Hornady SJ #3005, 100 gr.
Cast Bullets Used .... (Sized to .309" dia.)
*Gas Check Bullets .... *#311316, 112 gr.
*#311359, 115 gr.

TEST SPECIFICATIONS: (Velocity Only)

Firearm Used .... Thompson/Center Contender
Barrel Length .... 10"
Twist .... 1-10"
Groove Dia. .... .308"

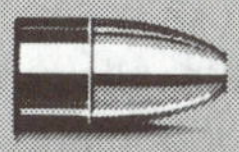

100 gr. Short Jacket
1.580" OAL

Powder	Sugg. Starting Grains	10" Bbl. Velocity fps	14" Bbl. Velocity fps	Max. Load Grains	10" Bbl. Velocity fps	14" Bbl. Velocity fps
SR-4756	4.6	899	—	5.8	1183	—
IMR-4227	7.7	829	—	9.7	1160	—
2400	7.5	933	—	9.5	1274	—
H-110	7.5	832	—	9.5	1208	—

#311316
112 gr. (#2 Alloy) 1.575" OAL

Powder	Sugg. Starting Grains	10" Bbl. Velocity fps	14" Bbl. Velocity fps	Max. Load Grains	10" Bbl. Velocity fps	14" Bbl. Velocity fps
SR-4756	4.2	934	—	5.5	1199	—
IMR-4227	7.0	854	—	9.0	1124	—
Unique	3.6	876	—	4.6	1074	—
2400	7.0	972	—	9.0	1276	—

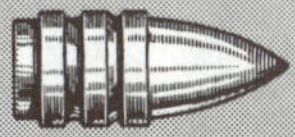

#311359
115 gr. (#2 Alloy) 1.675" OAL

Powder	Sugg. Starting Grains	10" Bbl. Velocity fps	14" Bbl. Velocity fps	Max. Load Grains	10" Bbl. Velocity fps	14" Bbl. Velocity fps
SR-4756	4.2	929	—	5.5	1181	—
IMR-4227	7.0	804	—	9.0	1098	—
Unique	3.6	868	—	4.6	1056	—
2400	7.0	960	—	9.0	1269	—

Note: Loads shown in shaded panels are maximum.

32 H & R Magnum

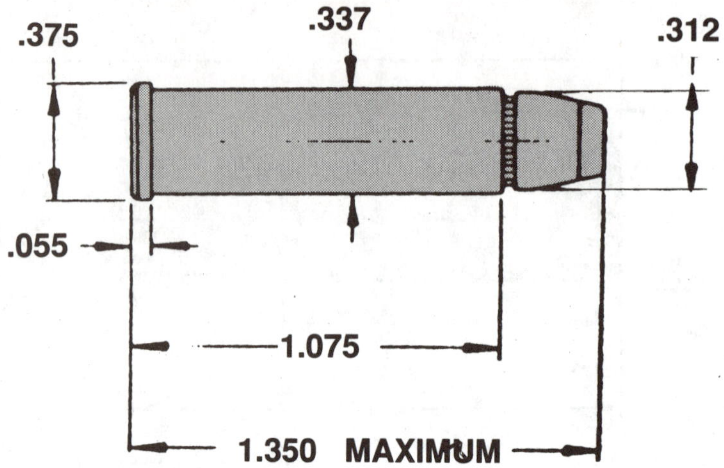

COMMENTS:

Contender barrels chambered for this cartridge use a smaller than normal groove diameter of 0.308 inch. This feature allows the reloader to select from a wide range of available 0.308" diameter bullets. The 110 grain round nose M1 Carbine bullets are a good choice for the relatively low velocities of this cartridge.

Not all sizing dies will reduce the diameter of this case sufficiently for 0.308" bullets. Check with the die manufacturer.

TEST COMPONENTS:

Cases .... Federal
Trim-to Length .... 1.070"
Primers .... Federal 100
Primer Size .... Small Pistol
Lyman Shell Holder .... No. 9
Jacketed Bullets Used .... Hornady SJ #3005, 100 gr.
(.308" dia.) Hornady RN #3015, 110 gr.
Sierra JSP #2120, 125 gr.
Cast Bullets Used .... (Sized to .310" dia.)
*Gas Check Bullets #313249, 85 gr.
*#313631, 100 gr.

TEST SPECIFICATIONS:
(Velocity Only)

Firearm Used .... Thompson/Center Contender
Barrel Length .... 10"
Twist .... 1-10"
Groove Dia. .... .308"

100 gr. Short Jacket
1.450" OAL

Powder	Sugg. Starting Grains	10" Bbl. Velocity fps	14" Bbl. Velocity fps	Max. Load Grains	10" Bbl. Velocity fps	14" Bbl. Velocity fps
Bullseye	2.5	838	—	3.7	1139	—
Unique	3.0	846	—	4.5	1219	—
Blue Dot	5.0	1015	—	6.5	1339	—
SR-4756	3.2	854	—	4.6	1231	—
800X	3.5	975	—	4.8	1249	—
231	2.5	792	—	3.8	1116	—
HP-38	2.0	694	—	3.4	1082	—
HS-6	4.7	1064	—	6.0	1320	—
AA 7	6.0	1152	—	7.2	1349	—

Note: Loads shown in shaded panels are maximum.

32 H & R Magnum

110 gr. Jacketed RN
1.450" OAL

Powder	Sugg. Starting Grains	10" Bbl. Velocity fps	14" Bbl. Velocity fps	Max. Load Grains	10" Bbl. Velocity fps	14" Bbl. Velocity fps
Bullseye	2.5	761	—	3.5	1000	—
Unique	3.0	799	—	4.4	1100	—
Blue Dot	5.0	965	—	6.1	1227	—
SR-4756	3.2	783	—	4.5	1130	—
800X	3.5	834	—	4.7	1171	—
231	2.5	693	—	3.7	1011	—
HP-38	2.0	563	—	3.2	941	—
HS-6	4.0	862	—	5.3	1114	—
AA 7	5.5	1010	—	6.7	1208	—

125 gr. Jacketed SP
1.700" OAL

Powder	Sugg. Starting Grains	10" Bbl. Velocity fps	14" Bbl. Velocity fps	Max. Load Grains	10" Bbl. Velocity fps	14" Bbl. Velocity fps
Bullseye	2.4	541	—	3.5	889	—
Unique	3.2	686	—	4.2	983	—
Blue Dot	5.0	860	—	6.0	1129	—
SR-4756	3.5	737	—	4.5	1007	—
800X	3.6	816	—	4.7	1071	—
231	2.5	445	—	3.7	903	—
HS-6	4.4	798	—	5.2	1066	—
AA 7	5.4	884	—	6.4	1071	—

#313249
85 gr. (#2 Alloy) 1.350" OAL

Powder	Sugg. Starting Grains	10" Bbl. Velocity fps	14" Bbl. Velocity fps	Max. Load Grains	10" Bbl. Velocity fps	14" Bbl. Velocity fps
Bullseye	2.5	1012	—	4.0	1363	—
231	2.5	959	—	4.1	1341	—
HP-38	2.0	895	—	3.5	1274	—
Red Dot	2.5	999	—	3.8	1330	—
Unique	3.0	1020	—	4.5	1369	—
SR-7625	2.7	939	—	4.3	1371	—
Herco	3.5	1071	—	4.9	1411	—
SR-4756	3.0	916	—	4.6	1393	—
800X	3.8	1170	—	5.2	1485	—
Blue Dot	5.0	1135	—	6.5	1503	—
AA 5	4.0	1111	—	5.4	1455	—

#313631
100 gr. (#2 Alloy) 1.325" OAL

Powder	Sugg. Starting Grains	10" Bbl. Velocity fps	14" Bbl. Velocity fps	Max. Load Grains	10" Bbl. Velocity fps	14" Bbl. Velocity fps
Bullseye	2.0	811	—	3.5	1190	—
231	2.2	824	—	3.7	1194	—
HP-38	1.8	754	—	3.3	1151	—
Red Dot	2.0	823	—	3.5	1204	—
Unique	2.7	852	—	4.2	1257	—
SR-7625	2.3	788	—	3.7	1176	—
Herco	2.8	844	—	4.2	1221	—
SR-4756	2.5	760	—	4.0	1210	—
800X	3.0	943	—	4.3	1254	—
Blue Dot	4.0	920	—	5.4	1247	—
AA 5	3.2	903	—	4.6	1269	—

Note: Loads shown in shaded panels are maximum.

357 Magnum

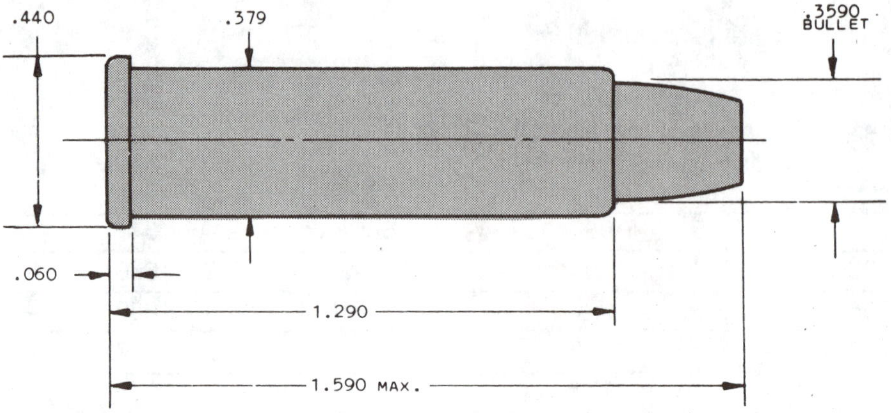

COMMENTS:

Lacking the cylinder gap of a revolver, the Contender pistol will deliver higher velocity from identical barrel length.

Do not use 357 Magnum Loads in 38 Special cases as very dangerous pressure will result.

Bullet #358156 is extremely popular for heavy loads. Bullet #358429 closely duplicates the factory 158 grain semi-wadcutter. This is the Elmer Keith design bullet and makes an excellent choice for hunting.

Heavy loads work well with a wide range of powders but you might want to try 2400 for your first accuracy tests. Bullets may be seated in tune with the individual Contenders rifling since there is no other constraint with regard to maximum overall cartridge length.

TEST COMPONENTS:

Cases .... Federal
Trim-to Length .... 1.280"
Primers .... CCI 550
Primer Size .... Small Pistol Magnum
Lyman Shell Holder .... No. 1
Jacketed Bullets Used .... Hornady HP #3570,110 gr.
Hornady HP #3571, 125 gr.
Sierra HC #8330, 150 gr.
Sierra SP #8340, 158 gr.
Hornady JTC SIL #3577, 180 gr.
Cast Bullets Used .... (Sized to .357" dia.)
*Gas Check Bullets #356242, 121 gr.
*#358156, 155 gr.
#358429, 168 gr.
*#358315, 204 gr.
*#358627, 215 gr.

TEST SPECIFICATIONS:
(Velocity Only)

Firearm Used .... Thompson/Center Contender
Barrel Length .... 10"
Twist .... 1-18"
Groove Dia. .... .356"

110 gr. Jacketed HP
1.590" OAL

Powder	Sugg. Starting Grains	10" Bbl. Velocity fps	14" Bbl. Velocity fps	Max. Load Grains	10" Bbl. Velocity fps	14" Bbl. Velocity fps
Unique	7.4	1334	—	10.0	1737	—
Blue Dot	10.0	1342	—	14.6	1968	—
2400	13.5	1427	—	20.3+	1952	—
IMR-4227	14.8	1446	—	19.5+	1733	—
AA 5	11.0	1832	—	13.0	2079	—

Note: Loads shown in shaded panels are maximum.
+ Designates a compressed powder charge.

125 gr. Jacketed HP
1.590" OAL

Powder	Sugg. Starting Grains	10" Bbl. Velocity fps	14" Bbl. Velocity fps	Max. Load Grains	10" Bbl. Velocity fps	14" Bbl. Velocity fps
Unique	7.0	1261	—	9.7	1643	—
Blue Dot	10.4	1355	—	13.3	1807	—
2400	13.0	1302	—	17.7	1778	—
IMR-4227	15.6	1435	—	19.5+	1749	—
AA 5	10.0	1648	—	12.0	1847	—

158 gr. Jacketed SP
1.590" OAL

Powder	Sugg. Starting Grains	10" Bbl. Velocity fps	14" Bbl. Velocity fps	Max. Load Grains	10" Bbl. Velocity fps	14" Bbl. Velocity fps
Unique	6.4	1102	—	8.3	1378	—
Blue Dot	9.6	1192	—	11.9	1580	—
2400	11.3	1093	—	14.9	1491	—
IMR-4227	12.2	1130	—	16.1	1430	—
AA 9	13.0	1471	—	16.0	1719	—

#358242
121 gr. (#2 Alloy) 1.640" OAL

Powder	Sugg. Starting Grains	10" Bbl. Velocity fps	14" Bbl. Velocity fps	Max. Load Grains	10" Bbl. Velocity fps	14" Bbl. Velocity fps
Unique	6.9	1241	—	9.1	1580	—
HS-7	10.0	1240	—	13.2	1590	—
2400	13.0	1381	—	16.4	1669	—
AA 5	8.6	1481	—	10.8	1771	—

150 gr. Jacketed HP
1.590" OAL

Powder	Sugg. Starting Grains	10" Bbl. Velocity fps	14" Bbl. Velocity fps	Max. Load Grains	10" Bbl. Velocity fps	14" Bbl. Velocity fps
Unique	6.5	1089	—	8.7	1451	—
2400	12.0	1199	—	15.6	1545	—
IMR-4227	12.5	1158	—	16.8	1500	—
AA 9	13.0	1477	—	16.0	1711	—

180 gr. JTC SIL
1.585" OAL

Powder	Sugg. Starting Grains	10" Bbl. Velocity fps	14" Bbl. Velocity fps	Max. Load Grains	10" Bbl. Velocity fps	14" Bbl. Velocity fps
HS-7	9.0	1139	—	11.2	1416	—
H-110	12.0	1231	—	15.0	1452	—
2400	10.0	1129	—	12.6	1412	—
IMR-4227	11.7	1093	—	14.6	1383	—
AA 9	11.2	1225	—	14.0	1491	—

#358156
155 gr. (#2 Alloy) 1.578" OAL

Powder	Sugg. Starting Grains	10" Bbl. Velocity fps	14" Bbl. Velocity fps	Max. Load Grains	10" Bbl. Velocity fps	14" Bbl. Velocity fps
Blue Dot	8.5	1187	—	10.5	1443	—
H-110	13.0	1430	—	15.7	1508	—
2400	10.6	1138	—	14.0	1466	—
AA 7	10.0	1333	—	12.7	1643	—
AA 9	12.2	1437	—	15.2	1672	—

Note: Loads shown in shaded panels are maximum.
\+ Designates a compressed powder charge.

357 Magnum

#358429
168 gr. (#2 Alloy) 1.647" OAL

Powder	Sugg. Starting Grains	10" Bbl. Velocity fps	14" Bbl. Velocity fps	Max. Load Grains	10" Bbl. Velocity fps	14" Bbl. Velocity fps
Blue Dot	8.3	1141	—	10.0	1368	—
H-110	13.0	1430	—	15.7	1503	—
2400	9.7	998	—	13.5	1422	—
AA 9	10.8	1301	—	13.5	1540	—

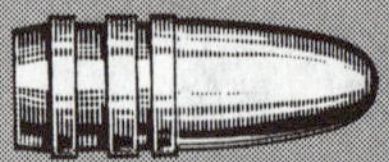

#358315
204 gr. (#2 Alloy) 1.880" OAL

Powder	Sugg. Starting Grains	10" Bbl. Velocity fps	14" Bbl. Velocity fps	Max. Load Grains	10" Bbl. Velocity fps	14" Bbl. Velocity fps
Blue Dot	9.0	1225	—	11.0	1410	—
H-110	13.0	1359	—	16.5	1515	—
2400	12.0	1318	—	14.5	1490	—

#358627
215 gr. (#2 Alloy) 1.680" OAL

Powder	Sugg. Starting Grains	10" Bbl. Velocity fps	14" Bbl. Velocity fps	Max. Load Grains	10" Bbl. Velocity fps	14" Bbl. Velocity fps
Blue Dot	7.0	1092	—	8.7	1263	—
H-110	10.3	1118	—	12.8	1322	—
2400	9.0	1028	—	11.2	1257	—

Note: Loads shown in shaded panels are maximum.

357 Remington Maximum

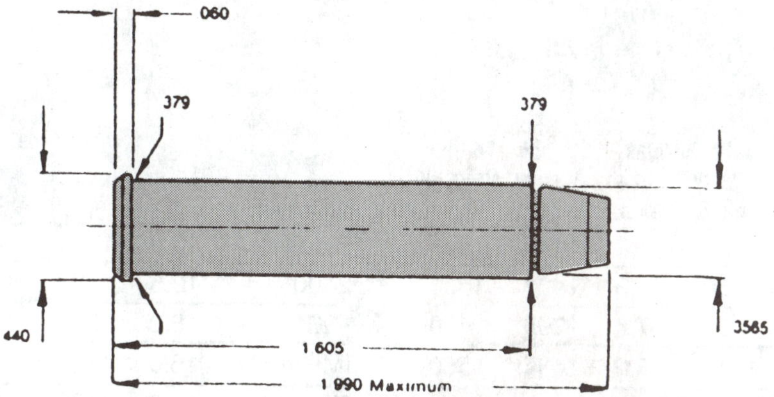

COMMENTS:

The 357 Remington Maximum was designed for silhouette shooting. Good results have been obtained in the T/C Contender.

Do not use 357 Remington Maximum loads in 357 Magnum cases as very dangerous pressure will result.

This cartridge will get all that can be had out of this basic case style with 0.357" handgun bullets.

TEST COMPONENTS:

Cases .... Remington
Trim-to Length .... 1.600"
Primers .... Remington 6½; CCI 450
Primer Size .... Small Rifle; Small Rifle Magnum
Lyman Shell Holder .... No. 1
Jacketed Bullets Used .... Hornady JHP #3575, 158 gr.
Sierra SIL FMJ #8350, 170 gr.
Hornady JTC SIL #3577, 180 gr.
Cast Bullets Used .... (Sized to .357" dia.)
*Gas Check Bullets
*#358156, 155 gr.
#358429, 168 gr.
*#358315, 204 gr.
*#358627, 215 gr.

TEST SPECIFICATIONS:
(Velocity Only)

Firearm Used .... Thompson/Center Contender
Barrel Length .... 10" and 14"
Twist .... 1-14"
Groove Dia. .... .357"

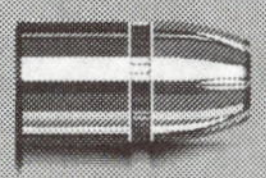

158 gr. Jacketed HP
1.910" OAL

Powder	Sugg. Starting Grains	10" Bbl. Velocity fps	14" Bbl. Velocity fps	Max. Load Grains	10" Bbl. Velocity fps	14" Bbl. Velocity fps
2400	14.5	1516	1554	17.2	1669	1754
RX7	20.0	1389	1477	24.5	1675	1765
IMR-4227	17.0	1493	1569	20.0	1630	1717
H-110	17.0	1519	1635	20.5	1715	1769

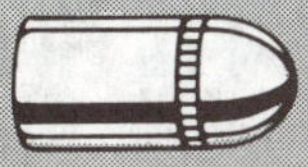

170 gr. Jacketed SIL FMJ
1.905" OAL

Powder	Sugg. Starting Grains	10" Bbl. Velocity fps	14" Bbl. Velocity fps	Max. Load Grains	10" Bbl. Velocity fps	14" Bbl. Velocity fps
2400	13.0	1306	1380	16.0	1603	1669
RX7	19.5	1310	1404	23.0	1657	1674
IMR-4227	16.0	1384	1375	19.0	1589	1651
H-110	16.0	1503	1542	19.0	1675	1736

Note: Loads shown in shaded panels are maximum.

357 Remington Maximum

180 gr. Jacketed SIL JTC
1.890" OAL

Powder	Sugg. Starting Grains	10" Bbl. Velocity fps	14" Bbl. Velocity fps	Max. Load Grains	10" Bbl. Velocity fps	14" Bbl. Velocity fps
2400	12.0	1130	1243	15.2	1449	1501
RX7	17.0	1106	1188	21.3	1424	1501
IMR-4227	14.0	1160	1288	17.0	1395	1470
H-110	15.0	1377	1411	18.0	1495	1560

#358156
155 gr. (#2 Alloy) 1.895" OAL

Powder	Sugg. Starting Grains	10" Bbl. Velocity fps	14" Bbl. Velocity fps	Max. Load Grains	10" Bbl. Velocity fps	14" Bbl. Velocity fps
Blue Dot	10.0	1261	1400	13.0	1598	1733
2400	12.5	1383	1520	16.0	1689	1794
RX7	18.0	1246	1402	22.5	1592	1715
IMR-4227	15.0	1385	1554	18.0	1544	1682
**H-110	16.0	1489	1603	19.0	1598	1714

#358429
168 gr. (#2 Alloy) 1.965" OAL

Powder	Sugg. Starting Grains	10" Bbl. Velocity fps	14" Bbl. Velocity fps	Max. Load Grains	10" Bbl. Velocity fps	14" Bbl. Velocity fps
Blue Dot	11.0	1390	1527	13.7	1642	1742
2400	13.0	1373	1526	16.0	1656	1749
RX7	18.0	1331	1457	22.5	1573	1715
IMR-4227	14.5	1353	1498	18.0	1577	1709
**H-110	16.0	1498	1631	18.7	1575	1690

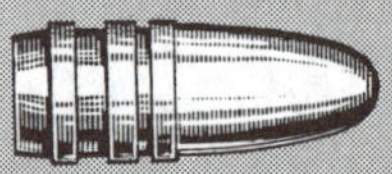

#358315
204 gr. (#2 Alloy) 2.155" OAL

Powder	Sugg. Starting Grains	10" Bbl. Velocity fps	14" Bbl. Velocity fps	Max. Load Grains	10" Bbl. Velocity fps	14" Bbl. Velocity fps
**2400	12.0	1238	1337	15.5	1501	1568
**RX7	18.0	1267	1404	22.5	1524	1637
**IMR-4227	14.5	1267	1378	18.0	1485	1587

#358627
215 gr. (#2 Alloy) 1.970" OAL

Powder	Sugg. Starting Grains	10" Bbl. Velocity fps	14" Bbl. Velocity fps	Max. Load Grains	10" Bbl. Velocity fps	14" Bbl. Velocity fps
**RX7	15.0	1073	1219	18.0	1304	1410
**IMR-4227	11.5	1089	1195	14.7	1292	1409

Note: Loads shown in shaded panels are maximum.
** Designates use of CCI 450 primers.

35 Remington

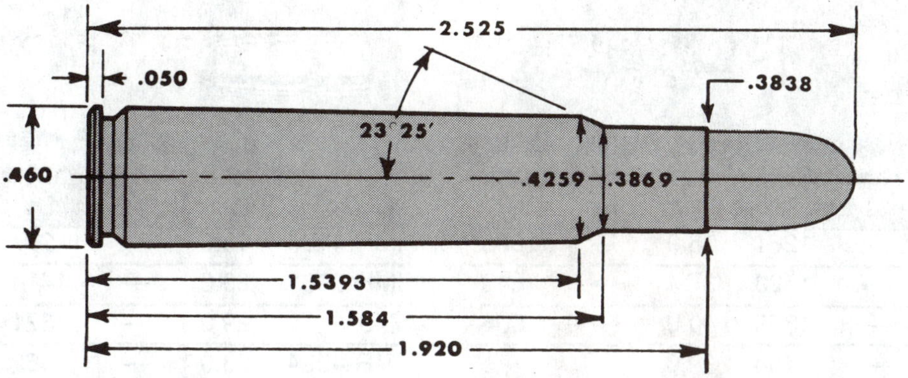

COMMENTS:

This cartridge is very popular in the Contender. It is, however, difficult to get reliable bullet expansion from most of the available 35 caliber rifle bullets at the velocities possible with this case. Powders faster than the common rifle powders such as IMR 4198 or IMR 3031 are required in this cartridge due to the shorter barrel length.

TEST COMPONENTS:

Cases .... Federal
Trim-to Length .... 1.910"
Primers .... Remington 9½
Primer Size .... Large Rifle
Lyman Shell Holder .... No. 2 or 8
Jacketed Bullets Used .... Sierra SP #8340, 158 gr.
Sierra FMJ #8350, 170 gr.
Sierra RN #2800, 200 gr.
Cast Bullets Used .... (Sized to .357" dia.)
*Gas Check Bullets .... #358430, 195 gr.
*#358315, 204 gr.

TEST SPECIFICATIONS: (Velocity Only)

Firearm Used .... Thompson/Center Contender
Barrel Length .... 14"
Twist .... 1-14"
Groove Dia. .... .357"

158 gr. Jacketed SP
2.185" OAL

Powder	Sugg. Starting Grains	10" BBL. Velocity fps	14" BBL. Velocity fps	Max. Load Grains	10" BBL. Velocity fps	14" BBL. Velocity fps
IMR-4198	23.7	—	1470	27.0	—	1757
IMR-3031	33.7	—	1649	38.4+	—	1956
748	34.0	—	1402	41.0	—	1673
IMR-4064	35.0	—	1580	40.0+	—	1881

Note: Loads shown in shaded panels are maximum.
+ Designates a compressed powder charge.

35 Remington

170 gr. Jacketed FMJ
2.205" OAL

Powder	Sugg. Starting Grains	10" BBL. Velocity fps	14" BBL. Velocity fps	Max. Load Grains	10" BBL. Velocity fps	14" BBL. Velocity fps
IMR-4198	20.5	—	1261	26.0	—	1603
IMR-3031	32.0	—	1523	37.4	—	1848
748	31.5	—	1359	40.0	—	1636
IMR-4064	33.5	—	1484	39.0	—	1775

200 gr. Jacketed RN
2.465" OAL

Powder	Sugg. Starting Grains	10" BBL. Velocity fps	14" BBL. Velocity fps	Max. Load Grains	10" BBL. Velocity fps	14" BBL. Velocity fps
IMR-4198	19.0	—	1217	24.5	—	1503
IMR-3031	29.0	—	1414	35.9	—	1767
748	29.0	—	1371	39.0	—	1671
IMR-4064	33.0	—	1482	39.0	—	1757

#358430
195 gr. (#2 Alloy) 2.215" OAL

Powder	Sugg. Starting Grains	10" BBL. Velocity fps	14" BBL. Velocity fps	Max. Load Grains	10" BBL. Velocity fps	14" BBL. Velocity fps
Unique	8.0	—	1143	12.0	—	1465
SR-4759	16.0	—	1317	19.0	—	1557
RX7	21.5	—	1312	23.0	—	1474

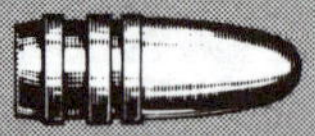

#358315
204 gr. (#2 Alloy) 2.500" OAL

Powder	Sugg. Starting Grains	10" BBL. Velocity fps	14" BBL. Velocity fps	Max. Load Grains	10" BBL. Velocity fps	14" BBL. Velocity fps
Unique	9.5	—	1203	12.0	—	1398
SR-4759	16.0	—	1245	18.5	—	1472
RX7	20.0	—	1262	22.8	—	1478

Note: Loads shown in shaded panels are maximum.

41 Remington Magnum

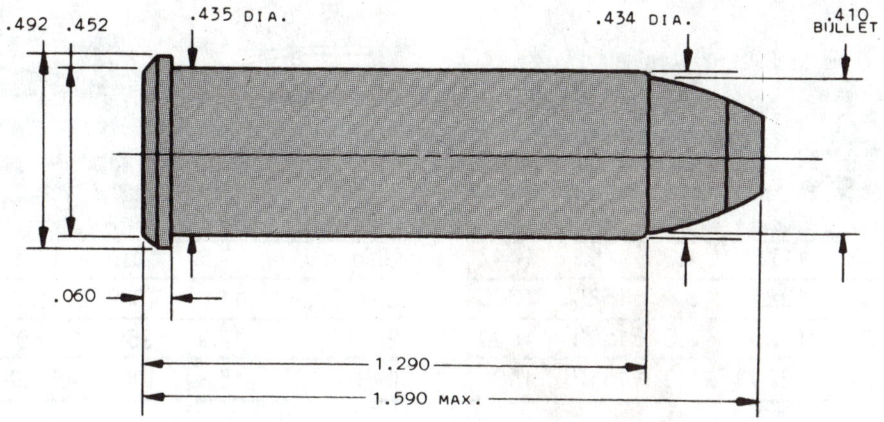

COMMENTS:

This cartridge has had a small but loyal following with hunters and silhouette shooters.

We suggest Hercules 2400 with all cast and jacketed bullets as being a good place to start your loading.

TEST COMPONENTS:

Cases .... Winchester
Trim-to Length .... 1.280"
Primers .... Remington 2½
Primer Size .... Large Pistol
Lyman Shell Holder .... No. 30
Jacketed Bullets Used .... Speer HP #4405, 200 gr.
Hornady HP #4100, 210 gr.
Sierra FMJ #8530, 220 gr.
Cast Bullets Used .... (Sized to .410" dia.)
*Gas Check Bullets *#410610, 215 gr.
#410459, 220 gr.

TEST SPECIFICATIONS:
(Velocity Only)

Firearm Used .... Thompson/Center Contender
Barrel Length .... 10" and 14"
Twist .... 1-20"
Groove Dia. .... .410"

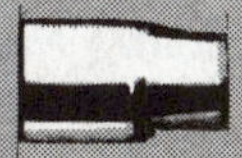

200 gr. Jacketed HP
1.590" OAL

Powder	Sugg. Starting Grains	10" Bbl. Velocity fps	14" Bbl. Velocity fps	Max. Load Grains	10" Bbl. Velocity fps	14" Bbl. Velocity fps
Unique	8.2	1127	1153	10.7	1429	1455
Blue Dot	12.5	1253	1267	15.4	1525	1576
2400	16.3	1376	1435	18.3	1514	1547
IMR-4227	18.0	1171	1235	21.0+	1415	1467

Note: Loads shown in shaded panels are maximum.
\+ Designates a compressed powder charge.

41 Remington Magnum

210 gr. Jacketed HP
1.585" OAL

Powder	Sugg. Starting Grains	10" Bbl. Velocity fps	14" Bbl. Velocity fps	Max. Load Grains	10" Bbl. Velocity fps	14" Bbl. Velocity fps
Unique	8.6	1145	1175	11.0	1432	1447
Blue Dot	12.6	1252	1321	15.8	1580	1606
2400	15.3	1166	1199	20.0	1505	1600
IMR-4227	17.8	1191	1220	21.0+	1412	1488

220 gr. Jacketed FMJ
1.590" OAL

Powder	Sugg. Starting Grains	10" Bbl. Velocity fps	14" Bbl. Velocity fps	Max. Load Grains	10" Bbl. Velocity fps	14" Bbl. Velocity fps
Unique	7.5	1010	1025	10.0	1278	1334
Blue Dot	11.0	1062	1128	14.2	1424	1456
2400	12.9	959	969	17.4	1325	1351
IMR-4227	15.3	1003	1029	19.7	1305	1355

#410610
215 gr. (#2 Alloy) 1.575" OAL

Powder	Sugg. Starting Grains	10" Bbl. Velocity fps	14" Bbl. Velocity fps	Max. Load Grains	10" Bbl. Velocity fps	14" Bbl. Velocity fps
Unique	7.8	1145	1197	10.4	1430	1470
Blue Dot	11.4	1209	1305	14.6	1542	1595
2400	14.2	1162	1233	18.8	1521	1589
IMR-4227	14.9	1047	1124	19.2	1351	1458

#410459
220 gr. (#2 Alloy) 1.700" OAL

Powder	Sugg. Starting Grains	10" Bbl. Velocity fps	14" Bbl. Velocity fps	Max. Load Grains	10" Bbl. Velocity fps	14" Bbl. Velocity fps
Unique	8.1	1165	1208	11.2	1459	1527
Blue Dot	13.0	1366	1449	16.0	1647	1727
2400	15.7	1306	1381	20.1	1609	1706
IMR-4227	17.3	1280	1371	21.5+	1527	1622

Note: Loads shown in shaded panels are maximum.
\+ Designates a compressed powder charge.

44 Remington Magnum

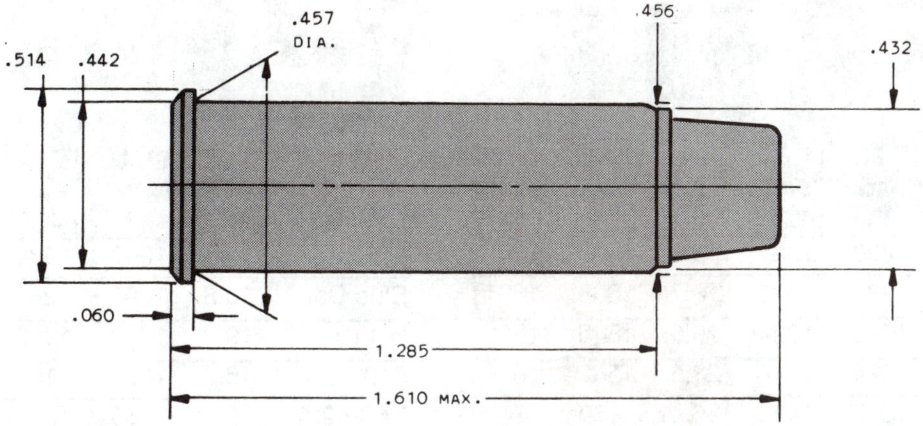

COMMENTS:

This is one of the more popular Contender cartridges and it will fill a broad range of applications.

Bullet #429215 is extremely popular. However, all cast bullet designs generally work well.

IMR 4227 and Hercules 2400 are suggested for combining maximum accuracy with full power loads.

TEST COMPONENTS:

Cases .... Remington
Trim-to Length .... 1.275"
Primers .... CCI 300
Primer Size .... Large Pistol
Lyman Shell Holder .... No. 7
Jacketed Bullets Used .... Sierra HC #8600, 180 gr.
Hornady HP #4410, 200 gr.
Hornady FMJ #4427, 240 gr.
Hornady FP #4300, 265 gr.
Cast Bullets Used .... (Sized to .430" dia.)
*Gas Check Bullets .... *#429215, 215 gr.
*#429244, 245 gr.

TEST SPECIFICATIONS:
(Velocity Only)

Firearm Used .... Thompson/Center Contender
Barrel Length .... 10" and 14"
Twist .... 1-22"
Groove Dia. .... .430"

180 gr. Jacketed HP
1.610" OAL

Powder	Sugg. Starting Grains	10" BBL. Velocity fps	14" BBL. Velocity fps	Max. Load Grains	10" BBL. Velocity fps	14" BBL. Velocity fps
Unique	12.0	1474	1505	13.6	1634	1679
Blue Dot	17.8	1570	1639	20.2	1770	1823
H-110	28.7	1757	1898	32.3+	1994	2135
2400	21.3	1424	1481	25.5	1683	1773
IMR-4227	24.2	1437	1503	27.0	1633	1741
AA 5	12.8	1448	1497	16.0	1744	1835

Note: Loads shown in shaded panels are maximum.
\+ Designates a compressed powder charge.

44 Remington Magnum

200 gr. Jacketed HP
1.610" OAL

Powder	Sugg. Starting Grains	10" BBL. Velocity fps	14" BBL. Velocity fps	Max. Load Grains	10" BBL. Velocity fps	14" BBL. Velocity fps
Unique	10.5	1280	1321	13.0	1498	1567
Blue Dot	15.7	1388	1421	18.7	1608	1691
H-110	24.5	1559	1624	29.9	1845	1986
2400	19.5	1324	1444	23.6	1562	1656
IMR-4227	22.8	1353	1446	26.4	1630	1709
800X	12.0	1412	1488	14.9	1646	1702
AA 7	18.0	1510	1639	21.0	1759	1831

240 gr. Jacketed FMJ
1.610" OAL

Powder	Sugg. Starting Grains	10" BBL. Velocity fps	14" BBL. Velocity fps	Max. Load Grains	10" BBL. Velocity fps	14" BBL. Velocity fps
Unique	9.0	1067	1088	12.1	1351	1412
Blue Dot	13.8	1206	1241	16.8	1447	1539
H-110	19.3	1250	1309	23.9	1532	1618
2400	16.7	1153	1162	21.0	1423	1474
IMR-4227	19.3	1160	1212	23.1	1413	1511
800X	11.0	1272	1319	13.7	1464	1538
AA 9	19.0	1411	1501	22.2	1575	1635

265 gr. Jacketed FP
1.650" OAL

Powder	Sugg. Starting Grains	10" BBL. Velocity fps	14" BBL. Velocity fps	Max. Load Grains	10" BBL. Velocity fps	14" BBL. Velocity fps
Unique	8.0	920	963	11.5	1255	1320
Blue Dot	12.5	1088	1101	16.0	1370	1395
2400	16.2	1072	1080	20.2	1363	1402
IMR-4227	17.5	1015	1037	22.0	1361	1387
800X	10.7	1181	1248	13.4	1375	1436
AA 9	18.0	1336	1377	21.3	1516	1583

#429215
215 gr. (#2 Alloy) 1.645" OAL

Powder	Sugg. Starting Grains	10" BBL. Velocity fps	14" BBL. Velocity fps	Max. Load Grains	10" BBL. Velocity fps	14" BBL. Velocity fps
Unique	10.0	1263	1323	13.2	1543	1624
Blue Dot	15.7	1397	1483	19.5	1677	1775
2400	20.0	1367	1488	25.0	1694	1787
IMR-4227	21.8	1314	1403	25.5+	1569	1680
800X	12.0	1426	1499	15.1	1639	1747
AA 7	16.0	1342	1451	20.2	1667	1797

#429244
245 gr. (#2 Alloy) 1.680" OAL

Powder	Sugg. Starting Grains	10" BBL. Velocity fps	14" BBL. Velocity fps	Max. Load Grains	10" BBL. Velocity fps	14" BBL. Velocity fps
Unique	8.9	1111	1155	12.1	1360	1434
Blue Dot	14.0	1255	1304	17.4	1494	1575
2400	18.2	1252	1327	22.2	1529	1605
IMR-4227	20.2	1227	1322	24.0+	1497	1575
800X	10.6	1282	1325	13.2	1488	1546
AA 9	17.5	1337	1430	21.7	1544	1645

Note: Loads shown in shaded panels are maximum.
+ Designates a compressed powder charge.

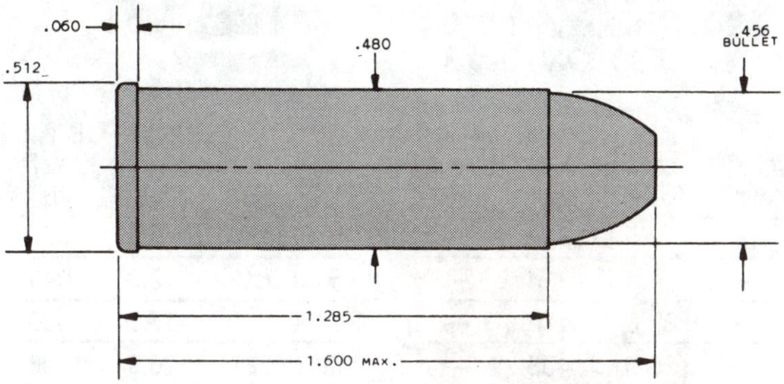

COMMENTS:

Care must be taken when sizing as the tiny rim of this cartridge does not afford a good purchase for the shell holder. If undue force is needed to withdraw the case, the rim may tear away leaving the case jammed in the die.

Unique is highly favored with all bullets for accuracy in this cartridge.

The first step of the expanding plug used must be 0.001" to 0.0015" under actual bullet diameter or bullets will not be tight in the case neck.

Bullet #452424, the Elmer Keith bullet, has replaced bullet #454424 in all our data. The bullets are identical except that #452424 is a few thousandths smaller in diameter. This is our most popular bullet in this caliber. Carefully control seating depths of cast bullets to the listed length.

TEST COMPONENTS:

Cases .... Winchester
Trim-to Length .... 1.275"
Primers .... Winchester 7-111
Primer Size .... Large Pistol
Lyman Shell Holder .... No. 11
Jacketed Bullets Used .... Speer HP #4477, 200 gr.
Sierra HP #8820, 240 gr.
Hornady HP #4520, 250 gr.
Cast Bullets Used .... (Sized to .452" dia.)
#454190, 250 gr.
#452424, 255 gr.
#454647, 285 gr.
#454 Round Ball, 142 gr.

TEST SPECIFICATIONS: (Velocity Only)

Firearm Used .... Thompson/Center Contender
Barrel Length .... 10"
Twist .... 1-24"
Groove Dia. .... .452"

200 gr. Jacketed HP
1.565" OAL

Powder	Sugg. Starting Grains	10" Bbl. Velocity fps	14" Bbl. Velocity fps	Max. Load Grains	10" Bbl. Velocity fps	14" Bbl. Velocity fps
Unique	11.2	1279	—	12.2	1391	—
Blue Dot	12.8	1030	—	15.8	1305	—
2400	19.0	1161	—	21.5	1366	—
IMR-4227	18.8	1080	—	23.5	1341	—

Note: Loads shown in shaded panels are maximum.

45 Colt

240 gr. Jacketed HP
1.580" OAL

Powder	Sugg. Starting Grains	10" Bbl. Velocity fps	14" Bbl. Velocity fps	Max. Load Grains	10" Bbl. Velocity fps	14" Bbl. Velocity fps
Unique	9.5	1096	—	12.5	1354	—
Blue Dot	11.8	973	—	14.8	1214	—
2400	18.5	1155	—	23.0	1472	—
IMR-4227	20.5	1179	—	25.0	1453	—

250 gr. Jacketed HP
1.586" OAL

Powder	Sugg. Starting Grains	10" Bbl. Velocity fps	14" Bbl. Velocity fps	Max. Load Grains	10" Bbl. Velocity fps	14" Bbl. Velocity fps
Unique	9.3	1027	—	12.2	1324	—
Blue Dot	12.0	946	—	14.5	1154	—
2400	17.5	1060	—	22.0	1410	—
IMR-4227	20.3	1196	—	24.0+	1362	—

#454190
250 gr. (#2 Alloy) 1.655" OAL

Powder	Sugg. Starting Grains	10" Bbl. Velocity fps	14" Bbl. Velocity fps	Max. Load Grains	10" Bbl. Velocity fps	14" Bbl. Velocity fps
Unique	9.5	1145	—	11.5	1278	—
Red Dot	6.5	985	—	9.4	1206	—
700X	6.5	966	—	9.0	1211	—
PB	8.0	1018	—	10.0	1204	—

#452424
255 gr. (#2 Alloy) 1.630" OAL

Powder	Sugg. Starting Grains	10" Bbl. Velocity fps	14" Bbl. Velocity fps	Max. Load Grains	10" Bbl. Velocity fps	14" Bbl. Velocity fps
Unique	8.5	1013	—	10.2	1162	—
Red Dot	6.5	917	—	8.3	1055	—
700X	6.3	955	—	8.0	1127	—
SR-7625	8.0	984	—	9.7	1153	—

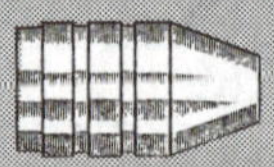

#454647
285 gr. (#2 Alloy) 1.700" OAL

Powder	Sugg. Starting Grains	10" Bbl. Velocity fps	14" Bbl. Velocity fps	Max. Load Grains	10" Bbl. Velocity fps	14" Bbl. Velocity fps
Unique	7.5	907	—	9.2	1056	—
Red Dot	5.0	716	—	7.3	961	—
700X	5.0	779	—	7.0	969	—
SR-7625	7.0	803	—	8.7	977	—

#454 Round Ball
142 gr. (#2 Alloy) 1.440" OAL

Powder	Sugg. Starting Grains	10" Bbl. Velocity fps	14" Bbl. Velocity fps	Max. Load Grains	10" Bbl. Velocity fps	14" Bbl. Velocity fps
Red Dot	3.0	530	—	4.3	701	—
Unique	3.5	409	—	6.5	732	—
SR-7625	5.0	550	—	7.2	774	—

Note: Loads shown in shaded panels are maximum.
+ Designates a compressed powder charge.

45 Winchester Magnum

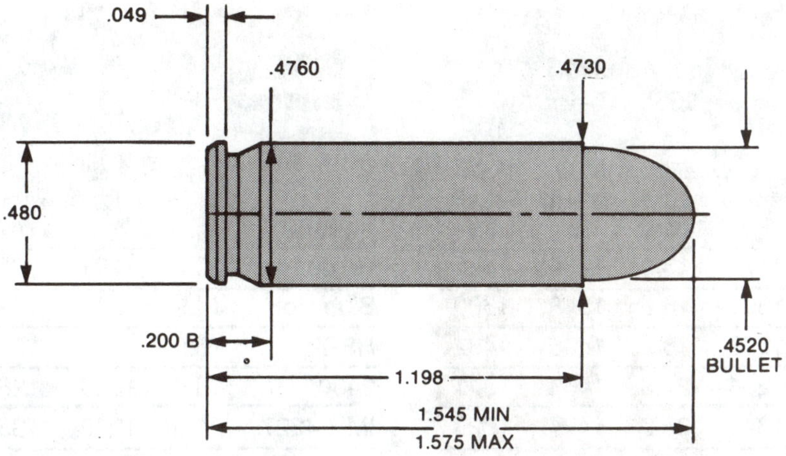

COMMENTS:

Never attempt to load 45 Magnum loads into the 45 Automatic cartridge as very dangerous pressure will result.

You may use most standard 45 Automatic dies to reload this cartridge since it is basically an elongated version of this round.

This cartridge headspaces on the case mouth. Adhere to the listed maximum and trim-to case lengths exactly. Do not roll crimp as doing so will alter headspacing. A taper crimp may be moderately applied if needed.

TEST COMPONENTS:

Cases .... Winchester
Trim-to Length .... 1.193"
Primers .... CCI 350
Primer Size .... Large Pistol, Magnum
Lyman Shell Holder .... No. 2
Jacketed Bullets Used .... Sierra HP #8800, 185 gr.
Speer HP #4477, 200 gr.
Hornady FMJ #4518, 230 gr.
Hornady HP #4520, 250 gr.
Cast Bullets Used .... (Sized to .452" dia.)
*Gas Check Bullets .... *#452484, 225 gr.
#454190, 250 gr.
#452424, 255 gr.

TEST SPECIFICATIONS:
(Velocity Only)

Firearm Used .... Thompson/Center Contender
Barrel Length .... 10" and 14"
Twist .... 1-16"
Groove Dia. .... .4515"

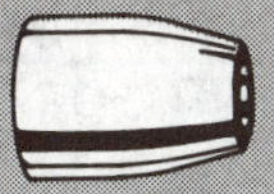

185 gr. Jacketed HP
1.525" OAL

Powder	Sugg. Starting Grains	10" BBL. Velocity fps	14" BBL. Velocity fps	Max. Load Grains	10" BBL. Velocity fps	14" BBL. Velocity fps
Unique	12.0	1429	1491	15.7	1771	1858
Blue Dot	17.5	1484	1634	21.0	1830	1903
HS-7	18.5	1547	1609	21.8	1775	1858
H-110	26.0	1669	1799	29.0	1792	1860
2400	25.0	1674	1704	28.7	1895	1948

200 gr. Jacketed HP
1.480" OAL

Powder	Sugg. Starting Grains	10" BBL. Velocity fps	14" BBL. Velocity fps	Max. Load Grains	10" BBL. Velocity fps	14" BBL. Velocity fps
Unique	10.0	1229	1237	13.2	1521	1607
Blue Dot	15.5	1348	1417	19.0	1733	1761
HS-7	16.5	1379	1421	20.0	1654	1713
H-110	24.5	1495	1613	27.0	1632	1719
2400	18.5	1287	1373	22.0	1497	1563

Note: Loads shown in shaded panels are maximum.

45 Winchester Magnum

230 gr. Jacketed FMJ
1.565" OAL

Powder	Sugg. Starting Grains	10" BBL. Velocity fps	14" BBL. Velocity fps	Max. Load Grains	10" BBL. Velocity fps	14" BBL. Velocity fps
Unique	10.0	1149	1172	13.0	1454	1475
Blue Dot	15.3	1305	1398	18.0	1548	1603
HS-7	15.5	1229	1297	18.0	1408	1482
2400	18.3	1220	1234	21.8	1468	1541
IMR-4227	22.0	1340	1346	25.0	1497	1523

250 gr. Jacketed HP
1.480" OAL

Powder	Sugg. Starting Grains	10" BBL. Velocity fps	14" BBL. Velocity fps	Max. Load Grains	10" BBL. Velocity fps	14" BBL. Velocity fps
Unique	10.0	1191	1210	11.5	1333	1355
Blue Dot	14.9	1317	1349	16.6	1488	1507
HS-7	15.0	1241	1275	17.0	1405	1428
2400	18.0	1246	1262	20.0	1392	1421
IMR-4227	21.0	1270	1359	24.5	1512	1585

#452484
225 gr. (#2 Alloy) 1.565" OAL

Powder	Sugg. Starting Grains	10" BBL. Velocity fps	14" BBL. Velocity fps	Max. Load Grains	10" BBL. Velocity fps	14" BBL. Velocity fps
Unique	9.5	1170	1207	12.5	1457	1475
Blue Dot	12.5	1102	1191	15.5	1435	1472
SR-7625	9.0	1116	1156	12.0	1381	1417
2400	18.0	1281	1319	21.0	1464	1510

#454190
250 gr. (#2 Alloy) 1.565" OAL

Powder	Sugg. Starting Grains	10" BBL. Velocity fps	14" BBL. Velocity fps	Max. Load Grains	10" BBL. Velocity fps	14" BBL. Velocity fps
Unique	8.5	1056	1109	11.3	1309	1375
Blue Dot	12.5	1137	1200	15.5	1351	1404
SR-7625	8.5	1074	1128	11.5	1339	1397
2400	17.0	1176	1241	20.0	1355	1490

#452424
255 gr. (#2 Alloy) 1.530" OAL

Powder	Sugg. Starting Grains	10" BBL. Velocity fps	14" BBL. Velocity fps	Max. Load Grains	10" BBL. Velocity fps	14" BBL. Velocity fps
Unique	8.0	1016	1078	11.0	1315	1349
Blue Dot	12.0	1093	1130	15.2	1383	1434
HS-6	9.7	997	1048	12.7	1228	1309
SR-7625	7.8	1034	1084	10.8	1312	1356
2400	15.0	1047	1157	18.0	1279	1315

Note: Loads shown in shaded panels are maximum.

45-70 Government

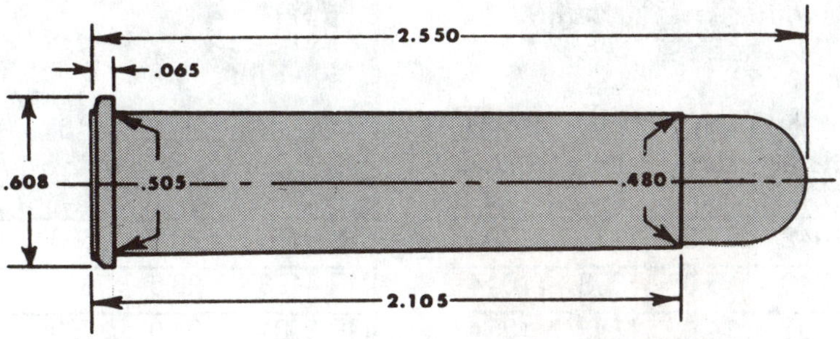

COMMENTS:

This is not a cartridge for handgunners of faint heart. Recoil is very severe when shot in the T/C Contender. Hercules 2400 and Reloder 7 are good choices.

TEST COMPONENTS:

Cases .... Remington
Trim-to Length .... 2.100"
Primers .... Remington 9½
Primer Size .... Large Rifle
Lyman Shell Holder .... No. 17
Jacketed Bullets Used .... Hornady HP #4500, 300 gr.
Speer SP #2479, 400 gr.
Cast Bullets Used .... (Sized to .459" dia.)
*Gas Check Bullets .... #457191, 292 gr.
#457193, 405 gr.

TEST SPECIFICATIONS: (Velocity Only)

Firearm Used .... Thompson/Center Contender
Barrel Length .... 10" and 16"
Twist .... 1-14"
Groove Dia. .... .460" in 10", .458" in 16"

300 gr. Jacketed HP
2.550" OAL

Powder	Sugg. Starting Grains	10" Bbl. Velocity fps	16" Bbl. Velocity fps	Max. Load Grains	10" Bbl. Velocity fps	16" Bbl. Velocity fps
IMR-4198	31.0	1008	1248	36.0	1297	1449
RX7	33.0	1173	1347	38.0	1342	1600
IMR-3031	38.0	960	1206	43.0	1299	1410
2400	24.0	1186	1346	30.0	1447	1685
IMR-4227	27.0	1062	1327	32.0	1307	1625

Note: Loads shown in shaded panels are maximum.

45-70 Government

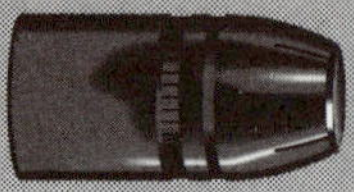

400 gr. Jacketed SP
2.550" OAL

Powder	Sugg. Starting Grains	10" Bbl. Velocity fps	16" Bbl. Velocity fps	Max. Load Grains	10" Bbl. Velocity fps	16" Bbl. Velocity fps
IMR-4198	28.0	844	1042	33.0	1116	1293
Unique	11.0	623	746	15.0	898	1014
RX7	29.0	897	1191	34.0	1255	1354
IMR-3031	34.0	750	1018	39.0	1073	1283
2400	22.0	954	1200	27.0	1287	1426
IMR-4227	25.0	960	1211	30.0	1324	1397

#457191
292 gr. (#2 Alloy) 2.550" OAL

Powder	Sugg. Starting Grains	10" Bbl. Velocity fps	16" Bbl. Velocity fps	Max. Load Grains	10" Bbl. Velocity fps	16" Bbl. Velocity fps
SR-4759	26.0	663	932	30.0	1187	1471
IMR-4198	28.5	811	1122	34.5	1194	1481
IMR-3031	34.0	758	961	44.0	1237	1433

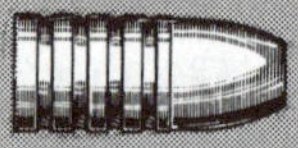

#457193
405 gr. (#2 Alloy) 2.550" OAL

Powder	Sugg. Starting Grains	10" Bbl. Velocity fps	16" Bbl. Velocity fps	Max. Load Grains	10" Bbl. Velocity fps	16" Bbl. Velocity fps
IMR-4198	21.5	626	733	28.5	1009	1195
RX7	33.0	1274	1386	41.0	1447	1578
IMR-3031	34.0	902	1061	38.5	1163	1304

Note: Loads shown in shaded panels are maximum.

8

REFERENCE SECTION

Appendix A
FACTORY AMMUNITION SPECIFICATIONS

Rifle Cartridge	Bullet Wt. Grs.	Type	Brand	Vel. Muzzle	Bbl. (in.)
17 Rem.	25	HP	Rem.	4040	24
22 Hornet	45	PSP	Rem., Win.	2690	24
22 Hornet	46	HP	Win.	2690	24
222 Rem.	50	PSP	Rem., Win., Fed.	3140	24
222 Rem.	55	HP	Rem.	3140	24
222 Rem.	55	FMJ	Win., Fed.	3020	24
223 Rem.	53	HP	Win.	3330	24
223 Rem.	55	PSP	Rem., Win., Fed.	3240	24
223 Rem.	55	HP	Rem., Fed.	3240	24
22-250	55	PSP	Rem., Win., Fed.	3680	24
220 Swift	50	PSP	Rem.	3780	24
224 Wea. Mag	55	PSP	Wea.	3650	26
243 Win.	80	PSP	Rem., Win., Fed.	3350	24
243 Win.	85	BTHP	Fed.	3320	24
243 Win.	100	PSP	Rem., Win., Fed.	2960	24
6mm Rem.	80	PSP	Rem., Win.	3470	24
6mm Rem.	100	PSP	Rem., Win., Fed.	3100	24
6mm BR Rem.	100	PSP	Rem.	2550	15
25-20 Win.	86	SP	Rem., Win.	1460	24
250 Sav.	100	ST, PSP	Rem., Win.	2820	24
257 Roberts	117	PSP	Rem.	2650	24
25-06 Rem.	100	PSP	Rem.	3230	24
25-06 Rem.	117	PSP	Fed.	2990	24
25-06 Rem.	120	PSP	Rem., Win.	2990	24
257 Wea. Mag	117	PSP	Wea.	3300	26
6.5x55 Swed.	140	PSP	Fed.	2850	24
264 Win. Mag	140	PSP	Rem., Win.	3030	24
270 Win.	100	PSP	Win.	3340	24
270 Win.	130	PSP	Rem., Win., Fed.	3060	24
270 Win.	150	PSP	Rem., Win., Fed.	2850	24
270 Wea. Mag	130	PSP	Wea.	3375	26
7mm BR Rem.	140	PSP	Rem.	2215	15
7-30 Waters	120	BTSP	Fed.	2700	24
7mm Mauser	140	PSP	Rem., Fed.	2660	24
7mm-08 Rem.	140	PSP	Rem.	2860	24
280 Rem.	140	PSP	Rem.	3000	24
280 Rem.	150	PSP	Rem., Fed.	2890	24
7mm Rem. Mag	150	PSP	Rem., Win., Fed.	3110	24
7mm Rem. Mag	165	BTSP	Fed.	2950	24
7mm Rem. Mag	175	PSP	Rem., Win.	2860	24
7mm Wea. Mag	175	PSP	Rem.	2910	24
30 Carbine	110	SP	Rem., Fed.	1990	20
30-30 Win.	150	SP	Rem., Win., Fed.	2390	24
30-30 Win.	170	SP	Rem., Win., Fed.	2200	24
300 Sav.	150	PSP	Rem., Win., Fed.	2630	24
308 Win.	150	PSP	Rem., Win., Fed.	2820	24
308 Win.	165	BTSP	Fed.	2700	24
308 Win.	180	PSP	Rem., Win., Fed.	2620	24
30-40 Krag	180	PSP	Rem., Win.	2430	24
30-06 Sprg.	125	PSP	Rem., Win., Fed.	3140	24
30-06 Sprg.	150	PSP	Rem., Win., Fed.	2910	24
30-06 Sprg.	180	PSP	Rem., Win., Fed.	2700	24
30-06 Sprg.	220	SP	Rem., Win., Fed.	2410	24
300 H&H Mag	180	PSP	Rem., Win., Fed.	2880	24
300 Win. Mag	150	PSP	Rem., Win.	3290	24
300 Win. Mag	180	PSP	Rem., Win., Fed.	2960	24
300 Wea. Mag	180	PSP	Rem.	3120	24
7.62x39mm	123	SP	Win.	2365	20
303 Brit.	180	PSP	Rem., Win., Fed.	2460	24
32-20 Win.	100	Lead	Rem., Win.	1210	24
32 Win. Sp.	170	SP	Rem., Win., Fed.	2250	24
8mm Mauser	170	PSP	Rem., Win., Fed.	2360	24
8mm Rem. Mag	180	PSP	Rem.	3080	24
338 Win. Mag	225	PSP	Rem., Fed.	2780	24
338 Win. Mag	250	PSP	Rem., Fed.	2660	24
340 Wea. Mag	250	PSP	Wea.	2850	26
348 Win.	200	ST	Win.	2520	24
35 Rem.	200	SP	Win., Fed.	2080	24
350 Rem. Mag	200	PSP	Rem.	2710	20
358 Win.	200	ST	Win.	2490	24
35 Whelen	200	PSP	Rem.	2675	24

Appendix A
FACTORY AMMUNITION SPECIFICATIONS

Rifle Cartridge	Bullet Wt. Grs.	Bullet Type	Brand	Vel. Muzzle	Bbl. (in.)
35 Whelen	250	PSP	Rem.	2400	24
375 Win.	250	PSP	Win.	1900	24
375 H&H Mag	270	SP	Rem., Win., Fed.	2690	24
38-55 Win.	255	SP	Win.	1320	24
38-40 Win.	180	SP	Win.	1160	24
416 Rem. Mag	400	PSP	Rem.	2400	24
416 Rigby	410	SP	Fed.	2370	24
44-40 Win.	200	SP	Rem., Win.	1190	24
444 Marlin	240	SP	Rem.	2350	24
45-70 Gov.	300	HP	Win., Fed.	1880	24
458 Win. Mag	500	FMJ	Rem., Win., Fed.	2040	24
460 Wea. Mag	500	FMJ	Wea.	2700	26

Note: Not all calibers or bullet weights are included in this chart.
Bullet Codes: PSP – pointed soft point; BTSP – boat tail soft point; SP – soft point; HP – hollow point; FMJ – full metal jacket; BTHP – boat tail hollow point; ST – Silver Tip
Rifle Brand Codes: Rem. – Remington; Win. – Winchester; Fed. – Federal; Wea. – Weatherby

Pistol and Revolver

Cartridge	Bullet Wt. Grs.	Bullet Type	Brand	Vel. Muzzle	Bbl. (in.)
221 Rem. Fireball	50	PSP	Rem.	2650	10
25 Auto	50	FMJ	Rem., Win., Fed.	760	2
30 Luger	93	FMJ	Win.	1220	4 1/2
32 Auto	71	FMJ	Rem., Win.	905	4
32 S&W	88	Lead	Rem.	680	3
32 S&W Long	98	Lead	Rem., Win., Fed.	705	4
32 H&R Mag	85	JHP	Fed.	1100	4 1/2
380 Auto	88	JHP	Rem.	990	4
380 Auto	95	FMJ	Rem., Win., Fed.	955	4
38 S&W	145	Lead	Win.	685	4
9mm Luger	88	JHP	Rem.	1500	4
9mm Luger	115	STHP	Win.	1225	4
9mm Luger	124	JHP	Fed.	1120	4
9mm Luger	147	STHP	Win.	1010	4
38 Super	130	FMJ	Win.	1215	5
38 Spec. +P	110	JHP	Rem., Win., Fed.	995	4
38 Spec. +P	125	JHP	Rem., Win., Fed.	945	4
38 Spec.	148	LWC	Rem.	710	4
38 Spec.	158	LSWC	Win., Fed.	755	4
38 Spec. +P	158	LSWC	Rem., Win., Fed.	890	4
357 Mag	110	JHP	Rem., Win., Fed.	1295	4
357 Mag	125	JHP	Rem., Win., Fed.	1450	4
357 Mag	140	JHP	Rem.	1360	4
357 Mag	158	JSP	Rem., Win., Fed.	1235	4
357 Mag	158	JHP	Rem., Win., Fed.	1235	4
40 S&W	155	JHP	Rem., Fed.	1140	4
40 S&W	155	STHP	Win.	1205	4
40 S&W	180	JHP	Rem., Win., Fed.	985	4
10mm Auto	155	JHP	Fed.	1325	5
10mm Auto	175	STHP	Win.	1290	5 1/2
10mm Auto	180	JHP	Fed.	1030	5
10mm Auto	200	FMJ	Rem.	1160	5
41 Rem. Mag	170	JHP	Rem.	1420	4
41 Rem. Mag	210	JSP	Rem., Win.	1300	4
44 S&W Spec.	246	Lead	Rem., Win.	755	6
44 Rem. Mag	180	JHP	Rem., Fed.	1610	4
44 Rem. Mag	210	STHP	Win.	1250	4
44 Rem. Mag	240	JHP	Rem., Win., Fed.	1180	4
45 Auto	185	JHP	Rem.	1000	5
45 Auto	230	FMJ	Rem., Win.	835	5
45 Colt	225	JHP	Fed.	900	5 1/2
45 Colt	255	Lead	Win.	860	5 1/2
45 Win. Mag	230	FMJ	Win.	1400	5

Note: Not all calibers or bullet weights are included in this chart.
Bullet Codes: JHP – jacketed hollow point; STHP – Silver Tip hollow point; FMJ – full metal jacket; JSP – jacketed soft point; LSWC – lead semi wad cutter; LWC – lead wad cutter; PSP – pointed soft point
Pistol and Revolver Brand Codes: Rem. – Remington; Win. – Winchester; Fed. – Federal

Appendix B
MAXIMUM CASE LENGTH & TRIM TO LENGTH GUIDE

Rifle Calibers	Maximum Case Length	Trim-to Length
17 Remington	1.796"	1.786"
22 Hornet	1.403"	1.393"
218 Bee	1.345"	1.335"
222 Remington	1.700"	1.690"
22 PPC	1.505"	1.495"
223 Remington	1.760"	1.750"
222 Remington Magnum	1.850"	1.840"
225 Winchester	1.930"	1.920"
224 Weatherby Magnum	1.925"	1.915"
22/250	1.912"	1.902"
220 Swift	2.205"	2.195"
22 Savage Hi-Power	2.050"	2.040"
6mm PPC	1.515"	1.505"
243 Winchester	2.045"	2.035"
6 M/M Remington	2.233"	2.223"
244 Remington	2.233"	2.223"
240 Weatherby Magnum	2.500"	2.490"
25-06 Remington	2.494"	2.484"
25/20 Winchester	1.330"	1.320"
256 Winchester	1.281"	1.275"
250/3000 Savage	1.912"	1.902"
257 Roberts	2.233"	2.223"
257 Weatherby Magnum	2.550"	2.540"
6.5 Japanese	1.988"	1.978"
6.5 Italian	2.065"	2.055"
6.5 Mann.-Schoe.	2.110"	2.100"
6.5 Swedish	2.160"	2.150"
6.5 x 55 Danish Krag	2.165"	2.155"
6.5 Remington Magnum	2.170"	2.160"
264 Remington Magnum	2.500"	2.490"
270 Winchester	2.540"	2.530"
270 Weatherby Magnum	2.550"	2.540"
7 MM Mauser	2.235"	2.225"
284 Winchester	2.170"	2.160"
280 Remington (7mm Express)	2.540"	2.530"
7 MM-08 Remington	2.035"	2.025"
7x61 MM Sharpe & Hart	2.402"	2.392"
7 MM Remington Magnum	2.500"	2.490"
7 MM Weatherby Magnum	2.550"	2.540"
7-30 Waters	2.040"	2.030"
7.35 Italian	2.01"	2.00"
30 MI Carbine	1.290"	1.286"
30 Remington	2.050"	2.040"
303 Savage	2.015"	2.010"
30/30 Winchester	2.039"	2.029"
300 Savage	1.871"	1.865"
307 Winchester	2.015"	2.005"
308 Winchester	2.015"	2.005"
30/40 Krag	2.314"	2.304"
30/06	2.494"	2.484"
300 H & H Magnum	2.850"	2.840"
300 Winchester Magnum	2.620"	2.610"
308 Norma Magnum	2.560"	2.550"
300 Weatherby Magnum	2.825"	2.815"
7.62 Russian Nagant	2.114"	2.105"
7.62 x 39 MM Russian	1.528"	1.515"
7.65 Argentine Mauser	2.105"	2.100"
303 British	2.222"	2.212"
7.7 Japanese	2.270"	2.260"
32/20 Winchester	1.315"	1.305"
32/40 Winchester	2.130"	2.120"
32 Remington	2.050"	2.045"
32 Winchester Special	2.040"	2.035"
8 MM Mauser	2.240"	2.235"
8 MM Remington Magnum	2.850"	2.840"
338 Winchester Magnum	2.500"	2.490"
340 Weatherby Magnum	2.825"	2.815"
348 Winchester	2.255"	2.245"
351 Winchester	1.380"	1.376"
35 Remington	1.920"	1.910"
356 Winchester	2.015"	2.005"
358 Winchester	2.015"	2.005"
350 Remington Magnum	2.170"	2.160"
35 Whelen	2.494"	2.484"
358 Norma Magnum	2.515"	2.505"
375 Winchester	2.020"	2.010"
375 H & H Magnum	2.850"	2.840"
378 Weatherby Magnum	2.913"	2.903"
38/55 Winchester	2.128"	2.118"
38/40 Winchester	1.305"	1.300"
416 Remington Magnum	2.850"	2.840"
416 Rigby	2.900"	2.890"
416 Weatherby Magnum	2.903"	2.898"
401 Winchester	1.500"	1.495"
44/40 Winchester	1.305"	1.300"
44 Remington Magnum	1.285"	1.280"
444 Marlin	2.225"	2.220"
45/70 Government	2.105"	2.100"
458 Winchester Magnum	2.500"	2.495"
460 Weatherby Magnum	2.913"	2.903"
50-70 Government	1.750"	1.740"

Pistol Calibers	Maximum Case Length	Trim-to Length
22 Remington Jet	1.288"	1.283"
221 Remington Fireball	1.400"	1.395"
25 A.C.P.	.615"	.610"
7mm T/CU	1.750"	1.740"
7mm BR	1.520"	1.510"
30 Luger	.850"	.845"
30 Mauser	.990"	.985"
32 A.C.P.	.680"	.672"
32 Smith & Wesson	.605"	.600"
32 Smith & Wesson Long	.920"	.915"
32 H & R Mag.	1.075"	1.070"
9 M/M Luger	.754	.751
357 Magnum	1.290"	1.285"
380 Auto	.680"	.677"
38 Super Auto	.900"	.895"
38 Special	1.155"	1.149"
38 Smith & Wesson	.775	.765"
357 Rem. Maximum	1.605"	1.600"
38/40 Winchester	1.305"	1.300"
40 S&W	.850"	.847"
10mm Auto	.992"	.989"
41 Magnum	1.290"	1.285"
44 Special	1.160"	1.152"
44 Remington Magnum	1.285"	1.280"
44/40 Winchester	1.305"	1.300"
45 A.C.P.	.898"	.895"
45 Auto Rim	.898"	.894"
45 Winchester Mag.	1.198"	1.193"
45 Colt	1.285"	1.280"

Appendix C
SHELLHOLDER CHARTS

Rifle Cartridges	Lyman	Hornady	RCBS	Lee	Redding	C-H
17 Remington	26	16	10	4	10	15
22 Hornet	4	3	12	7	14	HT
218 Bee	10	7	1	6	3	3
222 Remington	26	16	10	4	10	15
223 Remington	26	16	10	4	10	15
22 PPC	3	6	—	—	12	—
222 Remington Mag.	26	16	10	4	10	15
225 Winchester	5	4	11	10	4	4
224 Weatherby Mag.	3	1	—	—	12	28
22-250 Remington	2	1	3	2	1	1
220 Swift	5	4	11	10	4	4
5.6 x 50mm Mag.	26	16	6	—	12	—
6mm PPC	3	6	—	—	12	—
243 Winchester	2	1	3	2	1	1
6mm/244 Remington	2	1	3	2	1	1
240 Weatherby	2	1	3	2	1	1
25-20 Winchester	10	7	1	6	3	3
250-3000 Savage	2	1	3	2	1	1
257 Roberts	2 or 8	1	11	2	1	4
25-06 Remington	2	1	3	2	1	1
257 Weatherby	13	5	4	5	6	6
6.5mm Japanese	5	34	15	10	4	65
6.5mm Carcano	28	21	9	2	1	14
6.5x54mm Mau. Sch.	28	20	9	—	24	14
6.5x55 Swed. Mau.	27	19	2	3	1	2
6.5x57mm Mauser	2 or 8	1	3	2	1	—
6.5x57R Mauser	14B	—	—	—	—	—
6.5 Remington Mag.	13	5	4	5	6	6
264 Winchester Mag.	13	5	4 or 26	5	6	6
270 Winchester	2	1	3	2	1	1
270 Weatherby Mag.	13	5	4	5	6	6
7-30 Waters	6	2	2	3	2	2
7mm-08 Rem.	2	1	3	2	1	1
7mm Mauser	2	1	11 or 3	2	1	4
280 Remington	2	1	3	2	1	1
284 Winchester	2	1	3	2	1	1
7mm Remington Mag.	13	5	4 or 26	5	6	6
7mm Weatherby Mag.	13	5	4	5	6	6
7mm x 64 Brenneke	2	1	3	—	1	—
7mm x 65R Brenneke	14B	13	26	—	—	—
30 M1 Carbine	19	22	17	7	22	M1
30-30 Winchester	6	2	2	3	2	2
300 Savage	2	1	3	2	1	1
30-40 Krag	7	11	7	5	8	8B
308 Winchester	2	1	3	2	1	1
30-06 Springfield	2	1	3	2	1	1
300 H & H Magnum	13	5	4	5	6	6
300 Winchester Mag.	13	5	4 or 26	5	6	6
300 Norma Magnum	13	5	4	5	6	6
300 Weatherby Mag.	13	5	4	5	6	6
7.62 x 39 Russian	3	6	32	12	12	13
7.62 x 54R Russian	17	23	13	16	15	76
7.65mm Arg. Mauser	2	24	3	3	1	1
303 British	7	11	7	5	8	8B
32-20 Winchester	10	7	1	6	3	10
32 Winchester Spec.	6	2	2	3	2	2
8mm Mauser	2	1	3	2	1	1
8mm Remington Mag.	13	5	4	5	6	6
338 Winchester Mag.	13	5	4	5	6	6

Appendix C
SHELLHOLDER CHARTS

Rifle Cartridges	Lyman	Hornady	RCBS	Lee	Redding	C-H
340 Weatherby Mag.	13	5	4	5	6	6
348 Winchester	18	25	5	8	20	34
35 Remington	8 or 2	26	9	2	1	14
358 Winchester	2	1	3	2	1	1
350 Remington Mag.	13	5	4	5	6	6
35 Whelen	2	1	3	2	1	1
358 Norma Mag.	13	5	4	5	6	6
9.3 x 74mm R	13	13	4	—	6	6
38-55 Winchester	6	2	2	3	2	2
375 Winchester	6	2	2	3	2	2
375 H & H Magnum	13	5	4	5	6	6
378 Weatherby Mag.	17	14	14	—	18	—
416 Remington Mag.	13	5	—	—	6	—
416 Rigby	17	38	—	—	18	—
416 Weatherby	17	14	—	—	18	—
44-40 Winchester	14B	9	26 or 28	14	9	19
444 Marlin	14B	27	28	11	19	27
45-70 Government	17	14	14	8	18	47
458 Winchester	13	5	4	5	6	6
460 Weatherby	17	14	14	—	18	—
50-70 Government	22	—	31	—	—	57

Pistol Cartridge	Lyman	Hornady	RCBS	Lee	Redding	C-H
22 Jet	1	6	6	1	12	12
221 Fireball	26	16	10	4	10	15
25 ACP	32	37	29	15	27	—
256 Winchester Mag.	1	6	6	1	12	12
7mm T/C-U	26	16	10	4	10	15
7mm BR	2	1	3	2	1	1
30 Luger	12	8	16	19 or 6	13	—
30 Mauser	12	—	16	19 or 6	13	—
32 ACP	23	22	17	7	22	M1
32 STW Long	9	36	23	4	10	M1
32 H&R Magnum	9	36	23	4	10	M1
38 S&W	21	28	6	1	12	12
380 Auto	26	16	10	4	10	15
9mm Luger	12	8	16	19 or 6	13	Lug
38 Super Auto	12	8	1	19 or 6	5	3
38 Special	1	6	6	1	12	12
357 Magnum	1	6	6	1	12	12
357 Rem. Maximum	1	6	6	1	12	12
40 S&W	15	10	27	19	5	28
10mm Auto	15	10	27	19	5	28
38-40 Winchester	14B	9	26 or 28	14	9	19
41 Action Express	12	8	16	19 or 6	13	Lug
41 Remington Mag.	30	29	30	9	21	41
44 Special	7	30	18	11	19	8
44 Remington Mag.	7	30	18	11	19	8
45 ACP	2	1	3	2	1	1A
45 Auto Rim	14A	31	8	13	17	—
45 Winchester Mag.	2	1	3	2	7	1A
45 Colt	11	32	20	11	23	LC

Appendix D
RECOIL

When we consider bullet energy or muzzle energy in foot-pounds, we must realize that such force, or push, works equally in two directions.

The laws of science state that for every force there is an opposite and equal force. In the case of a firearm, the force (gases generated by the burning powder) which pushes the bullet down the barrel also pushes the rifle in the opposite direction. This force is, of course, equal in all directions so that if the firearm weighed exactly the same as the bullet, both would part company traveling at the same speed in different directions. Recoil calibrations are related to the bullet weight, firearm weight and weight of the powder charge.

The simplest means of measuring recoil is by the use of a pendulum. Such a device permits the measuring of both recoil velocity and recoil energy. Recoil energy (measured in foot pounds) may be reduced by increasing the weight of the firearm or by the use of one of the so-called "muzzle brakes." The idea behind the "muzzle brake" is to port some of the discharged gases to the rear. In theory, this creates another opposing force which tends to push the muzzle of the gun forward. The amount of recoil reduction accomplished by such devices varies (depending upon design) from almost none to as much as 25%.

Tables for recoil energy are far from infallible. The amount of recoil the shooter feels, or the sensation of recoil, is greatly influenced by stock design, butt plate design, how the rifle is held, etc. If the butt is held off the shoulder, then the effect of recoil is far more noticeable than if the butt is held firmly to the shoulder of a relaxed shooter. In the case of the relaxed shooter who shoulders the butt firmly, the gun recoils as if it were an extension of the shooter's body. The sensation of recoil is somewhat more moderate for he has added the weight of his upper torso to the weight of the rifle and, in a manner of speaking, has "rolled with the punch."

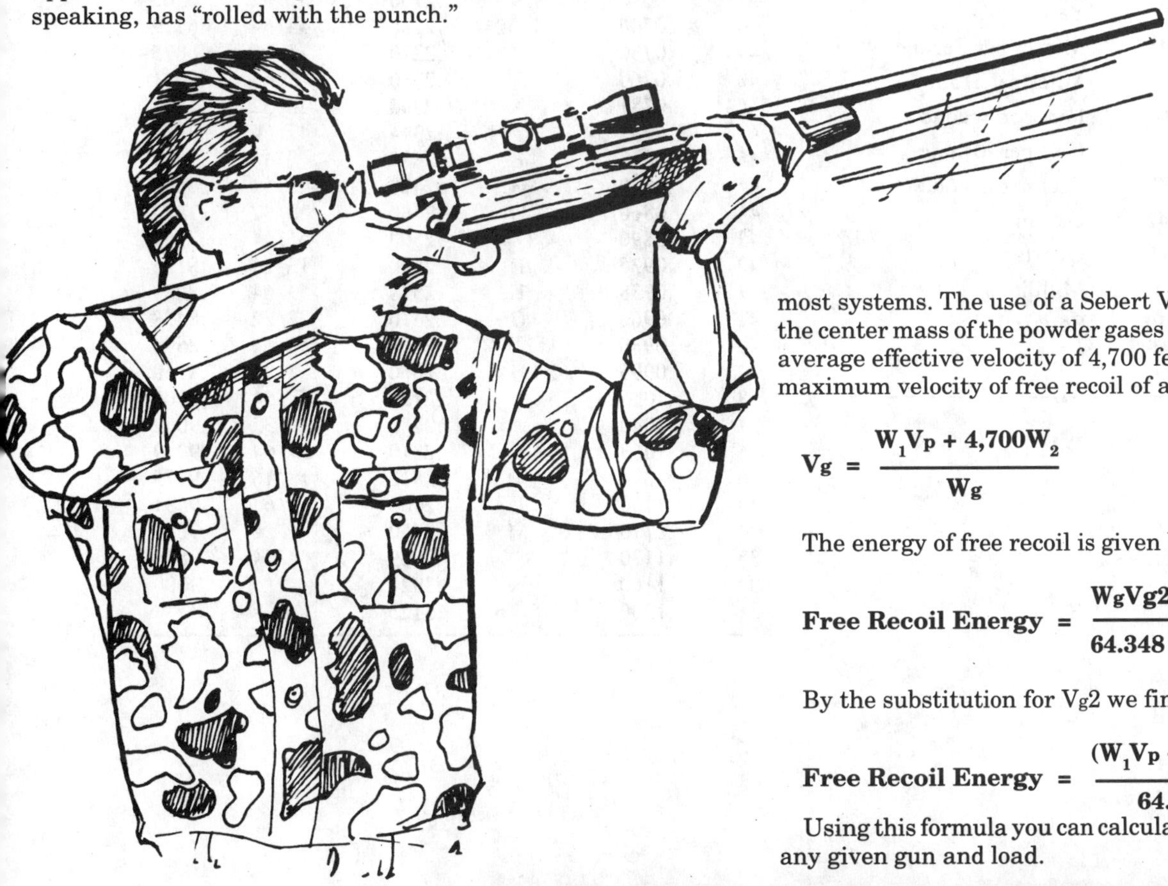

Technically speaking, recoil is based on the law of the Conservation of Momentum. As some readers may remember from physics class, this law states, "In a force and its reaction between two bodies (with no other forces present), equal and opposite changes in momentum will be given to the two bodies." Therefore, the momentum of a free-recoiling firearm is equal and opposite in direction to the momentum of the bullet and powder charge. Expressed mathematically, we have the following:

$$W_gV_g = W_1V_p + W_2V_c$$

A breakdown of this formula is as follows:

W_g = The weight of the gun in pounds.
W_1 = The weight of the bullet in pounds.
W_2 = The weight of the powder charge in pounds.
V_g = The recoiling velocity of the gun in feet per second.
V_p = The muzzle velocity of the bullet in feet per second.
V_c = Effective velocity of the powder charge.

The velocity of free recoil of a firearm is, therefore, as follows:

$$V_g = \frac{W_1V_p + W_2V_c}{W_g}$$

The weights of the gun, the bullet and the powder charge can be determined quite easily. By chronograph, we determine the muzzle velocity of the projectile. The effective velocity of the powder charge has been found to be generally consistent for most systems. The use of a Sebert Velocimeter indicates that the center mass of the powder gases leaves the muzzle with an average effective velocity of 4,700 feet per second. Hence, the maximum velocity of free recoil of a firearm becomes:

$$V_g = \frac{W_1V_p + 4{,}700W_2}{W_g}$$

The energy of free recoil is given by the relationship:

$$\text{Free Recoil Energy} = \frac{W_gV_g2}{64.348}$$

By the substitution for V_g2 we finally arrive at:

$$\text{Free Recoil Energy} = \frac{(W_1V_p + 4{,}700W_2)^2}{64.348W_g}$$

Using this formula you can calculate the amount of recoil for any given gun and load.

Appendix E

Conversion Factors

Multiply ×	By =	To Obtain
Atmospheres	14.70	Pounds per square inch
Kilograms per square centimeter	14.23	Pounds per square inch
Pounds per square inch	0.07032	Kilograms per cm^2
Drams	1.772	Grams
Drams	0.0625	Ounces
Grains (Troy)	1	Grains (Avoirdupois)
Grains	0.0648	Grams
Grams	15.43	Grains
Grams	0.03527	Ounces
Kilograms	1000	Grams
Kilograms	2.205	Pounds (Avoirdupois)
Ounces	16	Drams
Ounces	437.5	Grains
Ounces	0.0625	Pounds (Avoirdupois)
Ounces	28.35	Grams
Pounds	7000	Grains
Pounds	453.6	Grams
Pounds	16	Ounces
Centimeters	0.3937	Inches
Centimeters	0.01	Meters
Centimeters	10	Millimeters
Inches	2.540	Centimeters
Meters	100	Centimeters
Meters	3.281	Feet
Meters	39.37	Inches
Meters	1.094	Yards
Feet per Second	0.3048	Meters per Second
Feet per Second	0.6818	Miles per Hour
Meters per Second	3.281	Feet per Second
Miles per Hour	88	Feet per Minute
Miles per Hour	1.467	Feet per Second
Liters	0.2642	Gallons
Liters	1.057	Quarts
Liters	1000	Milliliters
Square Centimeters	0.1550	Square Inches
Square Inches	6.452	Square Centimeters

Decimal-Fraction Table

Drill Size	Decimal	Drill Size	Decimal	Drill Size	Decimal
80	.0135	1/8	.1250	O	.3160
79	.0145	30	.1285	P	.3230
1/64	.0156	29	.1360	21/64	.3281
78	.0160	28	.1405	Q	.3320
77	.0180	9/64	.1406	R	.3390
76	.0200	27	.1440	11/32	.3438
75	.0210	26	.1470	S	.3480
74	.0225	25	.1495	T	.3580
73	.0240	24	.1520	23/64	.3594
72	.0250	23	.1540	U	.3680
71	.0260	5/32	.1562	3/8	.3750
70	.0280	22	.1570	V	.3770
69	.0292	21	.1590	W	.3860
68	.0310	20	.1610	25/64	.3906
1/32	.0312	19	.1660	X	.3970
67	.0320	18	.1695	Y	.4040
66	.0330	11/64	.1719	13/32	.4062
65	.0350	17	.1730	Z	.4130
64	.0360	16	.1770	27/64	.4219
63	.0370	15	.1800	7/16	.4375
62	.0380	14	.1820	29/64	.4531
61	.0390	13	.1850	15/32	.4688
60	.0400	3/16	.1875	31/64	.4844
59	.0410	12	.1890	1/2	.500
58	.0420	11	.1910	33/64	.5156
57	.0430	10	.1935	17/32	.5312
56	.0465	9	.1960	35/64	.5469
3/64	.0469	8	.1990	9/16	.5625
55	.0520	7	.2010	37/64	.5781
54	.0550	13/64	.2031	19/32	.5938
53	.0595	6	.2040	39/64	.6094
1/16	.0625	5	.2055	5/8	.6250
52	.0635	4	.2090	41/64	.6406
51	.0670	3	.2130	21/32	.6562
50	.0700	7/32	.2188	43/64	.6719
49	.0730	2	.2210	11/16	.6875
48	.0760	1	.2280	45/64	.7031
5/64	.0781	A	.2340	23/32	.7188
47	.0785	15/64	.2344	47/64	.7344
46	.0810	B	.2380	3/4	.7500
45	.0820	C	.2420	49/64	.7656
44	.0860	D	.2460	25/32	.7812
43	.0890	1/4	.2500	51/64	.7969
42	.0935	E	.2500	13/16	.8125
3/32	.0938	F	.2570	53/64	.8281
41	.0960	G	.2610	27/32	.8438
40	.0980	17/64	.2656	55/64	.8594
39	.0995	H	.2660	7/8	.8750
38	.1015	I	.2720	57/64	.8906
37	.1040	J	.2770	29/32	.9062
36	.1065	K	.2810	59/64	.9219
7/64	.1094	9/32	.2812	15/16	.9375
35	.1100	L	.2900	61/64	.9531
34	.1110	M	.2950	31/32	.9688
33	.1130	19/64	.2969	63/64	.9844
32	.1160	N	.3020	1	1.000
31	.1200	5/16	.3125		

Appendix F
GLOSSARY OF TERMS

Anneal: The process of altering the structure of any metal so as to relieve its' working stresses and increase its' ductility.

Anvil: A metallic part of the primer. The blow from the firing pin crushes the priming mixture against the anvil causing ignition.

Ball: In military nomenclature this term refers to the bullet.

Ball Powder: This is a trade name for a double-base smokeless powder developed by Olin Industries. The grains have a spherical, or flattened spherical, shape.

Ballistics: The science of projectiles in motion.

Ballistic Coefficient: Ratio of the sectional density of a bullet to its coefficient of form. Represents the projectile's ability to overcome air resistence in flight.

Battery Cup: A type of primer used in shotshells which makes use of an outside cup to support it in the case.

Bearing Surface: That portion of the bullet surface which bears on, or touches, the bore.

Bell: To open the mouth of a case slightly in order to seat a bullet more easily. Also used in reference to a rifle barrel which is worn at the muzzle (belled muzzle).

Belted: A band which is formed into the head of some cases (magnum type) to strengthen the case and headspace the cartridge.

Berdan: A common European priming system which makes use of an anvil formed in the case itself.

Boattail: A bullet design having a tapered base.

Bore: The inside diameter of a barrel before the rifling is cut.

Boxer Primer: A type of primer used in all American rifle and pistol ammunition and featuring a self-contained anvil. It is fitted in the primer pocket in the head of centerfire cartridge cases.

Bullet: That portion of the cartridge which becomes a projectile when in flight.

Burning Rate: A term which is used to describe the rapidity with which a given powder will burn. The term itself is a relative one based on a comparison with other powders.

Caliber: The approximate diameter of a bullet, or bore, expressed in decimals of an inch in English. In Europe, where the metric system is used, this measurement is expressed in millimeters.

Cannelure: A groove (or grooves) cut around the circumference of a bullet. These grooves, usually one to a bullet, provide the best means of securely crimping the case mouth into the bullet.

Case: The metal, plastic or paper container which holds all of the components of a round of ammunition.

Case Neck Brush: The metal brush and handle used to clean the inside of case necks.

Case Trimmer: A device used to remove excess material from a case mouth. Metallic cases stretch after extensive reloading and firing because the brass flows forward. These cases must be trimmed back.

Case Trimmer Pilot: The pilot guides the cutting portion of a case trimmer by fitting inside the neck of the case to be trimmed.

Cast Bullet: Bullets for centerfire rifle or pistol which are cast from lead alloy. Muzzleloading projectiles and shotgun slugs are cast in pure lead.

Centerfire: Refers to a cartridge having a centrally-located primer in its' base.

Chamber: That part of the bore at the breech which is formed to accept and support the loaded cartridge.

Chamfer: To ream a taper on the inside of a case mouth.

Charge: The amount of powder used in the case at each loading. Also refers to the amount of shot used in a shotshell.

Choke: A constriction at the muzzle of a shotgun barrel designed to control the spreading of shot.

Chronograph: An instrument used to measure the velocity of a projectile.

Compressed Charge: A charge of powder which so nearly fills the case that it is compressed when the bullet is seated.

Components: Any of the various parts which go into the making of a cartridge.

Corrosion: The eating away of the bore by rust or chemical action.

Crimp: The bending inward of the case mouth to grip the bullet. With shotshells the term applies to the closure at the case mouth.

Crimped Primer: A forcing inward of the brass around the top of the primer pocket. This is frequently found on military cartridges and is done to prevent set-back of primers. The crimp must be removed before repriming the case.

Deburring Tool: The deburring tool removes burrs from the inside and outside of newly-trimmed case mouths.

Decap: To deprime a case by removing a spent primer from the primer pocket, most often accomplished by the decapping pin in a die during the resizing operation of reloading.

Die: A tool, in handloading, that reforms cartridge cases and seats bullets; or, in bullet manufacture, a tool that swages bullets or cores, extrudes lead wire or draws jackets.

Dram Equivalent: Term used to indicate the approximate velocity of a shot charge by a comparison.

Drift: Deviation of a projectile from the line of departure due to its' rotational spin or the force of wind.

Drop: The distance a projectile falls, calculated from the line of departure.

Double-Base Powder: Smokeless powder made with nitroglycerine and nitrocellulose base.

Duplex Load: The use of two different powders in loading the same cartridge.

Energy: A projectile's capacity for doing work at a given range, expressed in foot-pounds.

Erosion: The wearing away of the bore due to friction and/or gas cutting.

Fireform: Reforming or changing the shape of a cartridge case by firing it in a chamber of a desired shape.

Flash Hole: A hole leading from the primer pocket to the inside of the case.

Foot-Pound: A unit of kinetic energy equal to the effort required to raise one pound weight, to a height of one foot, against the normal pull of gravity.

Freebore: The distance in the barrel, if any, which the bullet travels before it contacts the rifling. Some barrels are purposely relieved to allow the bullet considerable free movement before it strikes rifling.

Appendix F
GLOSSARY OF TERMS

Gas Checks: A gilding metal cup which is used to protect the base of a cast lead alloy bullet from the effects of burning powder gases.

Gilding Metal: A copper-zinc alloy used for bullet jackets and gas checks.

Grain: A unit of weight measure. 437.5 grains equal one ounce; 7,000 grains equal one pound.

Grooves: Spiral cuts in the bore of a firearm which cause the bullet to spin as it moves through the barrel.

Hangfire: Delayed ignition.

Headspace: The distance from the breech face to that part of the chamber which acts as a stop and prevents the cartridge from moving forward. Also applies to the cartridge case.

Heel: The outer edge of the bullet base.

Hollow Point: A bullet design which features an axial hole at the point. The purpose of the hole is to aid expansion on impact.

Ignition: The action of setting the powder charge on fire.

IMR: Abbreviation for "Improved Military Rifle."

Jacket: The covering, or "skin", of a bullet.

Keyhole: The imprint of a bullet which struck sidewise on target, rather than point first.

Lands: The raised portion of the rifling.

Line of Sight: An imaginary straight line from the eye, through the sights, to the point of aim.

Loading Density: Ratio of the volume of the powder charge to the volume of the case. The higher the better in terms of accuracy.

Lube Dent: A dent in a cartridge case caused by using too much lubricant when resizing.

Lube Pad: A pad, impregnated with lubricant, on which cases are rolled before resizing.

Lubricant: Case sizing lubricant is used to reduce friction between the case and die during the resizing operation in reloading.

Mercuric Primer: A primer in which the priming mixture contains mercury.

Metal Case (MC): A type of bullet which, except for a small opening at the base, is completely encased in a jacket.

Mid-Range Trajectory: Refers to the distance the bullet rises above the line of sight. Mid-range trajectory is calculated halfway between the muzzle and the target.

Minute-of-Angle: A unit of angular deviation equal to 1/60th of a degree. For practical purposes, it is usually approximated as equal to one inch at 100 yards.

Misfire: The failure of a cartridge to fire after the primer is struck.

Muzzle: The front end of a barrel.

Muzzle Energy: See "Energy".

Neck: The upper section of a cartridge case that grips the bullet.

Neck-Size: To resize only the neck portion of a case.

Non-Corrosive: Usually refers to primers having a priming mixture which is free of corrosive compounds. Modern primers are non-corrosive.

Ogive: Refers to the nose shape of a bullet. The radius of the curve of a bullet nose.

Powder: The propellant used in most firearms. It produces a large volume of gas when ignited. There are two basic types—smokeless and black powder.

Powder Funnel: A helpful accessory that facilitates transfer of powder from a scale pan or measure to a cartridge case.

Powder Measure: An adjustable volumetric measure that meters out uniform charges of powder.

Powder Scale: A device to weigh charges of powder. Lyman's D-7 Scale is a beam scale where markers, called "poise", are moved along a weight-graduated beam, as the pan is filled with powder, until the balance point is reached.

Pressure: The force exerted by a burning charge of powder in the chamber of a firearm. Expressed normally in pounds per square inch.

Primer: A small metal cap containing the detonating mixture which is used to ignite the propellant charge.

Primer Pocket: The cavity in the base of a cartridge which receives and supports the primer.

Ram: The main shaft of a metallic cartridge reloading tool.

Resizing Die: The reloading die that resizes fired cartridge cases and removes spent primers by means of a decapping pin.

Round: A military term meaning one single cartridge.

Round Nose (RN): A bullet design which features a rounded nose.

SAAMI: Abbreviation for Sporting Arms and Ammunition Manufacturers Institute.

Seating Depth: The depth to which a bullet is seated below the case mouth.

Seating Die: The reloading die that presses the bullet into the neck of the cartridge case, crimping the case if so desired.

Sectional Density: A bullet's weight, in pounds, divided by the square of its' diameter in inches.

Shell Holder: This is attached to the top of the ram and holds the heads of cartridge cases as they are moved up and down, into and out of the die.

Swage: To pressure-form by forcing through or into a die.

Throat: That area of the bore immediately ahead of the chamber.

Trajectory: The path of a projectile in flight.

Trim-to-Length: The length a cartridge case should be trimmed to after it has stretched past its' "maximum case length".

Turret Press: A reloading press with a rotatable multi-station turret top for positioning dies and powder measure in their appropriate sequence.

Velocity: The speed at which a projectile travels. Usually measured in feet per second at a given range.

Vernier Caliper: A slide-type graduated instrument used to measure overall cartridge and case lengths precisely.

Wad Cutter (WC): A bullet design which features a sharp shoulder. Designed to cut a clean, round hole in a paper target.

Web: That portion of a cartridge case between the bottom of the primer pocket and the interior of the case.

Work-Harden: Brass becomes harder as it is worked. See "Anneal".

X-Ring: Center of a target. Bulls-eye.

Yaw: The action of a projectile spinning erratically around its' own axis.

Zero: The range at which a firearm will hit the point of aim.

LYMAN RELOADING RECORDS LOG (ONE)

Date Loaded	Lot No.	Caliber	Brass Brand Lot No.	Primer Brand Type Lot No.	Powder Brand Type Lot No. Charge	Times Fired	Times Trimmed	Bullet Brand Type Weight Lot No.
(sample entry)								
11/7/91	76	300 H&H	Rem. S21K6601	Win LR SML200G	IMR 4350 P84FE08A 48.0 6405	3	2	Nosler Part 180 XC1501

LYMAN RELOADING RECORDS LOG (TWO)

Date	My Lot No.	Comments
11/7/91	76	Case Length - 2.840" OAL w/bullet - 3.600"

LYMAN RELOADING RECORDS LOG (ONE)

Date Loaded	Lot No.	Caliber	Brass Brand Lot No.	Primer Brand Type Lot No.	Powder Brand Type Lot No. Charge	Times Fired	Times Trimmed	Bullet Brand Type Weight Lot No.
(sample entry)								
11/7/91	76	300 H&H	Rem. S21K6601	Win LR SML200G	IMR 4350 P84FE08A 6405 48.0	3	2	Nosler Part 180 XC1501

LYMAN RELOADING RECORDS LOG (TWO)

Date	My Lot No.	Comments
11/7/91	76	Case Length - 2.840" OAL w/bullet - 3.600"

LYMAN TARG-DOTS®

Instant Targets

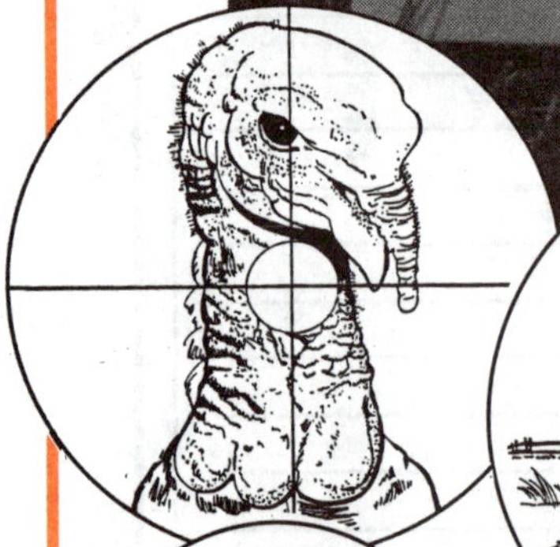

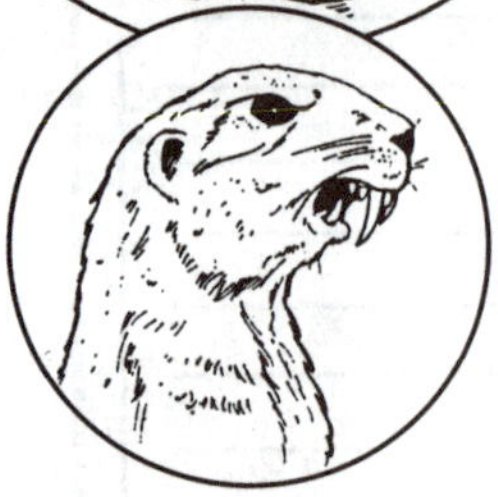

Fluorescent Animal Targets
6" Turkey (#2665)
6" Sabertooth Woodchuck (#2668)
3" Sabertooth Prairie Dog (#2679)

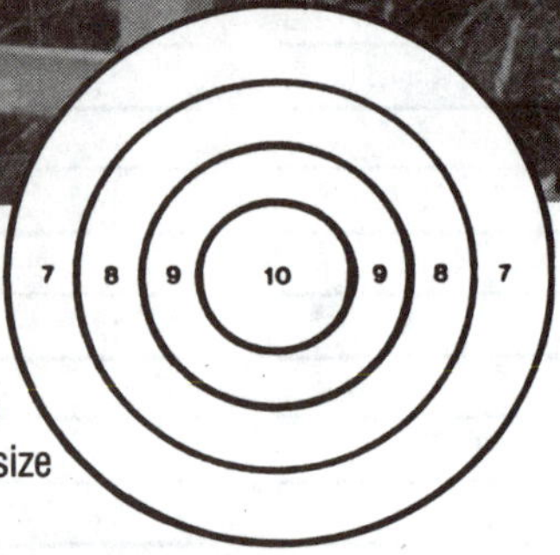

Hot Bullseye
1", 2" and 3" size

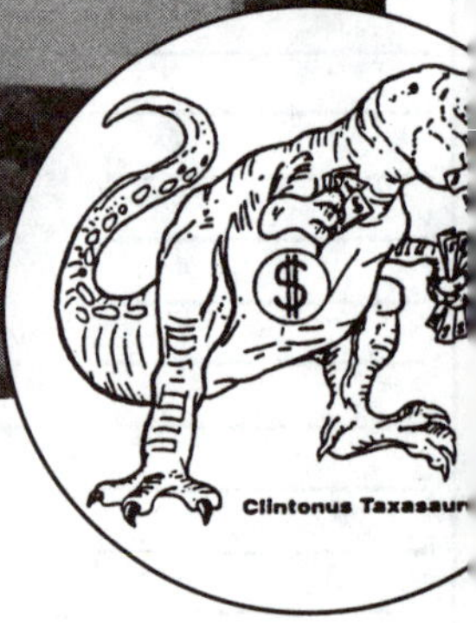

Clinton Taxasaurus
6"-10 pack (#2667)

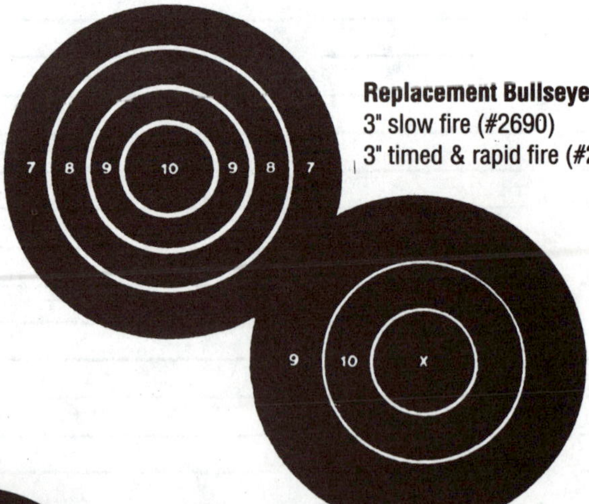

Replacement Bullseyes
3" slow fire (#2690)
3" timed & rapid fire (#2690A)

Accupoint Dots
1", 2" and 3" size

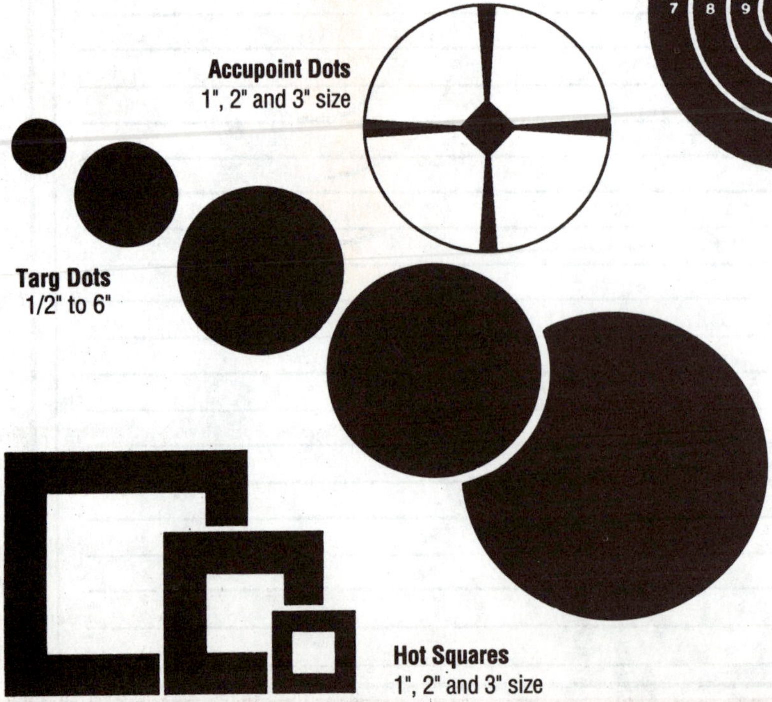

Targ Dots
1/2" to 6"

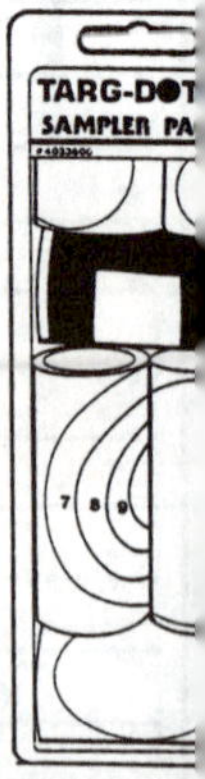

Sampler
Variety of 7 tar
Styles - 360
1/2" to 3" (#3

Hot Squares
1", 2" and 3" size

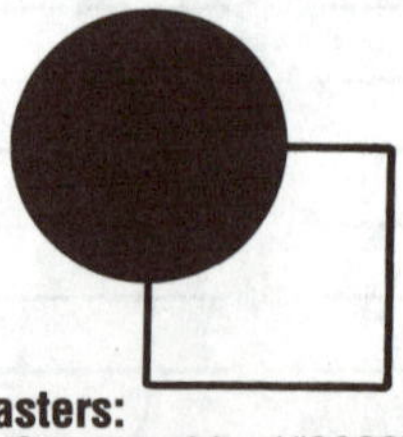

Pasters:
1" Square white (#2680)
1" Square Black (#2615)
1" Round Black (#2669)
(200p/pack)

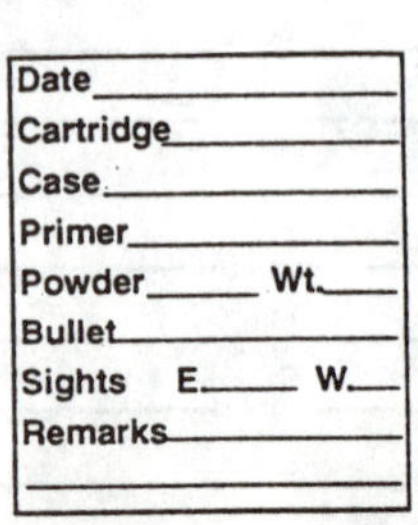

Fluorescent Green Reloaders Labels
1" x 2" (#2670)
2" x 2 1/2" (#26

TAKE THE TARG-DOT® CHALLENGE

"SHOOT AND WIN IN OUR EXPERT SHOOTER CONTEST"

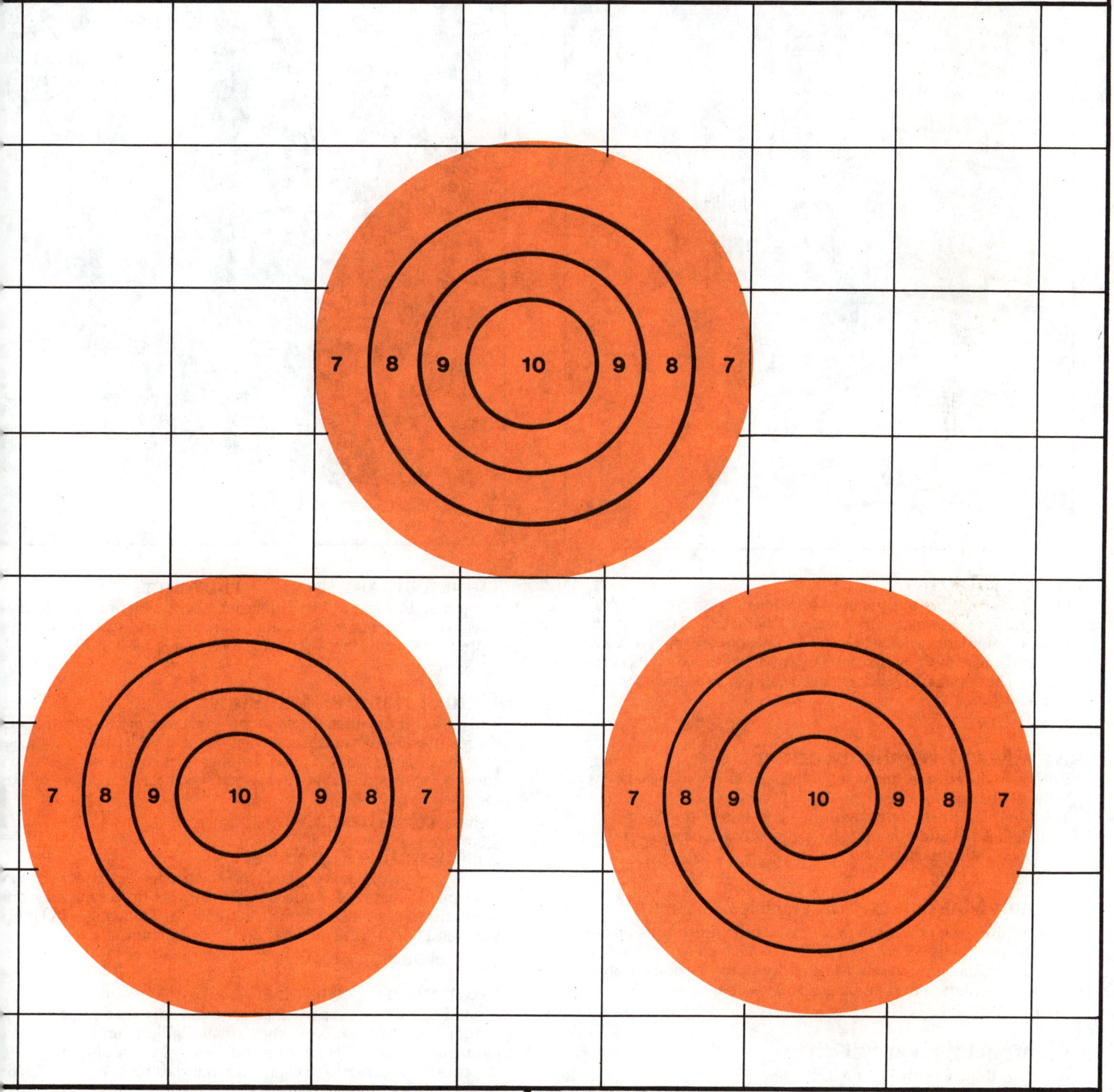

N AN "EXPERT SHOOTER" PATCH AND CERTIFICATE. SIMPLY FOLLOW E RULES BELOW.

'LE: Distance 100 yds. - 5 shots on any of these targets. A score of 45 or more qualifies for an expert award. Any *Centerfire Rifle* and scope may be used.

;TOL: Distance 50 feet - 5 shots on any of these targets. A score of 40 or more qualifies for an expert award. Any *Centerfire Pistol* and sight may be used.

ou do not qualify on these targets you may purchase Targ-Dot 3 inch "Hot Bulls" 50 foot slow fire replacement centers. Shoot until you qualify. Only Targ-Dots epted. Send your qualifying targets to:

Lyman Products 475 Smith St. Middletown, CT 06457

TE: To certify for award, dealer name and address must be included.

TARG-DOT DEALER

Dealer Name ______________________

Address ______________________

City ______________ **ST** ______ **Zip** ______

Shooter's Name ______________________

Address ______________________

City ______________ **ST** ______ **Zip** ______

Caliber: ________ **Type Rifle:** ________ **Type Pistol:** ________

Send to: Lyman® 475 Smith St. Middletown, CT 06457

Lyman Tools for Sportsmen & Gunsmiths

(see special offers on order card)

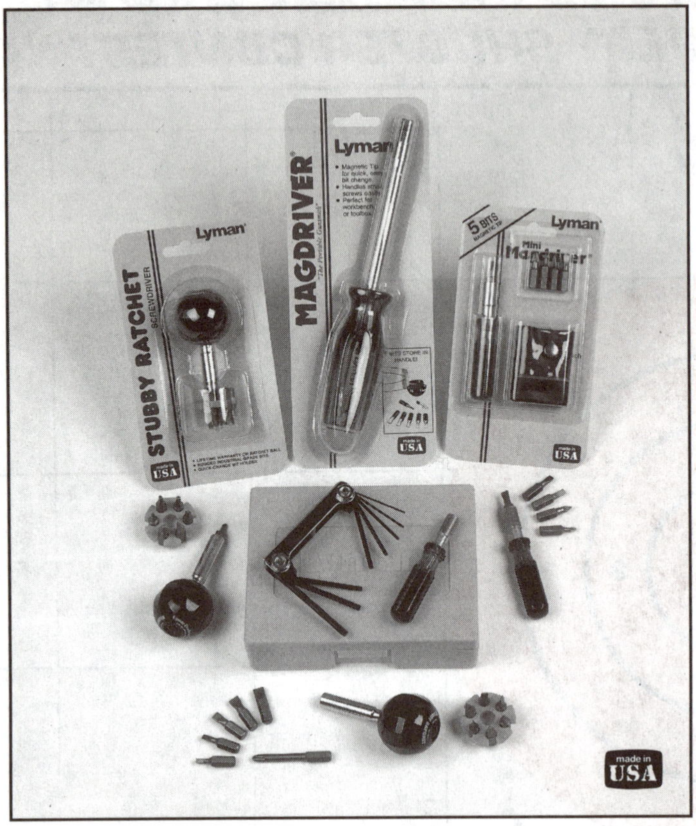

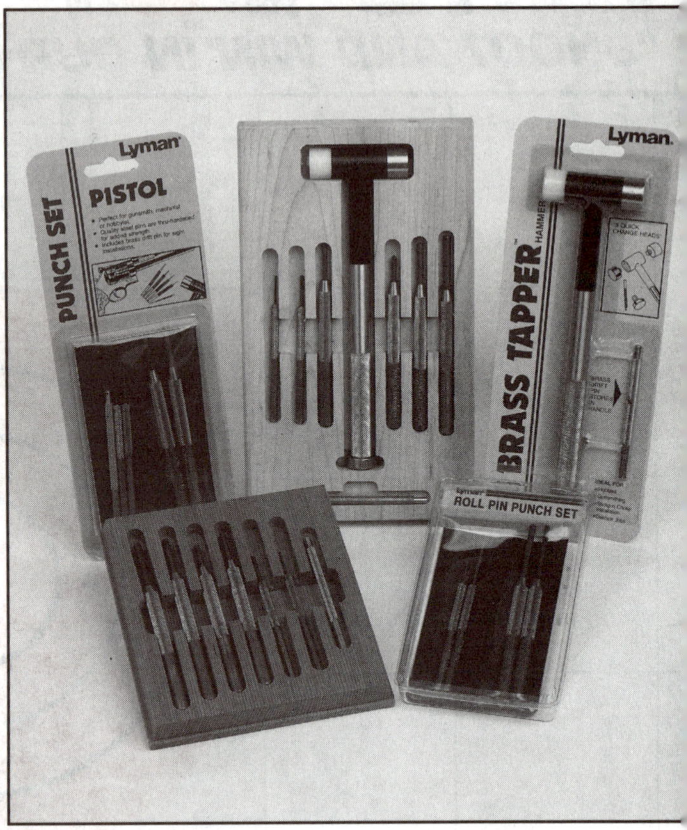

Stubby Ratchet

Our Stubby Ratchet fills the need for a small versatile high performance screwdriver that works in places that standard drivers won't fit at an affordable price. Each stubby includes a complete ratchet and bit caddy and 5 industrial hardened bits (three phillips, two blade). The reversible, quick change head lets you drive or retract with ease.

(4 oz.) (#7031275)

MiniMagdriver Screwdriver

With its reduced size, this little brother of our popular Magdriver is the perfect tool for those closeup, delicate jobs.

The MiniMagdriver comes with five bits - three slotted and two phillips heads - and a handy storage pouch. All bits are specially hardened, industrial grade.

(3.5 oz.) (#7631285)

Lyman Magdriver "The Portable Gunsmith"

A specially designed screwdriver to handle most gunsmithing applications. Unique storage handle contains all seven specially selected, hollow ground bits. Magnetic tip provides quick, easy bit changes and aids in the handling of small screws.

(6 oz.) (#7031249)

Compact Driver Kit

Ideal for Boat, Car and Home Use

The Compact Driver Kit includes our Stubby Ratchet Screwdriver, new Mini Magdriver Screwdriver and our fold-up Hex Key Set. Packaged in a rugged, latchable plastic storage kit. The kit services 20 heads that include 8 hex, 9 slotted and 3 phillips heads (one long-shank which is perfect for recoil pad use). All Lyman bits are specially hardened industrial grade and interchangeable with our popular Magdriver.

Compact Kit complete (1 lb.) (#7031283)

Hex Key Set only (4 oz.) (#7031295)

Archery Hex Set (not shown) (#7031296)

Lyman "Brass Tapper" Hammer

For gunsmithing and other delicate jobs, this hammer has three interchangeable heads including nylon, brass and steel. For delicate tapping, a knurled brass drift pin stores inside the handle.

(11 oz.) (#7031290)

Deluxe Hammer and Punch Set

The perfect companion for the home gunsmith or handyman. Each set features our complete selection of quality Gunsmith Punches plus the popular Brass Tapper Hammer. This unique set comes in a custom hardwood storage case which separates each punch by size for quick access.

Deluxe Set (3 lbs.) (#7031298)

Pistolsmith's Punch Set

For the specialized needs of pistol work, Lyman has put together a more economical assortment of punches. This basic set utilizes the same universal handle as the Gunsmith's Set with the following selection of punches: 1/16", 3/32", 1/8", 5/32" and brass drift punch.

Pistol Punch Set (8 oz.) (#7031271)

Gunsmith's Punch Set

A good selection of properly made and correctly sized punches is no longer hard to find. Now they're available from Lyman. The Gunsmith's Punch Set offers a complete selection of quality steel punches with hardened steel punch tips including 1/16", 3/32", 1/8", 5/32", 3/16", 1/8" center punch and a brass drift punch. Deluxe Set including the interchangeable head "Brass Tapper" hammer is also available. (see above)

Gunsmith Punch Set (12 oz.) (#7031273)

Pachmayr Shooters Accessories
(see special offers on order card)

’achmayr Shotgun Snap Caps and Dry Fire Dummy Rounds for Rifle, Pistol & Revolver

hotgun Snap Caps allow safe dry firing while protecting firing pin, sear and trigger. Excellent for firearms safety and storge. **Dry Fire Dummy Rounds** are made of tough plastic exterior with solid brass base. Ideal for training, working firearm ction and dry firing.

Part No.	Model	No. per pack	Price
04499	12GA	2	5.98
04493	16GA	2	5.98
04484	20GA	2	5.98
04500	.410GA	2	5.98
03201	.223	2	11.50
03203	.243	2	11.50
03205	.270	2	11.50
03209	30-06	2	11.50
03207	.308	2	11.50
03216	380	5	14.98
03231	9mm	5	14.98
03229	38/357	6	14.98
03228	40S&W	5	19.98
03218	44 Spec/mag	6	16.98
03230	45ACP	5	14.98
03220	45 Long Colt	6	16.98

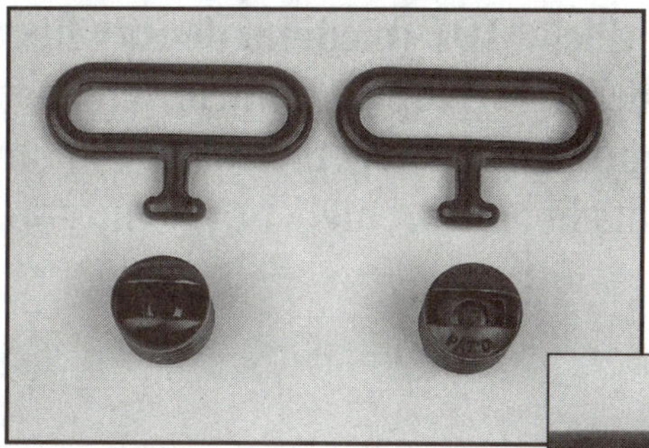

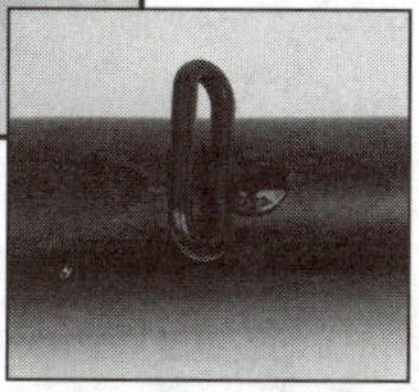

Flush Mount Sling Swivel Set

Back by popular demand the famous design preferred by custom gunsmiths for years. The unique, clean design that has been copied but never equaled in strength or clean appearance. Fully machined receptacles for strength. Simple “T” shaped sling loops lock positively and securely.

achmayr **Shooters Counter Mat** is ideal r cleaning or disassembling your firearm. rotects your firearm and furniture finish. olded grooves prevent losing screws and mall parts. Real 100% Pachmayr high uality rubber.

achmayr **Shooters Cap** features full nbroidered logo. One size fits all.

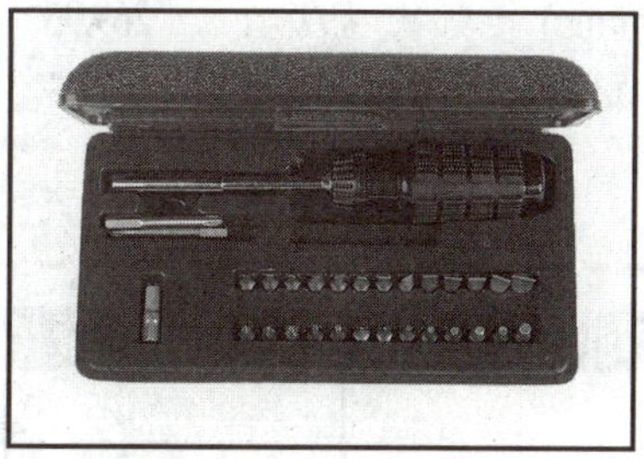

Gunsmith Tool Kit

Pachmayr’s 30 piece gunsmith screwdriver set contains the best selection of bits for working on all firearms. Gunsmith slotted bits for action screws, hex bits to fit the new scope mounting screws and Phillips bits including special long Phillips bits for butt plate and recoil pad removal. A must for the serious shooter.

’or complete information on Pachmayr products or to get a free Pachmayr catalog call

1-800-22-Lyman

Sturdy-Bench™

Save $30
Small Benches Reg. $239.95
Special $209.95

Save $10
Small Backboards Reg. $79.95
Special $69.95

SAVE!

The HOLUMBER™ STURDY-BENCH™ with its full-inch thick MDF (medium density fiberboard) top is ideal for mounting reloading machines and other bench top equipment. The posts, beams and diagonals are made from quality dried lumber and fastened together using 3/4" pipe size screw fasteners. This modular construction system results in an extremely heavy-duty workbench with exceptional rigidity.

Standard Bench- available in two work heights

36" High Bench (61" x 27" top)	#97336	$269.95	36" High Bench (49" x 24" top)	#97536	$239.95
41" High Bench (61" x 27" top)	#97341	269.95	41" High Bench (49" x 24" top)	#97541	239.95
Optional Backboard Kit, std. bench	#97342	89.95	Optional Backboard Kit, std. bench	#97542	79.95

Item #	Quantity	Cost*
	Total	

NAME ______________________

ADDRESS ______________________

CITY ____________ STATE ________ ZIP ________

Enclosed is my check/money order for $____________
made out to: LYMAN PRODUCTS CORPORATION
Or Please bill my:

❑VISA ❑MasterCard ❑Discover

Account number ______________ Exp. Date ______

Signature ______________________

All shipments will be freight prepaid in continental US. Prices valid through 1998. After, call Lyman for updated pricing.

***Note:Additional shipping and handling charge of $35.00 would apply to all shipments. All orders will be drop shipped from the manufacturer in Bellevue, WA. For technical questions feel free to call Lyman or to manufacturer direct at 1-800-375-8554.**

Lyman®

Lyman Products, 475 Smith Street, Middletown, CT 06457

Call toll-free 1-800-22-Lyman today for your free copy of our full-color catalog.

SEND TO COMPONENT MANUFACTURER

Gentlemen:

I have read about your products in Lyman's Reloading Handbook, 47th Ed. Please send me a current catalog showing your reloading components products.

Name: ______________________________

Address: ______________________________

____

Zip Code: ______________________________

47th Ed. 4

SPECIAL ORDER FORM

Congratulations! This Lyman purchase qualifies you to obtain selected products at special pricing below. Please complete the following.

Name: ______________________________

Address: ______________________________

City/State/Zip: ______________________________

Tel.#: (If we have a question) ______________________________

Method of Payment (Check one)

❑ C.O.D. ❑ VISA ❑ MasterCard ❑ Discover ❑ Check Enclosed

Minimum Order for Credit Card Use is $10.00

______________________________ /

Account number   Exp. Date

____

Signature

Description	Part No.	Qty.	Retail Price	Special Price	Total
Lyman Tools for Sportmen					
Stubby Ratchet	7031275		$7.95	$5.95	
Mini-Mag Driver	7631285		7.95	5.95	
MagDriver	7031249		16.50	12.95	
Compact Driver Kit	7031283		22.75	17.95	
Brass Tapper Hammer	7031290		13.75	11.95	
Deluxe Hammer and Punch Set	7031298		33.95	24.95	
Pistolsmith Punch Set	7031271		12.95	9.95	
Gunsmith Punch Set	7031273		17.50	14.95	
Pachmayr Shooters Accessories	Write in Ga or Cal.				
Shotshell Snap Caps (2 pk) (Specify Gauge)			5.98	4.50	
Rifle Dry Fire Dummy Rounds (Specify Caliber)(2 pk)			11.50	8.95	
Pistol Revolver Dry Fire Dummy Rounds (Specify Caliber)*			16.98	12.95	
Pachmayr Shooters Mat	02984		12.98	9.95	
Pachmayr Embroidered Cap	07089		14.98	9.95	
Pachmayr 31 Piece Gunsmith Tool Kit	03047		19.98	12.95	
Pachmayr Flush Mount Sling Swivel Set (Gunsmith Installation Recommended)	03291		17.98	12.95	
Other Popular Lyman Handbooks					
Shotshell Handbook 4th Edition	9827460		24.95	19.95	
Cast Bullet Handbook	9837249		19.95	14.95	
Pistol & Revolver Handbook 2nd Edition	9847200		19.95	14.95	
Reloaders Log	9847261		4.50	3.00	

* Revolver Calibers (6 p/pk) Pistol (5 p/pk)

Shipping/Handling Charge	**N/A**
CT Residents Only: (Add 6% Sales Tax)	
Total Amount	

Questions? Call Toll Free 1-800-22-LYMAN
All Shipments on above form are postpaid.
Offer valid only in U.S. and Canada. ***Good thru 2000***.

PLACE
STAMP
HERE

PRINT NAME AND ADDRESS OF MANUFACTURER BELOW

See **Special Order Form** on other side. Tear or cut off card, where shown. When folded and secured, card can be used as mailer by MC/Visa charge customers. For check or money order, use envelope. Any questions, please call toll-free **1-800-22-LYMAN** (except foreign customers, who may call (860) 632-2020). Hours: 9:00 a.m. -3:30 p.m. Eastern time, Monday-Friday.

Fold here and staple or tape order card enclosed. Do not send open!

PLACE
STAMP
HERE

LYMAN PRODUCTS CORP
DEPT 47 - HB OFFERS
475 SMITH STREET
MIDDLETOWN, CT 06457